Springer Texts in Statistics

Series editors

R. DeVeaux
S.E. Fienberg
I. Olkin

Springer Texts in Statistics (STS) includes advanced textbooks from 3rd- to 4th-year undergraduate courses to 1st- to 2nd-year graduate courses. Exercise sets should be included. The series editors are currently Stephen Fienberg and Richard D. De Veaux. George Casella and Ingram Olkin were editors of the series for many years.

More information about this series at http://www.springer.com/series/417

Douglas A. Wolfe • Grant Schneider

Intuitive Introductory Statistics

Douglas A. Wolfe
Department of Statistics
The Ohio State University
Columbus, OH, USA

Grant Schneider
Upstart Network
San Carlos, CA, USA

ISSN 1431-875X ISSN 2197-4136 (eBook)
Springer Texts in Statistics
ISBN 978-3-319-56070-0 ISBN 978-3-319-56072-4 (eBook)
DOI 10.1007/978-3-319-56072-4

Library of Congress Control Number: 2017950163

© Springer International Publishing AG 2017
This work is subject to copyright. All rights are reserved by the Publisher, whether the whole or part of the material is concerned, specifically the rights of translation, reprinting, reuse of illustrations, recitation, broadcasting, reproduction on microfilms or in any other physical way, and transmission or information storage and retrieval, electronic adaptation, computer software, or by similar or dissimilar methodology now known or hereafter developed.
The use of general descriptive names, registered names, trademarks, service marks, etc. in this publication does not imply, even in the absence of a specific statement, that such names are exempt from the relevant protective laws and regulations and therefore free for general use.
The publisher, the authors and the editors are safe to assume that the advice and information in this book are believed to be true and accurate at the date of publication. Neither the publisher nor the authors or the editors give a warranty, express or implied, with respect to the material contained herein or for any errors or omissions that may have been made. The publisher remains neutral with regard to jurisdictional claims in published maps and institutional affiliations.

Printed on acid-free paper

This Springer imprint is published by Springer Nature
The registered company is Springer International Publishing AG
The registered company address is: Gewerbestrasse 11, 6330 Cham, Switzerland

To our wives, Marilyn and Jingjing, for their patience and support through the lengthy preparation of this text

Preface

Understanding and interpreting data, both experimental data and the vast quantity of observational data that is now routinely available in the public domain, need not be linked solely to comprehension of a prescribed set of mathematical expressions. Introducing students to the beauty of statistics can be both motivational and interesting without being encumbered by equations. In fact, it is our view that the goal of an introductory statistics course should be to foster an appreciation for the role of statistics and associated data analysis approaches in our everyday lives rather than to prepare the students to be statistical analysts—if necessary, there will be time enough for that secondary emphasis as they begin to concentrate on their chosen fields of study. An introductory course should foster appreciation for the relevance and importance of using statistical methods for summarizing and interpreting data, but not at the expense of enjoying the process. While it is, of course, necessary to include common statistics such as the sample mean, standard deviation, correlation, etc. in an introductory statistics course, they do little to motivate and "grab" the students and can be introduced *after* a student has come to appreciate the nature of the basic information that is provided in collections of data. This desire leads us in this book to emphasize counting and ranking approaches as initial tools for eliciting information from data collections rather than as a fallback only when sample means and standard deviations are not effective. Not only can students easily understand such counting and ranking techniques, but they also find them to be quite valuable as they explain their analyses to others.

The second point of emphasis in this text is that examples should not be chosen simply to illustrate how a statistical procedure can be applied. Dull, contrived examples can make even interesting statistical analyses uninteresting. On the other hand, an example to which students easily relate can go a long way in both piquing and sustaining their interest in the associated statistical analysis. While they learn a statistical technique, they also unearth some information of interest to them in its own right. We all learn best when we see the relevance of a topic in our own lives. We have worked hard to motivate the statistical discussions in our text through experimental settings and data collections that we ourselves find interesting and that raise questions that we can use statistical tools to address. While we are the first to confess that our view of the world may not match up completely with that of high school juniors and seniors or college freshmen and sophomores, we have tried to include data sets in both the examples and the exercises that are both real (not just realistic) and of general relevance for the students.

A third new feature for this text is the way in which we have chosen to present the chapter exercises. In addition to the usual set of exercises at the end of each section (and there are many), we have included a substantial number of comprehensive exercises at the end of each chapter that include conceptual exercises, data analysis/computational exercises, exercises involving hands-on student activities, and exercises associated with a student's use of the Internet to access interesting and relevant data sets and statistical analyses. This provides instructors with a wide variety of exercises to challenge the individual interests of students in their classes.

The fourth and final unique feature of this introductory text is the inclusion of the necessary **R** functions to enable instructors and students to analyze data sets without the tedium often accompanying many statistical computations. Examples throughout the text still provide all of the details of the associated statistical calculations so that students are fully aware of how the various statistics elicit information from a data set(s). However, we also provide the appropriate **R** procedures that can be used to make the same calculations. It is easy enough for instructors to bypass the **R** procedures if they choose, but including them as part of the basic course would permit their students to apply the associated statistical procedures to large data sets where the direct numerical calculations by hand would be prohibitive. The use of these **R** programs also eliminates the need to include the normal, t, and χ^2 tables, as well as the relevant nonparametric null distribution tables as part of the text. We have also organized all of the R programs used in the text into a documented collection that is formally registered as an **R** package IIS specifically linked to this text.

The text is specifically designed for use in an AP statistics course for high school juniors or seniors or in a one-semester or two-quarter introductory precalculus statistics course for college freshmen or sophomores. How well we have succeeded in reaching these groups of students will clearly determine the impact of our not-just-one-more introductory statistics textbook!

Many friends and colleagues have helped with both the initial development and improvement of this text over the years. We owe a particular debt of gratitude to Brad Hartlaub for his invaluable help in initiating this project in the first place and his dedicated effort to move it forward over a number of succeeding years. We also appreciated his input from teaching much of this material in statistics classes at Kenyon College. Similarly, we owe thanks to Deborah Rumsey and Elizabeth Stasny for their feedback from using early drafts of the project in introductory statistics courses at The Ohio State

University. We also owe thanks to Jungwon Byun, Ben Chang, Neha Hebbar, Cindy Smith, and all of the attendees of the so-called "data party" at Upstart for providing a fertile testing ground for the material and concepts. Finally, we owe a special thank you to the computer support group in the Department of Statistics at The Ohio State University for their patience in helping us work through the various versions of MSWord that were confronted over the many years in preparation of the text.

Our editors Michael Penn and Hannah Bracken, who originally signed us to publish with Springer, and Rebekah McClure, who assumed our project when Hannah returned to school, were dedicated from the start of the project and provided tremendous support to see it through to publication. Our production manager Christina Oliver skillfully guided the manuscript through the production process.

To everyone who helped over the many years, our heartfelt thanks.

Columbus, OH, USA Douglas A. Wolfe
San Carlos, CA, USA Grant Schneider

Contents

1 Exploratory Data Analysis: Observing Patterns and Departures from Patterns

Data, data, everywhere and we are forced to look at it. We see it in the newspapers ("63% of the people polled support the president's decision to... "), on the news ("scientists at a major research university report that treatment of Parkinson's disease with a combination of drugs A and B has the potential to extend remission of the disease by an average of 2 years..."), from the government ("the nation's trade deficit narrowed last month, for the first time in... "), in sports ("over the past two years left-handed hitters have a batting average of .359 against ... "), in finance ("the Dow Jones Industrial Average rose again today to a new record high, marking the seventh consecutive day of record highs, but the broader market..."), and social settings ("despite the robust economy, the percentage of families living below the poverty line has not dropped substantially over the past six months..."), just to scratch the surface.

We are an information-oriented society–we demand facts in all aspects of our lives, but we are often overwhelmed by the magnitude of the response.

© Springer International Publishing AG 2017

D.A. Wolfe, G. Schneider, *Intuitive Introductory Statistics*, Springer Texts in Statistics,
DOI 10.1007/978-3-319-56072-4_1

However, it is vitally important that we are able to properly interpret the data we encounter, as well as to understand the circumstances behind the collection of the numbers. For example, what is it that is causing such a large increase in sales for a particular item produced by a major snack-food company? Which states are the safest to drive in and what issues are directly related to that safety? What percentage of college students are involved in "binge drinking" and are there consequences? What factors directly affect the growth of pine seedlings? What is a safe dosage for a new medication to treat Alzheimer's disease?

This chapter is about visualizing, summarizing, and interpreting existing collections of data. In later chapters we turn to the other set of issues regarding data collection and analysis themselves, including what, where, when, how, and from whom.

1.1 Interpreting Graphical Displays of Data Collections

Data are data are data–we all know what they are–it is as simple as that–right? Actually, nothing could be further from the truth. The word data itself is plural, referring to the fact that it corresponds to more than a single piece of information. The American Heritage College Dictionary (1993) defines the word data as "Factual information, esp. information organized for analysis." or "Values derived from scientific experiments." While one or the other of these definitions adequately describes the typical collection of data we might encounter, neither definition helps us understand or interpret a specific set of data.

The discipline of statistics offers an organized set of principles and procedures for addressing these many facets of data. One way to visualize the role of statistics in the collection and interpretation of data is to think of the interaction between data and statistics as an interlocking jigsaw puzzle – sometimes straightforward, sometimes quite intricate, in their connections.

In order to properly obtain and interpret a collection of data, it is important that we have taken the necessary steps to produce a fully complete puzzle. As we have all experienced, visualizing only portions of a jigsaw puzzle can produce misleading glimpses of the overall picture depicted in the puzzle. Thus, it is vital that we understand **all** of the various data-statistics connections so that every one of the puzzle pieces is in its proper place. This is the only way to ensure that a study or investigation leads to a correct representation and interpretation of the problem of interest. Figure 1.1 provides a visualization of this data-statistics jigsaw puzzle.

The types and varieties of data are as numerous as the studies that lead to their collection. As a result, the very first step in any study or investigation is to clearly understand **what type and how much data** should be obtained in order to address the question of interest. Once this decision has been reached, the next natural consideration is **how to properly collect** the appropriate data. These data-statistics connections relating to the design of an experiment and collection of the data are depicted as red pieces in the Fig. 1.1 jigsaw puzzle. Such issues will be discussed further in Chap. 3.

The second set of considerations in the data-statistics puzzle address the question of **how to summarize and describe a data collection** obtained from a properly designed study. How can the data be clustered into useful interpretive categories? What are the appropriate ways to summarize, describe, and compare the data? These types of data-statistics connections that relate to summarizing and describing a data collection are depicted as yellow pieces in Fig. 1.1. Issues related to the major data types and proper methods for summarizing and describing each type are the topics of Chaps. 1 and 2.

The remaining connections in the data-statistics puzzle deal **with appropriate methods for analyzing the various types of data and reaching proper conclusions from these analyses**. The data-statistics puzzle pieces associated with this aspect of data analysis and interpretation are colored blue in Fig. 1.1 and are discussed in detail in Chaps. 6, 7, 8, 9, 10, 11, and 12.

Fig. 1.1 Data-statistics jigsaw puzzle: an accurate picture only when we have made all of the proper data-statistics connections

We collect data because we are interested in the characteristics of some group or groups of people, places, things, or events. For example, we might want to know about temperatures in the month of June in Los Angeles, the intended functions of some clay artifacts excavated at an archaeological site, or the numbers of eggs for ruby-throated hummingbirds. In these contexts, **temperature**, **intended function**, and **number of eggs** are called **variables**.

Definition 1.1 A **variable** is an attribute of interest in a study. It can represent the outcomes of a designed experiment, the responses from a survey, or characteristics of a set of people, places, or things.

Variables assume different values across the group or groups we are studying. Thus, June temperatures in Los Angeles are different from day to day and from year to year, and different hummingbirds lay different numbers of eggs. To understand how a variable varies across a group, it is important to obtain a 'representative collection' of variable values from the group.

Definition 1.2 A particular measurement of an attribute is called an **observation**. The goal of a scientific study is to collect enough representative observations to tell us what we want to know about an attribute.

Each observation is a value of the variable of interest. For example, an observation at an archaeological site might be: "the clay artifact found in this location has a ceremonial function". On the other hand, an ornithologist might record: "this particular nest contains 3 eggs". Data sets encountered on a daily basis are nothing more than collections of observations on attributes of interest to the collectors of the data–but what types of variables make up these data collections? The first division that we make is between quantitative and categorical variables.

Definition 1.3 A **categorical variable** is one for which the associated observations are simply listings of physical characteristics or traits of the subjects or objects being studied. For example, eye color is a categorical variable, with categories brown, blue, green, etc. **Categorical data** then correspond to observed sample counts in each of the possible categories.

Definition 1.3 (continued)
In the case of eye color, the data would be the number of sample subjects with brown eyes, the number with blue eyes, the number with green eyes, etc.

Categorical variables abound in the way we *describe* people or objects to others. The classifications tall, young, poor, Hispanic, college graduate, athlete, male, and HIV positive are but a few of the traits that might be important in studying a human population.

For tabulation purposes, arbitrary numerical values are sometimes assigned to the classes of possible characteristics or traits for a categorical variable. However, arithmetic manipulations with such numerical labels are not meaningful. For example, although we might arbitrarily assign the number 1 to brown eyes, the number 2 to blue eyes, the number 3 to green eyes, etc., arithmetic computations with these labels would not make sense; that is, the sum of 4 for one brown-eyed and one green-eyed subject would not have the same interpretation as the sum of 4 for two blue-eyed subjects.

Definition 1.4 A **quantitative variable** is one for which the associated observations will be numerical and, therefore, such that the usual arithmetic manipulations make sense. **Quantitative data** then correspond to the measured numerical values of a quantitative variable.

We are very familiar with many quantitative variables as well, including weight, height, time between failures of an electronic device, blood pressure, number of eggs laid by a hummingbird, and percentage of tornadoes in a given year which caused at least one death. It makes sense to do arithmetic, such as adding and dividing by the number of observations to find an average value, on all of these variables.

In addition to variables that are clearly either categorical or quantitative, it is not uncommon to encounter 'borderline' variables. For example, consider an experiment in which each participant is asked to rank a set of five dish detergents, from most preferred to least preferred. For each participant the realized data (known as **ordinal data**) will be a ranking from 1 (best) to 5 (worst) for the five competing detergents. While such data clearly provide more than simple categorical information (e. g., the detergent labeled as 1 by a participant is actually preferred to and not just different from a detergent that he labels 3), they are not quite quantitative, as we have described it in Definition 1.4. One of the missing features is that numerical calculations are not valid on these ordinal rank data. Thus, for example, a particular dish detergent might receive two rankings of 1 and two rankings of 5 from four participants in the study, while a second detergent might receive all four rankings of 3 from the same participants. Even though both detergents received an average ranking of 3, the four participants clearly do not view the two detergents as equivalent. The averaging process is simply not a valid operation for such ranking data. While our major emphasis in this text will be on categorical and numerical data, we will at times also discuss how to deal with ordinal data.

Graphical Displays of Categorical Data The main purpose of some studies is to see how a set of data is distributed across a small set of *categories* or *classes*. If each observation falls into exactly one of the classes, we say that the classes *partition* the data collection. For example, the classes urban, suburban, and rural partition new housing construction, and the categories cats, dogs, and "other" partition domestic animals. The classes or categories in a partition are *exhaustive* and *exclusive*, meaning that they include every possible observation and they do not overlap, respectively.

As you learned in Definition 1.3, observations that can be placed into categories are called categorical data. Gender, hair color, and make of automobile are examples of categorical data. For such data we count how many observations fall into each of the categories. This collection of category *frequency counts* describes how the data collection is distributed across the partition.

Computation of the *relative frequencies*, corresponding to the frequency counts divided by the number of items in the sample, for each of the categories provides an adequate numerical summary measure for most collections of categorical data. However, there are two graphical techniques, *bar graphs* and *pie charts* that can be used to provide visual summarization of categorical data. Bar graphs can be used to display either the actual frequency counts or the relative frequencies (sample percentages) in a sample. Pie charts, on the other hand, are designed solely to visualize relative frequencies (sample percentages). Both techniques can be particularly effective when we are interested in comparing a number of different collections of data using the same partitioning categories. Examples 1.1 and 1.2 show how to construct bar graphs and relative-frequency pie charts for categorical data.

Example 1.1. Archaeological Excavations at Naco Valley Ed Schortman and Pat Urban, Professors in the Department of Anthropology and Sociology at Kenyon College in Gambier, Ohio, periodically organize and supervise expeditions for undergraduate students from a number of institutions to study the history of various regions in Central America. One such study involved the excavation of individual structures at Site 128 in the northwest portion of Naco Valley in Honduras. During the period AD 600–1000 (known as the Late Classic period) this site was the administrative home to a cadre of bureaucratic functionaries (lesser nobility) and their supporters. Among other things, the expedition group was interested in the types of artifacts that were being used by these inhabitants in order to find out if there were differences in such usage across different structures within Site 128 (possibly occupied by persons of differing levels of power or wealth, etc.). In one such archaeological study, the scientists and their students counted the numbers of the following types of objects found at a number of different structures within a particular excavation site:

1. Ocarinas (multi-note clay musical instruments)
2. Figurines (fired clay anthropomorphic and zoomorphic effigies)
3. Incensarios (incense burners)
4. Ground stone tools (used to process grain, primarily corn, into flour)
5. Stamps (for cloth decoration)
6. Sherd disks (possibly spindle whorls for processing thread)
7. Candelaros (for both practical lighting and ceremonial activities)
8. Jewelry (personal adornments made of clay).

The observed frequency counts for two of the structures at this site are presented in Table 1.1.

We are interested in the artifacts used by the inhabitants of these structures during the period of their existence, as partitioned into the eight noted artifact categories or classes. To construct bar charts for these categorical data, we label the horizontal baseline with the numbers as assigned to the artifact categories in Table 1.1. (Note that these labeling numbers are purely arbitrary, as they could have been assigned in any order to the eight categories.) Then we construct rectangles with common widths above each of the labels, with the height of a particular rectangle corresponding to the observed frequency count for the associated category. The bar graphs for Structures 13 and 17 are presented in Fig. 1.2a and b.

Several things are visually clear from the bar graphs. First, Categories 2 (figurines) and 3 (incensarios) dominate the types of artifacts found at both of the structures, perhaps indicating that each of them might have played some sort of ceremonial role for the inhabitants of the site. However, it is also clear from the bar graphs that the relative importance of Categories 2 and 3 are reversed at the two structures, suggesting that the associated ceremonial

Table 1.1 Artifacts recovered at two different structures during the excavation of Site 128 in Naco Valley, Honduras, 1996

Artifact	Frequency count at structure #13	Frequency count at structure #17
1. Ocarinas	14	21
2. Figurines	56	53
3. Incensarios	73	35
4. Ground stone tools	21	12
5. Candelaros	10	12
6. Stamps	5	9
7. Sherd disks	5	0
8. Jewelry	4	7

Source: Schortman and Urban (1998).

Fig. 1.2 Bar graphs for the Naco Valley artifact recovery data. (a) Structure 13. (**b**) Structure 17

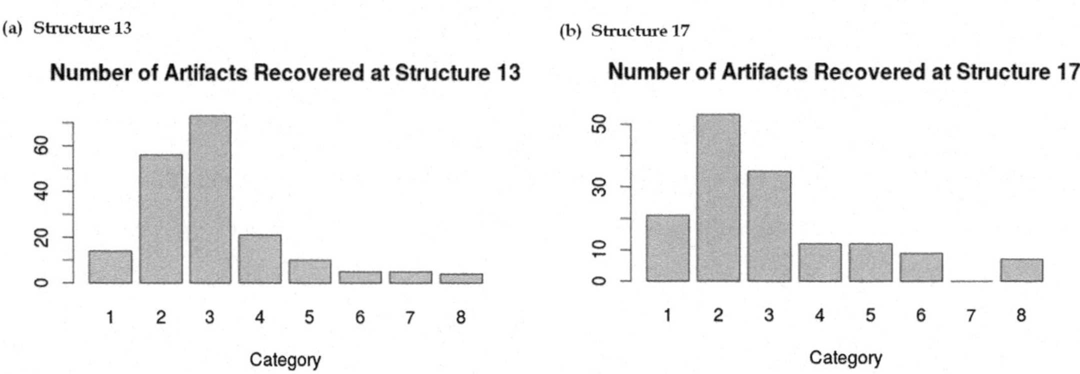

roles may have differed to some extent. (We will leave that discussion for the archaeologists!) Finally, there appears to be the possibility of differences in the use of stamps, sherd disks and jewelry at the two structures as well.

A second form of visual representation for categorical data is provided by the pie chart. It is most commonly used when we wish to pictorially display the sample relative frequencies, or percentages, rather than the raw frequencies for the various classes of the partition. It is particularly effective for displaying differences between two populations with respect to the same categories.

Example 1.2. Motor-Vehicle Deaths by Type of Accident In their annual report on accident statistics, the National Safety Council (1996) reported the breakdown of motor-vehicle deaths by type of accident for the period of years from 1913 through 1995. The population of deaths by motor-vehicle accident is partitioned into the following nine exhaustive categories:

1. Pedestrians
2. Other motor vehicles
3. Railroad trains
4. Streetcars
5. Pedalcycles
6. Animal-drawn vehicle or animal
7. Fixed objects
8. Noncollision accidents
9. Nontraffic deaths

Do you feel there would be a noticeable change in the relative percentages for these nine categories from, say, 1949 to 1985? One way to visually help us answer this question is to create pie charts for the data from those years. The dataset *traffic_accidents* contains the motor-vehicle accident data for the inclusive period 1924–1995. Accessing this file, we find the counts for the years 1949 and 1985 displayed in Table 1.2.

Summing, we find that the total number of reported motor-vehicle deaths in 1949 and 1985 are 32,536 and 47,019, respectively. To create relative frequencies (or percentages) for the nine categories for each of the years, we divide the observed frequencies by the total numbers of reported motor-vehicle deaths in the respective years. For example, the relative frequency (or percentage) for pedestrian deaths in 1949 is $8800/32{,}536 = .2705$, or slightly greater than 27%. These relative frequencies are then depicted visually in the two pie charts displayed in Fig. 1.3, where each relative frequency is

Table 1.2 Motor-vehicle deaths by type of accident in 1949 and 1985

Type of accident	1949	1985
Pedestrians	8800	8500
Other motor vehicles	10,500	19,900
Railroad trains	1452	538
Streetcars	56	2
Pedalcycles	550	1100
Animal-drawn vehicle or animal	140	100
Fixed objects	1100	3200
Noncollision accidents	9100	12,600
Nontraffic deaths	838	1079
Total	32,536	47,019

Source: National Safety Council (1996)

Fig. 1.3 Pie charts for the motor-vehicle deaths by types of accident data. (a) 1949. **(b)** 1985

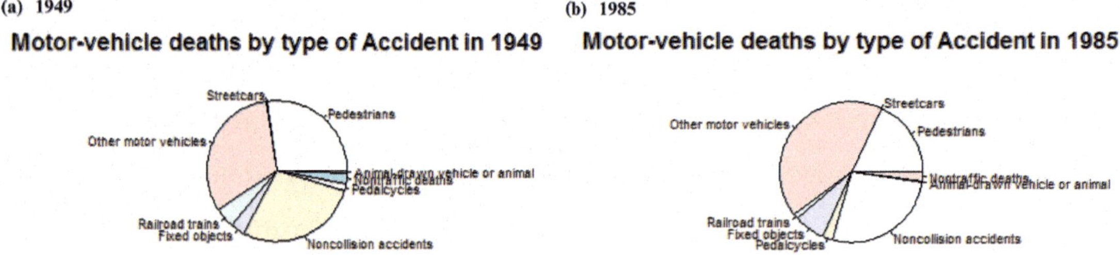

assigned that percentage of the area of the associated pie (circle). Note that for purposes of these pie charts, we have combined the figures for streetcars and animal-drawn vehicles or animals into one category (called others) because of the small relative frequencies associated with each of them individually. The main advantage from using pie charts as opposed to bar charts for presenting categorical data is the visual appeal of the 'pieces of the pie' associated with the various categories.

While much of the general pattern remains similar for the 2 years, it is clear from the pie charts that there have been substantial increases from 1949 to 1985 in the percentages of motor-vehicle deaths due to fixed objects and other motor vehicles, with corresponding decreases in such deaths for pedestrians and trains. Such findings do, of course, agree with our perceptions of the two different periods of time and the associated decreases in train traffic and increases in the number of automobiles and auto speed from 1949 to 1985. Do you think things have changed again since 1985? Explore this further in Exercise 1.D.11.

Graphical Displays of Quantitative Data While most data sets for categorical variables can be adequately summarized by simply listing either the numbers or percentages of outcomes in each of the possible categories (with visual aids, such as bar charts or pie charts), when we are faced with a collection of quantitative data of any size it is often difficult to make sense of the individual measurements directly. It is important to be able to summarize such data in meaningful and insightful ways. While computation of relevant statistics is a common means for obtaining such useful summaries, it is often the case that simple graphical ways of picturing or presenting the data can be just as important in getting a good feel for a data set and in providing the basis for an insightful interpretation of the data.

We consider three of the more useful graphical techniques, each of which has its own special merits for summarizing certain types of quantitative data sets. However, all three provide relevant information about important features of a data set, such as:

1. Is there *a single, dominant center* of the data set? Are most of the observations clustered around this one center or are there a number of secondary clusters of observations?
2. How *spread out* are the observations? Are they tightly clustered around the dominant center of the data or are they rather widely dispersed away from this center?

3. What is the general *shape or configuration* of the data collection? Do observations tend to be evenly distributed in both directions about the dominant center of the data set or are they more spread out in one direction than the other?
4. Are there unusual features of the data configuration? Are there scattered intervals, or *gaps*, containing no observations? Are there unusually large or small observations, called *outliers*, which are considerably outside the general pattern of the data?

Center, spread, shape, gaps, and outliers are all aspects of the *distribution* of a data collection.

Example 1.3. Cost of Engineering Drawings Large industrial complexes require a wide variety of mechanical devices and pieces of machinery as part of their manufacturing processes. A number of factors contribute to the overall cost of each piece of machinery or mechanical device, including those associated with the preparation of engineering drawings at various stages (pre-development, production of a prototype, final specification drawings for the purchaser, etc.) of development of the product. In Table 1.3 we present the

Table 1.3 Total engineering drawing hours contributing to the cost of pieces of machinery/mechanical devices for a large Ohio-based company

3	9	11	12	14	18	26	44
6	10	12	13	15	18	26	46
6	10	12	13	16	18	30	46
6	10	12	13	16	18	30	46
7	10	12	13	16	22	34	48
7	10	12	13	16	24	36	56
9	10	12	13	18	24	36	60
9	10	12	13	18	24	36	68
9	10	12	13	18	24	36	68
9	11	12	13	18	24	36	84
9	11	12	13	18	24	36	92
9	11	12	14	18	26	36	100

total number of engineering drawing hours that contributed to the cost of 96 pieces of machinery/mechanical devices for a major Ohio-based company.

While we can simply look at the 96 ordered total hour values in Table 1.3 and get some sense of the overall distribution of the observations, it is difficult in this raw form to detect any special features or patterns that might be present in the data. However, graphical displays can be very helpful in this regard and we describe three such techniques and use them to help provide informative 'pictures' of the total hours data.

Definition 1.5 A **dotplot** of a set of quantitative data is a technique for grouping observations that are equal. The horizontal axis is the scale of the variable being measured and a dot is placed above the value of each observation. Stacking the dots vertically above the outcome represents repeated values. This form of graphical display is only useful if there are a limited number of distinct outcomes among the sample data.

One possible dotplot (with observations grouped by twos) of the total engineering drawing hours data in Table 1.3 is presented in Fig. 1.4. We can see clearly that there is a clustering of drawings that required from 10 to 20 total engineering hours per drawing. The plot has a *dominant peak* or *visual center* somewhere between 10 and 12 h. While there is certainly some

Fig. 1.4 Dotplot for total engineering drawing hours

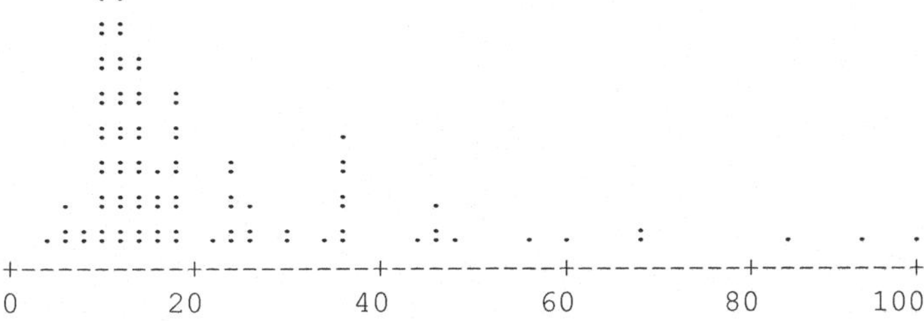

clustering of data around this visual center, many of the observations are quite far from the center. Moreover, the data configuration has a long *tail* extending to the right that includes values from 30 up to 100. We call such a configuration *skewed to the right*. Finally, there are a number of gaps in the data in the regions beyond 50 h, as well as three unusually large observations. We might call these three observations *outliers*, since they seem to fall outside the overall data pattern.

Definition 1.6 Another way to display quantitative data for which the number of observations is not too large is known as a **stemplot** (or **a stem and leaf display**). The **stem** usually corresponds to the first digit (or digits) in a number and the **leaf** then represents the final digit. To produce the stemplot for a given choice of stem and leaf, the stems are listed in a column from smallest (at the top) to the largest (at the bottom). Then the leaf for each observation is recorded to the right in the row of the display containing the observation's stem. For ease of interpretation, the leaves are also usually sorted from smallest to largest within a given stem.

The stemplot of the total engineering drawing hours data in Table 1.3 is presented in Fig. 1.5. Thus, for example, the eleven drawings that required 18 total engineering drawing hours are listed in the stemplot with a stem of 1 and leaves of 8. Similarly, the one drawing that required 100 total drawing hours is listed with a stem of 10 and a leaf of 0. As with the dotplot, the cluster of drawings requiring between 10 and 20 total engineering hours per drawing is clearly evident in the stemplot, as is the skewness of the data configuration toward the larger values. The three large outliers are perhaps a bit more evident in the stemplot since they follow the gap corresponding to the fact that there are no observations with drawing hours in the seventies.

Fig. 1.5 Stemplot of total engineering drawing hours

0	3666779999999
1	00000000111122222222222222333333333344566668888888888
2	2444444666
3	0046666666
4	46668
5	6
6	088
7	
8	4
9	2
10	0

Total Engineering Drawing Hours

Definition 1.7 The third and last graphical display that we discuss at this time is the **histogram**. It is the only one of the three presented thus far that involves an actual condensation or summary of the sample data. This is obtained by grouping the observations, something that is essential for effective graphical presentation of large data collections. Once the data are grouped, the histogram is a plot of either the numbers or relative frequencies (percentages) of observations in the grouping categories.

1.1.1 Construction of a Histogram

Here are the three steps for constructing a histogram:

Step 1: *Divide the range for the observed data values into a reasonable number of interval classes of equal width.* What constitutes a reasonable number of interval classes will become clear as you work with a given data collection. (While there is nothing to prevent histograms from being constructed with intervals of unequal width, experience has shown that it is often difficult to correctly interpret such histograms.)

Step 2: *Record the number of observations in each class, either as a straight count or as a percentage of the total number of observations in the data collection.* Thus, this step creates either a *frequency* or *relative frequency table* for the data collection and our particular choice of interval classes.

Step 3: *Graphically display the histogram.* The horizontal axis for this display corresponds to the units of measurement for our observations, divided into the interval classes specified in Step 1. Either frequency or relative frequency is plotted on the vertical axis, in the form of a bar over the corresponding interval class. The width of the base of the bar is the common width of our interval classes and the height is the class frequency or relative frequency. Unless there is an interval class which does not contain any observations (for which the vertical height would be zero), there are no gaps between adjacent bars in a histogram.

We illustrate the construction of a histogram with the total engineering drawing hours data in Table 1.3. Noting that the range of values for the data collection is from 3 to 100, we select six class intervals, each of length 20 h, centered at the values 0, 20, 40, 60, 80, and 100. While this choice for class intervals is surely arbitrary, we shall see that it leads to a reasonable condensation and summary of the drawing hours data. The frequency counts for each of these intervals for our data collection of 96 observations are then obtained from Table 1.3 to be:

Class interval	Frequency count
[−10, 10)	13
[10, 30)	61
[30, 50)	15
[50, 70)	4
[70, 90)	1
[90, 110)	2

Notice that we have constructed the class intervals so that they include the lower endpoints but not the upper endpoints. All of the drawings that took 10 h, for example, are counted in the **second** class interval, since 10 is part of this interval. We indicate this by using a bracket when the endpoint is included and a parenthesis when it is not. This provides us with a clearly defined rule for dealing with those observations with total engineering drawing hours exactly on one of the boundaries of the class intervals.

The histogram for the engineering drawing data with these class intervals is presented (courtesy of the **R** function *hist*( )) in Figure 1.6. The distribution of the engineering drawing hours data is clearly very concentrated in the interval [10, 30), its visual center, with slight concentrations in adjacent

Fig. 1.6 Histogram of total engineering drawing hours

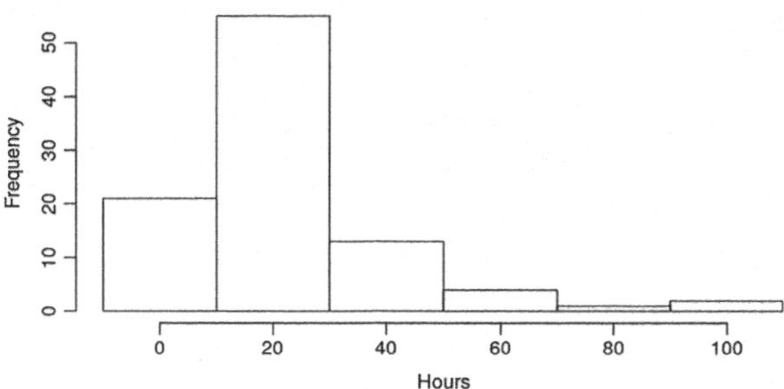

intervals. There are no complete gaps in the plot. However, there is a clear indication of heavy skewness to the right (i.e., the observations greater than the visual center are more spread out than those less than the visual center) and some evidence that at least the three observations in the final category [90, 110) are likely outliers for this data collection.

There is no hard and fast rule for the optimal number of interval classes to use in the construction of a histogram. Too few classes can lead to substantial loss of information from the data collection, while too many classes will do little to effectively summarize a large data collection. This is illustrated in Fig. 1.7 for the state-by-state percentage poverty data collection discussed in Exercise 1.1.14. Figure 1.7a corresponds to eight interval classes and provides an appropriately smooth graphical representation of the data collection. Figures 1.7b and c correspond to four and 25 interval classes, respectively, and both of these histograms provide misleading graphical representations of the data collection.

There is too much summarization of the data in (b). While the data collection is, indeed, a bit skewed to the right (as seen in (a)), it is not as badly skewed as suggested by (b). In addition, the dominant visual center for the data collection is completely masked in (b).

These two particular features of the data collection are not lost in (c) when we use a larger number of interval classes. However, the under-summarization that results from using such a large number of interval classes gives the

Fig. 1.7 Histograms of the state-by-state percentage poverty data collection. (**a**) Eight interval classes. (**b**) Four interval classes. (**c**) Twenty-five interval classes

(a)

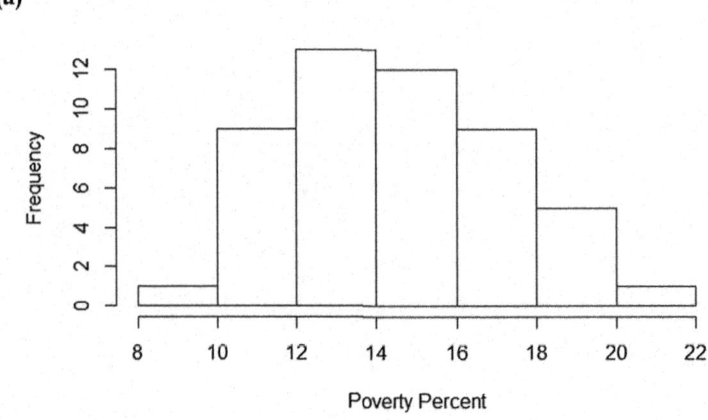

(b)

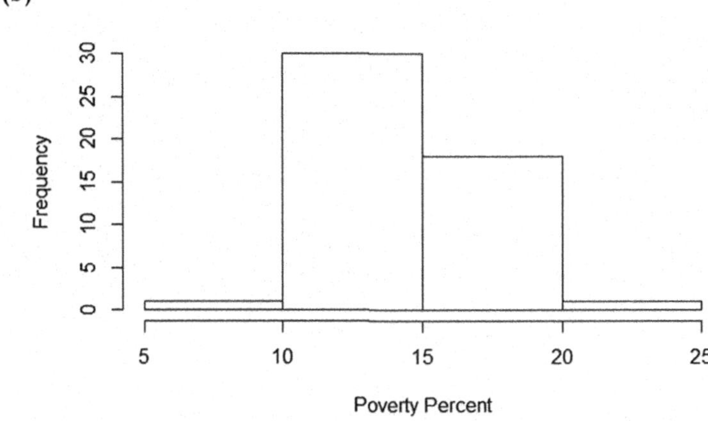

(c)

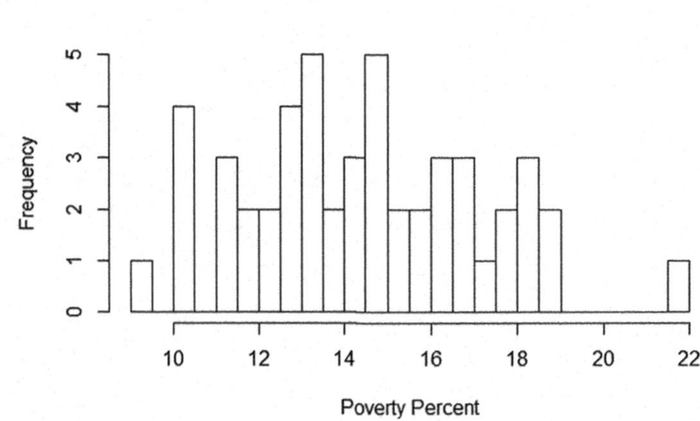

appearance that the data are relatively ragged in structure, especially to the right of the dominant visual center. This includes the apparent existence of a number of gaps in the data. This is also a misleading representation of this data collection, which is, in fact, a relatively smoothly structured set of data, as correctly displayed in (a) when eight interval classes are used.

There is nothing optimal about the particular use of eight interval classes, since using either seven or nine, for example, would yield histograms similar to the one in (a). However, it is clear that using as few as four or as many as 25 yields a histogram that distorts the true nature of this data collection. Software packages such as **R** typically allow the user to specify the number of endpoints for a histogram. When the user does not specify the endpoints, the software will use some criteria to choose the "best" endpoints. In the **R** function *hist*( ), the endpoints can be specified using the *breaks* argument.

Sometimes there are natural interval classes associated with the very nature of the data themselves. If not, it is often worthwhile to try several different choices for the number of interval classes and see which does the best at providing a good visual representation of the data. Using the **R** function *hist* ( ) makes this searching for the 'optimal' choice of number of class intervals a relatively easy task. Why don't you try it out on the total engineering drawing hours data collection (available in the dataset *engineering_drawing_hours*)?

When one of the purposes for construction of a histogram is to compare it with other histograms of related data collections, it is often better to plot relative frequencies (i. e., percentages) rather than frequencies. While the construction and general appearance of frequency and relative frequency histograms are similar for the same data collection, relative frequency histograms are more comparable across data collections, especially when the numbers of observations are quite different in the various collections.

In closing this section, we return to the idea of the general *shape* or *configuration* of a data collection. The histogram in Fig. 1.6 for the total engineering drawing hours once again clearly illustrates the fact that this data collection is more spread out to the right of its dominant visual center than it is to the left. In general, the overall configuration of a data collection is considered *skewed toward the larger values* (i.e., *skewed to the right*) if the observations greater than the dominant visual center of the data are more spread out (i.e., present a longer tail) than those less than the center. If the converse is true and the observations less than the dominant visual center are more spread out (i.e., present a longer tail) than those greater than the center, the overall configuration of the data collection is considered *skewed toward the smaller values* (i.e., *skewed to the left*). If neither of these extremes occurs so that the shape of the distribution for those observations greater than the dominant visual center of the data is similar to that for observations less than the center, we say that the overall configuration of the data collection is roughly *symmetric*. As noted, Fig. 1.6 provides a good example of a histogram for a data collection that is skewed to the right. In Figs. 1.8a and b, we present histograms that are representative of data collections that are skewed to the left and roughly symmetric, respectively.

Of course, other aspects of a data configuration can be ascertained from a well-constructed histogram. In (a)–(d) of Fig. 1.9, we present typical histograms for data collections that are, respectively, (a) heavily concentrated near a single dominant visual center, (b) considerably spread out from a single dominant visual center, (c) roughly uniform, and (d) bimodal (i.e., possessing two dominant visual centers) in shape.

Section 1.1 Practice Exercises

1.1.1. *American League Pitchers.* You have been asked to evaluate the effectiveness of pitchers in the American Baseball League. Describe at least five variables that would be relevant for such a study. Which of these variables are categorical and which are quantitative? For each of your variables, list some

Fig. 1.8 Histogram examples. (a) Histogram for a data collection that is skewed to the left. **(b)** Histogram for a data collection that is roughly symmetric

(a)

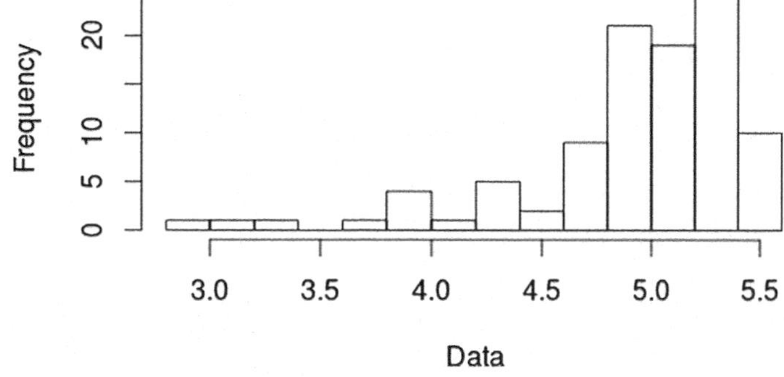

(b)

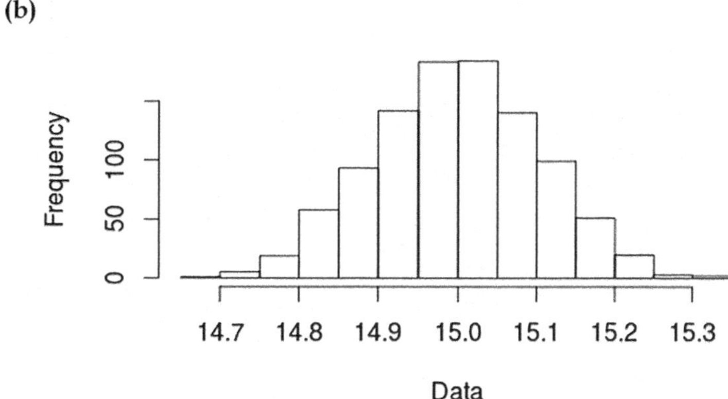

typical observations that might be obtained in an actual collection of data from American League pitchers.

1.1.2. *Eyewitnesses.* In a criminal court trial, eyewitnesses are often asked to describe the circumstances surrounding the crime that they allegedly witnessed. Describe at least five variables that would be relevant for such a description. Which of these variables are categorical and which are quantitative? For each of your variables, list some typical observations that might be part of such eyewitness testimony.

Fig. 1.9 Other typical histograms. (**a**) Histogram for a data collection that is heavily concentrated near a single dominant visual center. (**b**) Histogram for a data collection that has considerable spread around a single dominant visual center. (**c**) Histogram for a data collection that is roughly uniform. (**d**) Histogram for a data collection that has two dominant visual centers

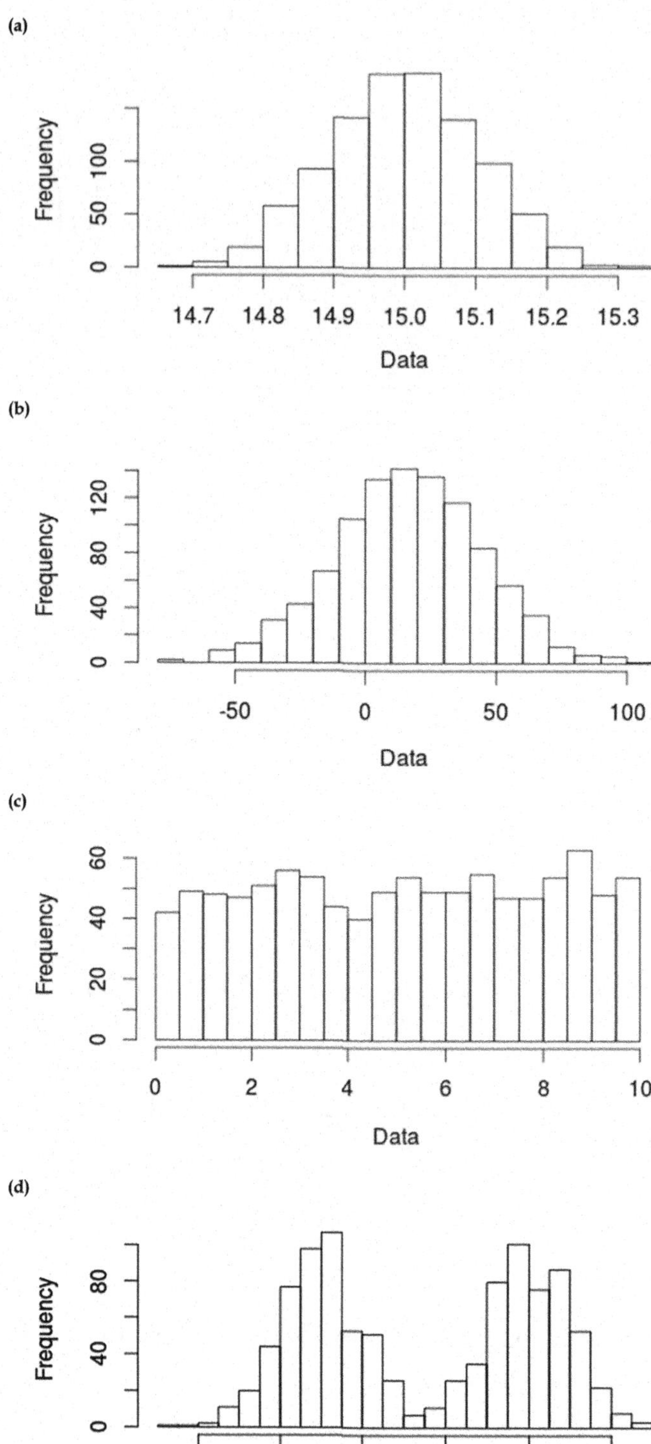

1.1.3. *National League Pitchers.*You have been asked to evaluate how well pitchers in the National Baseball League perform as hitters. What are some of the variables that would be relevant for such a study? Which of these variables are categorical and which are quantitative? For each of your variables, list some typical observations that might be obtained in an actual collection of data from National League pitchers.

1.1.4. *Best Friends.* Think of describing some of your best friends to someone who does not know them. What are some of the variables that you might use in these descriptions? Which of these variables are categorical and which are quantitative? Choose three particular best friends and list the specific observations of these variables that you would use to describe them.

1.1.5. *Course Evaluations.* You are asked at the end of each academic term to evaluate your course instructors. Describe at least five variables that you might choose to use in such an evaluation. Which of these variables are categorical and which are quantitative? List the specific observations that you would provide to describe the set of instructors that you had for your most recently completed academic term.

1.1.6. *Employment after Graduation.* After you have completed your academic career, you will likely seek full time employment in a position that utilizes what you have learned in college. Describe at least five variables that you will use in guiding your search for that ideal employment opportunity. Which of these variables are categorical and which are quantitative? List some typical observations that you might obtain in an evaluation of your first few employment opportunities.

1.1.7. *Adolescent Behavior*. In a study of adolescent sexual behavior, drug use, and violence, Turner et al. (1998) recorded the following social and demographic information for each of 1729 males ages 15–19 who were living in households in the continental United States between February and November

1995: Race-ethnicity, age, last year of school completed, parents' education levels, and marital status. What are the social and demographic variables for this study? Which of them are categorical and which are quantitative? What might be some typical observations in the data collection for the Turner *et al.* study?

1.1.8. *Adolescent Behavior.* For the Turner et al. (1998) study described in Exercise 1.1.7, describe at least three additional social and demographic variables that might be useful in studying sexual behavior, drug use, and violence in males ages 15–19 in the United States. Which of these variables are categorical and which are quantitative? What might be some typical observations for the group in the Turner *et al.* study?

1.1.9. *Public Restroom Lines.* In public places, the lines at women's restrooms are usually much longer than those at the corresponding men's restrooms. Is this due to fewer facilities available in women's restrooms or more women than men attending such events or simply that women are slower in such situations? Cornell engineering students studying toilet time behavior in airports, sports arenas, highway rest areas, and conference centers found that men averaged between 83.6 and 112.5 s in the rest rooms, while women spent between 152.5 and 180.6 s on average in the rest room. List at least five variables that might be measured in an attempt to assess why women are slower in such situations. Which of these variables are categorical and which are quantitative?

1.1.10. *Prostate Cancer.* Yoshizawa et al. (1998) investigated the conjecture that higher selenium intakes might reduce the risk of prostate cancer among men. In their study, they recorded the following information about 362 male subjects: age, body mass index (kg/m^2), prostate cancer family history, diabetes family history, smoking status (never, past, current), daily food and nutrient intake for lycopene (μg), calcium (mg), vitamin E with supplement

(International Units), saturated fat (*g*), and amount of selenium in toenails (parts per million). Which of these variables are categorical and which are quantitative?

1.1.11. *Residents in Poverty by States.* In addition to the well-known U.S. Census (dictated by the Constitution and conducted every 10 years), the United States Census Bureau also collects information annually through the American Community Survey on the population of the United States and its residents. One of these pieces of information concerns the number of individuals in each state whose annual income is below the established poverty level at the time. These state-by-state poverty level figures from 2013 are presented in Table 1.4 and in the dataset *state_poverty_levels_2013*.

(a) Using the following 29 class intervals to partition the total population data in Table 1.4, obtain frequency and relative frequency counts for the populations of the fifty states in 2013. (Notice that the last interval length is much longer than the other equal length intervals in order to capture the most populated states in a single interval and avoid a large number of empty intervals in the tabulation.)

[500,000, 1,000,000)	[1,000,000, 1,500,000)	[1,500,000, 2,000,000)
[2,000,000, 2,500,000)	[2,500,000, 3,200,000)	[3,200,000, 3,700,000)
[3,700,000, 4,200,000)	[4,200,000, 4,700,000)	[4,700,000, 5,200,000)
[5,200,000, 5,700,000)	[5,700,000, 6,200,000)	[6,200,000, 6,700,000)
[6,700,000, 7,200,000)	[7,200,000, 7,700,000)	[7,700,000, 8,200,000)
[8,200,000, 8,700,000)	[8,700,000, 9,200,000)	[9,200,000, 9,700,000)
[9,700,000, 10,200,000)	[10,200,000, 10,700,000)	[10,700,000, 11,200,000)
[11,200,000, 11,700,000)	[11,700,000, 12,200,000)	[12,200,000, 12,700,000)
[12,700,000, 13,200,000)	[13,200,000, 13,700,000)	[13,700,000, 14,200,000)
[14,200,000, 14,700,000)	[14,700,000, 15,200,000)	[15,200,000, 40,000,000)

(b) Partition the total population data into 15 equal length intervals (closed on the left and open on the right, as in part (a)) and once again obtain the frequency and relative frequency counts for the

Table 1.4 State population and poverty numbers in 2013

State	Total population[a]	Number at poverty level[b]
Alabama	4,833,996	883,371
Alaska	737,259	67,016
Arizona	6,634,997	1,206,460
Arkansas	2,958,765	565,469
California	38,431,393	6,328,824
Colorado	5,272,086	667,446
Connecticut	3,599,341	373,900
Delaware	925,240	111,327
Florida	19,600,311	3,253,333
Georgia	9,994,759	1,843,768
Hawaii	1,408,987	148,368
Idaho	1,612,843	246,550
Illinois	12,890,552	1,845,393
Indiana	6,570,713	1,015,127
Iowa	3,092,341	379,127
Kansas	2,895,801	393,358
Kentucky	4,399,583	800,635
Louisiana	4,629,284	888,019
Maine	1,328,702	180,639
Maryland	5,938,737	585,571
Massachusetts	6,708,874	770,513
Michigan	9,898,193	1,648,436
Minnesota	5,422,060	592,422
Mississippi	2,992,206	695,915
Missouri	6,044,917	931,066
Montana	1,014,864	163,637
Nebraska	1,868,969	239,433
Nevada	2,791,494	433,576
New Hampshire	1,322,616	111,495
New Jersey	8,911,502	998,549
New Mexico	2,086,895	448,461
New York	19,695,680	3,055,645
North Carolina	9,848,917	1,715,397
North Dakota	723,857	82,398
Ohio	11,572,005	1,796,942
Oklahoma	3,853,118	626,906
Oregon	3,928,068	642,138
Pennsylvania	12,781,296	1,690,405
Rhode Island	1,053,354	144,446

(continued)

Table 1.4 (continued)

State	Total population[a]	Number at poverty level[b]
South Carolina	4,771,929	860,380
South Dakota	845,510	115,454
Tennessee	6,497,269	1,126,772
Texas	26,505,637	4,530,039
Utah	2,902,787	361,181
Vermont	626,855	74,058
Virginia	8,270,345	938,733
Washington	6,973,742	967,282
West Virginia	1,853,595	332,347
Wisconsin	5,742,953	755,551
Wyoming	583,223	62,039

[a]*Source*: United States Census Bureau, Population Division (2014b)
[b]*Source*: United States Census Bureau, American Community Surveys (2014a)

populations of the fifty states in 2013. (If you wish, you may use a longer interval to capture the most populous states as is done in part (a).)

(c) Which of the two partitions in (a) and (b) do you feel provides a better summary of the relevant total population information in Table 1.4? What are your reasons for this preference?

1.1.12. *Residents in Poverty by States.* Consider the 2013 state-by-state poverty level numbers, as presented in Table 1.4. Select an appropriate partition and obtain frequency and relative frequency counts for the 2013 poverty level numbers in the 50 states.

1.1.13. *Residents in Poverty by States.* Consider the 2013 state-by-state poverty level numbers, as presented in Table 1.4.

(a) Compute the percentage of the population that had incomes below the poverty level in 2013 for each of the 50 states.

(b) Select an appropriate partition of the 2013 percentage poverty level numbers and obtain frequency and relative frequency counts for the fifty states.

(c) Compare and contrast the information summaries provided by the two sets of frequency counts discussed in part (b) and Exercise 1.1.12. Which do you prefer and why?

1.1.14. *Women and Poverty.* As dictated by the Constitution, every 10 years the United States Census Bureau conducts a census of the population of the United States and compiles information on its residents. One of these pieces of information concerns the percentage of individuals in each state whose annual income is below the established poverty level at the time. These state-by-state poverty level figures from the 2010 Census for women residents are presented in Table 1.5.

Table 1.5 State percentage poverty rates for women from the 2010 census

State	Percentage at poverty level
Alabama	18.7
Alaska	10.4
Arizona	16.2
Arkansas	18.1
California	15.0
Colorado	13.1
Connecticut	10.4
Delaware	11.3
Florida	15.8
Georgia	17.3
Hawaii	11.4
Idaho	16.0
Illinois	13.5
Indiana	14.8
Iowa	12.8
Kansas	13.8

(continued)

Table 1.5 (continued)

State	Percentage at poverty level
Kentucky	18.5
Louisiana	18.7
Maine	12.8
Maryland	10.1
Massachusetts	12.0
Michigan	16.2
Minnesota	11.7
Mississippi	21.6
Missouri	14.8
Montana	14.1
Nebraska	12.5
Nevada	14.3
New Hampshire	9.2
New Jersey	10.3
New Mexico	18.4
New York	14.8
North Carolina	16.8
North Dakota	14.3
Ohio	15.2
Oklahoma	16.1
Oregon	15.2
Pennsylvania	13.3
Rhode Island	13.6
South Carolina	17.6
South Dakota	15.0
Tennessee	16.8
Texas	16.9
Utah	13.1
Vermont	12.7
Virginia	11.5
Washington	13.4
West Virginia	17.8
Wisconsin	12.8
Wyoming	12.4

Source: National Women's Law Center (2011)

(a) Select an appropriate partition of these 2010 percentage poverty levels for women and obtain frequency and relative frequency counts for the 50 states.

(b) Compare and contrast the frequency and relative frequency counts obtained in part (a) for women in 2010 with the corresponding frequency and relative frequency counts for percentage poverty levels for all residents in 2013 obtained in part (b) of Exercise 1.1.13.

1.1.15. *Participation in Leisure Activities.* As part of its data collection process, the U.S. Census Bureau periodically reports the level of adult participation in selected leisure activities. Part of this collected information for Autumn 2010 is presented in Table 1.6.

Table 1.6 Frequency of adult participation in selected leisure activities in Autumn 2010

Activity	Two or more times a week	Once a week	Two or three times a month	Once a month
Adult education	3116	1973	762	1312
Attend auto shows	313	337	557	721
Attend art galleries/shows	78	215	879	2272
Attend classical music/ opera	99	65	409	900
Attend country music	67	125	239	458
Attend dance performances	122	162	335	403
Attend horse races	159	177	155	379
Attend other music	135	332	1120	2129
Attend rock music	187	173	730	1136
Backgammon	435	366	416	486
Baking	10394	8482	12482	9321
Barbecuing	12497	12939	18871	10473
Billiards/pool	975	1432	2125	2063
Bird watching	6101	1338	1169	876
Board games	2890	3134	6574	7759
Book clubs	285	234	419	2732
Chess	549	533	823	576

(continued)

Table 1.6 (continued)

Activity	Two or more times a week	Once a week	Two or three times a month	Once a month
Concerts on radio	1308	747	548	572
Cooking for fun	19162	7495	6795	4415
Crossword puzzles	12866	3136	2811	2674
Dance/go dancing	1636	2162	2728	2964
Dining out	20158	25173	26644	15686
Entertain at home	6976	9139	18565	19611
Fantasy sports league	2855	1559	372	330
Furniture refinishing	201	79	359	406
Go to bars/night clubs	3133	4846	7428	6430
Go to beach	3303	2018	4875	5428
Go to live theater	333	256	896	3331
Go to museums	121	198	1171	3317
Home decoration	890	977	1861	4178
Karaoke	460	401	665	904
Painting/drawing	2360	1288	1625	1609
Photo albums/scrap books	1237	743	1973	2332
Photography	4358	3310	5332	3508
Picnic	281	591	1672	3780
Play bingo	754	1095	811	1342
Play cards	5679	4969	6400	7567
Play musical instrument	7435	2096	1959	1211
Reading books	47483	8298	7513	6312
Reading comic books	1161	636	886	527
Sudoku puzzles	10265	2505	3159	2495
Trivia games	1891	1327	1397	1490
Woodworking	1714	965	1631	1443
Word games	7768	2709	2817	1899
Zoo attendance	189	239	632	2112

Source: United States Census Bureau (2012); GfK Mediamark Research & Intelligence (2010)
Entries are the numbers of adults (in thousands) participating in each frequency category

(a) Select an appropriate partition of the numbers of participants data and obtain frequency and relative frequency counts separately for each of the four frequencies of participation categories. Do the various activities appear in the same sections of the partition for the four frequencies of participation categories? Discuss your findings.
(b) Construct histograms for the numbers of participants separately in each of the four frequencies of participation categories. Compare and contrast the histograms. Are there any interesting features of these data collections?

1.1.16. *Bird Variety.* In a study designed to determine whether relationships exist between the numbers and types of bird species found at points along a river and the structure of the immediate surrounding forest, Groom (1999) recorded breeding-bird count data for riparian habitat along the Big and Little Darby Creeks in central Ohio. The data in Table 1.7 are the numbers of bird species detected in each of 39 distinct 5-min time intervals over the month of June, 1998.

(a) Use appropriate computer software (e.g., the **R** function *barplot*( )) to construct a labeled bar graph for this data collection.
(b) Comment on any unusual features of the data collection that are evident in this bar graph.

Table 1.7 Numbers of bird species detected in 5-min periods during June, 1998 at various riparian habitats along the Big and Little Darby Creeks in central Ohio

9	7	11	9	9	13	17	15	10	12	12	17
14	17	12	15	6	11	14	10	12	8	17	16
11	15	10	14	13	12	8	12	16	12	10	14
10	5	7									

Source: Groom (1999)

1.1.17. *Bird Variety.* Consider the bird species data presented in Table 1.7. Make a stemplot of these data.

1.1.18. *Bird Variety.* Consider the bird species data presented in Table 1.7. Make a dotplot of these data.

1.1.19. *Participation in Leisure Activities.* As part of its data collection process, the U.S. Census Bureau periodically reports the level of adult participation in selected leisure activities. Part of this collected information for Autumn 2010 is presented in Table 1.8.

Table 1.8 Frequency of adult participation in leisure activity in Autumn 2010

Activity	Participated at least once in the last 12 months
Adult education	16,640
Attend auto shows	19,346
Attend art galleries/shows	20,985
Attend classical music/opera	9715
Attend country music	11,266
Attend dance performances	10,010
Attend horse races	6654
Attend other music	26,536
Attend rock music	25,176
Backgammon	4234
Baking	57,703
Barbecuing	79,119
Billiards/pool	19,468
Bird watching	13,793
Board games	37,993
Book clubs	5747
Chess	6896
Concerts on radio	6441
Cooking for fun	50,243
Crossword puzzles	29,996
Dance/go dancing	20,995
Dining out	112,477

(continued)

Table 1.8 (continued)

Activity	**Participated at least once in the last 12 months**
Entertain at home	87,455
Fantasy sports league	8969
Furniture refinishing	6292
Go to bars/night clubs	43,513
Go to beach	58,670
Go to live theater	30,547
Go to museums	32,960
Home decoration	22,781
Karaoke	8186
Painting/drawing	13,791
Photo albums/scrap books	15,284
Photography	26,173
Picnic	26,321
Play bingo	10,271
Play cards	46,190
Play musical instrument	18,078
Reading books	86,540
Reading comic books	5557
Sudoku puzzles	26,540
Trivia games	11,872
Woodworking	10,202
Word games	22,147
Zoo attendance	28,148

Source: United States Census Bureau (2012); GfK Mediamark Research & Intelligence (2010)
Entries for the activity categories are the numbers of adults (in thousands) participating in the activity at least once in the past 12 months

(a) Make a stemplot and a dotplot for this collection of adult leisure activity participation data.

(b) Select an appropriate partition of the numbers of participants data and obtain frequency and relative frequency counts. Discuss your findings.

(c) Construct a histogram for the adult leisure activity participation data in Table 1.8. Are there any interesting features of the histogram?

(d) In Exercise 1.15 you were asked to construct histograms for adult leisure activity participation data broken into subcategories corresponding to frequency of participation. Compare and contrast the four histograms obtained in Exercise 1.15 with the overall activity histogram obtained in part (c) of this exercise. Are there any interesting similarities or differences?

1.1.20. *Firearms in the Home.* State health departments collaborate with the US Centers for Disease Control and Prevention to operate the Behavioral Risk Factor Surveillance System (BRFSS). Firearm storage questions were included in the 2002 BRFSS survey interviews. One of the questions was: "Are any firearms now kept in or around your home? Include those kept in a garage, outdoor storage area, car, truck, or other motor vehicle." The percentages of households (self reported respondents) with firearms for each of the fifty states at the time of the 2002 BRFSS survey are given in Table 1.9, as tabulated and discussed (among other topics) in Okoro et al. (2005).

Table 1.9 Percentages of state households that had firearms in or around their homes from the 2002 BRFSS survey interviews

State	Percentage of households with firearms
Alabama	57.2
Alaska	60.6
Arizona	36.2
Arkansas	58.3
California	19.5
Colorado	34.5
Connecticut	16.2
Delaware	26.7
Florida	26.0
Georgia	41.0
Hawaii	9.7
Idaho	56.8
Illinois	19.7

(continued)

Table 1.9 (continued)

State	Percentage of households with firearms
Indiana	39.0
Iowa	44.0
Kansas	43.7
Kentucky	48.0
Louisiana	45.6
Maine	41.1
Maryland	22.1
Massachusetts	12.8
Michigan	40.3
Minnesota	44.7
Mississippi	54.3
Missouri	45.4
Montana	61.4
Nebraska	42.1
Nevada	31.5
New Hampshire	30.5
New Jersey	11.3
New Mexico	39.6
New York	18.1
North Carolina	40.8
North Dakota	54.3
Ohio	32.1
Oklahoma	44.6
Oregon	39.8
Pennsylvania	36.5
Rhode Island	13.3
South Carolina	45.0
South Dakota	59.9
Tennessee	46.4
Texas	35.9
Utah	45.3
Vermont	45.5
Virginia	35.9
Washington	36.2
West Virginia	57.9
Wisconsin	44.3
Wyoming	62.8

Source: Centers for Disease Control and Prevention (2003), as tabulated and discussed in Okoro et al. (2005)

(a) Make a histogram of this data collection and describe any unusual features of the data that are evident in the histogram.

(b) Make a dotplot and stemplot for this data collection. Which do you prefer (histogram or dotplot/stemplot) for displaying the important features of these data and why?

1.1.21. *Global Warming—Natural or Man-Made?* CBS News/New York Times (2014) conducted a national poll between September 10 – September 14, 2014 in which they asked the following question of respondents: "Which statement comes closest to your view about global warming? Global warming is caused mostly by human activity such as burning fossil fuels. Global warming is caused mostly by natural patterns in the earth's environment. OR, Global warming does not exist." The results of this poll for Democrats, Republicans, and Independents are as follows:

	Mostly human activity	**Mostly natural patterns**	**It does not exist**	**Caused by both**	**Don't know what causes it/Unsure**
Republicans	35	42	18	4	1
Democrats	67	27	3	2	1
Independents	55	29	10	4	3

Using appropriate software, construct pie charts to graphically display these poll results separately for Republicans, Democrats, and Independents.

1.1.22. *Origin and Development of Humans.* The Pew Research Center (2009) conducted a poll that asked for respondents' views about the origins and development of living things. They were asked to select which of the following categories best reflected their views: (a) Humans and other living things have evolved over time through a natural selection process; (b) Humans and other living things have evolved over time with Supreme guidance; (c) Humans and other living things have existed in their present from since the beginning of time; or (d) Unsure/Don't know. The percentages for respondents in different age categories are as follows:

	Evolved over time			
Age range	**Natural selection process**	**Supreme guidance**	**Existed in present form from onset of life**	**Unsure/Don't know**
18–29	40	21	26	13
30–49	35	22	30	13
50–64	30	23	34	13
65+	23	19	35	23.

Using appropriate software, construct pie charts to graphically display these poll results for the four age categories. Why do you think there are such large percentages of Unsure/Don't know responses in all four age categories? Why so much larger for the age category 65+?

1.1.23. *Where is it Safest to Work?* The Bureau of Labor Statistics in the U.S. Department of Labor annually reports data on unintentional fatalities from accidents at work by industry. These data for accidental fatalities in 2013 are presented in Table 1.10.

(a) Use an appropriate graphical method to effectively display the raw number of fatalities by occupation/industry category.
(b) Use an appropriate graphical method to effectively display the number of fatalities per million hours worked by occupation/industry category.
(c) Compute the percentage of the total number of unintentional fatalities from accidents at work in 2013 for each of the occupation/industry categories. Use an appropriate graphical method to effectively display these percentages.
(d) Compute the percentages of total hours worked in 2013 for each of the occupation/industry categories and use an appropriate graphical method to effectively display these percentages.

Table 1.10 Unintentional fatalities from accidents at work, by industry, 2013

Occupation/Industry	Number of fatalities	Total hours worked (millions)
Management/Professional	638	106,269
Service	624	43,329
Sales/Office	281	59,770
Construction/Maintenance	1399	25,679
Production/Transportation/Material Moving	1394	33,122
Goods Producing	1733	53,929
Natural Resources	633	6746
Agriculture/Forestry Fishing/Hunting	479	4238
Mining/Quarrying/Oil and Gas Extraction	154	2508
Construction	796	16,972
Manufacturing	304	30,211
Wholesale Trade	190	7484
Retail Trade	253	27,936
Transportation/Warehousing	687	10,477
Utilities	23	1802
Information	39	5489
Financial Activities	84	18,889
Professional/Business Services	408	31,046
Educational/Health Services	131	39,936
Leisure/Hospitality	202	21,514
Other Services	179	12,429
Government	476	37,095

Source: United States Department of Labor (2014).

(e) Comment on the two graphical displays constructed in parts (c) and (d). Are there any problems with interpreting your results in part (c) without knowledge of the information obtained in part (d)? Can you think of a way to combine the information obtained in parts (c) and (d) to alleviate this problem?

1.1.24. *Sports Injuries.* In Table 1.11 we present estimates from the National Safety Council (2014), as reported by the Insurance Information Institute

Table 1.11 Estimated numbers of hospital-treated injuries sustained in 2012 during participation in sports

Sport	Number of injuries	Sport	Number of injuries
Archery	6055	Mountain biking	9176
Baseball	159,220	Mountain climbing	4446
Basketball	569,746	Racquetball/squash/ paddleball	5601
Bicycle riding	547,499	Roller skating	62,906
Billiards, pool	4983	Rugby	15,270
Bowling	18,685	Scuba diving	1437
Boxing	20,203	Skateboarding	114,120
Cheerleading	39,153	Snowboarding	38,805
Exercise	364,137	Snowmobiling	5633
Fishing	72,629	Soccer	231,447
Football	466,492	Softball	106,490
Golf	36,308	Swimming	213,464
Gymnastics	30,600	Tennis	24,224
Hockey (street/roller/field)	8243	Track and field	29,679
Horseback riding	66,543	Volleyball	61,495
Horseshoe pitching	1898	Waterskiing	7577
Ice hockey	18,962	Weight lifting	100,300
Ice skating	20,873	Wrestling	45,646
Martial arts	36,065		

Source: National Safety Council (2014), as reported in Insurance Information Institute (2015)

(2015), on the numbers of hospital-treated injuries sustained during participation in a variety of sports in 2012. Use a pie chart to graphically display these injuries data.

1.1.25. *Participation in Sports.* In Table 1.12 we present estimates provided by the U.S. Census Bureau (United States Census Bureau 2012) for the numbers of participants in a variety of sports in 2009. Consider those sports for which both the estimated sports participation data in Table 1.12 and the estimated hospital-treated sports injuries data in Table 1.11 are available. Use an appropriate graphical representation to effectively compare the relative safeties of participation in these sports.

Table 1.12 Estimated numbers of participants in a variety of sports, 2009

Sport	Number of participants	Sport	Number of participants
Archery	7,106,000	Paintball games	6,271,000
Baseball	11,507,000	Roller skating	7,874,000
Basketball	24,410,000	Skateboarding	8,418,000
Bicycle riding	38,139,000	Skiing	8,687,000
Billiards, pool	28,172,000	Snowboarding	6,189,000
Boating	23,959,000	Soccer	13,578,000
Bowling	44,972,000	Softball	11,829,000
Exercise	254,235,000	Swimming	50,226,000
Fishing	70,067,000	Table Tennis/Ping Pong	13,306,000
Football	8,890,000	Tennis	10,818,000
Golf	22,317,000	Volleyball	10,733,000
Hunting	25,003,000	Waterskiing	5,191,000
Ice hockey	3,057,000	Weightlifting	34,505,000
Mountain biking	8,368,000	Yoga	15,738,000

Source: United States Census Bureau (2012)

1.1.26. *Artifacts from Naco Valley, Honduras.* In their 1996 study of Site 128 in the Naco Valley, Honduras (see Example 1.1 for more details), Ed Schortman and Pat Urban, Professors in the Department of Anthropology and Sociology at Kenyon College in Gambier, Ohio, also collected information on the percentage of objects retrieved from individual excavated structures that were considered to be "elaborate imports". These "elaborate import" percentages for thirteen of the excavated structures at Site 128 in the Naco Valley are given in Table 1.13. Construct a stemplot for this data collection. What are some of the important features of the data?

1.1.27. *Artifacts from Naco Valley, Honduras.* In Example 1.1, we discussed the makeup of the artifacts found by Professors Schortman and Urban from Kenyon College for two different structures at Site 128 in Naco Valley, Honduras in 1996. In Table 1.14 we present similar data for Structures 4 and 23 at the same site.

Table 1.13 Percentage elaborate imports retrieved at selected excavation structures in Naco Valley, Honduras, 1996

Excavation structure number	Percentage elaborate imports
3	2.93%
4	4.79%
7	3.21%
12	2.00%
13	2.60%
17	4.36%
18	3.14%
19	5.63%
20	5.57%
21	4.00%
23	2.68%
24	3.42%
25	3.88%

Source: Schortman and Urban (1998)

Table 1.14 Artifacts recovered at structures 4 and 23 during the excavation of Site 128 in Naco Valley, Honduras, 1996

Artifact	Frequency count at structure #4	Frequency count at structure #23
1. Ocarinas	14	21
2. Figurines	56	53
3. Incensarios	73	35
4. Ground stone tools	21	12
5. Candelaros	10	12
6. Stamps	5	9
7. Sherd disks	5	0
8. Jewelry	4	7

Source: Schortman and Urban (1998)

Table 1.15 Math SAT scores for seniors graduating in 2013 or 2014 from a small private school

Males						Females				
660	510	680	750	510	570	460	400	470	520	740
480	390	680	660	610	430	360	620	550	470	530
700	350	490	400	450	650	580	490	570	350	500
600	510	600	460	570	730	580	560	650	400	390
570	530	650	500	520	670	600	670	570	680	590
600	520	710	550	550	400	510	510	440	520	410
630	650	590	630	570	720	480	440	640	510	500
500	540	440	460	490	590	490	490	620	660	740
740	520	710	580	620	590	670	600	430	570	470
640	730	670	470	480	650	510	680	570	390	570
550	590	680	630	550	500					
520	570	500	460	560	470					
490	460	530	510	700	540					
390										

Source: Depew (1999)

(a) Use appropriate computer software (e.g., the **R** function *barplot*( )) to construct labeled bar graphs for the data collections at the two structures.

(b) Comment on any unusual features of these data collections that are evident in these bar graphs.

1.1.28. *Gender and Math SAT Scores.* Math SAT scores were collected by Depew (1999) from seniors graduating in 2013 or 2014 from a small private school. These scores are presented in Table 1.15, separately for males and females. Make stemplots and dotplots separately for the collections of male and female SAT scores in Table 1.15.

1.1.29. *SAT Scores for Males.* Consider the SAT scores in Table 1.15 for the male students. Construct four separate histograms for these data, using five, ten, fifteen, and twenty equal length interval classes. Which of the four do you feel best depicts the important features of the data collection and why?

1.1.30. *Firearms in the Home.* Consider the percentages of state households that had firearms in 2002 as given in Table 1.9. Construct four separate histograms for these data, using four, eight, twelve, and sixteen equal length interval classes. Which of the four do you feel best represents the important features of the data collection and why?

1.1.31. *Histogram Shapes.* Pictured below are four histograms with a variety of shapes. Which of the following adjectives would you use to describe each of them: symmetric, skewed to the left, skewed to the right, single dominant visual center, more than one dominant visual center, uninformative, gappy, containing outliers, low spread, high spread?

(a)

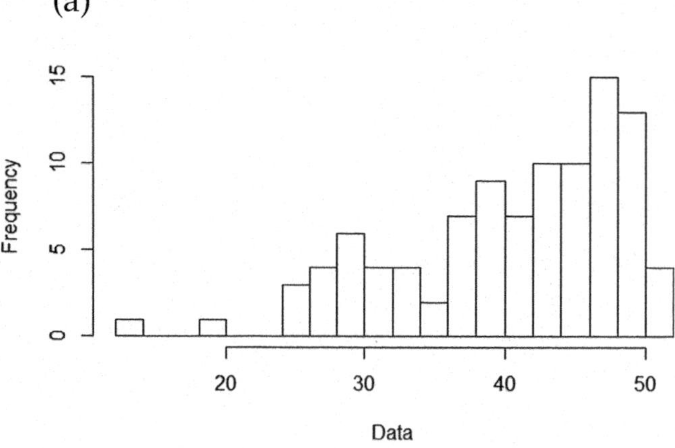

(b)

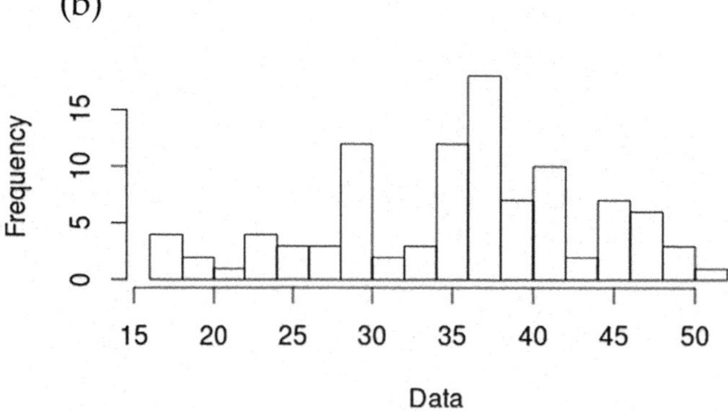

(c)

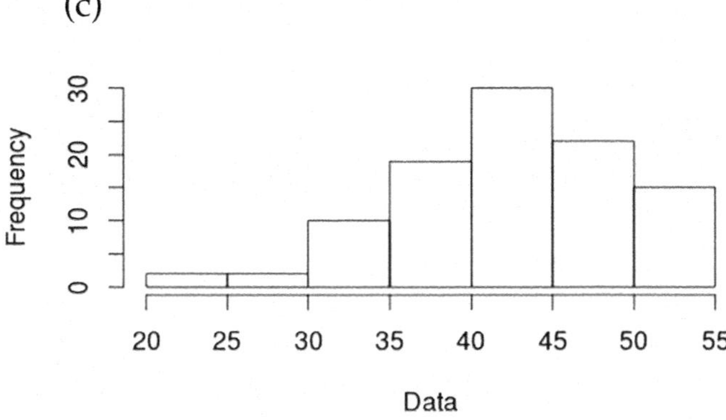

(d)

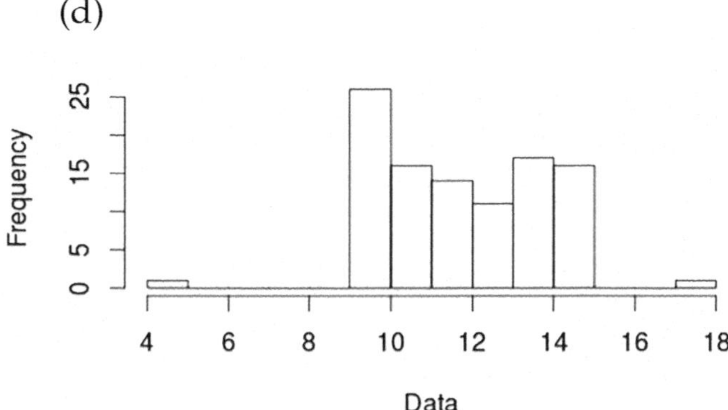

1.1.32. *More Histogram Shapes.* Pictured below are four histograms with a variety of shapes. Which of the following adjectives would you use to describe each of them: symmetric, skewed to the left, skewed to the right, single dominant visual center, more than one dominant visual center, uninformative, gappy, containing outliers, low spread, high spread?

(a)

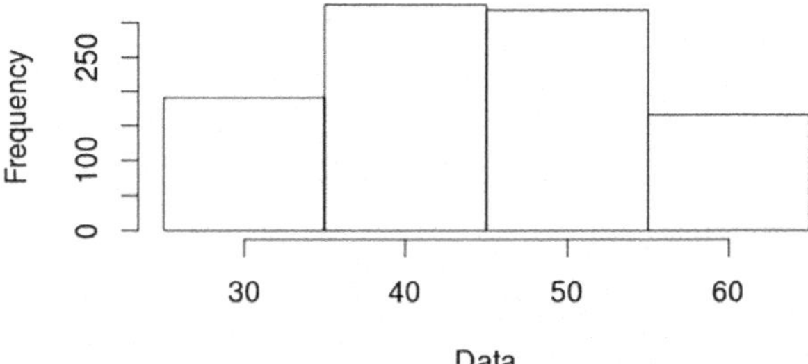

(b)

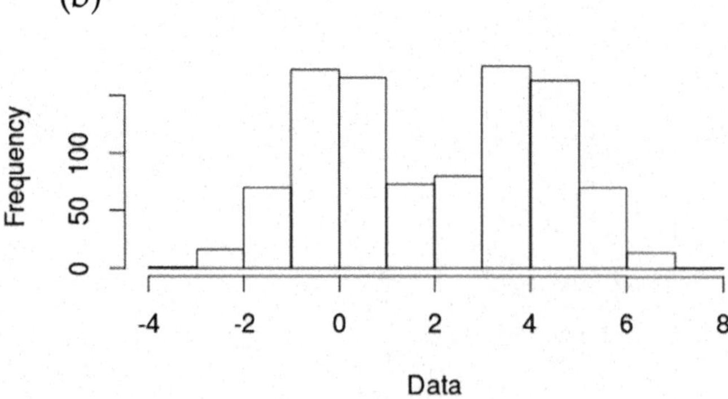

(c)

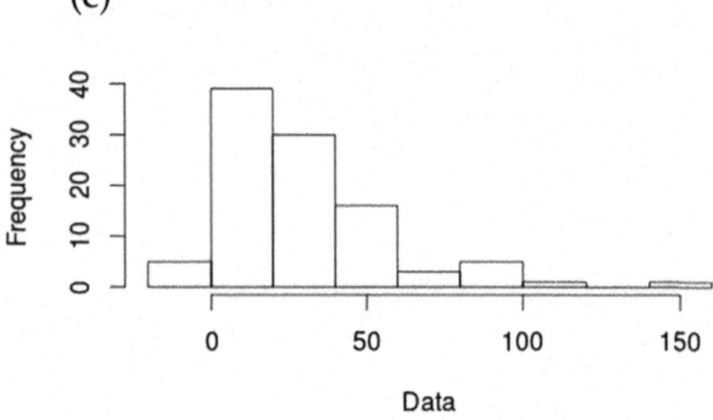

(d)

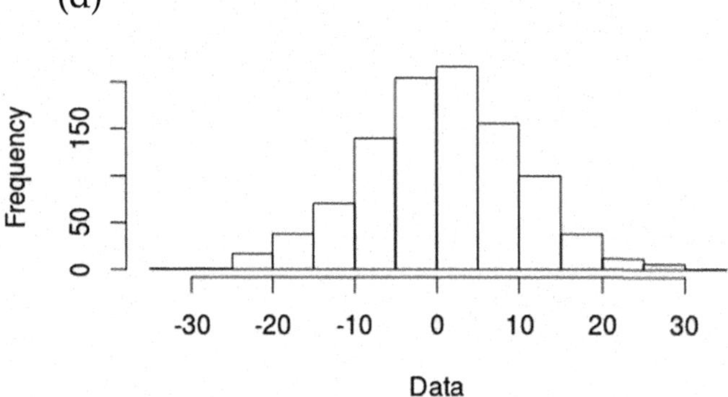

1.2 Numerically Summarizing One-Variable Data Collections

While each of the three graphical techniques discussed in Section 1.1 is useful in capturing a general sense of the important features in a one-variable data collection, none of them is designed to provide us with more numerically descriptive information about the data. This is especially true and troublesome if the data collection is rather large. We need better ways of *condensing* or summarizing data collections so that we can interpret and compare them more effectively. This often requires a number of different numerical data summaries.

Definition 1.8 A **statistic** S is a number computed from the observations in a data collection.

A statistic is a function solely of the observations in a data collection, so that given only the observations the value of the statistic can be determined. Hence, a statistic is by its very nature a summary of some attribute of the data collection.

It is easy to construct statistics S that can be computed for a given data collection. For example, S_1 = maximum observed value, S_2 = minimum observed value, S_3 = number of observations greater than 7.3, S_4 = sum of the squares of the observations, and S_5 = the number of observations between 3 and 9, inclusive, are all well defined statistics – the list could go on and on. (Can you think of five different statistics?) However, not all such statistics provide useful summaries for a data collection. In this section we concentrate on examples of statistics that can be used to numerically quantify the same important features of the data collection that we illustrated previously using graphical techniques; that is, the visual center of the data collection, spread of the observations around the visual center, shape of the data collection (i.e., symmetry versus skewness) and the presence of outliers and gaps in the data.

Measuring the Visual Center of the Data Collection: Mean, Median, and Trimmed Mean Each individual observation in a data collection provides information about the visual center of the data collection; that is, we have some (limited!) information about the center even if we have collected only a single observation. Moreover, it is quite natural to take the position that each observation contributes equally to our sample information about the visual center. Thus, we want a statistic that uses all of the observations and doesn't depend on the order in which they were collected. However, which particular statistic of this type we should use to measure the visual center of a data collection depends both on what we mean by the 'center' and on the observed data configuration.

Mean of the Observations If we interpret the visual center of a data collection to be the balance point where data values larger than the center are equally balanced by those that are smaller than the center, the numerical average or *mean* is a natural statistic for identifying and measuring the center. This is the case, for example, when our interpretation of the 'visual center' corresponds to a value for which the numerical contribution from data points that are greater than the 'center' is equally balanced by the numerical contribution from those that are less than it. In such settings, the appropriate statistic to measure this 'visual center' is naturally the average, or mean, of the collected observations.

Definition 1.9 The **mean** of the observations in a data collection is their numerical average. That is, the sum of the data values divided by the number of observations in the data collection. So, if $x_1, x_2, \ldots, x_n$ are the *n* observations in our data collection, their mean is

$$\bar{x} = \frac{1}{n}(x_1 + x_2 + \ldots + x_n).$$

A very convenient way to write the arithmetic average described in Definition 1.9 is with the **summation symbol Σ**. You will see it used throughout this book and you should become comfortable with this common notation. Using Σ, we can write the mean of our data collection as

$$\bar{x} = \frac{1}{n}\sum_{i=1}^{n} x_i. \tag{1.1}$$

You read this as follows: "x bar equals 1 over n times the sum of x sub i, for i equaling 1 to n". The expression says, "to find the mean, add up all the data values $x_1, x_2, \ldots, x_n$, and divide by n".

Example 1.4. Motor-Vehicle Fatalities in 2012 In their annual report on accident statistics, the National Highway Traffic Safety Administration (2013) reported figures for the number of motor-vehicle fatalities per 100 million vehicle miles traveled in each of the 50 states during the calendar year 2012. These figures are reproduced in Table 1.16.

Here we have $n = 50$ observations in our data collection and these observations can be represented by using the notation $x_1 = 1.33$ (Alabama), $x_2 = 1.23$ (Alaska), …, $x_{50} = 1.33$ (Wyoming). The national average of these statewide fatality rates is then

$$\begin{aligned}\bar{x} &= \frac{1}{50}\sum_{i-1}^{50} x_i = \frac{1}{50}[1.33 + 1.23 + 1.37 + \ldots + 1.04 + 1.33] \\ &= \frac{59.03}{50} = 1.18 \text{ fatalities per 100 million vehicles miles.}\end{aligned}$$

We note that this national average is relatively close to most of the statewide averages displayed in Table 1.16. However, there are a number of states that have motor-vehicle fatality rates that are considerably higher or lower than this national average; that is, some of the observations are rather widely dispersed from the national average. Are they to be viewed as 'unusually' high or low; that is, are they possibly data outliers? Are there reasons for these outcomes? We return to these questions later in this chapter.

Table 1.16 Total motor-vehicle fatalities per 100 million vehicle miles traveled in the 2012 calendar year

State	Fatality rate	State	Fatality rate
Alabama	1.33	Montana	1.72
Alaska	1.23	Nebraska	1.10
Arizona	1.37	Nevada	1.07
Arkansas	1.65	New Hampshire	0.84
California	0.88	New Jersey	0.79
Colorado	1.01	New Mexico	1.43
Connecticut	0.75	New York	0.91
Delaware	1.24	North Carolina	1.23
Florida	1.27	North Dakota	1.69
Georgia	1.11	Ohio	1.00
Hawaii	1.25	Oklahoma	1.48
Idaho	1.13	Oregon	1.01
Illinois	0.91	Pennsylvania	1.32
Indiana	0.99	Rhode Island	0.82
Iowa	1.16	South Carolina	1.76
Kansas	1.32	South Dakota	1.46
Kentucky	1.58	Tennessee	1.42
Louisiana	1.54	Texas	1.43
Maine	1.16	Utah	0.82
Maryland	0.89	Vermont	1.07
Massachusetts	0.62	Virginia	0.96
Michigan	0.99	Washington	0.78
Minnesota	0.69	West Virginia	1.76
Mississippi	1.51	Wisconsin	1.04
Missouri	1.21	Wyoming	1.33

Source: National Highway Traffic Safety Administration (2013)

For many data collections, the mean adequately locates the dominant visual center of the data. However, it is also clear that the mean, being a numerical average of all the observations, will be sensitive to either unusually large or unusually small observations in the collection of data, especially if the total number of observations is not large. For example, consider the data collection {1.9, 2.5, 3.6, 3.8, 18.2}. The mean for this data collection is $\bar{x} = \frac{(1.9 + 2.5 + 3.6 + 3.8 + 18.2\}}{5} = 6$, which is considerably larger than four of the five observations. The outlier 18.2 exerts a tremendous pull on the mean, moving it substantially away from the other four, more typical, data values. Thus, in the presence of such outliers the mean will not, in fact, provide a good representation for the dominant visual center or the typical value of a set of data.

Median of the Observations A measure of the center that is less sensitive to unusually large or small observations than the mean is provided by the median.

Definition 1.10 To find the **median,** $\tilde{x}$, of a data collection $x_1, \ldots, x_n$:

1. Sort the n data values in order from smallest to largest.
2. If n is odd, the median, $\tilde{x}$, is the single value in the middle of this ordered list.
   If n is even, there are two "middle values", and the median, $\tilde{x}$, is their average.

For example, to find the median of the data collection 2, 4, −1, 6, 5.1, we first sort the data values from smallest to largest:

$$\text{Sorted data values :} \quad -1, 2, 4, 5.1, 6.$$

Since there are $n = 5$ values in this data collection and 5 is odd, the median, $\tilde{x}$, is the middle number in this ordered list, namely,

$$-1, 2, \mathbf{4}, 5.1, 6$$
$$\uparrow$$

That is, the median for this data collection is $\tilde{x} = 4$.

To find the median for the data collection $-3.7, 8, -14, 6.3, 3, 6.3$, we again sort the data values from smallest to largest:

$$\text{Sorted data values}: \quad -14, -3.7, 3, 6.3, 6.3, 8.$$

Here $n = 6$, which is an even number, and there are two "middle values" in the ordered list, namely,

$$\begin{array}{c} -14, -3.7, \mathbf{3}, \mathbf{6.3}, 6.3, 8. \\ \uparrow\ \uparrow \end{array}$$

To find the median for this data collection, we average these two middle values to obtain $\tilde{x} = \frac{3 + 6.3}{2} = 4.65$.

Notice that for both of these data collections, the median divides the set of ordered data values in half. In the 5-element data collection, two values are larger than the median, 4, and two are smaller than it. In the 6-element data collection, three values are larger than the median, 4.65, and three are smaller.

We note that either a dotplot or a stemplot can be very useful in obtaining the value of a sample median for a collection of data. Both of these graphical representations involve the ordered observations and are thus naturally suited to yield the location of the median for the data.

If you are working with a large data collection or if you want to program a computer to find the median, it is convenient to have a rule that tells you which value or values to use from the ordered data collection. To write such a rule, we need to give names to the values in the ordered data. If $x_1, x_2, \ldots, x_n$ is your data collection, then the ordered data values are denoted by $x_{(1)}, x_{(2)}, \ldots, x_{(n)}$. The first value in the sorted list, $x_{(1)}$, is the smallest observation in the data collection. The second value, $x_{(2)}$, is the second smallest observation in the data collection, and so forth, until $x_{(n)}$ is the largest value in data collection. With this notation, we can now write our rule for computing the median of our data collection as follows:

$$\text{If } n \text{ is odd,} \quad \text{then} \quad \tilde{x} = x_{\left(\frac{n+1}{2}\right)}. \tag{1.2}$$

$$\text{If } n \text{ is even,} \quad \text{then} \quad \tilde{x} = \frac{x_{\left(\frac{n}{2}\right)} + x_{\left(\frac{n}{2}+1\right)}}{2}. \tag{1.3}$$

Notice how this works for our two previous examples. In the first, we have $n = 5$, so that $\frac{n+1}{2} = \frac{5+1}{2} = 3$. From (1.2), we see that the median for this odd number of observations is then element number 3 from the ordered data set; that is, $\tilde{x} = x_{(3)} = 4$. In the second example, we have $n = 6$, so that $\frac{n}{2} = \frac{6}{2} = 3$ and $\frac{n}{2} + 1 = 4$. Thus, for this odd number of observations, it follows from (1.3) that the median is obtained by averaging the 3rd and 4th ordered values; that is,

$$\tilde{x} = \frac{x_{(3)} + x_{(4)}}{2} = \frac{3 + 6.3}{2} = 4.65.$$

Example 1.5. Cost of Engineering Drawings (Continuation of Example 1.3)
In Table 1.3 we displayed the total engineering drawing hours for 96 pieces of machinery/mechanical devices for a major Ohio-based company. Here the number of observations, $n = 96$, is even, so that the median of this collection of data will be the average of the two "middle values". Since $\frac{n}{2} = \frac{96}{2} = 48$, it follows from (1.3) that the median of the collection of engineering drawing hours is $\tilde{x} = \frac{x_{(48)} + x_{(49)}}{2}$. Returning to the stemplot in Fig. 1.4, count up beginning with the smallest value $x_{(1)} = 3$ until we obtain $x_{(48)} = x_{(49)} = 14$ h, so that the median number of drawing hours for this collection of data is $\tilde{x} = (14 + 14)/2 = 14$ h. Note that it is clear from the way in which the median is calculated that the large observations 84, 92, and 100 do not directly influence its value; that is, these outliers could even be changed to 184, 192, and 200, respectively, and the value of the median would be unchanged. Because of this insensitivity to the effect of outliers, the median is often called *resistant* or *robust*. On the other hand, the mean does not have this feature.

First, the value of the mean for these 96 total drawing hour values is $\bar{x} = 22.2$ h. It is clear from the stemplot in Fig. 1.5 that the sample mean does not provide a very good measure of the visual center of this data collection. (Note that the mean is actually larger than about two-thirds (65 out of 96) of the data values.) Moreover, changing the outlier observations from their present values of 84, 92, and 100 to 184, 192, and 200, respectively, would also change the value of the data mean to $\bar{x} = 25.3$ h. This represents an increase of roughly 14% due simply to a substantial change in the three largest observations. This is not a desirable feature for a measure of the center of a collection of data if unusually large or small observations are likely to be present. In such settings the median is the preferred measure.

Example 1.6. Salary Figures for the New York Yankees in 2014 The dataset *american_league_salary_2014* contains the 2014 salaries (as of March 26, 2014) for all baseball players in the American League. In Table 1.17 we have recorded the salaries for the players on the New York Yankees, ordered from largest to smallest. Note that this ordering from largest to smallest, rather than smallest to largest, still enables us to compute the median. Since the number of observations here is 33, an odd number, the median 2014 salary for members of the New York Yankees baseball team will be $\tilde{x} = x_{((33+1)/2)} = x_{(17)}$, the unique middle ordered salary. From Table 1.17 we see that this median salary belongs to Brian Roberts and he was paid $\tilde{x} = x_{(17)} = \$2{,}000{,}000$ for 2014. How do you feel this 2014 median salary would compare with the mean salary for the 2014 New York Yankees? Check it out!

Trimmed Mean of the Observations Another estimate of the center of a collection of data that is also less sensitive to changes in the values of extreme observations than the mean, but which provides greater sensitivity to changes in the values of observations near the center of the data collection than the median, is a trimmed mean. For this measure, a selected portion of the

Table 1.17 Baseball salaries for members of the New York Yankees baseball team for 2014 (as of March 26, 2014)

Name of player	Total 2014 salary
CC Sabathia	24,285,714
Mark Teixeira	23,125,000
Masahiro Tanaka	22,000,000
Jacoby Ellsbury	21,142,857
Alfonso Soriano	19,000,000
Brian McCann	17,000,000
Hiroki Kuroda	16,000,000
Carlos Beltran	15,000,000
Derek Jeter	12,000,000
Ichiro Suzuki	6,500,000
Brett Gardner	5,600,000
David Robertson	5,215,000
Alex Rodriguez	3,868,852
Matt Thornton	3,500,000
Ivan Nova	3,300,000
Kelly Johnson	3,000,000
Brian Roberts	2,000,000
Brendan Ryan	2,000,000
Shawn Kelley	1,765,000
Francisco Cervelli	700,000
Eduardo Nunez	576,900
David Phelps	541,425
Michael Pineda	538,475
Adam Warren	527,400
Austin Romine	524,800
Preston Claiborne	511,325
Zoilo Almonte	511,300
Cesar Cabral	510,825
Vidal Nuno	504,500
John Ryan Murphy	502,700
Dellin Betances	502,100
Dean Anna	500,000
Shane Greene	500,000

Source: Petchesky (2014)

extreme values are deleted from each end of the ordered collection of data (this is called the 'trimming') and the average of the remaining observations (the 'trimmed mean') is used to provide information about the center of the data collection. This measure clearly is not influenced directly by the extreme observations (those trimmed), but is directly responsive to the untrimmed observations used in its calculation.

Definition 1.11 Let $x_1, x_2, \ldots, x_n$ be the n observations in a data collection and let d be an integer between 0 and $n/2$. Discard the d largest and d smallest values from the data collection and compute the average of the remaining n–$2d$ observations. This average is called the ***d*-th trimmed mean,** $\bar{x}_d$.

For most data collections where outliers are a problem, a trimmed mean based on trimming between 10% and 20% of the observations from each of the extremes will provide a representative estimate of the center of the observations. However, trimming proportions as high as 25–30% might be required if there are still outliers among the observations remaining after using an initial trimming proportion of 20%.

Once again, the summation notation provides us with a compact way of writing the definition for the d-th trimmed mean. If $x_{(1)}, x_{(2)}, \ldots, x_{(n)}$ are the ordered data values, then

$$\bar{x}_d = \frac{\left[x_{(d+1)} + \cdots + x_{(n-d)}\right]}{n-2d} = \frac{1}{n-2d}\sum_{i=d+1}^{n-d} x_{(i)}. \quad (1.4)$$

Example 1.7. Cost of Engineering Drawings (Continuation of Example 1.3) Consider once again the total engineering drawing hours presented in Table 1.3 for the 96 pieces of machinery/mechanical devices. Consider a trimming proportion of roughly 10%, corresponding to trimming 9 extreme observations from each end of the ordered data. Trimming the nine largest and smallest observations from the ordered total drawing hour data, we see that the value of the corresponding trimmed mean is

$$\bar{x}_9 = (9 + 9 + 9 + 9 + \ldots + 44 + 46 + 46)/78 = 18.55 \text{ h},$$

which is between the median, $\tilde{x} = 14$ h, and the mean, $\bar{x} = 22.2$ h, for these data, as we would expect in view of how the three statistics are calculated.

Measuring the Spread of the Data Collection: Range, Interquartile Range, and Standard Deviation In the previous section we discussed a number of statistics that can be used to measure the center of a data collection. Clearly, however, such measurements only provide partial information about the nature of the data collection. Consider, for example, the following two data collections of $n = 10$ observations:

Collection A: 3, 4, 5, 6, 7, 9, 10, 11, 12, 13
Collection B: 7.5, 7.6, 7.7, 7.8, 7.9, 8.1, 8.2, 8.3, 8.4, 8.5.

Find the mean, median and 10% trimmed mean for each of these data collections. Are you surprised that all statistics are equal in value? While this would probably only happen with such especially constructed sets of data, it is clear from these calculations that the visual center of both of these data collections is at the value 8. Does that mean that we can treat the two sets of data as virtually the same for all practical purposes? The answer is clearly no, as it is obvious from the following dotplots that these two collections do differ with respect to the spread (or what statisticians call *dispersion*) of their observations.

```
      .    .    .    .    .         .    .    .    .    .
----------|----------|----------|----------|----------|----------|  Collection A
          4          6          8          10         12         14


                              ..... .....
----------|----------|----------|----------|----------|----------|  Collection B
          4          6          8          10         12         14
```

This feature of a data collection is not captured in the previously discussed measures of center, as information about dispersion or variability is contained in the differences between observations, not in the individual values of the observations, as used in our measures of center. Thus, the pair of numbers (6, 8) has roughly the same amount of variability as does the pair of numbers (500, 502), but they certainly differ with respect to their centers. On the other hand, the pair of numbers (6, 8) has the same center as does (-94, 108), but they are associated with different amounts of variability.

There are a number of measures of variability that can be applied to data collections. The most obvious and simplest of these is the range of the data.

Definition 1.12 The **range, *R*,** of a data collection is the difference between the largest and the smallest observations in the collection.

Thus, using the notation we previously developed for the ordered values in a data collection $x_1, x_2, \ldots, x_n$, we can represent the range by

$$R = x_{(n)} - x_{(1)}. \tag{1.5}$$

The range for data collection *A* is $R_A = 13 - 3 = 10$ and that for data collection *B* is $R_B = 8.5 - 7.5 = 1$ and we have a clear indication of the differences in variability for these two data collections. However, the range is not an adequate measure of spread for most settings. For example, a third data collection *C*,

Collection *C*: 3, 7.6, 7.7, 7.8, 7.9, 8.1, 8.2, 8.3, 8.4, 13,

has the same measures of location and range as data collection *A*,

Collection *A*: 3, 4, 5, 6, 7, 9, 10, 11, 12, 13,

but the overall variability between the observations in collection C is clearly less than that for collection A. In order to develop a more differentiating measure of variability, we will need to take into account more than just the distance between the largest and smallest observations.

One way to provide a better measure of the variability in a data set is to use differences between observations in addition to the maximum and minimum.

Definition 1.13 Let $x_1, \ldots, x_n$ denote the n observations in a data collection. For a number q between 0 and 100, the ***q*th percentile** of the data collection is a value such that about q percent of the observations are less than or equal to it. Thus, for example, the median $\tilde{x}$ is the 50th percentile and the maximum $x_{(n)}$ is the 100th percentile. Two percentiles commonly used to describe sample variability are the first and third quartiles. The **first quartile, Q_1,** for the data collection (also the 25th percentile) is the median of the observations that are to the left of the sample median $\tilde{x}$ in the list of ordered data values. The **third quartile, Q_3,** for the data collection (also the 75th percentile) is the median of the set of observations that are to the right of the sample median $\tilde{x}$ in the list of ordered data values. The **interquartile range, *IQR*,** is the difference between the third and first quartiles, namely,

$$IQR = Q_3 - Q_1. \tag{1.6}$$

Note that IQR is similar to the range R (1.5). However, the range measures the maximum difference observed in the data, while the interquartile range measures the observed spread between more moderate, or typical, data values.

To illustrate the computation of the interquartile range for an even sample size, consider the data collection (3, 4, 14, −4, 9, −5, 10, 39, −40, 7, 3, 80). The ordered data collection is (−40, −5, −4, 3, 3, 4, 8, 9, 10, 14, 39, 80). Since $n = 12$, it follows from (1.3) that the median for this data collection is the average of the 6th and 7th ordered values; that is,

$$\tilde{x} = \frac{x_{(6)} + x_{(7)}}{2} = \frac{4+8}{2} = 6.$$

Hence, {−40, −5, −4, 3, 3, 4} and {8, 9, 10, 14, 39, 80} are the sets of observations that are to the left and right, respectively, of the sample median. Since each of these sets contains an even number (six) of observations, the median expression (1.3) is used once again to obtain both quartiles. The first quartile, Q_1, is the median of the six observations $\{-40, -5, -4, 3, 3, 4\}$ and we have $Q_1 = \frac{-4+3}{2} = -.5$. The third quartile, Q_3, is the median of the six observations {8, 9, 10, 14, 39, 80}, yielding $Q_3 = \frac{10+14}{2} = 12$. It then follows from (1.6) that the interquartile range is $IQR = 12 - (-.5) = 12.5$.

For the case of an odd sample size, consider the data collection {3, 6.6, 9.2, -.4, 7.7, -34, 27, 156, -4, -11, 99}. The ordered data collection is (−34, −11, −4, $-.4, 3, 6.6, 7.7, 9.2, 27, 99, 156$). Here, $n = 11$, and it follows from (1.2) that the median for this data collection is the 6th ordered value; that is, $\tilde{x} = x_{(6)} = 6.6$. Hence, {−34, −11, −4, −.4, 3} and {7.7, 9.2, 27, 99, 156} are the sets of observations that are to the left and right, respectively, of the sample median. Since each of these sets contains an odd number (five) of observations, the median expression (1.2) is used once again to obtain the two quartiles. The first quartile, Q_1, is the median of the five observations {−34, −11, −4, −.4, 3} and we have $Q_1 = -4$. The third quartile, Q_3, is the median of the five observations {7.7, 9.2, 27, 99, 156}, yielding $Q_3 = 27$. It then follows from (1.6) that the interquartile range is $IQR = 27 - (-4) = 31$.

In both of these examples, the same formula that was used to obtain the median of the entire sample was also used to obtain the two subset medians for Q_1 and Q_3; that is, for $n = 12$, expression (1.3) was used to obtain all three medians, while expression (1.2) yielded the three medians for $n = 11$. Do you think that this will always be the case; that is, whichever expression, (1.2) or (1.3), is used to obtain the median of the entire sample will also be used to obtain both Q_1 and Q_3? What if the sample size is $n = 13$?

Example 1.8. Cost of Engineering Drawings (Continuation of Example 1.3) We return to the total engineering drawing hours presented in Table 1.3 for the 96 pieces of machinery/mechanical devices. In Example 1.5 we found that the median for this data collection is $\tilde{x} = 14$ h. From Table 1.3 we see that the first quartile, Q_1, is the median of the 48 total drawing hours values that are to the left of the sample median $\tilde{x} = 14$ h (including one of the two observed drawings which required 14 total hours) and the third quartile, Q_3, is the median of the 48 total drawing hours that are to the right of the median $\tilde{x} = 14$ h (again including one of the two observed drawings which required 14 total hours). Since 48 is an even number, we see from (1.3) and Table 1.3 that $Q_1 = [x_{(24)} + x_{(25)}]/2 = [11 + 11]/2 = 11$ h and $Q_3 = [x_{(72)} + x_{(73)}]/2 = [26 + 26]/2 = 26$ h. It follows that the interquartile range for the total drawings data is $IQR = Q_3 - Q_1 = 26 - 11 = 15$ h.

For those data sets where outliers are not a major problem and there is no serious skewness in either direction, the mean provides a satisfactory measure of the center of the observations. For such settings a different approach based on individual distances of the observations from the mean can be used to assess the variability in the sample. If $\bar{x}$ is the mean for the n observations $x_1, x_2, \ldots, x_n$, then the *n deviations from the mean* are the differences $x_1 - \bar{x}, \ldots, x_n - \bar{x}$. For example, for the data collection $\{2, 3, 7, 8, 10\}$, the sample mean is $\bar{x} = 6$ and the $n = 5$ deviations from the mean are $2 - 6 = -4, 3 - 6 = -3, 7 - 6 = 1, 8 - 6 = 2$, and $10 - 6 = 4$. Each of the deviations from the mean provides a piece of information about the variability of the data about their center, as measured by the mean. In particular, the greater the magnitudes of the deviations from the mean, the greater the variability present in the data collection. Therefore, a natural first impulse to assess the variability about the center of the data collection would be to simply average these individual deviations from the mean. However, notice that the sum of the five deviations found above is (−4) + (−3) + 1 + 2 + 4 = 0, so that the average of this set of deviations from the mean is also 0. This is not an accident. The mean of a data collection has the property that the sum of the deviations from it is always zero; that is, $\sum_{i=1}^{n}(x_i - \bar{x}) = 0$ for every data collection. For this reason, neither the sum nor the average deviation of the observations from their mean is a useful measure of the variability in a data collection.

One way to circumvent this problem is to first square the deviations from the mean before summing. This continues to equally weight positive deviations and negative deviations of the same magnitude and the sum of the squared deviations will always be greater than zero (unless all the data values are the same, in which case the data quite naturally have zero variability). In fact, the greater the variability of the data around their mean, the greater will be this sum of squared deviations from the mean. For example, consider once again the two data collections:

Collection *A*: 3, 4, 5, 6, 7, 9, 10, 11, 12, 13
Collection *B*: 7.5, 7.6, 7.7, 7.8, 7.9, 8.1, 8.2, 8.3, 8.4, 8.5.

Each of these data collections has a mean value of $\bar{x} = 8$. However the sum of the squared deviations from this common mean is quite different for the two data collections. For data collection *A* we have

$$\begin{aligned}\sum_{i=1}^{10}(x_i - \bar{x})^2 &= \sum_{i=1}^{10}(x_i - 8)^2\\ &= (3-8)^2 + (4-8)^2 + (5-8)^2 + (6-8)^2 + (7-8)^2 + (9-8)^2 + (10-8)^2\\ &\quad + (11-8)^2 + (12-8)^2 + (13-8)^2\\ &= 25 + 16 + 9 + 4 + 1 + 1 + 4 + 9 + 16 + 25 = 110,\end{aligned}$$

while for data collection *B* we have

$$\sum_{i=1}^{10}(x_i - \bar{x})^2 = \sum_{i=1}^{10}(x_i - 8)^2$$
$$= (7.5-8)^2 + (7.6-8)^2 + (7.7-8)^2 + (7.8-8)^2 + (7.9-8)^2 + (8.1-8)^2$$
$$+ (8.2-8)^2 + (8.3-8)^2 + (8.4-8)^2 + (8.5-8)^2$$
$$= .25 + .16 + .09 + .04 + .01 + .01 + .04 + .09 + .16 + .25 = 1.10.$$

The sum of the squared deviations from the common mean $\bar{x} = 8$ for data collection A is 100 times that for data collection B! This provides good numerical justification for what can easily be observed directly from the two data collections themselves.

One natural measure of the variability in a data collection is the typical squared deviation, which we can find by averaging these n squared deviations.

Definition 1.14 Let $x_1, x_2, \ldots, x_n$ denote the n observations in a data collection. The **variance, s^2,** for these data is defined to be the sum of the squared deviations from the mean of the data divided by $n - 1$. Notationally, we have

$$s^2 = \frac{(x_1 - \bar{x})^2 + \cdots + (x_n - \bar{x})^2}{n-1} = \frac{\sum_{i=1}^{n}(x_i - \bar{x})^2}{n-1}. \tag{1.7}$$

The **standard deviation, s,** for the data collection is the square root of the variance s^2.

The units for the variance s^2 will be the square of the units of the measured data. However, the standard deviation s will have the same units as the data. For this reason (and others that will become apparent later), the standard deviation is generally the quantity of interest for most settings, rather than the variance. Computationally, of course, we must first calculate s^2 before we can obtain the value of s.[1]

Example 1.9. Asbestos Exposure and Lung Function Asbestos is a silicate mineral compound that is resistant to both heat and chemical reactions. These properties make it a natural insulator and it has been used for that purpose for a number of years. However, inhalation of asbestos dust can lead to asbestosis, a progressive disease characterized by chronic inflammation and congestion in the lungs, which leads to a loss in lung function. Employees of companies producing asbestos are particularly prone to development of this disease. Al Jarad et al. (1993) conducted a study in which they examined the decrease in lung function for a sample of asbestos workers who had not (yet) contracted asbestosis. The values in Table 1.18 are the percent decreases in lung function (as measured by a procedure known as FVC) over a period of slightly longer than 4 years for 20 asbestos workers who did not have asbestosis. The average percent decrease in lung function for these twenty asbestos workers is

$$\bar{x} = (10 + 2 + \ldots + 25 + 13)/20 = 232/20 = 11.6.$$

The 20 sample observations, their deviations from the mean, $\bar{x} = 11.6$, and the squared deviations are then:

[1] We note that the sum of the squared deviations from the mean is divided by n-1, not the seemingly more natural n, in the definition of the variance. While there are a number of statistical reasons for doing this, we note here only that the constraint implied by the fact that $\sum_{i=1}^{n} (x_i - \bar{x}) = 0$ is one of these reasons. Although we start out with n independent pieces of information, commonly known as *degrees of freedom*, from the sample observations $x_1, \ldots, x_n$, once we have computed the mean $\bar{x}$ and obtained the n deviations from the mean, we are left with only (n-1) degrees of freedom, since the constraint that $\sum_{i=1}^{n} (x_i - \bar{x}) = 0$ results in the loss of one degree of freedom.

Table 1.18 Percent decrease in lung function for asbestos workers who do not have asbestosis

Employee	Percent decrease in lung function
1	10
2	2
3	16
4	2
5	30
6	0
7	20
8	14
9	0
10	18
11	0
12	8
13	25
14	16
15	0
16	6
17	2
18	25
19	25
20	13

Source: Al Jarad et al. (1993)

x_i	$x_i - 11.6$	$(x_i - 11.6)^2$
10	−1.6	2.56
2	−9.6	92.16
16	4.4	19.36
2	−9.6	92.16
30	18.4	338.56
0	−11.6	134.56
20	8.4	70.56
14	2.4	5.76
0	−11.6	134.56

(continued)

(continued)

x_i	$x_i - 11.6$	$(x_i - 11.6)^2$
18	6.4	40.96
0	−11.6	134.56
8	−3.6	12.96
25	13.4	179.56
16	4.4	19.36
0	−11.6	134.56
6	−5.6	31.36
2	−9.6	92.16
25	13.4	179.56
25	13.4	179.56
13	1.4	1.96
——	——	——
232	0	1896.80

Summing the deviations from the mean, we obtain 0, as expected. To compute the variance, s^2, we sum the squares of the deviations from the mean and divide by $(n - 1) = 19$ to obtain

$$s^2 = (2.56 + 92.16 + \ldots + 179.56 + 1.96)/19 = 1896.8/19 = 99.832.$$

The standard deviation for these asbestos data is then $s = \sqrt{99.832} = 9.992$.

Almost all calculators and computer software packages will also compute the variance for a collection of data. In addition, a computationally simpler expression for s^2 than that given in (1.7) is discussed in Exercise 1.A.8.

Example 1.10. Motor-Vehicle Fatalities in 2012 (Continuation of Example 1.4) For the 50 statewide motor-vehicle fatality rates in Table 1.16, we have already seen in Example 1.4 that $\bar{x} = 1.18$ fatalities per 100 million vehicle miles. Using the **R** function *summary*( ) for the *"Population Death Rate"* column of the dataset *motor_vehicle_death_rate_2012* we obtain the following output.

```
> summary(motor_vehicle_death_rate_2012$"Population Death Rate")
 Min.    1st Qu.  Median   Mean   3rd Qu.  Max.
 0.6200  0.9675  1.1600   1.1810  1.4080  1.7600
```

Additionally, we can use the **R** function *sd*( ) to obtain s for the motor-vehicle death rates as follows.

```
> sd(motor_vehicle_death_rate_2012$"Population Death Rate")
[1] 0.2980303
```

As we will discuss more formally later in the text, the combination of mean and standard deviation for a data collection can also be used to describe a bit more of the structure for the data collection. For example, for most data collections roughly 95% of the observations in the collection can be found within two standard deviations of the mean; that is, in the interval $(\bar{x} - 2s, \ \bar{x} + 2s)$. For the 2012 motor-vehicle fatalities, we have $(\bar{x} - 2s, \ \bar{x} + 2s) = (1.18 - 2(0.298), 1.18 + 2(0.298)) = (0.5849, 1.777)$ motor-vehicle deaths per 100 million vehicle miles. Comparing the fatality rate data in Table 1.16 with this interval, we see that none of the states have a motor-vehicle death rate that falls outside this interval. Hence for these data, we actually have 100% (50 out of 50) of the observations in the collection falling inside the interval $(\bar{x} - 2s, \bar{x} + 2s)$.

For data sets for which the mean does not provide a satisfactory measure of the center of the observations, the standard deviation will also not be a good measure of the variability in the data. We can still use distances from a measure of the center of the data to assess variability, but we would need to use a more robust measure of the center, such as the median or a trimmed mean, as well as something other than the squares of the deviations, since squaring compounds the effects of any outliers. For more on this issue, see Exercise 1.A.2.

Summarizing Quantitative Data Collections Through Percentiles and Percentages There are two natural ways to describe a data collection in terms of the ordered values of the observations. Both are based on the concept of fenceposts.

> **Definition 1.15** A supply of **fenceposts** is a group of numerical values that divide a data collection into categories (i.e., partition the data collection). Fenceposts can be calculated from the data values themselves (**variable fenceposts**) or purposely selected before the data are collected (**fixed fenceposts**).

Although the two data summary methods we now describe are somewhat similar in their approach, one uses variable fenceposts and the other uses fixed fenceposts. Consequently, they lead to different types of information summaries of the data.

Measuring Position and Relative Positions: Percentiles, Standardized Scores, and Boxplots One approach to summarizing a data collection involves partitioning the data based only on the observations themselves. Previously we used percentiles of the data to define the interquartile range *IQR* (1.6) as a robust measure of variability in the data. However, when viewed as a package, percentiles provide more summary information about the data than just what is contained in *IQR*. In fact, a listing of the 10th, 20th, . . ., 80th, and 90th percentiles (known, collectively, as the nine sample deciles) provides us with a good sense as to how the data are distributed over the range of possible values for the variable being measured. We know that approximately 10% of the observations in the data collection fall in each of the ten categories created by these nine decile fenceposts. With this approach we are organizing the data by selecting the fenceposts in such a way that the number of observations between them is predetermined and the fenceposts themselves (i. e., the deciles) are the variables that provide us the summary information about the data collection; hence, the designation 'variable fenceposts' for this method of data summarization.

Example 1.11. Motor-Vehicle Fatalities in 2012 (Continuation of Example 1.4) The nine deciles for the 50 statewide motor-vehicle fatality rates in Table 1.16 correspond to values such that there are five of the observed statewide death rates between each adjacent pair of the deciles, as well as five death rates below the first decile and five above the ninth decile. Ordering the 50 statewide rates, we obtain the following ten partitioned sets of five observations each:

0.62	0.91	1.07	1.25	1.46
0.69	0.91	1.10	1.27	1.48
0.75	0.96	1.11	1.32	1.51
0.78	0.99	1.13	1.32	1.54
0.79	0.99	1.16	1.33	1.58
0.82	1.00	1.16	1.33	1.65
0.82	1.01	1.21	1.37	1.69
0.84	1.01	1.23	1.42	1.72
0.88	1.04	1.23	1.43	1.76
0.89	1.07	1.24	1.43	1.76.

Using these ordered values, we see that the nine deciles are:

$$d_1 = \frac{(0.79 + 0.82)}{2} = 0.805, \quad d_2 = \frac{(0.89 + 0.91)}{2} = 0.90, \quad d_3 = \frac{(0.99 + 1.00)}{2} = 0.995,$$

$$d_4 = \frac{(1.07 + 1.07)}{2} = 1.07, \quad d_5 = \frac{(1.16 + 1.16)}{2} = 1.16, \quad d_6 = \frac{(1.24 + 1.25)}{2} = 1.245,$$

$$d_7 = \frac{(1.33 + 1.33)}{2} = 1.33, \quad d_8 = \frac{(1.43 + 1.46)}{2} = 1.445, \quad d_9 = \frac{(1.58 + 1.65)}{2} = 1.615.$$

Including the minimum and maximum values, 0.62 and 1.76, respectively, with these nine deciles provides an informative summary of the statewide motor-vehicle fatality rates in 2012.

For many data collections, even fewer fenceposts will be needed to paint an adequate picture of the configuration of the observations.

Definition 1.16 It is often the case that important features of a data collection can be visualized through the summary provided by the minimum value $x_{(1)}$, the first quartile Q_1, the median $\tilde{x}$, the third quartile Q_3, and the maximum value $x_{(n)}$.This set of five fenceposts $x_{(1)}$, Q_1, $\tilde{x}$, Q_3, and $x_{(n)}$ is commonly known as the **five-number summary** of a data collection.

The **R** function *summary*( ) can be used to provide the five-number summary for a data collection. Such **R** output for the motor-vehicle death rate data in Table 1.16 was presented in Example 1.10. From that display, we see that the five-number summary for these data is given by $x_{(1)} = 0.62$, $Q_1 = 0.9675$, $\tilde{x} = 1.16$, $Q_3 = 1.408$, and $x_{(50)} = 1.76$.

A five-number summary is particularly useful for constructing a revealing graph called a boxplot. The center (as measured by the median), interior variability, and range of the values in the data collection are clearly apparent from such a boxplot representation.

Definition 1.17 The **boxplot** is a visual display of a five-number summary ($x_{(1)}$, Q_1, $\tilde{x}$, Q_3, and $x_{(n)}$) of a data collection and is constructed as follows:

1. The upper and lower ends of the box (called *hinges*) are supplied by the first and third quartiles Q_1 and Q_3, respectively. Thus, the length of the box is equal to the interquartile range $IQR = Q_3 - Q_1$.
2. The median, $\tilde{x}$, is indicated by a line within the box provided by the two quartiles.

Definition 1.17 (continued)

3. Extend two lines (called *whiskers*) outside the box to the minimum, $x_{(1)}$, and maximum, $x_{(n)}$.

We illustrate the details involved in the construction of a boxplot with the data collection (3, 4, 14, –4, 9, –5, 10, 30, –15, 8, 3, 34). With $n = 12$, the median for this data collection is $\tilde{x} = 6$, while the first and third quartiles are $Q_1 = -.5$ and $Q_3 = 12$, respectively. Combining these three values with the minimum $x_{(1)} = -15$ and maximum $x_{(12)} = 34$, we see that the five-number summary for this data collection consists of the five fenceposts $x_{(1)} = -15$, $Q_1 = -.5$, $\tilde{x} = 6$, $Q_3 = 12$, and $x_{(12)} = 34$. Following the guidelines in Definition 1.17, the boxplot for this data collection is then given by:

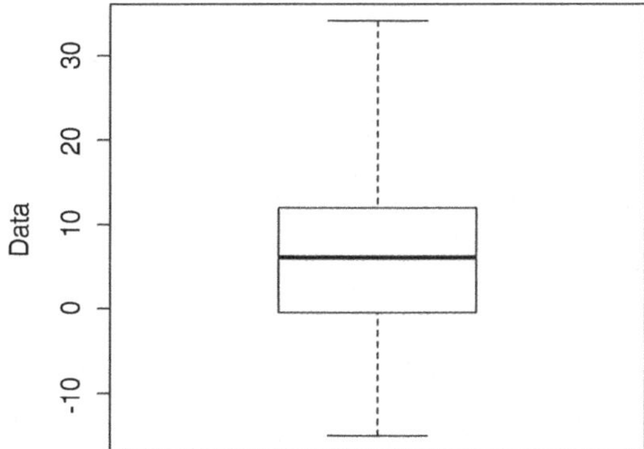

The centerline across the box is at the median, $\tilde{x} = 6$, and the lower and upper hinges (ends) of the box are at the first and third quartiles, $Q_1 = -.5$ and $Q_3 = 12$, respectively. The whiskers then extend from the box itself to the minimum, $x_{(1)} = -15$, and maximum, $x_{(12)} = 34$, values.

Example 1.12. Cost of Engineering Drawings (Continuation of Example 1.3) In Examples 1.5 and 1.8 we obtained the values of the median, $Q_1 = 11$ and $Q_3 = 26$, for the total engineering drawing hours presented in Table 1.3 for the 96 pieces of machinery/mechanical devices. Adding the minimum, $x_{(1)}$

Fig. 1.10 Boxplot for the total engineering drawing hours data

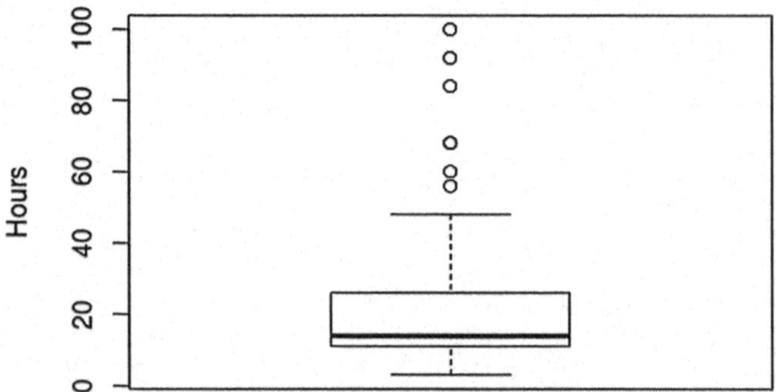

= 3, and maximum, $x_{(96)} = 100$, values provides the five-point summary for the data collection. A boxplot representation for this data collection (obtained using the **R** function *boxplot*( )) is presented in Figure 1.10. Notice that the **R** boxplot does not actually extend the whiskers to the smallest and largest observations in the data collection, as we recommend in Definition 1.17. Instead the **R** version of a boxplot extends the whiskers only to the largest and smallest observations within the region defined by the lower limit $Q_1 - 1.5\ IQR$ and the upper limit $Q_3 + 1.5\ IQR$. For this total hours data collection, $Q_1 - 1.5\ IQR = 11 - 1.5(26 - 11) = -11.5$ and $Q_3 + 1.5\ IQR = 26 + 1.5(26 - 11) = 48.5$. Thus, while the lower whisker in the **R** boxplot does, in fact, extend to the smallest observation, 3, in the data collection, the upper whisker extends only to 48, the largest observation in the data collection that is smaller than 48.5. The *outliers*, in this case, the seven observations larger than 48.5 (56, 60, 68, 68, 84, 92, and 100), are indicated by circles. (This example points out how important it is for you to be very familiar with your particular statistical software, especially with respect to graphical representations of a data collection, since many other programs will omit outliers by default.)

Several things are evident from this version of a boxplot of the total engineering drawing hours. The center of the data collection is at the median value, 14, and there is not a lot of variability in the middle portion of the data. The hinges are located at the quartiles, 11 and 26. However, the boxplot also provides a clear picture that the data collection is skewed to the right, since the upper whisker is much longer than the lower whisker. This indicates greater variability among those drawings with larger total hours than among those with smaller total hours, in complete agreement with the impression gathered by looking at the data collection itself.

Measuring Categorical Distribution of Quantitative Data: Fixed Fenceposts Another type of summary that can be used to describe a slightly different aspect of a data configuration is constructed by using fixed, prespecified fenceposts to divide the data into subsets of interest. The collection of fixed fencepost values used for this division of the data can vary from simply an equally spaced division of the expected range of measurements (similar to the approach taken in the construction of a histogram for graphical display of the data) to a more customized subset of potential data values that have particular meaning for a study.

Letting $f_1, \ldots, f_k$ denote such a collection of k fixed fencepost values, the data are then summarized by recording the numbers (or percentages) of observations falling between each of these fenceposts, as well as the numbers (or percentages) below f_1 and above f_k. **For convention, we choose to include values at the boundaries of these divisions (i.e., "on" the fenceposts) in the counts for the lower of the two involved intervals.**

Example 1.13. Cost of Engineering Drawings (Continuation of Example 1.3) A different picture of the engineering drawing hours data in Table 1.3 can be obtained by using the same number of fixed fenceposts equally spaced along the observed range of total hours from 0 to 100. Taking $k = 9$ and letting $f_i = 10 \times i$, for $i = 1, \ldots, 9$, we record the following numbers (percentages) of engineering drawings in the data collection with total hours either no more than 10, greater than 90, or between two of these fixed fenceposts:

Less than or equal to 10 h	21 (21.9%),
Greater than 10 but not more than 20 h	43 (44.8%),
Greater than 20 but not more than 30 h	12 (12.5%),
Greater than 30 but not more than 40 h	8 (8.3%),
Greater than 40 but not more than 50 h	5 (5.2%),
Greater than 50 but not more than 60 h	2 (2.1%),
Greater than 60 but not more than 70 h	2 (2.1%),
Greater than 70 but not more than 80 h	0 (0%),
Greater than 80 but not more than 90 h	1 (1.0%),
Greater than 90 h	2 (2.1%).

With these nine fixed fenceposts, we get a very clear picture that the bulk of the drawings in the collection required between 10 and 20 total hours to complete and that the median for the data is in this range as well. The right-skewness of the data is also evident in the strongly decreasing trend of the category percentages once past the 20-h boundary. Finally, since there are only three (3.1%) of the engineering drawings with total hours belonging to the final two categories created by these evenly spaced fenceposts, the case is strengthened for designation of these drawings as outliers for this data collection.

It is important to note that we have a lot of flexibility in the choice of these fixed fenceposts, both in their number and in where they are placed. For some studies, only one fencepost might be adequate to differentiate between 'large' and 'small' values or two fenceposts to describe 'minimal', 'moderate', and 'extensive' responses. Also, there is certainly no requirement to use fixed fenceposts that are equally spaced. For example, to summarize the 2014 American League baseball salary data in the dataset *american_league_salary_2014*, any set of fixed fenceposts would quite naturally include any preset boundaries established by union contracts with the major league teams, such as minimum salaries for starting and relief pitchers, for players in the starting lineups, for utility players, etc.

Interpretations of the Mean and Standard Deviation: Standardized Z-Scores The mean and standard deviation for a data collection can also be used to provide information about a particular observation's relative placement among the rest of the data.

> **Definition 1.18** Let $\bar{x}$ and s be the mean and standard deviation for a data collection. Then the **standardized z-score** for a particular observation x in the data set is
>
> $$\text{z-score for observation } x = \frac{x - \bar{x}}{s}. \tag{1.8}$$

The z-score describes the relative location of the observation x within the data collection by stating how many standard deviations it is away from the mean. The sign of the z-score indicates the direction from the mean – positive for those observations which are greater than the mean and negative for those which are less than the mean. In addition, the larger the magnitude of the z-score the further it is from the center (0).

Example 1.15. Motor-Vehicle Deaths in 2012 In Examples 1.4 and 1.10 we obtained the values $\bar{x} = 1.181$ and $s = 0.298$ to be the mean and standard deviation for the statewide motor-vehicle death rates for 2012, as listed in Table 1.16. The z-scores for the states of Massachusetts and Mississippi, for example, are then

$$\text{z-score}_{\text{Massachusetts}} = \frac{0.62 - 1.181}{0.298} = -1.88$$

and

$$z\text{-score}_{\text{Mississippi}} = \frac{1.51 - 1.181}{0.298} = 1.10,$$

respectively. Thus Massachusetts' motor-vehicle death rate is a good deal better (i. e., lower) than the average of the 50 states, while the motor-vehicle death rate in Mississippi is somewhat worse (i. e., higher) than this average.

Effect of Changing Units of Measurement on Summary Measures You have just finished preparing a variety of summary measures and graphs for the total drawing hours data in Table 1.3 when your boss tells you that you must convert all of your calculations to minutes and provide a new set of summary measures and graphs by the end of the day! Before you start thinking badly of your boss and dive headfirst into the entire process once again, you need to recall that the conversion of hours to minutes is simply a linear transformation of the form $y = a + bx$. For the data in Table 1.3, if x represents the total hours required for a drawing and y represents the total minutes required, then $y = 60x$, which is a linear transformation with $a = 0$ and $b = 60$. Moreover, fortunately for you, the effect of such linear transformations on most summary measures and associated graphs is completely determined by the form of the transformation and can be accounted for without redoing all of the calculations.

1.2.1 Effects of a Linear Transformation

Let $x_1, \ldots, x_n$ be a data collection and consider the linear transformation $y_i = a + bx_i$, $i = 1, \ldots, n$. This transformation scales every data value by a factor of b and then shifts it by a units. Then we have the following relationship between statistical summary measures and graphs associated with the two data sets $(x_1, \ldots, x_n)$ and $(y_1, \ldots, y_n)$:

1. The basic configuration or shape of the transformed data collection (y_1, ..., y_n) will be the same as for the original data collection (x_1, ..., x_n); that is, if the configuration of the x's is roughly symmetric (skewed to the right, skewed to the left), then so will be the configuration of the y's.
2. Every measure of the center considered in this text (mean, median, trimmed mean, and all percentiles) is changed by applying the linear transformation to the appropriate statistic for the (x_1, ..., x_n) data. For example, the median of (y_1, ..., y_n) becomes $\tilde{y} = a + b\tilde{x}$.
3. Measures of spread or variability are affected differently by linear transformations than are measures of the center. For example, the interquartile range is such that $IQR_y = |b| \times IQR_x$, which is the case with most measures of spread, including the standard deviation, s, and the range. (Note, however, that the variance s^2 is transformed by multiplication of b^2, rather than $|b|$, since it involves the squares of the deviations from the mean.) The fact that measures of spread are unaffected by changes in location (addition of constants) should agree completely with your intuition.

Based on this information, it is clear that you can easily comply with the last minute request by your boss. Multiplication of every location and spread or variability measure that you have already computed in hours by the factor 60 will immediately update your results to the new minute units. In addition, other information about the data set will carry over as well, including the general shape of the data configuration (symmetric or either right-skewed or left-skewed) and which of the observations are unusual or outliers.

Assessing Symmetry of a Data Collection: Triples and Percentile of the Mean One of the issues that is very important for the selection of appropriate statistics to summarize and analyze a data collection is whether or not the configuration of the data is generally symmetric. Earlier in this chapter we noted that a single observation provides some information about the center of data collection, but that it takes at least two observations to provide information about the variability in the data. How many observations are required to provide information about the symmetry of a data collection? If you answered three, you are correct.

Definition 1.19 Take any three data values from a data collection and order this **triple** of data values from smallest to largest. If the middle ordered item is closer to the smallest than to the largest, the triple is said to be a **right triple**. If the middle ordered item is closer to the largest, the triple is said to be a **left triple**. If the middle ordered value is exactly halfway between the other two, the triple is neither right nor left.

To measure the symmetry of a data set, we take all possible triples and classify them as right, left, or neither. A preponderance of right triples in a data collection is indicative of right-skewness, while a preponderance of left triples is indicative of left-skewness. If the data collection is roughly symmetrically configured, we would expect about an equal number of right and left triples.

Even for a small number of observations, the total number of triples can be quite large, making this evaluation of symmetry for a data collection computationally intensive. You may have learned in a previous math course that there are $\frac{n(n-1)(n-2)}{6}$ ways of choosing three items from a collection of n items. It follows that for a data collection containing $n = 10$ observations, there are 10 $(9)(8)/6 = 120$ triples to consider. An alternative method for determining whether a triple is a right or left triple involves a comparison between the average of the three observations and their median. If $\bar{x}$ and $\tilde{x}$ denote the mean and median, respectively, for a triple of observations (x_1, x_2, x_3), then that triple will be a right or left triple depending on whether $\bar{x}$ is greater than or less than $\tilde{x}$, respectively. (A triple for which the mean and median are equal is neither a right nor a left triple.)

Example 1.16. Where to Put the Microwave? The microwave oven is a virtual given in a modern kitchen, but where should it be placed in relation to other kitchen appliances, equipment, and work centers to optimize both convenience and efficiency? Yust (1982) reported on a study conducted in the Home Equipment Laboratory at the University of Minnesota that was designed to address this question. Twenty-four microwave oven owners (who did the majority of the meal preparations in their homes) were randomly selected to participate in the study. L-shaped laboratory kitchens were equipped with both range centers and mix centers and one of the following four microwave oven configurations:

(i) No microwave oven
(ii) Microwave oven located adjacent to the range center
(iii) Microwave oven located adjacent to the mix center
(iv) Microwave oven separated from both the mix and range centers.

Groups of six subjects were assigned to each of these experimental arrangements and asked to prepare the same complete meal. Two research assistants recorded the numbers of trips between work centers and the times spent at the various work centers for the 24 subjects participating in the study. The recorded numbers of trips between work centers for the six subjects working with the microwave separated from both the mix center and the range center are as follows:

$$x_1 = 91, \quad x_2 = 99, \quad x_3 = 114, \quad x_4 = 141, \quad x_5 = 171, \quad x_6 = 179.$$

To get some idea about the symmetrical nature of this small data set, we list the $6(5)(4)/6 = 20$ triples for these data, along with the mean and median for each triple and whether it is a right or left triple. Since there are 12 right triples and only 8 left triples, we have some indication that there is a slight skewness toward larger numbers of trips (i., e., to the right) for this set of data. Does this agree with a visual plot of the six data values?

Triple	Mean	Median	Right or left triple
(91, 99, 114)	101.33	99	Right
(91, 99, 141)	110.33	99	Right
(91, 99, 171)	120.33	99	Right
(91, 99, 179)	123	99	Right
(91, 114, 141)	115.33	114	Right
(91, 114, 171)	125.33	114	Right
(91, 114, 179)	128	114	Right
(91, 141, 171)	134.33	141	Left
(91, 141, 179)	137	141	Left
(91, 171, 179)	147	171	Left
(99, 114, 141)	118	114	Right
(99, 114, 171)	128	114	Right
(99, 114, 179)	130.67	114	Right
(99, 141, 171)	137	141	Left
(99, 141, 179)	139.67	141	Left
(99, 171, 179)	149.67	171	Left
(114, 141, 171)	142	141	Right
(114, 141, 179)	144.67	141	Right
(114, 171, 179)	154.67	171	Left
(141, 171, 179)	163.67	171	Left

Another quick indication of the symmetry or asymmetry of a collection of data can be provided by directly comparing the mean and the median of the collection. If the collection is skewed to the right, then the mean will be larger than the median, while the converse is true if the data are skewed to the left. If the data collection is roughly symmetric, then we would expect the median and mean to be relatively close in value. For the total engineering drawing hours of Table 1.3, for example, we recall from Example 1.5 that $\bar{x} = 22.2$ and $\tilde{x} = 14$, which is indicative of skewness to the right. This feature of the total engineering drawing hours data can also be detected by examining the set of data triples for the collection. However, for these data each of $\frac{96(95)(94)}{6} =$ 142,880 triples must be categorized as right, left, or neither. Although this would be a formidable task to complete by hand or even calculator, an **R** program can be used to easily obtain and classify these triples. For the total engineering drawing hours data collection you are asked in Exercise 1. B.6 to use this **R** program to show that the breakdown of the 142,880 triples is given by:

98,269 right triples
37,832 left triples
6779 triples that are neither right nor left.

This agrees with the previous evidence for skewness to the right provided by comparison of the mean and median of the data collection.

Finally we point out one additional statistic that provides information about the symmetry/asymmetry of a data collection. If a data collection is roughly symmetric, how many observations in the collection would you expect to be greater than the mean of the collection? Since the mean and median of a roughly symmetric data collection will be relatively close in value, then for such a set of data we should expect about one-half of the observations to be greater than the mean. What do you think would be the case for right-skewed or left-skewed data collections?

As an example, consider once again the data collection of total engineering drawing hours. Returning to Table 1.3, we see that only 31 of the 96 observations are greater than the mean $\bar{x} = 22.2$. Does this agree with your visual feel for these data provided by the dotplot and histogram in Figs. 1.3 and 1.5, respectively?

Section 1.2 Practice Exercises

1.2.1. *Heights of Classmates.* Consider the data collection of heights for all of your classmates (including you) in this course. Which of the following are statistics associated with this data collection?

(a) the tallest person in the class
(b) the average height for the students in the class
(c) the height of the shortest person in the class
(d) the number of persons in the class
(e) the building and room number where the class is held
(f) the number of students in the class who are over six feet tall
(g) the difference in the heights of the tallest and the shortest persons in the class
(h) the median grade point average for the students in this class
(i) the number of students in the class who are shorter than you
(j) the name of the class instructor
(k) the number of students who wore jeans to class today.

1.2.2. *Shooting Free Throws.* You shoot 65 free throws. Consider the data collection that simply records whether you make or miss each of these shots. Which of the following are statistics associated with this data collection?

(a) the total number of free throws that you made
(b) the shot on which you made your first free throw
(c) the number of misses for which you at least hit the rim
(d) the total number of free throws that you missed
(e) the number of free throws you shot left-handed
(f) the number of shots for which your foot was over the free-throw line
(g) the number of free throws you missed before you had made ten of them
(h) the percentage of made free throws
(i) the average amount of time between free throw shots

(j) the number of your last ten shots that you make

(k) the largest number of consecutively made free throws.

1.2.3. *Airline Passenger Complaints.* The U.S. Department of Transportation (2014) provided data on the numbers of complaints per 100,000 passengers in December 2013 for the following U.S. airlines:

Airline	**Number of complaints per 100,000 passengers**
Airtran	0.89
Alaska	0.60
American	1.99
American Eagle	1.83
Delta	0.53
Frontier	3.29
Hawaiian	0.61
JetBlue	0.63
Skywest	0.85
Southwest	0.36
United	1.89
US Airways	1.27

(a) Find the mean $\bar{x}$ and median $\tilde{x}$ of the airline complaint rates.

(b) What is the range R of the airline complaint rates?

(c) Describe any unusual features of this data collection.

1.2.4. Consider the data collection $\{1, 2, 6, 8, 9, 20, 24\}$.

(a) Find the mean $\bar{x}$ and median $\tilde{x}$ for this data collection.

(b) What happens to the mean and median if you add another observation 12 to the data collection? Justify your answers.

(c) What additional observation would have to be added to the original data collection in order to make the mean of the new data collection equal to 12? Justify your answer.

(d) Could **one** observation be added to the original data collection so that the median of the new data collection is equal to 12? equal to 8.25? Justify your answers.

(e) Which of the mean or median would be most affected by the addition of another observation 100 to the data collection? Justify your answer.

1.2.5. Consider the data collection {1, 2, 6, 8, 9, 20, 24}.

(a) Find the range R and standard deviation s for this data collection.

(b) What happens to the range and standard deviation if you add another observation 7 to the data collection? Justify your answers.

(c) What happens to the range and standard deviation if you add another observation 18? Justify your answers

(d) Could **one** observation be added to the original data collection so that the range of the new data collection is equal to 20? equal to 25? Justify your answers.

(e) Give an example of **one** observation value that could be added to the original data collection so that the range is affected more by its addition than is the standard deviation.

1.2.6. *Airline Passenger Complaints.* Consider the December 2013 airline complaint rates given in Exercise 1.2.3. Compute the value of the l-th trimmed mean $\bar{x}_1$ and compare it with the values of the mean $\bar{x}$ and median $\tilde{x}$ obtained in Exercise 1.2.3. What do these values suggest about the data collection?

1.2.7. *Gender and Math SAT Scores.* Consider the math SAT score data collections in Table 1.15.

(a) Find the mean $\bar{x}$ and median $\tilde{x}$ for the male students' math SAT scores.

(b) Find the 10-th trimmed mean $\bar{x}_{10}$ for the female students' math SAT scores.

(c) What are the separate values of the ranges for the male and female math SAT scores?

(d) Find the four quartiles and the interquartile range for the male students' math SAT scores.

(e) Find the standard deviation of the math SAT scores for the female students.

1.2.8. Consider the two data collections:

$$\{1, 3, 5, 7, 9, 11\} \quad \text{and} \quad \{9, 11, 13, 15, 17, 19\}.$$

Without formal calculation, which of the two data collections do you think will have the larger mean? Which do you think will have the larger standard deviation? Now formally justify your conjectures.

1.2.9. *Firearms in the Home.* Consider the state-by-state percentages of households with firearms in or around their homes, as given in Table 1.9.

(a) Find the five-number summary for this data collection.

(b) Use an appropriate software package to make a boxplot representation for the data collection.

(c) What important features of the data collection are observable from the five-number summary and boxplot?

1.2.10. *Bird Variety.* Consider the numbers of bird species data collection presented in Table 1.7.

(a) Find the mean $\bar{x}$ and median $\tilde{x}$ for these data.

(b) Find the variance s^2 and standard deviation s for these data.

1.2.11. *Bumped Airline Passengers.* The U.S. Department of Transportation (2014) provided data on the numbers of passengers bumped (i. e., involuntarily denied reserved seats on overbooked planes) during the calendar years 2012 and 2013 for the following major U.S. airlines:

Airline	Number of passengers bumped	
	2012	**2013**
Airtran Airways	2060	2302
Alaska	1103	714
American	5571	3233
American Eagle	1945	1923
Delta	5342	6070
Frontier	808	1272
Hawaiian	168	172
JetBlue	39	19
Skywest	5990	6768
Southwest	9490	12,221
United	14,394	9015
US Airways	3755	3531

(a) Find the five-number summaries separately for the 2012 and 2013 data.

(b) Use appropriate software to make separate boxplots for the 2012 and 2013 data.

(c) Compute the value of the *1*-th trimmed mean $\bar{x}_1$ and compare it with the values of the mean $\bar{x}$ and median $\tilde{x}$ for the 2013 numbers of bumped passengers. Does this comparison identify any important features of the 2013 data collection?

(d) Which airlines have improved their reservation operations from 2012 to 2013? Which have worsened? Justify your conclusions with appropriate statistics.

(e) Do you think your conclusions in part (d) might change if the data were percentages of passengers bumped, rather than the numbers of passengers bumped? Why or why not? (See Exercise 1.2.27.)

1.2.12. *Bumped Airline Passengers.* Consider the data presented in Exercise 1.2.11 for the numbers of bumped passengers in 2012 and 2013 for the 12 listed airlines.

(a) Find the mean change from 2012 to 2013 in the number of bumped passengers for the twelve airlines.

(b) Find the variance and standard deviation for the changes in the number of bumped passengers from 2012 to 2013 for the twelve airlines.

(c) Repeat parts (a) and (b) of this exercise without the United Airlines data. Comment on the reason for the observed differences between the means, variances, and standard deviations with and without the inclusion of United Airlines.

1.2.13. *Best Paid University Professors.* Which college/university professors are paid the most? Average 2011–2012 annual salaries (expressed in U.S. dollars) for academic employment as a new Assistant Professor, reported by the College and University Professional Association for Human Resources (2012), are presented in Table 1.19 for fourteen professional categories.

(a) Find the mean $\bar{x}$ and median $\tilde{x}$ for the average 2011–2012 academic salary for new assistant professors for the twenty-four disciplines.

(b) Find the variance s^2 and standard deviation s for the average 2011–2012 academic salary for new assistant professors for the 24 disciplines.

(c) What are the standardized z-scores for Engineering and Mathematics and Statistics?

(d) What are the standardized z-scores for Legal Professions and Visual and Performing Arts?

(e) Repeat parts (a)–(d) with the twenty-two professional categories obtained by eliminating the two categories of Business/

Table 1.19 Average annual salaries (U.S. dollars) for academic employment in 2011–2012 as a new assistant professor in twenty-four broad professional categories

Discipline	Average annual salary
Agriculture Related Sciences	68,999
Architecture and Related Services	63,386
Biological and Biomedical Sciences	57,249
Business, Management, and Marketing	98,212
Communication, Journalism and Related Programs	54,415
Computer and Information Sciences	74,563
Education	55,618
Engineering	78,650
English Language and Literature	52,405
Ethnic, Cultural, and Gender Studies	61,310
Family and Consumer Sciences	61,198
Foreign Languages, Literatures, and Linguistics	53,457
Health Professions	66,049
History	53,425
Legal Professions	96,955
Library Science	56,551
Mathematics and Statistics	58,266
Natural Resources and Conservation	62,926
Parks, Recreation, Leisure, and Fitness Studies	55,497
Philosophy and Religious Studies	54,340
Physical Sciences	58,786
Psychology	56,195
Social Sciences	60,240
Visual and Performing Arts	52,241

Source: College and University Professional Association for Human Resources (2012)

Management/Marketing and Legal Professions. Compare and contrast the results with your answers to parts (a)–(d).

1.2.14. *Most Employable Doctorates.* Which professional disciplines provide the best opportunities for employment for graduates with Ph. D. degrees? The data in Table 1.20 are the estimated percentage unemployed as of February 2013 for Ph. D. graduates in eight broad professional discipline categories, as reported by the National Science Foundation (2014).

Table 1.20 Percentage unemployed as of February 2013 for Ph. D. graduates in eight broad professional discipline categories

Discipline category	**Percentage unemployed**
Biological/Agricultural/Environmental Life Sciences	2.2
Computer/Information Sciences	1.8
Mathematics/Statistics	1.2
Physical Sciences	2.7
Psychology	1.6
Social Sciences	1.9
Engineering	1.9
Health	2.0

Source: National Science Foundation (2014).

(a) Find the mean $\bar{x}$ and median $\tilde{x}$ for the percentage unemployed Ph. D. graduates in February 2013 across the eight categories.

(b) Find the variance s^2 and standard deviation s for the percentage unemployed Ph. D. graduates in February 2013 across the eight categories.

(c) What are the standardized z-scores for Physical Sciences and for Biological/Agricultural/Environmental Life Sciences?

(d) What are the standardized z-scores for Computer/Information Sciences and for Mathematics/Statistics?

1.2.15. A data set contains four observations, $x_1 = 30$, $x_2 = 70$, $x_3 = 10$, and $x_4 = 90$. Find the values of $x_{(1)}$, $x_{(2)}$, $x_{(3)}$, and $x_{(4)}$.

1.2.16. A data collection $(x_1, \ldots, x_n)$ contains n observations.

(a) Give a formula for the median, $\tilde{x}$, of this data collection if $n = 77$.

(b) Give a formula for the median, $\tilde{x}$, of this data collection if $n = 100$.

1.2.17. *Engineering Drawing Hours.* Consider the total engineering drawing hours data displayed in Table 1.3. In Examples 1.5 and 1.8 we found the mean, median, and interquartile range for these data to be $\bar{x} = 22.2$ h, $\tilde{x} = 14$ h, and $IQR = 15$ h, respectively.

(a) If each of the 96 engineering drawing times is converted from hours to minutes, what will the corresponding values of the mean, median, and interquartile range be in minutes?

(b) If each of the 96 engineering drawing times is converted from hours to days, what will the corresponding values of the mean, median, and interquartile range be in days?

1.2.18. *Left and Right Triples.* Compute the numbers of left and right triples for the data collection { 1, 2, 3, 12, 15}. What do the triples indicate about symmetry versus skewness for the data collection?

1.2.19. *Bumped Airline Passengers.* Consider the data on numbers of passengers bumped (i. e., involuntarily denied reserved seats on overbooked planes) during the calendar year 2013, as given in Exercise 1.2.11 for twelve major U.S. airlines. Compute the numbers of left and right triples for these data. What do these numbers tell you about the symmetry/asymmetry of the data collection?

1.2.20. *Bumped Airline Passengers.* Consider the data on numbers of passengers bumped (i. e., involuntarily denied reserved seats on overbooked planes) during the calendar year 2013, as given in Exercise 1.2.11 for twelve major U.S. airlines. Compute the mean number of passengers bumped for these twelve airlines. How many observations in the data collection are less than this mean? What does this tell you about the symmetry/asymmetry of the data collection?

1.2.21. *Firearms in the Home.* Consider the state-by-state percentages of households with firearms in or around their homes, as given in Table 1.9. Find the mean for this data collection. How many observations in the data collection are less than this mean? What does this tell you about the symmetry/asymmetry of the data collection?

1.2.22. *Male Math SAT Scores.* Consider the data collection of male math SAT scores given in Table 1.15.

(a) Compute the nine deciles for this data collection.

(b) Select nine reasonable fixed fenceposts and summarize this data collection by recording the percentages of observations falling between each of the fenceposts.

1.2.23. *Firearms in the Home.* Consider the state-by-state percentages of households with firearms in or around their homes, as given in Table 1.9.

(a) Compute the nine deciles for this data collection.

(b) Select nine reasonable fixed fenceposts and summarize this data collection by recording the percentages of observations falling between each of the fenceposts.

1.2.24. *Residents in Poverty by States.* Use the five fixed fenceposts $f_1 = 250{,}000$, $f_2 = 500{,}000$, $f_3 = 750{,}000$, $f_4 = 1{,}500{,}000$, and $f_5 = 3{,}000{,}000$ to summarize the 2013 Census data in Table 1.4 on the number of individuals at poverty level in each of the fifty states.

1.2.25. *Women and Poverty.* Use eight fixed fenceposts of your choosing to summarize the 2010 Census data in Table 1.5 on percentage poverty rates for women in each of the fifty states.

1.2.26. *Bird Variety.* Consider the numbers of bird species data collection presented in Table 1.7.

(a) Find the five-number summary for this data collection.

(b) Use an appropriate software package to make a boxplot representation for the data collection.

1.2.27. *Which Airline Carries the Most Passengers?* The U.S. Department of Transportation (2014) provided data on the total numbers of passengers carried during the calendar years 2012 and 2013 for the following major U.S. airlines:

Airline	**Number of passengers carried**	
	2012	**2013**
Airtran Airways	21,744,193	17,832,245
Alaska	17,375,336	18,517,953
American	75,883,719	76,062,625
American Eagle	18,115,456	16,939,092
Delta	103,957,050	106,783,155
Frontier	10,324,099	10,361,896
Hawaiian	9,476,251	9,928,830
JetBlue	26,915,983	28,166,771
Skywest	25,867,287	26,518,312
Southwest	112,531,171	115,645,836
United	78,728,448	77,212,471
US Airways	55,237,069	57,834,693

(a) Find the five-number summaries separately for the 2012 and 2013 data.

(b) Use appropriate software to make separate boxplots for the 2012 and 2013 data.

(c) Compute the value of the *1*-th trimmed mean $\bar{x}_1$ and compare it with the values of the mean $\bar{x}$ and median $\tilde{x}$ for the 2013 numbers of passengers carried. Does this comparison identify any important features of the 2013 data collection?

(d) Which airlines have improved the scope of their operations from 2012 to 2013? Which have worsened? Justify your conclusions with appropriate statistics.

1.2.28. *Airline Passengers—Carried versus Bumped.* Consider the number of passengers carried and the number of bumped passengers for the major U.S. airlines, as given in Exercises 1.2.27 and 1.2.11, respectively. Create a new data collection for these U.S. airlines corresponding to the number of passengers bumped per 10,000 passengers carried for each of the years 2012 and 2013. Do the same analyses required in Exercise 1.2.11 for this new data collection. Compare and contrast your results with those obtained in Exercise 1.2.11 for the number of passengers bumped.

1.3 Comparing One-Variable Data Collections

In the previous two sections we emphasized various ways to describe and summarize a single one-variable data collection. Now we turn to the problem of comparison of two such data collections. This type of problem is very common in statistics as we attempt to use observations to decide if there are differences between two different groups or treatments or modes of operation.

In earlier sections of this chapter we presented a number of graphical and statistical methods for displaying and summarizing a single, quantitative data collection. It is then natural to compare two such quantitative data collections by looking at the graphical and statistical summaries for each collection.

For example, in Table 1.16 we presented the motor-vehicle fatality rates per 100 million vehicle miles traveled for each state during the calendar year 2012 and we illustrated the computation of several measures of center and variability for this data collection. Now let us see if we can find *regional* differences in death rates between the eastern and western halves of the country. We split the data into two collections, one for the eastern states and one for the western states, calling the eastern observations x and the western observations y.

We might want to know which region contains the states with the larger fatality rates per 100 million vehicle miles traveled. This is a question about the *centers* of the two data collections, so we will compute separate measures of center for each region and compare them. For example, if the difference in means, $\bar{y} - \bar{x}$, is positive, this will suggest that the western region contains the states with larger death rates. We could also look at the difference in medians, $\tilde{y} - \tilde{x}$, or the difference in trimmed means, $\bar{y}_d - \bar{x}_d$.

On the other hand, we might be interested in finding out which region has greater spread among its states' fatality rates. This is a question about *variability*, so we will compute statistics such as the range, interquartile range, or standard deviation for each region. Since measures of variability are affected by multiplicative (scale) changes in the data but not by changes in location (shift), to assess possible differences in variability between two data collections the natural comparisons are obtained by taking ratios rather than differences in the separate measures of variability. Thus we will look at ratios like R_y/R_x, IQR_y/IQR_x, or s_y/s_x.

Example 1.17. Regional Motor-Vehicle Fatality Rates (Extension of Example 1.4) In order to see if there are regional differences in these fatality rates, we divide the 50 states into two subcollections, corresponding to whether a state is east or west of the Mississippi River. We present the state fatality rates categorized in this fashion in Table 1.21. We can also create these two subcollections using the following **R** commands.

```
> east_states <- motor_vehicle_death_rate_2012[motor_vehicle_death_rate
_2012$Region == "East",]

> west_states <- motor_vehicle_death_rate_2012[motor_vehicle_death_rate
_2012$Region == "West",]
```

With this division, the eastern group consists of $m = 26$ states, while there are $n = 24$ states in the western group. As noted previously, we let x and y correspond to the eastern and western states, respectively. Computing the means, medians, ranges, interquartile ranges, and standard deviations for the two groups, we obtain:

Eastern States: $\bar{x} = 1.126$, $\tilde{x} = 1.055$, $R_x = 1.14$, $IQR_x = 0.398$, and $s_x = 0.303$
Western States: $\bar{y} = 1.240$, $\tilde{y} = 1.240$, $R_y = 1.030$, $IQR_y = 0.383$, and $s_y = 0.287$.

Table 1.21 Total motor-vehicle fatalities per 100 million vehicle miles traveled during calendar year 2012, categorized by location east or west of the Mississippi River

East of Mississippi		West of Mississippi	
State	**Fatalities**	**State**	**Fatalities**
Alabama	1.33	Alaska	1.23
Connecticut	0.75	Arizona	1.37
Delaware	1.24	Arkansas	1.65
Florida	1.27	California	0.88
Georgia	1.11	Colorado	1.01
Illinois	0.91	Hawaii	1.25
Indiana	0.99	Idaho	1.13
Kentucky	1.58	Iowa	1.16
Maine	1.16	Kansas	1.32
Maryland	0.89	Louisiana	1.54
Massachusetts	0.62	Minnesota	0.69
Michigan	0.99	Missouri	1.21
Mississippi	1.51	Montana	1.72
New Hampshire	0.84	Nebraska	1.10
New Jersey	0.79	Nevada	1.07
New York	0.91	New Mexico	1.43
North Carolina	1.23	North Dakota	1.69
Ohio	1.00	Oklahoma	1.48
Pennsylvania	1.32	Oregon	1.01
Rhode Island	0.82	South Dakota	1.46
South Carolina	1.76	Texas	1.43
Tennessee	1.42	Utah	0.82
Vermont	1.07	Washington	0.78
Virginia	0.96	Wyoming	1.33
West Virginia	1.76		
Wisconsin	1.04		

Source: National Highway Traffic Safety Administration (2013)

It is clear from these summary measures that there are, in fact, distinct differences between the motor-vehicle fatality rates in 2012 for these western and eastern groupings of states. The differences in mean and median death rates are $\bar{y} - \bar{x} = 1.240 - 1.126 = 0.114$ and $\tilde{y} - \tilde{x} = 1.240 - 1.055 = 0.185$, respectively, fatalities per 100 million vehicle miles, which indicates higher centers for the western grouping of states. The ratios of the ranges and standard deviations, namely, $R_x/R_y = 1.107$ and $s_x/s_y = 1.056$, respectively, indicate that there is slightly more variability present in the death rates for the eastern grouping of states. Why do you suppose that these differences exist for this particular division of the fifty states? Perhaps you can get more insight into these differences in the death rates by constructing separate histograms for the two groups. Are there other regional divisions of the United States that would make sense to examine as groups? Are there reasons other than geographical for comparing various groupings of the states with respect to motor-vehicle death rates? What about speed limits and seat-belt laws?

As illustrated in this example, many of the standard comparisons of two groups with respect to measures of center and variability do not involve any new forms of graphical or statistical summaries other than those discussed previously for a single data collection. However, statisticians have developed a number of graphical and statistical summaries that are specifically designed for comparison of two or more data collections and we devote the rest of this section to several of these approaches.

Graphical Displays to Compare Two Quantitative Data Collections: Back-to-Back Stemplots and Parallel Boxplots Two particularly nice ways to graphically display similarities or differences in two different data collections on the same measurement are provided by back-to-back stemplots and parallel boxplots. Since the underlying concepts are already familiar to us from previous work with a single data collection, we simply illustrate their adaptation to comparison of two groups via a pair of examples.

Example 1.18. Asbestosis and Lung Function (Extension of Example 1.9) In Example 1.9 we discussed the effect of exposure to asbestos dust on the potential loss of lung function. In that example, we considered a set of data gathered by Al Jarad et al. (1993) for a sample of twenty asbestos workers who had not (yet) contracted asbestosis. The same authors also presented similar data for a group of 30 such workers who had already been diagnosed with asbestosis. The percent decreases in lung function over the study period of slightly longer than 4 years for these thirty subjects with asbestosis are presented in Table 1.22.

We produce back-to-back stemplots in Fig. 1.12 for the two groups of decreases in lung function presented in Tables 1.18 and 1.22 for exposed workers without and with asbestosis, respectively. In this figure we have a common stem that provides the values of the tenths decimal place for the two data collections. The leaves then emanate in both directions from this common stem, to the left for workers with asbestosis and to the right for those workers free of the disease, and these leaves correspond to the hundredth decimal place for the observations in the two data collections.

Although no dramatic differences are apparent, the back-to-back stemplots clearly indicate a slightly higher average decrease in lung function over the 4 years in the study for those workers diagnosed with asbestosis. Why don't you see what other statistical techniques (graphical or summary) tell you about the decrease in lung function for these two groups?

To illustrate the use of parallel (also known as side-by-side) boxplots, we return to the regionally divided data on motor-vehicle death rates.

Example 1.19. Regional Motor-Vehicle Death Rates (Continuation of Example 1.17) A graphical summary of the motor-vehicle fatality rates in 2012 for the groupings of eastern (region 1) and western (region 2) states are presented in parallel boxplots in Figure 1.13. These side-by-side boxplots provide good visual verification of a number of the features we discovered with our numerical calculations in Example 1.17. For example, it is graphically clear from the parallel boxplots that the eastern grouping of states has a lower median fatality rate than the western grouping but that this lower median is accompanied by a slightly greater variability.

Table 1.22 Percent decrease in lung function for asbestos workers who have asbestosis

Subjects	Percent decrease in lung function
1	.02
2	.30
3	.08
4	.02
5	.08
6	.24
7	.30
8	.15
9	.02
10	.16
11	.23
12	.30
13	.15
14	.02
15	.25
16	.18
17	.08
18	.22
19	.23
20	.10
21	.11
22	.07
23	.15
24	.10
25	.50
26	.10
27	.15
28	.07
29	.15
30	.23

Source: Al Jarad et al. (1993)

Comparing Categorical Distributions of Quantitative Data: Fixed Fenceposts For many data collections a simple comparison by means of fixed fenceposts provides sufficient detail to highlight important similarities and contrasts. A particularly nice feature of this mode of comparison is that it

Fig. 1.12 Back-to-back stemplots for asbestos lung function loss data

Workers with Asbestosis		Workers without Asbestosis
888772222	.0	000022268
86555551000	.1	034668
543332	.2	0555
000	.3	0
	.4	
0	.5	

Fig. 1.13 Parallel boxplots for motor-vehicle death rates in 1995 for eastern and western groupings of states

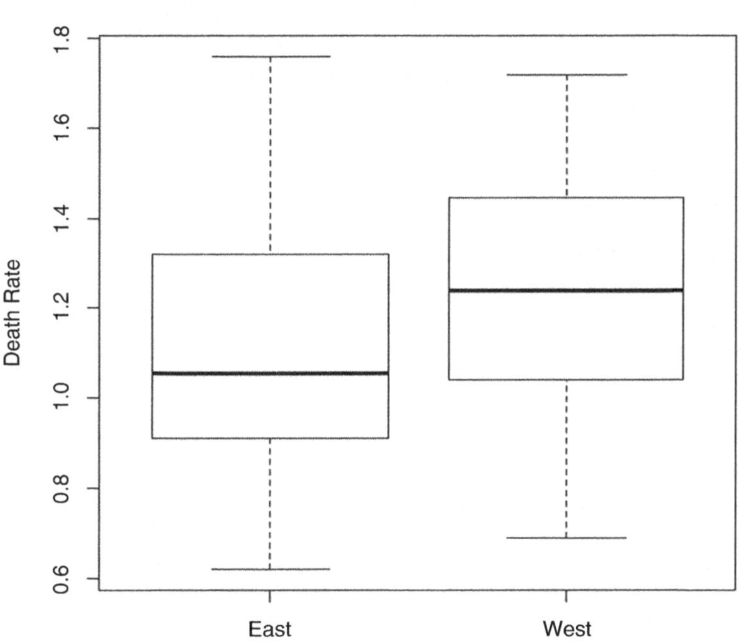

extends easily to more than two data collections (as does the use of parallel boxplots). We illustrate this use of fixed fenceposts in a comparison of major league baseball salary data for 2014.

Example 1.20. Comparison of 2014 Salary Figures for the New York Yankees and Cincinnati Reds The dataset *national_league_salary_2014* contains the 2014 salaries (as of March 26, 2014) for all baseball players in

Table 1.23 Baseball salaries for members of the Cincinnati Reds baseball team for 2014 (as of March 26, 2014)

Name of player	Total 2014 salary
Joey Votto	12,000,000
Brandon Phillips	11,083,333
Jay Bruce	10,041,667
Johnny Cueto	10,000,000
Homer Bailey	9,000,000
Ryan Ludwick	8,500,000
Aroldis Chapman	7,835,772
Mat Latos	7,250,000
Jonathan Broxton	7,000,000
Mike Leake	5,925,000
Sean Marshall	5,625,000
Manny Parra	2,000,000
Skip Schumaker	2,000,000
Chris Heisey	1,760,000
Alfredo Simon	1,500,000
Logan Ondrusek	1,425,000
Sam LeCure	1,200,000
Ramon Santiago	1,100,000
Jack Hannahan	1,000,000
Brayan Pena	875,000
Zack Cozart	600,000
Todd Frazier	600,000
Devin Mesoraco	525,000
J. J. Hoover	520,000
Pedro Beato	512,500
Tony Cingrani	512,500
Nick Christiani	500,000
Billy Hamilton	500,000
Brett Marshall	500,000
Neftali Soto	500,000

Source: Petchesky (2014)

the National League. In Table 1.23 we have recorded the salaries for the players on the Cincinnati Reds and we will use fixed fenceposts to compare this collection of salaries with those for the New York Yankees previously given in Table 1.17.

Using the $k = 4$ fixed fenceposts at $f_1 = 600{,}000$, $f_2 = 1{,}500{,}000$, $f_3 = 3{,}000{,}000$, and $f_4 = 10{,}000{,}000$ (and remembering that the square bracket indicates inclusion so that, for example, the category ($600,000, $1,500,000] includes a salary of $1,500,000), we obtain the following numbers (percentages) of New York Yankee and Cincinnati Reds baseball players with 2014 salaries in the five categories created by these four fenceposts:

Fencepost categories	New York Yankees	Cincinnati Reds
$\leq$ $600,000	13 (39.4%)	10 (33.3%)
($600,000, $1,500,000]	1 (3.0%)	6 (20%)
($1,500,000, $3,000,000]	4 (12.1%)	3 (10%)
($3,000,000, $10,000,000]	6 (18.2%)	8 (26.7%)
$>$ $10,000,000	9 (27.3%)	3 (10%)

If you are a major league ballplayer, does the use of these fixed fenceposts provide a clear picture of whether you would rather play for the Yankees or the Reds? How about as a baseball fan–which team would you rather root for?

Comparing Typical Observations from Two Data Collections: Using Individual Comparisons and Joint Ranks In many practical problems our primary interest is simply in a comparison of a typical observation from each of the two data collections. In such situations, there are more direct statistical approaches designed specifically for comparison of two data collections.

Suppose that x and y denote single observations randomly selected from data collections 1 and 2, respectively. If you were asked to use the pair (x, y) to evaluate the relative nature of typical observations from each of the two data collections, you would likely consider the difference $y - x$ as the primary source of information. However, it is important to note that the information in this difference consists of two components, namely, its sign and its magnitude. The sign provides evidence as to whether a typical observation from data collection 2 is larger or smaller than a typical observation from data collection 1. The magnitude $|y-x|$, on the other hand, helps assess the size of the difference between typical observations from the two data collections.

With this in mind, suppose we are asked to provide such an assessment of the similarity or difference between typical observations from the two data collections $(x_1, \ldots, x_m)$ and $(y_1, \ldots, y_n)$, but now we are permitted to use all of the data values from both collections. Based on our recent discussion, it is quite natural to base this assessment exclusively on the differences $d_{ij} = y_j - x_i$, for $i = 1, \ldots, m$ and $j = 1, \ldots, n$. How many such differences are available to us? For the data collections $(x_1, x_2, x_3) = (1, 4, 7)$ and $(y_1, y_2) = (3, 9)$, there are six differences:

$y_1 - x_1 = 3 - 1 = 2$	$y_2 - x_1 = 9 - 1 = 8$
$y_1 - x_2 = 3 - 4 = -1$	$y_2 - x_2 = 9 - 4 = 5$
$y_1 - x_3 = 3 - 7 = -4$	$y_2 - x_3 = 9 - 7 = 2.$

In this example we have 3×2 or 6 differences. In general, when there are m x's and n y's each x can be paired with each y and we have a total of mn differences at our disposal.

Virtually any statistical method for comparison of the typical observations from the two data collections will rely exclusively on these differences. Where the various statistical methods differ is in how they utilize this information.

One important piece of information is contained solely in the signs of the differences d_{ij}.

Definition 1.20 Let $(x_1, \ldots, x_m)$ and $(y_1, \ldots, y_n)$ be two data collections. The **counting statistic *U*** is the proportion of (x, y) pairs for which y is at least as large as x.

Clearly the statistic U provides us with partial information about the size relationship between typical observations from each of the two data collections. Values of U near 1/2 are indicative that the two data collections are similar in the sizes of their typical observations, since such values of U correspond to roughly one-half ($mn/2$) of the (x, y) pairs being such that y is at least as large as x, while for the other roughly one-half of the (x, y) pairs we have x at least as large as y. On the other hand, values of U near 1 provide rather strong evidence that the typical y observation is larger than the typical x observation. Conversely, values of U near 0 suggest that typical x's are larger than typical y's.

A convenient way to compute U for two given data collections is to look at all the differences d_{ij}. Count 1 if $d_{ij} \geq 0$ and count 0 if $d_{ij} < 0$. Then add up these counts and divide by mn (the total number of d_{ij}) to get U.[2]

In addition to the proportion statistic U that is based solely on the signs of the observed d_{ij} differences, their magnitudes can also be used to provide information about the magnitude of the typical $(y - x)$ difference. Here we rely on our 'old friends', the mean and median, to summarize the magnitude information contained in the individual d_{ij} differences.

First, consider the mean, $\bar{D}$, of the mn d_{ij} differences. If you work through the arithmetic, you will find that the mean of the d_{ij}'s is just $\bar{y} - \bar{x}$, the difference in the means of the two collections! Thus we do not obtain a truly

---

[2] Notationally, let h_{ij} be defined by

$$\begin{aligned} h_{ij} &= 1, \quad \text{if } d_{ij} \geq 0 \\ &= 0, \quad \text{if } d_{ij} < 0, \end{aligned}$$

for $i = 1, \ldots, m$ and $j = 1, \ldots, n$. Then

$$U = \frac{1}{mn} \sum_{i=1}^{m} \sum_{j=1}^{n} h_{ij}, \tag{1.10}$$

where the double sum says "add up the h's for all i and j combinations".

new comparison measure for the typical observations from the two data collections by averaging the pairwise d_{ij} differences.

What about the application of our median criterion to the individual d_{ij} differences? We denote the median of the differences by $\tilde{D}$. Is $\tilde{D}$ simply another representation for the difference, $\tilde{y} - \tilde{x}$, in the separate medians for the two data collections, just as $\bar{D}$ is for the separate means? The answer is no (as you are asked to verify in Exercise 1.A.6) and thus $\tilde{D}$ provides us with a statistic that is designed specifically to assess the magnitude of the difference between typical observations from *two* data collections. (We should point out that when both the $(x_1, \ldots, x_m)$ and $(y_1, \ldots, y_n)$ data collections are roughly symmetric in shape and neither contains a substantial number of outliers, the mean difference $\bar{D}$ and the median difference $\tilde{D}$ will generally be numerically close.)

Definition 1.21 Let $(x_1, \ldots, x_m)$ and $(y_1, \ldots, y_n)$ be two data collections. The **median difference statistic** is given by

$$\tilde{D} = \text{median}\left\{y_j - x_i\right\}, \tag{1.11}$$

where we consider all possible pairs that take one observation from each data collection.

We note that the counting statistic U (1.10) and the median difference statistic $\tilde{D}$ (1.11) provide quite different pieces of information about the relationship between a typical y observation and a typical x observation. Information about the *proportion* of y observations that are at least as large as x observations is provided by the statistic U, while the *size* of the difference between a typical y observation and a typical x observation is provided by $\tilde{D}$.

Example 1.21. Interstitial Lengths of Pine Species Habitat plays an important role in fish behavior, particularly feeding, spawning, and protection/

security. One of the modern methods of fisheries management is habitat modification in large, man-made reservoirs. Previous studies have shown that the type of structure introduced is an important factor in such habitat modifications. Of particular relevance in many settings is the size of openings or interstices in the introduced structure. The data in Table 1.24 represent a subset of that obtained by Kayle (1984) from Alum Creek Lake in Westerville, Ohio in a study to determine the relative effectiveness of blue spruce and white pine trees for habitat modification. The measurements in Table 1.24 are averages (*mm*) of interstitial lengths (distances between midpoints) of ten pairs of secondary branches for each of twelve blue spruce and twelve white pine trees.

Letting the blue spruce and white pine measurements correspond to the y and x data collections, respectively, we have $m = n = 12$ and there are 12 (12) = 144 differences y_j - x_i to compute. For example, the difference between the first y (blue spruce) observation and the first x (white pine) observation is $y_1 - x_1 = 46.7 - 75.2 =$ -28.5, which happens to be the smallest such difference for the two data sets. Using the **R** functions *outer*( ) and *sort*( ) for the dataset

Table 1.24 Mean interstitial lengths for blue spruce and white pine trees (*mm*)

Blue spruce (y)	White pine (x)
46.7	75.2
60.5	63.7
58.9	73.2
82.9	66.2
65.8	67.4
93.3	69.4
66.9	70.4
70.9	72.3
73.7	63.6
65.8	61.9
90.2	74.4
68.9	70.1

Source: Kayle (1984)

interstitial_lengths, we obtain the entire 144 d_{ij} differences, ordered from least to greatest, to be the following:

d_{ij}	d_{ij}	d_{ij}	d_{ij}
−28.5	−6.9	−0.5	10.0
−27.7	−6.5	−0.4	10.1
−26.5	−6.5	−0.4	10.6
−25.6	−6.3	0.5	11.8
−23.7	−6.3	0.5	12.5
−23.4	−5.7	0.7	12.8
−22.7	−5.5	0.8	13.5
−20.7	−5.4	1.4	15.0
−19.5	−4.8	1.5	15.5
−17.0	−4.7	1.5	15.8
−16.9	−4.6	2.1	16.7
−16.3	−4.6	2.1	17.0
−15.5	−4.3	2.2	17.9
−15.2	−4.3	2.2	18.1
−14.7	−4.3	2.7	18.9
−14.3	−4.3	3.2	19.2
−13.9	−3.6	3.3	19.3
−13.4	−3.6	3.3	19.8
−12.7	−3.5	3.5	20.1
−11.8	−3.5	3.6	20.1
−11.5	−3.4	3.9	20.8
−11.2	−3.2	3.9	21.0
−10.5	−3.2	4.3	21.0
−9.9	−3.1	4.7	22.8
−9.6	−3.0	5.0	22.9
−9.4	−2.5	5.2	23.2
−9.4	−2.3	5.3	23.9
−8.9	−1.6	6.3	24.0
−8.6	−1.6	7.0	25.9
−8.6	−1.5	7.2	26.5
−8.5	−1.5	7.3	26.6
−8.3	−1.4	7.5	27.1
−7.5	−1.4	7.7	28.3
−7.4	−1.2	8.5	29.6
−7.4	−0.7	9.0	29.7
−7.3	−0.5	9.7	31.4

The **R** command and output used to obtain this ordered listing follow.

```
> sort(outer(interstitial_lengths$"Blue Spruce", interstitial_lengths$"White Pine", "-"))
[1] -28.5 -27.7 -26.5 -25.6 -23.7 -23.4 -22.7 -20.7 -19.5 -17.0 -16.9 -16.3 -15.5 -15.2 -14.7 -14.3 -13.9 -13.4
[19] -12.7 -11.8 -11.5 -11.2 -10.5 -9.9 -9.6 -9.4 -9.4 -8.9 -8.6 -8.6 -8.5 -8.3 -7.5 -7.4 -7.4 -7.3
[37] -6.9 -6.5 -6.5 -6.3 -6.3 -5.7 -5.5 -5.4 -4.8 -4.7 -4.6 -4.6 -4.3 -4.3 -4.3 -4.3 -3.6 -3.6
[55] -3.5 -3.5 -3.4 -3.2 -3.2 -3.1 -3.0 -2.5 -2.3 -1.6 -1.6 -1.5 -1.5 -1.4 -1.4 -1.2 -0.7 -0.5
[73] -0.5 -0.4 -0.4 0.5 0.5 0.7 0.8 1.4 1.5 1.5 2.1 2.1 2.2 2.2 2.7 3.2 3.3 3.3
[91] 3.5 3.6 3.9 3.9 4.3 4.7 5.0 5.2 5.3 6.3 7.0 7.2 7.3 7.5 7.7 8.5 9.0 9.7
[109] 10.0 10.1 10.6 11.8 12.5 12.8 13.5 15.0 15.5 15.8 16.7 17.0 17.9 18.1 18.9 19.2 19.3 19.8
[127] 20.1 20.1 20.8 21.0 21.0 22.8 22.9 23.2 23.9 24.0 25.9 26.5 26.6 27.1 28.3 29.6 29.7 31.4
```

From this ordered listing we see that 69 of the 144 d_{ij} differences are positive, so that the proportion of (white pine, blue spruce) sample pairs for which the blue spruce interstitial length exceeds that for the white pine is $U = 69/144 = .479$. Moreover, since $mn = 144$ is an even number, the median difference statistic $\tilde{D}$ (1.11) is given by the average of the two middle ordered differences, corresponding to $\tilde{D} = (-0.5 - 0.5)/2 = -0.5$ *mm*. Since the sample value of U is close to 1/2 and the median of the mean interstitial length differences is only −0.5 mm, which is quite small relative to the interstitial length values, there is little evidence in the two data collections studied by Kayle to indicate any substantial difference in the typical mean interstitial lengths for blue spruce and white pine trees. (We note that the difference in means between the blue spruce and white pine trees is $\bar{D} = 1.39$ mm. Look once again at the mean interstitial length data in Table 1.24. Why do you think that the median difference $\tilde{D}$ (negative) and the mean difference (positive) have opposite signs for these two data collections?)

One additional way to provide a direct summary comparison for two data collections of a common measurement is to consider the relative positions of the observations from the two data collections. We first consider the relative positions for observations from a single data collection.

Definition 1.22 Let $(x_1, \ldots, x_n)$ be a collection of n distinct observations. The **rank of x_i, denoted Q_i**, is the number of observations in the collection which are less than or equal to x_i.

Example 1.22. Mean Interstitial Lengths for White Pine Trees Consider the data collection of mean interstitial lengths for the 12 white pine trees, as given in Table 1.24. The following table lists the ranks for these observations. Note that 75.2 is the largest observation, so it gets rank 12. Since three data values are less than or equal to 63.7, this observation gets rank 3. Proceeding similarly for the rest of the twelve observations leads to the ranks in the table.

Interstitial length	Rank
75.2	12
63.7	3
73.2	10
66.2	4
67.4	5
69.4	6
70.4	8
72.3	9
63.6	2
61.9	1
74.4	11
70.1	7

When the values in a data collection are not distinct (i. e., there are ties among the observations), we average the ranks associated with the tied values. Thus, for the data collection (3.4, 5.6, 3.4, 7.2, 5.6), the ranks are given by

Observation	Rank
3.4	1.5
5.6	3.5
3.4	1.5
7.2	5
5.6	3.5

The two smallest observations should get ranks 1 and 2. Since they are equal for this data collection, we give them both the average of these ranks, $(1 + 2)/2 = 1.5$. Similarly, the two 5.6 observations would have ranks 3 and 4, so they are both assigned the average rank, $3.5 = (3 + 4)/2$.

To use the idea of ranks to compare two data collections, we *jointly rank* (from least to greatest, with average ranks once again used for ties) the observations from both collections and then *separately* list the ranks associated with each of the data collections. Such a ranking provides an additional picture of the relative values of the observations in the two data collections. Often a further summary statistic is computed by obtaining the average rank associated with each data collection.

Example 1.23. Regional Motor-Vehicle Fatality Rates (Extension of Examples 1.4, 1.17, and 1.19) In Table 1.21 we presented the motor-vehicle fatality rates per 100 million vehicle miles traveled in calendar year 2012 for each of the 50 states categorized by its location east or west of the Mississippi River. We reproduce the data in Table 1.25, but this time with the joint ranking (from least to greatest) for each of the states given in bold type following its fatality rate. Note that an average rank of 6.5 has been assigned to the values for the two states of Rhode Island and Utah, each with a fatality rate of 0.82 per 100 million vehicle miles traveled. Similarly, average ranks of 11.5, 14.5, 17.5, 20.5, 25.5, 28.5, 33.5, 35.5, 39.5, and 49.5 were used for the fatality rates of the pairs of states (Illinois, New York), (Indiana, Michigan),

Table 1.25 Joint rankings of statewise motor-vehicle fatality rates per 100 million vehicle miles traveled during calendar year 2012, categorized by location east or west of the Mississippi River.

East of Mississippi			West of Mississippi		
State	**Fatality rate**	**Rank**	**State**	**Fatality Rate**	**Rank**
Alabama	1.33	**35.5**	Alaska	1.23	**28.5**
Connecticut	0.75	**3**	Arizona	1.37	**37**
Delaware	1.24	**30**	Arkansas	1.65	**46**
Florida	1.27	**32**	California	0.88	**9**
Georgia	1.11	**23**	Colorado	1.01	**17.5**
Illinois	0.91	**11.5**	Hawaii	1.25	**31**
Indiana	0.99	**14.5**	Idaho	1.13	**24**
Kentucky	1.58	**45**	Iowa	1.16	**25.5**
Maine	1.16	**25.5**	Kansas	1.32	**33.5**
Maryland	0.89	**10**	Louisiana	1.54	**44**
Massachusetts	0.62	**1**	Minnesota	0.69	**2**
Michigan	0.99	**14.5**	Missouri	1.21	**27**
Mississippi	1.51	**43**	Montana	1.72	**48**
New Hampshire	0.84	**8**	Nebraska	1.10	**22**
New Jersey	0.79	**5**	Nevada	1.07	**20.5**
New York	0.91	**11.5**	New Mexico	1.43	**39.5**
North Carolina	1.23	**28.5**	North Dakota	1.69	**47**
Ohio	1.00	**16**	Oklahoma	1.48	**42**
Pennsylvania	1.32	**33.5**	Oregon	1.01	**17.5**
Rhode Island	0.82	**6.5**	South Dakota	1.46	**41**
South Carolina	1.76	**49.5**	Texas	1.43	**39.5**
Tennessee	1.42	**38**	Utah	0.82	**6.5**
Vermont	1.07	**20.5**	Washington	0.78	**4**
Virginia	0.96	**13**	Wyoming	1.33	**35.5**
West Virginia	1.76	**49.5**			
Wisconsin	1.04	**19**			

Source: National Highway Traffic Safety Administration (2013)

(Colorado, Oregon), (Nevada, Vermont), (Iowa, Maine), (Alaska, North Carolina), (Kansas, Pennsylvania), (Alabama, Wyoming), (New Mexico, Texas), and (South Carolina, West Virginia), respectively.

An ordered listing of the joint ranks for states in the Eastern and Western collections, as well as the average rank for each group, follows:

Eastern Collection: 1, 3, 5, 6.5, 8, 10, 11.5, 11.5, 13, 14.5, 14.5, 16, 19, 20.5, 23, 25.5, 28.5, 30, 32, 33.5, 35.5, 38, 43, 45, 49.5, 49.5

Average joint rank for the Eastern collection $= \bar{Q}_{\text{Eastern}} = 587/26 = 22.58$

Western Collection: 2, 4, 6.5, 9, 17.5, 17.5, 20.5, 22, 24, 25.5, 27, 28.5, 31, 33.5, 35.5, 37, 39.5, 39.5, 41, 42, 44, 46, 47, 48

Average joint rank for the Western collection $= \bar{Q}_{\text{Western}} = 688/24 = 28.67$.

This joint ranking provides nice reinforcement for the observations previously made in Example 1.19 about the parallel boxplots for these data collections. The fact that the average rank is higher for the Western collection of states is in agreement with the observation from the parallel boxplots that the Western grouping has a higher median fatality rate.

This use of joint rankings to ascertain various patterns in more than one set of similar measurements can be a valuable tool for data analysis. We will return to such applications throughout the text.

Section 1.3 Practice Exercises

1.3.1. *Birthday Candles.* How long will birthday candles burn? Is it a function of how expensive they are? Koga (1999) conducted an experiment to address these questions. In particular, he was interested in how long it took two different brands of birthday candles to burn a prescribed length. The two brands he studied were: "Paper Art" (the cheaper candle) and "Party Express" (the expensive candle). The burning times (in seconds) data from his study are presented in Table 1.26.

(a) Compare the typical burning time for these two brands of birthday candles by computing the differences in the means, $\bar{x}_{PA}$ and $\bar{x}_{PE}$, and medians, $\tilde{x}_{PA}$ and $\tilde{x}_{PE}$, for the two data collections. Comment on your findings.

Table 1.26 Times (seconds) for birthday candles to burn a prescribed length

Paper art (y)	Party express (x)
543	385
606	739
623	728
634	749
749	427
770	452
760	424
812	473

Source: Koga (1999)

(b) Compare the variability in burning times for these two brands of birthday candles by computing the ratios of ranges, R_{PA} and R_{PE}, and standard deviations, s_{PA} and s_{PE}, for the two data collections. Comment on your findings.

1.3.2. *Firearms in the Home.* Apply the division of the fifty states into "Western States" and "Eastern States" used in Example 1.17 to the percentages of households that had firearms in or around their homes data in Table 1.9.

(a) Compare the typical percentage of households with firearms from the 2002 BRFSS survey interviews for these two state groupings by computing the differences in the means, $\bar{x}_{Western}$ and $\bar{x}_{Eastern}$, and medians, $\tilde{x}_{Western}$ and $\tilde{x}_{Eastern}$, for the two data collections.

(b) Compare the variability in percentage of households with firearms from the 2002 BRFSS survey interviews for these two state groupings by computing the ratios of ranges, R_{Western} and R_{Eastern}, and standard deviations, s_{Western} and s_{Eastern}, for the two data collections.

1.3.3. *Firearms in the Home.* Apply the division of the fifty states into "Western States" and "Eastern States" used in Example 1.17 to the percentages of households that had firearms in or around their homes data in Table 1.9.

Compare the percentages of households with firearms from the 2002 BRFSS survey interviews for these two state groupings by constructing

(a) back-to-back stemplots
(b) parallel boxplots.

1.3.4. *Motor-vehicle Fatalities.* Consider the total motor-vehicle fatalities per 100 million miles traveled during calendar year 2012 for each of the 50 states as presented in Table 1.16. Use some reasonable scheme to divide the 50 states into 25 "Northern" and 25 "Southern" states (as was done in Example 1.23 using the Mississippi River to produce "Eastern" and "Western" states). Discuss the similarities and differences in the motor-vehicle fatality rates for your groupings of "Northern" and "Southern" states by making use of

(a) appropriate summary statistics
(b) back-to-back stemplots and parallel boxplots.

1.3.5. Consider the two data collections

$$(x_1, x_2, x_3, x_4) = (1, 6, 9, 12) \quad \text{and} \quad (y_1, y_2, y_3) = (-3, 5, 15).$$

(a) How many $d_{ij} = y_j - x_i$ differences are there for these data collections?
(b) Compute all of these d_{ij} differences and obtain the value of $\tilde{D}$ (1.11). What does this tell us about the two data collections?

1.3.6. Consider the two data collections

$$(x_1, x_2, x_3, x_4) = (1, 6, 9, 12) \quad \text{and} \quad (y_1, y_2, y_3) = (-3, 5, 15).$$

(a) How many (x_i, y_j) pairs are there for these data collections?
(b) What proportion, U, of these (x, y) pairs are such that y is at least as large as x? What does this tell us about the two data collections?

1.3.7. *Baseball Salaries.* The datasets *national_league_salary_2014* and *american_league_salary_2014* contain the 2014 salaries (as of March 26, 2014) for all baseball players in the National and American Leagues, respectively. In Tables 1.27 and 1.28 we have recorded the salaries for the players on the St. Louis Cardinals and Baltimore Orioles baseball teams, respectively.

(a) Using the 4 fixed fenceposts at $f_1 = 600{,}000$, $f_2 = 1{,}500{,}000$, $f_3 = 3{,}000{,}000$, and $f_4 = 10{,}000{,}000$, compare the collections of 2014 salaries for the St. Louis Cardinals and Baltimore Orioles.

(b) Compare the results from part (a) with the similar comparison between the New York Yankees and the Cincinnati Reds discussed in Example 1.20. How would you combine these results to compare the 2014 salaries of all four of these baseball teams?

1.3.8. *Baseball Salaries.* Consider the 2014 salary data for the St. Louis Cardinals and Baltimore Orioles baseball teams, as given in Tables 1.27 and 1.28, respectively. Compare the salary data collections for the two baseball teams by constructing

(a) back-to-back stemplots

(b) parallel boxplots.

1.3.9. *Baseball Salaries.* Consider the 2014 salary data for the St. Louis Cardinals and Baltimore Orioles baseball teams, as given in Tables 1.27 and 1.28, respectively.

(a) Compare the typical 2014 salaries for these two baseball teams by computing the differences in the means, $\bar{x}_{SL}$ and $\bar{x}_B$, and medians, $\tilde{x}_{SL}$ and $\tilde{x}_B$, for the two data collections.

(b) Compare the variability in 2014 salaries for these two baseball teams by computing the ratios of ranges, R_{SL} and R_B, and standard deviations, s_{SL} and s_B, for the two data collections.

Table 1.27 Baseball salaries for members of the St. Louis Cardinals baseball team for 2014 (as of March 26, 2014)

Name of player	Total 2014 salary
Adam Wainwright	19,500,000
Matt Holliday	16,252,360
Jhonny Peralta	15,500,000
Yadier Molina	15,200,000
Jaime Garcia	7,875,000
Jason Motte	7,500,000
Mark Ellis	5,250,000
Jon Jay	3,250,000
Randy Choate	3,000,000
Allen Craig	2,750,000
Daniel Descalso	1,290,000
Matt Carpenter	1,250,000
Peter Bourjos	1,200,000
Lance Lynn	535,000
Joe Kelly	523,000
Arnoldi Cruz	521,000
Shelby Miller	521,000
Trevor Rosenthal	521,000
Shane Robinson	519,000
Pete Kozma	518,000
Matt Adams	516,000
Michael Wacha	510,000
Seth Maness	509,000
Carlos Martinez	505,000
Kevin Siegrist	505,000
Joey Butler	500,000
Jorge Rondon	500,000
Kolten Wong	500,000

Source: Petchesky (2014)

1.3.10. *Math SAT Scores.* The math SAT scores for seniors graduating in 2013 or 2014 from a small private school are given in Table 1.15. Compare the collections of these math SAT scores for males and females using:

Table 1.28 Baseball salaries for members of the Baltimore Orioles baseball team for 2014 (as of March 26, 2014)

Name of player	Total 2014 salary
Nick Markakis	15,350,000
Adam Jones	13,123,520
Ubaldo Jimenez	10,923,103
Chris Davis	10,350,000
Nelson Cruz	8,000,000
J. J. Hardy	7,916,667
Matt Wieters	7,700,000
Bud Norris	5,300,000
Wei-Yin Chen	4,155,333
Darren O'Day	3,200,000
Tommy Hunter	3,000,000
Brian Matusz	2,400,000
Ryan Webb	1,750,000
Troy Patton	1,100,820
Nolan Reimold	1, 025,000
Steve Pearce	850,000
Edgmer Escalona	550,000
Francisco Peguero	550,000
Chris Tillman	546,000
Miguel Gonzalez	529,000
Zach Britton	521,500
Manny Machado	519,000
Steve Lombardozzi	517,500
Jemile Weeks	515,000
Ryan Flaherty	512,500
David Lough	510,500
Brad Bach	509,500
Steve Johnson	506,000
T. J. McFarland	505,500
Steven Clevenger	505,000
Josh Stinson	504,000
Kevin Gausman	502,500
Henry Urrutia	501,500
Jonathan Schoop	500,500
Michael Almanzar	500,000

Source: Petchesky (2014)

(a) $k = 5$ fenceposts

(b) $k = 8$ fenceposts

(c) Which of the methods of comparison in (a) and (b) do you prefer and why?

1.3.11. *Full Professors' Salaries.* The average 2012–2013 salaries for full professors for 50 major universities in the United States are presented in Table 1.29.

(a) Label each university listed in Table 1.29 as either a public or private institution.

Discuss the similarities and differences in the average full professor salaries for these groupings of public and private institutions by making use of

(b) appropriate summary statistics

(c) back-to-back stemplots and parallel boxplots.

Table 1.29 Average full professor salaries for 50 major universities in the United States, 2012–2013

University	Average full professor salary
Duke	180,200
Vanderbilt	167,900
Washington University, St. Louis	175,800
Tulane	140,200
California Institute of Technology	179,200
Carnegie Mellon	146,500
Cornell	159,800
Virginia	143,200
Texas (Austin)	144,000
Rochester	138,600
Nebraska	116,000
Iowa	132,200
Stanford	207,300
Colorado	127,800
Penn	187,000
Michigan	148,700
Princeton	200,000

(continued)

Table 1.29 (continued)

University	Average full professor salary
Iowa State	119,300
Purdue	127,700
Chicago	203,600
Yale	186,300
Wisconsin	118,800
Penn State	138,700
California at Berkeley	158,900
Illinois (Urbana-Champaign)	141,700
Minnesota	134,300
Pittsburgh	135,900
Harvard	203,000
Northwestern	176,700
Missouri	117,200
Indiana	132,000
Florida	122,500
Case Western Reserve	132,300
Brown	160,800
MIT	178,700
Maryland (College Park)	138,100
Ohio State	136,900
North Carolina State	120,600
Syracuse	122,800
Michigan State	131,200
Southern California	160,500
Kansas	118,300
UCLA	167,000
Washington State	104,000
Oregon	110,900
California at San Diego	142,500
Florida State University	109,400
Tennessee	122,500
New York University	187,600
Columbia	212,300

Source: American Association of University Professors (2013)

1.3.12. *Math SAT Scores.* The math SAT scores for seniors graduating in 2013 or 2014 from a small private school are given in Table 1.15. Compare the collections of these math SAT scores for males and females by constructing

(a) back-to-back stemplots
(b) parallel boxplots.

1.3.13. *Lead-poisoned Geese.* March et al. (1976) examined the differences between healthy (normal) and lead-poisoned Canadian geese. One of the measures studied was plasma glucose ($mg/100ml$ plasma). The data March et al. obtained for eight healthy and seven lead-poisoned Canadian geese are given in Table 1.30.

(a) Compare the typical plasma glucose values for healthy and lead-poisoned Canadian geese by computing the differences in the means, $\bar{x}_H$ and $\bar{x}_{LP}$, and medians, $\tilde{x}_H$ and $\tilde{x}_{LP}$, for the two data collections.
(b) Compare the variability in the plasma glucose values for healthy and lead-poisoned Canadian geese by computing the ratios of ranges, R_{H} and R_{LP}, and standard deviations, s_{H} and s_{LP}, for the two data collections.

1.3.14. Consider the three data collections:

$A = (1, 3, 5, 7, 9, 11)$, B = (-6, -2, 2, 10, 14, 18), and C = (9, 11, 13, 15, 17, 19).

Compare and contrast the differences and similarities in these data collections by making use of

(a) appropriate summary statistics
(b) back-to-back stemplots and parallel boxplots.

Table 1.30 Plasma glucose values (*mg*/100*ml* plasma)

Healthy geese (y)	Lead-Poisoned geese (x)
297	293
340	291
325	289
227	430
277	510
337	353
250	318
290	

Source: March et al. (1976).

1.3.15. Consider the data collection

$(x_1, \ldots, x_{10}) = (1.6, -3.4, 5.5, 6.3, 14.9, 223.4, 55.8, -33.5, 20.4, 66.8)$. Compute the ranks ($Q_1, \ldots, Q_{10}$) of ($x_1, \ldots, x_{10}$).

1.3.16. Consider the two data collections

$$(x_1, x_2, x_3, x_4, x_5, x_6, x_7) = (-34.6, 20.0, 5.7, 16.6, -4.9, 147.9, 20.3)$$
$$\text{and } (y_1, y_2, y_3, y_4, y_5) = (-3.6, 16.6, 27.4, 543.8, 44.4).$$

(a) Compute the joint ranks of the x's and y's.

(b) What is the average joint rank for the x's? for the y's?

1.3.17. *Motor-Vehicle Fatalities by Region.* In Exercise 1.3.4 you were asked to divide the 50 American states into 25 "Northern" and 25 "Southern" states. For the statewise total motor-vehicle fatality rate data (Table 1.16), what is the average joint rank of the "Northern" states? of the "Southern" states? Discuss the implication of your findings.

1.3.18. *Baseball Salaries.* Consider the salary data for the players on the 2014 New York Yankees and Cincinnati Reds baseball teams, as given in Tables 1.17 and 1.23, respectively.

(a) What is the average joint rank of the Yankees players' salaries?
(b) How many possible pairings of one Yankees player's salary with one Reds player's salary are there?
(c) What proportion of the possible (Yankees player's salary, Reds player's salary) pairings are such that the Yankees player's salary is at least as large as the Reds player's salary?

1.3.19. *Baseball Salaries.* The datasets *national_league_salary_2014* and *american_league_salary_2014* contain the 2014 salaries (as of March 26, 2014) for all baseball players in the National and American Leagues, respectively. In Tables 1.17, 1.23, 1.27, and 1.28 we have recorded the salaries for the players on the New York Yankees, Cincinnati Reds, St. Louis Cardinals, and Baltimore Orioles baseball teams, respectively. Jointly rank the salaries of all four of the baseball teams and obtain the average rank for each of the teams. Discuss how these average ranks provide information about the relative payrolls for the four baseball teams.

1.3.20. *Full Professors' Salaries.* Consider the 2012–2013 average full professor salary data in Table 1.29. Use joint ranks to provide a comparison between public and private institutions with regard to the average salaries of their full professors in 2012–2013.

1.3.21. *Lead-poisoned Geese.* Consider the plasma glucose values for healthy (normal) and lead-poisoned Canadian geese given in Table 1.30. Compute the value of the median difference statistic $\tilde{D}$ for these data collections. What does this tell us about the data collections?

1.3.22. *Lead-poisoned Geese.* Consider the plasma glucose values for healthy (normal) and lead-poisoned Canadian geese given in Table 1.30.
(a) How many (x_i, y_j) pairs are there for these data collections?
(b) What proportion, U, of these (x, y) pairs are such that y is at least as large as x? What does this tell us about the data collections?

Chapter 1 Comprehensive Exercises

1.A. Conceptual

1.A.1. In Definition 1.14 the variance, s^2, for a data collection $(x_1, \ldots, x_n)$ is defined by

$$s^2 = \frac{\sum_{i=1}^{n} (x_i - \bar{x})^2}{n-1}.$$

However, the variance can also be computed without first calculating the value of the mean $\bar{x}$. Show that

$$s^2 = \frac{1}{n(n-1)} \sum_{1 \le i < j \le n} (x_i - x_j)^2$$

is an equivalent formula for computing s^2.

1.A.2. The variance s^2 (1.7) uses the sum of the squared differences of the observations from the mean, $\bar{x}$, to measure the variability associated with a data collection $(x_1, \ldots, x_n)$. An alternative statistic often used to measure the variability in a data collection is the mean absolute deviation (*MAD*) of the sample observations from $\bar{x}$, namely,

$$MAD = \frac{1}{n} \sum_{i=1}^{n} | x_i - \bar{x} | .$$

Compare and contrast these two measures of variability for a data collection. When do you think it might be more appropriate to use *MAD* rather than s^2 to assess the variability of a data collection?

1.A.3. Let (x_1, x_2, x_3) denote a triple of observations and let $\bar{x}$ and $\tilde{x}$ be the mean and median of these three observations. Show that (x_1, x_2, x_3) is a right triple if and only if $\bar{x} > \tilde{x}$.

1.A.4. Discuss why there are $n(n\text{-}1)(n\text{-}2)/6$ distinct triples of three observations each that can be constructed from the data collection $(x_1, \ldots, x_n)$.

1.A.5. Consider the two data collections $(x_1, \ldots, x_m)$ and $(y_1, \ldots, y_n)$. The median difference statistic $\tilde{D}$ (1.11) is the median of the $(y_j - x_i)$ differences for the mn possible (x_i, y_j) pairings, $i = 1, \ldots, m$ and $j = 1, \ldots, n$. Construct two specific data collections $(x_1, \ldots, x_m)$ and $(y_1, \ldots, y_n)$ that demonstrate that the median difference statistic $\tilde{D}$ is not equal to $\tilde{y} - \tilde{x}$, the difference in the separate medians for the two collections.

1.A.6. Let $\tilde{x}$ be the median for the data collection $(x_1, \ldots, x_n)$. Let $\bar{x}_{below}$ denote the mean of those observations in the data collection that are smaller than $\tilde{x}$ and let $\bar{x}_{above}$ denote the mean of those observations in the data collection that are larger than $\tilde{x}$.

(a) Discuss how the values of $\bar{x}_{below}$ and $\bar{x}_{above}$ can be used to provide information about the symmetry or lack of symmetry for the data collection.

(b) Compute $\bar{x}_{below}$ and $\bar{x}_{above}$ for the total engineering drawing hours data collection in Table 1.3. What do the values of $\bar{x}_{below}$ and $\bar{x}_{above}$ tell you about the symmetry/asymmetry of that data collection?

(c) Describe how percentiles other than (or in addition to) the median $\tilde{x}$ could be used in a similar fashion to help describe the symmetry/asymmetry characteristics of a data collection. Illustrate your proposal with the total engineering drawing hours data collection in Table 1.3.

1.A.7. In Section 1.2 we discussed how to use the number of observations that are smaller than the mean, $\bar{x}$, for a data collection to provide information about the symmetry/asymmetry of the data collection. In Exercise 1.A.6 we suggested that the mean, $\bar{x}_{below}$, of those observations in the data collection that are smaller than the median, $\tilde{x}$, and the mean, $\bar{x}_{above}$, of those observations in the data collection that are larger than $\tilde{x}$ can also be used to provide information about the symmetry/asymmetry characteristics of a data collection. Discuss how these two approaches to describing the symmetry/asymmetry characteristics of a data collection are related to the concepts of variable and fixed fenceposts, as discussed in Section 1.2.

1.A.8. Consider the data collection consisting of two observations (x_1, x_2). Compare the values of the median, $\tilde{x}$, and the mean, $\bar{x}$, for this data collection. What does this say about measuring asymmetry for a collection with fewer than 3 observations?

1.A.9. In Definition 1.14 the variance, s^2, for a data collection $(x_1, \ldots, x_n)$ is defined by

$$s^2 = \frac{\sum_{i=1}^{n} (x_i - \bar{x})^2}{n-1}.$$

Show that the variance can also be obtained using the computationally simpler formula

$$s^2 = \frac{1}{n-1}\left[\sum_{i=1}^{n} x_i^2 - \frac{1}{n}\left(\sum_{i=1}^{n} x_i\right)^2\right].$$

1.A.10. Consider the two data collections $(x_1, \ldots, x_m)$ and $(y_1, \ldots, y_n)$. Let U be the counting statistic (1.10) equal to the number of (x, y) pairs for which y is at least as large as x. Let $S_1, S_2, \ldots, S_n$ be the joint ranks for the observations $y_1, y_2, \ldots, y_n$, respectively, and set $V = S_1 + S_2 + \ldots + S_n$. If there are no tied values among the x and y observations, show that

$$V = U + \frac{n(n+1)}{2}.$$

1.A.11. *Government's Role in Social Nets.* Princeton Survey Research Associates of Princeton, New Jersey (1998) conducted an extensive series of surveys designed to assess American values about taking care of each other. Two of the survey questions were:

1. Do you think it is the government's responsibility to pay the health care expenses for all retired people?
2. Do you think the government should give people in poverty money to feed, clothe, and house their children under 18?

When these questions were asked of parents as a group and their adult children as a group, the percentages responding yes to questions 1 and 2 were:

	Parents	**Adult children**
Question 1	44%	48%
Question 2	61%	67%

However, it was also reported that only 50% of the adult children agreed with their parents with regard to Question 1 and that there was only slighter better 54% agreement between parents and their adult children for Question 2. How can the two groups (parents and adult children) agree so well on the issues in both Questions 1 and 2, yet there be such little agreement between individual parents and their own adult children on these two issues? What does this have to say about family influence on our attitude toward these issues?

1.A.12. *AIDS and Prisons.* The New York Times (1989) reported on the results of a study by Dr. Ford Brewer of The Johns Hopkins University School of Hygiene and Public Health that was designed to estimate the prevalence of AIDS among male prisoners. Dr. Brewer found that 476 of the 11,198 inmates in his nationwide study were infected with AIDS. In response to a subsequent query from the Columbus Dispatch about the prevalence of AIDS among male prisoners in Ohio, a spokesperson for the Ohio Department of

Rehabilitation and Correction (ODRC) indicated that Ohio does not test all male prisoners for AIDS. However, the spokesperson stated that ODRC projects the AIDS incidence rate for Ohio prisoners to be roughly 6 cases per 1000, "based on the incidence of AIDS among incoming prisoners, who can undergo voluntary testing for AIDS". Comment on whether these findings provide strong evidence that the incidence of AIDS among male Ohio prisoners is much lower than the national percentage for male prisoners.

1.B. Data Analysis/Computational

1.B.1. *Women's Shoes.* What type of shoes do women who work outside the home wear? The American Orthopedic Foot and Ankle Society (AOFAS) (1999) conducted a telephone survey of 531 women who identified themselves as working outside the home. The women were asked to describe the shoes they most commonly wear to work. Of the 531 women surveyed, 234 reported wearing "flats" (fashion shoes with heels less than 1 inch) at work, 159 wore athletic shoes, 90 wore low pumps (heels less than 2 1/4 inches), while 11 respondents reported regularly wearing shoes with a heel greater than 2 1/4 inches. The rest of the women surveyed did not indicate a clear, single preference for work footwear.

(a) How many of the surveyed women did not indicate a clear, single preference for work footwear?
(b) Use an appropriate graphical method to display these data.

1.B.2. *Government's Role in Health Care.* The Pew Research Center (2012) conducted separate surveys in 2009 and 2012 designed to assess American opinions about the government's role in health care, among other things. Two of the survey statements presented to the participants were:

1. I am concerned about the government becoming too involved in health care.
2. The government needs to do more to make health care affordable and accessible.

The participants were asked whether they agreed or disagreed with each of the statements. The percentages of individuals responding Agree, Disagree, or Don't Know to each of these statements in the 2009 and 2012 surveys are:

	2009		
	Agree	**Disagree**	**Don't Know**
Question 1	46%	50%	4%
Question 2	86%	12%	2%
	2012		
	Agree	**Disagree**	**Don't Know**
Question 1	59%	39%	2%
Question 2	82%	16%	2%

(a) Display these data in two well-labeled bar graphs, separately for Question 1 and Question 2.

(b) Use side-by-side pie charts to compare and contrast the results for Question 1 for the 2 years. Discuss the changes in public responses from 2009 to 2012.

(c) Use side-by-side pie charts to compare and contrast the results for Question 2 for the 2 years. Discuss the change in public responses from 2009 to 2012.

(d) Discuss anything you might find unusual in the public responses to the two statements in both years.

1.B.3. *Taking Care of Sick Parents.* Princeton Survey Research Associates of Princeton, New Jersey (1998) conducted an extensive series of surveys designed to assess American values about taking care of each other. One of the questions asked of the respondents in one of these surveys was:

Who should be responsible for taking care of parents if they become sick or disabled?

The respondents were also asked to self-classify themselves as Conservative, Moderate, or Liberal. The breakdown of the 1095 survey respondents with respect to both their political ideology and their answer to the stated question is:

	Numbers who gave this response		
Question response	**Conservatives**	**Moderates**	**Liberals**
People should feel entirely responsible	346	272	146
People should expect help from the government	339	409	221
It depends	60	61	28
Don't know/Refused to answer	8	15	0

(a) Use bar graphs to display these results.

(b) Use side-by-side pie charts to display these results.

1.B.4. *City Driving Gas Mileage.* The city driving gas mileage (miles per gallon) for some of the best (excluding electric vehicles) and worst mileage 2015 model automobiles are presented in Table 1.31.

(a) Find the five-number summaries separately for the nine best mileage and eleven worst mileage 2015 automobile models in Table 1.31.

(b) Use appropriate software to make separate boxplots for the nine best mileage and eleven worst mileage 2015 automobile models.

(c) Find the mean $\bar{x}$, median $\tilde{x}$, and standard deviation, s, separately for the nine best mileage and eleven worst mileage 2015 automobile models.

(d) Combine the nine best mileage and eleven worst mileage automobile models into one set of 20 automobile models and repeat parts (a), (b),

Table 1.31 City driving gas mileage for some of the best (excluding electric vehicles) and worst mileage 2015 model automobiles (miles/gallon)

Best mileage		Worst mileage	
Automobile model	**Gas mileage**	**Automobile model**	**Gas mileage**
Honda CR-Z	36	Bugatti Veyron	8
Scion iQ	36	Aston Martin D89	13
Audi A3 Diesel	31	Bentley Continental GT	9
Ford Fiesta SFE FWD	31	Maserati GranTurismo	13
Toyota Prius c Hybrid	53	Mercedes-Benz C63 AMG	13
Toyota Prius Hybrid	51	Chevrolet Camaro	12
Ford C-MAX-Hybrid FWD	42	Rolls-Royce Phantom	11
Honda Fit	33	Bentley Mulsanne	11
Toyota Prius v	44	Ferrari FF	11
		Infiniti QX50	17
		Mercedes-Benz E63 AMG S	15

Source: U.S. Department of Energy (2015).

and (c). Compare and contrast the results here with those previously obtained in parts (a)–(c).

1.B.5. *Statewise Motor-Vehicle Fatality Rates.* Use the computationally simpler formula for the variance, s^2, given in Exercise A.9 to compute the variance for the statewise motor-vehicle fatality rates data collection in Table 1.16.

1.B.6. *Engineering Drawing Hours.* Consider the total engineering drawing hours data collection in Table 1.3. We noted in Section 1.2 that there are $n(n\text{-}1)(n\text{-}2)/6 = 96(95)(94)/6 = 142{,}880$ possible triples involving these data. Use the **R** function *FindTriples*( ) for the dataset *engineering_drawing_hours* to show that 98,269 of these triples are right triples, 37,832 of them are left triples, and the remaining 6779 triples are neither right nor left.

1.B.7. *Stocking Game Fish.* To determine the number of game fish to stock in a given system and to set appropriate catch limits, it is important for fishery managers to be able to assess potential growth and survival of game fish in that system. Such growth and survival rates are closely related to the availability of appropriately sized prey. Young-of-year (YOY) gizzard shad (*Dorosoma cepedianum*) are the primary food source for game fish in many environments. However, because of their growth rate, YOY gizzard shad can become quickly too large for predators to swallow. Thus to be able to predict predator growth rates in such settings it is useful to know both the density and the size structure of the resident YOY shad population in a lake. With this in mind, Johnson (1984) sampled the YOY gizzard shad population in Kokosing Lake (Ohio) in summer 1984. The data in Table 1.32 are lengths (*mm*) for a subset of the YOY gizzard shad sampled by Johnson.

(a) Provide two different graphical representations of this data collection.
(b) Calculate a set of summary statistics for this data collection.
(c) Describe the most notable features of the data collection.

1.B.8. *Stretching a Hit into a Double.* The data in Table 1.33 were obtained by Woodward (1970) in a study of different methods of running to first base, with the goal of minimizing the time it would take to get from home plate to second base (i. e., get a double on a base hit). The times (in seconds) in Table 1.33 are averages of two runs from a point on the first base line 35 ft from home plate to a point 15 ft short of second base for the method of running known as "wide angle" for each of 22 different runners.

(a) Provide two different graphical representations of this data collection.

Table 1.32 Length of YOY gizzard shad from Kokosing Lake, Ohio, sampled in summer 1984 (*mm*)

46	41	42	58	38	28	31	25
28	42	60	27	33	26	30	25
46	45	32	51	26	27	27	24
37	38	42	42	25	27	29	27
32	44	45	52	28	27	30	30

Source: Johnson (1984).

Table 1.33 Times from home plate to second base using the "wide angle" method of running (*sec*)

5.55	5.45
5.75	5.45
5.50	4.95
5.40	5.40
5.70	5.50
5.60	5.35
5.35	5.55
5.35	5.25
5.00	5.40
5.70	5.55
5.10	6.25

Source: Woodward (1970)

(b) Calculate a set of summary statistics for this data collection.

(c) Describe the most notable features of the data collection.

(d) Comment on why you think Woodward chose to measure the time from a point on the first base line 35 ft from home plate to a point 15 ft short of second base, rather than from home plate to second base.

1.B.9. *Disciplining Physicians.* The Medical Board of California disciplined a total of 375 licensed physicians in the state of California between October 1995 and April 1997 (some for multiple offenses) and the State Medical Board of Ohio disciplined 340 licensed physicians in the state of Ohio between January 1997 and June 1999 (again some for multiple offenses). Morrison and Wickersham (1998) studied these discipline cases for California and Clay and Conatser (2003) compared the California data with those for Ohio. The numbers of offenses leading to the Boards' actions are provided in Table 1.34 for each of the two states.

(a) Use bar graphs to compare and contrast the California and Ohio data collections.

Table 1.34 Total numbers of various offenses leading to state medical board actions between October 1995 and April 1997 in California and between January 1997 and June 1999 in Ohio

	Number of occurrences	
Offense	**California**	**Ohio**
Negligence or incompetence	145	34
Inappropriate prescribing, drug possession	62	66
Alcohol and/or other drug use, impairment	56	100
Fraud, kickbacks, tax, worker's compensation	48	18
Sexual and/or inappropriate patient contact	40	17
Mental and/or physical impairment	21	21
Probation violation of previous action	18	71
Medical education or licensing violations	26	27
Misrepresenting credentials	9	26
Other crime	19	24
Other and/or miscellaneous	21	73

Source: Morrison and Wickersham (1998) and Clay and Conatser (2003).

(b) Use pie charts to compare and contrast the California and Ohio data collections.

1.B.10. *Disciplining Physicians.* The Medical Board of California disciplined a total of 375 licensed physicians in the state of California during the period between October 1995 and April 1997. Morrison and Wickersham (1998) studied these discipline cases and an accounting of the principal offenses leading to the Board actions are provided in Table 1.35, categorized by the genders of the doctors being disciplined (343 males and 32 females).

(a) Use bar graphs to display and compare these data collections.

(b) Use pie charts to display and compare these data collections.

1.B.11. *Disciplining Physicians.* Consider the California physician offense data discussed in Exercise 1.B.10. Morrison and Wickersham (1998) also gave an accounting of the type of disciplinary actions taken by the Board.

Table 1.35 Principal offenses leading to action by the medical board of California between October 1995 and April 1997, categorized by gender of the disciplined doctor

Principal offense	Number male	Number female
Negligence or incompetence	115	12
Inappropriate prescribing, drug possession	40	2
Alcohol and/or other drug use, impairment	45	7
Fraud, kickbacks, tax, worker's compensation	32	3
Sexual and/or inappropriate patient contact	36	1
Mental and/or physical impairment	16	4
Unlicensed assistant, poor supervision	7	1
Worked for unlicensed person or entity	8	0
Misrepresenting credentials	6	0
Other crime and/or miscellaneous	38	2

Source: Morrison and Wickersham (1998)

These actions are provided in Table 1.36, again categorized by the genders of the disciplined physicians.

(a) Recalling from Exercise 1.B.10 that there were 343 male and 32 female physicians who were given disciplinary actions, compute separately the percentages of male and female doctors who received each of the five types of disciplinary actions.

(b) Use bar graphs to display and compare the two groups of percentages obtained in (a).

(c) Could you also use pie charts to display the two groups of percentages obtained in (a)? Why or why not? [Hint: What do the percentages calculated in part (a) tell you about some of the physicians who were disciplined?]

1.B.12. *Plant Absorption of Soil Nutrients.* Vasicular-Arbuscular Mycorrhiza (VAM) is a fungus that affects the roots of plants and is found in almost all types of soil. VAM's relationship with plants is symbiotic, as it facilitates the absorption of nutrients from the soil and transfers them to the plants. In return, the plants supply the fungus with manufactured lipids. Brust (1984)

Table 1.36 Types of disciplinary actions taken by the medical board of California between October 1995 and April 1997, categorized by gender of the disciplined doctor

Type of discipline	Number male	Number female
Actual revocation of license	70	11
Actual suspension of license	42	7
Revocation or suspension, but action stayed	158	10
Letter of reprimand	73	4
Disciplined by reciprocity (other jurisdiction)	96	7

Source: Morrison and Wickersham (1998).

studied the effects that a variety of different tillage systems have on the ability of the VAM fungus to infect planted corn. In eight different samples of soil containing corn roots, Brust found the following percentages of roots infected with the VAM fungus:

20%	6%	38%	18%	60%	45%	53%	43% .

(a) How many possible triples can be constructed from these data?

(b) Use the **R** function *FindTriples*( ) to compute how many of these triples are right triples, how many are left triples, and how many are neither? What does this say about the data collection?

1.B.13. *Stocking Game Fish.* In Exercise 1.B.7, we considered length of YOY gizzard shad from Kokosing Lake, Ohio. There we treated the entire collection of 40 length measurements as if they were a single random sample from Kokosing Lake. In reality, Johnson (1984) collected these samples in groups of ten each from four different sites on the lake. The data collection, properly allocated to each of the four different sites, is reproduced in Table 1.37.

(a) Provide appropriate graphical representation(s) and a statistical summary for each of these four site data collections separately.

Table 1.37 Length of YOY gizzard shad, sampled from four different sites in Kokosing Lake, Ohio, in summer 1984 (mm)

Site 1	Site 2	Site 3	Site 4
46	42	38	31
28	60	33	30
46	32	26	27
37	42	25	29
32	45	28	30
41	58	28	25
42	27	26	25
45	51	27	24
38	42	27	27
44	52	27	30

Source: Johnson (1984)

(b) Comment on any similarities and differences between the four site data collections that are apparent from your statistical analyses in part (a).

1.B.14. *Stocking Game Fish.* In Exercise 1.B.13, we considered length of YOY gizzard shad from four different sites in Kokosing Lake, Ohio. In your separate statistical analyses of these four data collections in Exercise 1.B.13, you should have discovered that the lengths of YOY gizzard shad are quite similar at Sites 1 and 2, as well as being quite similar at Sites 3 and 4. Pool the ten observations from Sites 1 and 2 to constitute a single data collection of 20 observations from Site "A" and do the same for the ten observations from Sites 3 and 4 to constitute a single data collection of 20 observations from Site "B".

(a) What is the average joint rank of the YOY gizzard shad observations from Site "A"?

(b) How many possible pairings of one YOY gizzard shad length from Site "A" with one YOY gizzard shad length from Site "B" are there?

(c) What proportion of the possible (YOY gizzard shad length from Site "A", YOY gizzard shad length from Site "B") pairings are such that the YOY gizzard shad length from Site "A" is at least as large as the YOY gizzard shad length from Site "B"?

(d) Discuss the implications of your findings in parts (a)–(c).

1.B.15. *Stocking Game Fish.* Consider the YOY gizzard shad length data presented in Table 1.38.

(a) Compute the difference in means and the median difference statistic $\tilde{D}$ for the data collections from Sites 1 and 4. Do these two measures provide similar assessments of the difference in the "centers" of these data collections? Discuss.

(b) Compute the ratios of the ranges, R_1 and R_4, and standard deviations, s_1 and s_4, for these two data collections. Discuss your findings.

1.B.16. *Pine Tree Growth.* The Department of Biology at Kenyon College conducted a long-term experiment to study the growth of pine trees at a site located on a hill overlooking the Kokosing River Valley just south of Gambier, Ohio. In April 1990, student and faculty volunteers planted 1000 white pine (*pinus strobus*) seedlings on this site. A subset of the data collected between 1990 and 1997 by biology students at Kenyon College is contained in the dataset *pines_1997*. Consider the data collection corresponding to the tree heights of these white pines in 1996. Provide appropriate graphical representation(s) and a statistical summary for this data collection. What are the important features of the data collection?

1.C. Activities

1.C.1. *Basketball Attendance.* Go to your local athletic department and get the attendance figures for each of your home basketball games. Provide a graphical representation(s) and a statistical summary for this data collection.

1.C.2. *Hospital Admissions.* Go to a local hospital and obtain the numbers of patients admitted on each day in the past year.

(a) Provide graphical representation(s) and a statistical summary for this annual data collection.

(b) Separate your admission counts into the four seasons of winter, spring, summer, and autumn. Provide graphical and statistical summary comparisons for these four data collections.

(c) Separate your admission counts into the 12 months of the year. Provide graphical and statistical summary comparisons for these 12 data collections.

1.C.3. *How old are Our Coins?* Go to a bank and get 100 pennies, 100 nickels, and 100 dimes. Record the last digit in the year for each of the coins. Provide graphical and statistical summary comparisons for the data collections for each of the three coin denominations.

1.C.4. *Wall Street Profit.* Select 50 different stocks traded on the New York Stock Exchange and obtain their most recent selling prices, as well as their selling prices exactly 1 year ago. Record how much you would have made (or lost) per share in the past year for these 50 stocks. Provide an appropriate graphical representation(s) and a statistical summary for your "profit" data collection.

1.C.5. *How Much Sleep Do Students Need?* When do students go to bed and at what time do they get up in the morning? Ask each of your classmates in this course to write down for you when they typically go to bed and when they typically get up in the morning. Provide an appropriate graphical representation(s) for these two data collections. How would you use these data to provide information about the number of hours that students spend sleeping (at least at night!)?

1.D. Internet Archives

1.D.1. *U.S. Trade.* Search the Internet to find a site that provides the most recent itemized U.S. trade figures with different countries. Pick two countries of interest to you and obtain the year-to-date trade data by categories between the United States and each of these two countries. Provide appropriate graphical representation(s) to compare these two data collections. What are the important similarities/differences between our trade with these two countries?

1.D.2. *Gasoline Pump Prices.* Search the Internet to find a site that provides the most recent weekly gasoline pump prices for regular motor gasoline, auto diesel fuel, and auto propane fuel for various cities in the United States. Provide appropriate graphical representations and statistical summaries for each of these three data collections. What are the important features of the data collections separately and relative to one another?

1.D.3. *Cigarette Sales.* Search the Internet to find a site that provides the latest 12-month unit-volume comparisons for the leading U.S. cigarette brands for two recent years. Provide appropriate graphical representations to compare the cigarette share percentages of these leading brands for the 2 years.

1.D.4. *Division I College Basketball Attendance.* Search the Internet to find a site that provides the most recent men's college basketball average (per game) attendance figures for Division I institutions. Provide appropriate graphical representation(s) and a statistical summary for this data collection. What are the important features of the data collection?

Do the same for women's college basketball average (per game) attendance figures for Division I institutions.

Compare the results for men's and women's basketball games for Division I institutions.

1.D.5. *Division I Conference Basketball Attendance.* Search the Internet to find a site that provides the most recent men's college basketball average (per game) attendance figures for Division I conferences. Provide appropriate graphical representation(s) and a statistical summary for this data collection. What are the important features of the data collection?

Do the same for women's college basketball average (per game) attendance figures for Division I conferences.

Compare the results for men's and women's basketball games for Division I conferences.

1.D.6. *Weather Comparisons.* Search the Internet to find the most recent climatalogical data for large cities in the United States.

(a) Choose a winter month and provide appropriate graphical representation(s) and statistical summaries for both mean temperature and median precipitation for the listed cities. What are the important features of these two data collections?

(b) Choose a summer month and provide appropriate graphical representation(s) and statistical summaries for both mean temperature and median precipitation for the listed cities. What are the important features of these two data collections?

(c) Choose appropriate statistical methodology (graphical and/or summary statistics) to compare the mean temperatures and median precipitation figures for your chosen summer and winter months.

(d) Divide the mainland U.S. cities (not in Hawaii or Alaska) into four groups corresponding to the southwest, southeast, northwest, and northeast portions of the continental United States. Repeat parts (a) – (c) of this problem separately for each of these four groupings.

1.D.7. *National Football League Attendance.* Search the Internet to find a site that provides the most recent National Football League attendance report for each of the National Football League teams. Obtain appropriate graphical representation(s) and statistical summaries for both average home attendance and average road attendance for the teams. What are the important features of these data collections?

1.D.8. *Chronic Medical Conditions.* Search the Internet to find a site that provides the latest report on the number of chronic medical conditions per 1000 persons, by age, in the United States. Provide appropriate graphical representation(s) and statistical summaries to compare the numbers of chronic conditions per 1000 persons for five age groups (spanning 0 to 100 years) of your choosing. What are the important features of these data collections?

1.D.9. *EPA Mileage Ratings.* Search the Internet to find a site that provides the EPA mileage ratings for the latest car models. Pick two of the car categories and one truck category and provide appropriate graphical representation(s) and statistical summaries for both the city miles per gallon and the highway miles per gallon ratings for the listed vehicles in each of these categories. What are the important features of these data collections, both separately and relative to one another?

1.D.10. *Causes of Death in the United States.* Search the Internet to find a site that provides the most recent counts of the numbers of deaths resulting from the ten leading causes of death in the United States. Provide appropriate graphical representation(s) of this data collection.

1.D.11. *Motor Vehicle Fatalities by Type of Accident.* In Example 1.2, we compared the motor-vehicle deaths by types of accident for the 2 years 1949 and 1985. Search the Internet to find similar data for the year 2014 and analyze the data for 2014 in the same way as was done in Example 1.2 for the data from 1949 and 1985. Discuss the similarities and differences in the results for the 3 years.

Exploring Bivariate and Categorical Data 2

In Chapter 1 we focused on displaying and describing information on one variable at a time. In this chapter we consider graphical and numerical methods that can be used to investigate the relationship between two variables. Section 1 contains methods for exploring the relationship between two quantitative variables. Descriptive statistics for measuring the strength of association are provided in Sect. 2. Section 3 deals with relationships between two categorical variables.

2.1 Exploring the Relationship Between Two Quantitative Variables

The easiest method for exploring the relationship between two quantitative variables is a visual display of the pairs of observations. A quick look at this visual display, called a *scatterplot*, can help identify overall trends or patterns, deviations from the overall trends or patterns, clusters of observations, and unusual observations.

© Springer International Publishing AG 2017

D.A. Wolfe, G. Schneider, *Intuitive Introductory Statistics*, Springer Texts in Statistics,
DOI 10.1007/978-3-319-56072-4_2

Definition 2.1 A **scatterplot** is a two-dimensional plot with one variable on the horizontal axis and the other variable on the vertical axis. Each pair of values, one for each observation in the data set, is represented on the plot with a dot or some other plotting symbol.

Example 2.1. Looking for Association – Is There any Relationship Between the Median Weekly Earnings for Male and Female Employees? Data for median weekly earnings of male and female employees in service occupations, sales and office occupations, and construction and extraction occupations for each quarter from 2005 to 2015 are reported in Table 2.1 (and available in the **R** dataset *weekly_salaries*). Perhaps you can answer the question by simply looking at the raw data in Table 2.1. However, it is often easier to answer questions about relationships by constructing a visual display of the data. Figure 2.1 contains a scatterplot with median weekly earnings for women on the vertical axis and median weekly earnings for men on the horizontal axis. You can generate this plot for yourself using the **R** function *plot*( ). (We'll demonstrate this with an example soon.)

While an overall upward sloping pattern is evident, whatever relationship there might be, it does not necessarily appear to be linear; that is, many of the plotted data values would not be close to any straight line that you might choose to draw on the scatterplot.

2.1.1 Common Types of Relationships – No Association, Positive Association, Negative Association

If there is no apparent pattern in the scatterplot, then we say that the two variables are **not associated**. When no association exists between two variables the scatterplot will look like an unstructured collection of points.

Table 2.1 Median weekly earnings information for men and women 2005–2015

	Men in service occupations	Women in service occupations	Men in sales and office occupations	Women in sales and office occupations	Men in construction and extraction occupations	Women in construction and extraction occupations
Q1–2005	477	381	691	519	616	592
Q2–2005	473	371	684	523	599	406
Q3–2005	464	383	682	514	605	459
Q4–2005	493	383	707	524	606	501
Q1–2006	500	382	696	532	618	588
Q2–2006	492	389	669	534	611	542
Q3–2006	494	391	718	541	642	527
Q4–2006	488	397	704	544	619	515
Q1–2007	516	395	715	539	662	497
Q2–2007	521	404	722	546	635	566
Q3–2007	503	408	710	562	633	560
Q4–2007	520	415	708	552	661	622
Q1–2008	529	408	737	578	675	740
Q2–2008	539	416	730	575	676	733
Q3–2008	545	416	728	576	688	755
Q4–2008	539	441	737	583	714	769
Q1–2009	516	411	748	587	720	696
Q2–2009	520	419	726	589	704	629
Q3–2009	515	426	736	590	721	648
Q4–2009	566	418	737	593	738	687
Q1–2010	558	420	743	594	717	828
Q2–2010	533	433	720	595	698	587

(continued)

Table 2.1 (continued)

	Men in service occupations	Women in service occupations	Men in sales and office occupations	Women in sales and office occupations	Men in construction and extraction occupations	Women in construction and extraction occupations
Q3–2010	511	425	746	596	709	739
Q4–2010	585	421	732	602	713	626
Q1–2011	565	431	736	607	710	490
Q2–2011	544	439	739	606	721	688
Q3–2011	528	427	737	592	717	574
Q4–2011	578	440	740	605	726	671
Q1–2012	563	450	766	608	724	929
Q2–2012	529	435	764	615	766	647
Q3–2012	530	440	751	607	723	724
Q4–2012	550	420	801	609	749	670
Q1–2013	576	447	765	609	706	743
Q2–2013	534	461	750	615	747	668
Q3–2013	562	447	744	615	733	585
Q4–2013	546	452	766	621	765	620
Q1–2014	581	459	794	629	750	630
Q2–2014	580	452	759	623	729	699
Q3–2014	585	467	761	624	760	689
Q4–2014	588	470	753	605	786	790
Q1–2015	575	461	779	622	773	618
Q2–2015	587	457	765	632	742	700
Q3–2015	571	465	771	628	721	720
Q4–2015	607	471	799	626	777	729

Source: U.S. Department of Labor, Bureau of Labor Statistics (2016). Labor Force Statistics (CPS) [Accessed 04/18/2016]

Fig. 2.1 Scatterplot of median weekly earnings for men and women working in service, sales and office, and construction and extraction occupations

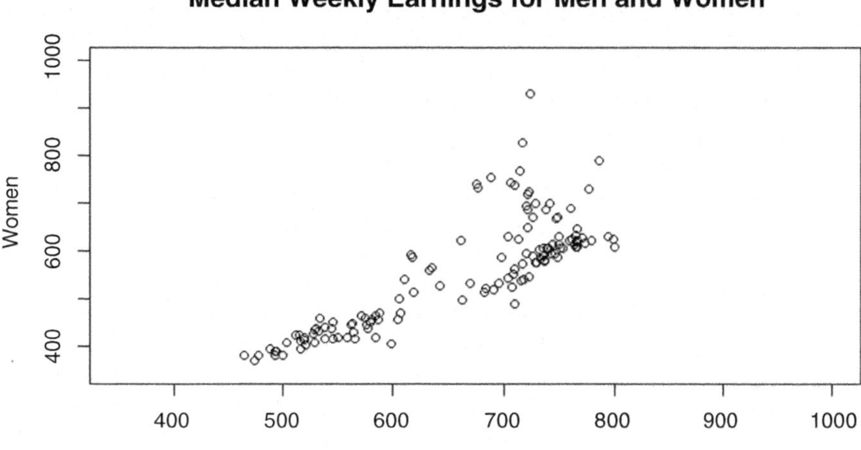

Example 2.2. Is There any Association Between the Length of Pine Needles and the Diameter of a Pine Tree? What About the Height and Diameter of Pine Trees? The Department of Biology at Kenyon College conducted an experiment to study the growth of pine trees at a site located just south of Gambier, Ohio, on a hill overlooking the Kokosing River valley. In April 1990, student and faculty volunteers planted 1000 White Pine (*pinus strobus*) seedlings on the Kenyon Center for Environmental Study (KCES). These seedlings were planted in two grids, distinguished by ten- and fifteen-foot spacings between the small trees. For a complete description of the design of the experiment and the measurements, see Example 3.4. A subset of the data collected by biology students at Kenyon College is contained in the dataset *pines_1997* (Table 2.2).

Figure 2.2 shows a scatterplot of the lengths of pine needles (Needles97) and the diameters of the pine trees (Diam97) at KCES in 1997. The plot shows that there is no obvious relationship between needle length and diameter for these trees. The most striking feature in Fig. 2.2 is the one unusual observation, with a needle length of approximately 170 *mm*. You can reproduce this

Table 2.2 Description of the dataset *pines_1997*

Column	Name	Contents
C1	Row	Row # in pine plantation
C2	Col	Column # in pine plantation
C3	Hgt90	Tree height at time of planting (*cm*)
C4	Hgt96	Tree height in September 1996 (*cm*)
C5	Diam96	Tree trunk diameter in September 1996 (*cm*)
C6	Grow96	Leader growth during 1996 (*cm*)
C7	Hgt97	Tree height in September 1997 (*cm*)
C8	Diam97	Tree trunk diameter in September 1997 (*cm*)
C9	Spread97	Widest lateral spread (*cm*)
C10	Needles97	Needle length in September 1997 (*mm*)
C11	Deer95	Type of deer damage: 1 = none, 2 = browsed
C12	Deer97	Type of deer damage: 1 = none, 2 = browsed
C13	Cover95	Amount of thorny cover (0 = none, 1 = <1/3, 2 = between 1/3 and 2/3, 3 = >2/3)
C14	Fert	Indicator for Fertilizer (0 = no, 1 = yes)
C15	Spacing	Distance (in feet) between trees (10 or 15)

Fig. 2.2 Scatterplot of needle length and tree diameter for pine trees at KCES in 1997

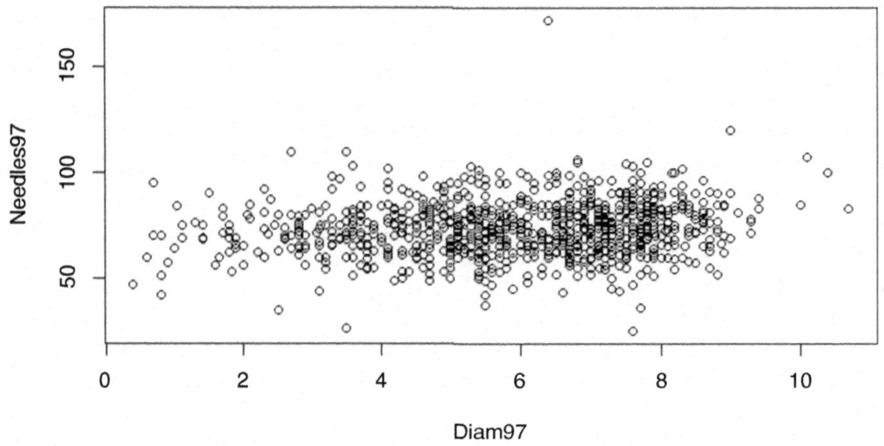

Fig. 2.3 Scatterplot of tree height and tree diameter for pines at KCES in 1997

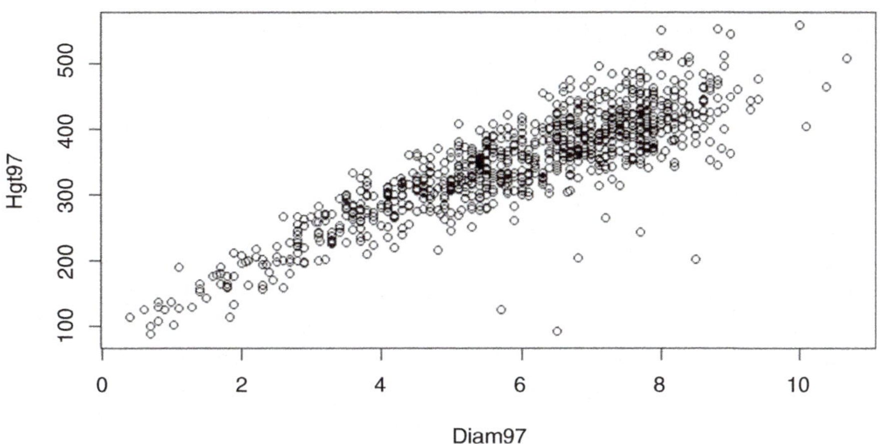

plot by calling the **R** function *plot*( ) as follows. The first two arguments specify the x and y values to be plotted, respectively. The other three are used to label the axes and give the plot a title. (These are only a few of the wide range of arguments that you can specify in the *plot*( ) function to make your plot as customized as you like!)

```
> plot(x = pines_1997$Diam97,
       y = pines_1997$Needles97,
       xlab = "Diam97",
       ylab = "Needles97",
       main = "Scatterplot of Needle Length and Tree Diameter")
```

A scatterplot of height (Hgt97) and diameter (Diam97) of the trees at KCES in 1997 is shown in Fig. 2.3. The scatterplot is again generated using the **R** function *plot*( ), as demonstrated by the following command.

```
> plot(x = pines_1997$Diam97,
       y = pines_1997$Hgt97,
       xlab = "Diam97",
       ylab = "Hgt97",
       main = "Scatterplot of Tree Height and Tree Diameter")
```

Our intuition that short trees have small diameters and tall trees have large diameters is clearly supported by this plot. When the nature of a relationship is of this type (i.e., small values of one variable occurring with small values of the second variable and large values with large values), we say that there is a *positive association* between the two variables.

Definition 2.2 Two variables are **positively associated** if large values of the variable on the horizontal axis occur with large values of the variable on the vertical axis and small values of the variable on the horizontal axis occur with small values of the variable on the vertical axis. When positive association exists between two variables the appearance of the scatterplot will be an upward sloping, egg-shaped or football-shaped, cloud of points.

Example 2.3. Domestic Greenhouse Gas Emissions of Air Pollutants Table 2.3 contains annual emissions data (in million metric tons of carbon dioxide equivalents) for four common air pollutants from 1990 to 2014 in the United States. Have methane emissions been reduced over this period of time? We can again use the **R** function *plot*( ) along with the dataset *emissions* to answer this question.

```
> plot(emissions$Year,
       emissions$Methane,
       xlab = "Year",
       ylab = "Methane",
       main = "Scatterplot of Methane Emissions and Year")
```

Figure 2.4 clearly shows such a decreasing pattern over time (although there appear to be multiple brief increases). As the year increases the amount of methane emissions tends to decrease. When the nature of a relationship is such that increases in one variable tend to occur with decreases in the other variable, we say that there is a *negative association*.

Table 2.3 Domestic greenhouse gas emissions of air pollutants, 1990–2014

Year	Carbon Dioxide	Methane	Nitrous Oxide	Fluorinated gases
1990	5115.095	773.8549	406.2285	101.966
1991	5064.88	777.0342	396.1137	92.91026
1992	5170.274	776.8698	404.0521	97.48091
1993	5284.759	764.0897	420.5032	97.06437
1994	5377.492	770.4504	402.4789	99.87502
1995	5441.599	767.9434	420.5857	118.4005
1996	5630.114	762.2038	428.9235	128.5969
1997	5704.997	747.1771	412.318	135.5071
1998	5744.672	737.8214	433.8726	150.4219
1999	5818.972	723.4549	401.1474	146.8679
2000	5992.438	717.4739	401.4002	147.6609
2001	5894.463	712.6768	399.3244	134.435
2002	5935.739	706.3302	400.9179	142.3412
2003	5982.289	708.5545	401.7753	132.355
2004	6096.978	704.8868	428.5542	139.5507
2005	6122.747	717.3562	397.5517	141.121
2006	6042.394	719.5807	410.0662	144.104
2007	6121.654	725.9746	418.9852	155.5943
2008	5923.201	738.8895	396.7771	157.5471
2009	5488.32	735.3587	399.5005	153.0503
2010	5688.756	722.4106	410.3142	163.9762
2011	5559.508	717.4237	416.5218	171.9448
2012	5349.221	714.4012	409.2856	170.1029
2013	5502.551	721.4751	403.3581	172.6038
2014	5556.007	730.8287	403.5098	180.1094

Source: U.S. Environmental Protection Agency (2016, May), Greenhouse Gas Inventory Data Explorer

Definition 2.3 Two variables are **negatively associated** if large values of the variable on the horizontal axis occur with small values of the variable on the vertical axis and small values of the variable on the horizontal axis occur with large values of the variable on the vertical axis. When negative association exists between two variables the appearance of the scatterplot will be a downward sloping, egg-shaped or football-shaped, cloud of points.

Fig. 2.4 Scatterplot of methane emissions versus year in the United States, 1990–2014

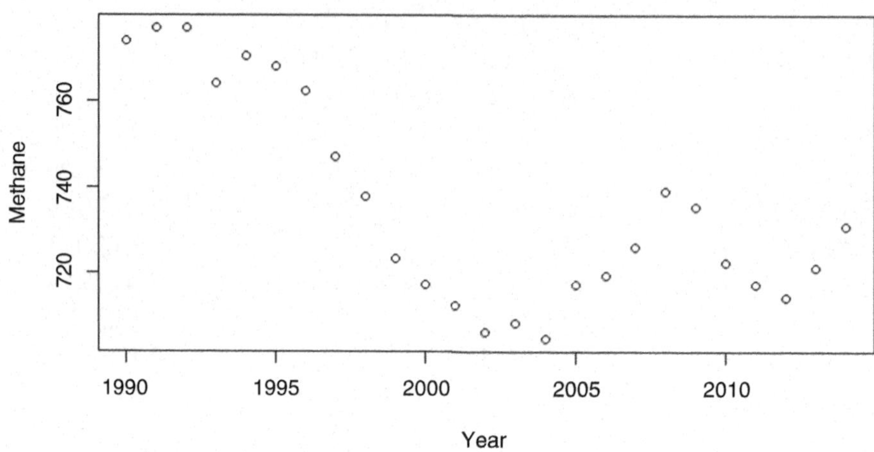

The scatterplot in Fig. 2.4 is a special type of plot, known as a time series plot. Researchers are often interested in looking at the pattern of variation in a variable over time and a simple time series plot (scatterplot with some unit of time on the horizontal axis) is the easiest way to visually display this variation.

Definition 2.4 Data collected on a variable over some interval of time (hours, days, weeks, months, years, etc.) is referred to as **time series data**. A useful graphical summary for data that is collected over time is called **a time series plot**. The plot is constructed by plotting the variable of interest on the vertical axis and time on the horizontal axis.

2.1.2 Scatterplot Smoothing

Although the pattern of variation in Fig. 2.4 is clear, scatterplot smoothing is a technique that is available in most modern statistical software packages and can be used to help the analyst interpret the overall trend in a scatterplot.

Fig. 2.5 Scatterplot Smoothing for methane emissions versus year in the United States, 1990–2014

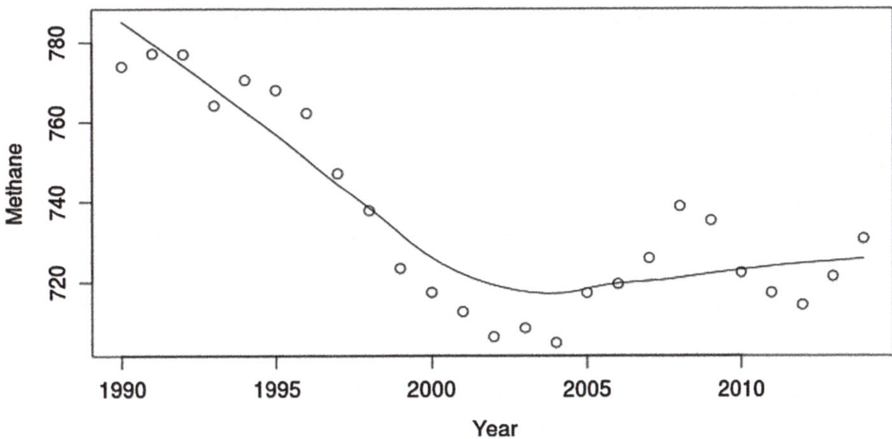

Figure 2.5 was obtained from **R** by simply replacing the previous call to *plot*( ) by a call to the function *scatter.smooth*( ) as follows.

```
> scatter.smooth(emissions$Year,
                 emissions$Methane,
                 xlab = "Year",
                 ylab = "Methane",
                 main = "Scatterplot of Methane Emissions and Year")
```

While the technical details of these computational smoothing methods are interesting, they take us beyond the scope of this text.

Example 2.5. Charge-off Rates Over Time Table 2.4 contains twenty-five years (1991 through 2015) of quarterly charge-off rates (dollars of loans expected to never be recovered as a proportion of the average dollars of total loans outstanding) for eight different types of loans as reported by the Federal Reserve. Each set of charge-off rates corresponds to the quarter which began on the date indicated by the Quarter column. The eight types of loans are as follows: Residential Real Estate, Commercial Real Estate, Farmland, Credit Cards, Other Consumer Loans, Leases, Commercial and Industrial Loans, and Agricultural Loans. The data in Table 2.4 can also be accessed in the **R** dataset *chargeoff_rates*.

Table 2.4 Charge-off data from the Federal Reserve for eight types of loans

Quarter	Resid. real estate	Comm. real estate	Farm-land	Credit cards	Other cons. loans	Leases	Comm. and indust.	Agri. loans
10/1/2015	0.21	0.05	0.05	2.91	0.8	0.23	0.36	0.08
7/1/2015	0.14	0.03	0.01	2.76	0.66	0.18	0.24	0.04
4/1/2015	0.2	0.04	0	3.03	0.58	0.13	0.21	0.06
1/1/2015	0.28	0.02	0	3.03	0.67	0.16	0.15	0.02
10/1/2014	0.27	0.08	0.03	2.98	0.82	0.11	0.26	0.07
7/1/2014	0.26	0.05	−0.01	2.89	0.76	0.13	0.2	0.02
4/1/2014	0.26	0.09	0.1	3.45	0.65	0.06	0.2	0.02
1/1/2014	0.36	0.06	0.03	3.32	0.83	0.02	0.2	0.04
10/1/2013	0.49	0.19	0.11	3.33	0.95	0.03	0.29	0.08
7/1/2013	0.47	0.21	0.02	3.19	0.88	0.02	0.27	0.11
4/1/2013	0.73	0.3	0.03	3.62	0.72	0.17	0.3	0.01
1/1/2013	0.9	0.35	0.14	3.78	0.89	0.38	0.33	−0.01
10/1/2012	1.07	0.67	0.27	3.78	1.13	0.17	0.39	0.35
7/1/2012	1.74	0.67	0.28	3.74	1.03	0.35	0.49	0.24
4/1/2012	1.23	0.76	0.33	4.15	0.95	0.22	0.52	0.29
1/1/2012	1.39	0.74	0.2	4.29	1	0.11	0.51	0.15
10/1/2011	1.4	1.35	0.36	4.53	1.33	0.32	0.77	0.22
7/1/2011	1.52	1.27	0.32	5.63	1.26	0.18	0.76	0.13
4/1/2011	1.67	1.44	0.34	5.58	1.3	0.12	0.82	0.27
1/1/2011	1.71	1.51	0.37	6.96	1.7	0.19	1.05	0.38
10/1/2010	1.99	2.46	0.47	7.7	1.93	0.74	1.45	0.9
7/1/2010	1.91	2.36	0.4	8.55	1.81	0.54	1.73	0.72
4/1/2010	2.14	2.36	0.44	10.97	2.05	0.72	1.76	0.52
1/1/2010	2.44	2.1	0.41	10.5	2.39	0.88	1.87	1.05
10/1/2009	2.78	3.27	0.54	10.19	3.03	1.43	2.65	0.64
7/1/2009	2.41	2.55	0.35	10.1	3.07	1.39	2.54	0.53
4/1/2009	2.32	2.26	0.27	9.77	3.1	1.53	2.31	0.41
1/1/2009	1.81	1.32	0.21	7.62	2.98	0.8	1.71	0.43
10/1/2008	1.62	2.25	0.21	6.3	3.03	0.77	1.55	0.34
7/1/2008	1.79	1.13	0.08	5.8	2.37	0.55	0.98	0.17
4/1/2008	1.16	0.96	0.05	5.47	2.01	0.48	0.79	0.13
1/1/2008	0.85	0.46	0.04	4.7	1.95	0.31	0.61	0.07
10/1/2007	0.47	0.43	0.04	4.18	2.03	0.36	0.82	0.19
7/1/2007	0.25	0.16	0.04	3.95	1.47	0.29	0.42	0.09
4/1/2007	0.18	0.12	0	3.85	1.35	0.14	0.4	0.08
1/1/2007	0.15	0.06	−0.02	3.93	1.38	0.15	0.31	0.06

(continued)

Table 2.4 (continued)

Quarter	Resid. real estate	Comm. real estate	Farm-land	Credit cards	Other cons. loans	Leases	Comm. and indust.	Agri. loans
10/1/2006	0.13	0.13	0.09	3.62	1.33	0.3	0.44	0.22
7/1/2006	0.1	0.04	0.01	3.87	1.08	0.09	0.27	0.09
4/1/2006	0.08	0.04	0.05	3.52	0.87	0.09	0.25	0.05
1/1/2006	0.09	0.02	0.02	3.12	0.92	0.19	0.18	0.08
10/1/2005	0.09	0.05	0.01	6.05	1.35	0.77	0.42	0.06
7/1/2005	0.07	0.05	0.03	4.33	2.04	1.04	0.19	0.03
4/1/2005	0.07	0.05	0.1	4.35	1	0.21	0.22	0.13
1/1/2005	0.08	0.04	0.02	4.6	1.11	0.32	0.22	0.05
10/1/2004	0.1	0.09	0.09	4.71	1.6	0.51	0.52	0.13
7/1/2004	0.09	0.06	0.04	4.4	1.19	0.36	0.4	0.16
4/1/2004	0.1	0.07	0.05	5.38	1.16	0.37	0.53	0.21
1/1/2004	0.12	0.06	0.01	5.38	1.28	0.43	0.67	0.26
10/1/2003	0.35	0.13	0.14	6.07	1.45	0.67	1.1	0.5
7/1/2003	0.12	0.15	0.08	5.4	1.36	0.79	1.17	0.36
4/1/2003	0.15	0.13	0.06	6.07	1.36	0.98	1.31	0.27
1/1/2003	0.14	0.09	0.04	5.77	1.41	1	1.38	0.29
10/1/2002	0.16	0.2	0.17	5.63	1.61	1.63	1.76	0.58
7/1/2002	0.15	0.14	0.05	5.94	1.46	0.86	2.03	0.48
4/1/2002	0.16	0.12	0.07	6.27	1.25	1.02	1.76	0.41
1/1/2002	0.15	0.14	0.09	7.8	1.45	0.8	1.44	0.34
10/1/2001	0.21	0.21	0.19	6.43	1.65	0.85	2.37	0.43
7/1/2001	0.45	0.12	0.06	5.23	1.27	0.79	1.29	1.07
4/1/2001	0.16	0.08	0.12	5.24	1.05	0.53	1.16	0.44
1/1/2001	0.13	0.09	0.12	4.7	1.12	0.46	0.87	0.57
10/1/2000	0.16	0.09	0.18	4.7	1.86	0.38	1.23	0.41
7/1/2000	0.12	0.06	0.1	4.31	0.9	0.3	0.66	0.3
4/1/2000	0.1	0.04	0.04	4.19	0.81	0.29	0.65	0.16
1/1/2000	0.11	0.03	−0.16	4.63	0.98	0.27	0.5	0.11
10/1/1999	0.16	0.06	0.11	4.62	1.16	0.4	0.75	0.35
7/1/1999	0.15	0.04	0.07	4.38	1.05	0.27	0.5	0.31
4/1/1999	0.11	0.02	0.04	4.36	0.83	0.31	0.5	0.34
1/1/1999	0.09	0.01	0	5.02	0.98	0.28	0.42	0.18
10/1/1998	0.09	0.06	0.07	5.26	1.16	0.27	0.57	0.45
7/1/1998	0.08	0.01	0	5.25	0.94	0.28	0.37	0.22
4/1/1998	0.08	−0.03	0.05	5.24	0.9	0.28	0.31	0.15
1/1/1998	0.08	0	−0.02	5.26	1	0.28	0.27	0.04

(continued)

Table 2.4 (continued)

Quarter	Resid. real estate	Comm. real estate	Farm-land	Credit cards	Other cons. loans	Leases	Comm. and indust.	Agri. loans
10/1/1997	0.11	0.04	0.24	5.49	1.12	0.39	0.35	0.32
7/1/1997	0.09	0.01	0.05	5.37	0.96	0.2	0.31	0.14
4/1/1997	0.1	0.01	0	5.39	0.92	0.25	0.21	0.19
1/1/1997	0.08	−0.02	−0.02	4.91	0.97	0.24	0.16	0.07
10/1/1996	0.11	0.1	0.09	4.73	1.07	0.26	0.26	0.37
7/1/1996	0.09	0.11	0.03	4.45	0.89	0.33	0.2	0.26
4/1/1996	0.1	0.14	0.03	4.53	0.77	0.22	0.26	0.28
1/1/1996	0.1	0.12	0	4.21	0.78	0.16	0.22	0.18
10/1/1995	0.14	0.32	0.09	4.02	0.89	0.19	0.35	0.37
7/1/1995	0.12	0.24	0.13	3.63	0.64	0.07	0.29	0.14
4/1/1995	0.13	0.51	0.05	3.33	0.56	0.22	0.11	0.15
1/1/1995	0.1	0.22	0	2.94	0.51	0.03	0.13	−0.01
10/1/1994	0.19	0.68	0.11	3.14	0.63	0.14	0.31	0.37
7/1/1994	0.12	0.51	0.06	2.93	0.47	0.14	0.17	0.15
4/1/1994	0.18	0.7	0.02	3.07	0.46	0.07	0.28	0.13
1/1/1994	0.14	0.64	0.02	3.29	0.45	0.04	0.21	0.11
10/1/1993	0.22	1.24	0.22	3.4	0.71	0.37	0.73	0.32
7/1/1993	0.2	1.23	0.14	3.58	0.6	0.35	0.6	0.06
4/1/1993	0.25	1.5	0.08	3.89	0.58	0.33	0.77	0.21
1/1/1993	0.17	1.3	0.04	4.08	0.63	0.47	0.65	0.1
10/1/1992	0.31	2.58	0.23	4.59	1	0.83	1.34	0.34
7/1/1992	0.28	2.57	0.34	4.31	0.83	0.53	1.28	0.38
4/1/1992	0.21	1.74	0.24	4.92	0.86	0.61	1.2	0.3
1/1/1992	0.16	1.69	0.16	4.9	1.01	0.72	1.37	0.19
10/1/1991	0.26	2.78	0.27	4.68	1.3	0.86	2.33	0.78
7/1/1991	0.21	2	0.41	4.73	1.03	0.77	1.79	0.46
4/1/1991	0.18	1.81	0.26	4.69	1.06	0.66	1.69	0.33
1/1/1991	0.16	1.34	0.12	4.16	1.18	0.91	1.05	0.17

Source: http://www.federalreserve.gov/releases/chargeoff/chgallnsa.htm [Accessed May 8, 2016]

We begin by looking at the overall variation in the quarterly charge-off rates over this twenty-five-year period. Figure 2.6 contains smoothed scatterplots for each of the eight loan types. The first of these plots is generated by the following **R** commands. (These differ from the earlier call

Fig. 2.6 Scatterplot smoothing for data from 1991 to 2015 for each of the eight loan types

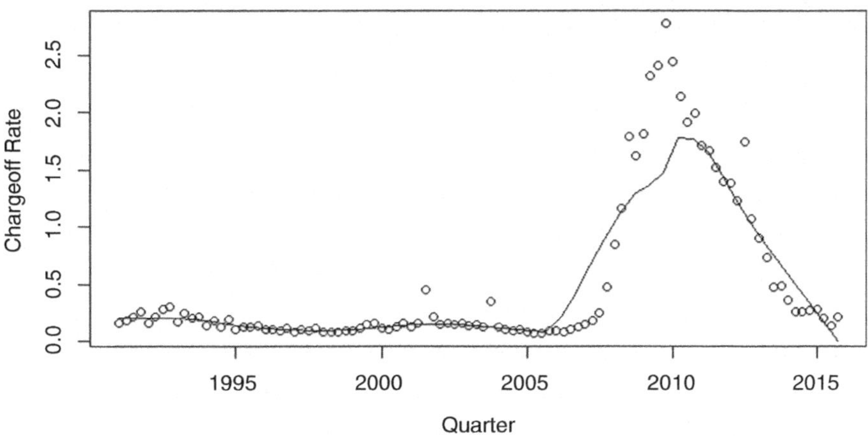

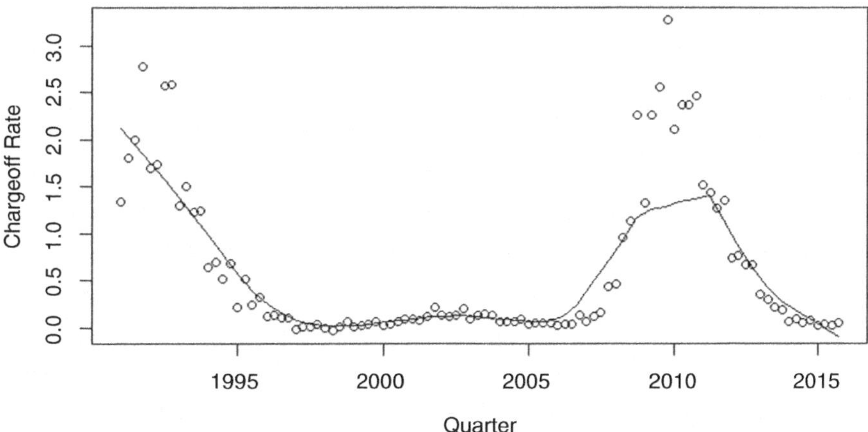

to the *scatter.smooth*( ) function only for aesthetic reasons. By default, the function does not handle the quarterly time points well and thus the horizontal axis can appear a bit strange. Try it for yourself to confirm!) Note that we have also specified the *span* argument to be equal to 1/5. This argument, whose technical details are beyond the scope of this book, controls how smooth or rough the smoothed line will be. Again, try a few values to see for yourself what it does!

Fig. 2.6 (continued)

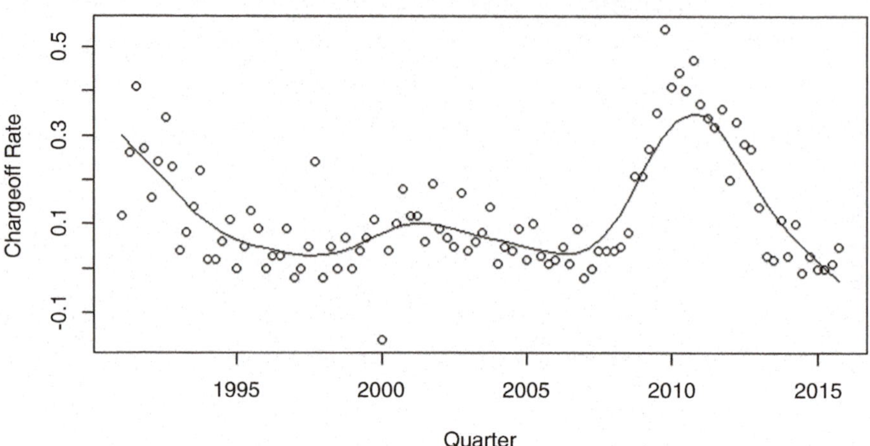

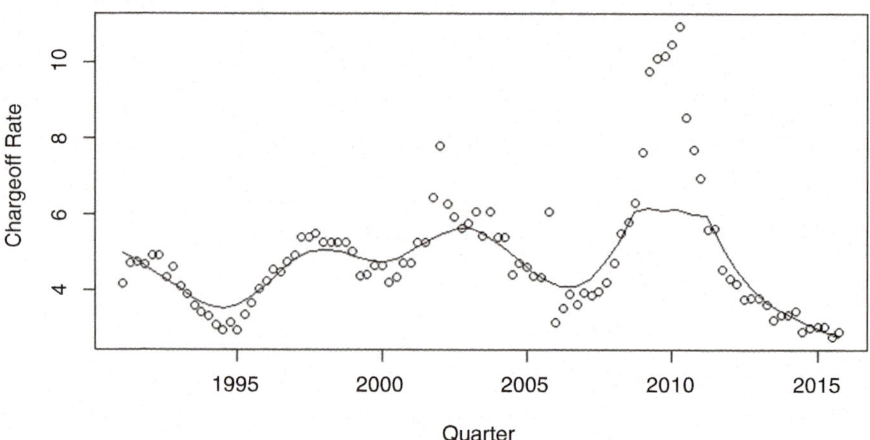


```
> plot(chargeoff_rates$Quarter,
       chargeoff_rates$ResidentialRE,
       xlab = "Quarter",
       ylab = "Chargeoff Rate",
       main = "Scatterplot of Residential Real Estate and Quarter")
> lines(loess.smooth(chargeoff_rates$Quarter,
                     chargeoff_rates$ResidentialRE,
                     span = 1/5))
```

The **R** code to generate the smoothed scatterplots for the other seven loan types is omitted, but you can replace the references above to "Residential Real Estate" and "ResidentialRE" with one of the other column names in the

Fig. 2.6 (continued)

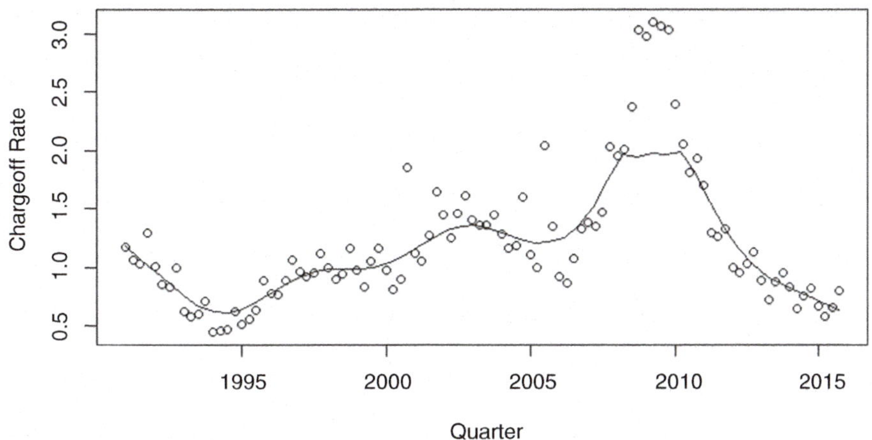

Scatterplot of Leases and Quarter

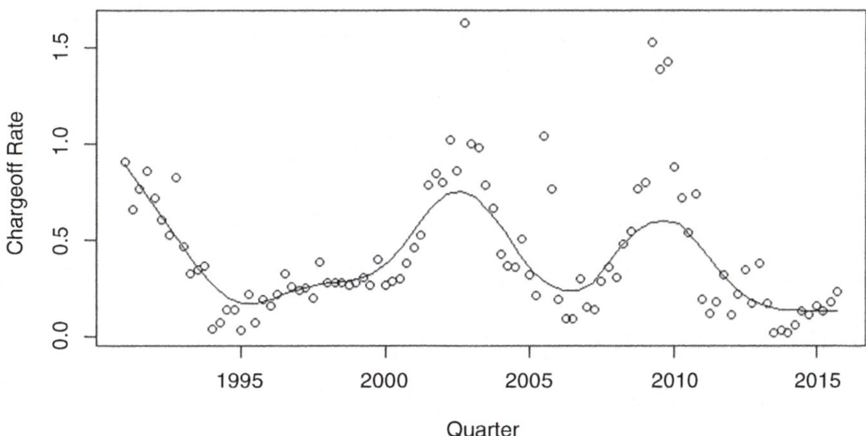

chargeoff_rates dataset. (Hint: You can use the command below to obtain all of the column names for the dataset.)

```
> colnames(chargeoff_rates)
```

Note that some of the charge-off rate values are negative, meaning that in these particular quarters, banks recovered more dollars on loans that were previously thought to be lost than actual losses. How well do you think that

Fig. 2.6 (continued)

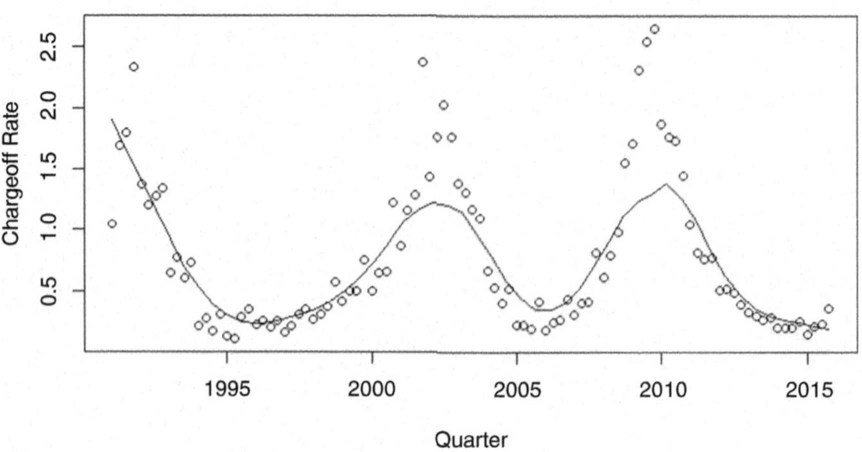

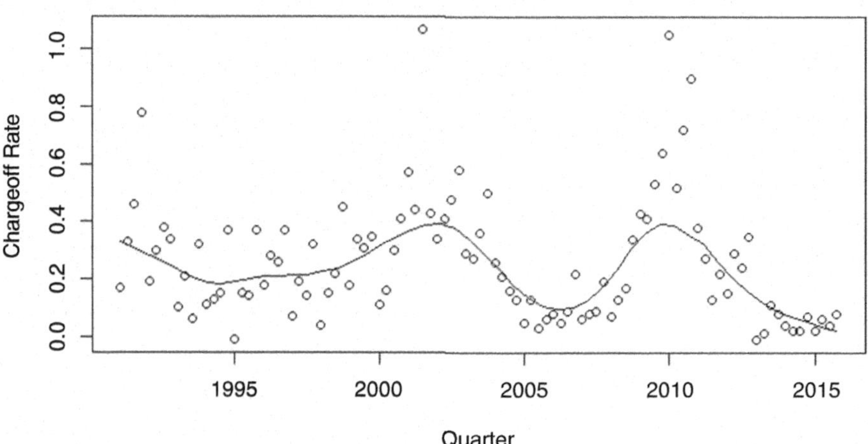

the smoothed lines fit the actual data? The real world is often very messy, as many people unfortunately discovered in the Great Recession of 2008! While most of the loan types have spikes in charge-off rates, the numbers, sizes, and timing of such spikes vary quite a bit. Also, note that the scales of the vertical axes are quite different for some of the loan types.

2.1.3 Including a Third Variable on Scatterplots

Example 2.6. Yearly Charge-Off Rates for Loans Used for Farmland Separated by Quarter Table 2.5 contains a subset of the charge-off rates on agricultural loans grouped by quarter for the years 1991 through 2015 (the rest of the data may be found in the **R** dataset *agricultural_chargeoff_rates_-by_quarter*). Note that this information is contained within Table 2.4, but is now presented in a different format. Instead of simply creating a time series plot, it is sometimes useful to create a scatterplot and then use a different plotting symbol to represent a third variable of interest. In this way three variables can be represented on a two-dimensional plot. Figure 2.7 contains a scatterplot of the second- and fourth-quarter charge-off rates for agricultural loans by year using quarter as the plotting symbol. We generate this by first selecting the indices of charge-off rates from the second and fourth quarters and then using these indices and the previous dataset to create a new variable named *q2_q4_agricultural_chargeoff_rates*.

Table 2.5 Charge-off rates for agricultural loans by year and quarter from 1991 to 2015

Year	Quarter	Agricultural loans
2015	4	0.08
2015	3	0.04
2015	2	0.06
2015	1	0.02
2014	4	0.07
...	...	...
...	...	...
1992	1	0.19
1991	4	0.78
1991	3	0.46
1991	2	0.33
1991	1	0.17

Source: http://www.federalreserve.gov/releases/chargeoff/chgallnsa.htm [Accessed May 8, 2016]

Fig. 2.7 Q2 and Q4 charge-off rates for agricultural loans by year from 1991 to 2015

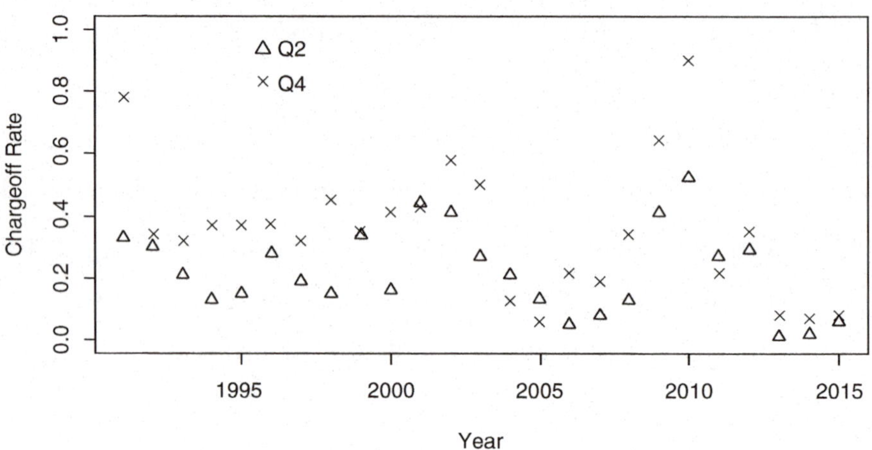


```
> second_and_fourth_indices <-
    agricultural_chargeoff_rates_by_quarter$Quarter %in% c(2,4)
> q2_q4_agricultural_chargeoff_rates <-
    agricultural_chargeoff_rates_by_quarter[second_and_fourth_indices,]
```

We generate the scatter plot as before, using the **R** function *plot*( ) on our newly defined dataset, but with the extra argument *pch* (which stands for "**p**lot **ch**aracters") specified. With this argument, we are instructing **R** to use the value stored in the *Quarter* column of our dataset to determine which symbol to use in the plot.

```
> plot(q2_q4_agricultural_chargeoff_rates$Year,
       q2_q4_agricultural_chargeoff_rates$AgriculturalLoans,
       xlab = "Year",
       ylab = "Chargeoff Rate",
       main = "Scatterplot of Agricultural Loans and Year, Grouped by
               Quarter",
       pch = q2_q4_agricultural_chargeoff_rates$Quarter)
```

Finally, to make the plot easier to interpret, we add a legend using the **R** function *legend*( ). An entire book could be (and many have been!) devoted to the rich capabilities of **R**'s plotting functions. We encourage you to explore this functionality, but in the interest of brevity, we simply provide the following function without discussion.

```
> legend("topleft",
         inset = c(0.2,0),
         bty = "n",
         legend = c("Q2", "Q4"),
         pch = c(2,4))
```

This modified scatterplot seems to indicate a higher percentage of agricultural loans charge off in Q4 than in Q2, at least for the 25-year period examined. Try producing the same scatterplot for all four quarters on your own, but be warned that the picture becomes much messier!

Section 2.1 Practice Exercises

2.1.1. *TIAA-CREF Account Performances.* Many teachers, professors, and other educational professionals have their retirement funds in TIAA-CREF accounts. In these accounts, individuals can make their own allocation decisions so it is important to track the performance of the various options. Table 2.6 contains a subset of historical unit values from January 2, 2015 to March 15, 2016 for seven TIAA and CREF variable annuities (the full table may be found in the **R** dataset *tiaa_cref*).

(a) Create a histogram for the unit values in the equity index account and comment on the overall shape of the data collection distribution.

(b) Comment on the pattern in the unit values for the Growth and Income account over this time period. What visual display do you think is most appropriate for demonstrating this pattern?

(c) The initial unit values for both the Equity Index and Real Estate Securities accounts were close to $15.50 on January 2, 2015. Compare these two sets of unit values over the time periods by constructing side-by-side boxplots and describe any differences.

(d) One way to compare accounts with different unit values is to create new variables that measure the percentage increases from January 2, 2015 to March 15, 2016. Explain how to create new variables that measure such percentage increases for the Global Natural Resources

Table 2.6 Selected rows of historical unit values of TIAA-CREF variable annuities from January 2, 2015 to March 15, 2016

Date	Bond index (TBIIX)	Equity index (TIEIX)	Global natural resources (TNRIX)	Growth & income (TIGRX)	International equity index (TCIEX)	Real estate securities (TIREX)	Social choice equity (TISCX)
1/2/2015	10.94	15.5	8.61	12.03	17.39	15.4	16.78
1/5/2015	10.97	15.23	8.38	11.81	16.99	15.45	16.46
1/6/2015	11.01	15.08	8.3	11.68	16.82	15.58	16.29
1/7/2015	11.01	15.26	8.38	11.84	16.92	15.82	16.49
1/8/2015	10.98	15.53	8.52	12.07	17.19	15.91	16.76
1/9/2015	11.01	15.4	8.52	11.98	17.1	15.92	16.61
1/12/2015	11.03	15.29	8.44	11.89	17.1	16.02	16.47
1/13/2015	11.03	15.26	8.38	11.87	17.19	15.97	16.44
...	...	...	...	...	...	...	...
...	...	...	...	...	...	...	...
3/7/2016	10.85	14.79	6.89	10.94	16.11	14.72	15.33
3/8/2016	10.89	14.59	6.58	10.82	15.93	14.57	15.12
3/9/2016	10.86	14.67	6.66	10.89	16	14.65	15.18
3/10/2016	10.85	14.66	6.65	10.9	15.99	14.61	15.17
3/11/2016	10.84	14.91	6.79	11.11	16.37	14.97	15.45
3/14/2016	10.85	14.89	6.74	11.1	16.36	14.93	15.42
3/15/2016	10.85	14.84	6.66	11.06	16.2	14.93	15.36

Source: www.tiaa-cref.org [Accessed 5/8/16]

and Social Choice Equity accounts. That is, identify the linear transformations that would be used to create the appropriate percentages.

(e) Is there any association between the units values for the Equity Index account and those for the Real Estate Securities account? If so, describe the nature of the association and comment on the strength of the association.

2.1.2. *TIAA-CREF Account Performances.* Use the TIAA-CREF variable annuities data in Table 2.6 (and the **R** dataset *tiaa_cref*) to answer the following questions.

(a) Create a time series plot for the set of historical unit values for each TIAA-CREF annuities account.

(b) Comment on the overall patterns and departures from those patterns.

(c) By simply looking at the graphs, which set of unit values appears to be the most variable?

(d) Calculate two numerical measures of variability for each set of unit values and compare them across the annuities accounts. Did you guess correctly in part (c)?

(e) Calculate the percentage change in each account from January 2, 2015 to March 15, 2016.

(f) If a teacher had invested $1000.00 in each of the 8 accounts on January 2, 2015, what would the value of each account have been on March 15, 2016?

2.1.3. *Median Weekly Earnings.* Use the weekly earnings data in Table 2.1 (and the **R** dataset *weekly_salaries*) to explore the relationships between the median weekly earnings for men and women in service occupations and construction and extraction occupations.

(a) Plot median weekly earnings for men who work in service occupations versus women who work in service occupations. Comment on the nature and strength of the association.

(b) Plot median weekly earnings for men who work in construction and extraction occupations versus women who work in construction and extraction occupations. Comment on the nature and strength of the association and compare your results to those in (a).

2.1.4. *Median Weekly Earnings.* Use the weekly earnings data in Table 2.1 (and the **R** dataset *weekly_salaries*) to analyze the time series for median weekly earnings for men and women in sales and office occupations.

(a) Construct a time series plot of median weekly earnings for men who work in sales and office occupations.

(b) Construct a time series plot of median weekly earnings for women who work in sales and office occupations.

(c) Do the plots show similar patterns? Are these patterns what you would have expected to see before you constructed the plots? Explain.

2.1.5. *Pine Tree Growth.* Use the pine data in the **R** dataset *pines_1997* to construct scatterplots and comment on the nature of the association for each pair of variables listed below. Would you be comfortable predicting the value of one variable if you were given information regarding the other variable?

(a) Spread97 and Hgt97

(b) Spread97 and Diam97

(c) Needles97 and Spread97

(d) Needles97 and Hgt97

(e) Diam96 and Diam97

(f) Grow96 and Spread97

2.1.6. *Pine Tree Growth.* Which of the following two relationships do you think will be stronger for the pine data? Why?

Hgt90 and Hgt96

or

Hgt96 and Hgt97

After you describe the rationale for your choice, construct appropriate scatterplots to check your conjecture.

2.1.7. *Greenhouse Gas Emissions.* Use the domestic greenhouse gas emissions of air pollutants data in Table 2.3 (and in the **R** dataset *emissions*) to evaluate the time series data on air pollutants from 1990 to 2014.

(a) Construct separate time series plots for carbon dioxide, nitrous oxide, and fluorinated gases.

(b) Use scatterplot smoothing methods on each of the three plots in part (a).

(c) Is the decreasing trend that was shown for methane (see Figs. 2.4 and 2.5) clearly visible for the other three air pollutants as well? If not, identify the pollutants that follow different trends and comment on the nature of those trends.

(d) How would you estimate the average yearly rate of change for a particular air pollutant from 1990 to 2014?

(e) Use your method from part (d) to estimate the average rate of change for each of the four air pollutants. Are the estimates close to one another? According to your estimates, which pollutant has decreased the most?

(f) Which pollutant was the most variable over the period from 1990 to 2014? Justify your answer by referring to the time series plots and computing appropriate measures of variability.

2.1.8. *Charge-off Rates.* Use the **R** dataset *chargeoff_rates* to answer the following questions.

Table 2.7 Kentucky Derby race statistics (1990–2012)

Date	Winner	Jockey	Net to winner	Time	Track
1990	Unbridled	C. Perret	581,000	2:02	Good
1991	Strike the Gold	C. Antley	655,800	2:03	Fast
1992	Lil E. Tee	P. Day	724,800	2:03	Fast
1993	Sea Hero	J. Bailey	735,900	2:02.4	Fast
1994	Go for Gin	C. McCarron	628,800	2:03.6	Sloppy
1995	Thunder Gulch	G. Stevens	707,400	2:01.2	Fast
1996	Grindstone	J. Bailey	869,800	2:01	Fast
1997	Silver Charm	G. Stevens	700,000	2:02.4	Fast
1998	Real Quiet	K. Desormeaux	738,800	2:02.2	Fast
1999	Charismatic	C. Antley	886,200	2:03.2	Fast
2000	Fusaichi Pegasus	K. Desormeaux	1,038,400	2:01	Fast
2001	Monarchos	J. Chavez	812,000	1:59.97	Fast
2002	War Emblem	V. Espinoza	1,875,000	2:01.13	Fast
2003	Funny Cide	J. Santos	800,200	2:01.19	Fast
2004	Smarty Jones	S. Elliot	5,854,800	2:04.06	Sloppy
2005	Giacomo	M. Smith	2,399,600	2:02.75	Fast
2006	Barbaro	E. Prado	2,000,000	2:01.36	Fast
2007	Street Sense	C. Borel	2,210,000	2:02.17	Fast
2008	Big Brown	K. Desormeaux	2,000,000	2:01.82	Fast
2009	Mine That Bird	C. Borel	2,000,000	2:02.66	Fast
2010	Super Saver	C. Borel	2,000,000	2:04.45	Fast
2011	Animal Kingdom	J. Velazquez	2,000,000	2:02.04	Fast
2012	I'll Have Another	M. Gutierrez	2,000,000	2:01.83	Fast

Source: www.kentuckyderby.ag [Accessed 5/10/16]

(a) Is there a positive association between the charge-off rates of agricultural loans and credit cards? If so, how strong is the association?

(b) Repeat part (a) for residential real estate loans and commercial real estate loans.

2.1.9. *Kentucky Derby Races.* Race statistics for the Kentucky Derby are presented in Table 2.7 (and are available in the **R** dataset *kentucky_derby_2012*) for 1990–2012.

(a) Construct a time series plot for the winning times and comment on any obvious patterns. (Note that you may need to change the Time column to be easier to work with and plot.)
(b) Construct a time series plot for the net amount of money paid to the winner of the race and comment on the pattern over time.
(c) Add a plotting symbol to the time series plot in part (a) that identifies the condition of the track. Does the condition of the track affect the winning times? Compute appropriate descriptive statistics to justify your response.
(d) Add a plotting symbol to the time series plot in part (b) that identifies the condition of the track. Does the condition of the track effect the net amount of money paid to the winner? Compute appropriate descriptive statistics to justify your response.
(e) Is there any association between winning time and the net amount of money paid to the winner? Justify your response.

2.2 Measuring the Strength of Association

As we have seen in Sect. 1, visualizing the raw data with scatterplots can be very helpful in getting a feel for the overall association between two variables. However, we would also like to have descriptive statistics to quantify the overall strength of the association. Two analysts may disagree in their subjective interpretations of a particular scatterplot so it is useful to include both visual displays and descriptive statistics in any complete analysis.

Our first measure of association is known as the Pearson correlation coefficient and it will be denoted by r. To formally define this statistic, we need to establish some mathematical notation. Suppose we refer to the variable on the horizontal axis as X and the variable on the vertical axis as Y.

Table 2.8 Understanding the direction of the association and the sign of the Pearson correlation coefficient

Deviation from x mean $(x_i - \bar{x})$	Deviation from y mean $(y_i - \bar{y})$	Product of deviations $(x_i - \bar{x})(y_i - \bar{y})$
Negative	Negative	Positive
Negative	Positive	Negative
Positive	Negative	Negative
Positive	Positive	Positive

Since we have n pairs of observations in our bivariate data set, we will refer to these pairs using subscript notation. The subscript may be written out for each pair using dot-dot-dot notation, $(x_1, y_1), \ldots, (x_n, y_n)$, or a generic subscript and its range may be provided for an arbitrary pair, (x_i, y_i), for $i = 1, \ldots, n$. To check whether large values of X occur with large or small values of Y, the deviation of x_i from its mean is multiplied by the deviation of y_i from its mean.

Table 2.8 shows the four nonzero possibilities for the deviations and the products of the deviations. If both deviations are negative (below average), then the product of the deviations will be positive. If both deviations are positive (above average), then the product of the deviations will again be positive. If one deviation is negative (below average) and the other deviation is positive (above average), then the product of the deviations will be negative.

This process of comparing deviations is completed for each pair of observations. Then, the results of the individual comparisons are combined by adding together all of the products. This process determines the sign of the correlation coefficient. The final step in computing the correlation coefficient is to standardize the statistic so that it will be on a unit-less scale that can be interpreted appropriately in an arbitrary setting. The required standardization is provided in Definition 2.5.

Definition 2.5 Suppose that n pairs of observations, (x_i, y_i) for $i = 1, \ldots, n$, are obtained on two quantitative variables. The **Pearson correlation coefficient** between x and y is

$$r = \frac{\sum_{i=1}^{n} (x_i - \bar{x})(y_i - \bar{y})}{\sqrt{\sum_{i=1}^{n} (x_i - \bar{x})^2 \sum_{j=1}^{n} (y_j - \bar{y})^2}} \tag{2.1}$$

Notice that the standardization relies on the sums of the squared deviations from the means. These sums of squared deviations were also fundamental in measuring variability about the mean in Chapter 1. In the exercises you will be asked to show that the Pearson correlation coefficient can be rewritten in terms of the standard deviations for x (s_x) and y (s_y) as

$$r = \frac{1}{n-1} \frac{\sum_{i=1}^{n} (x_i - \bar{x})(y_i - \bar{y})}{s_x s_y} \tag{2.2}$$

Example 2.7. On Time Airline Arrivals Table 2.9 (and the **R** dataset *airline_arrivals*) contains on time arrival records for U.S. flight carriers in 2015. Percentages of on time arrivals and ranks among the carriers are provided for all four quarters of the year 2015, the month of December, and the entire year 2015. Do you think there is any association between the percentages of flights that arrived on time during the first quarter and the percentages of flights that arrived on time during the second quarter for these airlines?

Figure 2.8 is a scatterplot of the on time percentages for the first quarter versus those for the second quarter across the thirteen airlines, which we again generate using the **R** function *plot*( ).

Table 2.9 On time arrival records for U.S. flight carriers, 2015

	Q1		Q2		Q3		Q4		December		2015	
Carrier	**%**	**Rank**	**%**	**Rank**	**%**	**Rank**	**%**	**Rank**	**%**	**Rank**	**%**	**Rank**
ALASKA	85.1	1	88.2	2	85.9	2	86.3	3	85.3	2	86.4	2
AMERICAN	75.9	7	78	7	82	4	82.9	6	79.2	5	80.3	4
DELTA	82.8	3	85.3	3	86.6	1	88.5	2	83.6	3	85.9	3
ENVOY	60.6	13	74.8	10	81.1	8	83.5	4	80.4	4	74.1	11
EXPRESSJET	73.6	9	76.1	9	81.6	5	80.7	8	77.3	7	77.9	9
FRONTIER	64	12	71.1	12	78	11	78.1	11	75	9	73.2	12
HAWAIIAN	85.1	2	91.3	1	84.8	3	92.5	1	93	1	88.4	1
JETBLUE	68	11	81	4	76.7	12	77.8	12	70.1	12	76	10
SKYWEST	76.8	6	80.8	5	81.1	7	80	9	72.9	10	79.7	6
SOUTHWEST	79	5	77.6	8	80	9	82.3	7	76.1	8	79.7	7
SPIRIT	70.5	10	61.8	13	69.6	13	73.9	13	68.7	13	69	13
UNITED	75.9	8	73.9	11	79.3	10	83.2	5	77.9	6	78.2	8
VIRGINAMERICA	79.2	4	79.9	6	81.2	6	79.2	10	71.1	11	79.9	5

Source: U.S. Department of Transportation https://www.transportation.gov/airconsumer [Accessed 5/12/16]

Fig. 2.8 Scatterplot of on time arrival percentages of U.S. flight carriers for the first and second quarters, 2015

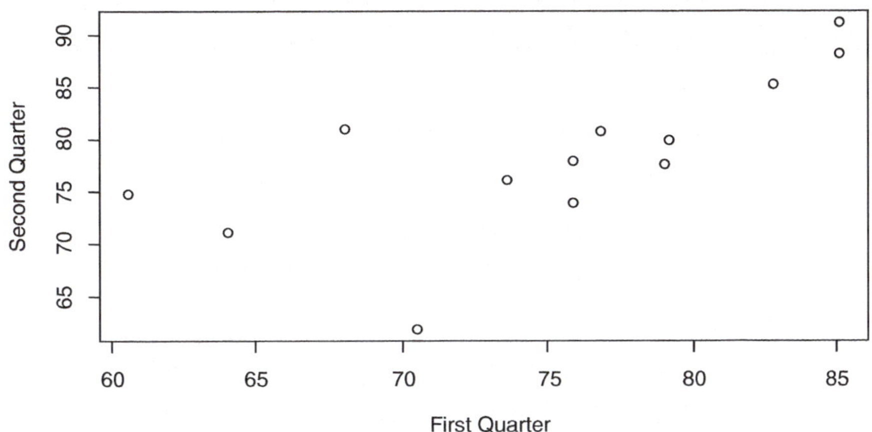


```
> plot(airline_arrivals$Q1.Percent,
       airline_arrivals$Q2.Percent,
       xlab = "First Quarter",
       ylab = "Second Quarter",
       main = "On Time Arrival Percentages in First and Second Quarters
              of 2015")
```

The scatterplot shows that there is positive association between the on time arrival percentages for these two quarters. (The association would be even stronger if it weren't for Spirit's nearly 9% drop in on time arrivals from Q1 to Q2!) Table 2.10 illustrates the steps involved in the computation of the correlation coefficient r for these data. Columns 3 and 4 contain the deviations from $\bar{x} = 75.11538$ and $\bar{y} = 78.44615$, respectively. Column 5 contains the product of the deviations, and Columns 6 and 7 contain the squared deviations.

To complete the computation of r, we add Columns 5, 6, and 7 and enter these sums into the formula in (2.1). The sums of Columns 5, 6, and 7 are, respectively,

Table 2.10 Computing the correlation coefficient between on time arrival percentages for U.S. flight carriers during the first and second quarters of 2015

Q1 Percent	Q2 Percent	x-dev $(x_i - \bar{x})$	y-dev $(y_i - \bar{y})$	Product $(x_i - \bar{x})(y_i - \bar{y})$	x-dev-sq. $(x_i - \bar{x})^2$	y-dev-sq. $(y_i - \bar{y})^2$
85.1	88.2	9.984615	9.753846	97.3884	7242.01	7779.24
75.9	78	0.784615	−0.44615	−0.35006	5760.81	6084
82.8	85.3	7.684615	6.853846	52.66917	6855.84	7276.09
60.6	74.8	−14.5154	−3.64615	52.92533	3672.36	5595.04
73.6	76.1	−1.51538	−2.34615	3.555325	5416.96	5791.21
64	71.1	−11.1154	−7.34615	81.65533	4096	5055.21
85.1	91.3	9.984615	12.85385	128.3407	7242.01	8335.69
68	81	−7.11538	2.553846	−18.1716	4624	6561
76.8	80.8	1.684615	2.353846	3.965325	5898.24	6528.64
79	77.6	3.884615	−0.84615	−3.28698	6241	6021.76
70.5	61.8	−4.61538	−16.6462	76.8284	4970.25	3819.24
75.9	73.9	0.784615	−4.54615	−3.56698	5760.81	5461.21
79.2	79.9	4.084615	1.453846	5.938402	6272.64	6384.01

$$\sum_{i=1}^{n}(x_i - \bar{x})(y_i - \bar{y}) = 477.8908$$

$$\sum_{i=1}^{n}(x_i - \bar{x})^2 = 702.7569$$

$$\sum_{i=1}^{n}(y_i - \bar{y})^2 = 692.9523$$

and the correlation coefficient r for these data is then

$$r = \frac{477.8908}{\sqrt{(702.7569)(692.9523)}} = 0.6848.$$

Having successfully computed the value of the correlation coefficient, how do we interpret $r = 0.6848$? Some properties of r will help us understand more about what we can and cannot conclude for a value of $r = 0.6848$. Figure 2.9 contains several scatterplots and the corresponding values of r.

2.2.1 Properties of *r*

Careful examination of Fig. 2.9 reveals several interesting properties of the correlation coefficient r. We list some of the properties below and then add some additional properties for you to think about and experiment with in the Sect. 2 Exercises.

The correlation coefficient r is always between -1 and 1, inclusive.

A positive value of r indicates positive association. The closer r gets to 1 the stronger the positive association. (Consider the first scatterplot in Fig. 2.9.)

A negative value of r indicates negative association. The closer r gets to -1 the stronger the negative association. (Consider the fourth scatterplot in Fig. 2.9.)

Perfect positive (negative) association is indicated by $r = 1$ ($r = -1$).

Fig. 2.9 Values of the correlation coefficient r for a variety of data scatterplots

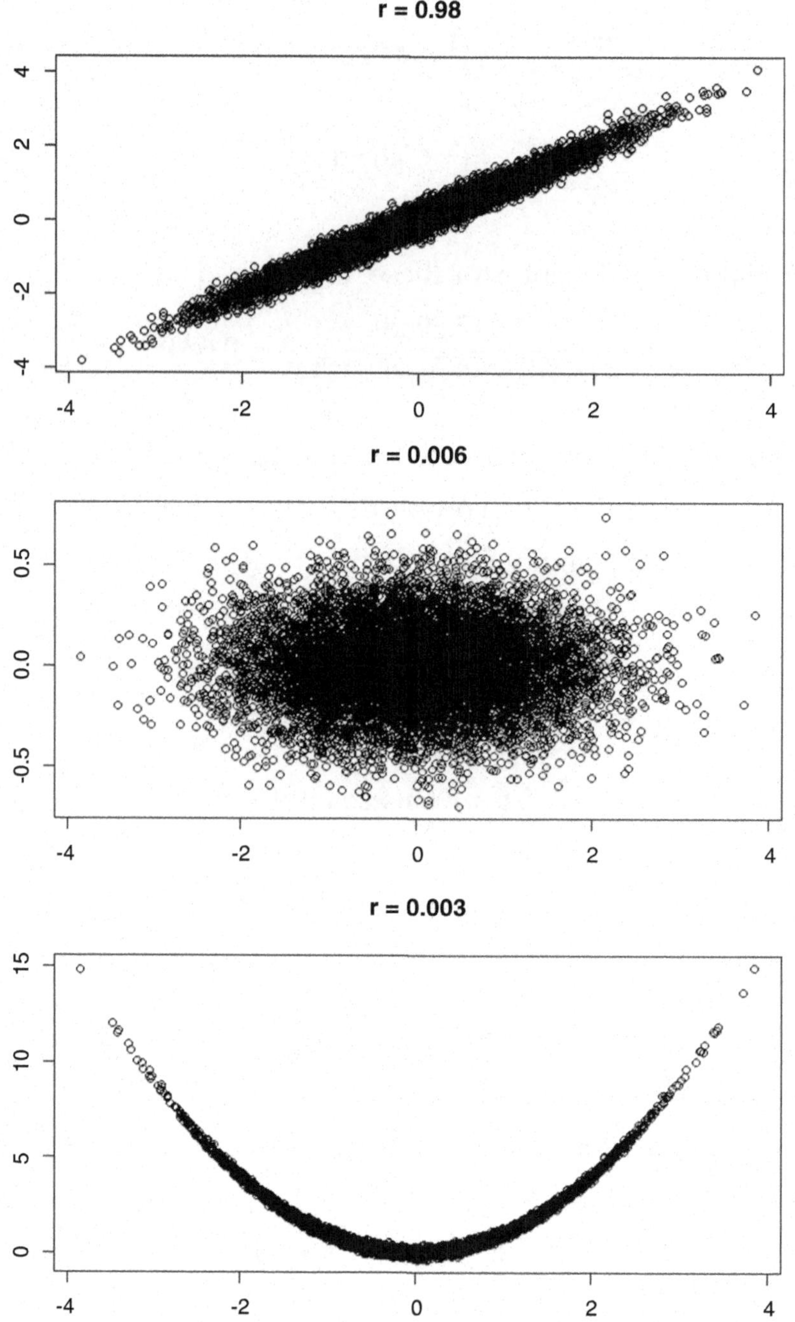

Fig. 2.9 (continued)

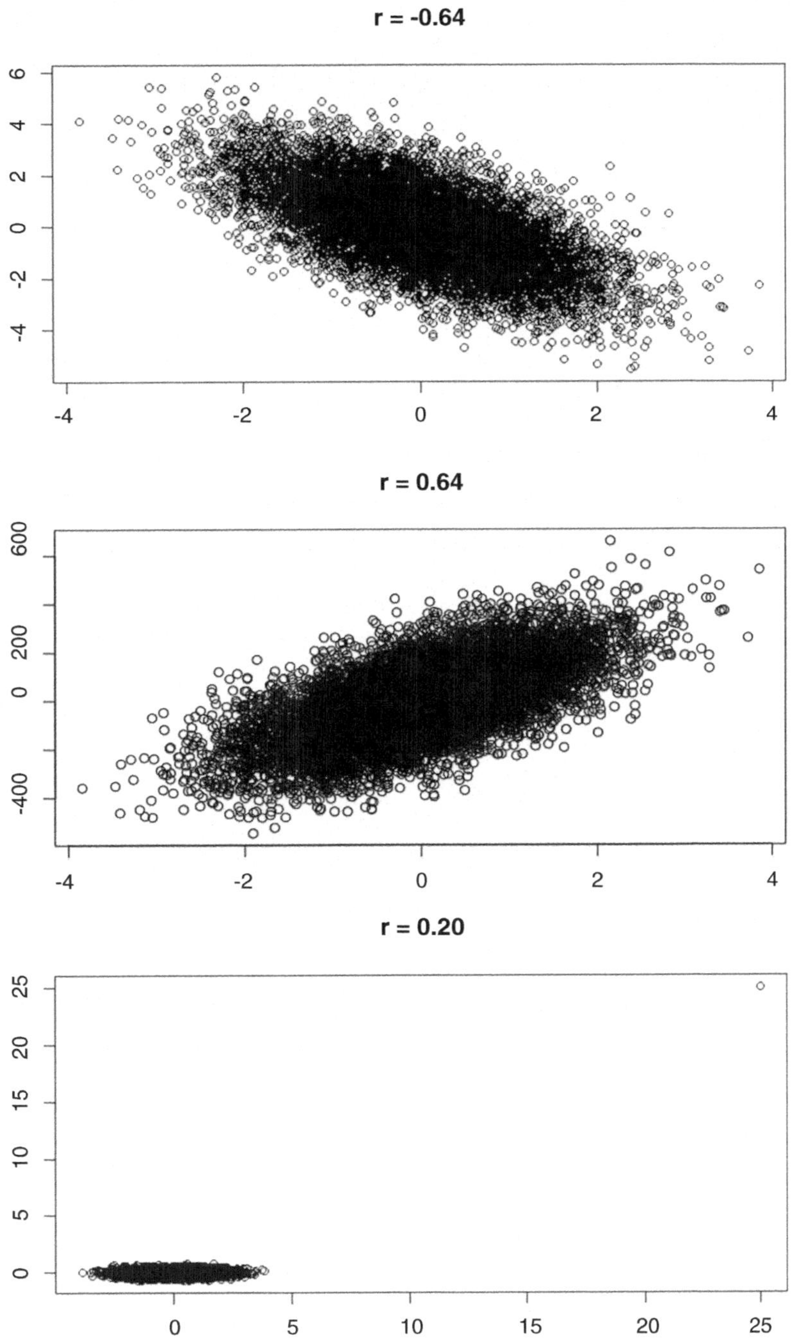

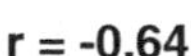

r = 0.64

r = 0.20

Values of r close to zero indicate very weak (minimal) association. (Consider the second scatterplot in Fig. 2.9.)

Interchanging the labels for x and y will not change the value of r.

Applying linear transformations to x, y, or both variables will not change the value of the correlation coefficient. The sign may change (if the slope parameters in the transformations have different signs), but the absolute value of the correlation coefficient will remain the same. (Consider the fourth and fifth scatterplots in Fig. 2.9.)

The value of r can be affected by a few unusual observations in the data set. (Consider the final scatterplot in Fig. 2.9.)

The correlation coefficient r measures only linear association. Two variables could be perfectly related by some nonlinear relationship (quadratic, periodic, etc.) and the value of r would be very close to zero. (Consider the third scatterplot in Fig. 2.9.)

The **R** function cor( ) can also be used to easily compute the value of r for a given dataset.

2.2.2 An Alternative Measure of Association

The fact that r measures only linear association and can be affected by unusual observations should raise at least some concern about the applicability of this statistic in some settings. Such concerns led to the development of an alternative measure of association, known as the **Spearman rank correlation**. Spearman's rank correlation coefficient, which we will denote by r_S, is based on separate ranks of the x's and y's. More specifically, let r_i be the rank of x_i among $x_1, \ldots, x_n$ and let s_i be the rank of y_i among $y_1, \ldots, y_n$. The Spearman rank correlation coefficient r_S, which is formally defined in Definition 2.6, is then simply the correlation coefficient from (2.1) computed on the ranks (r_i, s_i) instead of the original observations (x_i, y_i).

Definition 2.6 Suppose that n pairs of observations, (x_i, y_i) for $i = 1, \ldots, n$, are obtained on two quantitative variables. Let r_i denote the rank of x_i among $x_1, \ldots, x_n$ and let s_i denote the rank of y_i among $y_1, \ldots, y_n$. The **Spearman rank correlation coefficient** r_S for these data is

$$r_S = \frac{\sum_{i=1}^{n} (r_i - \bar{r})(s_i - \bar{s})}{\sqrt{\sum_{i=1}^{n} (r_i - \bar{r})^2 \sum_{j=1}^{n} (s_j - \bar{s})^2}} \tag{2.3}$$

Example 2.8. Computing Spearman's Rank Correlation Coefficient In Example 2.7 we found the Pearson correlation coefficient of $r = 0.6848$ between the first quarter and second quarter on time arrival percentages for U.S. flight carriers. To compute Spearman's rank correlation coefficient r_S, we return to Table 2.9 and notice that the necessary ranks have already been provided in Columns 3 and 5. Replacing the columns in Table 2.10 with the appropriate values for the ranks (or by again using the **R** function cor( ), but with the *method* argument specified to be "spearman"), we find $r_S = 0.741$. For these on time arrival percentages there appears to be only a small difference between Pearson's correlation coefficient and Spearman's correlation coefficient, something you can further investigate in Exercise 2.2.3.

Section 2.2 Practice Exercises

2.2.1. *On Time Flight Arrivals.* Use the on time arrival data for U.S. flight carriers in Table 2.9 to compute Pearson's correlation coefficient for:

(a) first quarter arrival percentages and December arrival percentages;

(b) third quarter arrival percentages and 12-month arrival percentages;

(c) fourth quarter arrival percentages and 12-month arrival percentages;

(d) 12-month arrival percentages and December arrival percentages.

2.2.2. *On Time Flight Arrivals.* Compute Spearman's rank correlation coefficient for each pair of variables in Exercise 2.2.1 and compare the two measures of association. Are the correlation coefficients roughly the same for all four pairs of variables?

2.2.3. *On Time Flight Arrivals.* Either by hand or using the **R** function cor( ), compute Pearson's correlation coefficient and Spearman's rank correlation coefficient for the first and second quarter on time arrival percentages as in Examples 2.7 and 2.8, **except** now remove Spirit Airlines from your calculations. Does this affect the coefficients? Why?

2.2.4. *Domestic Airline Flights.* Table 2.11 shows the aircraft departures, miles flown, and hours flown for domestic flights of large certified air carriers from 2000 to 2015.

Table 2.11 Number of aircraft departures, miles flown, and hours flown for domestic flights of large certified air carriers, 2000–2015

Year	Number of departures	Miles flown (thousands)	Hours flown
2000	7,895,860	5,066,796	12,422,456
2001	7,618,250	4,998,018	12,124,521
2002	8,079,009	5,024,072	12,341,520
2003	9,453,415	5,488,549	13,772,989
2004	9,962,389	5,942,197	14,772,895
2005	10,033,140	6,067,796	15,057,919
2006	9,707,992	5,967,798	14,765,098
2007	9,835,733	6,083,414	15,035,358
2008	9,375,728	5,834,031	14,458,611
2009	8,766,874	5,413,362	13,394,974
2010	8,700,353	5,455,511	13,362,769
2011	8,647,658	5,497,877	13,457,621
2012	8,445,486	5,460,376	13,250,921
2013	8,324,013	5,477,514	13,267,592
2014	8,109,302	5,450,627	13,145,456
2015	8,059,756	5,554,106	13,303,173

Source: "U.S. Air Carrier Traffic Statistics" – Bureau of Transportation Statistics www.rita.dot.gov/bts/acts [Accessed 5/15/16]

(a) Construct a time series plot to investigate the association between miles flown and hours flown. Comment on the direction and strength of the association.

(b) Find the two sets of ranks necessary for computing Spearman's rank correlation coefficient between year and miles flown. That is, identify $(r_1, \ldots, r_{16})$ and $(s_1, \ldots, s_{16})$.

(c) Compute the value of Spearman's rank correlation coefficient between year and miles flown.

(d) Compute the value of Pearson's correlation coefficient between year and miles flown. Is this value close to the value of Spearman's rank correlation coefficient obtained in part (c)?

(e) Suppose that the years were recoded to begin at 1 (2000) and end at 16 (2015) and the correlation coefficients in parts (c) and (d) were computed using the recoded values. Would the correlation coefficients remain the same or change? Explain.

2.2.5. *Domestic Airline Flights.* Use the data in Table 2.11 to explore the relationship between the number of hours flown and the number of departures for the large air carriers.

(a) Construct a scatterplot and compare the two statistics for measuring the strength of association between the number of hours flown and the number of departures for large air carriers.

(b) Which measure do you prefer for this pair of variables? Why?

2.2.6. A random number generator was used to produce the first column of data in Table 2.12. The second column was created from the first column by applying a transformation (not necessarily a linear transformation). The third and fourth columns show the ranks (r_i and s_i) of the values in columns one and two, respectively.

(a) Compute the value of Pearson's correlation coefficient between x and y.

Table 2.12 Randomly generated data and functions of randomly generated data

x	y	r	s
1.16342	1.8321	8	5
−0.49786	0.0614	7	1
−0.99839	0.9936	4	4
−0.81233	0.4354	5	3
−1.17106	1.8807	3	6
−1.34665	3.2887	2	7
−1.79934	10.4822	1	8
−0.68416	0.2191	6	2

(b) Compute the value of Spearman's rank correlation coefficient between x and y.

(c) Construct a scatterplot and examine the raw data to determine the transformation used to create the y-values in column 2 of Table 2.12.

(d) Which measure of association do you prefer for these data? Why?

2.3 Exploring the Relationship between Two Categorical Variables (Frequency Tables)

Example 2.9. Smoking During Pregnancy and Educational Level of Mother Is there any association between the educational level of a mother and whether or not she chooses to smoke during pregnancy? You might initially think that this question is an easy one to answer: simply collect the appropriate data and compute one of the two measures of association from Sect. 2. But there are some major problems with this thought. How are you going to measure educational level? What will you do with the yes or no responses from the pregnant mothers? When one or both of the variables of interest are categorical, there is no easy formula that can be used to measure

Table 2.13 Percentages of mothers who smoked during pregnancies by educational level, 2010–2014

Education level	2010	2011	2012	2013	2014
8th grade or less	4.24	4.75	4.75	4.65	5.06
9th through 12th grade with no diploma	19.85	20.07	20.20	20.09	20.51
High school graduate or GED completed	17.01	16.41	16.06	15.97	15.81
Some college, but not a degree	11.83	11.53	11.54	11.19	10.96
Associate degree	7.08	6.63	6.69	6.44	6.24
Bachelor's degree	1.46	1.35	1.33	1.27	1.20
Master's degree	0.57	0.58	0.54	0.52	0.49
Doctorate or Professional degree	0.42	0.38	0.34	0.33	0.32

Source: Centers for Disease Control and Prevention (CDC) "WONDER Online Database, February 2016. [Accessed at http://wonder.cdc.gov/natality-current.html on May 20, 2016]

Fig. 2.10 Bar graph of percentages of mothers who smoked during pregnancies in 2010

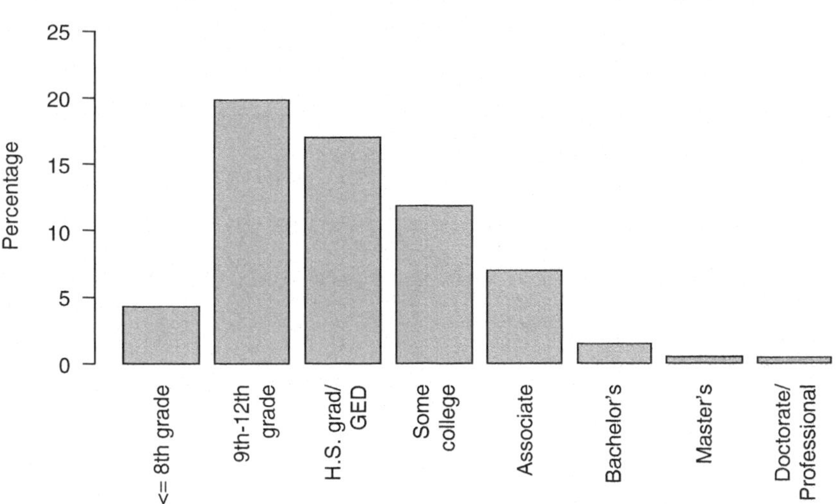

association. Instead, the analyst must compute appropriate percentages and use visual displays to answer the question.

Table 2.13 shows the percentages of mothers in the United States from 2010 to 2014 who smoked during pregnancy for various categories of educational level. Figure 2.10 is a bar graph of the percentages for the year 2010 generated using the **R** function *barplot*( ) and the dataset *mother_smoking_education_2010* as follows.

```
> barplot(mother_smoking_education_2010$Percentage,
          ylab = 'Percentage',
          names.arg = mother_smoking_2010$Education,
          las = 2,
          main = "Percentages of Mothers Who Smoked During Pregnancies
                  in 2010")
```

Note that the *barplot*( ) function takes many of the same arguments that we have seen previously when dealing with the *plot*( ) function. In addition to those we've encountered before, we use the *names.arg* argument to specify the names for the x axis and the *las* argument to specify the orientation of these names. (If we had left them horizontal rather than vertical, the bar graph would be much harder to read!) There appears to be a clear relationship between educational level and the percentage of mothers who smoked—as educational level increases the percentage of mothers who smoked during pregnancies tends to decrease, suggesting a negative association. This pattern appears to hold for all education levels except for "8th grade or less". Having checked the data on the CDC website, the rate of smoking during pregnancy for mothers who stop going to school in the 8th grade or earlier is consistently less than for mothers who finish 9th–12th grade but do not graduate across **all** age levels. Your authors' best guess is that mothers who drop out of school before high school avoid the associated peer pressure which may lead many of these mothers to start smoking in the first place. What do you think explains this phenomenon?

Do you think this association is the same for the other years 2011–2014? To investigate the association between these two categorical variables over time, we can create another bar graph by putting year on the *x*-axis and using different colors to identify the educational level of the mother. Figure 2.11 uses the **R** dataset *mother_smoking_education* to clearly show that the negative association is present for all 5 years. Note that the **R** function *barplot*( ) does not have an easy way to generate grouped bar graphs like the one in Fig. 2.11,

Fig. 2.11 Bar graph of percentages of mothers who smoked during pregnancies, 2010–2014

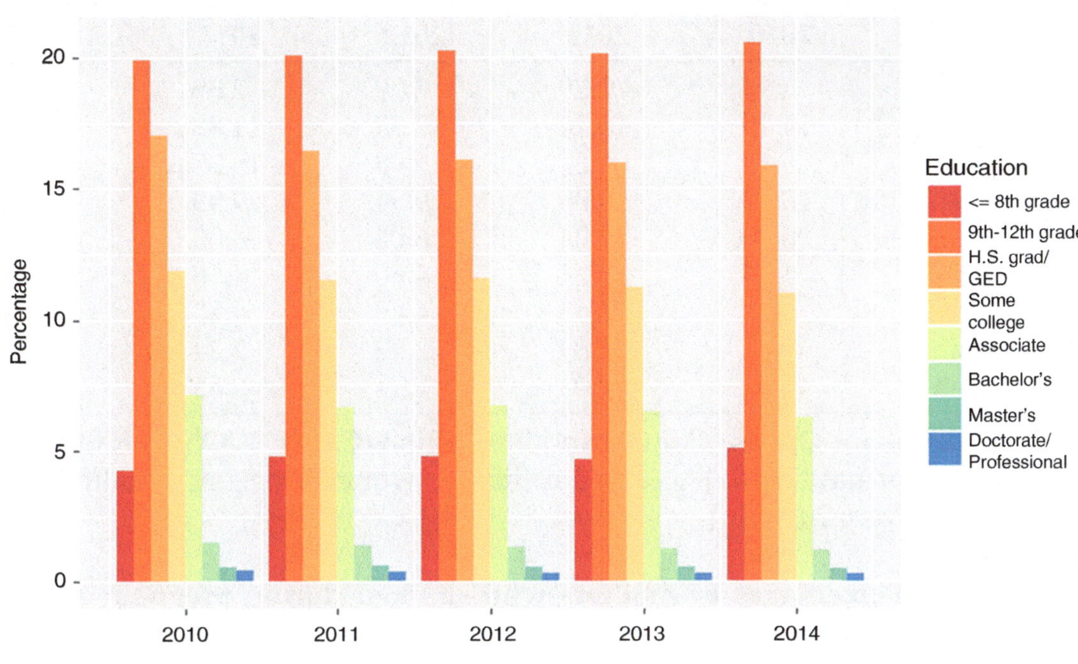

so the plot was generated using the *ggplot2* package, which is a very powerful (but also more complicated to learn!) plotting tool.

Section 2.3 Practice Exercises

2.3.1. *Age Related Smoking During Pregnancies.* Table 2.14 shows the percentage of mothers in the United States who smoked during pregnancies from 2010 to 2014 for various age categories. The data are also available in the **R** dataset *mother_smoking_age*.

(a) Create a bar graph to compare the percentages of mothers who smoked during pregnancies for each age category in 2010. Is there a strong association between age and whether or not a mother chooses to smoke during pregnancy?

Table 2.14 Percentages of mothers who smoked during pregnancies by age, 2010–2014

Age	2010	2011	2012	2013	2014
Under 15	2.94	2.29	2.99	2.59	2.68
15–19 years	14.19	13.24	12.53	11.88	11.37
20–24 years	16.88	16.37	15.98	15.37	14.92
25–29 years	10.23	10.00	9.91	9.83	9.91
30–34 years	6.00	6.04	6.02	6.03	6.05
35–39 years	4.74	4.71	4.60	4.58	4.80
40–45 years	4.87	4.63	4.48	4.27	4.31
45–49 years	3.11	2.89	2.72	2.24	2.68
50–54 years	1.5	2.04	1.21	0.85	1.95

Source: Centers for Disease Control and Prevention (CDC) "WONDER Online Database, February 2016. [Accessed at http://wonder.cdc.gov/natality-current.html on May 20, 2016]

Table 2.15 Percentages of mothers who smoked during pregnancies by educational level, 1989–1993

Educational level	1989	1990	1991	1992	1993
0–8 years	20.8	19.2	18.3	16.8	15.2
9–11 years	35	33.3	31.9	30.6	29
12 years	22.2	21.2	20.6	20.1	19.3
13–15 years	13.6	12.7	12.4	12	11.3
16 years or more	5	4.5	4.2	3.9	3.1

Source: Centers for Disease Control and Prevention, National Center for Health Statistics, Health, United States, 1995

(b) Similar to Fig. 2.11, create bar graphs grouped by year (using multiple plots if necessary) to compare the association for all 5 years. Do you see any consistent pattern?

(c) How do you think the data in Table 2.14 were obtained by the Centers for Disease Control?

2.3.2. *Educational Level and Smoking During Pregnancies.* Table 2.15 shows the percentages of mothers in the United States from 1989 to 1993 who smoked during pregnancy for various categories of educational level. The data are also available in the **R** dataset *mother_smoking_education_1989_1993*.

(a) Create a bar graph to compare the percentages of mothers who smoked during pregnancies for each educational level in 1990. In 1990, was there a strong association between age and whether or not a mother chooses to smoke during pregnancy? Compare this with the findings about the relationship between mothers' smoking habits and education in 2010 that were discussed in Example 2.9.

(b) Similar to Fig. 2.11, create bar graphs grouped by year (using multiple plots if necessary) to compare the association for all 5 years. Do you see any consistent pattern?

(c) What do you notice about the overall levels of smoking in Table 2.15 compared to those in Table 2.13? Do the changes in these levels seem to affect all education levels equally? What do you think the data will look like in 2031–2035?

2.3.3. *Does Where You Live Affect Your Health?* Table 2.16 shows a sample of "chronic disease indicators", as reported by the CDC's Division of Population Health, for the four states California, Michigan, Ohio, and West Virginia. These values represent the following indicators for the states:

1. Prevalence of obesity among adults 18 or older (age-adjusted, as of 2013)
2. Prevalence of heavy drinking among adults 18 or older (age-adjusted, as of 2013)

Table 2.16 U.S. Chronic Diseases Indicators (CDI)

State	Obesity	Heavy drinking	Life expectancy	Poverty
CA	24.10	6.30	78.80	17.00
MI	31.30	6.30	76.90	17.40
OH	30.20	5.50	76.49	16.30
WV	35.50	4.10	75.28	17.80

Source: Centers for Disease Control and Prevention http://catalog.data.gov/dataset/u-s-chronic-disease-indicators-cdi-ff843 [Accessed on May 21, 2016]

Table 2.17 Percentages of respondents who answered "yes" to various questions about science by political party

Political party	Global warming	Vaccines	Human astronauts	GMO
Republican	40.51	33.63	66.82	52.01
Democrat	75.79	20.82	62.15	37.70
Independent	62.57	31.26	60.22	41.19

Source: Pew Internet & American Life Project, Science Issues. http://www.pewinternet.com/datasets/2014-science-issues/ [Accessed on May 21, 2016]

3. Life expectancy at birth (as of 2001)
4. Prevalence of poverty (as of 2012)

The data are also available in the **R** dataset *state_cdi*.

(a) Create a bar graph to compare the prevalence of obesity for these states. Is the relationship between state and obesity what you would have expected?

(b) Create similar bar graphs for the four states and each of the other three indicators. Does the pattern you found for obesity hold as well for the other indicators?

(c) Create a scatterplot of the four pairs of observations for obesity and life expectancy. Does there appear to be any association between the two? Would you have expected this before seeing the data?

2.3.4. *Is There a Relationship between Political Affiliation and Views Toward Science?* Table 2.17 shows the percentages of people who answered "yes" to the following questions in a survey conducted by the Pew Research Center in 2014, grouped by political affiliation. The questions were:

1. From what you've heard or read, do scientists generally agree that the earth is getting warmer because of human activity?
2. Thinking about childhood diseases, such as measles, mumps, rubella, and polio, should parents be able to decide **NOT** to vaccinate their children?

3. The cost of sending human astronauts to space is considerably greater than the cost of using robotic machines for space exploration. As you think about the future of the U.S. space program, do you think it is essential to include the use of human astronauts in space?
4. Do you think it is generally safe to eat genetically modified foods?

These values are also available in the **R** dataset *pew_science_survey_data_by_party*.

(a) Create a bar graph to compare the percentages of respondents for each political party who answered "yes" that scientists generally agree that the earth is getting warmer because of human activity. Is there a strong association between political party and whether or not a person believes that scientists generally agree that global warming is caused by humans?

(b) Create bar graphs grouped by political party to see the association for the other three questions. Do you see any consistent pattern? Which questions produces the most disagreement and agreement, respectively?

(c) How do you think the data in Table 2.17 were obtained by the Pew Center? Do you think the manner in which the respondents are chosen would affect these numbers? What about the wording of the questions?

2.3.5. *Is There a Relationship between Age and Views toward Science?* Table 2.18 shows the percentages of people who answered "yes" to the Pew Research Center survey questions listed in Exercise 2.3.4. These values are also available in the **R** dataset *pew_science_survey_data_by_age_group*.

(a) Create a bar graph to compare the percentages of respondents for each age group who answered "yes" that scientists generally agree that the earth is getting warmer because of human activity. Is there a strong association between age and whether or not a person believes

Table 2.18 Percentages of respondents who answered "yes" to various questions about science by age group

Age	Global warming	Vaccines	Human astronauts	GMO
18–31	72.85	38.71	61.90	40.51
32–47	66.24	34.33	64.74	38.30
48–58	61.04	25.90	55.64	41.56
59–67	57.14	25.00	64.96	41.72
68+	49.86	18.65	64.78	47.03

Source: Pew Internet & American Life Project, Science Issues. http://www.pewinternet.com/datasets/2014-science-issues/ [Accessed on May 21, 2016]

that scientists generally agree that global warming is caused by humans?

(b) Create bar graphs grouped by age to see the association for the other three questions. Do you see any consistent pattern? Which questions produces the most disagreement and agreement, respectively?

(c) Which questions seem to get the most consistent responses across age groups? What about the most variable?

(d) Compare your answers to those you obtained in Exercise 2.3.4. Assuming the conventional wisdom that Republicans tend to be a bit older than Democrats on average, do the responses about global warming being caused by humans across age groups seem similar to those you observed across political party? What about the responses as to whether it is okay for parents not to vaccinate their children?

Chapter 2 Comprehensive Exercises

2.A. Conceptual

2.A.1. Use basic algebra to show that the correlation coefficient r in (2.1) can also be expressed by

$$r = \frac{1}{n-1} \frac{\sum_{i=1}^{n} (x_i - \bar{x})(y_i - \bar{y})}{s_x s_y}$$

as noted in (2.2).

2.A.2. An alternative form of Spearman's rank correlation coefficient r_S (2.3) is given by

$$r_S = \frac{12 \sum_{i=1}^{n} \left\{ \left(r_i - \frac{n+1}{2}\right)\left(s_i - \frac{n+1}{2}\right) \right\}}{n(n^2 - 1)}.$$

Compute this alternative version of r_S for the first and second quarter on time arrival data in Table 2.9 and verify that it matches the value given in Example 2.8. Compute $\bar{r} = \frac{1}{n}\sum_{i=1}^{n} r_i$ and $\bar{s} = \frac{1}{n}\sum_{i=1}^{n} s_i$ and compare them with the constants used to form the deviations in this version of the formula. What did you find?

2.A.3. Another alternative form of Spearman's rank correlation coefficient r_S (2.3) is given by

$$r_s = 1 - \frac{6 \sum_{i=1}^{n} d_i^2}{n(n^2 - 1)},$$

where $d_i = s_i - r_i$, for $i = 1, \ldots, n$; that is, d_i is the difference in the ranks for the ith pair.

(a) Compute the value of this alternative version of r_S for the first and second quarter on time arrival data in Table 2.9 and verify that it is equal to the value of r_S given in Example 2.8 (and computed in Exercise 2.A.2.).

(b) What are the d_i values, for $i = 1, \ldots, n$, if there is perfect positive association between x and y?

(c) What are the d_i values, for $i = 1, \ldots, n$, if there is perfect negative association between x and y?

2.A.4. Construct a set of 8 pairs of observations that are positively associated and compute the value of r for your data set.

(a) Apply a linear transformation of your choice to the x values only. Compute the value of r for the transformed data. Did the absolute value of r stay the same? Did it change sign? What would happen to the value of r if you change the sign of the slope parameter in your linear transformation?

(b) Compute r_S for the untransformed data (x_i, y_i) and for the linearly transformed data (x_i^*, y_i). Does the value of r_S remain the same?

(c) Do you think linear transformations will always affect r_S the same way that they affect r? Why or why not?

2.B. Data Analysis/Computational

2.B.1. *Quarterly Delinquency Rates and Charge-off Rates.* Consider the twenty-five years of quarterly delinquency rates for eight different types of loans, as reported by the Federal Reserve, presented in the **R** dataset *delinquency_rates.* Use **R** to construct smoothed scatterplots to compare the charge-off rates for each of the eight types of loans (see Example 2.5) with the corresponding delinquency rates. Does one set of rates appear more variable than the other? Do the two sets of rates appear to be related to each other over time?

2.B.2. *College Scorecard Comparisons.* Consider the College Scorecard Data reported by the U.S. Department of Education. A subset of these data for the year 2012 is presented in the **R** dataset *college_rankings_2012.* Your task is to prepare a report on U.S. educational institutions for a group of concerned parents. They are particularly interested in:

1. differences between public, private, and for-profit institutions;
2. relationships between faculty salaries and costs;
3. differences between schools of differing sizes.

(a) Compare in-state tuition for each of the three types of institution: public, private, and for-profit.
(b) Examine the relationship between average monthly faculty salaries and in-state tuition. Also examine the relationship between monthly faculty salaries and median graduate debt. Provide both graphical and numerical evidence to demonstrate any association you find.
(c) Examine the completion rates and average SAT scores of each of the three university types.
(d) Use enrollment numbers to group the institutions into "large", "medium", and "small" (using whatever thresholds you deem reasonable). Examine the relationship between school size and admission rates and that between school size and completion rates.
(e) Examine any other relationships that you think would be of interest to the concerned parents. Do you notice any patterns as to which types of schools seem more or less willing to provide data? (Missing values are denoted as NA in the dataset.)

2.B.3. *Population, Birth Rates, and Migration.* In this exercise, you will analyze population data provided by the U.S. Census Bureau at the state level. A subset of these data as of 2015 is presented in the **R** dataset *population_estimates_2015,* which contains population estimates, birth rates (per 1000 population), and net migration (per 1000 population) for each year 2011 through 2015.

(a) Begin by selecting the state you live in and any three other states that you would like to analyze. For each of these four states, produce a smoothed scatterplot of population estimates over time.

(b) For the four states you selected in (a), produce bar graphs of birth rates by year for each state. What patterns do you notice throughout time? What similarities and differences do you notice between the states you have selected?

(c) Produce scatterplots for population estimate and net migration for each of the 20 combinations of the 5 years and four states. What, if any, association do you see? Would you have expected this? Confirm your visual findings by computing two numerical measures of association.

2.B.4. *Population, Birth Rates, and Migration.* Repeat Exercise 2.B.3, but analyze the data for each of the four *regions* (rather than states) this time.

2.B.5. *Comparing NBA Teams.* In this exercise, you will analyze NBA teams' performance in the 2015–2016 season. A subset of the data made available at http://stats.nba.com/league/team/ is accessible in the **R** dataset *nba_2015_2016*, which contains information on various statistics measuring team performance.

(a) Ultimately, teams care about winning games. Using both graphical and numerical methods, examine the association between win percentage and each of the following statistics: rebounds, assists, turnovers, steals, and blocks. Based on these associations, if you were in charge of an NBA team, which area would you focus on most heavily?

(b) Without looking at the data, write down what you expect the association to be between each pair of the following three statistics: field-goal percent, three-point percent, and free-throw percent. Generate scatterplots and at least one numerical measure of these associations. Do the results agree with what you predicted?

(c) Produce bar graphs for the three percentages discussed in part (b) for each of the following four teams: Cleveland Cavaliers, Golden State Warriors, Los Angeles Lakers, and Philadelphia 76ers. Comment on any patterns you observe when comparing these teams.

2.C. Activities

2.C.1. You will need a standard measuring tape to collect information on a small group of people who are willing to participate in this activity. Collect the information described below for each individual. Construct an appropriate scatterplot for each pair of variables. Are the various pairs of variables associated? If so, explain the nature of the associations. Compute and compare two different measures of association for each pair of variables.

(a) Waist and neck sizes.
(b) Foot and forearm sizes.
(c) Shirt and shoe sizes.
(d) Inseam length and circumference of head.
(e) Height and distance from belly button to the floor.
(f) Height and weight.

2.C.2. *Who is better at rolling sixes?* Obtain a few standard six-sided dice and split participants into groups along a categorical variable (for example: hair color, height, gender, etc.). Each member of each group should roll a die five times and record the number of sixes that he or she obtains. Does group membership appear to be associated with ability to roll sixes? If so, explain the nature of the association. Repeat the experiment 5 times and construct a bar graph for the proportion of sixes for each group by experiment. Do your conclusions about each group's ability to roll sixes change when you analyze five repetitions of the experiment rather than one?

2.C.3. *Tossing Quarters.* Stand approximately 15 feet from a wall. Toss a U.S. quarter toward the wall and try to get it as close to the wall as you can. After some practice trials, record the distance from the edge of the wall to the closest edge of the coin for 15 consecutive tosses.

(a) Create a time series plot for the distance from the edge of the wall to the edge of the quarter.

(b) Describe any patterns or unusual observations you see on the plot. Did you get better or worse over time?

(c) Now try a different method of tossing the coin. After some practice trials, record the distance from the edge of the wall to the edge of the coin for 15 consecutive tosses. Create a time series plot for these distances and compare the plot with the one in part (a).

(d) Is one method clearly better than the other method for you? Justify your answer with appropriate descriptive statistics and graphical displays.

2.C.4. *Weather Forecasts.* Go online and find the 10-day weather forecasts for your local area, for Flint, Michigan, and for Tempe, Arizona. Record the temperatures for each of the 10 days at each of the three locations.

(a) Create a time series plot for the daily temperature at each location.

(b) What similarities and differences do you notice between the patterns of temperature at these locations?

(c) How do you think your answers to (a) and (b) would change if you were to repeat the exercise for *hourly* temperatures rather than daily temperatures?

(d) How do you think your answers to (a) and (b) would be affected if you were to repeat this activity in 6 months? (It may be helpful to find monthly average temperatures for each location.)

2.D. Internet Archives

2.D.1. Visit www.guessthecorrelation.com to check your intuition regarding the shape of a scatterplot and a statistic that measures the strength of linear association.

2.D.2. *Performance Statistics for the New York Stock Exchange.* Find a website that enables you to obtain recent performance statistics for stocks traded on the New York Stock Exchange (NYSE). Enter a ticker symbol for a stock of interest to you and obtain plots of performance statistics for your stock for each of the following time periods: intraday, 1-week, 1-month, 3-month, YTD (year-to-date), 1-year, 3-year, and 5-year. Do all of the plots show the same overall trend? If so, comment on that overall pattern. If not, what does this tell you about the importance of considering an appropriate scale for the horizontal axis?

2.D.3. *Kentucky Derby Races.* Find a website that provides data for all of the previous Kentucky Derby races (Derby charts, race statistics, etc.). Select a 25-year period of interest to you. Download the data from the 25-year period you selected and load it into **R**. (Hint: you can use the *read*.csv( ) function to load a .csv file into **R**.)

(a) Construct a time series plot for the winning times and comment on any obvious patterns.

(b) Construct a time series plot for the net amount of money paid to the winner of the race and comment on the pattern over time.

(c) Add a plotting symbol to the time series plot in part (a) that identifies the condition of the track at the time of the race. Does the condition of the track affect the winning times? Compute appropriate descriptive statistics to justify your response.

(d) Add a plotting symbol to the time series plot in part (b) that identifies the condition of the track at the time of the race. Does the condition of the track affect the net amount of money paid to the winner? Compute appropriate descriptive statistics to justify your response.

(e) Is there any association between winning time and the net amount of money paid to the winner? Justify your response.

2.D.4. *Stock Performances.* Visit YAHOO! Finance at finance.yahoo.com. Enter a ticker symbol for a company of interest to you and then click Go. (If you don't know the ticker symbol, you can do a quick search by typing a company name in the Quote Lookup.) After looking at the recent performance statistics, click the chart to see a time series plot of closing prices. Select 1-year under the time period options to get a different chart.

(a) Another method of smoothing time series data to search for trends is to compute moving averages of the prices. At the top of the chart you can add moving averages to your plot by clicking Indicator, then clicking Simple Moving Average, and finally entering the number of days you want to include in the moving average in the Period field. Select 25 and click the chart to add a 25-day moving average to your 1-year time series plot. Does the addition of the 25-day moving average help you see the overall trend?

(b) Now add a 50-day moving average to your 1-year time series plot. Do you see the same overall pattern with the 50-day moving average that you do with the 25-day moving average?

Designing a Survey or Experiment: Deciding What and How to Measure

3

Over the past few decades, storing data has become increasingly affordable. One measure of this, hard drive cost per gigabyte in U.S. Dollars, has fallen from over \$10,000 in 1990 to about 3 cents in 2014 according to data available at http://www.mkomo.com/cost-per-gigabyte-update. This steep decline has had a number of consequences, one of which is the explosion in popularity of so-called "big data". Technology companies are able to leverage this resource to conduct experiments at breakneck speeds and scales that would have been unthinkable at any other time in human history.

One of the more famous (or maybe infamous?) experimenters in the technology industry is the social network Facebook. According to the company's statistics page, as of June 2016, the site had 1.13 billion daily active users! One of the most important aspects of the social network is the "News Feed", which filters content according to Facebook's proprietary algorithms in order to deliver the most relevant content to each user. A highly publicized

© Springer International Publishing AG 2017
D.A. Wolfe, G. Schneider, *Intuitive Introductory Statistics*, Springer Texts in Statistics,
DOI 10.1007/978-3-319-56072-4_3

experiment was conducted in early 2012 to determine the effect that positive and negative content have on users' emotions.

The (generally negative) press focused on whether or not this sort of psychological experimentation was acceptable and whether the information gleaned from the study could actually serve any business purpose. However, beneath the heated debate about whether Facebook should have conducted the experiment lie many interesting and challenging questions which needed to be answered before the study could begin. What information should be collected? How can emotions be quantified? How can the researchers be sure that their personal biases about the importance of the News Feed do not affect the results? How many users' News Feeds would they need to alter? For each News Feed that they decided to alter, how many stories would they change?

These are some of the typical questions that all researchers must answer before they begin a research project or investigation, whether that investigation is conducted by observing the behavior of millions of people on the Internet, the medically-supervised reactions of dozens of people to an experimental drug, or countless other scenarios. Deciding what and how to measure are very often difficult questions to answer. However, without careful thought and planning before the data collection phase of a project, the entire investigation may be meaningless.

In Chaps. 1 and 2 we focused on displaying and describing data that were already available. In this chapter we step back and focus on the planning and data collection phases of a research project. Although there is a tendency to think that this is the "easiest" and "least important" of the numerous phases of a research project, nothing could be further from the truth. Devoting adequate time to careful planning in the early stages of a project is vital for its success. In Sect. 3.1 we describe several different methods of data collection. In Sect. 3.2 we focus on planning a survey or poll. In Sect. 3.3 we return to further examine the Facebook team's research question, as well as describe other types of designs.

3.1 Methods of Data Collection

If we are interested in a particular population, one way to obtain information on a set of traits or characteristics for this population is to gather and record information for every member of the population. This method of gathering information is referred to as a census.

Definition 3.1 A **census** is a complete enumeration or specification of characteristics for every member of a population of interest.

An admissions director may be interested in the SAT scores of all students attending her institution, a coach may be interested in the speed of the members on her team, a provost may be interested in the salaries for all full-time faculty members at her institution, or an analyst may be interested in the price of a particular stock for the last 30 days. In each of these situations the characteristic of interest can be collected for every member of the population and the techniques in Chaps. 1 and 2 can be used to interpret the data.

Example 3.1. Mathematics Faculty and Staff Salaries Table 3.1 (and the **R** dataset *osu_math_salaries_2015*) provides a complete list of the 2015 salaries of faculty and staff in the Mathematics Department at The Ohio State University. The population of interest is the faculty and staff of the Mathematics Department at The Ohio State University and the characteristic of interest is 2015 salary for these Individuals. Do any of the values surprise you?

Example 3.2. Student Report Cards As part of a major assessment effort the chancellor for high schools in the City of New York decided to start sending report cards home to parents. These report cards, called "School Quality Snapshots" did not contain student grades; they contained "grades" for the city school districts. The hope was that these report cards would help parents

Table 3.1 Salaries for all 226 math faculty and staff at The Ohio State University in 2015 (dollars)

237,980	105,924	80,433	53,768	42,240	24,530	12,800	4700
201,616	105,597	78,756	53,723	42,000	23,680	11,360	4322
187,594	104,537	78,390	52,730	40,368	23,163	11,080	3583
176,726	104,225	76,532	52,664	38,815	23,133	10,829	3341
174,460	103,342	71,438	51,762	38,540	22,895	10,800	3001
168,915	102,907	71,080	51,752	37,552	21,720	10,604	3001
167,775	102,808	69,940	50,733	37,539	21,456	10,604	3001
166,902	102,133	69,040	50,360	37,440	21,043	10,604	3001
165,638	101,774	66,872	50,360	37,331	21,003	9895	3001
162,352	100,788	66,554	49,998	36,368	21,003	9400	3001
146,111	99,468	66,138	49,739	36,340	20,883	9400	3001
140,696	98,703	63,816	48,540	36,186	20,883	9180	3001
140,548	96,525	63,688	48,449	35,616	20,619	9144	3001
137,076	96,448	63,360	48,396	33,589	20,300	8460	3001
131,473	95,628	62,736	47,718	33,133	20,180	8300	3001
128,465	94,716	62,016	47,416	32,883	18,820	7740	2940
128,319	93,204	62,016	46,956	31,988	18,258	7540	2870
127,463	92,251	57,912	46,757	31,676	18,140	7500	2477
122,436	92,248	57,726	46,600	31,604	16,620	7408	2090
119,020	91,968	56,296	45,924	31,440	16,360	7400	2030
118,183	90,436	56,296	45,585	31,007	16,280	6260	1820
116,016	90,388	56,176	45,009	30,743	15,990	6060	1501
114,954	89,240	56,072	44,336	30,723	15,080	6060	1109
114,349	89,088	55,832	44,336	30,300	15,040	5564	
113,136	88,840	54,988	44,136	28,775	14,960	5400	
108,610	86,978	54,852	44,060	28,542	14,780	5400	
108,328	85,104	54,036	42,741	25,120	14,600	5040	
106,588	84,360	53,858	42,443	25,036	13,872	5030	
106,180	82,820	53,812	42,280	25,013	13,720	5030	

Source: Cleveland.com (2016)

and students assess basic performance measures like teacher retention, graduation rates, math proficiency, reading proficiency, and standardized test scores. A subset of the provided characteristics is given in Table 3.2 and in the **R** dataset *school_report_cards_2014*. While the former displays a handful of schools for illustrative purposes, the latter contains data on all 484 high schools with data reported by the chancellor. (The additional characteristics

Table 3.2 Data for 484 New York City High Schools during the 2013–2014 school year

School	Enrollment	English	Math	QR Rating	SAT	Graduation	Attendance
Essex street academy	349	2.47	2.25	Proficient	1199	71.3	86.8
Pace high school	421	2.64	2.51	NA	1368	93.7	92.4
NYC iSchool	433	3.11	2.81	Well developed	1529	95.9	93.7
Kipp NYC college prep	NA	NA	NA	NA	1356	95.1	95.3
The Uft charter school	272	2.22	2.08	NA	1231	92	89.2

Source: New York City Department of Education (2016)

that we have omitted may be found by visiting http://schools.nyc.gov.) The population of interest contains the students attending high schools in New York City during the 2013–2014 school year. The characteristics provided for each city school are: name, enrollment, average grade 8 English proficiency (out of 4), average grade 8 math proficiency (out of 4), quality review rating (the school's overall grade), average SAT score, four-year graduation rate, and attendance rate.

Even though collecting data on particular characteristics for every member of a population of interest sounds like an easy task, imagine the amount of time and energy that are involved in this task as the size of the population increases. Observing, examining, questioning, or surveying every member of a population and recording the appropriate information become overwhelming tasks. The problem is not a lack of technical resources, since modern technology can handle the huge volume of data and other data management issues. Instead, the problem is the cost and time involved in completing a census.

The United States Census Bureau has been dealing with this problem for decades. Their mission is "to be the pre-eminent collector and provider of timely, relevant and quality data about the people and economy of the United States." As of July 1, 2015, the U.S. Census Bureau (www.census.gov) estimated that the size of the U.S. population was 321,418,820. Can you imagine trying to collect and record characteristics for every member of the U.S. population? That is exactly what the U.S. Census Bureau must do every 10 years, but there has been considerable debate in the United States Congress as to whether or not it is really necessary to conduct a complete census. Many people, including most statisticians, believe that the accuracy of the population estimates can actually be improved by collecting information from properly selected subsets of the population, called samples, and then using the information from these samples to provide estimates for the corresponding population characteristics.

Definition 3.2 A **sample** is a subset of the population.

Sampling is used to collect complete information on a subset of the population that is representative of the population. Thus, you are reducing the amount of information that must be obtained and the time involved in collecting it. The beauty of this method is that it cuts down on the required amount of time and effort, but it only works if the sample is representative of the population. How do we obtain a sample that is representative of the population? In Sect. 3.2 we will see that there are many plausible answers to this question, depending on the complexity of the situation being considered. There is, however, one thing that all reasonable solutions have in common; they rely on chance or randomization in some way.

Example 3.3. Teaching Evaluations Faculty members are evaluated by students in a variety of ways. Mandatory course and instructor evaluation forms, www.ratemyprofessor.com, informal word of mouth, reputation, and letter writing are just a few examples. Suppose the college policy at a particular institution requires the Provost to randomly select students from a faculty member's classes to participate in the evaluation process. The Provost decides that 5 students are to be selected from an introductory class of 25 students. How should she randomly select these 5 students? One method is to put the 25 names on 25 slips of paper, put the 25 slips of paper in a bag, shuffle the contents, and then pick 5 slips. The names appearing on the selected slips will be the 5 students who are asked to participate in the evaluation. Another method relies on the use of random numbers. Each of the 25 students is assigned a random number using a random number generator. The list of names is then sorted according to the random numbers and the students with the 5 smallest numbers are asked to participate in the evaluation. Table 3.3 contains a list of 25 students and random numbers assigned using the

Table 3.3 Class list and assigned random numbers

Name	Random number
Freddy	0.359234
Luke	0.940090
Molly	0.934925
JoAnne	0.809115
Elise	0.787670
Danny	0.154661
Joel	0.873139
Christian	0.812036
Monique	0.923244
Patricia	0.340109
Elizabeth	0.446112
Sara	0.482329
Jacob	0.927142
Chris	0.754434
Jed	0.521955
Jessica	0.403488
Olivia	0.102106
Mattie	0.197304
Whitney	0.480334
Vanessa	0.900897
Oliver	0.582211
George	0.323056
Chad	0.968394
Jimmy	0.485099
Laney	0.288986

R function *runif*( ), which will randomly generate numbers uniformly between 0 and 1. Danny, Mattie, Laney, Olivia, and George would be the five students asked to participate in the faculty review. Alternatively, this random selection can be done by defining a vector containing the 25 names and using the **R** function *sample*( ).

Another method of collecting data is experimentation. In an experiment, a new set of conditions or treatments is deliberately imposed on the subjects or experimental units by researchers. We are referring to treatments in a generic sense in this text. A treatment could refer to a typical medication like a cough suppressant or an antibiotic or it could refer to a new teaching technique, a new assembly method, a new fertilizer, etc. or to a new set of conditions such as limiting smoking, increased exercise, etc. Characteristics of the subjects or experimental units are recorded in order to investigate some research hypothesis.

Definition 3.3 An **experiment** is a research study where treatments are deliberately imposed on a set of experimental units or subjects. Characteristics of interest are measured and recorded after some prescribed event or time period (and often prior to the treatment to serve as baseline measurements).

One of the major differences between a sample and an experiment is the manner in which the data are collected. In sampling, the subjects or experimental units are randomly selected from a population and the characteristics of interest are recorded. In experimentation, there are typically only a limited number of appropriate subjects or experimental units available for the study. These subjects or experimental units are randomly assigned to one of the treatment groups and the characteristics of interest are recorded after a certain amount of time has passed or some prescribed event has taken place.

Example 3.4. Pine Tree Growth As discussed briefly in Example 2.2, the Department of Biology at Kenyon College conducted an experiment to study the growth of pine trees at a site located just south of Gambier, on a hill overlooking the Kokosing River valley. In April 1990, student and faculty volunteers planted 1000 white pine (*pinus strobus*) seedlings at the Kenyon Center for Environmental Study (KCES). These seedlings were planted in two grids, distinguished by 10 and 15 foot spacings between the small trees (see Fig. 3.1).

Fig. 3.1 Locations of pine trees planted in April 1990 at the Kenyon Environmental Center. Rows 1 through 23 are spaced at 15 feet intervals. Rows 24 through 44 were planted at 10 feet intervals. An X denotes a pine that is fertilized and an O denotes a pine that is not fertilized

	1	2	3	4	5	6	7	8	9	10	11	12	13	14	15	16	17	18	19	20	21	22	23	24	25	26	27	28	29	30
1	O	O	O	O	O	O	O	O	O	X	X	X	X	X	X	X	X	X												
2	O	O	O	O	O	O	O	O	O	X	X	X	X	X	X	X	X	X												
3	O	O	O	O	O	O	O	O	O	X	X	X	X	X	X	X	X	X												
4	O	O	O	O	O	O	O	O	O	X	X	X	X	X	X	X	X	X												
5	O	O	O	O	O	O	O	O	O	X	X	X	X	X	X	X	X	X												
6	O	O	O	O	O	O	O	O	O	X	X	X	X	X	X	X	X	X												
7	O	O	O	O	O	O	O	O	O	X	X	X	X	X	X	X	X	X												
8	O	O	O	O	O	O	O	O	O	X	X	X	X	X	X	X	X	X												
9	O	O	O	O	O	O	O	O	O	X	X	X	X	X	X	X	X	X												
10	O	O	O	O	O	O	O	O	O	X	X	X	X	X	X	X	X	X												
11	O	O	O	O	O	O	O	O	O	X	X	X	X	X	X	X	X	X												
12	O	O	O	O	O	O	O	O	O	X	X	X	X	X	X	X	X	X												
13	X	X	X	X	X	X	X	X	X	O	O	O	O	O	O	O	O	O												
14	X	X	X	X	X	X	X	X	X	O	O	O	O	O	O	O	O	O												
15	X	X	X	X	X	X	X	X	X	O	O	O	O	O	O	O	O	O												
16	X	X	X	X	X	X	X	X	X	O	O	O	O	O	O	O	O	O												
17	X	X	X	X	X	X	X	X	X	O	O	O	O	O	O	O	O	O												
18	X	X	X	X	X	X	X	X	X	O	O	O	O	O	O	O	O	O												
19	X	X	X	X	X	X	X	X	X	O	O	O	O	O	O	O	O	O												
20	X	X	X	X	X	X	X	X	X	O	O	O	O	O	O	O	O	O												
21	X	X	X	X	X	X	X	X	X	O	O	O	O	O	O	O	O	O												
22	X	X	X	X	X	X	X	X	X	O	O	O	O	O	O	O	O	O												
23	X	X	X	X	X	X	X	X	X	O	O	O	O	O	O	O	O	O												
24	O	O	O	O	O	O	O	O	O	O	O	O	O	X	X	X	X	X	X	X	X	X	X	X	X	X	X	X	X	X
25	O	O	O	O	O	O	O	O	O	O	O	O	O	X	X	X	X	X	X	X	X	X	X	X	X	X	X	X	X	X
26	O	O	O	O	O	O	O	O	O	O	O	O	O	X	X	X	X	X	X	X	X	X	X	X	X	X	X	X	X	X
27	O	O	O	O	O	O	O	O	O	O	O	O	O	X	X	X	X	X	X	X	X	X	X	X	X	X	X	X	X	X
28	O	O	O	O	O	O	O	O	O	O	O	O	O	X	X	X	X	X	X	X	X	X	X	X	X	X	X	X	X	X
29	O	O	O	O	O	O	O	O	O	O	O	O	O	X	X	X	X	X	X	X	X	X	X	X	X	X	X	X	X	
30	O	O	O	O	O	O	O	O	O	O	O	O	O	X	X	X	X	X	X	X	X	X	X	X	X	X	X	X		
31	O	O	O	O	O	O	O	O	O	O	O	O	O	X	X	X	X	X	X	X	X	X	X	X	X	X	X			
32	O	O	O	O	O	O	O	O	O	O	O	O	O	X	X	X	X	X	X	X	X	X	X	X	X	X	X			
33	O	O	O	O	O	O	O	O	O	O	O	O	O	X	X	X	X	X	X	X	X	X	X	X	X	X	X			
34	X	X	X	X	X	X	X	X	X	X	X	X	X	O	O	O	O	O	O	O	O	O	O	O	O	O	O			
35	X	X	X	X	X	X	X	X	X	X	X	X	X	O	O	O	O	O	O	O	O	O	O	O	O	O	O			
36	X	X	X	X	X	X	X	X	X	X	X	X	X	O	O	O	O	O	O	O	O	O	O	O	O	O	O			
37	X	X	X	X	X	X	X	X	X	X	X	X	X	O	O	O	O	O	O	O	O	O	O	O	O	O	O			
38	X	X	X	X	X	X	X	X	X	X	X	X	X	O	O	O	O	O	O	O	O	O	O	O	O	O	O			
39	X	X	X	X	X	X	X	X	X	X	X	X	X	O	O	O	O	O	O	O	O	O	O	O	O	O	O			
40	X	X	X	X	X	X	X	X	X	X	X	X	X	O	O	O	O	O	O	O	O	O	O	O	O	O	O			
41	X	X	X	X	X	X	X	X	X	X	X	X	X	O	O	O	O	O	O	O	O	O	O	O	O	O	O			
42	X	X	X	X	X	X	X	X	X	X	X	X	X	O	O	O	O	O	O	O	O	O	O	O	O	O	O			
43	X	X	X	X	X	X	X	X	X	X	X	X	X	O	O	O	O	O	O	O	O	O	O	O	O	O	O			
44	X	X	X	X	X	X	X	X	X	X	X	X	X	O	O	O	O	O	O	O	O	O	O	O	O	O	O			

W
S + N
E

The following events occurred between the time that the pine growth measurements were started in 1992 and the time that the measurements were concluded in 1997:

Fertilization –Spring 92, Fall 92, 93, 94, 95, 96;
Mowing between rows – Fall 93 (50 trees damaged or lost), Summer 95, 96, 97; and
Removing thorny vegetation in contact with trees – Summer 96.

Some notable environmental conditions that occurred between the beginning of the experiment in 1990 and the ending in 1997 are:

Rainfall

1992 – continuation of 5th year with low precipitation during the summer
1993 – record amounts of rainfall in July ends the period of low precipitation
1995 – heavy rainfall in June followed by very little precipitation
1996 – record amounts of rainfall in June

Winter Conditions

1994 – extreme cold combined with snow cover most of the winter
1995 – continuous snow cover during January and February
1996 –continuous snow cover during much of December and January
1997 – very little snow cover all winter; sixth coldest spring on record

The goal of this study was to determine which factors cause pines in the KCES to vary in growth rates. Potentially important factors that were examined include soil nutrients, the presence of competitors, and herbivores. A subset of the data collected by biology students at Kenyon College is contained in the **R** dataset *pines_1997*.

Many research studies are conducted by simply observing and comparing the characteristics or traits of interest for different subsets of the sample. The researchers do not deliberately impose the treatments – they just observe and record the information, which may or may not be obtained from randomly selected individuals. The major difference between such an observational study and an experiment is the manner in which the experimental units or subjects are assigned to the treatment groups. If the experimental unit or subject makes the choice and the researcher simply observes and records the characteristics of interest, then the investigation is an observational study. If the experimental units or subjects are randomly assigned to the treatment groups by the researcher, then it is an experiment.

Definition 3.4 An **observational study** is a research study where the experimenters do not decide which treatment group the subjects will be in and the researchers simply observe and record the characteristics of interest. Often the experimental units or subjects were born into or have made the choice to belong to their corresponding treatment group.

Example 3.5. Reading Habits How do reading habits vary by demographic and socio-economic categories? To answer questions related to this along with investigating the effect that the emergence of e-books is having on these habits, the Pew Research Center conducted a study in 2011 and released a report titled "The Rise of E-reading".

Notice that this was an observational study rather than an experiment, since treatment groups (for example: age, income, education level, etc.) were not assigned by researchers, but rather were predetermined. While the Pew researchers asked many other questions, we will restrict ourselves to considering the responses to the following two survey questions.

1. During the past 12 months about how many books did you read either all or part of the way through? Please include any print, electronic, or audiobooks you may have listened to.
2. Please tell me if you happen to have a handheld device made primarily for e-book reading, such as a Nook or Kindle e-reader.

Apparently the survey respondents included many avid readers, so the decision was made to censor responses to the first question to be a number between 0 and 96, inclusive, or simply "97 or more".

In addition to the responses to these questions of interest, we will consider responses to a few questions about demographics as well. Responses for a subset of the people interviewed by Pew are displayed in Table 3.4 and the responses of the full set of interviewees may be accessed in the **R** dataset *reading_habits_2011*.

Although a formal analysis of this study requires statistical inference methods which we will learn in Chaps. 7 and 8, we can use graphical methods from Chaps. 1 and 2 to make appropriate comparisons. For example, we can construct a bar graph of e-reader usage grouped by education level from the following **R** code. We first use the *aggregate*( ) function to compute the averages and store them in a data.frame.

```
> average_e_reader_use_by_education <-
  aggregate(e_reader=='Yes' ~ education,
      data = reading_habits_2011,
      mean)
> colnames(average_e_reader_use_by_education)[2] <-"e_reader_numeric"
```

We next use the *barplot*( ) function introduced in the previous chapter to construct the plot shown in Fig. 3.2.

```
> barplot(average_e_reader_use_by_education$e_reader_numeric * 100,
    ylab = 'Percentage Using E-readers',
    ylim = c(0,45),
    names.arg = average_e_reader_use_by_education$education,
    cex.names = .75, las =2,
    main = "Use of E-readers by Education Level in 2011")
```

Table 3.4 Reading habits for respondents to 2011 Pew Research Center study

Id	Sex	Education	Income	Internet	E_reader	Number_of_Books
100,014	Male	Some college	$20,000 – $30,000	Yes	No	10
100,015	Female	High school graduate	Don't know	No	No	36
100,016	Female	Some college	Don't know	Yes	No	1
100,017	Male	Some college	$10,000 – $20,000	Yes	No	5
100,019	Male	Some college	$100,000 –$150,000	Yes	No	1
100,025	Female	Post-graduate training	$150,000 +	Yes	No	12
100,028	Female	Some college	$100,000 –$150,000	Yes	Yes	48
211,280	Female	Some college	$100,000 – $150,000	Yes	Yes	1
211,306	Male	Some college	$50,000 – $75,000	Yes	No	30
211,319	Female	College graduate	$75,000 – $100,000	Yes	Yes	12
211,321	Female	High school graduate	$30,000 –$40,000	Yes	No	30
211,344	Male	Post-graduate training	$100,000 – $150,000	Yes	No	1

Source: Pew Internet & American Life Project: The rise of e-reading (2016)

Fig. 3.2 Barplot of e-reader usage by education, 2011

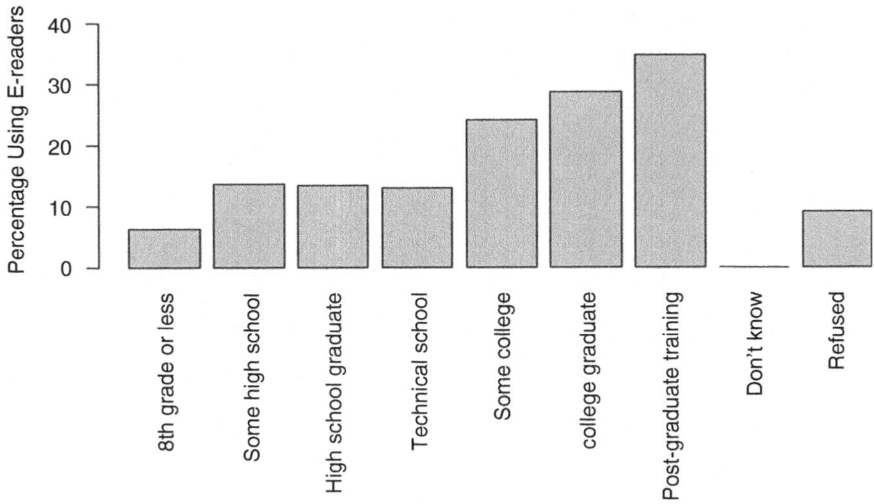

As you might have guessed, the rate of e-reader usage appears to increase with education level. Do you have any hypothesis about what the pattern will look like with respect to the respondents' incomes? With only a few small changes, we can replace "education" by "income" in the **R** code above to produce Fig. 3.3. This plot seems to indicate that a similar, possibly stronger relationship exists between income and e-reader usage. (In fact, you can download the raw data from the Pew Research Center Study and see the free-form responses given when participants who did not use e-readers were asked why they did not; one of the most common reasons was the cost of such devices.)

While census, sampling, experimentation, and observational studies are the most common and reliable methods of collecting data, there is one other method that is often used. Unfortunately, this method is not based on sound scientific principles. It relies on collecting information from isolated incidents, which are typically striking or alarming. Do you have a relative or friend who has smoked for years but is still in good health? Most people will respond yes

Fig. 3.3 Barplot of e-reader usage by income, 2011

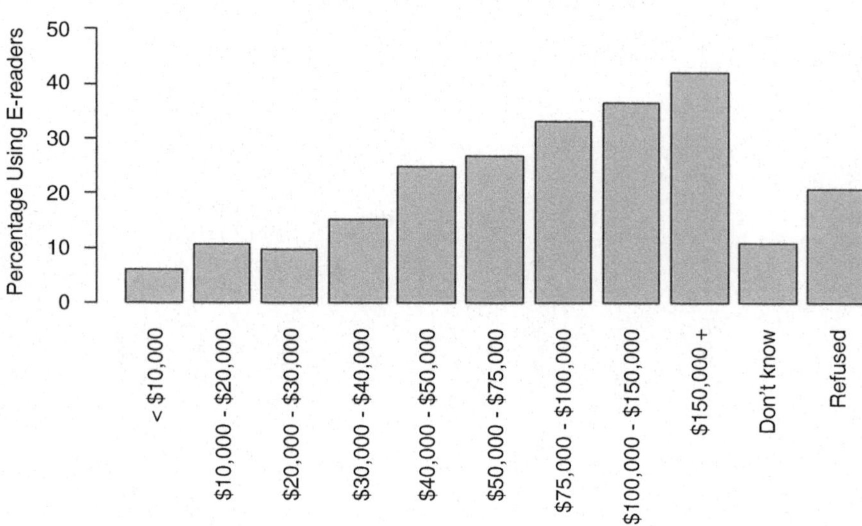

to this question, but that does not provide any useful information for a scientific assessment of the association between smoking and lung cancer. (Current scientific research clearly shows a link between smoking and lung cancer.) Anecdotal evidence gathered by looking at particular individuals or cases may be interesting and unusual, but it should certainly not be the basis for reaching valid scientific conclusions.

Definition 3.5 Anecdotal evidence is data collected on isolated individuals or cases because of their unusual or striking characteristics.

Although anecdotal evidence is the most widely accessible source of data, try to resist the temptation to draw inferences based on a few such isolated cases. If you do not have the time or the money to complete a census, as will often be the case in practice, inferences concerning a population should be based on information obtained from carefully collected samples and designed experiments.

Section 3.1 Practice Exercises

3.1.1. *Golf Handicaps*. The United States Golf Association (USGA) has developed a handicapping system that is based on the 20 most recent scores for a golfer. The handicapping system is supposed to enable two golfers of differing abilities to compete in a fair match. A golfer's handicap is based on the 10 best handicap differentials in the last 20 rounds of golf. (For more information, see www.usga.org.) Is a golfer's handicap based on a sample or census? Explain.

3.1.2. *Cardiovascular Exercise and Strength Training*. Does performing cardiovascular exercise and strength training together increase the effectiveness of both? Ho, Dhaliwal, Hills, and Pal (2012) studied the percent body fat and total weight lost by study participants over 12 weeks of participation in moderate-intensity aerobic, resistance, or combined exercise training. Do you think this is an example of an observational study or an experiment? Explain.

3.1.3. *Exercise and Pulse Rate*. In order to estimate the effect of short, quick spurts of exercise on pulse rates, an instructor asked the students in her class to take their pulses by counting the number of times their heart beat in 1 min. The students were then asked to do some form of exercise (push-ups, sit-ups, jumping jacks, jog-in-place, etc.) for 3 minutes. Immediately after the 3-minute exercise period, the students took their pulses again. Is this an observational study or an experiment? Explain.

3.1.4. A large family has been getting their cars serviced at the same garage for years. One of the family members notices that the tires he purchased from the garage were starting to show some serious signs of wear. After searching through his records he discovers that he only traveled 10,000 miles on these tires. He asks the other driving members of his family (parents, aunts, uncles, nieces, nephews, etc.) to check their records. After gathering the data, he discovers that two other members of the family have had problems with tire wear. Based on this evidence, he decides to contact a lawyer to inquire about filing a claim in court against the company. Would you consider the evidence collected to be the result of an observational study, the result of an experiment, or anecdotal data? Explain.

3.1.5. Provide a graphical and numerical summary for the salary data listed in Table 3.1.

3.1.6. *Teaching Evaluations*. To evaluate the effectiveness of its instructors, a university department distributes surveys to all students enrolled in one of its classes. Not all of the students have returned their surveys, but the department has created a database for the responses they have received. Do you think this database should be referred to as a census or a sample of the students?

3.1.7. *School Report Cards*. Use the report card data available in the **R** dataset *school_report_cards_2014* (and partially shown in Table 3.2) to answer the following questions.

(a) Which city school had the lowest graduation percentage.
(b) Create a scatterplot to examine the relationship between SAT scores and attendance rates for these high schools. Comment on the relationship you find.
(c) Create graphical displays to compare the four-year graduation rates for each of the possible quality review ratings. Note that some schools are missing this rating. Discuss the approach you decide to take in handling this missing data.
(d) Is there any association between enrollment numbers and average grade 8 English proficiency?

3.1.8. A youth soccer team contains 13 girls, but only 8 can play at a time. The names of the girls are Alison, Anna, Annie, Cindy, Jamie, Jenny, Johanna, Jungwon, Natalia, Neha, Ning, Victoria, and Xi. Use a random number generator (such as the **R** function *runif*( )) to explain how the coach should randomly decide who to start for the next game.

3.1.9. Repeat the experiment discussed in Example 3.3 using the **R** function *sample*( ). Provide the code you used to conduct the experiment and the results you obtained.

3.1.10. *Pine Tree Growth*. Use the **R** dataset *pines_1997* to answer the following questions.

(a) Did the trees that were fertilized grow better than those that were not fertilized? Use at least two different measures of center to make your comparison.

(b) Are the trees that are planted at 15 feet intervals taller than the trees that were planted at 10 feet intervals?

(c) Is there any difference in the variability of the tree heights for those trees that were fertilized and those trees that were not fertilized? Would you prefer S or IQR as your measure of spread for these tree heights? Provide a rationale for your choice.

3.1.11. *Reading Habits*. Analyze the number of books read in the past 12 months as presented in Table 3.4. That is, instead of looking at e-reader use as we did in Example 3.5, look at the number of books read. Note that you'll need to convert this column into a numeric vector. (Hint: use the **R** function *as.numeric*( ) but be careful handling the values of "97 or more"!) Obtain the average number of books read by educational level and create a plot similar to Fig. 3.3. Does your plot show a pattern similar to that in Fig. 3.3?

3.1.12. *Reading Habits—Part Two.* Repeat Exercise 3.1.11 for average number of books read in the past 12 months by income level.

3.1.13. *Reading Habits—Part Three.* Repeat Exercise 3.1.11 for average number of books read in the past 12 months by all possible combinations of sex and levels of Internet access.

3.2 Planning and Conducting Surveys or Polls

Time pressures and the costs associated with data collection often restrict our studies to subsets of a population. When this is the case, we are faced with an interesting and challenging problem. How do we select a subset of the population that is both manageable and representative of the population? One method that is used often in the popular media is self-selection.

Example 3.6. Homework and Family Stress In the 2015 study *Homework and Family Stress: With Consideration of Parents' Self Confidence, Educational Level, and Cultural Background*, 1173 parents who visited one of 27 pediatric offices in the Greater Providence area of Rhode Island answered the question: How many minutes does your child spend on homework per night? Among other things, researchers were interested in determining whether children are being assigned too much or too little homework, as compared to the "10 Minute Rule" recommended by the National Education Association, which states that the recommended nightly time spent on homework should be 10 min. multiplied by the child's grade level. For example, a second-grade student is recommended to be assigned 20 min. of nightly homework on average, while a ninth-grade student is recommended to be assigned 90 min. Table 3.5 provides a summary of the average responses by grade, as estimated from a plot in Pressman et al. (2015).

Notice that this survey was conducted while parents waited in doctors' offices. While this may be a convenient location to find many parents with

Table 3.5 Average minutes spent on homework per night

Grade	Average minutes spent on homework per night
Kindergarten	25
1st	29
2nd	29
3rd	34
4th	36
5th	35
6th	43
7th	51
8th	41
9th	50
10th	54
11th	53
12th	52

Source: R. M. Pressman, D. B. Sugarman, M. L. Nemon, J. Desjarlais, J. A. Owens, and A. Schettini-Evans (2015)

spare time to answer questions, it seems possible that the stress of having a sick child might cause parents to answer the question differently than if they were surveyed at home. Do you think it is safe to assume that the amount of homework assigned nightly in Providence is representative of the quantity assigned in the United States as a whole? Finally, note that the parents whose responses were recorded had chosen to take the time to fill out the survey. That is, there were many parents who chose not to voice their opinion on the amount of homework assigned. Would you expect there to be a relationship between the amount of homework a parent believes is being assigned and whether or not the parent **self-selects** himself or herself into the survey?

Definition 3.6 Self-selected samples are samples obtained by allowing subjects or experimental units to choose whether or not they want to participate in a survey or poll.

Surveys or polls based on self-selected samples often yield different results than those based on statistically designed samples. The reason for this is that self-selected samples tend to have different characteristics than the general population of interest.

> **Definition 3.7 Self-selection bias** is a bias that results from the use of voluntary or self-selected samples. The respondents in such samples typically favor one response over another because of a certain characteristic or strong feeling regarding a particular issue.

After CNN reported the study in a story with the headline "Kids have three times too much homework, study finds; what's the cost?", the Brookings Institute (a research organization) published an article titled "CNN's Misleading Story on Homework" that was quick to point out the flaws in this interpretation and potential sources of inaccuracy in the study itself.

The Internet has made it easier and cheaper to gauge public opinion via online surveys. This convenience comes at a cost, however. Many Internet users will only be willing to respond when they feel strongly about a subject, so you should be wary of such polling techniques because they are loaded with self-selection bias. As an analogy, think of Yelp reviews of a local restaurant. While there certainly exist some avid Yelp users who will provide a rating each time they dine at a restaurant (and similarly, there exist Internet users who are willing to give their opinion on every topic presented to them!), generally it's true that people are more likely to write a review when they have a very good or a very bad experience (in other words, a strong opinion).

Now we know what we should not be doing, but we still do not have an answer to our question regarding proper selection of a subset of the population that is both manageable and representative of the population. This answer relies on the use of chance or randomization in some planned way.

One natural method for selecting a proper sample is to use some chance or random mechanism that gives each element in the population an equal chance of being selected. Such a sample is called a simple random sample.

Definition 3.8 A simple random sample of size n is a subset of the population that is selected by using a chance or random mechanism that assigns an equal probability to every subset of n members of the population.

Example 3.7. Gallup Poll – Presidential Approval Ratings Since the 1930s the Gallup Organization has been evaluating presidents by conducting public opinion polls. Although their methods have changed over the years (from 1935 to the mid-1980s they used in-person interviews from across the country, while today they use telephone interviews), the idea has always been the same: randomly select around 1500 American adults to represent the opinions of the entire adult population in the United States. The samples are selected from a list of numbers generated using a random digit dialing procedure. Since approximately 30% of American households have unlisted phone numbers, random digit dialing is used to avoid a listing bias. Table 3.6 shows the results of such job approval polls for President Obama during the first 8 months of 2016.

The use of chance and randomization to ensure an equal probability of selection for all equal-sized subsets of the population is absolutely essential in order to obtain reliable data from surveys and polls. Not only does this randomization eliminate selection bias, but it also saves time and money by allowing us to focus on this small, representative portion of the population. However, there are also other sources of bias which we must be careful to avoid. Suppose that we are interested in the public opinion of The First Lady. How do we phrase a question regarding the President's spouse? Should we include both her title and her name, just her name, or just her title? While this

Table 3.6 Obama job approval ratings for January through August 2016

Do you approve or disapprove of the way Barack Obama is handling his job as president?

Date	Approve (%)	Disapprove (%)	No opinion (%)
2016 Aug 22–28	51	45	5
2016 Aug 15–21	51	44	4
2016 Aug 8–14	52	44	4
2016 Aug 1–7	52	45	4
2016 Jul 25–31	53	44	4
2016 Jul 18–24	49	47	4
2016 Jul 11–17	49	46	5
2016 Jul 4–10	51	45	4
2016 Jun 27-Jul 3	51	45	5
2016 Jun 20–26	50	46	5
2016 Jun 13–19	53	44	4
2016 Jun 6–12	53	43	4
2016 May 30-Jun 5	51	44	5
2016 May 23–29	52	44	4
2016 May 16–22	51	45	5
2016 May 9–15	51	45	4
2016 May 2–8	52	44	4
2016 Apr 25-May 1	51	46	4
2016 Apr 18–24	51	45	4
2016 Apr 11–17	48	47	4
2016 Apr 4–10	51	45	4
2016 Mar 28-Apr 3	51	45	4
2016 Mar 21–27	53	44	3
2016 Mar 14–20	50	46	4
2016 Mar 7–13	51	45	4
2016 Feb 29–Mar 6	50	46	4
2016 Feb 22–28	48	47	5
2016 Feb 15–21	48	48	4
2016 Feb 8–14	48	48	4
2016 Feb 1–7	47	50	3
2016 Jan 25–31	48	48	4
2016 Jan 18–24	48	47	5
2016 Jan 11–17	48	47	5
2016 Jan 4–10	47	49	4

Source: Gallup, Inc. (September 2016)

may seem like an unimportant detail, the choice we make could bias the responses in one direction or another. The wording of the questions may be the biggest source of bias in data obtained from surveys and polls.

Example 3.8. Another Opinion Poll – Congressional Approval The Gallup Organization also polls respondents about their opinion of Congress. The results of these opinion polls are shown in Table 3.7. How does the approval of Congress compare to the approval of the President? Note that the survey of Congressional approval is conducted much less frequently than that of Presidential approval. Does this surprise you?

The Gallup Organization carefully chooses its wording for survey questions and then keeps that same wording on such questions over time. If you are designing your own survey, how should you decide on the precise wording of a question? Perhaps the best starting point is to conduct a **pilot study** where you "field test" these questions on a small group of individuals and evaluate their reactions and responses to different wordings. For example, universities often gauge faculty opinions on a variety of work-related issues. One common topic of interest in surveys like these (and similar ones

Table 3.7 Congressional job approval ratings for January through August 2016

Do you approve or disapprove of the way Congress is handling its job?			
Date	**Approve (%)**	**Disapprove (%)**	**No opinion (%)**
2016 Aug 3–7	18	78	4
2016 Jul 13–17	13	83	4
2016 Jun 1–5	16	80	4
2016 May 4–8	18	78	4
2016 Apr 6–10	17	79	4
2016 Mar 2–6	13	84	3
2016 Feb 3–7	14	81	4
2016 Jan 6–10	16	80	4

Source: Gallup, Inc. (September 2016)

conducted by private employers) is faculty attitudes concerning salaries. One possible wording for this question is: Are you happy with your cost of living salary increases over the last 5 years? What if the word happy is replaced with unhappy to form an alternative wording of the question? Notice that changing one word in the question changes the entire tone. We could also replace happy with satisfied or unsatisfied, delete the cost of living qualifier, or restructure the entire question. The decisions you make regarding the precise wording of your survey questions could have a substantial impact on the responses and your final conclusions.

Example 3.9. Stratified Sampling As new policies are considered at a college or university, the administration often seeks feedback from various groups. Instead of asking for opinions from every member of every such group (that is, taking a census) or using randomization to select a simple random sample, it is sometimes reasonable to consider other factors. For example, the views of scientists may differ from those of artists, and the views of humanists may differ from those of social scientists, and the opinions of historians may differ from those of political scientists. To take into account these possible differences among the different specialties, a small number of people can be randomly selected from each specialty. This division of a population by specialty is known as stratification.

> **Definition 3.9 Stratification** is the division of a population according to some characteristic. Each division or subgroup of the population is referred to as a **stratum**.

Sampling within the different divisions or subsets of a population, known as stratified sampling, is often used to compare the opinions or attitudes of different groups. For example, political polls are commonly stratified by party affiliation, health surveys are conducted separately for men and women, and student surveys are stratified by class year. The major difference between stratified sampling and simple random sampling is that the randomization is carried out separately for each subgroup or stratum of the population.

Section 3.2 Practice Exercises

3.2.1. A zoning board in a small county with both urban and rural residents is interested in public opinion on the minimum lot size for a new subdivision of homes that has been proposed by a developer. Would you suggest simple random sampling or stratified sampling? Provide a rationale for your choice and describe how to implement your sampling plan. Your description should include the use of chance in some planned way.

3.2.2. A local community group is interested in findingout if residents are in favor of extending an existing bicycle/jogging path through their neighborhood. To estimate the appropriate population characteristic, the director of the community group uses the local telephone directory to randomly call residents and ask them their opinions on this issue. What population characteristic is the community group trying to estimate? Identify at least two sources of bias that the community group should be concerned about in their survey.

3.2.3. Do you think that residents who choose to have their phone number listed in the local phone directory are similar to those who keep their phone number unlisted? Identify at least one characteristic you would be interested in comparing for these two groups of residents. Explain how you would implement your sampling plan to collect appropriate data to respond to this question.

3.2.4. A Provost at a small liberal arts college is responsible for soliciting evaluations from students for faculty reviews. The faculty handbook suggests that these students should be randomly selected from the faculty member's courses since the last review. The Provost creates a list of students who have been taught by this faculty member since the last review, puts the list in alphabetical order, and then requests a letter from every 5th student on the ordered list. The Provost is using a sampling method known as systematic sampling. Although this sampling method can be very useful, especially in quality control programs for manufacturing processes, some faculty members were outraged when they heard about this practice. Why do you think some faculty members were so concerned and others were pleasantly surprised? Do you think this systematic sampling method will produce a representative sample of the student opinions?

3.2.5. Nielsen Media Research relies on statistical methods to create ratings and indices for television programs across the country. Do you think Nielsen Media Research uses simple random sampling or stratified sampling to create their ratings? Justify your response.

3.2.6. A steering committee is beginning to plan for the addition of new campus recreational facilities for students, faculty, and staff members. Existing recreational facilities are located on the south side of campus and the steering committee is interested in getting feedback on two different plans. Plan A would expand the existing facilities on the south side of campus and keep all recreational facilities in one location. Plan B specifies that the new recreational facility be placed on the north side of campus. Suggest a method for getting feedback from interested subgroups of students, faculty, and staff members. What subgroups would you be most interested in hearing from if you were on the steering committee? Is it reasonable to give all opinions equal weight in this decision?

3.2.7. An email has been sent to a random sample of all students to ask for their opinions on a new housing policy. Only 20% of all the randomly selected students responded to the message. If the Student Life Committee uses the information received from these responses to make inferences, what are they assuming about the responses for the other 80% of the randomly selected students who did not respond to the message?

3.3 Planning and Conducting Experiments

We begin our discussion of planning and conducting experiments by returning to the research question of Facebook. Recall that the researchers' primary interest was in determining the effect that positive and negative content have on users' emotions.

Example 3.10. Positive and Negative Facebook Posts—Emotional Contagion In order to test the hypothesis that emotional contagion exists, that is, that people who see more positive content are likely to produce more positive and fewer negative Facebook posts (and the opposite for people who see more negative content), the researchers began by dividing the 689,003 selected users into two experiments to be run in parallel: one to test positive contagion and another to test negative contagion. In each experiment, the treatment group received a reduced number of either positive or negative stories (one or the other, depending on which experiment). The actual reduction was determined randomly; depending on the user, a positive (or negative) story was suppressed with between 10 and 90% probability. In the control groups, posts were omitted at similar rates without consideration of whether they were positive, negative, or neutral. Posts were determined to be positive or negative using specialized software called "Linguistic Inquiry and Word Count", which is an example of work in an interesting branch of statistics known as "Natural Language Processing". While well beyond the scope of this book, we strongly encourage you to research this topic if you're interested!

Fig. 3.4 Effect of reducing negativity and positivity on status updates (*Source*: A. D. I. Kramer, J. E. Guillory, and J. T. Hancock (2014))

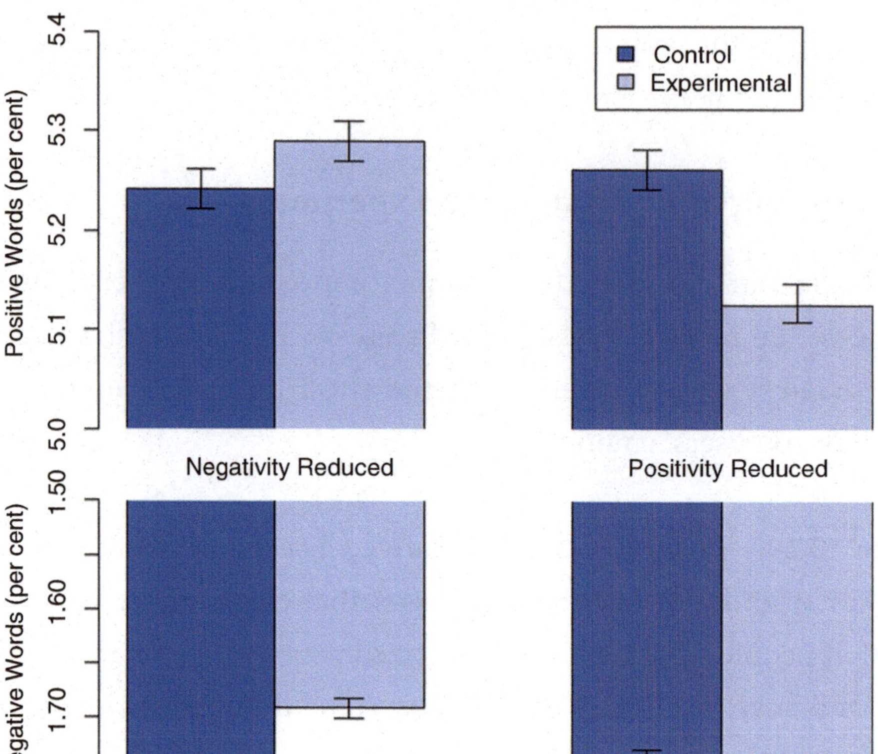

The researchers found evidence for emotional contagion, as users who had the amount of positive content in their Feeds reduced were less likely to post status updates with positive words and more likely to post them with negative words. The opposite appeared to happen when negativity was reduced. Although the sizes of the effects were small, the researchers point out that the amount of manipulation of the users was quite small as well. Their results are displayed graphically in Fig. 3.4.

The largest source of controversy concerning this story was the fact that users were not aware of their participation in the study. While Facebook claims that users agree to participate in such experiments as a condition of using the service, privacy advocates worried about the potential consequences of manipulating the emotions of unsuspecting users. Can you think of any issues that may have arisen if the researchers had asked participants for explicit consent and then only conducted the experiment on those who agreed?

Designing an experiment requires careful thought. A step-by-step approach to this planning process, as advocated by Dean and Voss (1999), is extremely helpful, but rarely does everything go as planned in experimentation. Some important factors may be overlooked, measurements on particular subjects may be difficult or impossible to obtain, equipment may break down during the experiment, etc. Scientific experimentation is an iterative process where the information and knowledge gained from one investigation is used to improve upon the next. Our goal in this section is to provide a basic introduction to the important planning process that should be part of every experiment. If the experiment is not properly designed, the objective of the experiment may be completely irrelevant. There is no magic formula statisticians can apply to fix design flaws in an experiment. Collecting relevant data in the proper way is absolutely essential for a successful experiment.

Step-by-Step Process for Planning an Experiment

Step 1. Identify your objective.

Step 2. Decide what information to collect.

Step 3. Decide how to collect the information.

Step 4. Decide how many experimental units are needed.

Step 5. Use randomization to assign experimental units to treatments.

Step 6. Collect the data.

Step 7. Analyze the data.

Example 3.11. Breaking Strengths for Mechanical Pencil Leads J. Ted Hunter (1997), a student in a design of experiments course, was searching for an idea for his course project. Like many other users of mechanical pencils, he was frustrated because the lead tip from his pencil kept breaking off while he was trying to write. After reading the promotional slogan on the side of his lead storage container, the student decided to test the manufacturer's claim that he was using the strongest lead in America. Testing a claim is a common objective in many experiments. More specifically, the student's objective was to compare the breaking strengths of pencil leads for different brands, thicknesses, and graphite densities. Brand, thickness, and graphite density are known as possible sources of variation in the breaking strength of pencil lead and are referred to as treatment factors.

Definition 3.10 Treatment Factors are controllable potential sources of variation in an experiment. Although the term treatment might make you think of medical or agricultural experiments, we will use it in a more generic sense throughout this text.

Three different brands (labeled A, B, and C), two different thicknesses (0.5 mm and 0.7 mm), and three different graphite densities (H, HB, and 2H) were used in this experiment. These specific settings or characteristics of the treatment factors are referred to as levels.

Definition 3.11 Levels are specific settings or characteristics of the treatment factors.

Moving to the second step in the planning process Ted decided to collect breaking strengths for sticks of pencil lead. The individual sticks of lead are referred to as experimental units.

> **Definition 3.12 Experimental units** are the subjects, individuals, or objects used in an experiment.

The third step presented an interesting challenge. How would breaking strength be measured? Ted decided to construct a small basket that could be attached to each stick of lead (see Fig. 3.5).

The ends of the sticks of lead were then placed on two even stacks of books and nickels were gently placed into the basket until the lead snapped. The response variable of interest, breaking strength of the lead, is the number of nickels that are placed in the basket before the lead snapped. There are obviously more sophisticated ways to measure the breaking strength of pencil lead, but this method is a reasonable approach if care is taken to avoid other possible sources of variation. For example, if different people placed the nickels in the basket or different methods of placement were used, the responses may differ because of these unwanted sources of variation. This would result in a problem known as **confounding**. We would not know if the

Fig. 3.5 Diagram of instrument used to measure breaking strengths of pencil lead

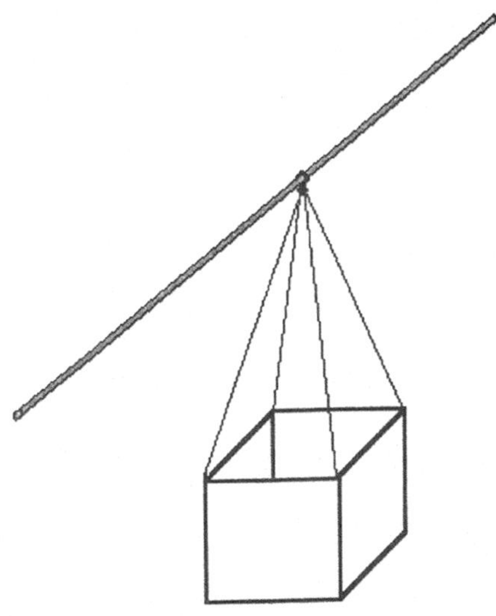

observed differences were due to the treatment factors or these unwanted sources of variation. Ideally, we would like to eliminate all possible sources of variation, except for the variability in the treatment factors we are interested in considering and random measurement error. One way to do this is called blocking.

Definition 3.13 Blocking is the formation of homogeneous subgroups of experimental units based on some characteristic or unwanted potential source of variation. The primary purpose of blocking is to create experimental conditions that are as identical as possible for the experimental units.

The next step in the process is to decide how many sticks of lead will be needed for the experiment. Since we have 3 brands, 2 thicknesses, and 3 graphite densities, we will need at least $3 \times 2 \times 3 = 18$ sticks of lead. However, using just 18 sticks of lead would allow us to measure the breaking strength for each treatment combination only one time. We would like to be able to repeat or replicate the experiment several times at each treatment combination to see if any patterns emerge. The problem is that this takes more time and typically requires more money. Thus, we have conflicting goals here. We want to replicate this experiment as often as we can so that we can be comfortable with our ability to detect treatment effects, but we also want to keep the number of replications as small as we can to help control the amount of time and money we spend on the experiment. A complete answer to the question of how many experimental units are needed requires more statistical concepts than we have mastered at this point, but the dilemma we are facing should be clear. Suppose that Ted decided to replicate this experiment 6 times for each treatment combination. Then he would need $6 \times 18 = 108$ sticks of lead.

Step 5 in the planning process is to randomly assign, using a table of random digits or a random number generator, the experimental units to the treatments. In this case, we have 108 sticks of lead with different treatment characteristics, so one way of completing this randomization is to label the sticks from 1 to 108 and then randomly decide on the order in which the breaking strengths for the 108 sticks will be measured.

The Importance of Randomization Randomization should always be used in Step 5 to eliminate potential bias and to "evenly distribute" unknown sources of variation.

The data collection process is often the place where some unknown sources of variation first become evident. For the moment, descriptive statistics and graphical displays from Chaps. 1 and 2 can be used to complete the data analysis step. More formal statistical inference procedures are presented in Chaps. 6, 7, 8, 9, 10, 11 and 12.

Example 3.12. The Placebo Effect As discussed in the 2010 NPR article "The Growing Power of the Sugar Pill", a new treatment for Parkinson's disease was being studied at the University of Colorado. The treatment involved major surgery, where cells from embryos were implanted in the brains of patients suffering from Parkinson's disease to replace the cells killed by the disease. George Doeschner, a patient suffering from Parkinson's disease, agreed to participate in a medical experiment designed to evaluate the effectiveness of this new surgical technique. George flew to Denver, was prepped for surgery, and was taken into the operating room like all other patients. However, George did not know exactly what was going to happen while he was in the operating room. There were two possibilities: either he was going to get the complete treatment or he was going to get a placebo (fake treatment). The surgeons actually drilled a hole in George's head and sewed it back up without doing anything else! You might be surprised to learn that this did not bother George a bit.

Many people are surprised when they first learn of the widespread use of such "fake" treatments in the medical research community. In fact, even though placebos have been advocated by statisticians for decades, the use of placebos can be controversial, and in some fields such as cancer clinical trials, still not the norm. Although it can be difficult to imagine giving a sugar pill to a person with a deadly disease when a promising experimental treatment is readily accessible, scientists believe that the benefits of using control groups for comparison purposes far outweigh the risks, even in major surgical operations. In fact, an article in TIME magazine (1999) claims that the National Institutes of Health rejects proposals from surgical researchers who do not employ placebo surgeries.

Interestingly, these fake treatments can have real effects. Kaptchuk et al. (2006) found that two different types of placebos (a sugar pill and fake acupuncture) both had the desired outcome of reducing arm pain. Furthermore, they both appeared to cause the side effects that participants were told the treatments could produce. The NPR article also cited a paper by Rief et al. (2009) that even suggests that the placebo effect has grown stronger over time as our expectations change!

Definition 3.14 A **placebo** is a fake treatment. A group of randomly selected experimental units are given a placebo to form a **control group**. This control group is used to measure **the placebo effect** and serve as a benchmark for comparison with the treatment group.

Even if the step-by-step process is followed and a randomly selected control group is used, other problems can destroy the experiment.

Example 3.13. Stopping Clinical Trials Early Due to Unanticipated Harm In 2007, a late stage clinical trial was being conducted in India and a handful of African countries to test the effectiveness of an HIV treatment called Ushercell, which was being developed by the nonprofit organization CONRAD. The compound, known as a "microbicide", was believed to reduce the rate of HIV infection in women who used it in gel form. However, researchers were shocked to learn that the gel actually seemed to *increase* the rate of HIV infection in the women who used it. (This contradicted nearly a dozen previous trials involving Ushercell.) Because of this increase compared to the placebo gel, the trial was immediately halted and the women in the treatment group were no longer given Ushercell. Honey (2007) discusses the steps taken by CONRAD and the researchers to better the science behind how exactly this could have happened.

Clinical trials can also be stopped early under more favorable conditions: for example, if the treatment appears to be so effective that the researchers feel that it would unethical to delay the development of the drug by waiting for the experiment to conclude. This is a somewhat controversial topic in the medical research community, however, and it is important to consider the conditions under which the trial would be stopped early *before* the experiment begins (rather than after seeing the data).

Designing a good experiment is difficult, some say impossible, if the researchers know which experimental units are in the control group and which experimental units are in the treatment group. To eliminate this potential source of bias, researchers should not be informed which patients are assigned to which group. Experiments where neither the patients nor the researchers know which group the experimental units have been assigned to are known as **double-blind experiments**.

Section 3.3 Practice Exercises

For each of the settings described in the following Exercises 3.2.1–3.2.6, complete Steps 1–4 in the step-by-step process for planning an experiment.

3.2.1. Design a taste test to compare consumer preferences for three different types of cola. Your design should include a detailed description of blocking factors and the randomization step.

3.2.2. Several consumers have filed complaints with a snack food company. The company produces a one-pound bag of chips, but the consumers feel that the bags of chips are starting to contain more air than chips. Design an experiment to test the company's claim that this product contains 1 pound of chips.

3.2.3. Is bottled water safer than tap water? Design an experiment to compare these two types of water.

3.2.4. Which brand of paper towels is the strongest? Design an experiment to compare the strengths of two different brands of paper towels.

3.2.5. Are certain pain relievers more effective than others? Design an experiment to compare Advil, Tylenol, and Aleve.

3.2.6. To compare the prices at supermarket A with those at supermarket B a student randomly selected 50 products from supermarket A and 50 products from supermarket B, computed the total bill at each store, and repeated these sampling procedures 28 times for each store. The student suggested that the average of the 28 bills for store A be compared with the average of the 28 bills for store B to decide which store offers the lowest prices, on average. Another student randomly selected 50 items from a list of products which the two stores have in common, computed the difference in the prices for each of the

50 products, and found the average of these 50 differences. She repeated this process 28 times and suggested that the average of the 28 average differences be used to decide which store offers the lowest prices on average. Which design do you prefer? Why?

3.2.7. In Example 3.11 the breaking strengths of the pencil leads are measured by counting the number of nickels that are placed in the basket before the lead snapped. Do you think it matters if the order of the nickels is the same for all sticks of lead? Would you prefer using randomization to decide on the order of the nickels for each stick of lead or using the same order for all sticks of lead? Explain.

Chapter 3 Comprehensive Exercises

3.A. Conceptual

3.A.1. A doctor wants to investigate the effect of a cholesterol-lowering drug on mental acuity, so she recruits 20 patients from the hospital she works at and conducts a series of tests to measure mental fitness. The patients are selected in such a way that 10 of them take the cholesterol-lowering drug and 10 of them do not. Is this an observational study or an experiment?

3.A.2. Suppose that the doctor in the previous exercise instead recruits 20 patients upon entry to the hospital and gives the drug to the first 10 patients and does not give the drug to the last 10 patients.

(a) Is this an observational study or an experiment? Explain.

(b) Are there any aspects of the doctor's implementation which you would change?

3.A.3. *Fish Oil and Epilepsy*. DeGiorgio et al. (2014) investigated the effects of fish oil on drug-resistant epilepsy. They were interested in testing both low dosage and high dosage amounts versus a placebo. However, due to practical limitations, they were only able to recruit 24 patients who participated throughout the study. Further complicating things was the fact that the baseline number of seizures varied widely (from as little as 3 per month to as many as 60 per month). Discuss how you would design an *ideal* study to examine these effects if the number of patients was not an issue and then discuss how you might work around this limitation.

3.B. Data Analysis/Computational

3.B.1. *Are Mac Users Charged More for Travel?* A 2012 Wall Street Journal article suggested that Mac users would pay \$20 to \$30 more per night for hotels on the online travel site Orbitz than PC users would. Partner with a classmate to design an experiment to test the validity of this claim. Collect and report the data from your experiment.

3.B.2. *Why Do Mac Users Pay More?* The article discussed in Exercise 3.B.1 was often misrepresented by third parties as claiming that Mac users would be charged more for the same hotel rooms. However, the article actually suggests that the Mac users would be steered to higher-end hotels, which end up costing \$20 to \$30 more per night. With this in mind, discuss any changes you would make to the design of your experiment in Exercise 3.B.1 and collect new data according to this updated experimental design.

3.B.3. *Police Body-Worn Cameras and Citizen Complaints*. Citizen complaints against police officers are often seen as being indicative of the level of compliance with police procedure and proper conduct. Ariel et al. (2016) conducted a study of the effect of police body-worn cameras on the number of complaints made by citizens. They randomly assigned officer shifts at seven police departments to either wear cameras or not wear them on a weekly basis. The results of their study are displayed in Table 3.8.

Table 3.8 Effect of body-worn cameras on citizen complaints against Police

Site	Complaints 12 months before	Complaints 12 months after	Number of complaints treatment shifts	Number of complaints control shifts	Treatment shifts	Control shifts	Rate per office (pre–treatment)	Rate per officer (post–treatment)
A	558	33	7	4	183	186	1.02	0.06
B	10	0	0	0	129	106	0.22	0
C	331	21	3	4	184	185	2.98	0.19
E	251	30	4	6	111	188	0.29	0.03
G	24	3	2	1	489	499	0.44	0.06
H	34	19	7	12	367	367	0.30	0.17
K	331	7	4	3	445	443	3.15	0.07

Source: B. Ariel, A. Sutherland, D. Henstock, J. Young, P. Drover, J. Sykes, S. Megicks, and R. Henderson (2016)

(a) Was this an experiment or an observational study?
(b) What is the treatment and what is the control? From looking at the data, do you think there is a strong treatment effect?
(c) Both the treatment and the control groups appear to have many fewer complaints than the same groups did before the experiment started. Why do you think this might be?
(d) The researchers initially reached out to 10 police departments and then reported the results based on the 7 sites which agreed to participate. What effect might this have on the results of the study?

3.C. Activities

3.C.1. *Effect of Salt on Ice.* You will need a tray of ice cubes and some table salt for this activity. The goal will be to determine whether salt causes ice to melt faster. Design an experiment to test this hypothesis. Describe your procedure, including what the experimental units are, what the outcome of interest is, what the treatment and control groups are, what process you used to assign experimental units to each group, and discuss any data that you collect.

3.C.2. *M & M Colors.* Do the colors of M&M's vary by which type of M&M they are? Although the company no longer posts the proportion of colors in each batch of M&M's, various sources on the Internet claim that there are, in fact, different color mixes for different types of M & M's. To test this, purchase two bags of M&M's, one each of two different types (for example, milk chocolate and peanut) and record the proportion of orange M&M's in each bag. Is there much difference in these proportions? Aggregate your results with those of your classmates to obtain a larger set of data. Is this an experiment or an observational study?

3.C.3. *Heart Rate and Exercise.* How do various activities affect your heart rate? Gather the necessary materials to measure your pulse and then measure it separately after performing 30 seconds of each of the following activities: jumping jacks, walking, sprinting, sitting in a chair, and lying on the ground.

(a) Is this an observational study or an experiment?
(b) What would you consider to be the placebo or baseline heart rate?
(c) Do you think the order in which these activities are performed will affect your results? If so, list a few steps that you can take to eliminate these effects.

3.D. Internet Archives

3.D.1. The polling company Reuters routinely conducts surveys on various topics of interest to Americans. You can view the results of recent polls in categories such as "Government & Policy", "Business & Finance", "Society & Lifestyle", and more by visiting http://polling.reuters.com/. Select a poll on a topic you find interesting or controversial.

(a) How many people participated in the poll?
(b) How were these people selected?
(c) Discuss any aspects of the poll that you might find surprising.

3.D.2. *NASA Experiments*. NASA makes available a list of all experiments that have been completed or are currently being conducted on the International Space Station (ISS) at http://www.nasa.gov/mission_pages/station/research/experiments/experiments_by_name.html. Find an experiment of interest to you and summarize it.

(a) How were the data collected?
(b) What were the treatment and control groups?
(c) What was the conclusion of the experiment and what practical use does this knowledge have?

4 Understanding Random Events: Producing Models Using Probability and Simulation

There are few things in life of which we are absolutely certain. Not knowing what is going to happen or how someone or something is going to react in a particular situation is what keeps life interesting. Let's take a moment to think about some of the things you do on a daily basis and list the events that have some chance or uncertainty associated with them. How do you wake up in the morning? Many people rely on an alarm clock, but what if the electricity is out or your alarm clock malfunctions or you forgot to set the alarm the night before? What are you going to wear? Usually you make this decision based on the weather forecast. Will it be warm or cold? Will it rain or not? Many of us have daily appointments and meetings that we must attend. Will you make it to your lunch meeting on time? Will this be one of those days filled with unforeseen problems? And after a long day at school or the office, will your team win the soccer game tonight? What other events can you think of that involve uncertainty?

© Springer International Publishing AG 2017
D.A. Wolfe, G. Schneider, *Intuitive Introductory Statistics*, Springer Texts in Statistics,
DOI 10.1007/978-3-319-56072-4_4

In this chapter we will develop your understanding of chance and uncertainty by focusing on some natural probability models for several basic settings. There are three common interpretations for the probability of an event.

Interpretation 1. Relative Frequency

The probability of an event A, say $P(A)$, is the long-run relative frequency of the event. That is, we independently repeat a random experiment n times and record the fraction of the experiments in which the event A occurs. The relative frequency interpretation of $P(A)$ is then the long-run proportion of independent experiments in which A occurs or, notationally, $P(A) = \text{limit}_{n\to\infty} \frac{\text{\# of experiments in which A occurs}}{\text{n}}$.

Interpretation 2. Logical Probability

If a random experiment can result in any one of n equally likely outcomes, then the probability of an event A, say $P(A)$, is equal to one over the number of equally likely outcomes, or notationally, $P(A) = \frac{1}{\text{n}}$.

Interpretation 3. Subjective Probability

The probability of an event A, say $P(A)$, is a number between 0 and 1, inclusive, which measures an individual's degree of belief in the event.

Although subjective probabilities are commonly used in everyday conversation, we will focus on the first two interpretations of probability in this text.

Example 4.1. Rolling a Pair of Fair Dice Consider the simple random experiment that consists of rolling a pair of fair dice and suppose we are interested in the probability of getting an even number. Using the relative frequency interpretation of probability, we would roll the dice over and over again and compute the fraction of times the event E = {even number} occurs. If you actually conduct this experiment with a pair of fair dice, you will find that this fraction converges to 1/2 as your number of trials increases. Try it!

Alternatively, we know that when we roll a pair of dice we are either going to get an even number or an odd number, each corresponding to 18 of the 36 possible combinations of numbers on the dice. Thus, there are two equally likely outcomes, even and odd, so the logical interpretation of probability implies that the probability of getting an even number is $P(E) = 1/2$.

We must be very careful in applying Interpretation 2. For example, even though the outcome from rolling a pair of dice is either an even number in the set {2, 4, 6, 8, 10, 12} or an odd number in the set {3, 5, 7, 9, 11}, it is not correct to conclude that $P(E) = 6/11$. When applying Interpretation 2 we are assuming that the basic outcomes are equally likely and the eleven outcomes {2, 3, 4, 5, 6, 7, 8, 9, 10, 11, 12} are not equally likely!

4.1 Probability as Relative Frequency: Law of Large Numbers

In this section we introduce the Law of Large Numbers, one of the most important, yet widely misunderstood, concepts in probability.

Example 4.2. Checking Your Intuition and Your Random Number Generator Suppose we generate two sets of 100 random numbers from the interval (0, 1). Then we create a new set of values corresponding to the differences between these two sets of random numbers. What should the collection of differences look like? Will it be symmetric? Where do you expect the center of the collection of differences to be located? What is the chance that a given number in the second column is greater than the corresponding number in the first column?

Fig. 4.1 Numerical and graphical summaries for differences between two sets of random numbers selected from the interval (0,1)

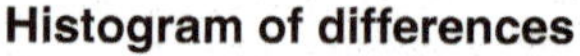

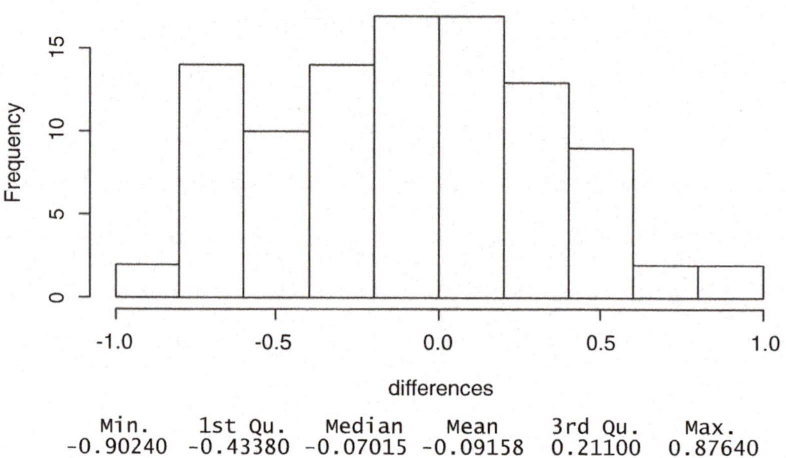


```
    Min.   1st Qu.    Median      Mean   3rd Qu.     Max.
-0.90240  -0.43380  -0.07015  -0.09158   0.21100  0.87640
```

We first generate the two sets of random numbers using two calls to the **R** function *runif*( ) while specifying the option $n = 100$. The following commands generate these two sets of 100 numbers and store them to local variables named *first_set* and *second_set*.

```
> first_set <- runif(n=100)
> second_set <- runif(n=100)
```

Next we compute and store the differences between the 100 pairs of numbers in a local variable named *differences*.

```
> differences <- first_set - second_set
```

Finally, we generate the numerical and graphical summaries displayed in Fig. 4.1 using the *summary*( ) and *hist*( ) functions.

```
> hist(differences)
> summary(differences)
```

Was your intuition correct? The collection of the differences is roughly symmetric and centered at zero. Now, let's use the random numbers in *first_set* and *second_set* to estimate the probability that the number in the second set will be greater than the number in the first set. We first create a new variable, *second_greater_than_first*, which will be a vector whose ith element is *TRUE* when the ith number in *second_set* is greater than the ith number in *first_set* and *FALSE* otherwise.

```
> second_greater_than_first <- second_set > first_set
```

Indicator variables like the one we have just created are often used to document whether an event of interest occurred. The *TRUE* and *FALSE* values (or, often, ones and zeroes) that are typically used for indicator variables make counting very easy. For example, if we count the number of *TRUE* values in *second_greater_than_first*, we will find out how many times the random number in the second set was greater than the random number in the first set. We can easily do this by using the **R** function *sum*( ), which will automatically convert the *TRUE* values to ones and the *FALSE* values to zeroes and then add them up for us.

```
> sum(second_greater_than_first)
[1] 57
```

For our simulated values the sum is 57 and so 57% of the time the number in the second column was greater than the number in the first column. Was your guess close to 50% or 0.5?

Another interesting way to view these simulation results is to keep a running total of the number of times the second number is larger than the first number. That is, suppose you generated two random numbers, compared them, and added one to a counter if the second number was greater than the first number. Do you think the percentage of times that the second number is greater than the first will always be close to 50%?

We can use the **R** function *cumsum*( ) to get the running total of the number of times that the second number is larger than the first and then divide that total by the number of trials to get the cumulative proportion. If we multiply

Fig. 4.2 Scatterplot of cumulative percentages versus trial number

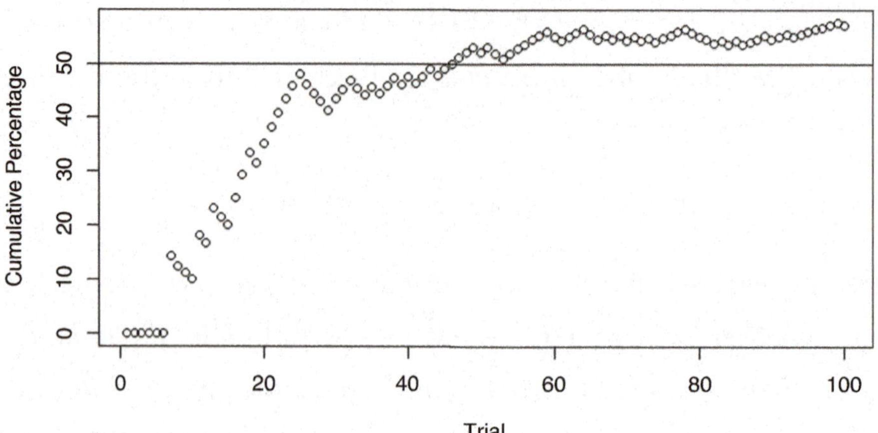

by 100, we'll have the cumulative percentages after each of the 100 trials. The following command does exactly that and stores the result in the variable *cumulative_percentages*.

```
> cumulative_percentages <- 100 * cumsum(second_greater_than_first) / 1:100
```

We use the following commands to create the scatterplot of these results (and to add a horizontal line at 50%) shown in Fig. 4.2.

```
> plot(1:100,
       cumulative_percentages,
       xlab = "Trial",
       ylab = "Cumulative Percentage")
> abline(h = 50)
```

Looking at Fig. 4.2 we notice that there is a considerable amount of variability in the cumulative percentages for the first few dozen trials. However, as we look at more and more pairs of random numbers (that is, the trial number increases), the estimates settle down and become less variable. This long-run stability is very important in the fields of probability and statistics. It leads to our relative frequency interpretation of probability and is known as the *Law of Large Numbers*.

Definition 4.1 The Law of Large Numbers says that if we repeat a random experiment a "large" number of times, then the cumulative fraction of times an event of interest occurs will converge to the probability of the event. Notationally, we have

$$\lim_{n\to\infty} \frac{\text{\# of experiments in which A occurs}}{\text{n}} = P(A).$$

Alternatively, the Law of Large Numbers says that the absolute value of the difference between the cumulative fraction and the probability of interest converges to zero; that is,

$$\lim_{n\to\infty} \left| \frac{\text{\# of experiments in which A occurs}}{\text{n}} - P(A) \right| = 0.$$

Far too often people think that the Law of Large Numbers is applicable to situations where the number of trials is small. In short, there is no such thing as the Law of Small Numbers. If you repeat the simulation described in Example 4.2 several times (as you're asked to do in Exercise 4.1.3), you will find that the regularity described by the Law of Large Numbers is only present in the long-run.

Many states now have lotteries and there are always people trying to take advantage of the common fallacy or belief in the "Law of Small Numbers." For example, several papers publish lists of so-called "hot" lottery numbers; that is, numbers that have occurred more frequently than others in recent drawings. What does the Law of Large Numbers say about "hot" and "cold" numbers if the lottery is fair?

Section 4.1 Practice Exercises

4.1.1. *Checking Your Intuition By Comparing Sequences of Events.* A sequence of nine coin tosses could result in the sequence HTTHHHTHH. One property of this sequence that may be of interest is the total number of heads, 6 in this case. Some other properties of interest in sequences of events are the number of runs and the lengths of those runs. A run is a string of outcomes that are the same. This sequence of 9 coin tosses has 2 runs of length 1, 2 runs of length 2, and 1 run of length 3. Now let's investigate these properties for some longer sequences.

(a) Write down a sequence of 50 outcomes (H or T) that you think could result from flipping a coin 50 times. (Don't actually flip the coin yet—just use your intuition to write down a sequence of 50 random outcomes.) How many heads are in the sequence? How many runs are in the sequence? Find out how long each run is by counting the number of H's or T's in the run and create a frequency table for run length. What is the length of the longest run?

(b) Flip a coin 50 times and record the sequence of 50 outcomes. How many heads are in the sequence? How many runs are in the sequence? Find out how long each run is by counting the number of H's or T's in the run and create a frequency table for run length. What is the length of the longest run? Compare your answers with your intuition in part (a).

(c) Use **R** (for example, the function *rbinom* ( ) may be helpful) to simulate 50 flips of a coin 30 separate times. For each of the 30 sequences, determine the number of heads and the number of runs in the sequence. Create graphical summaries for the number of runs and the number of heads. How do the values from your sequence in part (b) compare with those in the simulated distributions?

(d) Can you tell which the two sequences below was simulated and which was an intuitive guess?

Sequence R HHHHTHTHHTTHTHTHHHHTHHTTT
HHHTHHTHTHTTTHTHHHHTHHTHH

Sequence G HTHHHTTHHTTTHHHHTTTHTHTTH
HTTTTHHTTHHHTTTHTHTHHHTTT

4.1.2. *Functions of Random Numbers*. Suppose U is a random number between 0 and 1. Show what you need to do to U arithmetically to produce:

(a) a random number X between 1 and 10;

(b) a random number X between -1 and 1;

(c) a random integer X that is either 0 or 1 (consider a statement such as "if $U < 0.5$ then ...");

(d) a random integer in the set $\{1, 2, 3, 4, 5, 6\}$;

(e) Describe how you could use your random number U to make a random five letter word.

4.1.3. *Intuition, Random Number Generation, and the Law of Large Numbers*. Use a random number generator to create two separate sets of 100 random numbers each from the interval (0, 1). Call the sets *R1* and *R2*.

(a) Obtain the 100 differences $D = (R2 - R1)$ between the two sets of random numbers.

(b) What does the collection of differences D look like? Is it symmetric? Is the center of the collection of differences D located where you expected it to be? Explain.

(c) Create a new set of 100 values, say Q, that correspond to 1 if the number in *R1* is at least as large as the corresponding number in *R2* or 0 if the number in *R2* is larger.

(d) Find the sum of the values in Q. What does this statistic tell you? Explain how you would use this statistic to estimate the chance that one randomly generated number from the interval (0, 1) will be larger than a second randomly generated number from (0, 1).

(e) Compute the cumulative percentages of pairs for which *R2* is larger than *R1* as you move consecutively from the first to the 100th pair of random numbers.

(f) Create a scatterplot of the cumulative percentage of pairs for which *R2* is larger than *R1* against trial number.

(g) Describe the pattern in your scatterplot. How is this pattern related to the Law of Large Numbers?

4.1.4. A student is taking a standardized exam for the third time. When his parents ask him how he feels about his preparation, he responds with the following statement. "Well, I know that there are only three possibilities—I am either going to do the same, improve, or do worse. Using logical probability, I know that there is only a 1/3 chance that I will do worse, so I am feeling pretty good about the upcoming exam." Identify the flaw in the student's reasoning. Which interpretation of probability would you use to estimate his chance of doing worse on the exam? Explain.

4.1.5. Use the relative frequency interpretations of probability to identify the chance of getting a head when tossing a coin that is:

(a) balanced;
(b) unbalanced with heads being twice as likely as tails;
(c) unbalanced with heads being 5 times as likely as tails;
(d) unbalanced with heads being 1/4 as likely as tails.

4.1.6. Use the relative frequency interpretations of probability to identify the chance of getting an even number when rolling a die that is:

(a) balanced;
(b) weighted so that {1, 3, 5} are twice as likely as {2, 4, 6};
(c) weighted so that {1, 3, 5} are 1/3 as likely as {2, 4, 6}.

4.1.7. When listening to your local weather forecast, you often hear comments that involve probabilities. For example, there is a 60% chance of rain tomorrow or there is a 30% chance of snow. Do you think the meteorologists are using logical, relative frequency, or subjective probabilities? Explain.

4.1.8. A mutual fund manager invests \$1 million in a particular stock and then tells shareholders in the fund that she thinks there is a 75% chance that the company stock will increase in value by over 30% in the next 6 months. Do you think she is using the logical, relative frequency, or subjective interpretation of probability?

4.2 Some Basic Probability Rules

Definition 4.2 The **sample space** S of a random experiment is the set of all possible basic outcomes.

Example 4.3. Tossing Coins Consider the random experiment that consists of tossing three coins. The sample space for this experiment contains $2^3 = 8$ basic outcomes that we list as $S = \{$HHH, HHT, HTH, HTT, THH, THT, TTH, TTT$\}$. Figure 4.3 shows a convenient method of displaying this sample space.

Although tree diagrams like the one shown in Fig. 4.3 are useful for displaying sample spaces, often we are interested in particular characteristics of the basic outcomes rather than the basic outcomes themselves. For example, when tossing 3 coins we might be interested in the total number of heads, a random quantity determined from the basic outcomes of the random experiment, with its own derived sample space $H = \{0, 1, 2, 3\}$.

Fig. 4.3 Tree diagram for tossing three coins

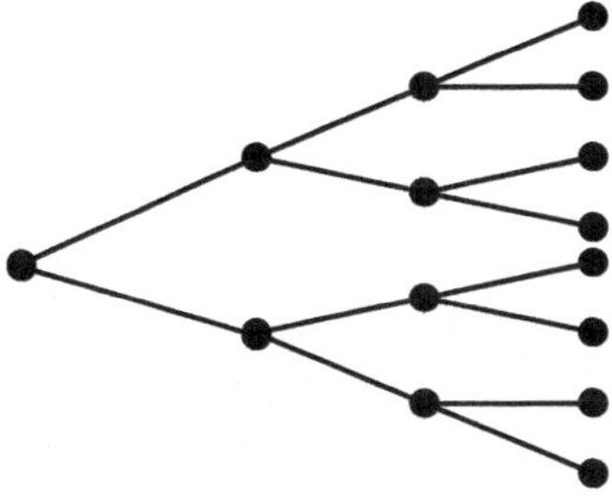

Definition 4.3 A **random variable** is a function with numerical values determined by the outcome of a random experiment. Random variables are typically denoted by capital letters.

We are often interested in calculating probabilities for events that can be written as functions (unions or intersections) of other events. When this is the case, we can usually calculate the probabilities of interest a number of different ways. For instance, in Example 4.1 we found that the probability of getting an even number, {2 or 4 or 6 or 8 or 10 or 12}, when we roll a pair of fair dice is 1/2. How would you calculate the probability of getting an odd number when rolling a pair of fair dice? You could go back to the original sample space and calculate the probability the same way we did in Example 4.1 or you could save time by using a simple relationship. Since each outcome is either even or odd, we know that $P(\text{even}) + P(\text{odd}) = 1$. We also know that the probability of getting an even number is 1/2, so $P(\text{odd}) = 1–1/2 = 1/2$.

The "trick" that we just used can be applied in many probability problems so we formally present it now.

Definition 4.4 If two events A and B do not have any basic outcomes in common (i.e., the intersection $A \cap B$ is the empty set) then we say that A and B are **disjoint events**.

Definition 4.5 Two events A and B are said to be **complementary** if the union $A \cup B = S$, the sample space, **and** they are disjoint events. The **complement** of an event A, denoted by A^c, is the set of all basic outcomes that are not in A.

Complement Rule The probability of the complement of an event A is $P(A^c) = 1 - P(A)$.

Example 4.4. Rolling Dice Many board games use dice to determine the number of spaces that a player will move when it is their turn. If you roll a pair of fair dice, what is the probability of getting something other than a sum of six on the upward facing sides? If we let $A = \{$sum of six on the roll of a pair of fair dice$\}$, then we want to find $P(A^c)$. Now, we can list all $6^2 = 36$ equally likely outcomes and calculate $P(A^c)$ by counting the number of equally likely outcomes that are not equal to six or we can count the number of equally likely outcomes that are equal to six and then use the complement rule. Since there are only five ways to get a six, namely, $\{(1, 5), (2, 4), (3, 3), (4, 2),$ and $(5, 1)\}$, $P(A) = 5/36$ and $P(A^c) = 1–5/36 = 31/36$.

Counting is a fundamental principle in probability. Often a probability problem can be reduced to counting the number of equally likely outcomes of interest. Although this sounds like a very simple task, careful attention to details in counting are required. We will return to some general counting techniques later in this chapter.

In the next few subsections we introduce some basic rules for calculating probabilities. While it is easy to fall into the trap of viewing these rules or formulas as something that must be memorized, we are hoping that you resist this temptation. Instead, try to view them as problem solving aids and continue to think about their intuitive nature.

4.2.1 Addition Rule

When an event can be written as the union of two or more events the addition rule can often provide a convenient way to calculate the probability of interest.

Addition Rule If A and B are two events of interest, then the probability that only A occurs or only B occurs or both A and B occur is

$$P(A \cup B) = P(A) + P(B) - P(A \cap B).$$

If A and B are disjoint events, then $P(A \cap B) = 0$ and the addition rule simplifies to

$$P(A \cup B) = P(A) + P(B).$$

That is, if A and B are disjoint, then the probability of their union is equal to the sum of the individual probabilities.

Example 4.5. Drawing a Card When drawing a card from a standard deck of 52 playing cards, what is the probability that the card is either red or a queen? Let $R = \{\text{red card is drawn from the deck}\}$ and $Q = \{\text{queen is drawn from the deck}\}$ denote the two events of interest. Using the addition rule, we find $P(R \cup Q) = P(R) + P(Q) - P(R \cap Q) = \frac{26}{52} + \frac{4}{52} - \frac{2}{52} = \frac{7}{13}$. An alternative method of solution works directly with the 52 equally likely cards. Since 26 of the cards are red and there are two black Queens, $P(R \cup Q) = \frac{28}{52} = \frac{7}{13}$.

The addition rule can be extended to the union of m events $A_1, A_2, \ldots, A_m$. If no two of the m events have any outcomes in common (they are mutually disjoint), then the probability of their union is equal to the sum of their probabilities. That is,

$$P(A_1 \cup A_2 \cup \cdots \cup A_m) = P\left(\bigcup_{i=1}^{m} A_i\right) = \sum_{i=1}^{m} P(A_i).$$

The setting where the m events are not mutually disjoint is discussed in Exercise 4.1.6.

4.2.2 Conditional Probability

What is the probability that a newborn baby will have a low birth weight? Researchers at Baystate Medical Center in Springfield, Massachusetts conducted a study in 1986 to estimate this probability and to investigate risk factors associated with low birth weight. A contingency table was constructed for birth weight (classified as low, medium, or high) and one of the risk factors, smoking. The output in Fig. 4.4 can be used to estimate the chance of having a baby with low birth weight.

Based on the summary information in the far right column, we would estimate the probability of having a baby with a low birth weight to be $59/189 = .3122$. However, if we consider the risk factor smoking, we would estimate the probability of having a baby with a low birth weight to be $30/74 = .4054$ for a smoking mother and $29/115 = .2522$ for a nonsmoking mother. Taking this risk factor into account is known in statistics as conditioning. That is, our estimate of the probability depends on whether or not the mother is a smoker. If the estimates would have been the same then there would be no need to consider the information on whether or not a mother is a smoker in assessing the probability of having a baby with low birth weight. In applied problems it is easy to overlook important conditional factors that may change your estimates.

Fig. 4.4 Contingency table for birth weights and smoking (rows are levels of weight, columns are levels of smoking)

	noSmoke	Smoke	total
High	42	13	55
Low	29	30	59
Med	44	31	75
total	115	74	189

The general form of the definition of conditional probability is given in Definition 4.6.

Definition 4.6 The conditional probability of an event A, given knowledge of the occurrence of the event B, is

$$P(\mathrm{A}|\mathrm{B}) = \frac{\mathrm{P}(\mathrm{A} \cap \mathrm{B})}{\mathrm{P}(\mathrm{B})}.$$

4.2.3 Multiplication Rule

If we multiply both sides of the equation in Definition 4.6 by $P(B)$, we obtain a probability rule for the intersection of two events.

Multiplication Rule If A and B are two events of interest, then the probability that both A and B occur is

$$P(A \cap B) = P(A \mid B)P(B).$$

An alternative version of the multiplication rule is obtained by simply reversing the roles of A and B. That is, the probability of the intersection of A and B can also be obtained from

$$P(B \cap A) = P(B \mid A)P(A),$$

if conditioning on A is more natural in a given problem.

Example 4.6. Low Birth Weight We can use the data in the contingency table in Fig. 4.4 to estimate the probability that a newborn baby will have a low birth weight and his/her mother is a smoker. Let L = {birth weight for a newborn baby is classified as low} and S = {mother is a smoker} denote the two events of interest. Using the multiplication rule, we estimate $P(S \cap L) = P(L|S)P(S) \approx \frac{30}{74} \times \frac{74}{189} = \frac{30}{189} = .1587$.

Fig. 4.5 General form of a contingency table for two categorical variables

		Column Variable				
		C_1	C_2	…	C_J	Row Totals
Row Variable	R_1	$P(R_1 \cap C_1)$	$P(R_1 \cap C_2)$	…	$P(R_1 \cap C_J)$	$P(R_1)$
	R_2	$P(R_2 \cap C_1)$	$P(R_2 \cap C_2)$	…	$P(R_2 \cap C_J)$	$P(R_2)$
	⋮	⋮	⋮	…	⋮	⋮
	R_I	$P(R_I \cap C_1)$	$P(R_I \cap C_2)$	…	$P(R_I \cap C_J)$	$P(R_I)$
	Column Totals	$P(C_1)$	$P(C_2)$	…	$P(C_J)$	1

Note that 30 is the table entry for the row labeled Low and the column labeled Smoke so the probability of the intersection is simply the cell entry divided by the total number of births. It is conventional to let the cells of a contingency table denote the counts (or probabilities) of the intersections. When presented as probabilities, the cell values are referred to as *joint probabilities* and the row and column totals are called *marginal probabilities*.

The general form of a contingency table (using probabilities) for two categorical variables is shown in Fig. 4.5. The row variable has m categories or levels and the column variable has n categories or levels. Notice that if you sum the row totals or the column totals or the $I \times J$ cell probabilities you will always get 1. These three different sets of probabilities denote three different probability distributions for discrete random variables. In Sect. 3 we will focus on a number of specific types of discrete random variables and their associated probability distributions.

In some settings we obtain special contingency tables where $P(R_i \cap C_j) = P(R_i) \times P(C_j)$ for all $i = 1, 2, \ldots, I$ and $j = 1, 2, \ldots, J$. When the joint probability of the intersection is equal to the product of the associated marginal probabilities for all cells in the two-way table, we say that the row variable and the column variable are *statistically independent*.

Definition 4.7 Events A and B are **independent** if knowledge about the occurrence of the event A does not change the probability of B and vice versa. Notationally, $P(A \mid B) = P(A)$ and $P(B \mid A) = P(B)$. Basically, the information given is not informative because the two events are unrelated or independent of one another.

If two events are independent then the multiplication rule states that the probability of their intersection is the product of their individual probabilities.

Multiplication Rule for Independent Events If A and B are independent events, then the probability that both A and B occur is

$$P(A \cap B) = P(A)P(B).$$

The multiplication rule can be extended to the intersection of m events A_1, $A_2, \ldots, A_m$. If the m events are mutually independent, then the probability of their intersection is equal to the product of their probabilities. That is,

$$P(A_1 \cap A_2 \cap \cdots \cap A_m) = P\left(\bigcap_{i=1}^{m} A_i\right) = \prod_{i=1}^{m} P(A_i).$$

The setting where the m events are not mutually independent is discussed in Exercise 4.A.7.

Section 4.2 Practice Exercises

4.2.1. *Eight-Sided Dice.* Consider the set of possible outcomes from rolling a pair of 8-sided dice.

(a) Identify the sample space S.

(b) Describe two events A and B that are disjoint and complementary.

(c) Describe two events C and D that are disjoint but not complementary.

4.2.2. *Five-Card Hands.* Consider the set of possible outcomes from dealing a hand of 5 cards from a standard deck of 52 playing cards (no jokers).

(a) Describe two events A and B that are disjoint and complementary.

(b) Describe two events C and D that are disjoint but not complementary.

4.2.3. A group of patients has been classified by gender and as having either high or low blood pressure. Let A be the event that the patient is female and B the event that the patient has high blood pressure. Are the events disjoint? Explain.

4.2.4. *Candy Colors.* Suppose a **large** bag of candy contains the following color distribution.

Color	Brown	Red	Yellow	Green	Orange
Proportion	0.3	0.2	0.2	0.2	0.1

Two pieces of candy are drawn randomly from this bag and we are interested in the color of each piece of candy.

(a) List the set of possible outcomes in the sample space S.

(b) Are the outcomes equally likely? Explain.

(c) Find the probability that the first piece of candy is red and the second piece of candy is orange.

(d) Find the probability that both pieces of candy are the same color.

4.2.5. *Smoke Detectors.* Most homes now have at least two smoke detectors. Suppose that the probability each smoke detector will function properly in the presence of smoke is .85 and that the smoke detectors function independently of one another.

(a) If you have two smoke detectors in your home, what is the probability that both of them will function properly during a fire?
(b) If you have three smoke detectors in your home, what is the probability that exactly one of them will function properly during a fire?
(c) If you have two smoke detectors in your home, what is the probability that at least one of them will NOT function properly during a fire?
(d) Repeat part (c) if you have three smoke detectors in your home and compare the two probabilities. How do you think this probability would change if you had four smoke detectors in your home?

4.2.6. *Venn Diagrams Can Be Helpful in Solving Probability Problems.* A *Venn diagram* is a graphical display where the sample space S is represented by a rectangular region and events are denoted by elliptical or circular subsets of the rectangular region. The following Venn diagram provides a visual representation for a sample space with two disjoint events.

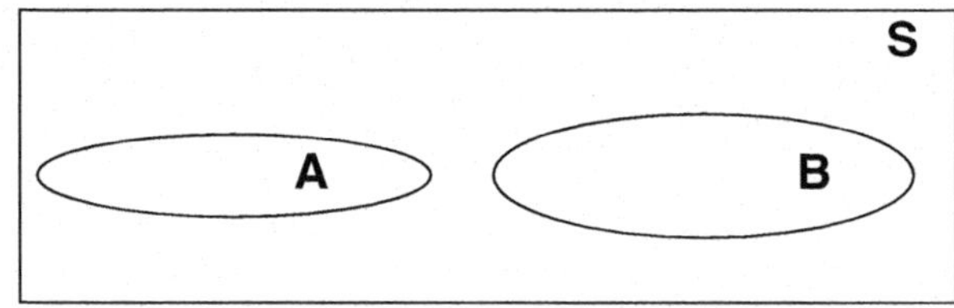

(a) Create a Venn diagram for one event A by removing the ellipse denoting the event B from the diagram above. Identify the complement of A on your diagram.
(b) Create a Venn diagram for two events A and B that are not disjoint. Identify the events $A \cap B$, $A \cap B^c$, and $A^c \cap B$ on your diagram.
(c) Create a Venn diagram for three events A, B, and C that are not disjoint. Identify the events $A \cap B \cap C$, $A^c \cap B \cap C$, $A \cap B^c \cap C$, $A \cap B \cap C^c$, $A^c \cap B^c \cap C$, $A^c \cap B \cap C^c$, $A \cap B^c \cap C^c$, and $A^c \cap B^c \cap C^c$.

4.2.7. *Daily Newspapers.* Suppose a particular city has two newspapers—one is delivered in the morning and one in the afternoon. If 75% of the households subscribe to the morning paper, 50% of the households subscribe to the afternoon paper, and 90% of the households subscribe to at least one of the papers, what proportion of the households subscribe to:

(a) both papers;
(b) the morning paper but not the afternoon paper;
(c) only the afternoon paper;
(d) neither paper.

4.2.8. *Dentist Visits.* The proportion of individuals visiting the dentist who have their teeth cleaned is 0.7; a cavity filled is 0.4; a tooth extracted is 0.05; their teeth cleaned and a cavity filled is 0.25; their teeth cleaned and a tooth extracted is 0.005; a cavity filled and a tooth extracted is 0.01; and their teeth cleaned, a cavity filled, and a tooth extracted is 0.0005.

(a) Find the proportion of individuals visiting the dentist who have none of these things done.
(b) What proportion of those individuals who had their teeth cleaned will also have a cavity filled?
(c) What proportion of those individuals who had a cavity filled will also have a tooth extracted?

4.2.9. *Sun around the Earth or Earth around the Sun?* National Public Radio (2014) reported on a number of results from a survey conducted by the National Science Foundation in the United States in 2012. One of the questions asked in the survey was "Does the earth revolve around the sun, or does the sun revolve around the earth?". Twenty six percent of the respondents said that the sun revolved around the earth! Suppose you randomly select ten individuals and ask them this question. If the results of the National Science Foundation survey are applicable:

(a) What is the probability that at least one of the individuals you interview will believe that the sun revolves around the earth?

(b) What is the probability that exactly half of the individuals you interview will believe that the sun revolves around the earth?

(c) What is the probability that more than half of the individuals you interview will believe that the sun revolves around the earth?

4.2.10. *Express Mail.* The staff members in a mailroom at a small company send 40% of their overnight packages via express mail service with Company A. Of these packages, 2% arrive after the guaranteed delivery time. Company B is used to send another 45% of the overnight packages and the remaining 15% are sent via Company C. Only 1% of the packages sent via Company B arrive late, while 5% of the packages handled by Company C are delivered late.

(a) If a mail record for an overnight delivery is randomly selected from the accounting database, what is the probability that the package went via Company A and was late?

(b) What is the probability that a randomly selected package arrived late?

4.2.11. *Dice and Board Games*. Many board games rely on the use of six-sided dice. For example, in "Dungeons and Dragons", a player rolls three, balanced six-sided dice and adds the spots on the upward facing sides to assign intelligence for a character. Yahtzee relies on five balanced six-sided dice.

(a) How many outcomes are in the sample space when rolling three balanced six-sided dice? Are these outcomes equally likely?

(b) List all of the possible values for the sum of the spots on the upward facing sides of three balanced six-sided dice? Are all of these values equally likely?

(c) Find the probability of getting a sum of 5 when rolling three balanced six-sided dice.

(d) How many outcomes are in the sample space when rolling five balanced six-sided dice? Are these outcomes equally likely?

(e) Find the probability of getting a sum of 30 when rolling five balanced six-sided dice.

4.2.12. *Pizza Pies.* You work at Mike's Pizza shop. There are 7 pizzas in the oven and you know the following information about the pizzas: 3 pizzas have thick crust, 4 have regular crust, one thick crust pizza and two regular crust pizzas have only sausage, and two thick crust and two regular crust pizzas have only mushrooms.

(a) Suppose a pizza is randomly chosen from the oven. Are the events getting a thick crust pizza and getting a pizza with mushrooms independent? Explain.

(b) Suppose an 8th pizza is added to the oven. This pizza has thick crust with cheese only. Now are the events getting a thick crust pizza and getting a pizza with mushrooms independent? Explain.

4.2.13. *Economy and Voting.* A survey organization asked respondents their views on the likely future direction of the economy and whether they had voted for the President in the last election. The two-way table below shows the proportion of responses in each category

		View on Economy		
		Optimistic	Pessimistic	Neutral
Voting behavior	For President	0.2	0.1	0.1
	Against President	0.1	0.15	0.05
	Did not vote	0.05	0.1	0.15

What is the probability that a randomly selected respondent:

(a) voted against the President;

(b) is pessimistic about the future of the economy;

(c) voted for the President and is pessimistic about the future of the economy;

(d) voted for the President but is not pessimistic about the future of the economy?

(e) Are the respondents' views on the economy and voting behavior independent? Explain.

4.2.14. *Space Missions.* Prior to the fatal 1986 explosion of the space shuttle Challenger, many government officials believed that the space shuttle program would never have a fatal failure. Within NASA, estimates of the probability of a mission failure had ranged from 1 in 100,000 (by management) to 1 in 100 (by engineers). The Challenger explosion came on its 26th mission.

(a) What is the probability of 25 successes in the first 25 missions if the probability of failure on each mission was 1%? 4%? 10%?

(b) What is the probability of at least one failure in the first 26 missions if the probability of failure on each mission was 1%? 4%? 10%?

4.2.15. *Diagnostic Tests.* A diagnostic test for the AIDS virus has probability 0.005 of yielding a false positive; i.e., indicating the presence of the AIDS virus when it is not present. If the 140 employees of a medical clinic are all free of the AIDS virus and they each take this diagnostic test, what is the probability that there will be at least one false positive outcome?

4.2.16. *Treating Childhood Diseases.* Parents of children diagnosed with a certain disease must decide whether or not to allow their children to be given a prescribed treatment. The treatment consists of a series of five independent shots and the children must receive all five shots. Preliminary studies of this treatment have shown that the probability of death or serious side effects associated with each shot in the treatment is 1/1750. A medical doctor claims that the probability of death or serious side effects from the complete five shot treatment is 1/1750. Is the medical doctor correct? Explain.

4.3 Discrete Random Variables and Their Probability Distributions

Discrete random variables are categorical or qualitative variables that can assume only a fixed number of values. To specify a discrete probability distribution the possible values of the random variable are listed with their corresponding probabilities, typically in tabular form.

Example 4.7. Rebooting Personal Computers Let X be the random variable that counts the number of times a personal computing system must be rebooted or restarted in an 8-h workday because of system errors. For a particular system the probability distribution for X is

Number of Reboots (x)	0	1	2	3	4	5
Probability $P(X = x)$	.53	.22	.13	.07	.03	.02

To find the probability that the system will need to be rebooted at least once during the next 8-h workday, we can simply add the probabilities corresponding to 1, 2, 3, 4 and 5 reboots or use the complement rule. Either way, there is a 47% chance that the computer will need to be rebooted at least once.

In the next two sub-sections we will focus on two general probability distributions that occur frequently in practice. In both cases we are interested in studying random experiments with only two possible outcomes classified as "success" or "failure."

Definition 4.8 A **Bernoulli trial** is a random experiment that can result in one of only two possible outcomes. One outcome is classified as a "success" and the other outcome is classified as a "failure." The classification labels "success" and "failure" need not correspond to any rational interpretation.

In the first setting involving Bernoulli trials, we are interested in counting the random number of successes in a fixed number, n, of independent Bernoulli trials. In the second setting we are interested in counting the random number of Bernoulli trials until the first success occurs.

4.3.1 Binomial Distribution

We are interested in finding the probability distribution for the total number of successes in n independent Bernoulli trials with the same probability of success, p, on each trial. Let X_i be an indicator variable that records the outcome of the ith Bernoulli trial, for $i = 1, \ldots, n$. That is, $X_i = 1$ if a "success" occurs on the ith trial and $X_i = 0$ if a "failure" occurs. Let B denote the total number of successes in the n independent Bernoulli trials. That is,

$$B = \sum_{i=1}^{n} X_i. \tag{4.1}$$

To find the probability distribution of B (called its *sampling distribution*) we will list all 2^n possible outcomes of the Bernoulli trials, compute the probability of each outcome, identify the value of B for each outcome, and tally the possible values of B and their associated probabilities.

Example 4.8. Binomial Distribution Suppose that four patients are given an experimental medication to treat an illness. Unfortunately, there is a problem with side effects for this treatment. The pharmaceutical manufacturer estimates that there is a 50% chance that a particular patient will experience side effects under normal usage of the medication. How many of the four patients will experience side effects with this medication? We don't know, of course, but we can use the information provided to determine the probability distribution for the total number of patients (out of four) who will experience side effects. With $p = ½$, the probability of each of the possible outcomes is $(1/2)^4 = 1/16$. The list of possible outcomes (0 indicates no side effects and 1 indicates side effects), the common probability 1/16 for each possible outcome and the value of B for that outcome are shown in Table 4.1.

To complete our derivation of the probability distribution of B we combine the probabilities for each of the possible values of B. The resulting tallies are

Table 4.1 Possible outcomes and probabilities for four independent Bernoulli trials when the probability of success is $p = 1/2$

Possible outcome from Bernoulli trials	Probability of possible outcome	Value of B
0000	1/16	0
0001	1/16	1
0010	1/16	1
0011	1/16	2
0100	1/16	1
0101	1/16	2
0110	1/16	2
0111	1/16	3
1000	1/16	1
1001	1/16	2
1010	1/16	2
1011	1/16	3
1100	1/16	2
1101	1/16	3
1110	1/16	3
1111	1/16	4

Table 4.2 Probability distribution for B when $n = 4$ and $p = 1/2$

Value of B	0	1	2	3	4
Probability	1/16	1/4	5/8	1/4	1/16

provided in Table 4.2. Notice that the probability distribution of B shown in Table 4.2 is symmetric about $b = 2$; that is, $P(B = 4) = P(B = 0)$ and $P(B = 3) = P(B = 1)$.

Example 4.9. Continuation of Example 4.8. Suppose that the pharmaceutical manufacturer improves the medication so that there is only a 10% chance (i.e., $p = .1$) that a particular patient will experience side effects under normal usage. How does that change the probability distribution for the number of patients who will experience side effects from this medication?

Table 4.3 Possible outcomes and probabilities for four independent Bernoulli trials when the probability of success is $p = .1$

Possible outcome from Bernoulli trials	Probability of possible outcome	Value of B
0000	$(.1)^0(.9)^4 = .6561$	0
0001	$(.1)^1(.9)^3 = .0729$	1
0010	$(.1)^1(.9)^3 = .0729$	1
0011	$(.1)^2(.9)^2 = .0081$	2
0100	$(.1)^1(.9)^3 = .0729$	1
0101	$(.1)^2(.9)^2 = .0081$	2
0110	$(.1)^2(.9)^2 = .0081$	2
0111	$(.1)^3(.9)^1 = .0009$	3
1000	$(.1)^1(.9)^3 = .0729$	1
1001	$(.1)^2(.9)^2 = .0081$	2
1010	$(.1)^2(.9)^2 = .0081$	2
1011	$(.1)^3(.9)^1 = .0009$	3
1100	$(.1)^2(.9)^2 = .0081$	2
1101	$(.1)^3(.9)^1 = .0009$	3
1110	$(.1)^3(.9)^1 = .0009$	3
1111	$(.1)^4(.9)^0 = .0001$	4

Table 4.4 Probability distribution for B when $n = 4$ and $p = .1$

Value of B	0	1	2	3	4
Probability	.6561	.2916	.0486	.0036	.0001

The possible outcomes and values of B listed in Table 4.1 will remain the same, but the probability of each possible outcome must be changed to reflect the change in the probability of success for each trial. Table 4.3 shows the necessary changes and Table 4.4 shows the new probability distribution for B. Notice that the probability distribution of B in Table 4.4 for $p = .1$ is not symmetric.

As Examples 4.8 and 4.9 illustrate, once the number of trials (n) and the probability of success (p) are known, computing the probability distribution for B basically becomes a counting problem. The notation $B(n, p)$ will be used to denote the probability distribution for the binomial random variable corresponding to the number of successes from n independent Bernoulli trials with probability of success p. We will write $B \sim B(n, p)$ to indicate that the random variable B is distributed as a binomial random variable with n independent trials and probability of success on each trial equal to p. The general form of the $B(n, p)$ probability distribution can be expressed as follows.

General Form of a Binomial Probability Distribution If B is the number of successes in n independent Bernoulli trials, each with probability of success p, then

$$P(B = b) = \binom{n}{b} p^b (1 - p)^{n-b}, \quad \text{for } b = 0, 1, \ldots, n,$$

where $\binom{n}{b} = \frac{n!}{b!(n-b)!}$ and $n\,! = n \times (n - 1) \times (n - 2) \times \cdots \times 2 \times 1$.

The coefficient $\binom{n}{b}$ counts how many of the 2^n possible outcomes for the n trials contain exactly b successes and $(n\text{-}b)$ failures.

Example 4.10. If $B \sim B(15, 0.70)$, what is $P(B = 10)$?

$$P(B = 10) = \binom{15}{10} (.7)^{10} (.3)^{15-10} = \frac{15!}{10!5!} (.7)^{10} (.3)^5 = .2061.$$

You can easily compute this quantity for yourself by using the **R** function *dbinom*( ) and specifying the *x, size,* and *prob.* arguments, which represent the number of successes, number of trials, and probability of success, respectively, as follows.

```
> dbinom(x = 10, size = 15, prob = 0.7)
[1] 0.2061304
```

4.3.2 Geometric Distribution

Instead of counting the number of successes in a fixed number, n, of independent Bernoulli trials, sometimes we are interested in how many trials, G, must be completed until we observe the first success. The random variable, G, that counts the number of independent Bernoulli trials necessary to obtain the first success is called a *geometric random variable.*

Example 4.11. Suppose you were hired by a small telemarketing company to sell a very expensive product. Company research shows that only 3% of all calls will result in a successful sale. **If** we can assume that the results of your calls to different individuals are independent **and** that the probability of a sale is the same for each individual contacted, then we can use the Multiplication Rule for Independent Events to determine the probabilities of interest. For example, if $G =$ [the number of calls necessary to get the first sale], then the probability that we will make our first sale on the third phone call is $P(G = 3) = (.97)^2 \times .03 = .0282$. Do you think the assumptions in this telemarketing example are reasonable?

The general form of the probability distribution for the geometric random variable G, corresponding to the number of independent Bernoulli trials with common probability of success p needed to obtain the first success, is as follows.

General Form of the Geometric Probability Distribution If G is the number of independent Bernoulli trials, each with probability of success p, necessary to obtain the first success, then

$$P(G = g) = (1 - p)^{g-1}p, \text{ for } g = 1, 2, \ldots$$

You can also use the **R** function *dgeom*( ) to calculate the value of the probability distribution for a geometric random variable by specifying the arguments x and *prob*, which are the number of failures (not the number of trials!) and probability of success, respectively. Running the following command verifies the result we obtained in Example 4.11.

```
> dgeom(x = 2, prob = 0.03)
[1] 0.028227
```

Notice that the geometric probability distribution assigns positive probability to an infinite number of values, since there is no limit to the number of trials it MIGHT take to obtain the first success.

Section 4.3 Practice Exercises

4.3.1. *On-the-Job-Stress*. In a recent poll, Gallup (November, 2012) found that 33% of employees in the United States were totally dissatisfied with the amount of stress in their jobs. Suppose you randomly select five employed individuals and ask them how they felt about the stress in their jobs. Determine the probability distribution for the number of interviewed individuals who indicate that they are totally dissatisfied with the amount of stress in their jobs. How would the probability distribution change if the proportion of employees in the United States who are dissatisfied with the level of stress in their jobs were 50% instead of 33%?

4.3.2. *Landlines Versus Cell Phones*. According to a recent survey by GfK (2015a), 44% of adults in the United States live in households with cell phones but no landline. Suppose you place calls to 12 randomly chosen telephone numbers.

(a) What is the probability that exactly 4 of your calls will be to households without a landline?
(b) What is the probability that at least 4 of your calls will be to households without a landline?
(c) What is the probability that at most 4 of your calls will be to households without a landline?

(d) What is the probability that more than 4 of your calls will be to households without a landline?

(e) What is the probability that fewer than 4 of your calls will be to households without a landline?

4.3.3. Is the binomial distribution $B(n, p)$ always symmetric (regardless of the number of trials) when the probability of success is 1/2? Does this intuitively make sense? Explain. Use the **R** functions *dbinom*( ) and *plot*( ) for a few different numbers of trials to graphically support your answer.

4.3.4. *Does Lefty-Lefty Mean Lefty Children*? There is a 26% probability that a child of two left-handed parents will also be left-handed. If a couple who are both left-handed have four children, find the probability distribution for the number of their children who are also left-handed. Assume independence of this trait between births of the children.

4.3.5. A manager at a restaurant is training her new hostess and during the discussion the new employee asks about the typical size of a group that needs to be seated. The manager says company research indicates that one of the following three probability models is appropriate. Identify the correct probability model and explain the problems with the other two models.

Group Size	**1**	**2**	**3**	**4**	**5**	**6**	**7**	**≥8**
Model 1	0.15	0.35	0.10	0.15	0.04	0.06	0.01	0.10
Model 2	0.10	0.40	0.05	0.25	0.04	0.06	0.01	0.09
Model 3	0.15	0.35	0.10	0.20	0.05	0.03	0.01	0.20

4.3.6. A box contains four slips of paper marked 1, 2, 3, and 4. Two slips are selected with replacement between selections. Find the probability distribution for the random variable that records the difference (second number minus first number). How does the probability distribution change if the first drawn slip of paper is NOT replaced before the second slip is drawn?

4.3.7. Consider the random variable that counts the number of heads in 10 flips of a fair coin. Find the probability of getting

(a) exactly 5 heads;

(b) at most 5 heads;

(c) at least 5 heads;

(d) less than 5 heads;

(e) more than 5 heads.

4.3.8. Find the probability distribution for B = [the number of heads obtained in 10 flips of an unbalanced coin with probability 0.4 of getting a head on each of the independent trials].

4.3.9. Suppose that $B \sim B(16, .75)$. Use the **R** functions *dbinom*( ) and *pbinom*( ) to find the following probabilities.

(a) $P(B = 12)$

(b) $P(B \leq 12)$

(c) $P(B < 12)$

(d) $P(B > 12)$

(e) $P(B \geq 12)$

(f) $P(B < 9)$

(g) $P(B > 10)$

(h) $P(B \leq 8)$

4.3.10. Suppose that $B \sim B(20, .4)$. Use **R** to find the complete probability distribution for B. Graph this probability distribution and comment on the shape and at least one other feature of the distribution.

4.3.11. *Is God on Your Side?* USA Today (2013) reported on a number of results from a survey conducted by the Public Religion Research Institute in the United States in 2013. One of the questions asked in the survey was "True or False: God plays a role in determining which team wins a sporting event.".

Twenty seven percent of the respondents answered "True", indicating that they believe that God does play a role in determining winners of sporting events. Suppose you randomly select 20 individuals and ask them this question. Let the random variable B denote the number of selected individuals who answer "True" to the question. If the results of the Public Religion Research Institute survey are applicable and $b \in \{0, 1, 2, \ldots, 20\}$, what is the expression for

(a) $P(B = b)$?
(b) $P(B < b)$?
(c) $P(B \geq b)$?

4.3.12. *Does God Reward Good Athletes?* USA Today (2013) reported on a number of results from a survey conducted by the Public Religion Research Institute in the United States in 2013. One of the questions asked in the survey was "True or False: God rewards athletes who have faith with good health and success.". Fifty three percent of the respondents answered "True", indicating that they believe that God does reward athletes who have faith with good health and success. Suppose you randomly select 30 individuals and ask them this question. Let the random variable B denote the number of selected individuals who answer "True" to the question. If the results of the Public Religion Research Institute survey are applicable and $b \in \{0, 1, 2, \ldots, 30\}$, what is the expression for

(a) $P(B = b)$?
(b) $P(B > b)$?
(c) $P(B \leq b)$?

4.3.13. *Big Bang or Not?* National Public Radio (2014) reported on a number of results from a survey conducted by the National Science Foundation in the United States in 2012. One of the questions asked in the survey was "True or False: The universe began with a huge explosion.". Sixty one percent of the respondents answered False, indicating that they did believe that the universe began with a huge explosion. Suppose you randomly select 15 individuals and ask them this question. If the results of the National Science Foundation survey are applicable, what is the probability that:

(a) exactly 8 of the selected individuals do not believe that the universe began with a big explosion;

(b) at most 8 of the selected individuals do not believe that the universe began with a big explosion;

(c) at least 8 of the selected individuals do not believe that the universe began with a big explosion;

(d) exactly 8 of the selected individuals believe that the universe began with a big explosion;

(e) less than half of the selected individuals believe that the universe began with a big explosion.

4.3.14. A potato chip manufacturer decides whether to purchase a truckload of potatoes by selecting samples and inspecting them to see if they meet the company standards. Suppose that 20 potatoes are randomly selected and the company policy is to purchase the load of potatoes if less than 2 of the potatoes are deemed unsatisfactory. Find the probability of purchasing a load of potatoes for which:

(a) 5% are rotten;

(b) 10% are rotten.

4.3.15. *From Whence We Came?* National Public Radio (2014) reported on a number of results from a survey conducted by the National Science Foundation in the United States in 2012. One of the questions asked in the survey was "True or False: Human beings, as we know them today, developed from earlier species of animals.". Forty eight percent of the respondents answered True, indicating that they did believe that human beings did develop from earlier species of animals. Suppose you randomly select 30 individuals and ask them this question. If the results of the National Science Foundation survey are applicable, what is the probability that:

(a) less than half of the selected individuals indicated that they believe that human beings developed from earlier species of animals?

(b) more than 20 of the selected individuals indicated that they believe that human beings developed from earlier species of animals?

(c) at least five of the selected individuals indicated that they do not believe that human beings developed from earlier species of animals?

4.3.16. *Tail-gaiting*. The AAA Foundation of Public Safety reported (July, 2016) that 50.8% of drivers acknowledged having aggressively tail-gated another vehicle to express displeasure at least once during 2014. Consider a sequence of cars following you on a major highway. Assuming independence between individual cars behind you on the highway, what is the probability of being tail-gated for the first time:

(a) by the second car following you?

(b) by the fourth car following you?

(c) before the sixth car following you?

4.3.17. *Monopoly*. In the game of Monopoly a player must stay in jail until they roll doubles with a pair of balanced dice (if they do not have the Get Out of Jail Free card). What is the probability that the player will get out of jail by rolling doubles:

(a) on their first try;

(b) on their second try;

(c) on their third try;

(d) before their sixth try?

4.3.18. *Will You Be More Educated than Your Significant Other*? In a 2007 survey, the Pew Research Center (January, 2010) found that 53% of spouses had the same education level, 19% of husbands had more education than their wives, and 28% of wives had more education than their husbands. Suppose you are in a class with ten men and ten women (including yourself). Assume that all 20 of you will marry (at least once) someone of the opposite gender.

(a) What is the probability that more than half of the men in your class will marry (for the first time) a woman with more education?

(b) What is the probability that more than half of the women in your class will marry (for the first time) a man with more education?

(c) (More difficult) What is the probability that more than half of your class (men and women combined) will marry (for the first time) someone of the opposite gender with the same education level?

4.4 Simulating Probability Distributions

Suppose you are interested in collecting a set of n different items, such as baseball cards, figurines, rare coins, or dolls. If you can purchase the entire set at once or over the course of several weeks then there is no chance involved. However, we are interested in the more intriguing situation where we are not certain which one of the items we will receive when we obtain it. This famous problem is known as the collector's problem.

Example 4.12. Collector's Problem Each box of a particular brand of cereal contains one out of a set of n different prizes. Suppose that the prize in each box is equally likely to be any one of the n possible prizes. How many cereal boxes do you think the collector will have to purchase in order to obtain the complete set of n prizes? The answer will vary randomly from collector to collector. That is, one collector may be lucky enough to get the complete set of n prizes in the first n boxes, although, as we shall see, this is highly unlikely if n is larger than 2. It may take another collector $2n$ boxes to get the complete set. Yet another collector may need to purchase n^2 boxes of cereal. In general, we would like to know how many boxes a typical collector must buy to get the complete set. In other words we want to find the center of the probability distribution for the number of purchased boxes required to complete the collection.

When you are faced with a practical problem like this, one approach is to use simulation to help you develop a better understanding of the underlying variability in the process. Suppose there are 6 prizes in the set. You can simulate the collection process for the set of 6 prizes by:

1. Repeatedly rolling a fair six-sided die and counting how many rolls it takes until each number appears at least once.
2. Writing the numbers 1 through 6 on identical pieces of paper, placing them in a container, drawing (with replacement between draws) one piece of paper at a time, and counting how many draws you must make until each number is chosen at least once. (With replacement means that once you observe the number on the piece of paper you have chosen, you put that piece of paper back into the container and mix thoroughly before making your next selection.)
3. Using a random number generator on your calculator or computer and counting how many random numbers you must generate until each of the 6 numbers {1,2,3,4,5,6} (or {0,1,2,3,4,5}) appears at least once as the first number to the right of the decimal point.

Alternatively, we can use **R** to simulate the number of selections necessary. We begin by defining a variable n which represents the number of different prizes (in this case, 6) and a variable *prize_collected* which is a vector of length n, where the ith element is *TRUE* if the ith prize has been collected and *FALSE* otherwise.

```
> n <- 6
> prize_collected <- logical(n)
> number_of_boxes_purchased <- 0
```

The following three lines represent the process of drawing one of the 6 prizes and indicating that the prize has been collected. The first line will randomly select an integer between 1 and 6 and store that as *new_prize*, while the second line will set the value of *prize_collected* corresponding to *new_prize* to be *TRUE*. The third line will handle the bookkeeping of updating the number of cereal boxes that have been purchased in total.

```
> new_prize <- sample(n, 1)
> prize_collected[new_prize] <- TRUE
> number_of_boxes_purchased <- number_of_boxes_purchased + 1
```

We now introduce the concept of a "while loop" in **R**, which will allow us to repeat the commands above *while* a certain condition is true. In this case, we want to continue purchasing boxes and drawing prizes until we have collected all the prizes. That is, (in clumsier English that more closely matches the **R** code) "while we have not collected all of the prizes". The *all*( ) function will allow us to check whether every element of *prize_collected* is *TRUE* or not, which will indicate whether we've collected all the prizes. Combining all of the pieces together, we end up with the following code which can be used to simulate the number of boxes purchased by any single collector. It has been defined inside a function named *CollectorsProblem* so that it can be easily called a repeated number of times.

```
> CollectorsProblem <- function(n){
    prize_collected <- logical(n)
    number_of_boxes_purchased <- 0

    while(!all(prize_collected)){
      new_prize <- sample(n, 1)
      prize_collected[new_prize] <- TRUE
      number_of_boxes_purchased <- number_of_boxes_purchased + 1
    }

    number_of_boxes_purchased
  }
```

Finally, we simulate the number of boxes selected for 100 different collectors, store the results in *number_of_boxes_purchased_by_collector* and display a subset of the results in Table 4.5. Numerical and graphical summaries of the full results can be found in Fig. 4.6.

Table 4.5 Simulation of collector's problem for $n = 6$

Collector	Number of boxes
1	7
2	10
3	19
4	19
5	6
6	26
...	...
95	12
96	13
97	17
98	19
99	26
100	22

Fig. 4.6 Descriptive statistics and graphical summary for collector's problem when there are $n = 6$ prizes

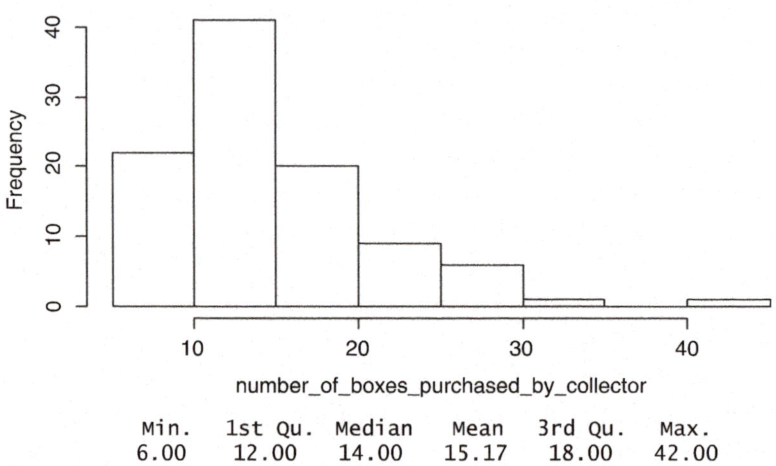


```
  Min.  1st Qu.  Median   Mean  3rd Qu.   Max.
  6.00   12.00    14.00  15.17   18.00   42.00
```

```
> number_of_collectors <- 100
> number_of_boxes_purchased_by_collector <- numeric(number_of_collectors)
> for(i in 1:number_of_collectors){
    number_of_boxes_purchased_by_collector [i] <- CollectorsProblem(n = 6)
  }
> hist(number_of_boxes_purchased_by_collector)
> summary(number_of_boxes_purchased_by_collector)
```

Based on this simulation we would estimate the center of the probability distribution to be around 14. Notice that we are using the median as our estimate of the center, since the distribution of the number of boxes purchased is clearly skewed to the right. The estimate of the center of the probability distribution based on the average is 15.17. Thus, if we had to answer the original question, we would conclude that the typical collector must purchase around 14 boxes of cereal to get the complete set of $n = 6$ prizes.

In Exercise 4.B.3 you will be asked to use these simulation techniques to estimate the number of purchases required when we increase the number of prizes to $n = 10$. You can use the same simulation methods, but you must make sure that each prize is equally likely to be chosen for each selection.

The power of simulation cannot be overstated. However, some situations require precise mathematical solutions. In Sect. 6 we will take a more mathematical approach to solving the collector's problem.

Section 4.4 Practice Exercises

4.4.1. *Collector's Problem With Twelve Prizes.* How would you use simulation to solve the collector's problem when there are 12 different prizes?

(a) Explain how you would use a random number generator that produces numbers between 0 and 1 to carry out the simulation.

(b) Can you repeatedly roll a pair of fair six-sided dice to carry out the simulation? Explain why or why not.

4.4.2. *Collector's Problem With Four Prizes.* Use simulation to estimate the number of purchases required by a collector who is interested in obtaining a complete set of $n = 4$ prizes.

4.4.3. *Stock Prices and Random Walks.* A stock will close either higher or lower when the market trading ends on a given day (ignore the possibility that the stock will end the day unchanged). Some economists believe that this day-today variation in stock prices follows what is known as a random walk, where it is equally likely that a stock will increase or decrease on a given day and the days' changes are independent.

(a) Describe how you would simulate this day-to-day behavior in stock prices by using a random number generator.
(b) Carry out your simulation for a month with 30 days.
(c) Did your stock close higher or lower at the end of the 30 days?

4.4.4. *Movement of Particles and Two-Dimensional Random Walks.* Some physicists believe that under certain conditions particle movements can be viewed as a random walk (see Exercise 4.4.3) in two dimensions. Imagine placing a particle in the middle of a grid that looks like a checkerboard. Describe a simulation study to investigate the movement of a particle on the grid when:

(a) the particle is equally likely to move to any of the 8 squares surrounding it;
(b) the particle can only move north, south, east, or west with equal probability;
(c) the particle can only move north, south, east, or west and the north and east directions are twice as likely as the south and west directions.

4.4.5. *Streaky Behavior in Sports.*

(a) Describe how you would simulate the next 10 at bats for a baseball player with a batting average of 0.295.
(b) Describe how you would simulate the next 20 free throws for a basketball player with a successful free throw shooting percentage of 88%.
(c) Describe how you would simulate the number of one-putts in a round of golf (18 holes) for a golfer who averages 6 one-putts per round.

4.5 Expected Values and Standard Deviations for Random Variables

The expected value or mean of a random variable measures the center of a probability distribution in a manner similar to the way $\bar{x}$ measures the center of a collection of data.

Definition 4.9 If X is a discrete random variable assuming the values x_1, $x_2, \ldots, x_k$ with probabilities $p_1, p_2, \ldots, p_k$, respectively, then the **mean or expected value of X**, denoted by either $E(X)$ or μ_X (we will use both), is

$$E(X) = \mu_X = (x_1 \times p_1) + (x_2 \times p_2) + \cdots + (x_k \times p_k) = \sum_{i=1}^{k} (x_i \times p_i).$$

Notice the similarity between $\bar{x}$ and μ_X. Recall that in computing $\bar{x}$, each of the n observations has an equal weight of $1/n$. In a similar manner, μ_X is a weighted average of the possible values of X with each value being weighted by its probability of occurrence. (If the k values of X happen to be equally likely, then the computation of μ_X is identical to the computation of $\bar{x}$.)

Example 4.13. Expected Number of Reboots Consider the probability distribution for the number of reboots in an 8-h workday, as discussed in Example 4.7. To find the expected number of times the computer must be rebooted in an 8-h workday, we calculate.

$$E(X) = \mu_X = (0 \times .53) + (1 \times .22) + (2 \times .13) + (3 \times .07) + (4 \times .03) + (5 \times .02)$$
$$= .91.$$

Now the question is, how do we interpret $\mu_X = .91$? A common misinterpretation is to say that there is a 91% chance that we will have to reboot the computer in a given workday. In fact, we found in Example 4.7 that the probability we will have to reboot the computer at least once in a given workday is only .47, not .91. A correct interpretation of μ_X is that .91 is the center of the probability distribution for the number of reboots in an 8-h workday. However, another common interpretation of μ_X is based on the Law of Averages. It says that on average we expect to reboot the computer .91 times in an 8-h workday.

Definition 4.10 Let $\bar{x}$ be the average of n independent sample observations from the same probability distribution with mean μ_X. Then the **Law of Averages** states that $\lim_{n \to \infty} \bar{x} = \mu_X$; that is, as n gets larger, the sample mean, $\bar{x}$, converges (i.e., gets closer) to the mean, μ_X, of the probability distribution.

Example 4.14. Bernoulli Random Variable Consider a single Bernoulli trial with probability of success p, and let $X = 1$ or 0, depending on whether the trial is a success or failure, respectively, be the associated Bernoulli random variable. Then the expected value of X is given by

$$E(X) = \mu_X = 1(p) + 0(1 - p) = p.$$

Example 4.15. Geometric Random Variable Suppose that G has a geometric probability distribution with probability of success p. What is the expected number of Bernoulli trials that will be required to obtain the first success? Using Definition 4.9, we see that

$$E(G) = \mu_G = \sum_{g=1}^{\infty} g \times (1 - p)^{g-1} \times p.$$

Making use of properties of a geometric series (which is beyond the scope of this course), it can be shown that this yields the result

$$E(G) = \frac{1}{p}.$$

Although the proof of this result makes use of properties of a geometric series that are beyond the scope of this course, we want to focus on the intuitive nature of the result for Bernoulli trials. We saw in Example 4.14 that the average number of successes per trial is p, so that is it not too surprising that the average number of trials per success is $1/p$.

Knowing the mean μ_X provides us with information about the center of a probability distribution, but it does not tell us anything about the associated variability for the random variable X. To measure this variability, we form the deviation of each possible value for X from the expected value μ_X and weight the square of each deviation by its associated probability. Summing these weighted squared deviations leads us to the variance of the probability distribution.

Definition 4.11 If X is a discrete random variable assuming the values $x_1, x_2, \ldots, x_k$ with probabilities $p_1, p_2, \ldots, p_k$, respectively, then the **variance of X** is

$$Var(X) = \sigma_X^2 = \left[(x_1 - \mu_X)^2 \times p_1\right] + \left[(x_2 - \mu_X)^2 \times p_2\right] + \cdots + \left[(x_k - \mu_X)^2 \times p_k\right] = \sum_{i=1}^{k} (x_i - \mu_X)^2 \times p_i.$$

The **standard deviation of X**, σ_X, is then the square root of the variance, namely, $\sigma_X = \sqrt{\sigma_X^2}$.

Example 4.16. Variance and Standard Deviation of the Number of Reboots Consider the number of times, X, that a computer must be rebooted in an 8-h workday, as discussed in Example 4.7. In Example 4.13

we found that $\mu_X = .91$. To find the variance and standard deviation for X, we use Definition 4.11 directly and obtain

$$\begin{aligned}\sigma_X^2 &= \left[(0-.91)^2 \times .53\right] + \left[(1-.91)^2 \times .22\right] + \left[(2-.91)^2 \times .13\right] \\ &\quad + \left[(3-.91)^2 \times .07\right] + \left[(4-.91)^2 \times .03\right] + \left[(5-.91)^2 \times .02\right] \\ &= 1.5219.\end{aligned}$$

It follows that $\sigma_X = \sqrt{\sigma_X^2} = \sqrt{1.5219} = 1.2337.$

Example 4.17. Variance and Standard Deviation of a Bernoulli Variable - Consider a single Bernoulli trial with probability of success p, and let $X = 1$ or 0, depending on whether the trial is a success or failure, respectively, be the associated Bernoulli random variable. In Example 4.14 we found that $E(X) = \mu_X = p$, so that

$$\sigma_X^2 = p(1-p)^2 + (1-p)(0-p)^2 = p(1-p).$$

It follows that $\sigma_X = \sqrt{p(1-p)}$.

Finally, we note that linear transformations affect the expected value and standard deviation of random variables in the same way that they affect the descriptive measures of center and spread we discussed in Chap. 1. In Exercises 4.A.8 and 4.A.9 you will be asked to verify that

$$\mu_{(a+bX)} = E(a+bX) = a + b\mu_X$$

and

$$\sigma_{(a+bX)}^2 = Var(a+bX) = b^2\sigma_X^2,$$

for all possible values of a and b.

Section 4.5 Practice Exercises

4.5.1. Find the expected number of patients who will experience side effects for each of the probability distributions in Exercise 4.3.1.

4.5.2. Using the correct probability model for the group size for patrons at a restaurant in Exercise 4.3.5, find the expected group size. Interpret this value in language that the new hostess will understand. Is the expected group size the same as the most likely group size? Explain.

4.5.3. Let X be the outcome when a fair die is rolled once.

(a) Verify that the expected value of X is 3.5.

(b) Suppose that a friend offers to pay you back the \$10 they borrowed or the amount \$$(35/X)$ after rolling a fair die. Which choice has the higher expected payback? Explain.

4.5.4. Attorneys often must decide whether to charge a client a flat rate or take a percentage of the settlement. Suppose that an attorney has a case where she could charge \$7500 or take 25% of any settlement. The attorney believes that there is a 35% chance of losing the case. She also believes that, if they win the case, there is a 60% chance that the judge will award a settlement of \$50,000 and a 40% chance that the judge will award a settlement of \$100,000.

(a) Specify the probability distribution for the attorney's fees if she decides to take a percentage of the settlement. Don't forget that there are three possible values for her fees in this case.

(b) What is the expected fee if she decides to take a percentage of the settlement?

(c) Would you recommend that the attorney charge a flat rate or take a percentage of the settlement?

4.5.5. *Dungeons and Dragons.* In Dungeons and Dragons, a character's intelligence is assigned by rolling three balanced dice. If X is the total number of dots on the upward facing sides of the three dice, find:

(a) the expected value of X;

(b) the variance of X;

(c) the standard deviation of X.

4.5.6. A firm estimates its future profit (with loss expressed as negative profit) through the probability distribution provided below.

Profit (P) in millions of $	−1	0	1	2	3
Probability	0.1	0.1	0.2	0.4	0.2

(a) Find the probability that the firm's profit is at least $2 million.

(b) Find the mean and standard deviation for the profit.

(c) The firm does not retain all of its profit because it will pay $1 million in dividends to shareholders and the executives will share 10% of the profit as a bonus (or hardship if the profit is negative). The amount the firm retains is $Y = 0.9P - 1$ (in million of dollars). Find the mean and standard deviation of Y.

4.5.7. *Keno.* Consider the game of Keno where balls numbered from 1 to 100 are placed in a cylinder. Players bet $1 on a number of their choice and the payoff is $3 if that number is one of the 25 randomly selected balls (without replacement between draws). Assume that you have been hired to do a few calculations for a casino.

(a) What is the probability distribution for the payoff?

(b) What is the mean or expected payoff?

(c) Interpret this expected payoff in language that the manager of the casino would understand. Would you recommend that they add this game at the casino?

(d) What is the variance of the payoff?

(e) What is the standard deviation of the payoff?

4.5.8. *Keno*. Consider the game of Keno described in Exercise 4.5.7, but now we want to view the game from the gambler's perspective. The gambler's winnings (W) can be written as a linear transformation of the payoff (P), namely, $W = P - 1$.

(a) Find the expected value of the gambler's winnings, $E(W)$.

(b) Explain the relationship between $E(W)$ and $E(P)$.

(c) Find the standard deviation of the gambler's winnings.

4.6 Combining Random Variables

In Chap. 1 we pointed out how simple linear transformations of one variable can be very useful in many settings. In this section we extend that discussion to include situations where we are combining several variables, typically by summing or averaging. For example, a farmer planting corn on six plots of land at six different locations would be interested in the total yield from his farm (all six plots of planted land) during a particular growing season. If a teacher wishes to assess a new instructional method using an instrument designed to measure the performance of the students, the teacher may be interested in comparing class averages as well as individual scores. With these types of problems in mind, we consider properties of sums and averages of random variables.

Expected Value of a Sum If $T = \sum_{i=1}^{n} X_i$ is the sum of n random variables $X_1, X_2, \ldots, X_n$, then the **expected sum** is

$$\mu_T = E\left(\sum_{i=1}^{n} X_i\right) = \sum_{i=1}^{n} E(X_i) = \sum_{i=1}^{n} \mu_{X_i}.$$

In words, the expected value (mean) of a sum is the sum of the expected values (means).

Example 4.18. How many successes, B, do we expect to obtain in n independent Bernoulli trials with common probability of success p? Since B is the sum of n independent Bernoulli trials $X_1, X_2, \ldots, X_n$, we can use our result about the expected value of a sum and the fact that the mean for a Bernoulli random variable is p (see Example 4.14) to conclude that

$$E(B) = E\left(\sum_{i=1}^{n} X_i\right) = \sum_{i=1}^{n} E(X_i) = n \times p.$$

Since $B \sim B(n, p)$, we now know that the expected value for a $B(n, p)$ random variable is $\mu_B = np$.

The result in Example 4.18 for Bernoulli trials can be generalized to the sum, T, of **any** n independent and identically distributed (i.i.d.) random variables. If $X_1, X_2, \ldots, X_n$ are i.i.d. random variables and $T = \sum_{i=1}^{n} X_i$, then

$$\mu_T = E\left(\sum_{i=1}^{n} X_i\right) = n \times E(X_1).$$

Variance of a Sum of Independent Random Variables If $T = \sum_{i=1}^{n} X_i$ is the sum of n **independent** random variables $X_1, X_2, \ldots, X_n$ with variances $\sigma^2_{X_1}, \ldots, \sigma^2_{X_n}$, respectively, then the **variance of the sum** is.

$$\sigma^2_T = Var\left(\sum_{i=1}^{n} X_i\right) = \sum_{i=1}^{n} Var(X_i) = \sum_{i=1}^{n} \sigma^2_{X_i}.$$

In words, the variance of the sum is the sum of the variances. Note that this is not the case if the random variables are dependent.

Example 4.19. Binomial Random Variable What is the variance of the number of successes, B, in n independent Bernoulli trials with constant probability of success p? Recall that B can be viewed as the sum of n independent Bernoulli trials $X_1, X_2, \ldots, X_n$, so that we have

$$\sigma_B^2 = Var\left(\sum_{i=1}^{n} X_i\right) = \sum_{i=1}^{n} Var(X_i) = n \times p \times (1-p).$$

The result in Example 4.19 for Bernoulli trials can be generalized to the sum, T, of **any** n independent and identically distributed (i.i.d.) random variables. If $X_1, X_2, \ldots, X_n$ are i.i.d. random variables and $T = \sum_{i=1}^{n} X_i$, then

$$\sigma_T^2 = Var\left(\sum_{i=1}^{n} X_i\right) = n \times Var(X_1).$$

Example 4.20. Collector's Problem—A Mathematical Approach We return to the collector's problem discussed in Example 4.12 and apply some of our new results so that we can compare the simulation estimates with theoretical results. Let T_n denote the number of cereal boxes required to get a complete set of n prizes. We are interested in finding the expected value of T_n, say $\mu_n = E[T_n]$. Let X_i denote the number of additional boxes required to get the ith new prize once $(i - 1)$ different prizes have already been obtained. Then we can express T_n as

$$T_n = \sum_{i=1}^{n} X_i \qquad (4.2)$$

The collector gets one of the n prizes in the first box, so $X_1 = 1$. Each subsequent box contains a different prize with probability $p = (n - 1)/n$ and the same prize with probability $1/n$. Thus, X_2 is a geometric random variable with parameter $p = (n - 1)/n$. Using the property of geometric random variables obtained in Example 4.15, we find that $E[X_2] = 1/p = n/(n-1)$. Once two different prizes have been obtained, each subsequent box contains a new prize with probability $p = (n-2)/n$ and one of the same prizes already obtained with probability $2/n$. Thus, X_3 has a geometric distribution with parameter $p = (n-2)/n$ and $E[X_3] = n/(n-2)$. We can continue this process until we have collected $(n-1)$ different prizes. Once $(n-1)$ different prizes have been obtained, each subsequent box contains the final new prize with probability $p = 1/n$ and X_n has a geometric distribution with parameter $p = 1/n$ and $E[X_n] = n$.

Using the property for the expected value of a sum of random variables, we find that the expected value (long run average) for the number of boxes required to get the complete set of n prizes is

$$\begin{aligned}\mu_n &= E[T_n] = E\left[\sum_{i=1}^{n} X_i\right] = 1 + \frac{n}{n-1} + \frac{n}{n-2} + \cdots + \frac{n}{2} + n \\ &= n\left(\frac{1}{n} + \frac{1}{n-1} + \cdots + \frac{1}{2} + 1\right).\end{aligned}$$

For $n = 6$, for example, we find $\mu_6 = 14.7$. Now we can compare the mathematical result $E[T_6] = 14.7$ with the simulation estimate we obtained in Sect. 4.4 for 35 collectors. Recall that we obtained a simulated average of 15.17 (not too bad!) and a simulated median of 14. The fact that the expected value is closer to the average than the median should be no surprise since the expected value of a random variable corresponds to the long run average.

Section 4.6 Practice Exercises

4.6.1. *Tossing Coins.* Use the **R** function *rbinom*( ) to simulate the tossing of 15 unbalanced coins with common probability 0.4 of getting a head.

(a) How many heads did you get?

(b) How many heads should you expect to get?

(c) Repeat this entire simulation 100 separate times and record the number of heads obtained for each simulation.

(d) Create a frequency histogram for your data.

4.6.2. *Tossing Different Coins.* Repeat Exercise 4.6.1 for unbalanced coins with common probability 0.1 of getting a head.

4.6.3. *Simulating Sales.* A salesperson has been very successful in getting new customers. If she can get people to watch her demonstration of the product, she has an 85% chance of completing the sale.

(a) Describe how to simulate the total number of sales for the next 20 people who watch her demonstration.

(b) Carry out your simulation and record the total number of sales.

(c) Repeat part (b) 100 times. On average, how many sales did she make? How many should she have expected to make?

4.6.4. *Expected Keno Payoff.* Suppose that a gambler makes a $1 bet on two separate games of Keno, as described in Exercises 4.5.7 and 4.5.8. Let T denote the total payoff after the two independent games are completed.

(a) What is the expected total payoff, $E(T)$?

(b) What is the variance of the total payoff?

(c) What is the standard deviation of the total payoff?

(d) If you add the standard deviations of the payoff for each game, do you get the standard deviation for the total payoff in part d? Why or why not?

4.7 Normal Distributions

Normal distributions are the most important and widely used probability distributions in statistics. They are also referred to as bell curves because of their general shape, as illustrated in Figs. 4.7, 4.8, and 4.9. Notice that in each case the distribution is single-peaked and symmetric about a central point.

Fig. 4.7 A normal density curve centered at 0

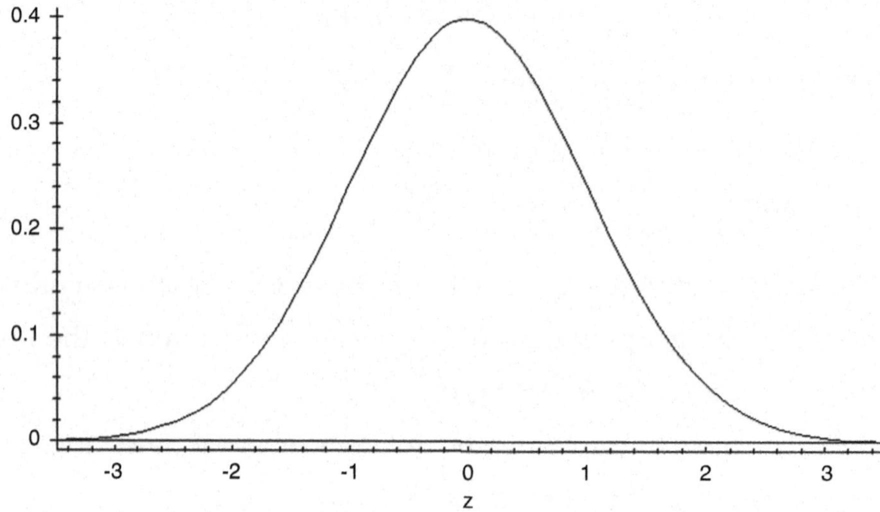

Fig. 4.8 A normal density curve centered at −10

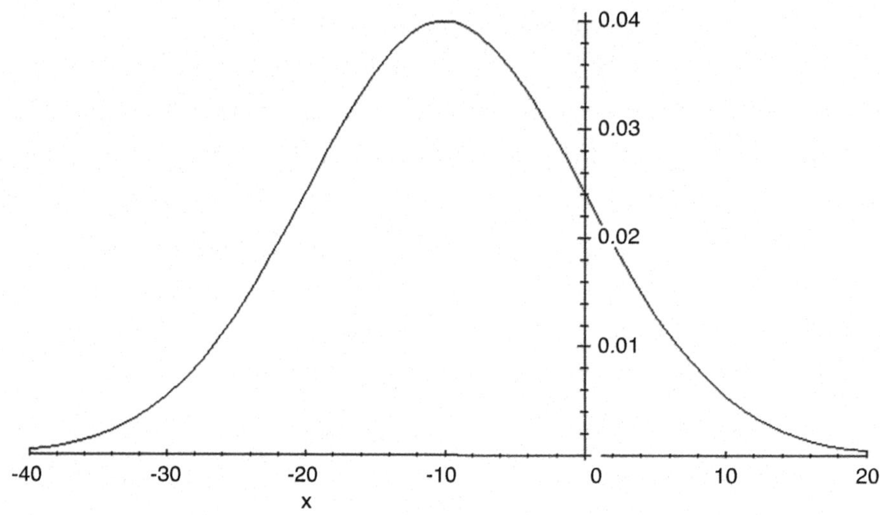

The point of symmetry is referred to as the visual center or mean of the distribution. Although the visual center changes from 0 in Fig. 4.7 to −10 in Fig. 4.8 and to 100 in Fig. 4.9, all normal distributions have the same general shape. The two characteristics that change for different normal distributions are the visual center, which we will denote by μ, and the standard deviation (or spread), which we will denote by σ.

Fig. 4.9 Two normal density curves centered at 100

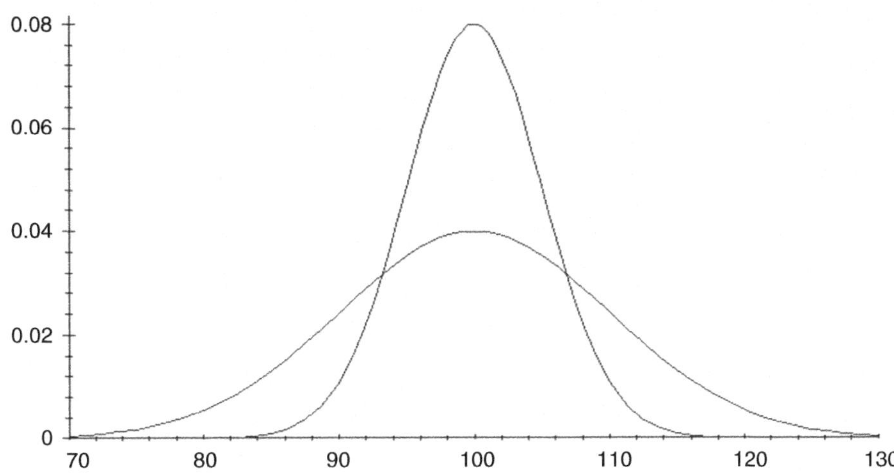

A normal density curve is a function of μ and σ. Once μ and σ are specified, the normal density curve is completely determined.

> **Definition 4.12** A normal density curve is completely determined by μ and σ. The precise mathematical function for the density is $f(x) = \frac{1}{\sqrt{2\pi\sigma^2}} e^{-\frac{(x-\mu)^2}{2\sigma^2}}$. Although we will not work directly with this function, we will use the shorthand notation $N(\mu, \sigma)$ to identify a normal distribution with mean μ and standard deviation σ.

The two normal distributions in Fig. 4.9 have a common visual center at $\mu = 100$, but the spread in the two density curves is clearly different. One of the curves is a $N(100, 10)$ density curve and the other curve is a $N(100, 5)$ density curve. Can you correctly label the two curves?

Bell-shaped curves like the ones shown in Figs. 4.7, 4.8, and 4.9 are commonly used as models for probability distributions of data and to calculate or approximate probabilities of interest. To be useful as a realistic model these curves must satisfy certain constraints or conditions.

> **Definition 4.13** A **probability density curve** is a nonnegative function for which the area under the curve is equal to one. Density curves are often useful as models for data and areas under such curves correspond to relative frequencies. Thus, the probability of a region of interest can be obtained by simply computing the corresponding area under the density curve.

Figure 4.10 shows a probability density curve that is skewed toward higher values (we refer to this as skewed "to the right"). We might use a model like this for variables that are known to have unusually large possibilities without corresponding unusually small possibilities. To estimate the probability of getting an observation in the interval from 10 to 15, we would find the value of the shaded area for this model.

In the next section we focus on calculating probabilities for normal distributions.

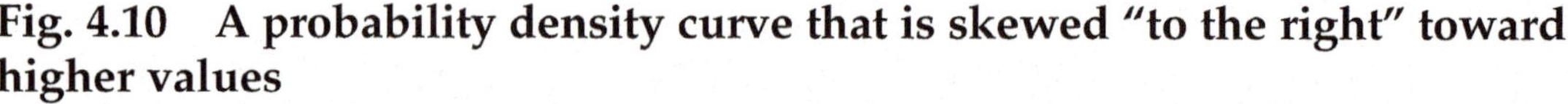

Fig. 4.10 A probability density curve that is skewed "to the right" toward higher values

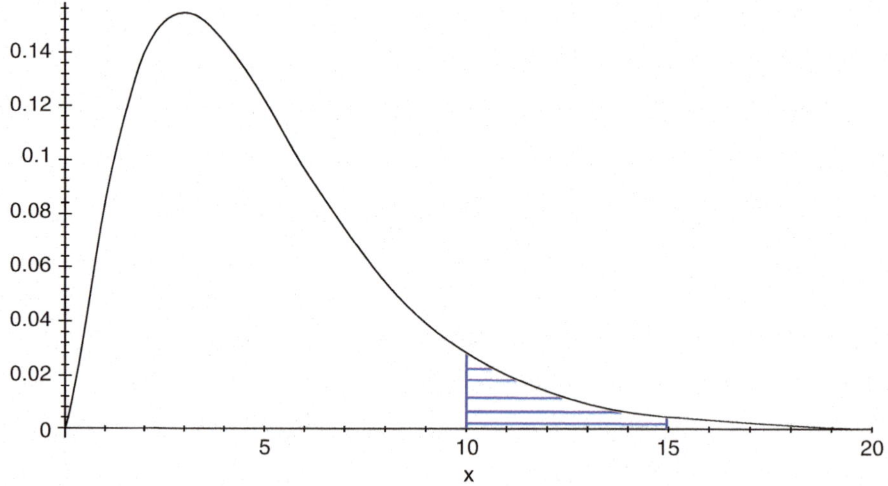

4.7.1 Probability Calculations for Normal Distributions

Fortunately, all normal distributions have the same general shape and we can use one special normal distribution, called the *standard normal distribution* and denoted by $N(0, 1)$, to calculate probabilities for ***any*** normal distribution. Thus we concentrate first on calculating probabilities for regions of the $N(0, 1)$ probability distribution. In every problem we suggest that you sketch the normal distribution and shade the appropriate region of interest before beginning your calculations. The basic types of calculations that will be useful for normal distributions are illustrated through a series of examples.

Example 4.21. Find the area to the left of -1 for the $N(0, 1)$ distribution.

Step 1. Sketch the $N(0, 1)$ distribution and shade the region of interest.

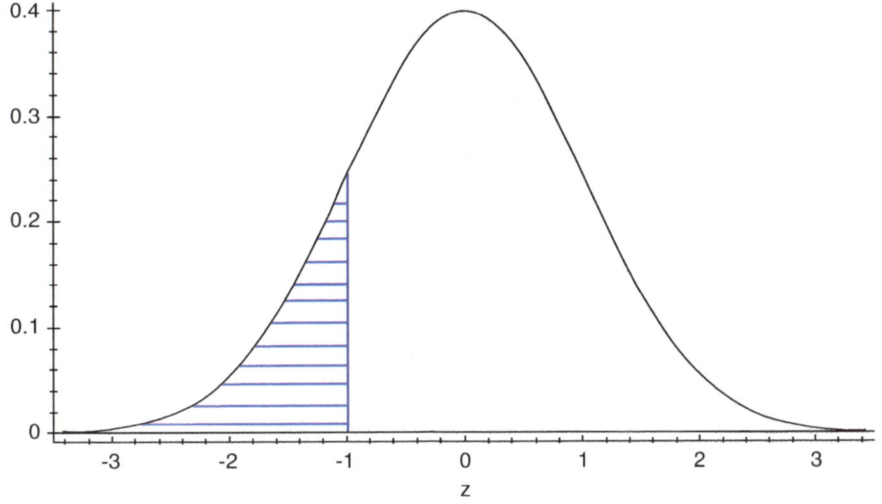

Step 2. Use the **R** function *pnorm*( ) (or any of the hundreds of online results you get when you search "normal distribution calculator") with the arguments q and *lower.tail* specified to be -1 and *TRUE*, respectively. (Note that *lower.tail* = *TRUE* is the default value, so the result would be the same if only q were specified, but it's good to get in the habit of explicitly specifying the region of interest for these calculations.)

```
> pnorm(q = -1, lower.tail = TRUE)
[1] 0.1586553
```

Notice that we are simply accumulating all of the area under the normal curve up to the point −1. This area can be viewed as a probability for the N(0,1) model. That is, if we let Z denote a random variable having the N(0,1) probability distribution, then the shaded area corresponds to $P(Z < -1)$ and is equal to roughly 0.1587. If we were using this curve to model a population, we would say that the proportion of observations below −1 is 0.1587.

Before moving to our next example, we want to formally introduce the idea of cumulating the area under a curve at or below a point. This is a common calculation of interest and leads to a new function that is quite useful in statistics.

> **Definition 4.14** The **cumulative distribution function (c.d.f.)** is a function that corresponds to the area (or probability) at or below a prescribed value. We use the notation $F(x) = P(X \leq x)$ to denote the c.d.f. of the random variable X evaluated at x.

We use $\Phi(z)$ to denote the c.d.f. of the standard normal distribution evaluated at z. As demonstrated in Example 4.21, the **R** function *pnorm*( ) can be used to provide values of $\Phi(z)$ by specifying the q argument to be equal to the z value of interest.

Example 4.22. Find the area under the $N(0, 1)$ curve between .6 and 1.4.

Step 1. Sketch the $N(0,1)$ curve and shade the region of interest.

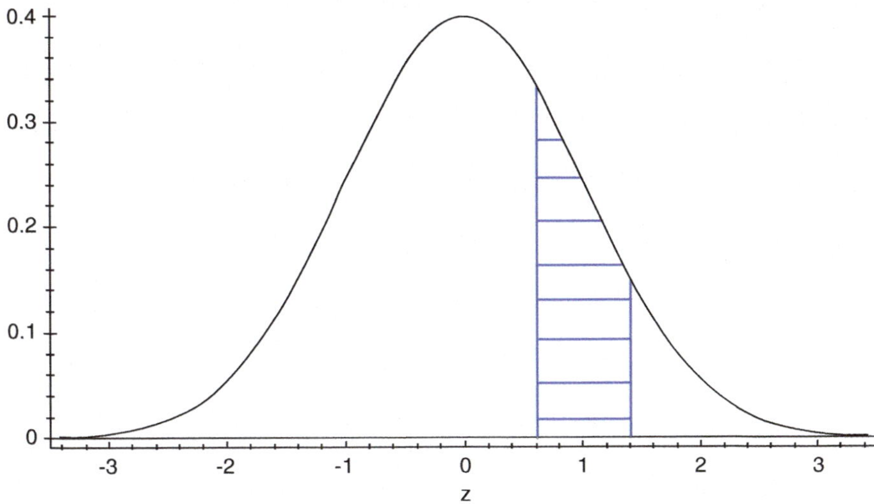

Step 2. Find the shaded area of interest by calculating the area to the left of 1.4 and then subtracting the area to the left of 0.6. This can be achieved by calling *pnorm*( ) for each of the two values and taking the difference.

```
> pnorm(q = 1.4, lower.tail = TRUE) - pnorm(q = 0.6, lower.tail = TRUE)
[1] 0.1934965
```

That is, the probability that a standard normal random variable will be between 0.6 and 1.4 is $\Phi(1.4) - \Phi(0.6){=}0.1934965$.

In general, the probability $P(a \leq Z \leq b)$ that the observed outcome of a standard normal random variable Z falls in the interval $[a, b]$ can be computed by finding the value of the c.d.f. at the larger number b and subtracting from it the value of the c.d.f. at the smaller number a, yielding $P(a \leq Z \leq b) = \Phi(b) - \Phi(a)$.

Example 4.23. Find the area under the $N(0, 1)$ curve above 1.4. In Example 4.22, the first call to *pnorm*( ) gives us $P(Z \leq 1.4)$. We can either use the complement rule to find the area above 1.4 to be $P(Z > 1.4) = 1 - P(Z \leq 1.4)$ or we can specify the *lower.tail* option to be *FALSE*. Both methods lead us to the result that $P(Z > 1.4) = 0.0808$, as demonstrated below.

```
> 1 - pnorm(q = 1.4, lower.tail = TRUE)
[1] 0.08075666

> pnorm(q = 1.4, lower.tail = FALSE)
[1] 0.08075666
```

In general, the area to the right of a number z for the $N(0, 1)$ curve is equal to $1 - \Phi(z)$. Using the symmetry of the $N(0, 1)$ distribution, we also notice that the area to the right of any number z is equal to the area to the left of $-z$. Notationally, we have $[1 - \Phi(z)] = \Phi(-z)$. You will likely find this relationship to be most useful when you have already calculated one of these areas and are asked about the other, say, for example in the exercises at the end of this chapter.

Example 4.24. Find the value z for the standard normal random variable Z that has area .25 above it.

Step 1. Sketch the $N(0, 1)$ distribution and identify the area of interest.

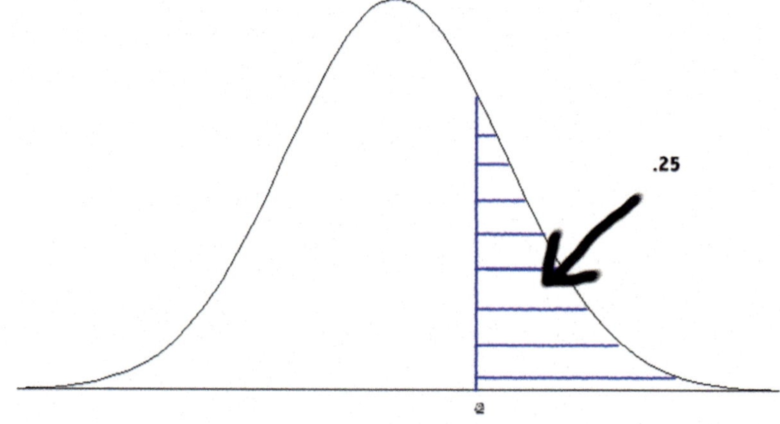

Step 2. Here instead of calculating the area, we are given the area and are asked to find the value of z that corresponds to it. This is the inverse (or reverse) of the problem that we have previously been solving and so it is natural to use the inverse of the cumulative distribution function to find z. Since the c.d.f. $\Phi(z)$ corresponds to the area to the left of z, we rewrite our problem in terms of the un-shaded area (0.75). Thus we want to find the value

of z such that $\Phi(z) = 0.75$. The solution is $z = \Phi^{-1}(0.75)$. We can do this using the *qnorm*( ) function in either of the following two ways. (Does it make sense that they produce the same answer?)

```
> qnorm(p = 0.75, lower.tail = TRUE)
[1] 0.6744898
> qnorm(p = 0.25, lower.tail = FALSE)
[1] 0.6744898
```

Both methods tell us that z = 0.6745. Even though the notation is different, notice that we have just calculated the third quartile, Q3, for the N(0,1) distribution.

To calculate probabilities or percentiles for a random variable X that has an arbitrary $N(\mu, \sigma)$ probability distribution, we can make use of the standardizing transformation $Z = \frac{X-\mu}{\sigma}$. In Exercise 4.A.10 you will be asked to use properties of expected value to show that $E[Z] = 0$ and $Var[Z] = 1$. It can also be shown that the standardized variable Z will always have a standard normal probability distribution.

Example 4.25. Suppose that the sum of exam scores for a large class roughly follows the $N(275, 43)$ distribution. What is the probability that a randomly selected student from the class will have an exam total above 200?

Step 1. Sketch the $N(275, 43)$ distribution and identify the problem of interest.

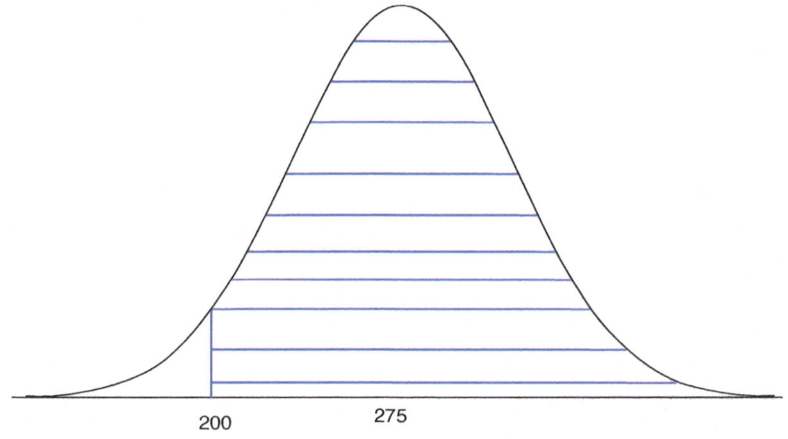

Step 2. There are several different ways to calculate the shaded area of interest. One of the easiest ways is to find the value of the c.d.f. at 200 and then subtract that value from 1 (which **R** will do for us automatically if we specify *lower.tail* to be *FALSE*). That is, if we let S denote the exam total of the randomly selected student, then we can use the **R** function *pnorm*( ) similar to what we did in Exercise 4.23, but now with the *mean* and *sd* arguments specified as 275 and 43, respectively, to find that $P(S > 200) = 0.9594$.

```
> pnorm(q = 200, mean = 275, sd = 43, lower.tail = FALSE)
[1] 0.9594367
```

Alternatively, we can use the standardizing transformation $z = \frac{200-275}{43} \approx -1.7442$ and compute the probability that a $N(0, 1)$ random variable is greater than this standardized value. We store the standardized value as the variable *z_value* and then use the **R** function *pnorm*( ) to get $P(Z > z\_value) = P(S > 200) = 0.9594$.

```
> z_value <- (200 - 275) / 43
> pnorm(z_value, lower.tail = FALSE)
[1] 0.9594367
```

Using the standardization technique may be necessary sometimes, for example, in a classroom setting where access to software is restricted. Be wary of rounding errors that may creep into your calculations, though!

Example 4.26. Suppose that the sum of exam scores for a large class roughly follows the $N(430, 100)$ distribution. How high must a student score on her exams to be in the top 5%?

Step 1. Sketch the $N(430, 100)$ distribution and identify the problem of interest.

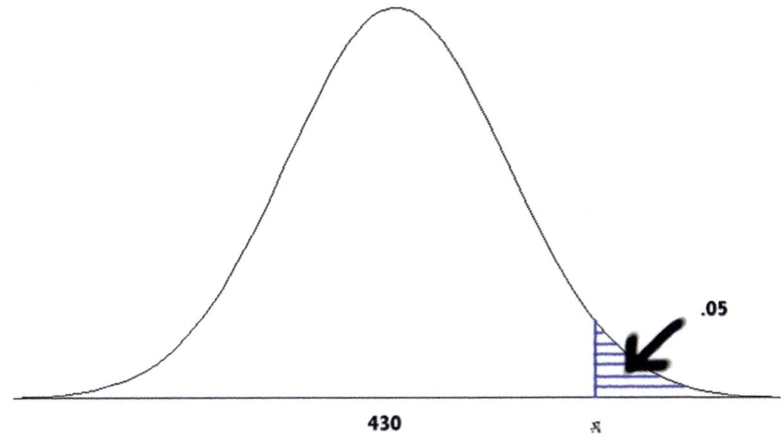

Step 2. Similar to Example 4.24, here we are given the area and we need to find the value of x that has area .95 under the $N(430, 100)$ probability density curve to the left of x. In other words, we need to find the value of the inverse cumulative distribution function for the $N(430, 100)$ distribution at $.1-.05 = .95$. We proceed very similarly to Example 4.24, except with the argument p updated to be the new percentile of interest and now specifying the *mean* and *sd* arguments. As we saw in Example 4.24, the two following commands will give us the same result: a student must score above 594.4854 to be in the top 5%.

```
> qnorm(p = 0.95, mean = 430, sd = 100, lower.tail = TRUE)
[1] 594.4854
> qnorm(p = 0.05, mean = 430, sd = 100, lower.tail = FALSE)
[1] 594.4854
```

Note that we can also use the standardization technique shown in Example 4.25 and a bit of algebra to arrive at the same result. We first compute and store the 95th percentile of the $N(0,1)$ distribution as the variable *z_95* using the following command.

```
> z_95 <- qnorm(p = 0.05, lower.tail = FALSE)
```

We then set $(x - 430)/100$ equal to *z_95* and solve for x. Again, we find that the 95th percentile for the sum of the exam scores is 594.4854.

```
> 100 * z_95 + 430
[1] 594.4854
```

4.7.2 Using Normal Distributions as Models for Measurements

Normal distributions can be very useful in providing estimates of probabilities and making inferences for populations for many settings. However, if the normal distribution is inappropriate as a model for a given set of measurements, then the associated estimates and inferences can be useless. To help decide whether or not the normal distribution is appropriate for a given data collection, we recommend two quick checks.

The 1–2–3 standard deviation rule can be used to compare the observed percentages of observations within 1, 2, and 3 standard deviations of the sample mean with the expected percentages of observations within 1, 2, and 3 standard deviations of the population mean for a normal model.

Definition 4.15 The **1–2-3 standard deviation rule** states that for any normal model, approximately 68% of the observations will fall within one standard deviation of the mean, approximately 95% of the observations will fall within two standard deviations of the mean, and approximately 99.7% of the observations will fall within three standard deviations of the mean. Thus, a crude check of normality for a data collection can be obtained by computing the percentages of observed values that fall in the intervals, $(\bar{x} - s, \bar{x} + s)$, $(\bar{x} - 2s, \bar{x} + 2s)$, and $(\bar{x} - 3s, \bar{x} + 3s)$, where $\bar{x}$ and s are the sample mean and standard deviation, respectively, for the data collection, and then comparing these percentages with 68%, 95%, and 99.7%, respectively.

A second approach is our recommended, more rigorous, method for checking normality. A normal probability plot is constructed to compare the empirical percentiles of the data collection with the theoretical percentiles for the normal model. The empirical percentiles for a collection of n data observations are the values of the order statistics $X_{(1)}, X_{(2)}, \ldots, X_{(n)}$. Loosely speaking, the corresponding theoretical percentiles are the expected values of the n order statistics for a random sample of size n from a standard normal model. We can generate these percentiles by using the **R** function *ppoints*( ) and properly specifying the argument n. Alternatively, we can use the **R** function *qqnorm*( ) to automatically generate these theoretical percentiles and plot them along with the observed percentiles. A linear trend in a normal probability plot indicates that the empirical percentiles are matching up well with the corresponding theoretical percentiles and therefore it is not unreasonable to use a normal distribution as a model for these measurements. On the other hand, curvature in a normal probability plot indicates that there are some discrepancies in the comparisons of the empirical and theoretical percentiles, indicating that the normal model may not be appropriate.

A major problem with this approach is that interpreting probability plots can be rather subjective. How do we know what to look for in a probability plot? What types of curvature should we worry about? Will the plot be perfectly straight if the data are normal? Good judgement comes with experience so we will use simulation to help develop your understanding of normal probability plots. We will generate 500 random observations from three different distributions: the $N(0,1)$ distribution, the Uniform(-3,3) distribution (whose density curve is constant at $1/6$ from -3 to 3 and zero elsewhere), and the t-distribution with 1 degree of freedom (also known as the Cauchy distribution), whose pdf at first glance may look similar to the pdf of a normal distribution, but in fact has much larger amounts of probability assigned to the tails. The following three lines generate the three sets of values.

```
> normal_sample <- rnorm(n = 500)
> uniform_sample <- runif(n = 500, min = -3, max = 3)
> t_sample <- rt(n = 500, df = 1)
```

Fig. 4.11 Normal probability plot for N(0, 1) data

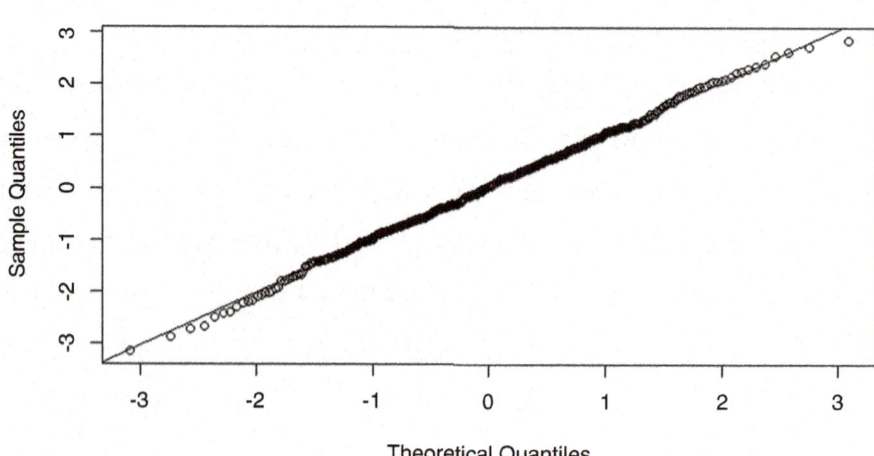

We now use the **R** function *qqnorm*( ) to generate the plots discussed earlier for each of the three samples. To help make the patterns more discernible, we also use the **R** function *qqline*( ) to add the line y = x to the plots.

First, we have the normal probability plot for the $N(0,1)$ data shown in Fig. 4.11. Although the plot shows some slight departures from a linear trend, they are nothing to be concerned about. We are not looking for a perfectly straight line, just no systematic departures from an overall linear trend.

```
> qqnorm(normal_sample)
> qqline(normal_sample)
```

The normal probability plot in Fig. 4.12 for the Uniform(−3, 3) data clearly shows some systematic departures from linearity in the two tails. Notice the sharp turn downward in the lower tail and the sharp turn upward in the upper tail. These departures from linearity in the tails of the normal probability plot are evident because the Uniform(−3, 3) distribution does not have a smooth decline in the tails like a normal distribution. As mentioned earlier, its density curve drops to zero at −3 and 3.

```
> qqnorm(uniform_sample)
> qqline(uniform_sample
```

Fig. 4.12 Normal probability plot for Uniform(−3, 3) data

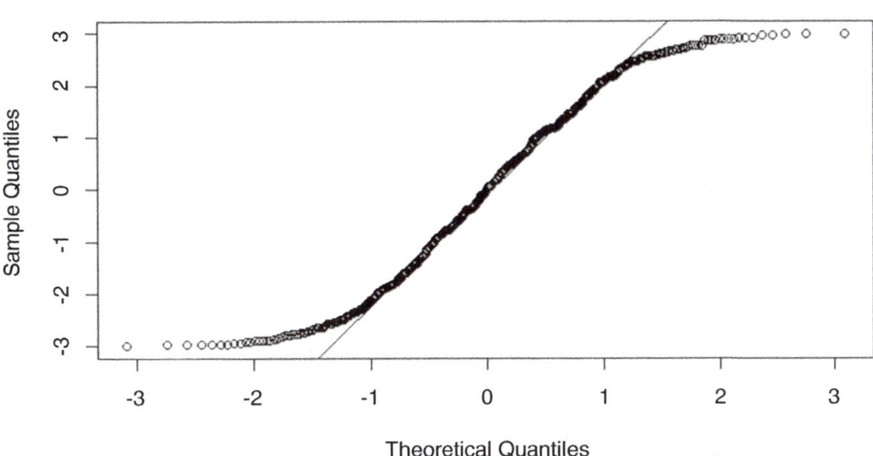

Fig. 4.13 Normal probability plot for t(1) data

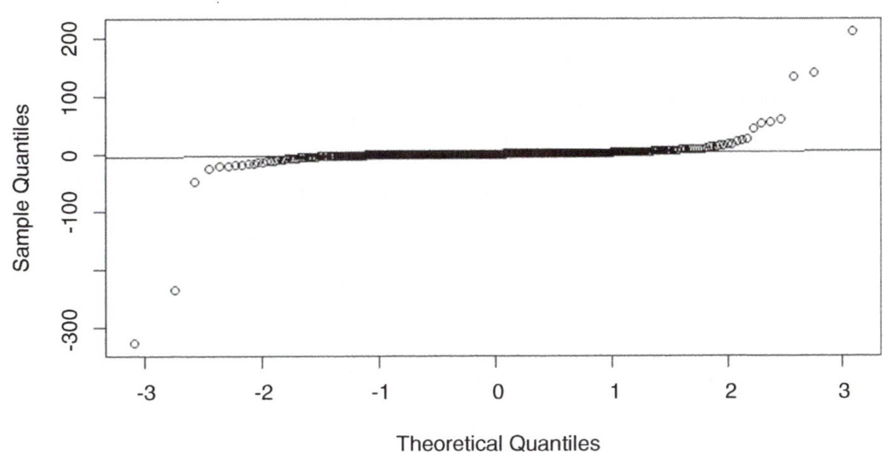

The probability plot for the t(1) data in Fig. 4.13 also shows some clear departures from linearity in the tails. The S-shape indicates that we observed several values that are much smaller than we would expect from a normal model, as well as several values that are much larger than we would expect. The only major difference between the t(1) density curve and a normal density curve is the aforementioned heavier tails of the t(1) distribution.

```
> qqnorm(t_sample)
> qqline(t_sample)
```

Section 4.7 Practice Exercises

4.7.1. Suppose that Z has a standard normal distribution. Find the following probabilities:

(a) $P(Z < 1.5)$;

(b) $P(Z > 1.5)$;

(c) $P(0 < Z < 1.5)$;

(d) $P(-1.5 < Z < 1.5)$.

4.7.2. The 1–2-3 standard deviation rule is also commonly referred to as the 68–95-99.7 rule. To understand why, calculate the area under the standard normal curve for the following intervals:

(a) $(-1, 1)$;

(b) $(-2, 2)$;

(c) $(-3, 3)$.

4.7.3. Find the following percentiles for the standard normal distribution.

(a) 10th

(b) 25th

(c) 50th

(d) 75th

(e) 90th

(f) Explain the relationship between the 10th and 90th and 25th and 75th percentiles. Identify the general form of this relationship for an arbitrary percentile, say z_p, for p between 0 and 100 and explain why it holds.

(g) What is the value of the interquartile range (IQR) for the standard normal model?

4.7.4. Suppose that X follows the normal distribution with mean 85 and standard deviation 5; that is, $X \sim N(85, 5)$. Find:

(a) $P(X < 78)$;

(b) $P(X > 88)$;

(c) $P(78 < X < 88)$;

(d) the value $x_{0.8}$ such that$P(X \leq x_{0.8}) = 0.8$;

(e) the value $x_{0.05}$ such that$P(X \leq x_{0.05}) = 0.05$.

4.7.5. Suppose the time taken by a computer on a local area network to establish a connection to a remote site follows the normal distribution with a mean of 15 s and a standard deviation of 3 s.

(a) What proportion of the connections will occur in less than 12.2 s?

(b) What proportion of the connections will occur between 12.2 and 16.5 s?

(c) What proportion of the connections will take longer than 10 s?

(d) 90% of the connections will occur in less than how many seconds?

(e) 75% of the connections will take longer than how many seconds?

4.7.6. Suppose the wrapper of a candy bar lists its weight as 8 ounces. The actual weights, however, of individual candy bars naturally vary to some extent. Suppose that these actual weights vary according to the normal distribution with mean $\mu = 8.3$ ounces and standard deviation $\sigma = 0.125$ ounces.

(a) What proportion of the candy bars weigh less than the advertised 8 ounces?

(b) What proportion of the candy bars weigh more than 8.5 ounces?

(c) What proportion of the candy bars weigh between 8.1 and 8.4 ounces?

(d) What is the weight such that only 1 candy bar in 1000 weighs less than that amount?

(e) What is the weight such that 60% of the candy bars weigh more than this amount?

4.7.7. Suppose high school grade point averages (gpas) for a class of 2000 students are normally distributed with mean 3.3 and standard deviation 0.3.

(a) What percent of the students in the class have gpas below 2.7?

(b) How high must a student's gpa be for her to place in the top 5% of the class?

(c) What percent of the students in the class have gpas above 3.5?

(d) How low must a student's gpa be for her to place in the bottom 20% of the class?

(e) What percent of the students in the class have gpas between 2.85 and 3.65?

4.7.8. The annual sales revenues of a new product are normally distributed with a yearly mean of $45,000 and standard deviation of $3500. The cost of producing this product is $38,000.

(a) What is the chance that next year's revenues will cover the cost of production?

(b) Determine the revenues that mark the 10th and 90th percentiles.

4.7.9. Two companies are bidding on a contract to supply personal computers to a school district. The first company claims that the mean life of its computers is 1825 days with a standard deviation of 180 days. The second company claims that the mean life of its computers is 1770 days with a standard deviation of 90 days. Assuming no differences in cost and that computer life is normally distributed, which company would you recommend if the computers need to last at least 1645 days? Justify your choice with appropriate calculations.

4.7.10. An automobile manufacturer claims that the fuel consumption for a certain make and model of car should average 28 miles per gallon. After driving such a car for over 6 months (long enough to get through the typical break-in period), you notice that you are only getting 20 miles per gallon. During a phone conversation with a customer service representative for the company you are told that the standard deviation of the fuel consumption for your car is 3 miles per gallon. Assuming that the company's claims are valid, what is the chance that your car would perform as badly as it is or worse?

4.7.11. A soup company claims that one of its products contains 480 *mg* of sodium per serving with a standard deviation of 6 *mg*. Assuming that the amount of sodium in a particular serving varies according to the normal distribution, what is the chance that a particular serving contains:

(a) fewer than 475 *mg* of sodium;

(b) at most 490 *mg* of sodium;

(c) at least 475 *mg* of sodium;

(d) more than 490 *mg* of sodium.

(e) Find the interval of sodium levels that contains 80% of the values for the amount of sodium in a particular serving.

4.7.12. Use statistical software to generate 100 random numbers that are uniformly distributed over the following intervals.

(a) (0, 1)

(b) (−1, 1)

(c) (0, 5)

(d) (90, 100)

Construct normal probability plots for each set of random numbers. Do the plots show the same overall pattern as Fig. 4.11? Construct histograms for each set of random numbers. Describe the patterns in these histograms.

4.7.13. *Are 2014 Major League Baseball Players "Normal"?* The datasets *american_league_salary_2014* and *national_league_salary_2014* contain the 2014 salaries (as of March 26, 2014) for all baseball players in the American and National Leagues, respectively. Construct normal probability plots for each dataset. Would you be willing to use the normal model for salaries of either American or National League baseball players? Why or why not?

4.7.14. *Are Pine Tree Heights Normally Distributed?* The dataset *pines_1997* contains a subset of the data on pine trees collected between 1990 and 1997 by biology students at the Kenyon College Environmental Center. A complete description of this experiment can be found in Chap. 3.

(a) Construct a normal probability plot for the initial planting heights (hgt90) of the trees. Are there any obvious departures from a linear trend on the plot?

(b) Construct a histogram for the initial planting heights (hgt90) of the trees.

(c) Based on your plots from (a) and (b), would you be willing to use the normal model for these initial tree heights?

(d) Repeat parts (a), (b), and (c) for the tree heights in 1997 (hgt97).

(e) Repeat parts (a), (b), and (c) for the tree diameters in 1997 (diam97).

4.7.15. A group of 25 students were asked how much they paid for their last haircut. The responses were: 20, 18, 25, 25, 0, 20, 60, 0, 20, 10, 10, 20, 40, 26, 40, 12, 16, 16, 36, 38, 21, 15, 13, 10, and 10.

(a) Find the percentage of responses that fall within one standard deviation of the mean.

(b) Find the percentage of responses that fall within two standard deviations of the mean.

(c) Find the percentage of responses that fall within three standard deviations of the mean.

(d) Based on the 1–2–3 standard deviation rule, would you be willing to use the normal model for these haircut prices? Why or why not?

(e) Construct a normal probability plot for the haircut prices and comment on whether or not you would be willing to use the normal model based on the pattern in this plot.

Chapter 4 Comprehensive Exercises

4.A. Conceptual

4.A.1. Is the $B(n, ½)$ probability distribution always symmetric (regardless of the number of trials)? Does this make intuitive sense? Explain.

4.A.2. What happens to the $B(n, p)$ probability distribution as the number of trials n increases and the probability of success p remains fixed?

4.A.3. What happens to the $B(n, p)$ probability distribution as the probability of success p gets closer to zero or one and the number of trials n remains fixed?

4.A.4. Suppose five patients are given an experimental medication with potential side effects to treat an illness. The pharmaceutical manufacturer estimates that 50% of the patients who take this medication will experience side effects. Use this information to determine the probability distribution for the total number of patients (out of five) who take this medication and experience side effects. How would the probability distribution change if the chance of experiencing serious side effects with use of the medication increased to .75?

4.A.5. Describe several disjoint events. Describe several complementary events. Describe several events that are disjoint but not complementary.

4.A.6. Display a Venn diagram for an experiment with three events, say A, B, and C, that are not mutually disjoint. Carefully label the events A, B, C, $A \cap B$, $A \cap C$, $B \cap C$, and $A \cap B \cap C$ on the Venn diagram. Describe how the probabilities of these events should be combined to form an addition rule for $m = 3$ events. Repeat this exercise for $m = 4$ events. Describe a general addition rule that can be used for an arbitrary number (m) of events.

4.A.7. Developing a general multiplication rule is an iterative process. To find $P(A_1 \cap A_2 \cap A_3)$ we first condition on the event $A_1 \cap A_2$ and then apply the multiplication rule with $A = A_3$ and $B = A_1 \cap A_2$. After the first application of the multiplication rule we have $P(A_1 \cap A_2 \cap A_3) = P(A_3 | A_1 \cap A_2) \times P(A_1 \cap A_2)$. Now, we apply the multiplication rule again to the second term on the right hand side of our equation to obtain $P(A_1 \cap A_2 \cap A_3) = P(A_3 | A_1 \cap A_2) \times P(A_2 | A_1) \times P(A_1)$.

(a) Apply the multiplication rule three times to derive a multiplication rule for $m = 4$ events.

(b) Apply the multiplication rule (m - 1) times to derive a general multiplication rule for m events.

4.A.8. Verify that the expected value of a linear transformation of a random variable is equal to the value obtained by applying the linear transformation to the expected value of the random variable. In other words, show that $\mu_{(a + bX)} = E(a + bX) = a + b\mu_X$, for all possible values of a and b.

4.A.9. Verify that the variance of a linear transformation of a random variable is equal to the square of the slope coefficient times the variance of the random variable. In other words, show that $\sigma^2_{(a+bX)} = Var(a + bX) = b^2\sigma^2_X$, for all possible values of a and b. Carefully explain why the value of the intercept a does not play a role in computing the variance of the linear transformation of a random variable.

4.A.10. If $X \sim N(\mu, \sigma)$ and the standardizing transformation is applied to create $Z = \frac{X-\mu}{\sigma}$, show that:

(a) Z is a linear transformation of X by identifying the slope and intercept constants;

(b) $E(Z) = 0$ and $Var(Z) = 1$. [Hint: Recall how linear transformations affect the mean and variance (and, hence, the standard deviation).]

4.B. Data Analysis/ Computational

4.B.1. Using the **R** function *table*( ) on the "Cover95" and "Deer95" columns of the *pines_1997* dataset produces the following contingency table. (The *dnn* argument is used only to label the dimensions of the table for clarity and can be any vector of strings.) Use this output to answer the following questions.

```
> table(pines_1997$Cover95, pines_1997$Deer95, dnn = c("Cover95", "Deer95"))

          Deer95
Cover95     0   1
        0 151  60
        1 158  76
        2 177  44
        3 176  29
```

(a) Estimate the overall probability that a randomly selected tree was browsed by deer in 1995.

(b) Consider the amount of thorny cover on the tree ($0 =$ no cover, $1 =$ less than $1/3$, $2 =$ between $1/3$ and $2/3$, and $3 =$ greater than $2/3$). Separately for each different level of thorny cover, estimate the probability that deer browsed a randomly selected tree in 1995. Does knowledge about the amount of thorny cover on the tree affect your estimate of the probability?

(c) Thorny vegetation in contact with the pine trees was removed during the summer of 1996. Use the variable "Deer97" to estimate the probability that deer browsed a randomly selected tree in 1997. How does this estimate compare with your estimates in parts (a) and (b)?

4.B.2. Describe how to solve the collector's problem with simulation when the set contains 12 different prizes. Can you repeatedly roll a pair of six-sided dice to carry out the simulation? Explain.

4.B.3. Use simulation to estimate the number of purchases required by a collector who is interested in obtaining a complete set of $n = 10$ prizes? Find $E[T_{10}]$ and compare it with your simulation estimate.

4.B.4. For large n, approximate values of $\mu_n = E[T_n]$ can be obtained using Euler's approximation for the harmonic series. That is,

$$1 + \frac{1}{2} + \frac{1}{3} + \cdots + \frac{1}{n} \approx \ln(n) + \gamma + \frac{1}{2n},$$

where $\gamma = 0.57721566490153286060\ldots$ is Euler's constant. Approximate μ_n for $n = 6$ and $n = 10$ and compare these approximations with the exact and simulated estimates obtained in Example 4.20 and Exercise 4.B.3, respectively.

4.B.5. Use the **R** functions *qqplot*( ) and *qqline*( ) to see if the normal distribution would be an appropriate model for the variables in the datasets below.

(a) 2014 salaries for American League baseball players (*american_league_salary_2014*)
(b) 2015 salaries for faculty and staff in the Mathematics Department at The Ohio State University (*osu_math_salaries_2015*)
(c) Unit values on TIAA and CREF variable annuities (*tiaa_cref*)
(d) Measurements on pine trees at the Kenyon College Environmental Center (*pines_1997*)
(e) Amount of money paid to the winner of the Kentucky Derby from 1990 to 2012 (*kentucky_derby_2012*).

4.B.6. *Romantic Relationships in High School—To Have or Not To Have.* The Pew Research Center (2015b) reported on topics related to teens and the use of technology in establishing or maintaining romantic relationships. In a national survey of 1060 teens ages 13–17, they found that 64% of the respondents had never been in a romantic relationship. Suppose we select a random sample of $n = 30$ teens ages 13–17 and ask them if they have ever been in a romantic relationship.

(a) What is the probability that more than 50% of our sample have never been in a romantic relationship?

(b) What is the probability that less than 10 students in our sample have been in a romantic relationship?

(c) How many of the students in our sample should we expect to have been in a romantic relationship?

4.B.7. *Romantic Relationships—How To Make Them Happen.* The Pew Research Center (2015b) reported on topics related to teens and the use of technology in establishing or maintaining romantic relationships. In a national survey of 1060 teens ages 13–17, they found that 50% of the respondents had showed a romantic interest in someone by sharing something funny/interesting with them online. Suppose we select a random sample of $n = 20$ teens ages 13–17 and ask them if they had ever showed a romantic interest in someone by sharing something funny/interesting with them online.

(a) How many of the students in our sample should we expect to have showed a romantic interest in someone by sharing something funny/interesting with them online?

(b) What is the probability that no more than 5 students in our sample have showed a romantic interest in someone by sharing something funny/interesting with them online?

(c) What is the probability that exactly half of the students in our sample have showed a romantic interest in someone by sharing something funny/interesting with them online?

4.B.8. *Romantic Relationships—Maintenance.* The Pew Research Center (2015b) reported on topics related to teens and the use of technology in establishing or maintaining romantic relationships. In a national survey of 1060 teens ages 13–17, they found that among those respondents who had been or are currently in some kind of romantic relationship, 11% expected to hear from their partner on an hourly basis! Suppose we select a random sample of $n = 40$ teens ages 13–17 who are currently in some kind of romantic relationship and ask them how often they expect to hear from their romantic partners.

(a) How many of the students in our sample should we expect to say that they expected to hear from their romantic partners on an hourly basis?

(b) What is the probability that no students in our sample expected to hear from their romantic partner on an hourly basis?

(c) What is the probability that more than 30 of the students in our sample expected to hear from their romantic partners on an hourly basis?

4.B.9. *Romantic Relationships—Control.* The Pew Research Center (2015b) reported on topics related to teens and the use of technology in establishing or maintaining romantic relationships. In a national survey of 1060 teens ages 13–17, they found that among those respondents who had been or are currently in some kind of romantic relationship, 26% said that their romantic partner checked up with them multiple times per day asking where they were, who they were with, or what they were doing. Suppose we select a random sample of $n = 40$ teens ages 13–17 who are currently in some kind of romantic relationship and ask them if their romantic partner exhibited this controlling behavior.

(a) How many of the students in our sample should we expect to say that their romantic partner behaved in this manner?

(b) What is the probability that no less than 30% of the students in our sample say that their romantic partner behaved in this manner?

(c) What is the probability that more than 70% of the students in our sample say that their romantic partner behaved in this manner?

4.B.10. *Romantic Relationships—After Shocks.* The Pew Research Center (2015b) reported on topics related to teens and the use of technology in establishing or maintaining romantic relationships. In a national survey of

1060 teens ages 13–17, they found that among those respondents who had previously been in a romantic relationship that had now ended, 8% said that their ex-romantic partner called them names, put them down or said really mean things to them on the Internet or their cellphone. Suppose we select a random sample of $n = 100$ teens ages 13–17 who had previously been in a romantic relationship that had now ended and ask them if their ex-romantic partner called them names, put them down or said really mean things to them on the Internet or their cellphone.

(a) What is the probability that exactly 8 of the students in our sample say that their ex-romantic partner called them names, put them down or said really mean things to them on the Internet or their cellphone?
(b) What is the probability that more than 8 of the students in our sample say that their ex-romantic partner called them names, put them down or said really mean things to them on the Internet or their cellphone?
(c) What is the probability that less than 8 of the students in our sample say that their ex-romantic partner called them names, put them down or said really mean things to them on the Internet or their cellphone?
(d) How many of the students in our sample should we expect to say that their ex-romantic partner called them names, put them down or said really mean things to them on the Internet or their cellphone?

4.B.11. *Laughing Online.* In a recent Facebook Blog Post, Weinsberg et al. (2015) presented the results of their analysis of de-identified posts and comments posted on Facebook in the last week of May, 2015. They found that 15% of the posts or comments during that period of time included characters identifiable as laughter. Suppose we select a random sample of $n = 75$ recent posts or comments on Facebook and record whether or not they include laughter characters.

(a) What is the probability that more than 15 of the posts or comments include laughter characters?

(b) What is the probability that fewer than 5 of the posts or comments include laughter characters?

(c) How many of the posts or comments should we expect to include laughter characters?

4.B.12. *Laughing Online—LOL, HaHa, Hehe, or Emoji?* In a recent Facebook Blog Post, Weinsberg et al. (2015) presented the results of their analysis of those de-identified posts and comments posted on Facebook in the last week of May, 2015 that contained "at least one string of characters matching laughter". They reported that 51.4% of these posts/comments used "haha", 13.1% used "hehe", 33.7% used "emoji', and only 1.9% used "lol" to indicate laughter. Suppose we select a random sample of $n = 60$ recent posts or comments on Facebook that contained "at least one string of characters matching laughter".

(a) How many of these sample posts/comments should we expect to have used "haha"? "hehe"?, "emoji"?, "lol"?

(b) How many of these sample posts/comments should we expect to have used something other than "emoji"?

(c) What is the probability that more than 50 of our sample posts/ comments used "emoji" to indicate laughter?

(d) What is the probability that less than 5 of our sample posts/ comments used "lol" to indicate laughter?

(e) What is the probability that more than 30% of our sample posts/ comments used either "haha" or "hehe" to indicate laughter?

(f) What is the probability that less than 40% of our sample posts/ comments did not use "haha" to indicate laughter?

4.B.13. *Beware of the Gigabyte!* The Los Angeles Times (2014) reported on a number of results from a survey conducted by Vouchercloud.net, a coupons website. One of the items in the survey asked respondents to select which of a number of possible choices best defined a "gigabyte". Twenty seven percent of the respondents incorrectly selected the answer that identified a "gigabyte" as an insect commonly found in South America. Suppose you randomly select individuals and ask them this question until you find the first respondent that identifies a "gigabyte" as an insect commonly found in South America. If the results of the Vouchercloud.net survey are applicable:

(a) What is the probability that you will interview more than six people before you obtain the first respondent who identifies a "gigabyte" as an insect commonly found in South America?

(b) What is the probability that you will find the first respondent who identifies a "gigabyte" as an insect commonly found in South America prior to your fourth interview?

(c) How many individuals would you expect to interview before you find the first respondent who identifies a "gigabyte" as an insect commonly found in South America?

4.B.14. *Church Attendance and Torture.* Does attending church services on a regular basis (at least once a week) increase your approval of using torture against suspected terrorists? The Cable News Network (2009) reported on a number of results obtained in a survey conducted April 14–21, 2009 by the Pew Research Center. One of the questions asked in the survey was "Do you think the use of torture against suspected terrorists in order to gain important information can often be justified, sometimes be justified, rarely be justified, or never be justified?" Forty two percent of the respondents who "seldom or never" attend church services agreed that the use of torture against suspected terrorists is "often" or "sometimes" justified. On the other hand, 54% of the respondents who attend church services at least once a week agreed that the use of torture against suspected terrorists is "often" or "sometimes" justified. Suppose that the results of the Pew Research Center are applicable.

(a) If you select a random sample of 50 individuals who "seldom or never" attend church services, what is the probability that less than half of the individuals you interview will agree that the use of torture against suspected terrorists is "often" or "sometimes" justified? How many of the 50 would you expect to agree that the use of torture against suspected terrorists is "often" or "sometimes" justified?

(b) If you select a random sample of 50 individuals who attend church services on a regular basis (at least once a week), what is the probability that more than half of the individuals you interview will agree that the use of torture against suspected terrorists is "often" or "sometimes" justified? How many of the 50 would you expect to agree that the use of torture against suspected terrorists is "often" or "sometimes" justified?

4.B.15. *Stormy Weather and Cloud Computing*. The Business Insider (2012) reported on a number of results from a national survey conducted by Wakefield Research (commissioned by Citrix) in August 2012. One of the questions asked in the survey was "Can stormy weather interfere with cloud computing?". Fifty one percent of the respondents (including a majority of Millennials) agreed that stormy weather can interfere with cloud computing! Suppose you randomly select 30 individuals and ask them this same question. If the results of the Wakefield Research survey are applicable:

(a) What is the probability that no more than ten of the individuals you interview will believe that stormy weather can interfere with cloud computing?
(b) What is the probability that less than 20 of the individuals you interview will believe that stormy weather can interfere with cloud computing?
(c) How many of the 30 individuals that you interview would you expect to believe that stormy weather can interfere with cloud computing?
(d) P.S. In the Wakefield Research survey, when asked what "the cloud" is, the participants' responses included: "a fluffy white thing", toilet paper, cyberspace, and pillow.

4.C. Activities

4.C.1. *Checking Your Intuition Using a Random Number Generator—Revisited.*

(a) Generate two sets of 10 random numbers each from the same probability distribution. (You may use any probability distribution you want, just make sure you use the same one for both sets.)

(b) Create a new set of 100 values corresponding to the differences between each value in the first set and each value in the second set; that is, subtract the first value in the second set from all ten values in the first set. Then subtract the second value in the second set from all ten values in the first set. Continue this process until you subtract the tenth value in the second set from all ten values in the first set.

(c) Follow the steps described in Example 4.2 to create the cumulative proportion of times the second number was larger than the first number and display the results in a scatterplot like the one in Fig. 4.2.

(d) Comment on the similarities and differences in the two scatterplots.

(e) If you repeated steps (a)–(d) with a different probability distribution would the overall pattern be the same or different for each distribution you consider? Explain your answer in language that your friends would understand. [Hint: You may want to repeat (a)-(d) with a few different distributions before you formulate your explanation.]

4.C.2. *Long Run Relative Frequencies.*

(a) Simulate 100 trials of a random experiment with two outcomes, labeled "success" and "failure", under the assumptions that the trials are independent and that a "success" is three times more likely than a "failure" on each trial.

(b) For each trial, record the outcome, the cumulative number of successes thus far, and the cumulative proportion of successes thus far.

(c) Construct a scatterplot with cumulative number of successes on the y-axis and trial number on the x-axis.

(d) Construct a scatterplot with cumulative proportion of successes on the y-axis and trial number on the x-axis.

(e) Repeat parts (a)-(d) several times. Do you notice any similarities in the plots? Is there any regularity in the plots for the first ten trials? What happens as the number of trials grows?

4.C.3. *Checking Your Intuition By Comparing Sequences of Events.*

(a) Write down a sequence of 50 outcomes (H or T) that you think could result from flipping a fair coin 50 times. (Don't actually flip the coin just yet—use only your intuition to write down the sequence.) How many heads are in the sequence? How many runs are in the sequence? Determine the length of each run by counting the number of H's or T's in the run and create a frequency table for the set of run lengths. What is the length of the longest run?

(b) Flip a fair coin 50 times and record the sequence of 50 outcomes. How many heads are in the sequence? How many runs are in the sequence? Determine the length of each run by counting the number of H's or T's in the run and create a frequency table for the observed run lengths. What is the length of the longest run? Compare these answers with those from your intuitive sequence in part (a).

(c) Use appropriate statistical software to simulate 30 sets of sequences, each of which corresponds to flipping a fair coin 50 times. For each of the observed 30 sequences, determine how many heads and how many runs are in the sequence. Create graphical summaries for the number of runs and the number of heads in the 30 sequences. How do the values from your sequences in parts (a) and (b) compare with those in the simulated sequences?

(d) One of the following two sequences was simulated as in part (c) and the other was just an intuitive guess as in part (a). Which is which? Why?

Sequence 1 HHHHTHTHHTTHTHTHHHHTHHTTT
HHHTHHTHTHTTTHTHHHHTHHTHH

Sequence 2 HTHHHTTHHTTTHHHHTTTHTHTTH
HTTTTHHTTHHHTTTHTHTHHHTTT

4.C.4. *Hitting Streaks.*

(a) Find the most recent batting average for your favorite baseball player. If you don't have a favorite player then select a player from the Major League Baseball team that is closest to your hometown. (If baseball is not in season, you can use the batting average from the end of the previous season.)

(b) Use statistical software to simulate five at bats for the player you have selected. How many hits did he get? How many hits would you expect him to get in five at bats?

(c) Repeat part (b) 25 times and provide numerical and graphical summaries for the number of hits and the percentage of hits.

(d) Use statistical software to simulate 20 at bats for the player you have selected. How many hits did he get? How many hits would you expect him to get in 20 at bats?

(e) Repeat part (d) 25 times and provide numerical and graphical summaries for the number of hits and the percentage of hits.

(f) Compare your simulation results for parts (c) and (e). Describe the similarities and differences.

4.C.5. *Estimating Probabilities.* Carefully describe your method of estimation in each of the following situations.

(a) Estimate the probability of randomly selecting a red M&M from a small bag of M&M's.

(b) Estimate the probability of randomly selecting an orange candy from a bag of Reese's Pieces.
(c) Estimate the probability of randomly selecting a green piece of cereal from a box of Froot Loops.
(d) Estimate the probability of randomly selecting a yellow bear from a bag of gummy bears.
(e) Did you enjoy your simulations?

4.D. Internet Archives

4.D.1. *Baseball Players' Bonuses.* Search the Internet to find data sets containing the bonuses received by all National League and, separately, all American League baseball players at the end of the most recently completed baseball season. Construct normal probability plots for each of these two datasets. Would you be willing to use the normal model for season bonuses for either American or National League baseball players? Why or why not?

4.D.2. *Poker Hands.* Search the Internet to find a site that provides the exact probabilities for each of the ten possible five card poker hands (nothing, one pair, two pairs, three of a kind, straight, flush, full house, four of a kind, straight flush, and royal flush).

(a) Deal yourself 100 poker hands (each time thoroughly shuffling the deck between hands).
(b) What is the expected number of times each of the possible poker outcomes should occur in your 100 hands? Compare these expected numbers with the observed numbers in your 100 hands. Would these two sets of numbers get closer if we were to deal 5000 hands instead of 100? Why?
(c) Compare the observed percentage of times each of the possible poker outcomes occurred in your 100 hands with the exact probability for

the outcome. Would these two sets of percentages get closer if we were to deal 5000 hands instead of 100? Why?

(d) Can you mathematically verify the probabilities for each of the ten possible poker hands? If not, describe how you could use simulation to arrive at good empirical evidence that the probabilities are correct.

4.D.3. *Roulette.* Search the Internet to find a site that provides the basic rules behind the American version of roulette.

(a) Describe the makeup of the roulette wheel for this version.

(b) What are some of the possible betting schemes for playing the game?

(c) Consider the simple choice of betting on a red color. What is the probability that you will win with this bet? How much should you expect to "win" if you played this red color bet 100 times for $1 each bet?

(d) Consider the choice of betting that the roulette ball will rest in one of the even numbers (0 and 00 are NOT considered even or odd numbers). What is the probability that you will win with this bet? How much should you expect to "win" if you played this even number bet 100 times for $1 each bet?

(e) How could you use random number simulation to support your conclusions in parts (c) and (d)?

(f) Find an Internet site that allows you to simulate playing roulette with an electronic version of a roulette wheel—no betting involved! Have fun spending some time providing additional support to your conclusions in parts (c) and (d).

4.D.4. *Monty Hall Problem.* Suppose you are on a game show and you are given the choice of one of three doors. Behind one door there is a new car—behind the other two doors are goats. You pick a door, say number 3, and the host of the show (who knows what is behind each of the doors), opens one of the other two doors, say number 1, to show you that it concealed one of the two goats. He then asks you if you would like to switch from your original choice of door number 3 to door number 2. You win whatever "prize" is behind your final choice.

(a) Should you make the switch from door number 3 to door number 2? Does it even matter?

(b) Search the Internet to find a website that allows you to electronically play this game. (See, for example, www.math.ucsd.edu.) Play the game 100 times, each time choosing to stick with your originally chosen door rather than switching after one of the goats is revealed. What percentage of times did you end up with the car?

(c) Now play the game 100 more times, each time choosing to switch doors from your original choice after one of the goats is revealed. What percentage of times did you end up with the car?

(d) Do you want to change your answer to part (a) in view of the simulation results in parts (b) and (c)?

(e) Can you mathematically show that switching doors after one of the goats is revealed greatly enhances your probability of winning the car? How much greater is it?

4.D.5. *Simulating From a Large Population.* Search the Internet to find a large dataset that describes characteristics or attributes of the American public that are of interest to you.

(a) Use a random number generator to select a random sample of size $n = 100$ from the population and summarize the data of interest to you in your sample. How do your sample results compare to the same features of the entire population?

(b) Repeat part (a) for a random sample of size $n = 500$. Compare the results from this larger sample with those for the sample size 100 in part (a).

5 Sampling Distributions and Approximations

Now that we have built a solid foundation based on exploratory data analysis techniques, proper design of experiments, and basic probability, we will put the finishing touches on this foundation by studying sampling distributions, the most important part of statistical inference. Before we can use the information we have collected and analyzed from our sample to make inferences about the population of interest, we must make sure that we understand how the statistic we have computed varies with repeated sampling. Our goal here is to describe what might happen if we repeat the entire sampling process and computation of the desired statistic again and again. Do you think that if you take a different sample you will get exactly the same value for the statistic? While it is certainly possible that this could happen, in most practical settings it is very unlikely that you will get exactly the same value of the statistic. At first, it might appear that this would be a major problem for the field of statistics. If we collect different samples and they usually give us different results, how can we make any inferences? The fact is that even though the

© Springer International Publishing AG 2017
D.A. Wolfe, G. Schneider, *Intuitive Introductory Statistics*, Springer Texts in Statistics,
DOI 10.1007/978-3-319-56072-4_5

values of the statistic are likely to differ from sample to sample, they will follow a pattern. This pattern of variation in repeated sampling is described by the sampling distribution of the statistic.

Definition 5.1 The **sampling distribution of a statistic** is the probability distribution for the set of possible values that can be assumed by the statistic.

Our goal in this chapter, then, is to learn some results that will help us understand and describe the nice patterns that are typically encountered in sampling distributions. In Chap. 6 our understanding of these patterns will pay off by showing us how to draw conclusions from samples. When we are interested in some feature of a large population we usually can't examine every member of the population, so we take a random sample and use this incomplete information to make reasonable guesses about the population. These reasonable guesses based on sampling are what we've been calling *statistical inferences*. They are expressed using statements like:

> "Based on the results of a clinical trial with 30 patients, we have convincing evidence to assert that this drug is an effective treatment;"
>
> OR
>
> "Based on a random sample of 1000 voters, we are quite confident that somewhere between 45% and 49% of all voters favor Proposition 11."

If we are going to act on these inferences, by administering the drug to more patients or campaigning for (or against) Proposition 11, we would like to know how reliable they are. How surprised would we be if the drug turns out, in fact, to be ineffective? How surprised would we be if 55% of all voters favor Proposition 11? Sampling distributions will help us answer these questions. We can't eliminate uncertainty – it is a natural consequence of sampling – but we can learn how to understand and assess it. Our statements will still be fuzzy, but we will be able to say *precisely how fuzzy* they are.

In Sect. 1 we study the sampling distribution for a sample average. In Sect. 2 we consider the sampling distribution for a sample proportion. The *Central Limit Theorem*, perhaps the most important and surprising result in statistics, is explored in Sect. 3, and it is used to approximate sampling distributions. Finally, in Sect. 4, we discuss simulation and resampling techniques to provide additional ways to approximate sampling distributions of important statistics.

5.1 The Sampling Distribution for a Sample Average

Consider Fig. 5.1 depicting the selection of a random sample $X_1, \ldots, X_n$ of size n from a normal population with mean μ and standard deviation σ. We will use the mean of our sample,

$$\bar{X} = \frac{X_1 + X_2 + \cdots + X_n}{n},$$

as our estimator for the population mean, μ, which is the value we are interested in. However, if we took a number of different samples, each of size n, from this population, we would get different values of the sample average for each of these samples. Before we start drawing conclusions from such samples, we need to know if the mean of a random sample from a

Fig. 5.1 Selecting a random sample of size *n* from a *N*(μ, σ) population

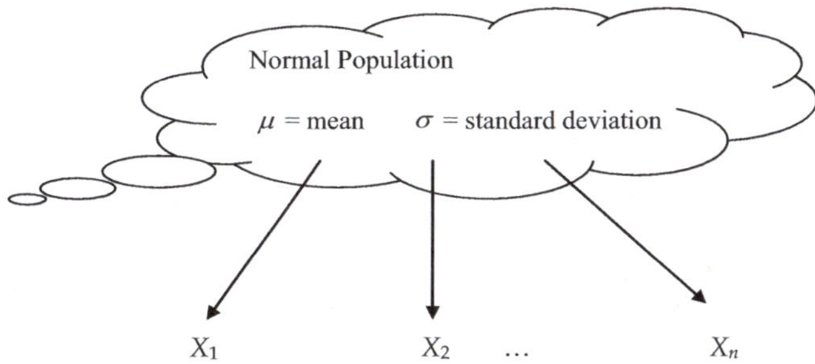

population is a good estimator for the mean of the population. This concern can be expressed in two important questions:

1. If we took many samples, would their means average out to the true population mean?
2. How close to μ will $\bar{X}$ typically be? That is, how variable is the mean of a random sample?

These are both questions about the *sampling distribution* of $\bar{X}$: "Where is it centered?" and "How does the variability of $\bar{X}$ across samples compare to the variability, σ, in the population?"

Since we will be using $\bar{X}$ to estimate μ we would hope that the sampling distribution of $\bar{X}$ is centered at μ. To show that this is indeed the case, we use the properties of expectation developed in Sects. 4.5 and 4.6. To compute $\bar{X}$ we add the n random variables $X_1, X_2, \ldots, X_n$ and divide by n. We learned in Chap. 4 that the mean of a sum of random variables is the sum of their means, and that scaling a random variable by a constant scales the mean by the same constant. Thus,

$$E\left(\frac{X_1 + X_2 + \cdots + X_n}{n}\right) = \frac{1}{n}(E(X_1) + E(X_2) + \cdots + E(X_n)).$$

Since each member of our random sample comes from a population with mean μ, the expected value of each X_i is μ, so $\frac{1}{n}(E(X_1) + E(X_2) + \cdots + E(X_n)) = \frac{1}{n}(\mu + \mu + \cdots + \mu) = \mu$. Putting it all together, we have $E(\bar{X}) = \mu$.

So the answer to our first question is YES! The sampling distribution of the sample mean is centered at the population mean μ. Since the center of the sampling distribution of $\bar{X}$ is equal to μ, we say that the statistic $\bar{X}$ is an *unbiased estimator* of μ.

We assess the variability of $\bar{X}$ by its variance, $Var(\bar{X})$. The variability in $\bar{X}$ comes from two sources, the randomness in our sample and the variability in the population we're sampling from. We would expect a larger sample to give us better information, so $Var(\bar{X})$ should decrease as the sample size n increases. In addition, we would expect a population with a smaller σ to deliver less variable samples. For example, a sample of the weights of college sophomores who wear a size 10 will cluster more tightly around the mean than a sample from all women between the ages of 16 and 50.

Using the properties of variances from Sects. 4.5 and 4.6, we can show that this intuition is correct: $Var(\bar{X}) = \frac{\sigma^2}{n}$. Notice that the variability in the sampling distribution of $\bar{X}$ is always smaller than the variability in the population. By taking a sample of size n and computing the sample average, we are reducing the variability by a factor of $1/n$. In other words, the variability in one random observation from this population is n times larger than the variability of a sample average of n observations from the population. This is precisely what allows us to control the fuzziness in statistical inferences. Even if the population has a large σ, we can get reliable estimates by taking large enough samples.

We now know how to describe the center and spread of the sampling distribution of $\bar{X}$, but what about its shape? Theorem 5.1 formally states that the overall shape of the sampling distribution of the sample average will resemble that of the underlying bell-shaped normal population.

Theorem 5.1 If $\bar{X}$ is the mean of a random sample of size n from a population that is normal with mean μ and standard deviation σ, then $\bar{X}$ is also normal with mean μ and reduced standard deviation $\frac{\sigma}{\sqrt{n}}$. Using our $\sim$ notation we can write this compactly as $\bar{X} \sim N\left(\mu, \frac{\sigma}{\sqrt{n}}\right)$.

This is a particular case of a much more general result that says linear combintations of normal random variables are normal. Although this general theorem is beyond the scope of this text, it is very natural to think that linear combinations of normal, bell-shaped curves will remain normal and bell-shaped. We will often apply another special case of this general result when comparing the sum or difference of two variables with possibly different normal distributions. Theorem 5.2 states this special case:

Theorem 5.2 If X and Y are independent normal variables with means μ_X and μ_Y and variances σ_X^2 and σ_Y^2, respectively, then the variable $aX + bY$ is also normal with mean $a\mu_X + b\mu_Y$ and variance $a^2\sigma_X^2 + b^2\sigma_Y^2$. In particular,

$$X + Y \text{ has mean } \mu_X + \mu_Y \text{ and variance } \sigma_X^2 + \sigma_Y^2,$$

and

$$X - Y \text{ has mean } \mu_X - \mu_Y \text{ and variance } \sigma_X^2 + \sigma_Y^2.$$

Notice that both the sum $X + Y$ and difference $X - Y$ have the same variance, which is the sum of the variances of the two random variables.

Example 5.1. Lecture Lengths Suppose that the class periods at a particular university are exactly 50 min in length. However, the actual lecture time on a particular day for Professor Staub varies according to the normal distribution with mean 52 min and standard deviation 2 min. Suppose that the lengths of different lectures are independent of one another and Professor Staub gives 36 lectures during the semester.

1. What is the probability that Professor Staub's lecture on Monday will be less than 50 minutes in length?

If we let L denote the length of Professor Staub's lecture on Monday, then we want to find $P(L < 50)$. Standardizing both sides and finding the appropriate area under the standard normal distribution yields

$$P(L < 50) = P\left(Z < \frac{50 - 52}{2}\right) = P(Z < -1) = .1587.$$

We can verify this calculation in **R** using the *pnorm*( ) function, as we did in Chap. 4, in either of the following two ways.

```
> pnorm(q = 50, mean = 52, sd = 2)
[1] 0.1586553
> pnorm(q = -1)
[1] 0.1586553
```

2. What is the probability that Professor Staub's average lecture time over the entire semester will be less than 50 min?

Let $\bar{L}$ denote Professor Staub's average lecture time over the 36 lectures in the semester. From Theorem 5.1, we know that $\bar{L}$ is normally distributed with a mean of 50 min and a standard deviation of $2/\sqrt{36}$. Thus, we find

$$P(\bar{L} < 50) = P\left(Z < \frac{50 - 52}{2/\sqrt{36}}\right) = P(Z < -6) \approx 0.$$

Again, we can verify this calculation with *pnorm*( ).

```
> pnorm(q = 50, mean = 52, sd = 2 / sqrt(36))
[1] 9.865876e-10
> pnorm(q = -6)
[1] 9.865876e-10
```

Thus, even though there is approximately a 16% chance that Professor Staub will dismiss the class early on any particular day, it is highly unlikely that Professor Staub will lecture for less than 50 min on average for a semester. The actual probability is about 10^{-9}, so although it is not at all unlikely for a particular lecture to be less than 50 min, there is virtually no chance that the average of 36 lectures will be this far below 52 min.

Example 5.2. Soccer Competition Two soccer players with equal ability enter a friendly competition. They will each run one-half mile and if one player beats the other by 10 s or more, the loser will buy ice cream for the winner. Suppose that the time for each player to complete one-half mile varies normally with a mean of 3 min and a standard deviation of 15 s and the two times are independent. What is the probability that one of the players will have to buy ice cream for her competitor after the run?

Notice that this is a question about the sampling distribution of $X - Y$, where X is the time for the first player and Y is the time for the second player. Using the results from Theorem 5.2, we know that the difference in the times, $X - Y$, is normally distributed with a mean of 0 s and a standard deviation of $\sqrt{15^2 + 15^2} = 21.2132$ seconds. Thus, the probability that one player will have to buy ice cream for the other player is $P(|X - Y| > 10)$. Using symmetry of the normal distribution and software to calculate areas under normal curves, we find

$$\begin{aligned} P(|X - Y| > 10) &= 2 \times P(X - Y < -10) \\ &= 2 \times P\left(Z < \frac{-10}{21.2132}\right) \\ &= 2 \times P(Z < -.4174) \\ &= 2 \times .3187 \\ &= .6374. \end{aligned}$$

We again provide the **R** command for verification of this number.

```
> 2 * pnorm(q = -10, mean = 0, sd = sqrt(15^2 + 15^2))
[1] 0.6373519
```

Thus, there is approximately a 64% chance that one of the players will have to buy ice cream for the other player.

5.1.1 Comparing Two Averages

We now turn to the setting where we want to compare two sample averages. For example, we might want to compare the average height for men with the average height for women, the average strength for athletes with the average strength for non-athletes, the average response time for a treatment group with the average response time for a control group, etc. We will still assume that both samples are from normal populations.

Let $\bar{X}$ denote the sample average for a random sample of size m from a normal population with mean μ_X and variance σ_X^2 and let $\bar{Y}$ denote the sample average for a second random sample of size n from a normal population with mean μ_Y and variance σ_Y^2. Suppose that the two random samples are also independent. One way to compare these sample averages is through the sampling distribution of the difference $\bar{X} - \bar{Y}$. Using properties of expectations and variances, we can easily describe the centers and spreads of the two separate sampling distributions. More specfically, the centers and spreads for the sampling distributions of $\bar{X}$ and $\bar{Y}$ are, respectively,

$$E[\bar{X}] = \mu_X \quad Var[\bar{X}] = \frac{\sigma_X^2}{m};$$

$$E[\bar{Y}] = \mu_Y \quad Var[\bar{Y}] = \frac{\sigma_Y^2}{n}.$$

To find the center and spread for the sampling distribution of the difference $\bar{X} - \bar{Y}$ we again make use of the results in Sect. 4.6. Since the expected value of a difference is the difference of the expected values, we find that

$$E[\bar{X} - \bar{Y}] = \mu_X - \mu_Y.$$

Moreover, since the two samples are independent, we recall that the variance of a difference is the sum of the variances, so that

$$Var\left[\bar{X} - \bar{Y}\right] = \frac{\sigma_X^2}{m} + \frac{\sigma_Y^2}{n}.$$

But Theorems 5.1 and 5.2 tell us even more, namely, that the sampling distribution of $\bar{X} - \bar{Y}$ is also normal with mean and variance as specified above. We formally state this as Theorem 5.3.

Theorem 5.3 If $X_1, X_2, \ldots, X_m$ is a random sample from a $N(\mu_X, \sigma_X)$ population and $Y_1, Y_2, \ldots, Y_n$ is a second independent random sample from a $N(\mu_Y, \sigma_Y)$ population, then the sampling distribution of $\bar{X} - \bar{Y}$ follows the normal distribution with mean $\mu_X - \mu_Y$ and variance $\frac{\sigma_X^2}{m} + \frac{\sigma_Y^2}{n}$; that is, $\bar{X} - \bar{Y} \sim N\left(\mu_X - \mu_Y, \sqrt{\frac{\sigma_X^2}{m} + \frac{\sigma_Y^2}{n}}\right)$.

Example 5.3. College Placement Exams Suppose that over the past 5 years the scores on a college placement exam have varied according to a $N(75, 10)$ distribution for women and a $N(72, 15)$ distribution for men. If 20 men and 30 women take the placement exam next fall, what is the chance that the average score for the men on this test is at least as high as the average score for the women?

Let $\bar{Y}$ and $\bar{X}$ denote the average scores for the 30 women and 20 men, respectively. Since it is reasonable to assume that the scores of the men and women will be independent of one another, Theorem 5.3 implies that the sampling distribution of $\bar{X} - \bar{Y}$ follows the normal distribution with mean $72 - 75 = -3$ and standard deviation $\sqrt{\frac{15^2}{20} + \frac{10^2}{30}} = 3.8188$. The probability that the average score for men with be at least as large as the average score for women is then

$$P(\bar{X} \geq \bar{Y}) = P(\bar{X} - \bar{Y} \geq 0) = P\left(Z \geq \frac{0 - (-3)}{3.8188}\right) = P(Z \geq .7856) = .2161.$$

The following command verifies this calculation using the **R** function *pnorm*( ).

```
> pnorm(q = 0, mean = -3, sd = sqrt(15^2 / 20 + 10^2 / 30),
        lower.tail = FALSE)
[1] 0.2160555
```

We note that even though the men perform slightly worse than the women on average, there is still slightly more than a 20% chance that on this particular exam the men will do as well as or better than the women. This might initially seem somewhat surprising, since you may have been tempted to think that women would *always* do better than men because their average score over the past 5 exams had been 3 points higher—intuition must be quantified!

Section 5.1 Practice Exercises

5.1.1. *Statistics GPAs.* The grade point average of all students taking introductory statistics in the United States varies according to the $N(3, .1)$ distribution.

(a) What is the sampling distribution of the sample average for a class of 5 students?

(b) What is the sampling distribution of the sample average for a class of 40 students?

(c) What is the sampling distribution of the sample average for a class of 100 students?

5.1.2. *Pokemon Cards.* A group of young children is wild about Pokemon cards. The monthly amount of time spent trading these cards by members of this group is normally distributed with a mean of 40 h and a standard deviation of 5 h.

(a) What is the probability that a randomly selected child from this group spends more than 45 h trading Pokemon cards during the month of August?

(b) If 40 children are randomly selected from this group, what is the sampling distribution of the average amount of time they spend monthly trading Pokemon cards?

(c) What is the probability that a randomly selected group of 40 children will average more than 45 h of trading Pokemon cards in a particular month?

(d) How many children would have to be randomly selected from the group to reduce the standard deviation to 2.5 for the average amount of time they spend monthly trading Pokemon cards?

(e) How many children would have to be randomly selected to reduce the standard deviation by a factor of 10 for the average amount of time they spend monthly trading Pokemon cards?

5.1.3. *Statistics Exams.* Thirty-two students, 17 women and 15 men, enroll for a basic statistics course. Assume that the scores on the first exam follow the $N(75, 10)$ distribution.

(a) What is the sampling distribution of the sample average for all 32 students?

(b) Find the probability that the sample average for all 32 students will be more than one standard deviation away from its mean.

(c) What is the sampling distribution of the sample average for the 17 women?

(d) Find the probability that the sample average for the 17 women is greater than 70.

(e) What is the sampling distribution of the sample average for the 15 men?

(f) Find the probability that the sample average for the 15 men is less than 60.

(g) Find the probability that the sample average for the 17 women will be at least five points higher than the sample average for the 15 men.

5.1.4. *Sales Commissions.* A salesperson earns a 10% commission on all sales of a particular product. The weekly sales of this product varies according to the normal distribution with a mean of $3700 and a standard deviation of $100.

(a) What is the probability that the salesperson will earn more than $350 in any particular week?

(b) What is the probability that she will average more than $350 per week in a month that contains 4 weeks?

5.1.5. *Beverage Content.* The content of a bottle of a popular beverage varies according to the normal distribution with a mean of 20 ounces and a standard deviation of .5 ounces.

(a) Find the 90th percentile for the content of a typical bottle of this beverage.

(b) Find the 90th percentile for the average content in a six-pack of this beverage.

(c) Find the 90th percentile for the average content in a 12-pack of this beverage.

(d) Comment on the relationship between the percentiles in parts (a), (b), and (c).

5.1.6. *Soda Content.* The beverage content in soda cans produced by a bottling company varies according to the normal distribution with a mean of 12 ounces and a standard deviation of .25 ounces.

(a) What is the chance of getting some extra soda (more than the stated 12 ounces) for free in a single can?

(b) What is the chance of getting more than .25 ounces of free soda in a single can?

(c) What is the chance of getting .25 free ounces in a six-pack of this soda?

(d) What is the chance of getting .25 free ounces in a 12-pack of this soda?

5.1.7. *Pocket Change.* The amount of change carried by a group of students varies normally with a mean of 70 cents and a standard deviation of 10 cents. Two students are randomly selected and asked to count how much change each of them has.

(a) What is the expected difference in the amounts?

(b) What is the standard deviation of the difference in the amounts?

(c) What is the probability that one student has at least 10 cents more than the other student?

(d) Let T be the total amount of change carried by three students riding in a car. Find the mean and standard deviation of T.

(e) Let Y be the average amount of change carried by three students riding in a car. Find the mean and standard deviation of Y. Compare this with the result in part (d).

5.1.8. *Learning New Manufacturing Process.* The amount of time it takes for an employee at a manufacturing plant to learn a new packaging process varies normally with a mean of 30 h and a standard deviation of 10 h. Two employees are randomly selected from all the employees at the plant.

(a) What is the expected difference in the amount of time it takes the two employees to learn the new process?

(b) What is the standard deviation for the difference in the amount of time it takes the two employees to learn the new process?

(c) What is the chance that it will take one employee at least 5 h longer than the other employee to learn the new process?

5.1.9. *Delivering Sunday Papers.* A student earns extra money by delivering Sunday newspapers. The collections from her customers vary from week to week, with a weekly mean of $50 and a weekly standard deviation of $5. The student assumes that her weekly collections can be represented by independent random variables, each with a $N(50, 5)$ distribution. She is interested in the total amount of money she will make during a particular month with five Sundays.

(a) Find the expected value for the total collections from these five Sundays.

(b) Find the standard deviation for this total.

(c) What is the chance that the student collects a total of at least $250 during this month?

(d) What is the chance that the student collects a total of at most $225 during this month?

(e) How many Sundays will she have to deliver newspapers for the expected value of her total collections to be $1000?

5.1.10. *Weight of Chips.* A snack food company claims there are 16 ounces of chips in a bag of their product. Actually the contents vary according to the normal distribution with a mean of 16.05 ounces and a standard deviation of .1 ounce.

(a) What is the chance that a particular bag of chips contains less than 15.9 ounces?

(b) What is the chance that a case of these chips (a case contains 12 bags) will have an average bag content of less than 15.9 ounces?

(c) If each of two customers purchase a bag of these chips, what is the chance that one customer will get 1 ounce more chips than the other customer?

5.1.11. *Average Temperatures.* Based on historical records since 1974, as obtained from Intellicast (2016), the average high temperature for September in Columbus, Ohio has been 77 degrees Fahrenheit. In September 2015 the average high temperature in Columbus was 80.2° Fahrenheit, which is 3.2° above the historical average. What is the probability that the average for the 30 days in September 2015 would be 3.2° above normal if daily high temperatures in September in Columbus vary according to the normal distribution with the historical average high temperature and a standard deviation of 5° Fahrenheit?

5.1.12. *Average Precipitation.* The long-term average amount of precipitation for a particular city during the combined months of May and June is 4 inches. In 2015 this city received 1.80 inches of precipitation in May and 0.65 inches in June. From the information provided, can you determine if the combined May and June precipitation in 2015 was unusually low? If so, explain how. If not, identify any other information you would need.

5.2 Sampling Distributions for Proportions and Counts

Suppose that we are sampling from a Bernoulli population with only two possible outcomes, one labeled "success" and the other labeled "failure", and we are interested in the proportion p of successes in the population. For example, we might be interested in the proportion of patients who experience side effects with a particular drug, the proportion of voters who are in favor of re-electing their senator, the proportion of workers who test positive for drug use, etc. The setting is basically the same as the one described in Sect. 4.3, but here we are interested in the proportion of successes, p, rather than just the number of successes, B. Thus it is natural to use $\hat{p} = \frac{B}{n}$, the proportion of successes in our sample, as our estimate of p.

Example 5.4. Experimental Drug Consider the setting where four patients are given an experimental drug to treat an illness. If the chance of

Table 5.1 Sampling distribution for $\hat{p} = \frac{B}{n}$ when $n = 4$ and $p = 1/2$

Value of $\hat{p} = B/n$	0	.25	.5	.75	1
Probability	$\frac{1}{16}$	$\frac{1}{4}$	$\frac{3}{8}$	$\frac{1}{4}$	$\frac{1}{16}$

experiencing side effects with this medication is .5 for each patient and the patients react to the medication independently, what is the sampling distribution for the proportion of patients who will experience side effects?

If this drug is repeatedly given to samples of size four,[1] then the proportion of patients experiencing side effects will always be one of the values 0/4, 1/4, 2/4, 3/4, or 4/4. The set of possible Bernoulli outcomes and the probability of each outcome will be the same as those listed in Table 4.1. The only difference here is that instead of keeping track of the total number of successes B (the last column in Table 4.1) we need to tally the proportion of successes, B/n. The sampling distribution of the proportion of patients who will experience side effects is almost identical to the one shown in Table 4.2. The only difference here is the value of the statistic of interest: the numbers of successes 0, 1, 2, 3, and 4 in the last column of Table 4.2 need to be replaced by the proportions of successes 0/4, 1/4, 2/4, 3/4, and 4/4, respectively. Table 5.1 shows the sampling distribution for the proportion of patients (out of four) who will experience serious side effects.

The result of Example 5.4 can easily be generalized to any sample size n and probability of success p. Since the proportion of successes in a sample is simply the total number of successes divided by the sample size, the sampling distribution for the sample proportion can always be obtained from the appropriate binomial distribution for the number of successes B.

[1] We have made the sample size very small here to simplify the calculations, but small sample sizes are often typical in clinical trials of treatments for rare illnesses.

Theorem 5.4 The sampling distribution of the proportion of successes, $\hat{p} = B/n$, in n independent Bernoulli trials with common probability of success p is given by

$$P\left(\hat{p} = \frac{b}{n}\right) = \frac{n!}{b!(n-b)!} p^b (1-p)^{n-b}, \quad \text{for } b = 0, 1, 2, \ldots, n.$$

Notice that the sampling distribution of $\hat{p}$ in Theorem 5.4 is equivalent to the $B(n, p)$ binomial distribution discussed in Sect. 4.3. That is, $P(\hat{p} = b/n) = P(B = b)$, for $b = 0, 1, \ldots, n$. If we take a moment to stop and think about what this actually says, the result makes sense. The only way the proportion of successes can be b/n is if the number of successes is equal to b, so we are just describing the same event in two different ways. Since the events are the same, the probabilities must be the same.

Example 5.5. Effect of Increasing Sample Size What happens to the distribution of $\hat{p} = B/n$ as the number of patients increases and the probability of experiencing side effects remains fixed at $p = 0.5$? To answer this question we use the **R** function *dbinom*( ) to examine the exact sampling distribution of B/n for $n = 10$, 25, 50, and 100. Dotplots displaying these sampling distributions are shown in Figs. 5.2, 5.3, 5.4, and 5.5, respectively. The following **R** code generates the plot shown in Fig. 5.2. The other three plots can be obtained by simply changing the value of n on the first line.

```
> n <- 10
> prob_vals <- dbinom(x = 0:n, size = n, prob = 0.5)
> plot((0:n)/n,
       prob_vals,
       main = paste0("Distribution of B/", n),
       xlab = paste0("B/",n),
       ylab = "Probability")
```

Fig. 5.2 Exact sampling distribution of $B/10$ when $p = .5$ and $n = 10$

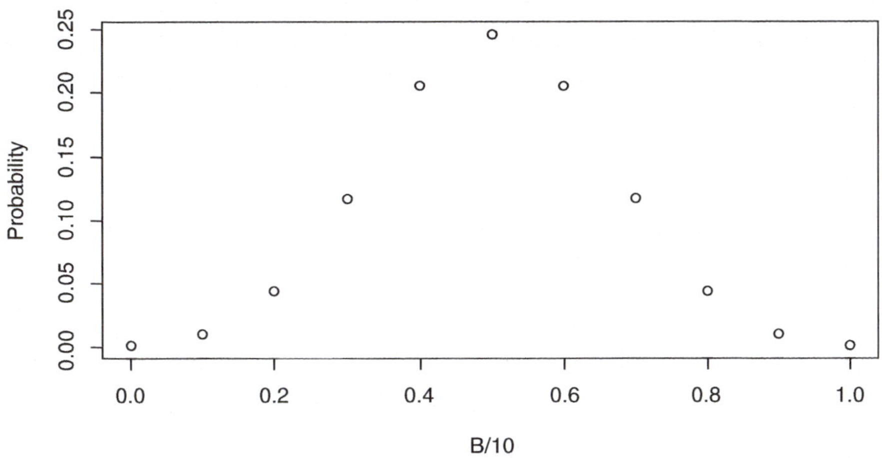

Fig. 5.3 Exact sampling distribution of $B/25$ when $p = .5$ and $n = 25$

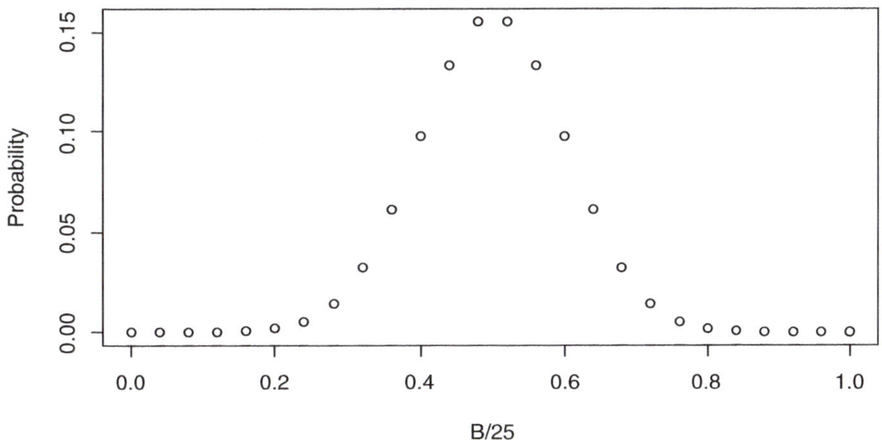

The smooth bell-shaped pattern that emerges is common for many sampling distributions and it reminds us of the normal distributions discussed in Sect. 4.7. In Sect. 3 we will use this bell-shaped pattern to develop normal approximations for a number of sampling distributions. These plots also show the same behavior we saw for sample averages for samples from a normal population: Increasing the sample size decreases the variability. Here

Fig. 5.4 **Exact sampling distribution of $B/50$ when $p = .5$ and $n = 50$**

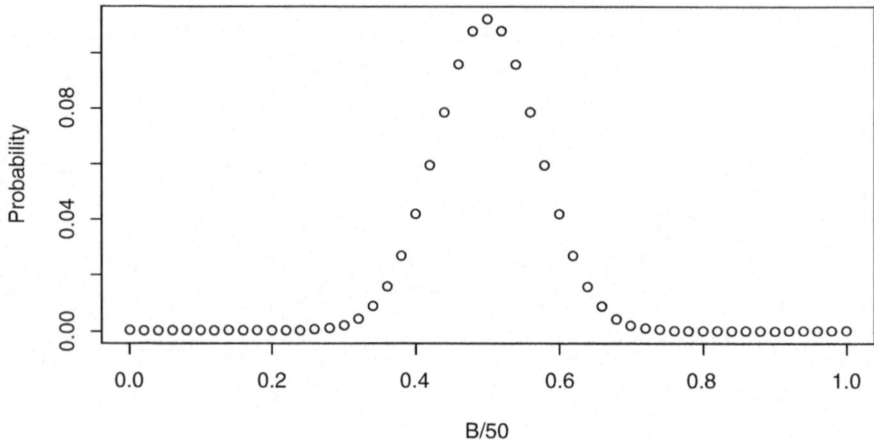

Fig. 5.5 **Exact sampling distribution of $B/100$ when $p = .5$ and $n = 100$**

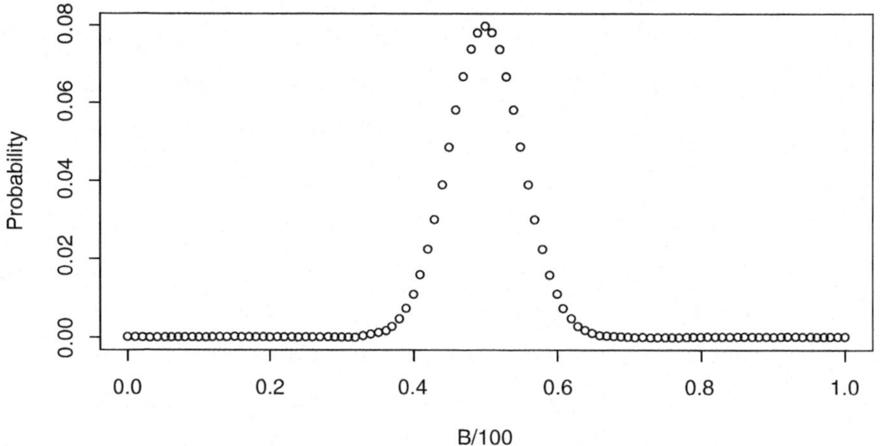

you can see that as n increases from 10 to 100, the values of $\frac{B}{n}$ cluster more closely around the population mean of 0.5.

5.2.1 Comparing Two Proportions

Suppose we are interested in comparing two population proportions. For example, on December 19, 1998 the United States House of Representatives voted to impeach President William Jefferson Clinton. If we wanted to determine whether Republicans and Democrats differed on this controversial and politically charged issue, we could have collected two separate random samples, one from Republicans and one from Democrats, and compared the sample proportions $\hat{p}_R$ and $\hat{p}_D$ of individuals in the separate samples who supported the impeachment decision. If we were to repeatedly take such samples (many news organizations at the time were conducting these surveys on a daily basis), what is the sampling distribution of the difference $\hat{p}_R - \hat{p}_D$ between the two independent sample proportions?

Suppose we sample m Republicans and n Democrats. Using Theorem 5.4, we know that the sampling distribution of $B_R = m\hat{p}_R$ is $B(m, p_R)$ and the sampling distribution of $B_D = n\hat{p}_D$ is $B(n, p_D)$, where p_R and p_D are the proportions of all Republicans and Democrats, respectively, who supported the impeachment decision. Thus, the exact form of the sampling distribution of the difference $\hat{p}_R - \hat{p}_D$ can be obtained by using the two separate binomial distributions. The possible values of $\hat{p}_R - \hat{p}_D$ range from $\frac{0}{m} - \frac{n}{n} = -1$ to $\frac{m}{m} - \frac{0}{n} = 1$ and the computations involved in calculating these probabilities are tedious. We will not get into the specific details here. Instead we simply point out the fact that for most practical settings the sampling distribution of $\hat{p}_R - \hat{p}_D$ also follows the nice bell-shaped pattern we observed in Figs. 5.2, 5.3, 5.4, and 5.5 and then develop an approximation to this sampling distribution later in Sect. 3. Even though computation of the exact probabilities for the values of $\hat{p}_R - \hat{p}_D$ is tedious, it is easy to obtain the center and spread of the sampling distribution of $\hat{p}_R - \hat{p}_D$ using properties of expectations and variances from Sect. 4.6. The expected value of $\hat{p}_R - \hat{p}_D$ is $E[\hat{p}_R - \hat{p}_D] = p_R - p_D$. Since the two samples are independent, it follows that the variance of $\hat{p}_R - \hat{p}_D$ is given by

$$Var[\hat{p}_R - \hat{p}_D] = \frac{p_R(1 - p_R)}{m} + \frac{p_D(1 - p_D)}{n}.$$

In general, for independent random samples of sizes m and n from *Bernoulli*(p_1) and *Bernoulli*(p_2) populations, respectively, the center of the sampling distribution of the difference in sample proportions $\hat{p}_1 - \hat{p}_2$ is $p_1 - p_2$ and the variance of the sampling distribution of $\hat{p}_1 - \hat{p}_2$ is given by $Var[\hat{p}_1 - \hat{p}_2] = \frac{p_1(1-p_1)}{m} + \frac{p_2(1-p_2)}{n}$. Once again, the exact form of this sampling distribution can be obtained by using the $B(m, p_1)$ and $B(n, p_2)$ binomial distributions, but the calculations are quite tedious. In Sects. 3 and 4 we will examine methods for approximating the sampling distributions for a sample proportion and for the difference between two independent sample proportions. Figure 5.6 illustrates the exact sampling distribution of $\hat{p}_1 - \hat{p}_2$ when $m = 10$, $p_1 = .4$, $n = 12$, and $p_2 = .8$. Notice that the distribution is centered at $.4 - .8 = -.4$ and the overall shape is approximately normal.

5.2.2 Comparing Several Proportions

Consider sampling from a population where each member of the population belongs to one and only one of k distinct categories. (Basically we are

Fig. 5.6 Exact sampling distribution of $\hat{p}_1 - \hat{p}_2$ when $m = 10$, $p_1 = .4$, $n = 12$, and $p_2 = .8$

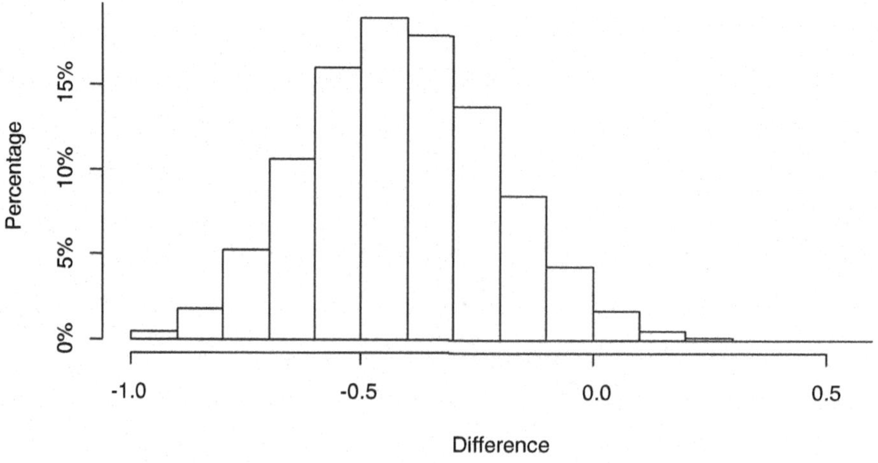

extending the Bernoulli population, which has only two categories, to a population that has k distinct categories.) For example, we might be interested in the hair color of a randomly selected person, the model of an observed automobile, the categorized weight (light, medium, or heavy) of an object, etc. What will be the form of the sampling distribution for sample proportions of such categorical data?

Example 5.6. M & M Colors Mars, Inc. claims that the color ratio in plain M&M's is 30% brown, 20% yellow, 20% red, 10% orange, 10% green, and 10% blue. If 10 M&M's are randomly selected from a **large** bag of plain M&M's, what is the probability of getting 3 brown, 2 yellow, 2 red, 1 orange, 1 green, and 1 blue M&M's, corresponding exactly to the proportions of these colors in the overall population of M&M's?

We extend the logic that we used to compute binomial probabilities in Sect. 4.3 to categorical populations with more than two categories. Since this is a **large** bag of M&M's and we are selecting them at random, it is reasonable to assume that the color of any M&M chosen will be (at last approximately) independent of the color of any other chosen M&M. Using the multiplication rule for independent events and a counting formula, we find that the probability of selecting a sample of 3 brown, 2 yellow, 2 red, 1 orange, 1 green and 1 blue M&M's is

$$\begin{aligned} &P(Br = 3,\ Y = 2,\ R = 2,\ O = 1,\ G = 1,\ Bl = 1) \\ &= \frac{10!}{3!2!2!1!1!1!}\ .3^3.2^2.2^2.1^1.1^1.1^1 = .0065. \end{aligned}$$

We can also calculate this probability in **R** using the *dmultinom*( ) function as follows.

```
> dmultinom(x = c(3,2,2,1,1,1), prob = c(0.3,0.2,0.2,0.1,0.1,0.1))
[1] 0.00653184
```

The product of the individual probabilities, $.3^3.2^2.2^2.1^1.1^1.1^1$, represents the probability that any particular selection sequence will have the desired color specification. The coefficient, $\frac{10!}{3!2!2!1!1!1!}$, in front of the product of the individual probability is the number of ways we can pick 3 of the 10 M&M's to be brown, 2 from the remaining 7 M&M's to be yellow, 2 from the remaining 5 M&M's to be red, 1 from the remaining 3 M&M's to be orange, 1 from the remaining 2 M&M's to be green and have the last one M&M be blue. That is,

$$\frac{10!}{3!2!2!1!1!1!} = \binom{10}{3}\binom{7}{2}\binom{5}{2}\binom{3}{1}\binom{2}{1}\binom{1}{1} = 151{,}200$$

is the number of arrangements that give us our desired color specification.

As previously noted, the sample outcome for which we have just calculated the probability is special: it matches the population proportions perfectly. If you try different values of the argument x in the *dmultinom*( ) function in order to experiment with sample outcomes that are close to but not perfect matches with the color proportions claimed by Mars Inc. (check it out!), you will discover that the outcome in Example 5.6 is, indeed, the **most likely** outcome. However, as we have seen, even this perfect match is extremely unlikely to occur. For comparison, an outcome with 3 brown, 2 yellow, 3 red, 0 orange, 1 green, and 1 blue M&M's has .0044 chance of occurring, and the probability of getting a sample of 10 M&M's with none of the "rare" colors (orange, green, and blue) is about .0065. To put these small probabilities in perspective, think a bit about just how many different possible outcomes there are.

5.2.3 Using Ranks and Counts to Compare Two Samples

Suppose we are interested in comparing a random sample of m values, say X_1, $X_2, \ldots, X_m$, from some continuous population with an independent random sample of n values, say $Y_1, Y_2, \ldots, Y_n$, from a second continuous population to

Fig. 5.7 Possible age distribution for individuals enrolled in an adult education program

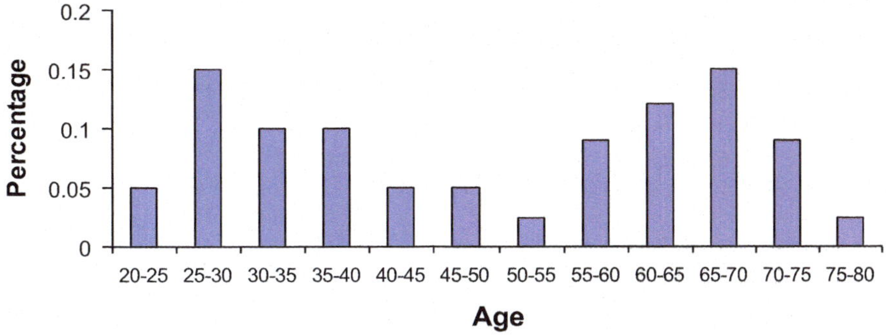

see if the Y's are generally bigger than, smaller than, or about the same as the X's. For example, suppose we are operating an adult education program. In preparing our course offerings for the program, we have been assuming that the age distributions for men and women enrolled in adult education are similar, but we would like to test this assumption. Although we do not know the actual age distributions, we do know they are not bell-shaped and normal. Each of the distributions will likely have two peaks, one for adults a few years out of college and another for retirees, perhaps looking something like the relative frequency distribution in Fig. 5.7.

Since such a distribution is clearly not normal, we can not use the sampling distribution for the difference in normal means that we used in Example 5.3 to test our assumption that the age distributions for men and women enrolled in adult education are similar. We now describe a technique for making this comparison that does not require the inappropriate assumption of normality.

We take independent random samples of $m = 10$ women and $n = 11$ men enrolled in our program. We rank all 21 of these individuals in order of age from youngest to oldest, using a label X for the women and Y for the men. Our labeled list of ranks might look like this:

$$XYXXXYXXYXXXYYYYXYYYY.$$

As we look at the sequence we notice that most of the Y's are larger than most of the X's. But how can we quantitatively measure or describe this outcome? One way to compare the relative locations of the X's and the Y's is to compare the average of the X ranks (ordered placements) with the average of the Y ranks (ordered placements). Doing so with the sequence of observed X's and Y's above, we find that the average of the X ranks, $(1 + 3 + 4 + 5 + 7 + 8 + 10 + 11 + 12 + 17)/10 = 7.8$, is much smaller than the average of the Y ranks, $(2 + 6 + 9 + 13 + 14 + 15 + 16 + 18 + 19 + 20 + 21)/11 \approx 13.9$. The difference (average rank for men) – (average rank for women) is $13.9 - 7.8 = 6.1$, which clearly indicates that the men tend to be older.

Another way to compare the X and Y samples is to pair each of the 10 women with each of the 11 men, giving us a total of $mn = 110$ pairs. For each pair, we count 1 if the woman is younger and 0 if the man is younger. Our comparison statistic is then the number of female/male pairs out of 110 for which the woman is younger. Obtaining these counts is tedious, but statistical software can do it easily. For example, the **R** function *wilcox.test*( ) will quickly tell us that the woman is older in 23 of the 110 pairs for the rank sequence above. This comparison also seems to indicate that the men and women have different age distributions, with the men tending to be older.

Thus, we have identified two statistics that can be used to compare the X's and the Y's even when the populations are not bell-shaped and normal:

$$W = \text{average of the Y ranks} - \text{average of the X ranks}$$

and

$$U = \text{the number of } (X_i, Y_j) \text{ pairs for which } X_i < Y_j, i = 1, \ldots, m \text{ and } j = 1, \ldots, n.$$

For our samples of 10 women and 11 men, we have $W = 6.1$ and $U = 87$.

Note that the **R** function *wilcox.test*( ) reports the number of pairs for which the X value is *larger* than the Y value, which we will call U', rather than the number of pairs for which the X value is *smaller* than the Y value. Since we know that there are a total of mn pairs, we can use $U = mn - U'$ to determine the number of pairs for which $X_i < Y_j$. We found that the woman is older than the man in 23 of the 110 pairs above; equivalently the man is older than the woman in $110 - 23 = 87$ of the pairs.

But simply obtaining numerical values for these statistics is not sufficient. What we need to know is whether or not outcomes like $W = 6.1$ and $U = 87$ are typical for random samples of 10 women and 11 men from identical populations, or do they instead indicate that we should question our basic assumption that men and women participants in adult education have about the same age distribution? To reach an informed conclusion, we need to learn more about the sampling distributions of W and U.

What are the sampling distributions for the statistics W and U if, in fact, the X's and Y's are actually coming from a common age population? If that were the case we would expect the X's and Y's to be "evenly mixed", since then every one of the $(m + n)!$ possible ordered arrangements of the m X's and n Y's is equally likely to occur. Using this fact and the fact that there are actually only $\frac{(m+n)!}{m!n!}$ different possible arrangements of the X's and Y's, we can use straightforward enumeration to find the sampling distributions for these two statistics.

We illustrate how to obtain these sampling distributions with a small-scale example where it is easier to follow the details of the calculations.

Example 5.7. Sampling Distributions for W and U Suppose that $m = 3$ and $n = 2$. Then there are $\frac{5!}{3!2!} = 10$ possible ordered arrangements of the X's and Y's and each one has the same chance $\left(\frac{1}{10}\right)$ of occurring when the X's and Y's have the same distribution. These ten ordered arrangements, associated X and Y ranks, common probability for each arrangement, and the corresponding values of W and U are shown in Table 5.2.

Thus, the sampling distributions of W and U for $m = 3$ and $n = 2$ are given by:

$$P(W = -2.5) = P(W = 2.5) = P\left(W = -1\frac{2}{3}\right) = P\left(W = 1\frac{2}{3}\right) = 0.1$$
$$P\left(W = -\frac{5}{6}\right) = P\left(W = \frac{5}{6}\right) = P(W = 0) = 0.2,$$

and

$$P(U = 0) = P(U = 6) = P(U = 1) = P(U = 5) = 0.1$$
$$P(U = 2) = P(U = 3) = P(U = 4) = 0.2,$$

respectively. Notice that the sampling distribution of W is symmetric about 0 and the sampling distribution of U is symmetric about 3.

For our samples of $m = 10$ women and $n = 11$ men, a table like Table 5.2 would be huge—in fact it would have $\frac{21!}{10!11!} = 352,716$ rows! However, using a simulation method that we will develop in Sect. 4 (or the **R** functions *pwilcox* ( ) and *qwilcox*( )) we can get a very good idea of what the sampling distributions of W and U look like in this case with $m = 10$ and $n = 11$. The

Table 5.2 Possible ordered arrangements and values of W and U for $m = 3$ and $n = 2$

Arrangement	*Y* Ranks	*X* Ranks	Probability	Value of *W*	Value of *U*
YYXXX	1,2	3, 4, 5	0.1	-2 ½	0
YXYXX	1,3	2, 4, 5	0.1	−1 ⅔	1
YXXYX	1,4	2, 3, 5	0.1	−5/6	2
YXXXY	1,5	2, 3, 4	0.1	0	3
XYYXX	2,3	1, 4, 5	0.1	−5/6	2
XYXYX	2,4	1, 3, 5	0.1	0	3
XYXXY	2,5	1, 3, 4	0.1	5/6	4
XXYYX	3,4	1, 2, 5	0.1	5/6	4
XXYXY	3,5	1, 2, 4	0.1	1 ⅔	5
XXXYY	4,5	1, 2, 3	0.1	2 ½	6

Fig. 5.8 Sampling distributions of W and U for samples of $m = 10$ X's (women) and $n = 11$ Y's (men)

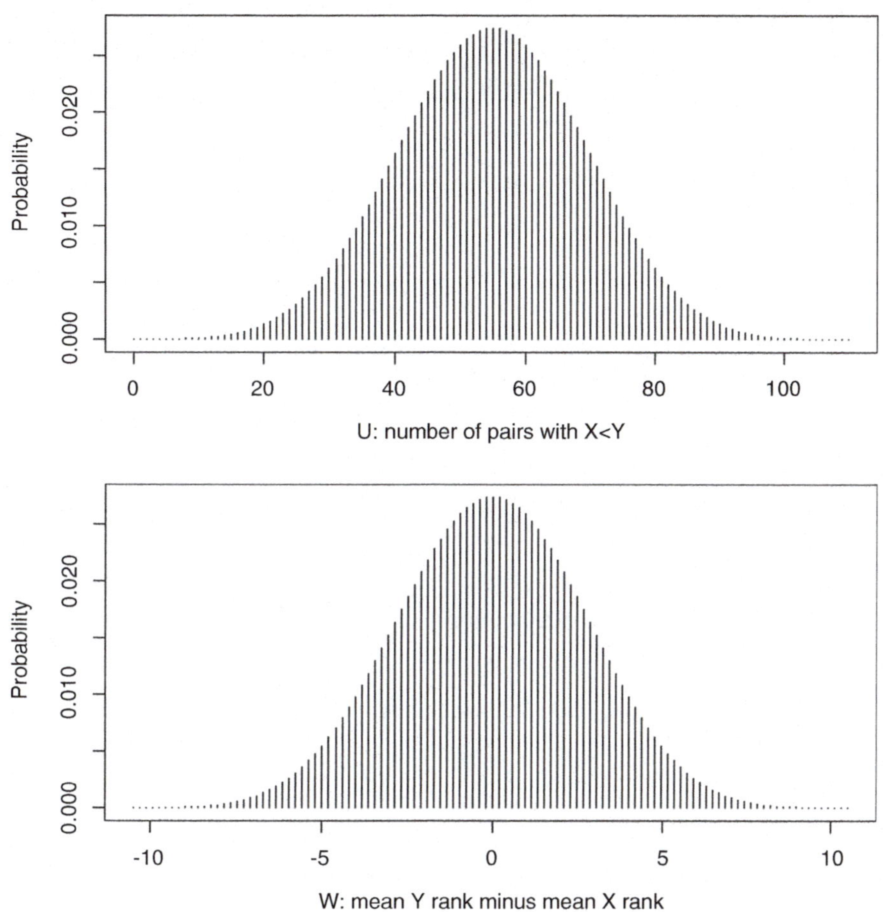

histograms in Fig. 5.8 show the sampling distributions for W and U in this setting.

It is clear from these sampling distribution histograms that our observed samples with $W = 6.1$ and $U = 87$ would be very unusual if women and men actually had identical age distributions. In fact, if the ages were equally distributed the chance of a sample this unusual is roughly 0.012. We can verify this using the **R** function *pwilcox*( ) as follows, keeping in mind that we need to specify $U' = 23$ instead of $U = 87$.

```
> pwilcox(q = 23, m = 10, n = 11, lower.tail = TRUE)
[1] 0.01207771
```

Thus, we conclude that it is very likely that the age distributions for men and women in adult education are, indeed, different.

Section 5.2 Practice Exercises

5.2.1. *Sampling Distribution of* $\hat{p}$. Obtain the sampling distribution of $\hat{p} = \frac{B}{n}$ by listing all the possible values of the sample proportion and calculating the appropriate probabilities for each of the following settings.

(a) $n = 5, p = .5$.

(b) $n = 5, p = .9$.

(c) $n = 5, p = .1$.

(d) Discuss the relationship between the distributions in parts (b) and (c).

5.2.2. *Sampling Distribution of* $\hat{p}$. Obtain the sampling distribution of $\hat{p} = \frac{B}{n}$ by listing all the possible values of the sample proportion and calculating the appropriate probabilities for each of the following settings.

(a) $n = 8, p = .5$.

(b) $n = 8, p = .75$.

(c) $n = 8, p = .25$.

(d) Discuss the relationship between the distributions in parts (b) and (c).

5.2.3. Values of $\hat{p}_1 - \hat{p}_2$. List all possible values of $\hat{p}_1 - \hat{p}_2$ when $10\hat{p}_1 \sim B$ (10, .5) and $5\hat{p}_2 \sim B(5, .5)$.

5.2.4. *Sampling Distributions of U and W*. Obtain the sampling distributions of the two statistics U and W when $m = n = 3$.

5.2.4. *Sampling Distributions of U and W*. Obtain the sampling distributions of the two statistics U and W when $m = 4$ and $n = 2$.

5.2.5. *Sun around the Earth or Earth around the Sun?* National Public Radio (2014) reported on a number of results from a survey conducted by the National Science Foundation in the United States in 2012. One of the questions asked in the survey was "Does the earth revolve around the sun, or does the sun revolve around the earth?". Twenty six percent of the respondents said that the sun revolves around the earth! Suppose you randomly select 60 individuals and ask them this question. If the results of the National Science Foundation survey are applicable:

(a) What is the probability that exactly 1/3 of your respondents will say that the sun revolves around the earth?

(b) What is the probability that at least 1/3 of your respondents will say that the sun revolves around the earth?

(c) What is the probability that at most 1/3 of your respondents will say that the sun revolves around the earth?

(d) What is the probability that more than 1/3 of your respondents will say that the sun revolves around the earth?

(e) What is the probability that less than 1/3 of your respondents will say that the sun revolves around the earth?

(f) How many of the respondents would you expect to say that the sun revolves around the earth?

5.2.6. *Moving Aging Parents in to Live with You—Mom or Dad?* In a national survey of 1118 individuals aged 40 and older with both living parents, Visiting Angels (2013), one of the largest in-home senior care companies in the United States, asked respondents the following question: "If you had to choose only one of your aging parents to move in and live with you, would you choose your mom or your dad? 745 of the respondents said they would choose mom over dad.

Suppose you randomly select 60 individuals aged 40 and older with both living parents and ask them this question. If the results of the Visiting Angels survey are applicable:

(a) How many of your interviewees would you expect to choose mom over dad?

(b) What is the probability that exactly 2/3 of your interviewees would choose mom over dad?

(c) What is the probability that more than 2/3 your interviewees would choose mom over dad?

(d) What is the probability that at most 2/3 of your interviewees would choose dad over mom?

(e) Explain why the probabilities in (a), (b), and (c) do not sum to one.

5.2.7. *Coin Tosses.* When tossing a balanced coin ten times, what is the chance that exactly 50% of the outcomes will have heads facing up?

5.2.8. *Rolling a Fair Die.* When rolling a balanced six-sided die 12 times, what is the chance that more than 40% of the rolls will result in even numbers?

5.2.9. *Multiple Choice Exam.* An exam contains 25 multiple choice questions with 5 possible responses for each question. Assume that a student did not have time to study and is simply guessing at the correct answer for each question.

(a) What percentage of the questions would you expect the student to answer correctly?

(b) What is the standard deviation of the percentage of questions the student will answer correctly?

(c) Repeat parts (a) and (b) for a student who knows the correct answers for 13 questions and must simply guess for the remaining 12 questions.

5.2.10. *University Seating Policy.* Suppose that 40% of the women at a university are supportive of a new seating policy in the dining halls, but only 25% of the men are supportive of the policy. If we survey 20 men and 20 women at random, where would you expect the sampling distribution of the difference in the sample proportions, $\hat{p}_{women} - \hat{p}_{men}$, to be centered? Identify the variance of the difference $\hat{p}_{women} - \hat{p}_{men}$.

5.2.11. *Peanut M&M's.* According to Mars, Inc., the color ratio for peanut M&M's is 20% brown, 20% yellow, 20% red, 10% orange, 10% green, and 20% blue. If six peanut M&M's are selected randomly from a **large** bag of peanut M&Ms., what is the chance that 1/3 are brown, 1/3 are yellow, and 1/3 are red?

5.2.12. *Class Sizes.* Suppose there are roughly the same numbers of students in each of the four classes (Freshman, Sophomore, Junior, and Senior) at a small university. If a random sample of 12 students is selected from this university, what is the chance of getting exactly 3 students from each class? Do you think this is the most likely configuration of students in the sample? Explain.

5.2.13. *Heart Rates.* Suppose we are interested in comparing heart rates for athletes and non-athletes at your school. If we select 15 athletes and 15 non-athletes at random, how many pairs of individuals would we have to compare to determine the number of athletes that have lower heart rates than non-athletes in these two samples?

5.2.14. *Do Children Know Where Household Firearms Are Stored?* In a survey of parents and children in households with firearms, Baxley and Miller (2006) collected information from 201 such households in rural Alabama. The parents in 60 of those households reported that their children did not know the storage place for the firearms. Suppose you randomly select 30 additional households with firearms in rural Alabama and ask them whether their children know the storage place for the firearms. If the results of the Baxley-Miller survey are applicable:

(a) In how many of the 30 households in your survey would you expect that the parents indicate that their children know the storage place for the firearms?

(b) What is the probability that the parents say their children know the storage place for the firearms in more that ten of your 30 surveyed households?

(c) What is the probability that the parents say their children know the storage place for the firearms in less than five of your 30 surveyed households?

5.2.15. *What Parents Don't Know!* Consider the survey of parents and children in households with firearms discussed in Exercise 5.2.14. Baxley and Miller followed up their initial survey with a second set of questions for the children without their parents being present. For those 60 households where the parents reported that their children did not know the storage place of the firearms, Baxley and Miller found that children from 23 of these households did, in fact, know where the firearms were stored!

(a) What percentage of the total 201 households surveyed by Baxley and Miller were such that the parents were not aware that their children knew of the storage place for their firearms?

As in Exercise 5.2.14, suppose you randomly select 30 additional households with firearms in rural Alabama.

(b) In how many of the 30 households in your survey would you expect that the parents were not aware that their children knew the storage place for the firearms?

(c) What is the probability that there were no households in your survey where the parents were not aware that their children knew the storage pale for the firearms?

(d) Would it be surprising if you found that ten of the 30 households in your survey were such that the parents were not aware that their children knew the storage plae for the firearms? Justify your answer.

5.2.16. *Does God Reward Good Athletes?* USA Today (2013) reported on a number of results from a survey conducted by the Public Religion Research Institute in the United States in 2013. One of the questions asked in the survey was "True or False: God rewards athletes who have faith with good health and success." Fifty three percent of the respondents answered "True", indicating that they believe that God does reward athletes who have faith with good health and success. Suppose you randomly select 30 individuals and ask them the same question. If the results of the Public Religion Research Institute's Survey are applicable:

(a) How many of your respondents would you expect to answer "True"?

(b) What is the probability that more than 20 of your respondents answer "True"?

(c) What is the probability that less than 15 of your respondents answer "True"?

(d) What is the probability that exactly 15 of your respondents answer "True"?

(e) What is the probability that between 12 and 18, inclusive, of your respondents answer "True"?

5.2.17. *Multinomial Distribution.* Consider the general setting where we have a population with k distinct categories $C_1, \ldots, C_k$, and let p_i denote the proportion of the population that belongs to category C_i, for $i = 1, \ldots, k$. (Note that $\sum_{i=1}^{k} p_i = 1$.) Suppose we collect a random sample of size n from this population and let B_i denote the number of sample items that belong to category C_i, for $i = 1, \ldots, k$. (Note that $\sum_{i=1}^{k} B_i = n$.) (You may assume that the population is large enough so that $(p_1, \ldots, p_k)$ remain constant throughout the sampling process.) Using counting techniques similar to those used in the discussion in Example 5.6, find a general expression for the joint sampling distribution of $(B_1, \ldots, B_k)$.

5.3 Approximating Sampling Distributions

Exact sampling distributions for a variety of statistics were presented in Sects. 1 and 2. While the sampling distributions are absolutely essential for making inferences based on those statistics, the computations involved in obtaining exact probabilities for the discrete distributions in Sect. 2 can become tedious and time consuming as the sample size(s) increases. Even using **R** as discussed in the previous sections, there will always be a point where exact calculations for these distributions are just not practical. This is precisely the case in the following example.

Example 5.8. Binomial Distribution for a Large Number of Bernoulli Trials Suppose we want to compute the probability that the proportion of successes is exactly equal to .5 when sampling 10,000 observations from a *Bernoulli*(p = .5) population. In other words, we want the probability of getting exactly 5000 successes in 10,000 such Bernoulli trials. As we did in Example 5.5, we can use the **R** function *dbinom*( ) to compute this probability. Behind the scenes, however, the function is not calculating the value in Theorem 5.4 directly. In fact, some other statistical software programs will produce an error rather than the number. To see why, consider the first term of Theorem 5.4, known as "n choose b". We can try to compute this with n = 10,000 and b = 5000 in **R** using the arguments *n* and *k* in the *choose*( ) function as follows.

```
> choose(n = 10000, k = 5000)
[1] Inf
```

But certainly this probability can't be infinite! The point of this example is not to illustrate a limitation of the **R** functions, but rather to increase our awareness of what we are asking the software to do. Computing the quantity $\frac{10000!}{5000!5000!}$ is difficult even for the most sophisticated technology. Note that, because the *dbinom*( ) is such a commonly used distribution, the authors have implemented a clever workaround to avoid needing to calculate these factorials. (The details, which you can read by running *?dbinom*, are well beyond the scope of this book.) However, for general distributions that we may encounter, there is no guarantee that such clever workarounds will always exist and will have already been implemented.

We are thus faced with a difficult situation. We need sampling distributions to carry out proper statistical inferences, but it is problematic for even the latest available technology to do all of the exact computations that are often required. What can we do in such situations? The answer is one of the most surprising, yet most important results, in the entire field of statistics. Students have been heard to say that this result is shocking, amazing, counter intuitive, impossible, etc. No matter what your initial reaction is, think about the upcoming result very carefully because it is the cornerstone of many statistical inference procedures.

Looking closely at the sampling distributions presented in Sects. 1 and 2, we notice that every one of the statistics considered there can be viewed as a sum or an average of individual pieces of information. We might be averaging 0's and 1's, adding counts, averaging normal observations, or adding ranks, but in each case we are adding or averaging. When a statistic can be viewed as a sum or average of many individual pieces of information, the sampling distribution of this statistic will often be approximated well by an appropriate normal distribution. Yes, it is true!! No matter what distribution the individual pieces of information follow -- different pieces can even follow different distributions -- the combination of this information through adding or averaging will generally lead to a sampling distribution that can be approximated well by an appropriate bell-shaped normal distribution!

Formal statements of these types of results are collectively known as *Central Limit Theorems*. Theorem 5.5 details a version that applies in the setting where the n pieces of information, $X_1, X_2, \ldots, X_n$, are independent and have identical distributions. In Exercise 5.3.1, you are asked to consider components of a version of the Central Limit Theorem for the setting where the pieces of information, X_i, are independent, but could have different distributions. There are also Central Limit Theorems that are appropriate even when the pieces of information are not independent, but such settings are beyond the scope of this text.

Theorem 5.5 A Central Limit Theorem If $X_1, X_2, \ldots, X_n$ are a random sample from a population with mean μ and standard deviation σ, then as the sample size n increases the sampling distributions of

$$Z_{sum} = \frac{\sum_{i=1}^{n} X_i - n\mu}{\sigma\sqrt{n}} \quad \text{and} \quad Z_{average} = \frac{\bar{X} - \mu}{\sigma/\sqrt{n}}$$

can both be well approximated by the standard normal distribution with mean 0 and standard deviation 1.

Before applying Theorem 5.5 to some of the count and average statistics presented earlier in this chapter, a few general remarks are in order. Central Limit Theorems are mathematical results that deal with sampling distributions of statistics as the sample size, n, approaches infinity. The accuracy of the approximation will differ from one application to another. While we will provide some general rules of thumb regarding the use of Central Limit Theorems, you must always remain aware that these are only approximations to the exact distributions. As we will see, in many cases the approximations are very, very good, but there are cases, especially for small and moderate sample sizes (values of n), where the approximations are inappropriate or can be improved upon by making some adjustments.

It is also important to remember that you do not need the Central Limit Theorem to tell you about the mean and standard deviation of the average of any sample of size n. An argument exactly like the one we used for a normal population in Sect. 1 shows that for a random sample $X_1, X_2, \ldots, X_n$ from any population with mean μ and standard deviation σ, the sample average $\bar{X}$ has mean μ and standard deviation $\frac{\sigma}{\sqrt{n}}$. So even without the Central Limit Theorem we know that as the sample size increases the variability in the sampling distribution of $\bar{X}$ will decrease. In other words, the larger the sample size, the more concentrated the distribution will be around its center, μ.

What the Central Limit Theorem adds to this discussion is the fact that the distribution of $\bar{X}$ also becomes approximately normal as the sample size increases. Theorem 5.5 tells us how to use the standardized statistics Z_{sum} and $Z_{average}$ and the standard normal distribution to obtain approximate detailed information about the sampling distributions of $\sum_{i=1}^{n} X_i$ and $\bar{X}$.

In Exercise 5.3.2 you are asked to show that the center and spread of the sampling distribution for the sum of the observations in a random sample depend on the sample size. The center of the sampling distribution for the sum will not be the same for all sample sizes, unless $\mu = 0$. As we sum different amounts of information, the center of the sampling distribution will typically depend on how many pieces of information we are including in our sum.

Now let's apply the Central Limit Theorem to the settings we have previously discussed. Hopefully, as you see more and more such applications you will begin to appreciate the magnitude of this result. The easiest application is to the sample average presented in Sect. 1. There we assumed that the samples were from normal populations. Now, we remove the normal assumption and stipulate that the sample sizes are large. The implications of this change are relatively straightforward. The computations remain the same, but the results are now only approximately correct.

The application of Theorem 5.5 to binomial counts B requires some additional attention. Recall that B is the total number of successes in n independent *Bernoulli*(p) trials. Since we have n independent pieces of information and B is a sum, the only remaining issue is identifying μ and σ in Theorem 5.5. Since the theorem is written in terms of the individual pieces of information, we need to recall that the mean and variance of the individual *Bernoulli*(p) random variables are $\mu = p$ and $\sigma^2 = p(1 - p)$, respectively. Thus, Theorem 5.5 implies that the standardized count

$$Z_{sum} = \frac{B - np}{\sqrt{np(1-p)}} \quad \text{is approximately N}(0,\ 1).$$

Note that this is equivalent to saying that B is approximately normal with mean np and variance $np(1\text{-}p)$ for large n.

Example 5.9. Normal Approximation for Binomial Counts Consider the sampling distribution of the number of successes in $n = 30$ independent *Bernoulli*($p = .6$) trials. In order to check the accuracy of the Central Limit Theorem in this situation, we compute the probability of getting at most 25 successes using both the exact sampling distribution and the normal approximation.

Let B denote the number of successes in the 30 Bernoulli trials. We know that $B \sim B(n = 30, p = .6)$, so that the probability of getting at most 25 successes is exactly equal to

$$P(B \leq 25) = \sum_{i=1}^{25} P(B = i) = \sum_{i=1}^{25} \binom{30}{i} (.6)^i (1 - .6)^{30-i} = .9985.$$

Using the normal approximation for this probability, we find

$$P(B \leq 25) = P\left(Z_{sum} \leq \frac{25 - 30(.6)}{\sqrt{30(.6)(1 - .6)}}\right) = P(Z_{sum} \leq 2.6087) \approx \Phi(2.6087)$$
$$= .9955.$$

Since the difference in these two probabilities, namely, $.9985 - .9955 = .003$, is close to zero, the normal approximation does a good job in this situation.

Finally, we need to decide when it is reasonable to use the normal approximation to the sampling distribution of B. In short, the answer depends on p. If $p = .5$, the sampling distribution of B is symmetric and the approach to normality is relatively fast. If p is close to 0 or 1, however, the sampling distribution of B is skewed and the approach to normality is slower. A general rule of thumb is that if np and $n(1 - p)$ are both at least 10 then it is reasonable to use the normal approximation. As we will see in Example 5.10, the accuracy of the normal approximation varies from application to application.

Normal Approximation for Binomial Counts If B is the number of successes in n independent *Bernoulli*(p) trials and n is reasonably large (i.e., $np \geq 10$ and $n(1 - p) \geq 10$), then the sampling distribution of B is approximately $N(np, \sqrt{np(1-p)})$. Thus, under these conditions, the probability that B is in the interval $[a, b]$, where a and b are integers such that $0 \leq a < b \leq n$, is approximately

$$\begin{aligned} P(a \leq B \leq b) &= P\left(\frac{a - np}{\sqrt{np(1-p)}} \leq Z_{sum} \leq \frac{b - np}{\sqrt{np(1-p)}}\right) \\ &\approx \Phi\left(\frac{b - np}{\sqrt{np(1-p)}}\right) - \Phi\left(\frac{a - np}{\sqrt{np(1-p)}}\right). \end{aligned}$$

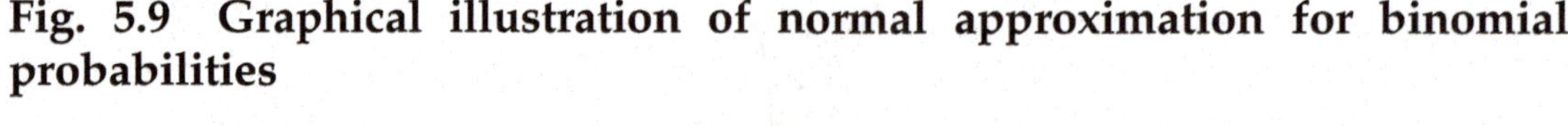

Fig. 5.9 Graphical illustration of normal approximation for binomial probabilities

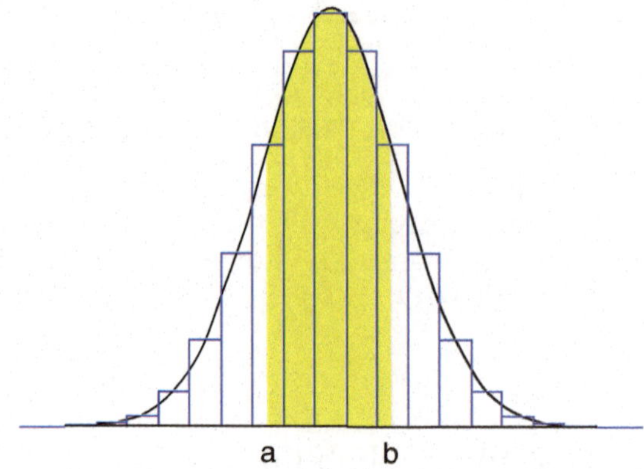

Figure 5.9 provides a graphical illustration of the normal approximation for the sampling distribution of binomial counts. The bars of the histogram represent the exact binomial probabilities. The shaded area under the normal curve between a and b provides an approximation for the probability of getting at least a but no more than b successes.

Theorem 5.4 specified how the exact sampling distribution for a proportion is directly related to a binomial distribution. Thus, the application of the Central Limit Theorem to sampling distributions for proportions will be very similar to the result above. The only major difference is that the sample proportion, $\hat{p}$, is an average. Thus, we will apply Theorem 5.5 with $\mu = p$ and $\sigma = \sqrt{p(1-p)}$.

Normal Approximation for Proportions If $\hat{p}$ is the proportion of successes in n independent *Bernoulli*(p) trials and n is reasonably large (i.e., $np \geq 10$ and $n(1-p) \geq 10$), then the sampling distribution of $\hat{p}$ is approximately $N\left(p, \sqrt{\frac{p(1-p)}{n}}\right)$. The probability that $\hat{p}$ is in the interval $[a/n,\ b/n]$, where a and b are integers such that $0 \leq a < b \leq n$, is approximately

$$P\left(\frac{a}{n} \leq \hat{p} \leq \frac{b}{n}\right) = P\left(\frac{\frac{a}{n}-p}{\sqrt{\frac{p(1-p)}{n}}} \leq Z_{average} \leq \frac{\frac{b}{n}-p}{\sqrt{\frac{p(1-p)}{n}}}\right)$$
$$\approx \Phi\left(\frac{\frac{b}{n}-p}{\sqrt{\frac{p(1-p)}{n}}}\right) - \Phi\left(\frac{\frac{a}{n}-p}{\sqrt{\frac{p(1-p)}{n}}}\right).$$

Example 5.10. Normal Approximation for Binomial Counts and Percentages Let's compare the normal approximation for binomial counts with the normal approximation for proportions and the exact binomial probability for a particular event of interest. Consider a $B(40, 0.3)$ random variable for which we want to find the probability of getting at most 30% successes. Since 30% of 40 equals 12, this problem can be rewritten as $P(B \leq 12)$, where $B \sim B(40, 0.3)$. Using the *pbinom( )* function in **R**, we find that $P(B \leq 12) = .5772$.

```
> pbinom(q = 12, size = 40, prob = 0.3)
[1] 0.5771809
```

Figure 5.10 shows the normal approximation for $P(B \leq 12)$ based on counts. Since the shaded area under the $N\left(12, \sqrt{40(.3)(1-.3)}\right)$ curve to the left of 12 accounts for half of the total area under this normal distribution, we have $P(B \leq 12) \approx 0.5$. The difference between the exact probability and the approximate probability is $0.5772 - 0.5 = 0.0772$. Even though our rule of thumb is satisfied, since $40(.3) = 12 \geq 10$ and $40(.7) = 28 \geq 10$, this approximation is not as accurate as the one in Example 5.9. The reason for this loss of accuracy is the slight skewness in the shape of the exact $B(40, 0.3)$ distribution, since $p = 0.3$ is further from 0.5. Although the normal distribution does not fit quite as well in this situation, it still provides a reasonably good approximation.

Fig. 5.10 Normal approximation for $P(B \leq 12)$ when $B \sim B(40,0.3)$

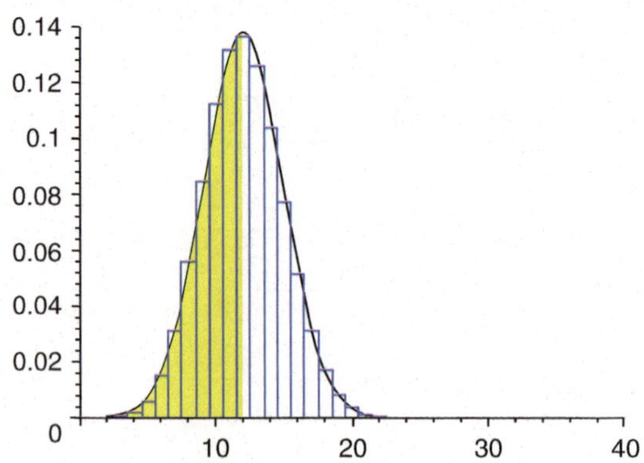

To apply the normal approximation for proportions we need to use the $N\left(.3, \sqrt{\frac{(.3)(1-.3)}{40}}\right)$ distribution to find $P(\hat{p} \leq .3)$. Since half the area under the $N\left(.3, \sqrt{\frac{(.3)(1-.3)}{40}}\right)$ curve will be to the left of .3, we get the same answer using the normal approximation for proportions, namely, $P(\hat{p} \leq .3) \approx 0.5$, as we did when we used the normal approximation for counts. The bottom line is that it doesn't matter if you use the approximation for counts or the approximation for proportions—you will always get the same answer. If you are more comfortable with one or the other, please take that approach to solving problems—just be sure to use the mean and standard deviation that are appropriate for your choice.

Section 5.3 Practice Exercises

5.3.1. *Normal Approximations.* Suppose $X_1, \ldots, X_n$ are independent random variables with means $\mu_1, \ldots, \mu_n$ and standard deviations $\sigma_1, \ldots, \sigma_n$, respectively. In most practical applications the sampling distributions for $\sum_{i=1}^{n} X_i$ and $\bar{X}$ will become approximately normal as n increases.

(a) Find the means of the approximate sampling distributions for $\sum_{i=1}^{n} X_i$ and $\bar{X}$.

(b) Find the standard deviations of the approximate sampling distributions for $\sum_{i=1}^{n} X_i$ and $\bar{X}$.

(c) Identify the standardized variables Z_{sum} and Z_{average} for this setting, where we are summing or averaging independent but not identically distributed random variables.

5.3.2. *Approximate Sampling Distribution.* Rewrite the statement in Theorem 5.5 in terms of an approximate sampling distribution for $\sum_{i=1}^{n} X_i$ and comment on the center and spread of this approximate sampling distribution. Does the center of the sampling distribution depend on n? What happens in the special case where $\mu = 0$?

5.3.3. *Unbalanced Coin Tosses.* Consider the experiment where you toss an unbalanced coin with probability .4 for obtaining a head. Calculate the probability of getting at least 20 heads in 40 tosses using:

(a) the binomial distribution;

(b) the normal approximation;

(c) Comment on the accuracy of the approximation.

5.3.4. *Bernoulli Trials.* Consider the sampling distribution for the number of successes in 20 independent Bernoulli trials with $p = .1$. Calculate the probability of getting less than 3 successes using:

(a) the binomial distribution;

(b) the normal approximation;

(c) Comment on the accuracy of the approximation.

5.3.5. *Baseball and Beer.* Baseball is the American pastime, but what goes with watching a baseball game? The well-known song says peanuts and crackerjack, but how about some beer to wash those snacks down? Wolfe et al. (1998) conducted a study to see just how much beer and baseball had become synonymous. Male spectators of drinking age were sampled over a three-game period—on a Friday night, a Saturday afternoon, and a Monday night—during the 1993 season at two major ballparks. Wolfe et al. found that 65 out of 166 sampled spectators in the age group 20–35 had consumed alcohol immediately prior to entering the ballpark.

(a) What percentage of their interviewees had consumed alcohol immediately prior to entering the ballpark?

Assume that this percentage is applicable to all male spectators at baseball games. Suppose that you interview your own sample of 75 male baseball spectators in the age group 20–35 at the next baseball game you attend.

(b) How many of your 75 interviewees would you expect to have consumed alcohol immediately prior to entering the ballpark?

(c) What is the approximate probability that the sample percentage of your interviewees who had consumed alcohol immediately prior to entering the ballpark is greater than 50%?

(d) What is the approximate probability that fewer than 20 of your interviewees had consumed alcohol immediately prior to entering the ballpark?

(e) Do you think the results obtained by Wolfe et al. remain valid today? How do you think they might have changed? What about female spectators at baseball games? Are their alcohol consumption habits similar to male spectators? How would you support your opinions? Check it out!

5.3.6. *College Alumni Surveys.* Past experience indicates that about 40% of all college alumni who receive a survey will take the time to complete and return it. A self-study committee at a small liberal arts college mails a survey to 4000 randomly selected alumni.

(a) What are the mean and standard deviation for the random number of alumni who complete and return the survey?

(b) Approximate the probability that at least 1650 alumni complete and return the survey.

5.3.7. *Multiple Choice Exams.* Your instructor has decided to construct a 250 question multiple choice final exam for the course, with five possible choices for each question. Suppose that you get so busy that you do not find time to study for this final exam. As a result, you decide to simply guess (select one of the five choices in a completely random fashion) on each question.

(a) What is the sampling distribution for your number of correct responses on the exam?

(b) What is your expected score on the exam?

(c) Compute the variance and standard deviation for your number of correct responses.

(d) Is it likely that you will score over 50% on this exam? Justify your response.

5.3.8. *Art and the Color Purple.* Wypijewski (1997) reported on the results of a comprehensive scientific poll of American tastes in art, as commissioned by Vitaly Komar and Alexander Melamid in conjunction with the National Institute, a nonprofit offshoot of *The Nation* magazine. A random sample of 526 females were aked to name their favorite color. Thirty-seven females indicated that purple was their favorite color.

(a) What percentage of the respondents named purple as their favorite color?

Suppose you were able to interview a random sample of 125 of the participants in this study.

(b) How many of your 125 interviewees would you expect to name purple as their favorite color?

(c) What is the expected value of the sample proportion of your interviewees who name purple as their favorite color?

(d) What is the standard deviation of the sample proportion?

(e) What is the approximate probability that more than 10% of your interviewees name purple as their favorite color?

(f) What is the approximate probability that fewer than 10 of your interviewees name purple as their favorite color?

(g) What is the **exact** probability that more than 40 of your interviewees name purple as their favorite color? Justify your answer.

5.3.9. *Crates of Potatoes.* The weight of a crate of potatoes averages 1200 pounds with a standard deviation of 25 pounds. At the end of the day, 20 crates of potatoes are still sitting in a storage area. Approximate the probability that more than 25,000 pounds of potatoes are in the storage area. What assumptions are you making in your calculations?

5.3.10. *Grocery Bills.* Suppose the average weekly grocery bill for a family of four is $140 with a standard deviation of $10.

(a) Estimate the amount of money that this family will spend on grocery bills in 1 year.

(b) Will the standard deviation of the annual grocery expense be larger or smaller than $10? Explain.

(c) Will the standard deviation of the average weekly grocery expense for 1 year be larger or smaller than $10? Explain.

(d) Approximate the probability that the average weekly grocery expense for 1 year is less than $135.

5.3.11. *Stress.* Suppose 75% of the American people feel more stress this year than theydid last year. A random sample of 150 individuals is selected.

(a) What is the approximate probability that 105 or more of the individuals in the sample feel more stress this year than they did last year?

(b) What is the approximate probability that 70% or more of the selected individuals feel more stress this year than they did last year?

5.3.12. *Airline Overbookings.* An airline would like to fill an airplane with 107 seats. Since some individuals do not show up for their flights (even though they purchase tickets), the airline typically overbooks flights.

(a) Suppose the airline sells 110 tickets for this flight. If the chance that an individual passenger will not show up for the flight is .01 (independently of all the other passengers), what is the probability that the plane will not be full when it takes off?

(b) If the airline sells 110 tickets for this flight and the chance that an individual passenger will not show up for the flight is .03 (independently of all the other passengers), what is the probability that the plane will not be full when it takes off?

5.3.13. *Family Vacation Times.* A selective program would like to have 250 students participate each week of its summer institute. The director of the program knows that July 4 is a popular vacation time for families so he admits 300 students for the week of July 4. If the chance of a student accepting the offer and attending the institute is .8 (independently of all the other students), what is the probability that the director will get more participants than he wants during the week of July 4? Will this probability increase or decrease if the chance of a student accepting the offer and attending the institute drops to .7?

5.4 Simulating Sampling Distributions

In this section we take a different approach to approximating sampling distributions. Technology has changed the way the world operates and statistical methodology and practice is no exception. The techniques in this section make heavy use of modern technology, but they are also very intuitive and easy to understand. The methods we discuss, simulation and bootstrapping/resampling, represent two of the most important developments in the field of statistics in the last few decades and their use will only continue to expand over the years to come.

The ideas are very simple. To understand the features (shape, center, spread, etc.) of a sampling distribution for a particular statistic, we simply simulate samples again and again from the appropriate population or a representative sample, compute the value of the statistic of interest for each simulated sample, and combine these values to form a simulated sampling distribution for the statistic. This simulated sampling distribution can then be used to approximate the shape, center, spread, or other characteristics of interest, as well as relevant probabilities, for the true sampling distribution of the statistic.

Figure 5.11 illustrates the first of two common simulation methods that will be used repeatedly throughout this text. It is applicable whenever we know everything about the distribution of the population from which we are sampling or when we are willing to make an assumption about the form of the underlying population. Our goal is to learn about a sampling distribution that we cannot explicitly obtain analytically and is very difficult and tedious to compute directly. We have already used this method twice before in Sect. 2. To construct Fig. 5.6, samples were repeatedly obtained from $B(10, 0.4)$ and $B(12, 0.8)$ distributions, the corresponding sample proportions $\hat{p}_1$ and $\hat{p}_2$ were created, and then the differences $\hat{p}_1 - \hat{p}_2$ were formed. Figure 5.8 was created by repeatedly selecting 10 integers at random from the set of possible ranks

Fig. 5.11 Basic idea behind the use of simulation to approximate sampling distributions

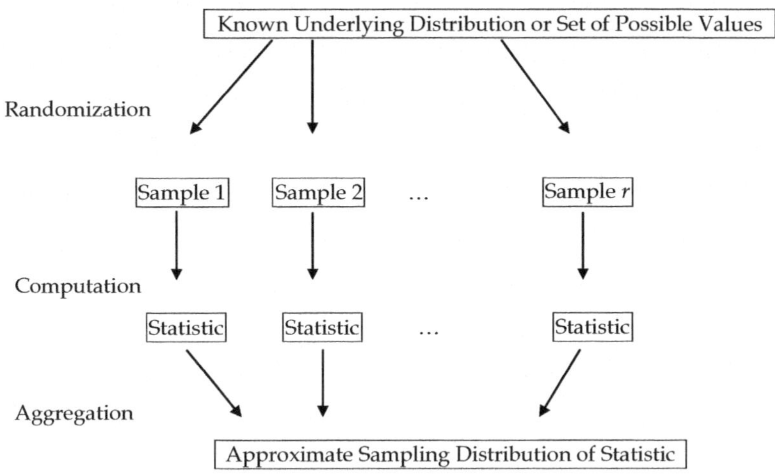

{1, 2,. . ., 21} to assign as ranks for the women, with the remaining 11 ranks assigned to the men. Exact sampling distributions can be obtained in both settings, but the calculations involved are very tedious. By repeatedly using random sampling to obtain different samples from either a known population or the complete set of possible values, we can obtain reliable approximate sampling distributions. After computing the statistic of interest (e.g., mean, median, IQR, sample percentage, difference in rank averages, standard deviation, etc.) for each random sample and aggregating these values, we can use the techniques from Chap. 1 to display the approximate sampling distribution for the statistic.

The more repetitive samples we take (i.e., increasing *r*), the better the approximation will become. This technique is used in a wide variety of statistical and practical settings.

Example 5.11. A Classroom Activity to Simulate the Sampling Distribution for Making Inferences About the Proportions in a Categorical Population Mars, Inc. claims that the color ratio in plain M&M's is 30%

Table 5.3 Color distribution for a bag of 400 plain M&M's

Color	Brown	Yellow	Red	Orange	Green	Blue
Observed count	104	71	93	33	38	61
Expected count	120	80	80	40	40	40

brown, 20% yellow, 20% red, 10% orange, 10% green, and 10% blue. Furthermore, the company claims that each large production batch is blended precisely to the ratios specified above and mixed thoroughly. Suppose we select a large bag of plain M&M's from a grocery store shelf and count how many of each color are in the bag. For illustrative purposes only, we assume there are 400 plain M&M's in the bag and the color distribution for our bag is given in Table 5.3. Do you think the observed counts in Table 5.3 are generally in line with the color distribution stipulated by Mars, Inc. or does there appear to be some concern with their claim based on this one sample?

If the company's claim is true then we would expect to get $400(.3) = 120$ brown, $400(.2) = 80$ each of yellow and red, and $400(.1) = 40$ each of orange, green, and blue M&M's. These expected counts are also shown in Table 5.3. Now, the question is what statistic should we use to measure the overall difference between what we observed and what we expected to observe if the Mars, Inc. claim is true? A natural choice is known as the *Pearson goodness of fit statistic* and is given by

$$G = \sum_{\text{all categories}} \frac{(\text{observed count} - \text{expected count})^2}{\text{expected count}}.$$

Note that this is just one possible statistic that can be used to compare the observed counts with the expected counts. Perhaps you might be interested in averaging the absolute value of the differences or looking at some other function of the differences. The beauty of this simulation approach is that you can investigate whichever statistics you think are reasonable! Of course,

Table 5.4 Calculating the Pearson goodness of fit statistic G for our bag of M & M's

Color	$O - E$	$(O - E)^2$	$(O - E)^2/E$
Brown	104–120 = −16	256	256/120 = 2.1333
Yellow	71–80 = −9	81	81/80 = 1.0125
Red	93–80 = 13	169	169/80 = 2.1125
Orange	33–40 = −7	49	49/40 = 1.225
Green	38–40 = −2	4	4/40 = 0.1
Blue	61–40 = 21	441	441/40 = 11.025
			$G = 17.60833$

you should choose wisely, as you would still need to convince others that your statistic is appropriate for evaluating the color distribution claim.

Table 5.4 shows the calculations necessary to compute G. After finding $G = 17.60833$ for our particular bag of plain M&M's, we still need to have a way to assess whether or not this observed value of the statistic is in reasonable agreement with the Mars, Inc. claim. To make this decision, we simulate the sampling distribution of G by selecting $r = 100$ independent samples, each of size $n = 400$, from a multinomial distribution with $p_1 = .3$, $p_2 = .2$, $p_3 = .2$, $p_4 = .1$, $p_5 = .1$, and $p_6 = .1$, corresponding to the categorical color proportions claimed by Mars, Inc. We can use the **R** functions *rmultinom*( ) and *chisq.test*( ) to generate samples and to calculate the goodness of fit statistic, respectively. Using the following code, we generated 100 independent samples of size 400 each, calculated G for each of the samples, and stored the results in the local variable *g_results*.

```
> color_proportions <- c(0.3, 0.2, 0.2, 0.1, 0.1, 0.1)
> n_samples <- 100
>
> samples <- rmultinom(n = n_samples, size = 400, prob = color_proportions)
> g_results <- numeric(n_samples)
> for(i in 1:n_samples){
   g_results[i] <- chisq.test(samples[,i], p=color_proportions)$statistic
 }
```

The 100 simulated values of G are listed in Table 5.5 (from smallest to largest) and displayed graphically with a histogram in Fig. 5.12. This Figure

Table 5.5 Ordered simulated values ($r = 100$) of the Pearson goodness of fit statistic G

0.9208	1.1375	1.2375	1.2833	1.3208	1.3500	1.3500	1.5083	1.6708	1.7250	1.7375	1.8083
1.9875	1.9875	2.0375	2.3875	2.4208	2.4750	2.4875	2.6208	2.7208	2.7208	2.7333	3.0000
3.0833	3.2000	3.2750	3.3500	3.3583	3.4208	3.5375	3.6500	3.6583	3.7208	3.7583	4.0083
4.0208	4.1750	4.2000	4.3375	4.5833	4.6333	4.7000	4.8083	4.8083	4.8208	4.8375	4.8375
4.9083	4.9208	5.1375	5.2833	5.3083	5.4000	5.4083	5.4208	5.5333	5.6500	5.6583	5.7083
5.7583	5.8000	5.8708	5.9208	5.9500	6.1333	6.1333	6.2083	6.3750	6.4000	6.5875	6.8333
6.8375	6.8833	7.0208	7.0583	7.2000	7.2708	7.2833	7.4208	7.5000	7.6833	7.8750	8.0583
8.3708	8.4375	8.5875	8.6500	9.0875	9.2833	9.3375	9.9375	9.9833	10.1083	10.8833	
10.9333	11.0208	13.6708	14.3333	17.6083							

Fig. 5.12 Histograms for simulated sampling distribution for Pearson goodness of fit statistic G for $r = 100$ and $r = 1000$ random samples

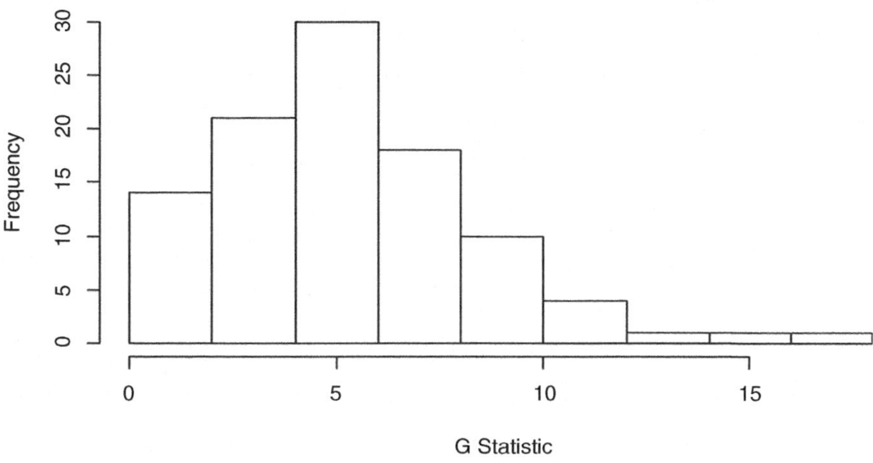

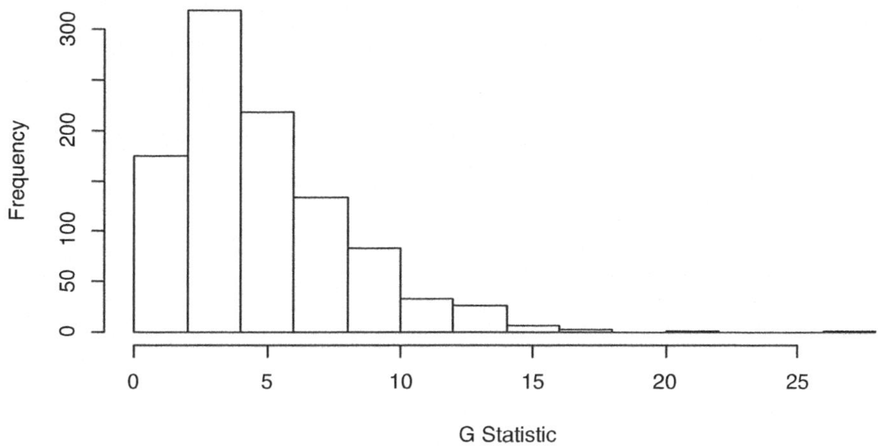

also shows the results of a simulation for 1000 values of G (we use the same code as above, but change *n_samples* to be 1000). These histograms provide approximations for the sampling distribution of G (the approximation is more accurate, of course, using $r = 1000$ samples than for using only $r = 100$) and can be used to describe the shape, center, and spread in the sampling distribution, as well as compute approximate probabilities for G.

Notice that the value $G = 17.60833$ that we obtained for our one sample bag of plain M&M's is equal to the largest value of G observed in the 100 simulated samples. This is not a coincidence. In fact, to make a point we picked our counts in Table 5.3 so that they would yield the largest value of G from our simulation. Now YOU take a sample of 400 pieces from a real bag of plain M&M's, count the colors, and compute the value of G for your sample. (Of course, enjoy the candy, too!) Where is the value of the statistic G for your sample located in the ordered list in Table 5.5? Where does it appear on the histogram? Is it somewhere in the middle or is it more toward one of the tails? What does this tell you about the color distribution claim by Mars, Inc.? If the value of your test statistic happens to be larger than 17.60833, what does that tell you about the claim?

When sampling from a categorical population with k categories, the statistic G is often used to compare observed counts with expected counts. A formal description of the Pearson goodness of fit statistic is provided in Definition 5.2. Although the simulated sampling distributions in Fig. 5.12 provide reasonable approximations to the exact distribution of G, some readers may be uncomfortable relying completely on simulation and not having a mathematical approximation to the sampling distribution of G. For a large number of sample observations, n, a mathematical approximation to the sampling distribution of G is provided by the chi-square distribution with $(k - 1)$ degrees of freedom. (We point out that the simulated sampling distribution for G given in Fig. 5.12 for $r = 1000$ random samples looks very much like the corresponding chi-square distribution with $k - 1 = 5$ degrees of freedom.) While the mathematical function for a chi-square distribution is not of interest to us, percentiles for chi-square distributions can be found using the *pchisq*( ) and *qchisq*( ) functions in **R** with the relevant degrees of freedom specified.

Definition 5.2 Consider a categorical population with k distinct categories $C_1, \ldots, C_k$ and proportions p_i in category C_i, for $i = 1, \ldots, k$. When collecting a random sample of size n from such a population, the Pearson goodness of fit statistic G for comparing observed (B_i) and expected (np_i) counts, is given by $G = \sum_{i=1}^{k} \frac{(B_i - np_i)^2}{np_i}$. The approximate ($n$ large) sampling distribution for G is the chi-square distribution with $(k - 1)$ degrees of freedom.

We will also make use of a second simulation method known as *bootstrapping* or *resampling*. In this setting very little is known about the distribution of the population. In fact, the only information we have about the population is contained in a single random sample from it. We use randomization to *resample* from this one set of data and assume that these samples, known as bootstrap[2] samples, are representative of what we might see if we were to sample repeatedly from the whole population, even though they are all based solely on the single observed random sample. We can bootstrap to estimate parameters for the population, investigate the overall shape of the sampling distribution for a statistic, or make inferences about the population. The size of each bootstrap sample is often taken to be equal to the size of the single observed sample, so the bootstrap sampling is usually done with replacement, as will be the case in our examples. Impact of the choice for how many bootstrap samples to obtain when resampling is explored in the next example.

[2] The term bootstrap originates from an old expression that encouraged individuals to improve matters by lifting themselves up by their own bootstraps. Real life bootstraps are not very common now, but the process is similar to the more common concept of rebooting a computer from a core set of instructions. When all else fails, reboot. In statistical settings, when the calculations are tedious or the situation is difficult because you cannot remember the detailed formulas for a distribution, try bootstrapping.

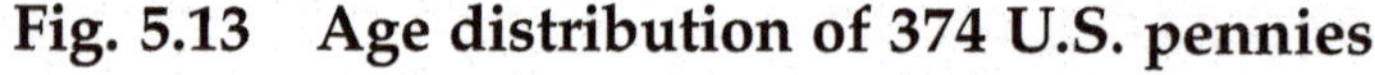

Fig. 5.13 **Age distribution of 374 U.S. pennies**

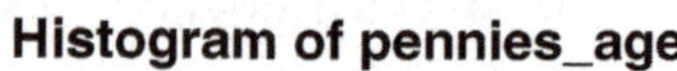

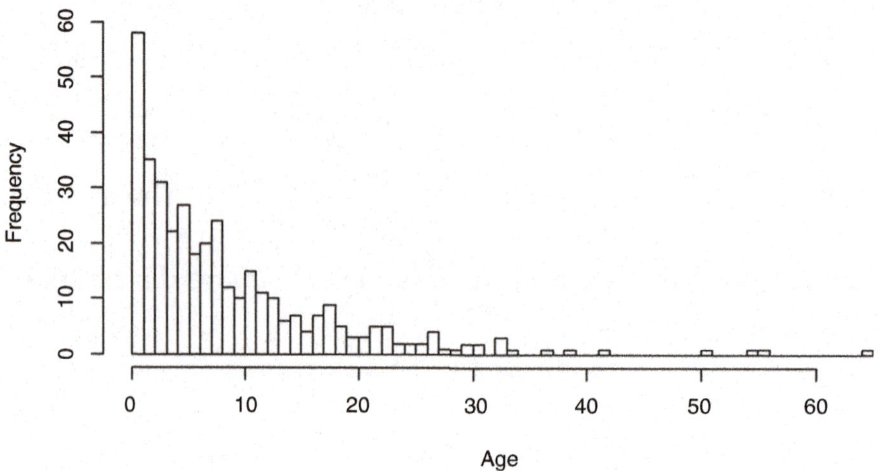

Example 5.12. Bootstrapping the Age Distribution of U.S. Pennies An introductory class of 25 students collected 374 United States pennies and calculated the ages of the pennies by subtracting the minting dates printed on the pennies from the current year.

The age distribution of the 374 pennies, which can be accessed via the **R** dataset *pennies_age*, is shown in Fig. 5.13 and some descriptive statistics are provided below.

```
> summary(pennies_age)
   Min. 1st Qu.  Median    Mean 3rd Qu.    Max.
  0.000   3.000   6.000   9.053  12.000  65.000
> sd(pennies_age)
[1] 9.439653

> hist(pennies_age,
         breaks = 50,
          xlab = "Age")
```

Although the 374 pennies were not randomly selected, we will assume that they are representative of the age of U.S. pennies currently in circulation. Not surprisingly, the age distribution is clearly skewed to the right. To

Fig. 5.14 Approximate sampling distribution for sample average age of 5 U.S. pennies

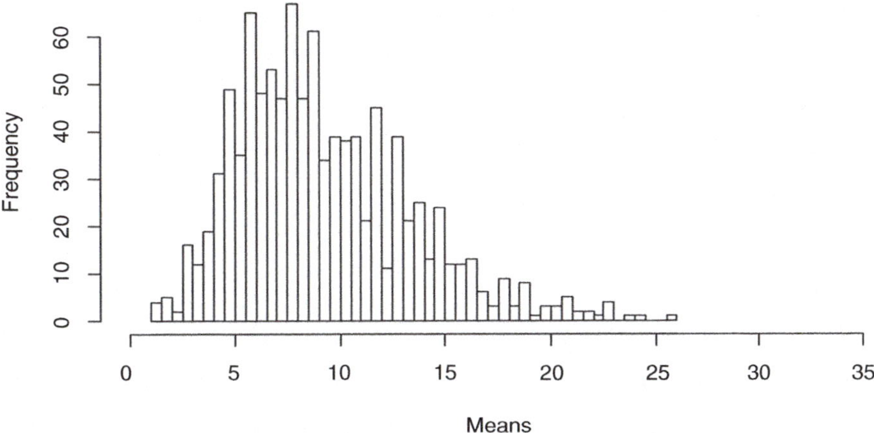

investigate the shape (and other properties) of the sampling distribution of the average age, $\bar{X}$, for the pennies, we take 1000 individual samples of size 5 each, 1000 individual samples of size 10 each, 1000 individual samples of size 20 each, and 1000 individual samples of size 40 each from the 374 observed ages, all sampling done, of course, with replacement. For each of the 4000 samples we compute the average age, yielding 1000 sample average ages for size 5 samples, 1000 sample average ages for size 10 samples, 1000 sample average ages for size 20 samples, and 1000 sample average ages for size 40 samples. Figures 5.14, 5.15, 5.16, and 5.17 present the approximate sampling distributions of the sample average obtained in this manner for sample sizes 5, 10, 20, and 40, respectively. The **R** code to generate the plot for 1000 sample averages for size 5 samples, which uses the *replicate*( ) and *sample* ( ) functions and specifies a handful of arguments for the *hist*( ) function for aesthetics, is provided below and can be modified for the other sample sizes.

```
> hist(x = replicate(1000, mean(sample(pennies_age, 5, replace = TRUE))),
       breaks = 50,
       xlim = c(0,35),
       xlab = "Means",
       main = "Histogram of 1000 Replicates of Mean for 5 Sampled Pennies ")
```

Fig. 5.15 Approximate sampling distribution for sample average age of 10 U.S. pennies

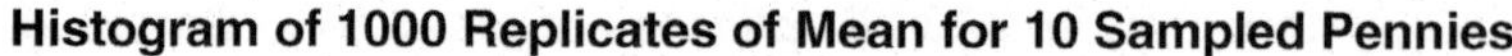

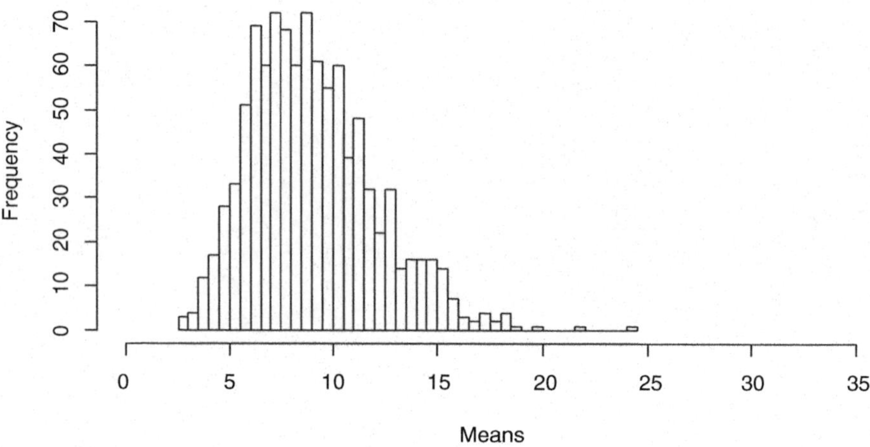

Fig. 5.16 Approximate sampling distribution for sample average age of 20 U.S. pennies

Histogram of 1000 Replicates of Mean for 20 Sampled Pennies

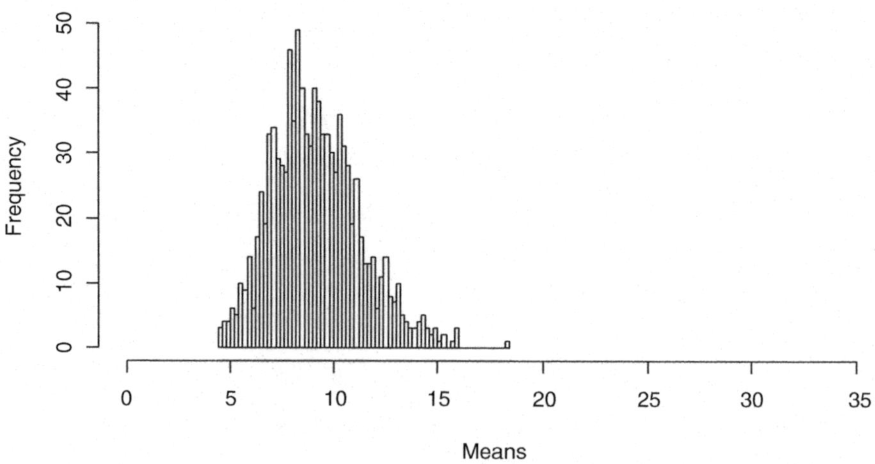

We are using the information contained in our one observed sample of 374 U. S. pennies to estimate what the sampling distribution of $\bar{X}$ would look like for samples of various sizes (5, 10, 20, and 40) drawn from the entire population of all U. S. pennies in circulation. We are not really interested, for

Fig. 5.17 Approximate Sampling Distribution for Sample Average Age of 40 U.S. pennies.

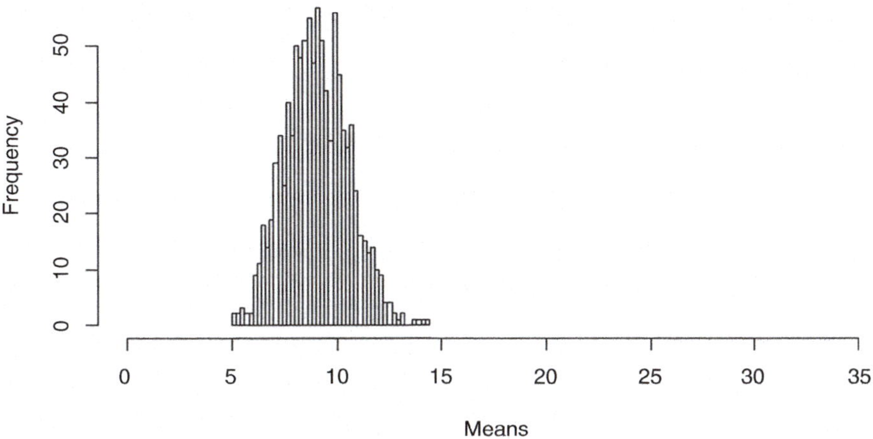

example, in the average ages for the individual 1000 samples of size 5 selected from our 374 pennies, but, instead, we are interested in what these 1000 sample averages of size 5 are able to tell us about the sampling distribution for the sample average of a single sample of size 5 selected from the entire population of all U. S. pennies in circulation. We don't want to collect additional pennies from the full population, so we resample from the ones we've already collected, letting them serve as representatives of the entire population. We are simply trying to extract as much information as we can from this one random sample of 374 pennies that we have already collected.

Notice that the approximate sampling distributions for the various sample averages are NOT skewed to the right like the original age distribution for the 374 pennies shown in Fig. 5.13. In fact, they are roughly bell-shaped and symmetric! The centers of all four approximate sampling distributions appear to be somewhere between 8 and 10 years, agreeing roughly with the average age of 9.053 for all 374 pennies. As the sample size increases from 5 to 40, the approximate sampling distributions become more bell-shaped and the variability clearly decreases. While the main thrust of this example has been to illustrate how to use the bootstrapping technique to simulate the sampling distribution of $\bar{X}$ for sample sizes $n = 5$, 10, 20, and 40, the histograms also provide bonus empirical evidence to support the Central Limit Theorem presented in Sect. 3. As the sample size n increases from 5 to 40, the true sampling distribution of $\bar{X}$ looks more and more like a $N\left(9.053,\ \frac{9.439653}{\sqrt{n}}\right)$, as formally prescribed by the Central Limit Theorem.

This bootstrapping technique can be used for <u>any</u> statistic, not just the sample mean $\bar{X}$. For example, instead of computing the sample average for all 4000 samples we could have computed the sample median, range, standard deviation, IQR, etc. Figures 5.18, 5.19, 5.20, and 5.21 depict the bootstrap approximate sampling distributions of the sample median for samples of size 5, 10, 20, and 40, respectively. Notice that the approximate sampling distributions for the sample median do NOT look like either the approximate sampling distributions for the sample average OR the original age distribution for all 374 pennies. Although the exact form of the sampling distribution for the sample median depends on the distribution of the ordered sample

Fig. 5.18 Approximate sampling distribution for sample median age of 5 U.S. pennies

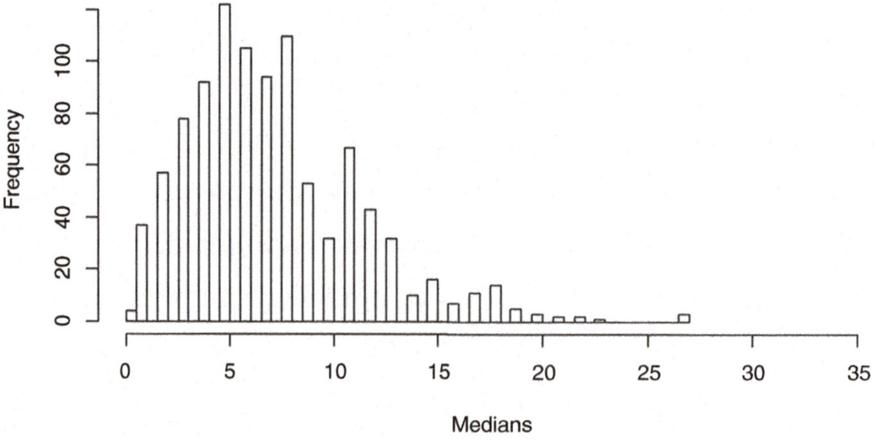

Fig. 5.19 Approximate sampling distribution for sample median age of 10 U.S. pennies

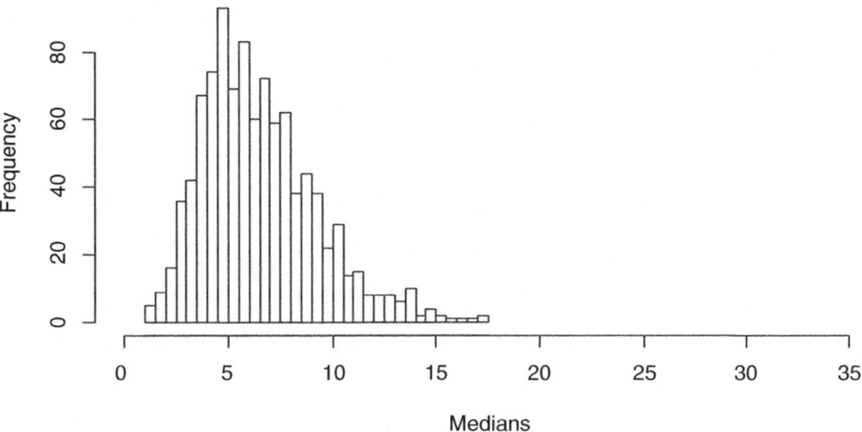

Fig. 5.20 Approximate sampling distribution for sample median age of 20 U.S. pennies

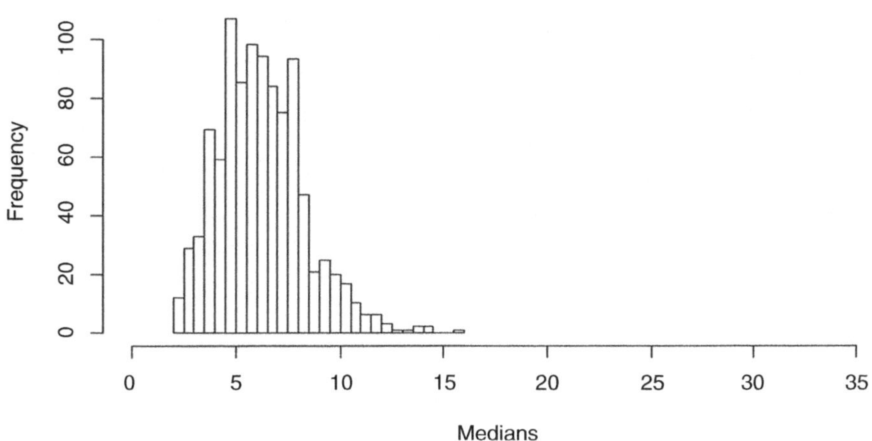

items (i.e., order statistics) and is beyond the scope of this course, we can clearly use bootstrapping to approximate it.

Bootstrapping can be used in a wide variety of settings. The power, utility, and simplicity of bootstrapping make it one of the most important recent

Fig. 5.21 Approximate sampling distribution for sample median age of 40 U.S. pennies

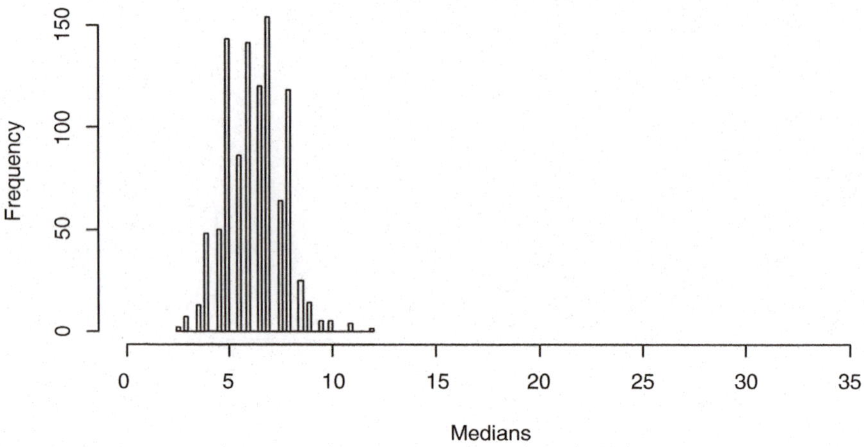

Fig. 5.22 Summary of bootstrapping/resampling technique for approximating sampling distributions

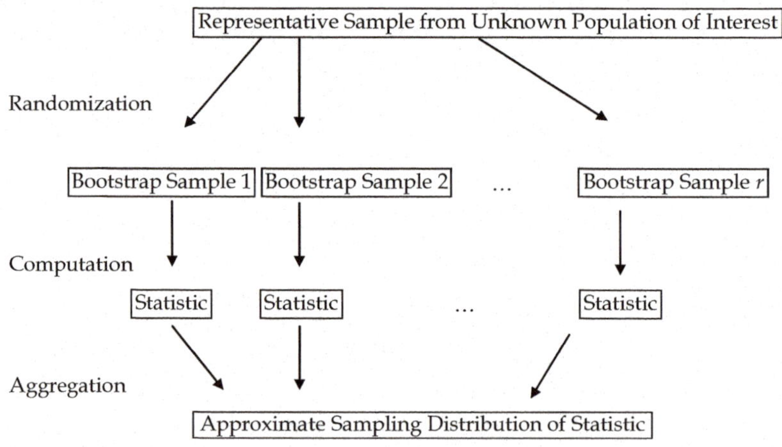

developments in the field of statistics. Figure 5.22 summarizes the major steps involved in using bootstrapping/resampling methods to approximate a sampling distribution. One remaining question that we need to address is: How many bootstrap samples are needed? The short (and a bit amazing) answer is that, for <u>any</u> original sample size n, as we increase the number of bootstrap samples (i.e., the number of runs in our simulation) the simulated sampling distribution will look more and more like the true sampling distribution for the statistic of interest. Using 500 or more bootstrap samples will almost always be enough to get a reasonable approximation to a sampling distribution. In fact, there are many situations where using 100 or fewer bootstrap samples will yield rough approximations that are sufficient to address the questions of interest.

Section 5.4 Practice Exercises

5.4.1. *Random Numbers.* Use the **R** function *runif*( ) to simulate $r = 5$ samples of size $n = 10$ each chosen randomly from the interval (0,1).

(a) Create histograms for each of the 5 samples.
(b) Do the histograms vary from sample to sample?
(c) Are there similarities in the histograms?
(d) Sketch the distribution that these histograms are approximating.
(e) Repeat part (a) for $r = 5$ samples of size $n = 20$ each from the interval (0,1).
(f) Repeat part (a) for r $= 5$ samples of size $n = 30$ each from the interval (0,1).
(g) What happens to the approximating histograms as n increases?

5.4.2. *Random Numbers.* Simulate $r = 50$ samples of size $n = 10$ each chosen randomly from the interval (0, 1) using the **R** function *runif*( ). Find the sample mean for each of the 50 samples. Display the approximate sampling distribution for the sample mean and comment on its shape. Repeat the simulation. Does the approximate sampling distribution from your second simulation look exactly like the one obtained in your first simulation?

5.4.3. *Random Numbers.* Simulate $r = 1000$ samples of size $n = 10$ each chosen randomly from the interval (0, 1) using the **R** function *runif*( ). Find the sample mean for each of the 1000 samples. Display the approximate sampling distribution for the sample mean. Repeat the simulation for samples of size $n = 20$. What happens to the center and spread of the approximate sampling distributions as n increases from 10 to 20?

5.4.4. *Random Numbers.* Simulate $r = 1000$ samples of size $n = 10$ each chosen randomly from the interval (0, 1) using the **R** function *runif*( ). Find the sample median for each of the 1000 samples. Display the approximate sampling distribution for the sample median and comment on its main features.

5.4.5. *Random Numbers.* Simulate $r = 1000$ samples of size $n = 10$ each chosen randomly from the interval (0, 1) using the **R** function *runif*( ). Find the sample standard deviation for each of the 1000 samples. Display the approximate sampling distribution for the sample standard deviation and comment on its main features.

5.4.6. *Pay Increases.* Suppose the average pay increase for 52 employees at a company was 3.4% in 2004. A particular employee was unhappy with the size of his pay increase and wanted to learn more about the minimum raise offered by the company. He has asked you to help him simulate $r = 500$ samples of size $n = 52$ each from a normal distribution with mean 3.4 and standard deviation 0.5 using the **R** function *rnorm*( ). For each of the random samples he wants you to compute and save the smallest pay increase for that random sample. Display the approximate distribution for the sample minimum (i.e., smallest pay increase). If the employee received a 1% raise, do you think his unhappiness is justified?

5.4.7. *Bernoulli Simulation.* Simulate $r = 100$ random samples of size $n = 25$ each from a Bernoulli population with $p = 0.5$ using the **R** function *rbinom*( ). Display the approximate sampling distribution of the sample percentage $\hat{p}$. Repeat the simulation for $p = 0.1$ and $p = 0.8$. Do the shapes of the sampling distributions for $\hat{p}$ appear the same for all three simulations? Where are the sampling distributions centered? What about the variability in the sampling distributions?

5.4.8. *Coin Flips.* Simulate 25 flips of a balanced coin 100 separate times and compute the percentage $\hat{p}$ of heads for each of the 100 iterations using the **R** function *rbinom*( ). Display the approximate sampling distribution of $\hat{p}$. Repeat your simulation for 100 and 400 tosses of a balanced coin. Display the three approximate sampling distributions of $\hat{p}$, using the same horizontal and vertical scale. What happens to the center and spread of the sampling distributions as the number of flips increases?

5.4.9. *Age of Pennies.* The dataset *pennies_age* contains the ages of 374 U.S. pennies. Simulate $r = 1000$ bootstrap samples of size $n = 10$ each using the **R** function *sample*( ) and find the sample standard deviation for each sample. Display the approximate bootstrap sampling distribution for the sample standard deviation for samples of size 10.

5.4.10. *Home Values.* The dataset *homes_prices* contains the total assessed values for a sample of homes in Wake County, North Carolina. Simulate $r = 100$ bootstrap samples of size $n = 25$ each using the **R** function *sample*( ) and compute the sample mean and sample median for each sample. Display the approximate sampling distributions for the sample mean and sample median. Comment on the similarities and differences in these two approximate sampling distributions.

Chapter 5 Comprehensive Exercises

5.A. Conceptual

5.A.1. *Continuity Correction for the Normal Approximation to a Binomial Distribution.* The accuracy of the normal approximation for binomial counts can be improved by making one simple modification. Since B is a count of the total number of successes, it can assume only integer values. When we consider the sampling distribution of B to be approximately normal, however, we are also

Fig. 5.23. Normal approximation for binomial counts using the continuity correction

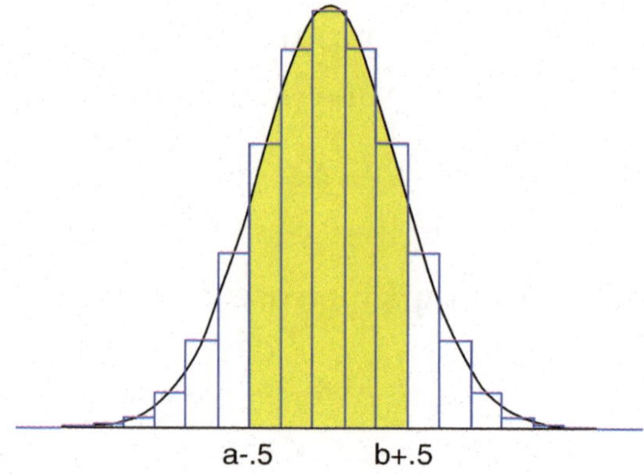

permitting B to assume values between the integers. To correct for the fact that we are using a continuous distribution to approximate a distribution that can assume only integer values we can adjust the process by acting as if the values of B are obtained by rounding to the nearest integers. This amounts to adding or subtracting 1/2 to the number of successes and is referred to as the *continuity correction*. Figure 5.23 illustrates the use of this continuity correction for the probability

$$\begin{aligned} P(a \leq B \leq b) &= P\left(\frac{a - \frac{1}{2} - np}{\sqrt{np(1-p)}} \leq Z_{sum} \leq \frac{b + \frac{1}{2} - np}{\sqrt{np(1-p)}}\right) \\ &\approx \Phi\left(\frac{b + \frac{1}{2} - np}{\sqrt{np(1-p)}}\right) - \Phi\left(\frac{a - \frac{1}{2} - np}{\sqrt{np(1-p)}}\right). \end{aligned}$$

Compare Fig. 5.23 with Fig. 5.9 and then apply the normal approximation with continuity correction to approximate $P(B \leq 25)$ when $B \sim B(30, .6)$ using the **R** function *pbinom*( ). Compare your answer with that obtained in Example 5.9 using the normal approximation without the continuity correction.

5.A.2. *Continuity Correction for the Normal Approximation to a Binomial Distribution*. Consider the sampling distribution for the number of successes in 20 independent Bernoulli trials when $p = 0.5$. Calculate the probability of getting less than 8 successes using:

(a) the exact binomial distribution;

(b) the normal approximation without the continuity correction;

(c) the normal approximation with the continuity correction.

5.A.3. *Continuity Correction for the Normal Approximation to a Binomial Distribution*. Consider the sampling distribution for the number of successes in 40 independent Bernoulli trials when $p = 0.4$. Calculate the probability of getting at most 20 successes using:

(a) the exact binomial distribution;

(b) the normal approximation without the continuity correction;

(c) the normal approximation with the continuity correction.

5.A.4. *Continuity Correction for the Normal Approximation to a Binomial Distribution*. Consider the sampling distribution for the number of successes in 50 independent Bernoulli trials when $p = 0.45$. Calculate the probability of getting at least 20 successes using:

(a) the exact binomial distribution;

(b) the normal approximation without the continuity correction;

(c) the normal approximation with the continuity correction.

5.A.5. *Continuity Correction for the Normal Approximation to a Binomial Distribution*. Consider the sampling distribution for the number of successes in 100 independent Bernoulli trials when $p = 0.8$. Calculate the probability of getting more than 82 successes using:

(a) the exact binomial distribution;

(b) the normal approximation without the continuity correction;

(c) the normal approximation with the continuity correction.

5.A.6. *Continuity Correction for the Normal Approximation to the Distribution of a Sample Proportion.* Modify the normal approximation to the sampling distribution for a proportion (corresponding to counts of Bernoulli random variables) to adjust for the fact that we are approximating a discrete distribution with a continuous normal curve. That is, specify the appropriate continuity correction for using the normal distribution to approximate the sampling distribution for a proportion.

5.B. Data Analysis/Computational

5.B.1. *Football Yardage.* A football team averages 3.5 yards per run with a standard deviation of 2 yards. If the team calls 30 running plays per game and the number of yards gained per rush is approximately normally distributed, what is the probability that they will rush for over 120 yards in a game?

5.B.2. *Loss of Smell?—Might Not Bode Well.* In a study investigating a link between the loss of our sense of smell and increased risk of dying, Devanand et al. (2015) collected information from 1169 adults in New York City. At their initial evaluations, participants took a "scratch and sniff" test (known as UPSIT) in which they were asked to identify 40 common odors. In a follow up study several years later, the authors found that 45.36% of the participants with the lowest UPSIT scores in the range [0, 20] had died during the follow up period. Suppose we selected a random sample of $n = 100$ of the participants in this study with UPSIT scores in the range [0, 20].

(a) How many of the participants in our sample would we expect to still be alive at the end of the follow up period?

(b) Use the **R** function *pbinom*( ) to determine the exact probability that at least half of the participants in our sample were still alive at the end of the follow up period.

(c) Use the **R** function *pnorm*( ) to approximate the probability that at least half of the participants in our sample were still alive at the end of the follow up period. Compare this result with your finding in part (a).

(d) What is the approximate probability that more than 40 of the participants in our sample died during the follow up period?

(e) What is the approximate probability that fewer than 20 of the participants in our sample were still alive at the end of the follow up period?

5.B.3. *Smelling Fine?—Enjoy the Wine*. Consider the olfactory study by Devanand et al. (2015) described in Exercise 5.B.2. The authors also reported that only 18.39% of the participants with the highest UPSIT scores in the range (31, 40] had died during the follow up period. Suppose we selected a random sample of $n = 100$ of the participants in this study with UPSIT scores in the range (31, 40].

(a) How many of the participants in our sample would we expect to still be alive at the end of the follow up period?

(b) Use the **R** function *pbinom*( ) to determine the exact probability that at least half of the participants in our sample were still alive at the end of the follow up period.

(c) Use the **R** function *pnorm*( ) to approximate the probability that at least half of the participants in our sample were still alive at the end of the follow up period. Compare this result with your finding in part (a).

(d) What is the approximate probability that more than 40 of the participants in our sample died during the follow up period?

(e) What is the approximate probability that fewer than 20 of the participants in our sample were still alive at the end of the follow up period?

(f) Compare your findings for these participants with high UPSIT scores with what you obtained in Exercise 5.B.2 for participants with low UPSIT scores.

5.B.4. *Having Heart Surgery?—Get Married First.* Neuman and Werner (2015) used data from the University of Michigan Health and Retirement Study (http://hrsonline,isr.umich.edu), a longitudinal panel survey that has enrolled 29,053 adults 50 years of age or older since 1998, to study the postoperative function characteristics of married and unmarried patients undergoing cardiac surgery between 2002 and 2010. Of the 1026 married cardiac surgery patients, 199 died or developed new ADL (Activities of Daily Living) dependencies following surgery prior to their first scheduled postoperative interview. Of the 550 unmarried (separated, divorced, widowed, or never married) cardiac surgery patients, 172 died or developed new ADL dependencies following surgery prior to their first scheduled postoperative interview. Suppose we select a random sample of $n = 100$ of the married cardiac surgery patients in this study and an independent random sample of $m = 50$ of the unmarried cardiac surgery patients in the study.

(a) How many of the married cardiac surgery patients in our sample would we expect to have either died or developed new ADL dependencies following surgery prior to their first scheduled postoperative interview?

(b) Use the normal distribution to approximate the probability that fewer than 20% of the married cardiac surgery patients in our sample had either died or developed new ADL dependencies following surgery prior to their first scheduled postoperative interview.

(c) What is the approximate probability that more than 15% of the married cardiac surgery patients in our sample had either died or developed new ADL dependencies following surgery prior to their first scheduled postoperative interview?

(d) Repeat parts (a) – (c) for the unmarried cardiac surgery patients in our second sample. Compare with your findings in parts (a) – (c).

(e) Let $\hat{p}_M$ and $\hat{p}_U$ denote the percentages of the 100 married and 50 unmarried cardiac surgery patients, respectively, in our random samples who either died or developed new ADL dependencies following surgery prior to their first scheduled postoperative interview. What is the approximate sampling distribution of $\hat{p}_U - \hat{p}_M$?

5.B.5. *One-Party Government?—Depends on When You Ask.* Gallup, Inc. (2015d) conducted telephone interviews September 9-13, 2015 with a random sample of 1025 adults, aged 18 and older, living in all 50 U.S. states and the District of Columbia. The participants were asked to state their political party preference (Democrat, Independent, or Republican) and whether or not they favored one party control of both Congress and the Presidency. Among those participants whose party preference was Republican, 40% had a preference that the same party should control both Congress and the Presidency. Suppose we collect a random sample of $n = 80$ from the Republican participants in these telephone interviews.

(a) How many individuals in our random sample would we expect to have a preference that the same party should control both Congress and the Presidency?

(b) What is the exact probability that more than half of the Republicans in our random sample have a preference that the same party should control both Congress and the Presidency?

(c) What is the approximate probability that less than 30 of the Republicans in our random sample have a preference that the same party should control both Congress and the Presidency?

(d) Gallup, Inc. had conducted a similar poll in 2014 prior to the November 2014 elections and found that only 24% of Republicans in that poll had a preference that the same party should control both Congress and the Presidency. Suppose we were able to collect a random sample of $n = 80$ Republican participants in the 2014 poll. Answer parts (a) – (c) for this random sample of 2014 poll participants.

(e) Compare your answers in parts (a)–(c) with those obtained in part (d). Can you think of possible reasons that might have led to such a sudden change in governance principles among Republicans from 2014 to 2015?

5.B.6. *Company Branding—Do Employees Get It?* Gallup, Inc. (2012a) asked more than 3000 randomly selected workers whether they agreed with the statement: "I know what my company stands for and what makes our brand(s) different from our competitors." Only 41% of the respondents strongly agreed with this statement. Suppose we were able to collect a random sample of $n = 200$ from the workers who participated in this survey.

(a) How many individuals in our random sample would we expect to have strongly agreed with the statement?

(b) What is the probability that exactly 41% of the individuals in our sample strongly agreed with the statement?

(c) What is the approximate probability that less than 41% of the individuals in our sample strongly agreed with the statement?

(d) What is the approximate probability that between 80 and 120, inclusive, of the individuals in our sample strongly agreed with the statement?

5.B.7. *Smoking in Public Places.* In recent years there has been a strong push by states and cities to ban cigarette smoking in public places (such as restaurants and bars) to protect non-smokers from secondhand smoke. How does the American public feel about such laws? As part of their annual Consumption Habits survey during the period July 7-10, 2014, Gallup, Inc. (2014) conducted telephone interviews with a random sample of 1013 adults, aged 18 and older, living in the U. S. states and the District of Columbia. They found that 56% of the respondents supported a ban on smoking in public places. Suppose we were able to collect a random sample of $n = 150$ from among the participants in these telephone interviews.

(a) How many individuals in our random sample would we expect to have supported the ban on smoking in public places?

(b) What is the approximate probability that more than 2/3 of the individuals in our random sample supported the ban on smoking in public places?

(c) What is the approximate probability that less than half of the individuals in our random sample supported the ban on smoking in public places?

(d) Do you think the public opinion on this issue might differ among different age groups? When do you think the public opinion changed from supporting smoking in public places to banning it? Do you think the public supports a total ban on smoking? Go to Gallup, Inc. (2014) and find out!

5.B.8. *Teenagers and Sports.* National Public Radio (2015), in conjunction with the Robert Wood Johnson Foundation and the Harvard T. H. Chan School of Public Health, supported a major study about sports and health in America. They polled 2506 adults during the period January 29-March 8, 2015. In particular, they asked 437 parents with children currently attending middle school, junior high school, or high school and participating in a sport to name the sport their child participated in MOST OFTEN during the previous year. The results of their poll are as follows:

Sport Participated in MOST OFTEN	**Percentage of Children**
Basketball	16
Soccer	14
Baseball/softball	11
Football	9
Running/jogging/trail running/track	7
Volleyball	6
Swimming	5
Others	32

Suppose we select a random sample of $n = 8$ children from the group represented in this poll. Use the **R** functions *dmultinom*( ) and *pbinom*( ) to answer the following questions.

(a) What is the probability that the sport participated in most often by the children in our sample was basketball(3), soccer(2), baseball/softball (2), and other(1)?

(b) What is the probability that more than half of the children in our sample participated most often in basketball?

(c) What is the probability that all of the eight children participated most often in "other" sports?

5.B.9. *Adults and Sports/Moderate/Vigorous Exercise.* National Public Radio (2015), in conjunction with the Robert Wood Johnson Foundation and the Harvard T. H. Chan School of Public Health, supported a major study about sports and health in America. They polled 2506 adults during the period January 29-March 8, 2015 and asked them about their sports/exercise participation during the previous year. For those 690 respondents who indicated that they DID NOT play sports or do vigorous- or moderate-intensity exercise in the previous year, 47% gave health-related reasons, 38% gave time availability/cost /lack of opportunity reasons, and the remaining 15% cited a lack of interest. Suppose we select a random sample of $n = 30$ of the adults in this study who DID NOT play sports or do vigorous- or moderate-intensity exercise in the previous year. Use the **R** functions *dmultinom*( ) and *pbinom*( ) to answer the following questions.

(a) What is the probability that 10 adults in our sample gave health-related reasons, 10 of them gave time availability/cost/lack of opportunity reasons, and 10 of them cited lack of interest.

(b) What is the probability that more than half of the adults in our sample gave health related reasons?

(c) What is the probability that none of the adults in our sample cited lack of interest?

(d) What is the probability that none of the adults in our sample cited health-related reasons and twice as many of them gave time/availability/cost/lack of opportunity reasons as gave lack of interest as the reason?

5.B.10. *Math SAT Scores.* Consider the Math SAT scores for seniors graduating in 2013 or 2014 from a small private school, as presented in Table 1.15.

(a) Find the sample average and sample standard deviation for the 79 male graduates.

(b) Find the sample average and sample standard deviation for the 50 female graduates.

Assume that these sample averages and standard deviations can be used as reasonable surrogates for the corresponding population means and standard deviations for all graduating seniors from small private schools. Suppose you collect additional random samples of $m = 60$ male and $n = 50$ female seniors graduating from other small private schools.

(c) What is the probability that the average SAT score for your sample of 60 male graduating seniors will be less than 600?

(d) What is the probability that the average SAT score for your sample of 50 female graduating seniors will exceed 550?

(e) What is the probability that the average SAT score for your sample of 60 male graduating seniors will be larger than the average SAT score for your sample of 50 female graduating seniors?

5.B.11. *Stretching a Hit into a Double.* Woodward (1970) conducted a study of different methods of running to first base, with the goal of minimizing the time it would take to get from home plate to second base (i.e., get a double on a base hit). The times (in seconds) given in Table 1.33 are averages of two runs from a point on the first base line 35 ft from home plate to a point 15 ft short of second base for the method of running known as "wide angle" for each of 22 different runners.

(a) Find the sample average and sample deviation for the 22 runners.

Assume that this sample average and sample standard deviation can be used as reasonable surrogates for the corresponding population mean and population standard deviation for all baseball players similar in caliber to the sampled runners. Suppose you collect an additional random sample of $n = 40$ baseball players and measure the average of two "wide-angle" runs from home plate to second base for each of them.

(b) What is the probability that the "wide-angle" time for any one of your ballplayers is between 5.35 and 5.45 seconds?

(c) What is the probability that the average "wide-angle" time for your 40 ballplayers is between 5.35 and 5.45 seconds?

(d) What is the probability that the average "wide-angle" time for your 40 ballplayers is less than 5.20 seconds?

(e) How many of your ballplayers would you expect to have a "wide-angle" time between 5.35 and 5.45 seconds?

(f) What is the probability that all of your 40 ballplayers record "wide-angle" times less than 5.20 seconds?

5.B.12. *How Long Are Movies?* The *Movie and Video Guide* is a ratings and information guide to movies that had been prepared annually by Leonard Maltin. Moore (2006) selected a random sample of 100 movies from the 1996 edition of the *Guide.* He compiled the dataset *movie_facts* containing relevant information about the selected movies. One of the pieces of information provided is the running length of the movies, in minutes.

(a) Find the mean and standard deviation for the running length of this random sample of 100 movies.

Suppose you collect a new random sample of running length times for 50 current movies. Assume that the distribution of running lengths for current movies is similar to those produced in 1996.

(b) What are the mean and standard deviation of the average running length for your sample of 50 current movies?

(c) What is the probability that the average running length for your sample of 50 current movies is less than 2 h?

(d) What is the probability that the average running length for your sample of 50 current movies is between 1 h 50 min and 2 h 10 min?

(e) How many of the current movies in your sample would you expect to have running lengths of at least 2 h?

(f) What is the probability that none of your movies has a running length of more than 2 h 15 min?

5.B.13. *Gender Differences in Body Temperature.* Mackowiak et al. (1992) collected body temperature data from 148 individuals aged 18 through 40 years. The dataset *body_temperature_and_heart_rate* contains body temperature values (artificially generated by Shoemaker 1996, to closely recreate the original data obtained by Mackowiak et al.) for 65 male and 65 female subjects.

(a) Obtain the mean and standard deviation for the body temperatures of the 65 male subjects in this dataset.

(b) Obtain the mean and standard deviation for the body temperatures of the 65 female subjects in this dataset.

Suppose you now collect additional random samples of 50 female and 50 male subjects and measure their body temperatures. Assume that the current populations from which you selected your random samples are similar to the populations that led to the random samples in the Mackowiak et al. study.

(c) What is the probability that the average body temperature for your sample of 50 female subjects is greater than 98.6 degrees Fahrenheit?

(d) What is the probability that the average body temperature for your sample of 50 male subjects is between 98 and 99 degrees Fahrenheit?

(e) What is the probability that the average body temperature for your sample of 50 male subjects will be greater than the average body temperature for your sample of 50 female subjects?

(f) How many of your 50 male subjects would you expect to have body temperatures between 98 and 99 degrees Fahrenheit?

5.B.14. *Gender Differences in Heart Rate.* Mackowiak et al. (1992) collected heart rate data from 148 individuals aged 18 through 40 years. The dataset *body_temperature_and_heart_rate* contains heart rate values (artificially generated by Shoemaker, 1996, to closely recreate the original data obtained by Mackowiak et al.) for 65 male and 65 female subjects.

(a) Use the **R** function *sample*( ) to simulate $r = 750$ bootstrap samples of size $n = 15$ each from the heart rates for the 65 male subjects and find the sample average for each sample. Display the approximate bootstrap sampling distribution for the sample average for samples of size 15.

(b) Simulate r = 500 bootstrap samples of size n = 12 each from the heart rates for the 65 female subjects and find the sample standard deviation for each sample. Display the approximate bootstrap sampling distribution for the sample standard deviation for samples of size 12.

(c) Simulate $r = 1000$ bootstrap samples of size $n = 20$ each separately from the male and female subjects. Find the difference in the sample medians for the male and female subjects for each of the 1000 pairs of samples. Display the approximate bootstrap sampling distribution for the difference in sample medians for common sample sizes of 20 each.

5.B.15. *How Much Do Euros Weigh?* The Euro is the common currency coin for the countries comprising the European Union. According to information from the "National Bank of Belgium", the 1 Euro coin is stipulated to weigh 7.5 g. Shkedy et al. (2006) obtained eight separate packages of 250 Euros each from a Belgian bank and their assistants Sofie Bogaerts and Saskia Litière individually weighed each of these 2000 coins using a weighing scale of the type Sartorius BP310, which provided an accurate reading up to one thousandth of a gram. These 2000 weights, indexed by package number, are provided in the dataset *weight_of_Euros*.

(a) Using only the 250 coins from package number 2, simulate $r = 500$ bootstrap samples of size $n = 25$ each and find the minimum Euro weight for each of the 500 samples. Display the approximate bootstrap sampling distribution for the minimum Euro weight for samples of size 25.

(b) Using only the 250 coins from package number 4, simulate $r = 1000$ bootstrap samples of size $n = 30$ each and find the average Euro weight for each of the 1000 samples. Display the approximate bootstrap sampling distribution for the sample average for samples of size 30.

(c) Using only the 250 coins from package number 5, simulate $r = 750$ bootstrap samples of size $n = 40$ each and find the range of the Euro weights for each of the 750 samples. Display the approximate bootstrap sampling distribution for the sample range for samples of size 40.

(d) Combining the 500 coins from packages numbered 4 and 5, simulate $r = 1000$ bootstrap samples of size $n = 30$ each and find the average Euro weight for each of the 1000 samples. Display the approximate bootstrap sampling distribution for the sample average for samples of size 30. Compare your results with those obtained in part (b).

5.B.16. An automobile manufacturer claims that the fuel consumption for a certain make and model of car averages 28 miles per gallon with a standard deviation of 3 miles per gallon. Suppose you test a random sample of $n = 25$ cars of this make and model.

(a) What is the probability that the average fuel consumption for your random sample of 25 cars will be greater than 30 miles per gallon?

(b) What is the probability that the average fuel consumption for your random sample of 25 cars will be between 26 and 31 miles per gallon?

(c) How many of your cars would you expect to have fuel consumptions between 26 and 31 miles per gallon?

(d) What is the probability that none of your random sample of 25 cars has fuel consumption greater than 28 miles per gallon?

5.B.17. *How Well Does Your Beer Hold Its Foam?* Two features of bottled beer that are important to beer consumers are the amount of initial head formation when a beer is poured and how long the head lasts. Ault et al. (1967) measured the height of the initial head formation upon pouring, the percentage adhesion of the head, and the percentage collapse of the head 4 min after pouring for 20 bottles selected from two different production lots of the beer. The dataset *beer_head* contains the results of their study.

(a) Find the sample average and standard deviation for the maximum head formation for the sample of 20 bottles of beer from the first production lot.

(b) If you were to collect another random sample of $n = 30$ bottles of beer from the first production lot, what is the probability that the average maximum head formation for your sample of 30 bottles would be greater than 175?

(c) Find the sample average and standard deviation for the percentage collapse of the head 4 min after pouring for the sample of 20 bottles of beer from the second production lot.

(d) If you were to collect another random sample of $n = 40$ bottles of beer from the second production lot, what is the probability that the average percentage collapse for your sample of 40 bottles would be less than 80 percent?

(e) Find the sample averages and standard deviations for the percentage adhesion of the head separately for the two production lots.

(f) If you were to collect additional random samples of $n = 20$ bottles of beer from each of the two production lots, what is the probability that the average percentage adhesion for the sample from the second production lot would exceed the average percentage adhesion for the sample from the first production lot?

5.C. Activities

5.C.1. *Age of U.S. Dimes.* Collect a large sample of U.S. dimes and find the ages of the dimes (in years). Simulate $r = 1000$ bootstrap samples of size 5 each, 1000 bootstrap samples of size 10 each, 1000 bootstrap samples of size 20 each, and 1000 bootstrap samples of size 40 each from this large sample of dimes and compute the sample mean for each of these 4000 samples. Display the approximate sampling distributions for sample averages of sizes 5, 10, 20, and 40 from the age distribution of all dimes in circulation at the time. Compare your average age distributions with those for U.S. pennies, as displayed in Figs. 5.14, 5.15, 5.16, and 5.17.

5.C.2. *Age of U.S. Quarters.* Repeat Exercise 5.C.1 for U.S. quarters.

5.C.3. *Peanut M&Ms.* Collect a sample of Peanut M&Ms. and compute the Pearson goodness of fit statistic for the sample relative to the claim by Mars, Inc. that the color combination in M&M's Peanuts is 20% brown, 20% yellow, 20% red, 10% orange, 10% green, and 20% blue. Use simulation to assess their claim.

5.C.4. *Peanut Butter and Almond M&Ms.* Collect a sample of Peanut Butter or Almond M&M's and compute the Pearson goodness of fit statistic for the sample relative to the claim by Mars, Inc. that the color combination in both M&M's Peanut Butter and M&M's Almond is 20% brown, 20% yellow, 20% red, 20% green, and 20% blue. Use simulation to assess their claim.

5.C.5. *Reese's Pieces.* Collect a sample of Reese's Pieces and compute the Pearson goodness of fit statistic relative to the claim by Mars, Inc. that Reese's Pieces are evenly mixed among two colors, orange and brown. Use simulation to assess their claim.

5.D. Internet Archives

5.D.1. *Country Characteristics/Attributes.* Search the Internet to find a dataset that contains the characteristics or attributes for a sample of individuals residing in a particular country of interest to you. Use these data and bootstrap simulation to obtain the approximate sampling distribution for the sample average of 30 individuals from this country for one of these characteristics or attributes.

5.D.2. *Drug Studies/Medical Treatment.* Search the Internet to find an article that provides details for the results of a clinical trial studying the effectiveness of a new drug or medical procedure. Use these data and bootstrap simulation to obtain the approximate sampling distribution for the average effect of the new drug or medical procedure on a sample of 20 additional individuals who might use it.

5.D.3. *Church Attendance.* Search the Internet to find the results of a survey dealing with weekly church attendance in the United States. Using these results and bootstrap simulation, obtain the approximate sampling distribution for the percentage of all Americans who attend church weekly.

5.D.4. *Political Survey.* Search the Internet to find the results of a political survey that includes questions of interest to you. Select a question with at least five possible responses indexed by a categorical variable with at least four categories. Use the responses of the survey participants to this question and bootstrap simulation to obtain the approximate sampling distribution for the Pearson goodness of fit statistic G for a sample of size $n = 200$ from the population of interest.

5.D.5. *Time to Degree.* Search the Internet for a study that provides information about times to degree for a collection of undergraduate students. Use the results of this study to simulate the sampling distribution for the average time to degree for a random sample of 100 undergraduate students.

6 Statistical Inference: Estimating Probabilities and Testing and Confirming Models

In the previous three chapters we have discussed various aspects of the following three important topics:

(a) What data are required to address a question of interest and how should these data best be obtained?

(b) What models are appropriate for the relevant data and which properties of these models are important for the question of interest?

(c) Using the data models to help us understand the proper data collection process, how do we then best summarize the information contained in the resulting observations?

All of these topics relate most directly to the preliminary planning stage for an investigation of the question of interest and to the initial summary of the amassed sample data. For some studies, such as population censuses or preliminary investigations designed to help in planning more comprehensive later studies, dealing effectively with these three issues might be sufficient.

© Springer International Publishing AG 2017

D.A. Wolfe, G. Schneider, *Intuitive Introductory Statistics*, Springer Texts in Statistics,
DOI 10.1007/978-3-319-56072-4_6

Fig. 6.1 Relationships between population and sample

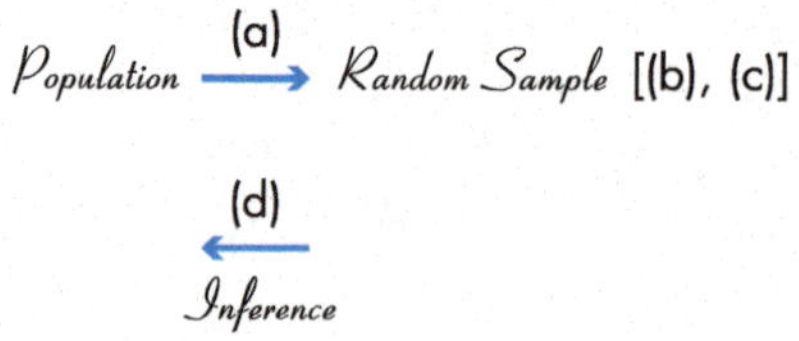

However, for most scientific investigations we take a sample to learn more about the population from which the sample data were obtained. This still involves choosing a model and summarizing the data we collect (questions (b) and (c)), but now we want to go further and use our data to draw conclusions about the underlying population. These conclusions about a population based on statistical evidence from the sample are called **inferences**. We illustrate these relationships between the population and sample in Fig. 6.1.

For example, what can be said about the general effectiveness of a medication designed to halt or reverse the balding of men based on the results of the use of this medication on a properly chosen sample of 24 men who are showing signs of losing their hair? What conclusions can be drawn about college students, in general, based on the information that 17 out of a sample of 52 first-year college students worked more than 20 h per week, while 34 out of a sample of 73 college seniors did the same? Using data collected from such samples of a population of college graduates, what can be said, in general, about the health benefits or detriments from having participated in intercollegiate athletics versus either an active involvement in intramural athletics or virtually no athletic participation?

In this chapter we begin the discussion of using sample data to make inferences about properties of the population(s) from which the data were obtained. Such inferences will generally take one or more of the following three forms for important attributes of the population(s):

(a) simple numerical estimation of the attributes;
(b) interval estimation of the attributes, with some indication of both the accuracy and reliability of the intervals;
(c) tests of hypotheses about the attributes, with some indication as to the effectiveness of the test procedures.

We introduce these three modes of inference through discussion of how to use a random sample from a single population to make inferences about the relative frequency of an event A in the population. This relative frequency is also the probability $P(A)$ that a single random observation drawn from the population will result in occurrence of the event A.

Example 6.1. Impact of Community Nature on Flower Color Ecologists are often interested in the results of natural selection. Ostler and Harper (1978) studied the evolution of flower parts and colors in relation to community structures, particularly open (i.e., communities such as meadows and grasslands which lack woody vegetation) and closed (i.e., communities forested with several species of trees). One interesting attribute of a community structure is its relative distribution of white versus colored flowers. Meadows, for example, are usually full of color in spring and summer, and it is possible that white flowers do not stand out in sharp contrast to their surroundings in such a setting. In a forest, however, white flowers may have an advantage over colors because they are pale against a dark background, thereby attracting pollinators and ensuring continuation of the species.

In this ecological study, there are several questions we might ask about the floral colorings of open and closed communities. For example, is there a difference in the percentage of white flowers for the two types of communities? What is the probability that a randomly selected flower from an open community will be white? from a closed community? For newly created communities of either type with initially balanced percentages of white and colored flowers, is there a tendency over time to move toward dominance of one or the other color category?

All these questions deal with the particular attribute 'percentage of white flowers' in a community. Ostler and Harper collected sample data to address these questions and make inferences based on these data for the general populations of closed and open communities. We will return to this example to illustrate a number of the statistical concepts and procedures discussed later in this chapter.

6.1 Point Estimation

General Concept Let X_1, ..., X_n denote the collected measurements for a random sample from an underlying population that may be either continuous or discrete. A **population parameter** is a numerical attribute of a population that is often unknown. Common examples of parameters include the population mean, median, and variance, as well as the probability that a randomly selected member of the population will have a particular characteristic. However, it is often the case that the attributes of a probability distribution that are most of interest are not adequately expressed by these four parameters. Thus, for example, we might be interested in whether the underlying population has a probability distribution that is symmetric or in how likely it is that we might experience unusually large or small observations (commonly called 'outliers') when sampling from this population. Other attributes of interest (parameters) can be represented by categorical probabilities. For example, what is the probability that a randomly selected observation from the population will exceed an important threshold value (e.g., that the lifetime of a part in our new automobile will exceed the manufacturer's warranty or that a randomly selected individual from a diseased population will respond better to an innovative new medical treatment than to the current standard treatment for the disease). In this text, we will often denote parameters of interest by lower case Greek letters.

Whatever the form of a parameter, say θ, of interest for a given population, there are a variety of criteria that statisticians have used to develop natural and effective methods for estimation of θ based on sample data X_1, ..., X_n obtained from the population. One such criterion is to estimate the parameter θ by its natural statistical counterpart in the sample. Thus, for example, we would estimate the population mean μ by its sample analog, the sample mean $\bar{X} = (X_1 + \ldots + X_n)/n$, the population median θ by its sample analog $\tilde{X} = \text{median}\{X_1, \ldots, X_n\}$, and the percentage, p, of the population with a certain characteristic by the percentage of the sample observations that have this characteristic.

In Sect. 5.2 we used the notation $\hat{p}$ to denote the sample proportion of successes used to estimate the population proportion of successes, p. In general in this text when we use the sample analog to estimate a population parameter we will use the 'hat' (^) notation to distinguish the sample estimator. Thus $\widehat{\theta}$ represents a sample estimator (based on the sample data X_1, ..., X_n) for the population parameter θ.

Probability of an Event Let A denote an event of interest in the underlying population and let p denote the parameter representing the probability that this event occurs when we sample a single observation from our population. For this population parameter p, the natural sample analog associated with a random sample of size n from the population is simply the proportion of sample observations for which the event A occurs.

For example, suppose you are driving a long distance at night and decide to pass the time by counting the number of cars you meet and keeping track of how many of them have only one working headlight. Here, A corresponds to the event that a car has only one working headlight and p is the proportion of cars that have only a single working headlight in the area through which you are traveling. If 73 of the 896 cars you encountered that evening had only one working headlight, you would naturally estimate p by your observed proportion, $73/896 = .0815$, of one-headlight cars.

In general, if we let B denote the number of times the event A occurs among the sample observations $X_1, \ldots, X_n$, then our point estimator for p will be the sample statistic:

$$\hat{p} = [\text{number of times the event } A \text{ occurs in the sample}]/n = \frac{B}{n}. \qquad (6.1)$$

We call $\hat{p}$ a *point estimator* because it estimates p with a single number. Later in this chapter we will see how to estimate p more reliably by providing an interval in which we expect p to lie.

For your headlight data, $p = \{$proportion of cars that have only one working headlight in the area through which you are traveling$\}$, $n = 896$, and $B = 73$. The natural point estimate for p is then $\hat{p} = \frac{73}{896} = .0815$.

We illustrate the use of this estimator for the probability of an event in two different settings, one where the sample data themselves simply record the occurrence or non-occurrence of the event A, and a second where information in addition to the occurrence or non-occurrence of A is contained in the collected sample data.

Example 6.2. Estimation of Percentage White Flowers (Continuation of Example 6.1) In their study to assess the effect that community structure (open or closed) has on the coloration of flowers in the community, Ostler and Harper (1978) collected floral characteristic data from 14 open and 11 closed plant communities in the Wasatch Mountains of northern Utah and adjacent Idaho. At each site, the authors conducted a complete census of the flowers in the community and reported the percentage of white flowers present. The resulting data are presented in Table 6.1.

Table 6.1 Percentage white flowers in open and closed communities in the Wasatch Mountains of northern Utah and adjacent Idaho

Community type	Percentage white flowers			
Open communities	52.6	42.6	43.4	39.4
	15.3	34.9	38.9	45.6
	32.7	27.9	23.8	17.6
	24.4	29.7		
Closed communities	54.8	75.8	60.1	55.8
	55.7	44.9	45.8	58.6
	62.6	70.8	69.2	

Source: Ostler and Harper (1978)

The authors were concerned with a number of general differences between open and closed communities. One such comparison of interest would be the proportions of these two types of communities that contain more than 50% white flowers. The associated events are {a randomly selected open community contains more than 50% white flowers} and {a randomly selected closed community contains more than 50% white flowers} and the related parameters are p_{Open} and p_{Closed}, respectively. Viewing the data collected by Ostler and Harper as a representative random sample of $m = 14$ open and $n = 11$ closed communities, we see that only one of the sample open communities exhibited more than 50% white flowers, while nine of the closed communities had a majority white flowers. Once again from Eq. (6.1) we obtain the point estimates $\hat{p}_{\text{Open}} = 1/14 = .0714$ and $\hat{p}_{\text{Closed}} = 9/11 = .8182$, indicating a considerable estimated difference between the two types of communities with respect to the predominance of white versus colored flowers.

For our second example concerned with estimation of the probability of an event, we consider the results of a formally designed medical study to assess the effectiveness of a possibly improved treatment for patients with Alzheimer's disease. Here the subjects in the study serve as their own controls and we are interested in evaluating whether or not the new treatment regime leads to improvements in cognitive awareness for the subjects.

Example 6.3. Medical Improvements for Subjects with Alzheimer's Disease As the population of the industrial world continues to increase in age, we are faced with ever increasing physical and mental health needs of the elderly. One of the most serious of these needs is that associated with people who suffer from Alzheimer's disease and their caregiving families. One accepted form of medical treatment for Alzheimer patients has been the use of cholinesterase inhibitors (such as *tacrine* and *physostigmine*), designed to inhibit enzymatic degradation at nerve connections, thereby enabling improved nerve-to-nerve communications. In an attempt to see if the positive effects of such cholinesterase inhibitors can be enhanced by supplementing the treatment with low doses of *l-deprenyl* (previously shown to be effective in animal studies), Schneider et al. (1993) undertook a study with ten outpatients with probable Alzheimer's disease living in the Los Angeles area. The six women and four men who were part of this study had previously completed multicenter clinical trials of either *tacrine* or sustained-release *physostigmine salicylate* and had continued receiving the medication after that clinical trial ended. These ten patients were randomly assigned to receive either a 4-week trial of oral *l-deprenyl* (5 *mg* b.i.d.) followed by 4 weeks of placebo or a 4-week trial of placebo followed by 4 weeks of *l-deprenyl* (5 *mg* b.i.d.). The *l-deprenyl* and placebo were administered as identical-appearing tablets. The study was also double-blinded in the sense that neither the subjects nor the persons administering the dosages knew which of the participants received a particular order of the placebo and *l-deprenyl* (see Chap. 3 for more discussion of double-blinded experiments). Assessment of the effects of the placebo and *l-deprenyl* was made at the end of each 4-week period by a psychometrician who was also blinded to the treatment assignments. The cognitive subscale of the Alzheimer's Disease Assessment Scale was used to assess the state of the subjects at each of these times. Values of these subscale data for both periods of measurement and each of the ten subjects are given in Table 6.2, where lower scores indicate preferable behavior patterns.

Table 6.2 Scores on the cognitive subscale of the Alzheimer's disease assessment scale for 10 patients with Alzheimer's disease under treatment with *Tacrine* or *Physostigmine Salicylate* plus *l-deprenyl*

		Cognitive subscale scores			
Subject	Sex	Placebo	*l-deprenyl*	"Improvement"	Positive?
1	F	29	26	3	Yes
2	M	40	32	8	Yes
3	M	42	36	6	Yes
4	M	45	41	4	Yes
5	F	41	40	1	Yes
6	F	24	27	−3	No
7	M	32	34	−2	No
8	F	53	50	3	Yes
9	F	38	32	6	Yes
10	F	28	29	−1	No

Source: Schneider et al. (1993)

To assess the enhancement effect of *l-deprenyl,* we look at the subject differences between cognitive subscale scores for the 4-week placebo period and those for the 4-week period of treatment with *l-deprenyl*. One aspect of interest in this study relates to the population event, D, that a randomly selected Alzheimer patient would show improvement in cognate subscale score from the combined use of *l-deprenyl* and a *cholinesterase inhibitor* relative to the latter in combination with a placebo. Let X and Y denote the *l-deprenyl* and the placebo cognitive subscale scores, respectively, for a randomly selected Alzheimer patient and set $Z = Y - X$ to be the "improvement" (i.e., lowering of cognitive subscale scores) associated with the use of *l-deprenyl*. Then the probability of the event D is $p = P(Z > 0)$, corresponding to the proportion of all Alzheimer patients for whom this treatment would lead to improved cognitive subscale scores. To estimate p using our sample data, we let X_i and Y_i denote the cognitive subscale score for the ith subject after the 4-week period of treatment with *l-deprenyl* and after the 4 weeks of placebo, respectively. The observed "improvement" in cognitive subscale score for the

i^{th} subject associated with the use of *l-deprenyl* is then represented by $Z_i = Y_i - X_i$, for $i = 1, ..., 10$. For the event $\{Z > 0\}$, we then use Eq. (6.1) to provide us with the point estimate $\hat{p}$ = (sample percentage of subjects who show an improvement with the use of *l-deprenyl*) = # $\{Z_i\text{'s} > 0\}/10 = 0.70$.

Thus, based on these sample data, we estimate that the probability is .70 that a randomly chosen Alzheimer patient being treated with a *cholinesterase inhibitor* will show additional improvement in behavior patterns if *l-deprenyl* is added to the treatment regime. Comparing this .70 estimate with the baseline value of $p = .50$ (corresponding to no effect from the treatment and only a 50–50 chance that an Alzheimer patient would have greater cognitive sub-scale scores after receiving the treatment) provides some indication of the impact that the treatment has on Alzheimer patients. (The magnitude of the expected improvement in behavior patterns from the use of *l-deprenyl* is also likely to be of interest in this study. Can you think of an intuitive way to estimate this magnitude effect using the collected data? We return to this example to formally address that issue in Chap. 8.)

In Examples 6.2 and 6.3, respectively, we used sample data to estimate that 81.82% of all closed communities will contain a majority of white flowers and that 70% of Alzheimer's patients will show improvement with the new treatment. We know that these are both just rough estimates, but "how rough" are they? Since our estimate of 81.82% is based on a sample of only 11 closed communities, would we be too surprised to find out that the true percentage of closed communities with a majority of white flowers is 75%? In view of the fact that we used data from a random sample of only 10 Alzheimer's patients to obtain our 70% point estimate, how unusual would it be to find out that only 62% of all Alzheimer patients actually experience improvement with *l-deprenyl*? One way to answer these questions is to make interval estimates rather than point estimates, and in the next section we look at the process of building such reliable interval estimates.

Section 6.1 Practice Exercises

6.1.1. *Grass Greener on the Other Side of the Fence?* Jobvite (2014) presented the results of a nationwide survey about the social, mobile job seeker. They reported that **51%** of the survey respondents who had employment were actively seeking or open to a new job. Is the boldface number a statistic or parameter? Describe the population and population parameter of interest in this survey.

6.1.2. *Snoring Intensity.* Wilson et al. (1999) used polysomnographic testing to measure the sound intensity of snoring for **1139** patients who had been referred to the sleep laboratory of a primary care hospital. One of the measures of snoring intensity studied by the authors is L_{10}, recorded in *dBA* units (a sound pressure measurement in decibels that employs the *A*-weighting network). The average L_{10} for the patients in the study was **48.8** *dBA* units. Which of the boldface numbers are statistics? Describe the population and a population parameter of interest in this portion of their study.

6.1.3. *Snoring Intensity and Noise Restrictions.* Regulations of the Minnesota Pollution Control Agency (MPCA) restrict the acceptable outdoor nighttime noise level to an L_{10} value no higher than **55** *dBA*. In their investigation (see Exercise 6.1.2) of the sound intensity of snoring, Wilson et al. (1999) found that **12.3%** of the **1139** patients in their study had snoring levels higher than 55 *dBA*. Which of the boldface numbers are statistics? Describe the population and a population parameter of interest in this portion of their study.

6.1.4. The average grade on an examination for a large lecture introductory statistics course is **78.5**. You attend a particular **8:30** am recitation section for this course, and you want to find out if the **30** students in your section did better on this examination than the average. The morning the examination is handed back, only **15** of the students in your recitation show up. The average score on the examination for these 15 students is **83.5**. Which of the boldface numbers are statistics? Describe the population and a population parameter of interest.

6.1.5. A manufacturing process produces light bulbs with an average burning life of **2000** h. Over the course of a year, you purchase **40** of their light bulbs and find that the average burn life for these bulbs is **1570** h. Which of the boldface numbers are statistics? Describe the population and population parameter of interest.

6.1.6. *Depression among Twins.* The Virginia Twin Study of Adolescent Behavioral Development is a prospective study of more than 1400 male and female juvenile twin pairs. Silberg et al. (1999) used the data from this study to compare the trajectory of depressive symptoms among boys and girls from childhood to adolescence. The Child and Adolescent Psychiatric Assessment (CAPA), a semistructured, investigator-based psychiatric interview administered by separate skilled interviewers to a pair of twins and at least one of the twins' parents, provided the data used to assess psychopathology. The study included 338 monozygotic (MZ – identical) and 156 dizygotic (DZ – fraternal) female twin pairs and 252 MZ and 158 DZ male twin pairs. Describe a relevant population for this study and list at least four population events (and associated probabilities) that would be of interest to the investigators.

6.1.7. *Cigarette Demand among Adolescents.* Pierce et al. (1998) were interested in the proposition that tobacco industry advertising and promotional activities were contributing causal agents in the stimulation of demand for cigarettes among adolescents. A total of 1752 adolescent never smokers who were not susceptible to smoking when first interviewed in 1993 in a telephone survey in California were reinterviewed in 1996. More than half the sample named a favorite cigarette advertisement in 1993 (with Joe Camel advertisements being the most popular). Describe a relevant population for this study and list at least two population events (and associated probabilities) that would be of interest.

6.1.8. *Coffee and Stress.* Lane (1999) measured the blood pressures and stress hormone levels for 72 habitual coffee drinkers (those who typically drink four or five cups of coffee in a morning) on days when they drank coffee and days when they voluntarily abstained from coffee. Describe a relevant population for this study and list at least two population events (and associated probabilities) that would be of interest in this setting.

6.1.9. *Stress and Coronary Heart Disease.* Cardiovascular reactivity (CVR) has been hypothesized as one possible mechanism linking stress to risk for coronary heart disease (CHD). To demonstrate the feasibility of the reactivity hypothesis, investigators must evaluate the cardiovascular activities of individuals during common and naturally occurring activities, including those that are stressful as well as those that do not produce stress. Brondolo et al. (1999) conducted one such study involving 115 New York City traffic enforcement officers (TEAs). Measures of systolic blood pressure (SBP), diastolic blood pressure (DBP), and heart rate (HR) were collected throughout a workday via the Suntech Accutracker II, using a blood pressure cuff attached to the non-dominant arm and three standard electrocardiogram chest electrodes. One question of interest in the study was "Do traffic enforcement agents have higher levels of blood pressure and heart rate when they are in stress-related situations than when they are involved in non-stressful daily activities?" What are the population and the population event of interest here? What is the point estimate for the probability of this event if 87 of the 115 subject TEAs exhibited higher systolic blood pressure when talking to motorists than when they talked to coworkers or supervisors?

6.1.10. *Women's Shoe Preferences.* The American Orthopedic Foot and Ankle Society (1998) conducted a telephone interview of 531 women who worked outside the home.

(a) They reported that 234 of the women interviewed indicated that they typically wore "flats" (fashion shoes with heels less than 1 inch) at work. What are the population and event of interest here? What is the sample estimate for the probability of this event?

(b) They also reported that 58% of the women in their survey with more than 4 years of college typically wore flats to work, 46% of the women with 4 years of college typically wore flats to work, and 37% of the women in the survey who completed the 12th grade or less typically wore flats to work. What are the relevant populations and events of interest in this portion of the study? What are the sample estimates for the probabilities of these events?

6.1.11. *Who are More Stubborn—Women or Men?* Men have long had a reputation for being more stubborn than women when it comes to asking for directions when they are lost while driving. In a study, digital map developer Navigation Technologies (1999) conducted a nationwide telephone survey with 503 men and 502 women between the ages of 18 and 65. They asked the respondents what they did when they were lost while driving.

(a) They reported that 400 of the male respondents indicated they would stop and ask for directions or consult a map, while 306 of the female respondents indicated they would do the same. What are the relevant populations and events of interest in this portion of their study? What are the sample estimates for the probabilities of these events?

(b) The authors also reported that 756 of all (male and female) respondents indicated that the main emotion felt when they were lost while driving was that of frustration. What is the relevant population and event of interest in this portion of the study? What is the sample estimate for the probability of this event?

(c) Finally, the authors reported that 11.9% of older Americans (ages 54-64) in their survey reported that they had experienced being lost when driving, while 23.1% of the younger drivers (ages 18-24) surveyed reported the same. What are the relevant populations and events of interest in this portion of the study? What are the sample estimates for the probabilities of these events?

6.1.12. *Connection between Psychological and Physical Health.* Can addressing a patient's psychological needs produce both psychological and physical health benefits? In particular, can writing about stressful life experiences affect a patient's disease status? Smyth et al. (1999) conducted a study to address this question for patients with asthma or rheumatoid arthritis. In the study, 107 patients with asthma or rheumatoid arthritis were assigned either to a control group (37 patients) or an experimental group (70 patients). Patients in the control group were asked to write about emotionally neutral topics, while those in the experimental group were asked to write about the most stressful events in their lives. Using standard quantitative measures of medical status, the authors found that 33 of the patients in the experimental group had an improved disease status following the writing exercise, while 9 of the control group patients showed such an improvement. What are the relevant populations and events of interest in this study? What are the sample estimates for the probabilities of these events?

6.1.13. *American Tastes in Art.* What do Americans want to see in their art, particularly paintings? Komar et al. (1997) presented the results of an extensive poll regarding American tastes in art. A group of 1001 randomly selected subjects answered a list of 102 individual questions about their art preferences. Of the 1001 respondents, 681 indicated that they preferred the colors in a painting to be blended into each other, rather than kept separate.

What is the relevant population and event of interest in this question from their study? What is the sample estimate for the probability of this event?

6.1.14. *Do Paintings Need to Have Meaning?* In the art opinion poll discussed in Exercise 6.1.13, the authors reported that 751 of the respondents indicated that paintings do not necessarily have to teach us any lessons to be appreciated. What is the event of interest in this question from their study? What is the sample estimate for the probability of this event?

6.2 Interval Estimation

General Concept In the previous section, discussion centered around how to use data in a random sample $X_1, \ldots, X_n$ to provide a point estimator $\widehat{\theta}$ for a population parameter θ. While the particular value of this estimator for our observed sample data does, indeed, yield information about the unknown value of θ, we must also recognize the fact that it has been obtained in the process of random sampling from the underlying probability distribution. Thus the value of the estimator $\widehat{\theta}$ will vary from random sample to random sample; that is, $\widehat{\theta}$ is a sample statistic with its own probability distribution, known as the *sampling distribution* of $\widehat{\theta}$. As we discussed previously in Chap. 5, this sampling distribution of $\widehat{\theta}$ contains important information about the chance variation that accompanies the fact that our estimate is based on the realization of a specific, single random sample. The sampling distribution can be used to provide us with estimates of the margin of error that accompanies the use of $\widehat{\theta}$ to estimate the parameter θ, thereby enabling us to develop not only a plausible interval of values for θ but also to attach a specific confidence that the true value of θ will, in fact, be captured by this interval.

We encounter information of this type on a daily basis in the media, whether it is newspapers, magazines, radio, television, or on the Internet. For example, the results of an opinion poll regarding how well Americans think the President is doing his job might be reported (in any of the aforementioned media) as a 57% favorable job approval rating with a stated margin of error of ± 4%. The 57% approval rating obtained in the poll provides the sample estimate, $\hat{p} = .57$, of $p =$ {proportion of all Americans who approved of the way the President was doing his job at the time of the opinion poll}. On the other hand, the ± 4% margin of error results from the fact that the observed sample percentage of respondents who believe that the President is doing a good job (57% for this particular sample) would, in all likelihood, be something other than 57% if we interviewed a different set of randomly selected subjects in a second random sample; that is, the percentage who approve of the way the President is doing his job will vary from sample to sample. The stated margin of error (± 4% for this example) is a direct consequence of this sampling variability and the confidence we wish to have that the observed interval (53%–61% for this example) does, indeed, contain the true percentage of all Americans who approve of the way the President is doing his job. Survey results such as these are usually stated with a margin of error that provides us with 95% confidence that the obtained interval does indeed contain the true percentage.

Using this preliminary discussion of typical poll results to help motivate the concept and guide us through its various components, we consider now the formal definition of a confidence interval for a population parameter θ based on a random sample from the population.

Definition 6.1 Let $X_1, \ldots, X_n$ be a random sample from a probability distribution for which θ is a parameter of interest and let CL be a number between 0 and 1. A **level *CL* confidence interval for the parameter $\boldsymbol{\theta}$** is an interval that depends on the sample $X_1, \ldots, X_n$ in such a way that our sampling process has probability CL of producing an observed interval that contains the unknown value of θ.

We commonly let $(\widehat{\theta}_L, \widehat{\theta}_U)$ represent a level *CL* confidence interval for θ, where $\widehat{\theta}_L$ and $\widehat{\theta}_U$ are a pair of sample statistics. Then the sampling process is such that we have probability *CL* of obtaining sample values $X_1 = x_1, \ldots, X_n = x_n$ for which the observed interval $(\widehat{\theta}_L, \widehat{\theta}_U)$ contains the unknown value of θ; that is, for which

$$\widehat{\theta}_L < \theta < \widehat{\theta}_U. \tag{6.2}$$

Once our sample values $x_1, \ldots, x_n$ have been observed, we say that we are **100*CL*% confident** that the true value of θ is in the attained interval $(\widehat{\theta}_L, \widehat{\theta}_U)$. The level *CL*, which represents our confidence in the observed interval, is almost always taken to be at least .90 (most commonly, .95) and only rarely would one ever use a level smaller than .80.

The percent confidence that an interval will contain the true unknown value of θ is usually related to the variability from one sample to another of a point estimator $\widehat{\theta}$ for θ. The more variable is the estimator $\widehat{\theta}$, the larger will be the required number of observations in our sample in order to attain a desired confidence level.

As discussed in Chap. 5, such information about the sampling variability of an estimator $\widehat{\theta}$ is provided directly by its sampling distribution. We shall see explicitly in later examples in this section how to use the sampling distribution for a point estimator $\widehat{\theta}$ to establish an appropriate confidence level *CL* prior to collecting our random sample $X_1, \ldots, X_n$. On the other hand, computation of the values of the statistics $\widehat{\theta}_L$ and $\widehat{\theta}_U$ for a specific set of sample data and the resulting determination of the confidence interval for θ are functions solely of the single, observed values, $x_1, \ldots, x_n$ of our random sample.

It is important to understand why the interval estimates for a parameter θ are called "confidence intervals" rather than "probability intervals", which might at first impression seem more natural. The distinction relates to the fact that the randomness involved in developing the interval of plausible values for θ is associated entirely with the process of obtaining the random sample, not with the parameter itself. Thus, for example, if we have a pair of statistics $\widehat{\theta}_L$ and $\widehat{\theta}_U$ such that $(\widehat{\theta}_L, \widehat{\theta}_U)$ is a 95% confidence interval for θ, the proper interpretation is that 95% of all possible random samples that could be chosen from the underlying population will yield data for which the computed value of $(\widehat{\theta}_L, \widehat{\theta}_U)$ will, in fact, contain the unknown value of the population parameter θ. In terms of a probability statement, this corresponds to

$$P\left\{\text{we select a random sample } X_1, \ldots, X_n \text{ such that } \widehat{\theta}_L < \theta < \widehat{\theta}_U\right\} = .95.$$

On the other hand, once we actually select our random sample and obtain the sample data $X_1 = x_1, \ldots, X_n = x_n$, the corresponding observed confidence interval $(\widehat{\theta}_L, \widehat{\theta}_U)$ will, with certainty, either contain the true (unknown) value of θ or it will not. There is no longer any probability (i.e., randomness) associated with the statement that we are 95% confident that $\widehat{\theta}_L < \theta < \widehat{\theta}_U$. We are either right or wrong and we have no way to know which. However, we have high confidence in the *procedure* we used to construct our interval. In fact, we know that 95% of the confidence intervals we could build using this procedure do, in fact, contain the true value of the parameter θ. We return at the end of the section to illustrate with a specific example this important relationship between the *pre-sampling determination of the level* of a confidence interval and the *observed confidence intervals* produced by different individual samples.

As an illustrative example of the concepts involved in the construction of a confidence interval for a parameter, consider the estimation of the percentage white flowers discussed in Examples 6.1 and 6.2. One of the quantities of interest in these examples was the proportion of closed community structures that contain more than 50% white flowers, denoted by p_{Closed}. From the data collected by Ostler and Harper (1978), recall that nine of the eleven closed communities studied had a majority of white flowers, yielding the point estimate $\hat{p}_{\text{Closed}} = 9/11 = .8182$ for p_{Closed}. However, we also know that if we were to examine other randomly selected groups of eleven closed communities, we might very well observe that eight or ten (or, in fact, any number from zero to eleven) of them have a majority of white flowers. This phenomenon represents the variability inherent in the sampling process and information about this uncertainty is provided by the sampling distribution of $\hat{p}_{\text{Closed}}$ = [percentage of eleven randomly selected, closed communities that have a majority white flowers]. While our best estimate of the true value of p_{Closed} for our obtained data is, in fact, .8182, we also know that we COULD have obtained nine out of eleven closed communities having a majority white flowers even if the true value of p_{Closed} is .7 or .9 or, in fact, any value strictly between 0 and 1. Intuitively, we feel that the observed outcome $\hat{p}_{\text{Closed}} = 9/11$ is more likely to have occurred if the true value of p_{Closed} is in the interval (.7, .9), as opposed to being in the interval (.1, .3), for example.

Now, the sampling distribution of $\hat{p}_{\text{Closed}}$ contains information about the variability of $\hat{p}_{\text{Closed}}$ associated with the sampling process. In fact, we can use this sampling distribution to obtain estimates of the random error of the estimator $\hat{p}_{\text{Closed}}$ and use this error estimate to provide an interval of plausible values for p_{Closed}, say $(\hat{p}_L, \hat{p}_U)$, such that a prescribed percentage, say 90%, of all possible random samples of eleven closed communities would provide sample observations for which the true value of p_{Closed} belongs to the observed confidence interval $(\hat{p}_L, \hat{p}_U)$. Then we can state that $(\hat{p}_L, \hat{p}_U)$ is a 90% confidence interval for p_{Closed}, with the interpretation that 90% of all possible random samples of eleven closed communities would produce values of $\hat{p}_L$ and $\hat{p}_U$ for which p_{Closed} is where we say it is, namely, between $\hat{p}_L$ and $\hat{p}_U$.

As in this example about p_{Closed}, it is often the case that a confidence interval for a parameter θ has a very intuitive representation. Such an interval is commonly centered at a point estimate $\widehat{\theta}$ for the parameter and the length of the interval is related to both the desired percentage confidence in the interval and the sampling variability associated with the estimate $\widehat{\theta}$; that is, the confidence interval for θ is of the form

$$\widehat{\theta} \pm \text{margin of error for } \widehat{\theta} \text{ as an estimator of } \theta, \tag{6.3}$$

where the margin of error increases as a function of both the stipulated confidence level *CL* and the sampling variability of the estimator $\widehat{\theta}$. A confidence interval is thus typically centered at our best estimate of θ, and the degree of uncertainty we have in that estimate is directly reflected in the width of the confidence interval. In Figs. 6.2 and 6.3 we present two possible sampling distributions for a point estimator $\widehat{\theta}$. It is clear from the figures that both sampling distributions are centered at the true parameter value θ = 5. However, it is also clear that the sampling variability associated with the sampling distribution of $\widehat{\theta}$ is considerably greater for Fig. 6.3 than it is for Fig. 6.2. Confidence intervals of the form (6.3) for θ would directly reflect

Fig. 6.2 Sampling distribution of $\widehat{\theta}$ with small sampling variability

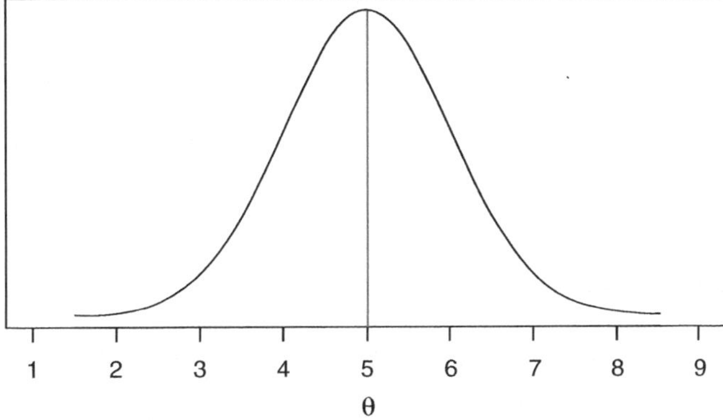

Fig. 6.3 Sampling distribution of $\widehat{\theta}$ with large sampling variability

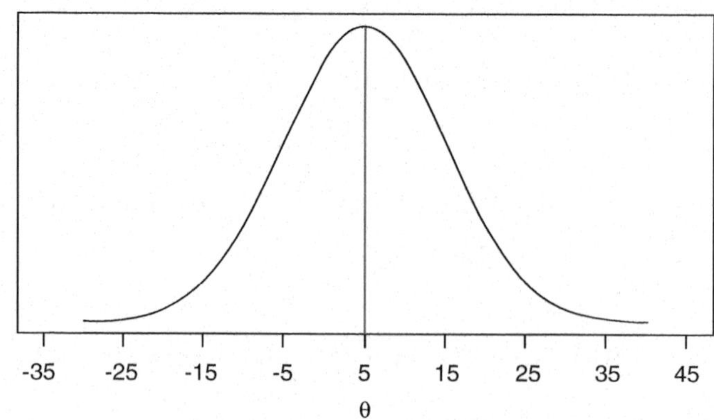

this fact in the sense that the margin of error for $\widehat{\theta}$ will be much larger for the sampling distribution in Fig. 6.3 than for the one in Fig. 6.2. This would result in wider confidence intervals for θ under the sampling distribution conditions of Fig. 6.3.

Example 6.4. Survey Data–the Buckeye State Poll To illustrate the confidence interval expression displayed in Eq. (6.3), consider the results of the Buckeye State Poll (conducted by The Ohio State University College of Social and Behavioral Sciences by telephone during the period February 17–26, 1998) reported in the April 9, 1998 issue of the Columbus Dispatch newspaper. That poll involved 936 women in central Ohio and a number of questions in the poll dealt with issues related to whether standards governing sex and violence on television should be tightened. For one such question, 431 of the 936 respondents indicated that they considered the television industry to be a greater danger to society than government restriction on what appears on television. Based on these sample data, it was estimated that $(431/936)\% = 46\%$ of all women in central Ohio consider the television industry to be a greater danger to society than government restrictions on what can appear on television, with a margin of error of $\pm$ 3.2%. As we shall see later, this margin of error is not only a function of the estimated percentage (46%), but also the number of women included in the poll and the stated confidence level (95%) for the interval. The end result is that (42.8%, 49.2%) is an approximate 95% confidence interval for the proportion of all central Ohio women who consider the television industry to be a greater danger to society than government restrictions on what can appear on television.

Many, but not all, of the confidence intervals that we study in this text will be of the form given in (6.3), namely, $\widehat{\theta} \pm$ margin of error for $\widehat{\theta}$. However, even those that we consider which do not formally match the structure in (6.3) will have the same flavor; namely, the center of the interval will be roughly associated with a point estimate of the parameter and the length of the interval will be directly related to the variability associated with the sampling distribution of the point estimator.

Probability of an Event Let A denote an event of interest with respect to the underlying population and let p denote the probability that this event occurs when we randomly sample a single observation from our population (i. e., p is the proportion of the population that belongs to event A). For the Buckeye State Poll data, for example, we are interested in the population of central Ohio women and the event A corresponds to the statement that the television industry is a greater danger than government restrictions on television. The parameter p is then the proportion of central Ohio women who view the television industry to be the greater danger.

As discussed in Sect. 1, the natural estimator for p is $\hat{p}$ (6.1), the proportion of the sample observations for which the event A occurs. Labeling a response that rates the television industry as a greater danger than government regulation to be a success, the survey data collected in the Buckeye State Poll recorded the number of successes (431) obtained in the polling of the sample of 936 central Ohio women. This led to the point estimate of $\hat{p} = 431/936 = .46$.

The estimator $\hat{p}$ provides us with a natural place to start in our search for a confidence interval for p. However, in order to determine the appropriate upper and lower endpoints for a confidence interval for p, we also need to have additional information about the variability of $\hat{p}$ from random sample to random sample. This leads us naturally to the sampling distribution of $\hat{p}$ or, equivalently, that of the random variable $B = n\hat{p}$ = [number of times the event A occurs in a random sample of size n]. Thus, for the Buckeye State Poll data it is necessary to know the sampling variability associated with the numbers of 'successes' that might be obtained in interviews of different random samples of 936 central Ohio women. As we shall see later, this is the additional information that was used by the Buckeye State Poll to construct the 95% confidence interval of 42.8% to 49.2% for p.

Previously, in Sect. 4.3, we discussed the fact that a statistic corresponding to the number of successes observed in n independent Bernoulli trials with constant probability of success p has a sampling distribution that is binomial with parameters n and p. Since the statistic B is such a count of independent Bernoulli successes, its sampling distribution will be binomial with parameters n and p. In particular, this means that $\hat{p}$ is an unbiased estimator for p (i.e., that $\mathrm{E}\left[\hat{p}\right] = p$) and that the standard deviation of $\hat{p}$ (measuring the variability of the estimator from random sample to random sample) is $\left[\frac{p(1-p)}{n}\right]^{1/2}$. These facts can be used along with the **R** function *binom.test*( ) to obtain exact confidence intervals for p even for small sample sizes n. See, for example, Section 2.3 of Hollander et al. (2014). However, in our discussion here we will concentrate on the use of the large sample approximation to the sampling distribution of $\hat{p}$ (or, equivalently, B) to provide more intuitive approximate confidence intervals for p. These large sample approximations will be quite adequate for most settings where the sample size n is at least 50. First we note that the standard deviation of $\hat{p}$, namely, $\left[\frac{p(1-p)}{n}\right]^{1/2}$, depends on the unknown value of p. However, an intuitive sample estimator for this standard deviation is provided by simply replacing p by its sample estimator, $\hat{p}$, yielding $\left[\frac{\hat{p}(1-\hat{p})}{n}\right]^{1/2}$. This estimated standard deviation is called **the standard error of the estimator** $\hat{p}$ and we denote this standard error by S.E.($\hat{p}$). It is then quite reasonable to construct confidence intervals for the parameter p that are of the form

$$\hat{p} \pm k\text{S.E.}(\hat{p}) = \hat{p} \ \pm k\left[\frac{\hat{p}(1-\hat{p})}{n}\right]^{1/2}, \tag{6.4}$$

where k is an appropriate constant chosen to be the proper number of standard errors we are required to add and subtract from $\hat{p}$ to achieve the approximate desired confidence in the resulting interval. Thus, while the point estimator $\hat{p}$ provides us with the center of the interval, we must rely on the standard error of $\hat{p}$ to help determine the width of the interval. In addition, the constant k is an increasing function of the confidence associated with the interval. In order to increase our confidence in the interval (6.4) for a fixed value of the point estimator $\hat{p}$, we must increase our value of k, which, in turn, results in a wider confidence interval.

One last piece of the puzzle remains to be completed. For an interval of the form (6.4) we need to know how to compute the approximate confidence level CL for a given k and how to find the value of k that will yield a desired approximate confidence level CL. Once again, this requires the sampling distribution of $\hat{p}$ (or, equivalently, of $B = n\hat{p}$). In Sect. 5.3, we pointed out that the sampling distribution of the standardized variable

$$Z_{\hat{p}} \ = \ [\hat{p} - p]/[\hat{p}(1-\hat{p})/n]^{1/2}$$

can be approximated by the $N(0, 1)$ probability distribution for a large number of observations n. Thus, for large n, we have

Fig. 6.4 Standard normal distribution with central probability *CL* between –*k* and *k*

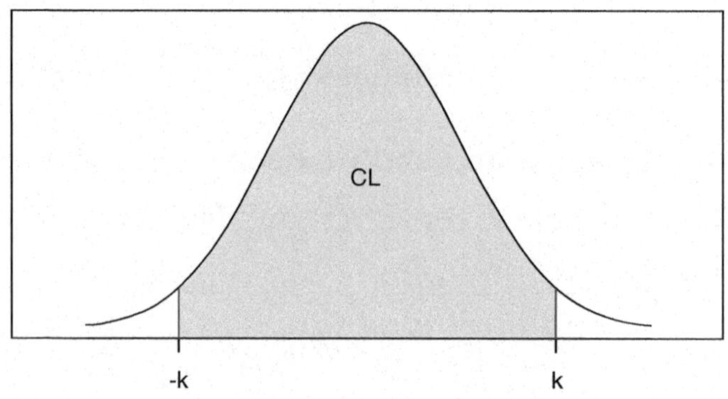

$$P\{-k < Z_{\hat{p}} < k\} \approx P\{Z < k\} - P\{Z < -k\}, \tag{6.5}$$

where $Z \sim N(0, 1)$. This central probability CL for a standard normal distribution is illustrated using **R** in Fig. 6.4.

Because we want the probability between $-k$ and k to be CL, the symmetry of the standard normal distribution implies that the probability to the right of k and the probability to the left of $-k$ must both be equal to (1-CL)/2. This is satisfied if we choose k to be the upper (1-CL)/2th percentile for the standard normal distribution, $z_{(1-CL)/2}$, since then

$$\begin{aligned} &P\left(-z_{(1-CL)/2} < Z_{\hat{p}} < z_{(1-CL)/2}\right) \approx P\left(Z < z_{(1-CL)/2}\right) - P\left(Z < -z_{(1-CL)/2}\right) \\ &= \left(1 - \frac{1-CL}{2}\right) - \frac{1-CL}{2} = CL. \end{aligned} \tag{6.6}$$

It follows that

$$\begin{aligned} &P\left(-z_{(1-CL)/2} < Z_{\hat{p}} < z_{(1-CL)/2}\right) \\ &\quad = P\left\{\hat{p} - z_{(1-CL)/2} S.E.(\hat{p}) < p < \hat{p} + z_{(1-CL)/2} S.E.(\hat{p})\right\}, \end{aligned}$$

and Eq. (6.6) implies that

$$P\{\hat{p} - z_{(1-CL)/2}S.E.(\hat{p}) < p < \hat{p} + z_{(1-CL)/2}S.E.(\hat{p})\} \approx CL.$$

Hence, it follows that **an approximate 100*CL*% confidence interval for the parameter *p*** that is appropriate for large n is given by

$$\begin{aligned}(\hat{p}_{\mathrm{L}}, \hat{p}_{\mathrm{U}}) &= (\hat{p} - z_{(1-CL)/2}S.E.(\hat{p}), \hat{p} + z_{(1-CL)/2}S.E.(\hat{p})) \\ &= \hat{p} \pm z_{(1-CL)/2}S.E.(\hat{p}).\end{aligned} \tag{6.7}$$

Thus we see that getting a confidence interval for p with approximate level .05 (that is, an approximate 95% confidence interval) involves using the upper 2.5% point for the standard normal distribution, while an approximate 90% confidence interval for p uses the upper 5% point for the standard normal distribution. In general, getting a confidence interval for p with approximate level CL uses the upper (1-CL)/2% point, $z_{(1-CL)/2}$, for the standard normal distribution.

Example 6.5. Confidence Intervals for Open and Closed Communities (Continuation of Examples 6.1 and 6.2) In Example 6.2 we saw that the point estimates of p_{Open} = [proportion of open communities which contain a majority white flowers] and p_{Closed} = [proportion of closed communities which contain a majority white flowers] based on the Ostler-Harper data are $\hat{p}_{\mathrm{Open}} = .0714$ and $\hat{p}_{\mathrm{Closed}} = .8182$. Moreover, these estimates are based on samples of $n_{\mathrm{Open}} = 14$ and $n_{\mathrm{Closed}} = 11$ open and closed communities, respectively. Note that both of these sample sizes are much smaller than the recommended size of 50 for best use of the large sample approximation in Eq. (6.6). However, we choose to illustrate the confidence interval in (6.7) with these small sample sizes in order to point out difficulties that can occur in such applications. (For discussion of an alternative approach to obtaining confidence intervals for a population probability when we have a small sample size, see Exercises 6.A.19, 6.B.20, and 6.B.22.)

To compute an approximate 95% confidence interval for p_{Open}, we set *CL* = .95 and find $z_{(1\text{-CL})/2} = z_{.05/2} = z_{.025} = 1.96$. We can also find this upper tail probability by using the R function *qnorm*( ) with the *lower.tail* argument specified to be *FALSE* as follows.

```
> qnorm(0.025, lower.tail=FALSE)
[1] 1.959964
```

From (6.7), the approximate 95% confidence interval for p_{Open} is then given by

$$\begin{aligned}\hat{p}_{\text{Open}} \pm z_{.025}\, S.E.(\hat{p}) &= .0714 \pm 1.96\,[.0714(1-.0714)/14]^{1/2}\\ &= .0714 \pm .1349 = (-.0635, .2063).\end{aligned}$$

Since p_{Open} is a probability, it must be strictly between 0 and 1, inclusive. Hence, the lower bound of -.0635 provided in the approximate confidence interval for p_{Open} is not reasonable. This is a direct consequence of the fact that the large sample approximation used in construction of the approximate 95% confidence interval for p_{Open} is not sufficiently accurate for a sample size as small as $n = 14$. In such cases, we simply truncate the lower limit of the confidence interval at 0 and state that we are approximately 95% confident that no more than 20.63% of all open communities will have a majority of white flowers.

On the other hand, we see that a corresponding approximate 95% confidence interval for p_{Closed} is given by

$$\begin{aligned}\hat{p}_{\text{Closed}} \pm z_{.025}\text{S.E.}(\hat{p}) &= .8182 \pm 1.96\,[.8182(1-.8182)/11]^{1/2}\\ &= .8182 \pm .2279 = (.5903, 1.0461).\end{aligned}$$

Here we see that the inaccuracy of the normal approximation for the small sample size $n = 11$ manifests itself in an unreasonable upper bound of 1.0461 for the probability p_{Closed}. Once again the proper approach is to truncate the upper limit for the confidence interval at 1 and state that we are approximately 95% confident that no less than 59.03% of all closed communities will have a majority white flowers.

We note that the approximate confidence intervals for p_{Open} and p_{Closed} do not contain any values in common, providing some evidence that there likely is a difference between the percentages of all open and closed communities that have a majority of white flowers. We will return to formally assess this conjecture in Sect. 3.

We point out that the form of the approximate $100CL\%$ confidence interval for p in (6.7) implies that we are roughly $100CL\%$ confident that our margin of error in using $\hat{p}$ to estimate p will not be more than $\pm z_{(1-\text{CL})/2}\,\text{S.E.}(\hat{p})$. Thus from the approximate 95% confidence interval for p_{Closed} discussed in Example 6.4, we are roughly 95% confident that our estimate of p_{Closed}, namely, $\hat{p}_{\text{Closed}} = .8182$, is not in error by more than $\pm z_{.025}\,\text{S.E.}(\hat{p}) = \pm .2279$.

A similar discussion applies to the Buckeye State Poll data reported in the April 9, 1998 issue of the Columbus Dispatch and discussed previously in Example 6.4. There we noted that a survey of 936 women in central Ohio led to the approximate 95% confidence interval of (42.8%, 49.2%) for the proportion, p, of all central Ohio women who consider the television industry to be a greater danger to society than government restrictions on what can appear on television. Thus we estimate the proportion p to be 46% and we are roughly 95% confident that our margin of error with this estimate is no more than $\pm$ 3.2%, as noted in the Dispatch article. Now we know where this 3.2% comes from, since it is the margin of error in (6.7); that is, for this particular question from the February 1998 Buckeye State Poll survey, we have (with 95% confidence) that the margin of error associated with our estimate $\hat{p} = 46\%$ is

$$\pm z_{.025}\text{S.E.}(\hat{p}) = \pm 1.96\,[(.46)(1-.46)/963]^{1/2} = \pm .0315,$$

in agreement (up to possible rounding error) with the margin of error of $\pm$ 3.2% quoted in the Dispatch article.

What Should One Expect of the Approximate Confidence Interval for *p* in (6.7)? When we introduced the general concept of confidence intervals in Sect. 2 we made a special point to emphasize that the randomness associated with a given confidence interval is in the sampling process to obtain the data. The population parameter θ for which we are seeking an interval estimate is a constant that does not vary from sample to sample. Thus if $(\widehat{\theta}_L, \widehat{\theta}_U)$ is a 95% confidence interval for the parameter θ, the proper interpretation is that 95% of all possible random samples will yield observed confidence limits $\widehat{\theta}_L$ and $\widehat{\theta}_U$ for which the interval $(\widehat{\theta}_L, \widehat{\theta}_U)$ contains θ, and the other 5% of the random samples will yield $\widehat{\theta}_L$ and $\widehat{\theta}_U$ values for which $(\widehat{\theta}_L, \widehat{\theta}_U)$ fails to contain θ, that is, either $\theta > \widehat{\theta}_U$ or $\theta < \widehat{\theta}_L$. Any specific confidence interval computed from a particular random sample either will or will not contain θ. Moreover, since the value of θ is unknown, we would have no way to deduce whether the **single** confidence interval resulting from our **one** random sample has, in fact, successfully covered θ . However, we remain 95% confident that the process of collecting our random sample has provided us with a set of data for which this is the case.

We illustrate the process of simulation using **R** and the approximate confidence interval for the parameter $p = P(\text{event } A \text{ occurs})$ given in Eq. (6.7). Taking 1 - $CL = .05$ and obtaining the upper .025 percentile for the N(0,1) distribution, $z_{.025} = 1.96$, we see that the approximate 95% confidence interval for p has the form $\hat{p} \pm 1.96\text{S.E.}(\hat{p})$. For our simulation study we consider the setting where $p = .6$ and our sample size is $n = 100$. We obtain the results of one such sample in **R** as follows.

```
> rbinom(n = 1, size = 100, prob = 0.6)
[1] 61
```

Note that in the *rbinom*( ) function, the argument *n* refers to the *number* of samples, and the argument size refers to the sample size (which we denote in this text by n). If we had specified $n = 100$ and $size = 1$, we would have ended up with 100 samples of size 1 (which could then be added together to obtain 1 sample of size 100). For our randomly generated sample, we can use the **R** function *binconf*( ) to construct the approximate 95% confidence interval for p.

```
> binconf(x = 61, n = 100, method = 'asymptotic')

 PointEst     Lower     Upper
     0.61 0.5144028 0.7055972
```

Note that in the *binconf*( ) function, the argument n refers to the sample size (which agrees with our notation). By specifying the *method* argument to be *asymptotic*, we obtain the 95% confidence interval of the form $\hat{p} \pm 1.96\,\text{S.E.}(\hat{p})$ (rather than one of the alternative forms; details can be found using the command ?*binconf*).

We see that for the above sample, the approximate confidence interval contains the true value $p = 0.6$ using the following commands to compare the true value to the second and third numbers output by the *binconf*( ) function (since the first number output is $\hat{p}$).

```
> binconf(x = 61, n = 100, method = 'asymptotic')[2] < 0.6
[1] TRUE
> binconf(x = 61, n = 100, method = 'asymptotic')[3] > 0.6
[1] TRUE
```

To get some idea of the variability from random sample to random sample (remember that is the only variability associated with the confidence interval), we can repeat this entire sampling process $M = 50$ times, in each case constructing the approximate 95% confidence interval for p and recording whether it contains the true value .6. With our understanding of the proper interpretation of a 95% confidence interval, we should expect that roughly .95 (50) = 47.5 of the confidence intervals so constructed will, in fact, contain 0.6.

We first obtain the results of the $M = 50$ simulations by setting the argument $n = 50$ in the function *rbinom*( ) and storing the results in the variable *binomial_ci_results*.

```
> binomial_ci_results = rbinom(n = 50, size = 100, prob = 0.6)
```

For each result stored in *binomial_ci_results*, we compute the approximate 95% confidence interval and check to see whether or not it contains 0.6. (This is achieved using a *for loop*.) The results of the simulation are presented in Table 6.3, along with the indication whether each of the fifty observed confidence intervals actually contains the true probability .6.

Table 6.3 Fifty approximate 95% confidence intervals for p, each based on 100 randomly generated independent Bernoulli variables with constant probability of success p = .6

Sample number	Number of occurrences of event A	$\hat{p}$	Lower CB	Upper CB	Does the observed interval contain the true value p = .6?
1	62	0.62	0.524864	0.715136	Yes
2	56	0.56	0.462708	0.657292	Yes
3	55	0.55	0.452491	0.647509	Yes
4	49	0.49	0.392020	0.587980	No
5	62	0.62	0.524864	0.715136	Yes
6	63	0.63	0.535370	0.724630	Yes
7	58	0.58	0.483263	0.676737	Yes
8	59	0.59	0.493601	0.686399	Yes
9	57	0.57	0.472965	0.667035	Yes
10	57	0.57	0.472965	0.667035	Yes
11	64	0.64	0.545920	0.734080	Yes
12	59	0.59	0.493601	0.686399	Yes
13	56	0.56	0.462708	0.657292	Yes
14	64	0.64	0.545920	0.734080	Yes
15	61	0.61	0.514401	0.705599	Yes
16	58	0.58	0.483263	0.676737	Yes
17	60	0.60	0.503980	0.696020	Yes
18	69	0.69	0.599351	0.780649	Yes

(continued)

Table 6.3 (continued)

Sample number	Number of occurrences of event A	$\hat{p}$	Lower CB	Upper CB	Does the observed interval contain the true value $p = .6$?
19	58	0.58	0.483263	0.676737	Yes
20	64	0.64	0.545920	0.734080	Yes
21	70	0.70	0.610182	0.789818	No
22	62	0.62	0.524864	0.715136	Yes
23	60	0.60	0.503980	0.696020	Yes
24	50	0.50	0.402000	0.598000	No
25	58	0.58	0.483263	0.676737	Yes
26	54	0.54	0.442314	0.637686	Yes
27	54	0.54	0.442314	0.637686	Yes
28	71	0.71	0.621063	0.798937	No
29	62	0.62	0.524864	0.715136	Yes
30	62	0.62	0.524864	0.715136	Yes
31	67	0.67	0.577838	0.762162	Yes
32	65	0.65	0.556514	0.743486	Yes
33	62	0.62	0.524864	0.715136	Yes
34	64	0.64	0.545920	0.734080	Yes
35	56	0.56	0.462708	0.657292	Yes
36	55	0.55	0.452491	0.647509	Yes
37	61	0.61	0.514401	0.705599	Yes
38	64	0.64	0.545920	0.734080	Yes
39	63	0.63	0.535370	0.724630	Yes
40	54	0.54	0.442314	0.637686	Yes
41	56	0.56	0.462708	0.657292	Yes
42	57	0.57	0.472965	0.667035	Yes
43	66	0.66	0.567153	0.752847	Yes
44	52	0.52	0.422078	0.617922	Yes
45	56	0.56	0.462708	0.657292	Yes
46	52	0.52	0.422078	0.617922	Yes
47	58	0.58	0.483263	0.676737	Yes
48	68	0.68	0.588571	0.771429	Yes
49	59	0.59	0.493601	0.686399	Yes
50	60	0.60	0.503980	0.696020	Yes

We see that 46 of the 50 (92%) constructed confidence intervals correctly contain the true value $p = .6$. While this is slightly below the 'expected' number of 47.5 (95%) successful confidence intervals, it is certainly not unreasonable in view of what we have learned about sampling variation. We note that two of the four observed confidence intervals which did not contain .6 were such that the upper confidence bound was less than .6 while the other two unsuccessful intervals were such that the lower confidence bound exceeded .6. This is also not unexpected in view of the symmetric nature of our confidence interval for p.

To further add to our understanding of the proper interpretation of confidence intervals, we use **R** to generate a second set of 50 approximate 95% confidence intervals for p based on random samples of size 100 with probability .6 for occurrence of the event A. The results of this second simulation are presented in Table 6.4.

Table 6.4 A second set of fifty approximate 95% confidence intervals for p, each based on 100 randomly generated independent Bernoulli variables with constant probability of success $p = .6$

Sample number	Number of occurrences of event A	$\hat{p}$	Lower CB	Upper CB	Does the observed interval contain the true value $p = .6$?
1	59	0.59	0.493601	0.686399	Yes
2	63	0.63	0.535370	0.724630	Yes
3	57	0.57	0.472965	0.667035	Yes
4	60	0.60	0.503980	0.696020	Yes
5	55	0.55	0.452491	0.647509	Yes
6	54	0.54	0.442314	0.637686	Yes
7	58	0.58	0.483263	0.676737	Yes
8	61	0.61	0.514401	0.705599	Yes
9	62	0.62	0.524864	0.715136	Yes
10	64	0.64	0.545920	0.734080	Yes
11	55	0.55	0.452491	0.647509	Yes
12	63	0.63	0.535370	0.724630	Yes
13	60	0.60	0.503980	0.696020	Yes

(continued)

Table 6.4 (continued)

Sample number	Number of occurrences of event A	$\hat{p}$	Lower CB	Upper CB	Does the observed interval contain the true value $p = .6$?
14	61	0.61	0.514401	0.705599	Yes
15	59	0.59	0.493601	0.686399	Yes
16	60	0.60	0.503980	0.696020	Yes
17	61	0.61	0.514401	0.705599	Yes
18	64	0.64	0.545920	0.734080	Yes
19	57	0.57	0.472965	0.667035	Yes
20	59	0.59	0.493601	0.686399	Yes
21	59	0.59	0.493601	0.686399	Yes
22	59	0.59	0.493601	0.686399	Yes
23	58	0.58	0.483263	0.676737	Yes
24	67	0.67	0.577838	0.762162	Yes
25	72	0.72	0.631996	0.808004	No
26	59	0.59	0.493601	0.686399	Yes
27	57	0.57	0.472965	0.667035	Yes
28	57	0.57	0.472965	0.667035	Yes
29	59	0.59	0.493601	0.686399	Yes
30	55	0.55	0.452491	0.647509	Yes
31	54	0.54	0.442314	0.637686	Yes
32	55	0.55	0.452491	0.647509	Yes
33	61	0.61	0.514401	0.705599	Yes
34	66	0.66	0.567153	0.752847	Yes
35	63	0.63	0.535370	0.724630	Yes
36	63	0.63	0.535370	0.724630	Yes
37	57	0.57	0.472965	0.667035	Yes
38	59	0.59	0.493601	0.686399	Yes
39	65	0.65	0.556514	0.743486	Yes
40	58	0.58	0.483263	0.676737	Yes
41	65	0.65	0.556514	0.743486	Yes
42	62	0.62	0.524864	0.715136	Yes
43	61	0.61	0.514401	0.705599	Yes
44	59	0.59	0.493601	0.686399	Yes
45	65	0.65	0.556514	0.743486	Yes
46	60	0.60	0.503980	0.696020	Yes
47	65	0.65	0.556514	0.743486	Yes
48	59	0.59	0.493601	0.686399	Yes
49	53	0.53	0.432177	0.627823	Yes
50	53	0.53	0.432177	0.627823	Yes

For this second set of simulated results, we see that 49 of the 50 (98%) constructed confidence intervals correctly contain the true value $p = .6$. Thus in this case we slightly exceed the 'expected' number of 47.5 (95%) successful confidence intervals. However, we are certainly still in rough agreement with the 95% confidence that we place in the randomness of the sampling process. (We note in passing that if the two sets of 50 confidence intervals are combined to make a single set of 100 confidence intervals, we would, in fact, have exactly 95 of them which contain the true value .6, in perfect agreement with the 'expected' number of successful confidence intervals! Of course, this is simply the luck of the draw, as another randomly generated set of 100 such confidence intervals could easily have 92 or 97 or even 88 intervals that correctly contain .6.)

In practice, of course, we would collect only one sample of 100 observations (rather than the 50 samples of 100 observations in each of the two simulations) and it would lead to a single approximate 95% confidence interval for p. Moreover, since p is unknown we would have no way to know whether or not it belonged to the one observed confidence interval. For the case of $p = .6$, for example, we might obtain an interval that does, indeed, contain .6, such as one of the 95 successful intervals in our simulation study. On the other hand, we might be unfortunate and obtain an interval (such as the interval (.610182, .789818) from the first set of 50 simulated samples, for example) that does not correctly contain .6. Once we have collected our single random sample and computed our one approximate 95% confidence interval for p, we cannot know whether or not we are correct in our assertion that the observed interval contains p. However, what we can say is that we are 95% confident that our random sampling and interval construction process did, in fact, produce an interval that contains p. Thus the proper interpretation of our 95% confidence interval for p is that if we always compute the interval in the prescribed fashion in every situation, then 95% of the intervals obtained will contain the true parameter value p.

Selecting the Sample Size In constructing the approximate 95% confidence interval for p_{Closed}, the proportion of closed communities that contain a majority white flowers, in Example 6.4, we found that the computed upper limit for the interval was 1.0461. Since we knew that p_{Closed} can not exceed 1, we truncated the upper limit to be 1 to obtain the approximate 95% confidence interval (.5903, 1) for p_{Closed}. This problem with the computed upper confidence limit is due to the fact that the accuracy of the approximation used in construction of the confidence interval for p given in (6.7) is directly related to the number of observations in the sample. The approximation is simply not very accurate for the Ostler and Harper sample size $n = 11$. The approximation would, of course, have been much better if we had been able to collect a greater number of sample observations. In particular, for the approximate 95% confidence interval for p_{Closed}, suppose for purposes of illustration that Ostler and Harper had included $n = 110$ closed communities in their study (instead of the eleven that they examined) and found that 90 of these closed communities had a majority white flowers. For this larger data set, we would still have $\hat{p}_{\text{Closed}} = 90/110 = .8182$. However, now the approximate confidence interval for p_{Closed} would be given by

$$\begin{aligned} .8182 \pm 1.96\,[.8182(1-.8182)/110]^{1/2} &= .8182 \pm .0721 \\ &= (.7461, .8903). \end{aligned}$$

With these "additional 99 observations" our new interval is still centered at the point estimate $\hat{p}_{\text{Closed}} = .8182$. However, we no longer have a problem with the upper limit (.8903) exceeding the known boundary (1) for p_{Closed}. In addition, we notice that the length of the approximate 95% confidence interval has decreased from 1 - .5903 $=$ .4097 for $n = 11$ closed communities to .8903 - .7461 $=$ .1442 for the hypothetical $n = 110$ closed communities. Thus we would be able to make a much stronger statement (with the same approximate 95% confidence and observed point estimate .8182) about the unknown value of p_{Closed} had we observed 110 observations than we can with the

11 observations collected by Ostler and Harper. This is no accident and it should not be a surprise. If we were able to consider larger random samples, we would expect less variability in our point estimator from sample to sample. This fact translates into a smaller margin of error (i. e., more accuracy) for our point estimator and thus shorter confidence intervals from the particular sample observed.

This naturally raises the question about whether we can decide before we collect our sample how large a sample size is necessary in order for our achieved confidence interval to have a margin of error no greater than some prespecified level with which we are comfortable. To proceed along these lines, we note that the margin of error for the approximate 100CL% confidence interval for p given in (6.7) is $\pm z_{(1-\text{CL})/2}\text{S.E.}(\hat{p}) = \pm z_{(1-\text{CL})/2}[\hat{p}(1-\hat{p})/n]^{1/2}$, which depends on the value of the point estimate $\hat{p}$ from our sample. Fortunately, it can be shown mathematically that the quantity $\hat{p}(1-\hat{p})$ can be no larger than $\frac{1}{2}\left(1-\frac{1}{2}\right) = \frac{1}{4}$; that is, the greatest margin of error for a fixed sample size n will occur when the estimate $\hat{p}$ is equal to $1/2$. This is not too surprising, since it says that the most variability in our observed sample percentages will be associated with those events A that are least predictable (i.e., events A for which there is a 50-50 chance of occurrence), while those events that are virtually certain to occur (p near 1) or virtually certain not to occur (p near 0) will yield sample percentages that vary little from sample to sample.

Using this upper bound of $1/4$ for the quantity$\hat{p}(1-\hat{p})$, we see that the margin of error for the approximate 100CL% confidence interval for p given in (6.7) is never any greater than $\pm z_{(1-\text{CL})/2}\left[\frac{1}{4n}\right]^{1/2}$. Hence if we desire an approximate 100CL% confidence interval for p with margin of error (either plus or minus) no greater than a pre-specified value d, it suffices to choose a sample size n such that

$$z_{(1-CL)/2}\left[\frac{1}{4n}\right]^{1/2} = \frac{z_{(1-CL)/2}}{2\sqrt{n}} \leq d$$

or, equivalently, such that

$$\sqrt{n} \geq \frac{z_{(1-CL)/2}}{2d}.$$

Squaring both sides yields the required inequality

$$n \geq z^2_{(1-CL)/2}/4d^2$$

for the necessary sample size.

Sample Size Adequate to Provide Desired Margin of Error In order to obtain an approximate $100CL\%$ confidence interval for the probability, p, of an event A, that has margin of error no greater than $\pm\, d$, it suffices to collect at least

$$n \geq z^2_{(1-CL)/2}/4d^2 \tag{6.8}$$

sample observations and estimate the probability p by the corresponding sample percentage $\hat{p} =$ [number of times the event A occurs in the sample]$/n$.

Example 6.6. Sample Survey Design You have been hired to design and conduct a sample survey in the state of Tennessee to determine the strength of support among eligible voters for legalized casino gambling in the state. Let p denote the proportion of eligible voters in Tennessee who favor legalizing casino gambling in the state. It is important to your employers that you obtain accurate information regarding the statewide sentiment on this issue. Consequently, they have asked you to provide them with an estimate of p for which you are confident that the margin of error is not greater than $\pm$ 2%. Since you will be using the data from a random sample to obtain your estimate, you cannot, of course, guarantee such a margin of error with certainty. However, you can do so with any desired approximate confidence. For example, if 90% confidence is taken to be adequate for the study, we see from Eq. (6.8) that to satisfy the prescribed requirements, you will need to interview at least

$$n \geq z_{.05}^{2}/4d^{2} = (1.645)^{2}/4(.02)^{2} = 1691.3$$

randomly selected eligible state voters. Rounding up to maintain the desired confidence and maximum margin of error, you decide to interview 1692 eligible state voters. With this size random sample, you can be approximately 90% confident that the observed percentage of your sample in favor of legalizing casino gambling will not differ from the unknown p by more than $\pm$ 2%. (We point out that the actual process of obtaining a random sample of 1692 eligible voters in the state of Tennessee to interview on this question is not necessarily an easy task. We refer you to Chap. 3 for some of the possible approaches for creating such a random sample, as well as discussion of some of the potential hazards to avoid in the sampling process.)

We note that if one has additional knowledge (for instance, from a preliminary study) that the true value of the probability p for an event A is greater than or equal (or less than or equal) to some known value $p^{*} \neq .5$, the necessary sample size to have prescribed confidence in an upper bound for the margin of error of the sample estimator $\hat{p}$ can sometimes be substantially lower than the worst case scenario (using $p = .5$) provided by the conservative approach in Eq. (6.8).

Section 6.2 Practice Exercises

6.2.1. Below are three statements about a 95% confidence interval $(\widehat{\theta}_L, \widehat{\theta}_U)$ for the population parameter θ. Which of them are true statements? For each of the false statements, give a brief explanation why it is false.

(a) If the true value of θ is 2, then the probability is .95 that we will obtain a sample for which the observed value of the statistic $\widehat{\theta}_L$ is less than or equal to 2 and the observed value of the statistic $\widehat{\theta}_U$ is greater than or equal to 2.

(b) If the observed values of the statistics $\widehat{\theta}_L$ and $\widehat{\theta}_U$ for our sample are $\widehat{\theta}_L = 3.68$ and $\widehat{\theta}_U = 8.44$, then the probability is .95 that θ is between 3.68 and 8.44.

(c) If we were to collect 1000 independent samples and numerically compute the values of the confidence interval $(\widehat{\theta}_L, \widehat{\theta}_U)$ for each of them, we would expect that the true value of θ would be contained in approximately 950 of these intervals.

6.2.2. Below are three statements about a 90% confidence interval $(\widehat{\theta}_L, \widehat{\theta}_U)$ for the population parameter θ. Which of them are true statements? For each of the false statements, give a brief explanation why it is false.

(a) Once we have collected our sample and numerically computed the value of the confidence interval, we will be able to tell whether it contains the true value of θ.

(b) We have collected ten separate random samples of data and numerically computed the value of the confidence interval for each of them. If each of the first nine of these intervals does, indeed, contain the true value of θ, then the tenth confidence interval will not contain θ.

(c) Since the endpoints of the interval $\widehat{\theta}_L$ and $\widehat{\theta}_U$ are statistics, their values will vary from sample to sample. However, the probability is .95 that the random sample we collect will produce values of $\widehat{\theta}_L$ and $\widehat{\theta}_U$ such that the true value of θ is between them.

6.2.3. If a level *CL* confidence interval $(\widehat{\theta}_L, \widehat{\theta}_U)$ for the population parameter θ is based on a sample estimator $\widehat{\theta}$, how does the sampling variability of $\widehat{\theta}$ affect the interval? How does the choice of confidence level *CL* affect the interval?

6.2.4. *Stress and Blood Pressure.* Consider the blood pressure study discussed in Exercise 6.1.9. What is the standard error for the point estimate of

$p = P$(a person will have higher blood pressure in stress-related situations than during non-stressful daily activities)?

6.2.5. *Women's Shoe Preferences.* Consider the shoe preference data in Exercise 6.1.10. What is the standard error for the point estimate of the proportion of those women who work outside the home who typically wear "flats" to work?

6.2.6. *Who Are More Stubborn—Women or Men?* Consider the results of the survey presented in Exercise 6.1.11. What is the standard error for the point estimate of the proportion of men who would not stop and ask for directions or consult a map if they were lost?

6.2.7. *Stress and Blood Pressure.* Using the blood pressure data in Exercise 6.1.9, find an approximate 92% confidence interval for

$p = P$(a person will have higher blood pressure in stress-related situations than during non-stressful daily activities).

6.2.8. *Women's Shoe Preferences.* Consider the shoe preference data in Exercise 6.1.10.

(a) Find an approximate 95% confidence interval for the proportion of those women who work outside the home who typically wear "flats" to work.

(b) Find an approximate 99% confidence interval for the proportion of those women who work outside the home who typically wear "flats" to work.

(c) Discuss the differences between the two confidence intervals obtained in parts (a) and (b).

6.2.9. *Women's Shoe Preferences.* Consider the shoe preference data for women with more than 4 years of college who work outside the home, as presented in part (b) of Exercise 6.1.10. Can you use these data to find an approximate 98% confidence interval for the proportion of women with more than 4 years of college who work outside the home and who typically wear "flats" to work? If yes, find the confidence interval. If no, explain why not.

6.2.10. *Lost While Driving.* Consider the results of the survey presented in Exercise 6.1.11. Can you use these data to find an approximate 90% confidence interval for the proportion of younger drivers (ages 18-24) who have experienced being lost while driving? If yes, find the confidence interval. If no, explain why not.

6.2.11. Using an appropriate software package (such as **R**), simulate 50 random samples, each with sample size $n = 100$, from a population with $p = P$ (event A occurs) $= .9$. For each of these 50 random samples, compute an approximate 95% confidence interval for p and record whether or not .9 is in the interval (i.e., whether or not the observed confidence interval does, in fact, contain the true value of p).

(a) How many of the 50 confidence intervals should we expect to contain the value $p = .9$?

(b) How many of the 50 simulated confidence intervals actually did contain the value $p = .9$?

(c) Repeat the entire process another 50 times and answer parts (a) and (b) once again for this next set of random samples. Comment on your findings from the two sets of 50 random samples.

(d) Compare these results with the corresponding outcomes discussed in the text when $p = .6$. In particular, what are the average center and average length for the one hundred 95% confidence intervals when $p = .6$? What are the corresponding average center and average length for the one hundred 95% confidence intervals obtained in this exercise when $p = .9$? Discuss similarities or differences between these summary statistics for the two sets of simulations.

6.2.12. Using an appropriate software package (such as **R**), simulate 50 random samples, each with sample size $n = 100$, from a population with $p = P$ (event A occurs) $= .9$. For each of these 50 random samples, compute an approximate 90% confidence interval for p and record whether or not .9 is in the interval (i.e., whether or not the observed confidence interval does, in fact, contain the true value of p).

(a) How many of the 50 confidence intervals should we expect to contain the value $p = .9$?

(b) How many of the 50 observed confidence intervals actually did contain the value $p = .9$?

(c) Repeat the entire process another 50 times and answer parts (a) and (b) once again for this next set of random samples. Comment on your findings from the two sets of 50 random samples.

(d) Compare these results with the corresponding outcomes for the 95% confidence intervals discussed in Exercise 6.2.11 when $p = .9$. In particular, what are the average center and average length for the one hundred 95% confidence intervals when $p = .9$? What are the corresponding average center and average length for the one hundred 90% confidence intervals obtained in this exercise when $p = .9$? Discuss similarities or differences between these summary statistics for the two sets of simulations.

6.2.13. Using an appropriate software package (such as **R**), simulate 50 random samples, each with sample size $n = 100$, from a population with $p = P$ (event A occurs) $= .6$. For each of these 50 random samples, compute an approximate 99% confidence interval for p and record whether or not .6 is in the interval (i.e., whether or not the observed confidence interval does, in fact, contain the true value of p)

(a) How many of the 50 confidence intervals should we expect to contain the value $p = .6$?

(b) How many of the 50 observed confidence intervals actually did contain the value $p = .6$?

(c) Repeat the entire process another 50 times and answer parts (a) and (b) once again for this next set of random samples. Comment on your findings from the two sets of 50 random samples.

(d) Compare these results with the corresponding outcomes for the 95% confidence interval discussed in the text when $p = .6$. In particular, what are the average center and average length for the one hundred 95% confidence intervals when $p = .6$? What are the corresponding average center and average length for the one hundred 99% confidence intervals obtained in this exercise when $p = .6$? Discuss similarities or differences between these summary statistics for the two sets of simulations.

6.2.14. Let p be the probability of the event A in a population. Suppose that A occurs 18 times in a random sample of 25 observations from the population. What approximate confidence do we have that p is contained in the interval (.544, .896)?

6.2.15. Let p be the probability of the event A in a population. If (.1788, .3212) is an approximate 90% confidence interval for p based on a random sample of size n from the population, in what percentage of the sample observations did the event A occur? How many observations were there in the random sample?

6.2.16. Let p be the probability of the event A in a population. Suppose that (.1877, .4123) is a confidence interval for p based on a random sample of size n from the population.

(a) In what percentage of the sample observations did the event A occur?
(b) What approximate confidence do we have in the interval (.1877, .4123) if the sample size is $n = 64$?
(c) If we have approximate confidence 98.58% in the interval (.1877, .4123), what is the sample size n?
(d) Discuss the implications of your answers to (b) and (c).

6.2.17. Let p be the probability of the event A in a population. Suppose that A occurs in 37 out of 82 observations from the population. Compute the approximate 90%, 95%, and 99% confidence intervals for p. Discuss the similarities and differences in these three intervals.

6.2.18. Let p be the probability of the event A in a population. Suppose that A occurs in 30 out of 50 observations from the population.

(a) Compute an approximate 92% confidence interval for p.
(b) What approximate confidence do you have in the interval obtained in part (a) if the sample size had been 100 (with 60 event A occurrences) instead of 50?
(c) What approximate confidence do you have in the interval obtained in part (a) if the sample size had been 25 (with 15 event A occurrences) instead of 50?
(d) Discuss your findings in parts (a) – (c).

6.2.19. Let p denote the probability of the event A in a population. Consider the approximate 95% confidence interval for p based on a random sample of size 100 from the population, as given in expression (6.7). Suppose we collect 10 independent samples, each of sample size $n = 100$, from the population and compute the 95% confidence interval for each of these samples. Let B denote the number of these ten confidence intervals that contain the true value of p.

(a) How many of the ten confidence intervals should we expect to contain the true value of p?

(b) What is the sampling distribution of the statistic B?

(c) If the true value of p is .75, what is the probability that B is greater than 7?

(d) Does your answer to part (c) change if the true value of p is .34? Explain.

6.2.20. Let p denote the probability of the event A in a population. You wish to use a random sample of size n from the population to construct an approximate 95% confidence interval for p.

(a) How large should n be if you want your confidence interval to have a margin of error no greater than $\pm$ .10?

(b) How large should n be if you want your confidence interval to have a margin of error no greater than $\pm$ .01?

6.2.21. Let p denote the probability of the event A in a population. You wish to use a random sample of size n from the population to construct an approximate confidence interval for p that will have a margin of error no greater than .05.

(a) How large should n be if you want your approximate confidence in the interval to be 95%?

(b) How large should n be if you want your approximate confidence in the interval to be 90%?

6.2.22. *Women's Shoe Preferences.* Consider the shoe preference data for women with more than 4 years of college who work outside the home, as presented in part (b) of Exercise 6.1.10. Let p denote the proportion of this group of women who typically wear 'flats' to work. How many women with this educational background would have to have been part of the telephone interviewing process in order to guarantee that the approximate 95% confidence interval for p would not be in error by more than $\pm$ .04?

6.2.23. *Lost While Driving.* Consider the study discussed in Exercise 6.1.11 regarding a number of issues about being lost while driving. In part (c) of that exercise, it was noted that 11.9% of older Americans (ages 54-64) in the survey reported that they had experienced being lost while driving. Let p denote the proportion of all Americans in this age group who have experienced being lost while driving. How many Americans would have to have been part of the survey in order to guarantee that the approximate 90% confidence interval for p would not be in error by more than $\pm$ .25?

6.2.24. *Whose Rights Do the Courts Protect?* In a telephone interview survey conducted on February 2, 1994, ABC News found that 447 out of 520 interviewed adults agreed with the following statement: "The court system in the United States does too much to protect the rights of people who are accused of crimes, and not enough to protect the rights of crime victims."

(a) Let p denote the percentage of all adults who agree with the given statement. Using the ABC News data, find an approximate 94% confidence interval for p.

(b) Can we conclude from the findings of the ABC News telephone interviews that a large majority of adults feel that the court system in the United States does too much to protect the rights of people who are accused of crimes? Or that a large majority of adults feel that the court system in the United States does not do enough to protect the rights of crime victims? Comment on the wording of the statement used by ABC News in their telephone interviews.

6.2.25. *Men and Candy*. In a study of 7841 men, Lee and Paffenbarger (1998) found that 3312 of the respondents indicated that they 'almost never' ate candy. This was taken to be the 'non-consumer' group. The remaining 4529 men made up the 'candy consumer' group.

(a) Let p denote the percentage of all men who almost never eat candy. Viewing the 7841 men in the Lee and Paffenbarger study as a random sample from the population of all men, find an approximate 96% confidence interval for p.

(b) In reality, the 7841 men in the Lee and Paffenbarger (1998) study were Harvard alumni who had been undergraduates at Harvard between 1916 and 1950. Does this affect the interpretation of the confidence interval for p obtained in part (a)? If so, how?

6.2.26. *Men and Candy*. Consider the candy-consumer survey data discussed in Exercise 6.2.25. Lee and Paffenbarger (1998) also found that 1565 of the non-consumers of candy and 1987 of the candy-consumers in their sample took vitamin or mineral supplements on a regular basis.

(a) Viewing the 3312 male non-consumers in the Lee-Paffenbarger sample as representative of all male non-consumers of candy, find an approximate 95% confidence interval for the percentage of male non-consumers of candy who take vitamin or mineral supplements on a regular basis.

(b) Viewing the 4529 male candy-consumers in the Lee-Paffenbarger sample as representative of all male candy-consumers, find an approximate 95% confidence interval for the percentage of male candy-consumers who take vitamin or mineral supplements on a regular basis.

(c) Compare the two confidence intervals obtained in parts (a) and (b). What can be said about the vitamin and mineral supplements habits of the male non-consumer and candy-consumer groups?

6.3 Hypothesis Testing

The General Concept Thus far in our study of statistical inference we have concentrated directly on point and interval estimation of a population parameter θ of interest. In some settings such estimation is in itself the most appropriate way to address the important questions in a study and additional methods of inference are not necessary. However, in many settings we are also interested in using the sample data to assess the validity of some claim (hypothesis) about the population. In such cases, we need statistical procedures, called hypothesis tests, that will enable us to use the evidence available in the collected sample observations to reach a logical and defensible conclusion about the validity of the *a priori* ("from the earlier" in Latin) hypothesis. Often the appropriate hypothesis tests are themselves based on point or interval estimators of a relevant population parameter. However, the statistical methodology for testing hypotheses and the associated procedural steps needed to reach a conclusion regarding the hypotheses are generally more complicated than those associated with either point or interval estimation. We introduce the structure associated with hypothesis testing with the following example.

A 'good' friend of yours has asked you to spend a nice sunny afternoon playing a 'friendly' game of chance out on the college oval. She suggests that you flip a coin and she will pay you a dime for every tail that occurs and you must pay her a dime for every head that occurs. Since it is a beautiful day and you like your friend's company, you agree to meet her on the oval at 2 p.m. However, as the day moves along you begin to have a few concerns about the game. After all, you have heard that several students created biased coins with $P(\text{heads}) = .9$ in an industrial arts class and your friend is in that class. You would not be so worried except for the fact that she said she would bring the coin and that she really felt that heads would be lucky for her today!! This leaves you with a dilemma--you certainly do not want to unfairly accuse your friend of trying to cheat you, but you also do not want to lose the money you worked so hard to earn.

Such a setting is ideally suited for hypothesis testing. Once you meet on the oval you convince your friend that you would like to 'warm up' for the game by practice flipping the coin eight times, to which she agrees. Unknown to your friend, you have decided to base your decision about whether to flip the coin for money solely on the outcome of these eight flips. There are two very different hypotheses or claims involved in this situation. One, which we will call the null hypothesis and denote by H_0, is that your friend is playing with a fair coin and simply wants to enjoy the sunny afternoon with you. Since she is your friend, this is the hypothesis that you are willing to believe (or accept) if, in fact, you were not able to collect any data at all. The second, more sinister, hypothesis is that your friend has chosen to use her weighted coin to take a bit of the 'chance' out of your game of chance. We refer to this as the alternative hypothesis and denote it by H_A. It is a conjecture about what your friend MIGHT be doing and requires sufficient sample evidence before you are willing to reject H_0 in favor of this conjecture. Thus for this setting we are interested in testing the null hypothesis

$$H_0 : [\text{the coin is fair; that is, } P(\text{head}) = .5]$$

versus the alternative hypothesis

$$H_A : \left[\text{the coin is weighted such that} P(\text{head}) > .5\right].$$

The next step in constructing our hypothesis test is to decide which possible outcomes for our eight practice flips will provide sufficient evidence to reject H_0 in favor of H_A. Letting B denote the number of heads obtained in our eight practice flips, we know that the set of possible outcomes for B (i. e., the sample space) is $S = \{0, 1, \ldots., 7, 8\}$. Clearly large values of B are going to be in better agreement with the alternative hypothesis H_A than with the null hypothesis H_0, while the opposite is true for small values of B. In order to make our decision totally objective (and not be influenced by any glare that you might be getting from your friend), we need to divide the total sample space into two regions, one set of outcomes which will cause you to reject H_0 and refuse to play the game with your friend and the complementary set of outcomes which will not lead you to reject H_0 and proceed to play the game with your friend. One such division of the sample space S would be to reject H_0 and refuse to play the game only if you obtain at least 6 heads in your 8 flips; that is,

$$\text{Reject } H_0 \text{ if and only if } B \geq 6.$$

Thus, we choose to reject H_0 only if the 'evidence' provided by our 8 coin flips is sufficiently unlikely (i. e., we obtain at least 6 heads) to have resulted from a fair coin. (We note that the choice of 6 as 'extreme' is up to you and other options for this 'critical value' would certainly be reasonable. We return to this issue later and explore the effect that other possibilities, such as 5 or 7, for the critical value have on the decision process.)

With this example in mind, we state a few fundamental definitions for the methodology of hypothesis testing.

Definition 6.2 The **null hypothesis** is a statement of interest that we wish to evaluate in light of collected sample evidence. We denote a null hypothesis by $\boldsymbol{H_0}$ **(pronounced *H* naught)** and it usually corresponds to the hypothesis of "no change" or "no effect" from a new procedure under investigation.

A null hypothesis can often be viewed as a default statement if there were no collected data. An individual or organization might make such a statement about a standard product or treatment. With this in mind, failure to reject H_0 on the basis of the collected sample data generally leads to at least temporary validity for H_0 and a continuation of doing things as they were being done prior to the experimental study.

For the coin tossing quandary with our friend, we would certainly believe that the null hypothesis H_0 of a fair coin is true if we are not permitted to flip the coin eight times. There would be no reason to accuse our friend of trying to cheat without sufficient supporting evidence.

Definition 6.3 The **alternative hypothesis is denoted by H_A** and corresponds to the claim that is being made for the problem of interest. This claim is almost always the primary reason for conducting the experiment to collect the data in the first place. Often it is also what the experimenter hopes or strongly believes to be true, but is not willing or able to act on unless the evidence in the sample is sufficiently convincing of the claim.

The alternative hypothesis is the statistical analogue of the 'guilty' verdict in a court of law. A person on trial is presumed innocent (analogue of the null hypothesis) unless the evidence and the lawyers (analogues of sample data and statisticians) convince the jury of the person's guilt (corresponding to the alternative hypothesis) beyond a reasonable doubt (a concept for which a statistical analogue will be defined later in this section).

In the case of the projected game with our friend, the alternative hypothesis H_A corresponds to the claim that she is using an unfairly weighted coin. While this is certainly not what we hope to be the truth for this particular setting, it definitely remains the case that we do not want to act on the claim and accuse our friend of trying to cheat unless the evidence (i. e., the number of heads) obtained from the eight flips is sufficiently unlikely to have resulted from flipping a fair coin.

Definition 6.4 A hypothesis test of a null hypothesis H_0 versus an alternative hypothesis H_A is a rule that, on the basis of appropriate collected sample observations, leads to a decision whether or not to reject H_0. Such a rule corresponds to a division of the sample space into two subsets. One of these subsets contains those sample outcomes that provide sufficient critical evidence supporting H_A to warrant rejection of H_0. This subset is called the **critical (rejection) region for the test**. The second complementary subset contains those sample outcomes that are not considered sufficiently incompatible with H_0 to warrant its rejection. Often the division of the sample space into these two subsets is based on one or two numerical values, known as **critical value(s)**, of a sample statistic, called the **test statistic**.

For the experiment with your friend's coin, the hypothesis test is the previously stated decision rule

$$\text{Reject } H_0\text{: [coin is fair] if and only if } B \geq 6,$$

where here the test statistic B is the number of heads in our eight practice flips. The associated sample space $S = \{0, 1, ..., 7, 8\}$ is divided into the two subsets $C = \{6, 7, 8\}$, the critical region, and its complement $C' = \{0, 1, ..., 5\}$. The complement corresponds to those outcomes that are not sufficiently supportive of the alternative hypothesis $P(\text{head}) > .5$ to warrant rejecting the null hypothesis that the coin is fair. (Note that while the final decision regarding H_0 depends on the eventual observed value of the test statistic B, the formal construction of the critical region C for the test does not. Its construction is based solely on the sampling distribution of the test statistic B under the null hypothesis condition that the coin is fair and how willing we are to incorrectly reject this fairness when it is true.)

The description of this hypothesis test consists of three separate pieces. First there is the test statistic, B, used to assess the information about the coin's fairness that is available from our eight sample flips. The second piece of the hypothesis test is the 'direction' of the critical region. The 'direction' of this test is to reject the fairness of the coin for large ($\geq$) values of B. The final component of the test is stipulation of the degree of sample evidence that is sufficient to reject H_0. This requires specifying the boundary of the critical region. For the above hypothesis test, this is equivalent to answering the question 'How many heads in the eight sample flips is enough to warrant rejection of fairness for the coin?' The answer in the stated hypothesis test is '6 or more'.

This three-component breakdown is true, in general, for any hypothesis test. That is, a statistical test of a null hypothesis H_0 versus an alternative H_A consists of specifying:

- the appropriate test statistic, say T,
- the direction of the rejection region (i. e., do we reject H_0 for large values of T, small values of T, or both large and small values of T),

and

- the critical value(s) that delineates the boundary of the critical region and corresponds to the degree of sample evidence in favor of the alternative hypothesis H_A that we require in order to reject H_0.

This construction is illustrated in Fig. 6.5, where t_1, t_2, t_3, and t_4 are the critical values that define the boundaries of the various critical regions.

Fig. 6.5 Three components of a test of hypothesis

TEST STATISTIC	DIRECTION OF REJECTION	CRITICAL VALUE(S)
	↗ Reject for **large values** of T	$T \geq t_1$ **(critical value)**
T	→ Reject for **small values** of T	$T \leq t_2$ **(critical value)**
	↘ Reject for **either small or large values** of T	$T \leq t_3$ **(critical value)** or $T \geq t_4$ **(critical value)**

The particular problem of interest and the type of data being collected generally suggest an appropriate test statistic T. The direction of the rejection region corresponds to those possible values for T that are most indicative that H_A is a more reasonable conclusion than H_0. Finally, both the sampling distribution of the test statistic T under the condition that the null hypothesis H_0 is true and the degree of our willingness to incorrectly reject such a true H_0 play roles in determination of the critical value boundary. (We will return for a more detailed discussion of this latter aspect later in this chapter.)

For the coin tossing experiment, it is clear that large values of B (i.e., obtaining a large number of heads in our eight tosses of the coin) are more likely to occur if the alternative H_A is true than if the null hypothesis H_0 is true. Hence, the natural direction for the rejection region in this case is to reject H_0 for large values of B. (A hypothesis test with a rejection region of this form is often referred to as an **upper-tailed hypothesis test**, because of the nature of the values of the test statistic that lead to rejection of H_0.) The final piece of the test requires the selection of a critical value, corresponding to the minimum number of heads that we need to obtain in our eight flips before we can no longer have faith in the fairness of the coin (i.e., H_0). The particular critical region selected for our example was $C = \{6, 7, 8\}$. (Why you might choose this particular critical value 6--and not 5 or 7, for example--will be discussed later.)

Type I and Type II Errors When we use sample data to conduct a test of the null hypothesis H_0 versus the alternative hypothesis H_A, we have no way of knowing for certain whether or not our test procedure has led to a correct decision. Therefore, it is important to be aware of the possibility that our decision could be in error, and to provide ways to assess the likelihood of this possibility in different situations.

Whether or not the decision indicated by a test procedure for a specific set of sample data is correct depends on the true state of the population providing the sample data. If the population is such that the null hypothesis, H_0, is true, then we will make a correct decision if the collected sample data do not lead to rejection of H_0. If, on the other hand, the null hypothesis is true but we just happen, by chance, to obtain a sample that leads to rejection of H_0, our decision is clearly in error. Such a mistake (i. e., rejecting a true null hypothesis) is called a Type I error. In our coin-tossing example, a Type I error would occur if the coin were actually fair but we obtained, by chance, 6 or more heads in our 8 flips.

Conversely, if the true state of our population is that the null hypothesis is false, we make a correct decision when our collected sample data lead to rejection of H_0 by our hypothesis test. Of course if we happen to obtain a sample that leads us not to reject H_0 when it is false, we will have reached yet a different incorrect conclusion, namely, failing to reject H_0 when, in fact, it is false. This type of mistake is called a Type II error.

Definition 6.5 A statistical test of the null hypothesis H_0 against the alternative hypothesis H_A is said to result in a **Type I error** if it rejects H_0 (and thus, at least in effect, does not reject H_A) when, in fact, H_0 is true. If, on the other hand, a test fails to reject a null hypothesis H_0 when, in fact, it is false, we say that a **Type II error** has occurred.

The way we state things may seem a bit strange as you first become acquainted with the language of hypothesis testing. For example, we are being especially careful to make the statement "fail to reject H_0" rather than the seemingly more natural statement "accept H_0". Simply because the sample evidence is consistent with the null hypothesis (so that we do not want to reject it) does not prove that H_0 is true. In fact, the same sample evidence is usually consistent with a number of different hypotheses about the population other than the stated null hypothesis H_0, so that "accepting H_0" is too strong a conclusion to reach from the results of a hypothesis test.

Suppose, for example, that you conduct the 8 'test' tosses with your friend's coin and you observe 5 heads. As this outcome is not in our prescribed critical region, we would not reject H_0. This is, of course, a reasonable conclusion, since observing 5 heads in 8 tosses is not surprising if the coin is fair. However, such an outcome is also not surprising if $P(\text{Heads}) = 0.47$ or, in fact, if $P(\text{Heads}) = \frac{5}{8} = 0.625$! There are, in fact, a large number of possible values for $P(\text{Heads})$ that are consistent with the observed outcome of 5 heads in 8 tosses. Thus, while it is reasonable to conclude that $P(\text{Heads}) = 0.5$ should not be rejected based on the sample outcome, the evidence is not sufficiently strong for us to conclude that it is the true description of our friend's coin. In fact, it may very well not be true that $P(\text{Heads}) = 0.5$, but we simply have not obtained enough evidence in our eight tosses to refute it.

Take your time becoming familiar with these hypothesis testing concepts and related terminology. Study a number of examples until you become comfortable with the twists and turns of the specialized language.

The different scenarios relating the possible outcomes of a hypothesis test with the truth of the hypotheses can be represented in the following simple diagram:

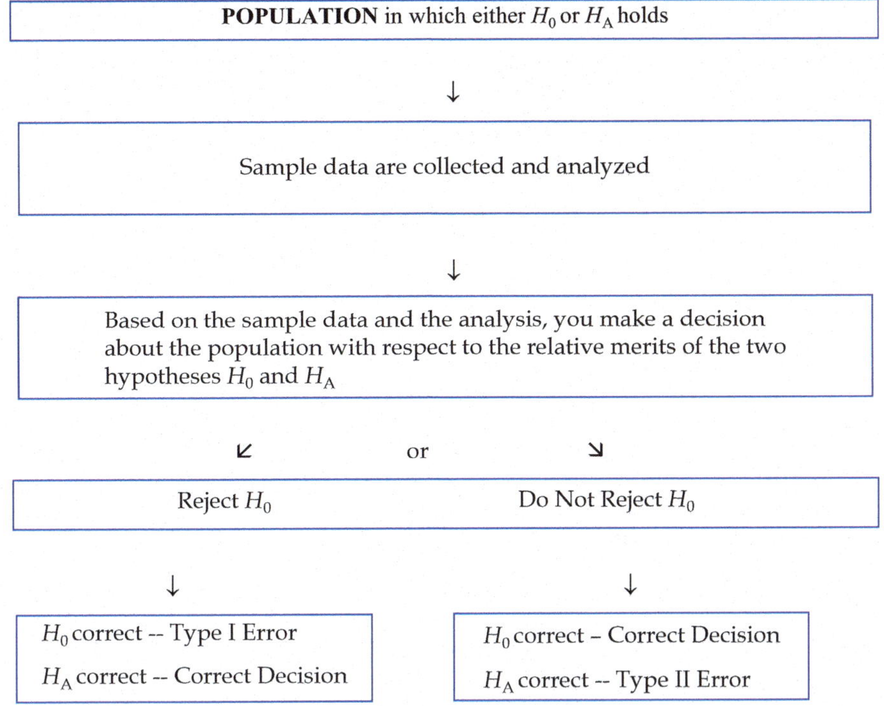

Unless we miraculously ascertain whether or not the null hypothesis is actually true, we will never be able to determine if an error (either Type I or Type II) has occurred with respect to a specific decision from a hypothesis test. However, it is clear that both errors cannot occur simultaneously. For any specific set of sample data, a hypothesis test can result in a Type I error only if the null hypothesis is true or a Type II error only if the alternative hypothesis is correct. In addition, for a given problem the two errors can be of varying importance and lead to quite different practical implications should they occur. For instance, consider the eight practice flips with your friend's coin. If the coin is, indeed, fair (i.e., H_0 is true), but our eight flips simply happen by chance to result in 6 or more heads, we will end up committing a Type I error. The practical implication of this combination of fair coin with an unusually large number of heads is that you incorrectly accuse your 'friend' of trying to use an unfair coin to cheat you out of your money--the net result is probably at least a strained relationship. On the other hand, if your 'friend' is truly trying to take advantage of you with the weighted coin for which $P(\text{heads}) = .9$, but simply from the random nature of the experiment you end with fewer than 6 heads in your 8 practice flips, you are likely to have a long, financially costly afternoon with your 'friend'--the net result is a 'friend' for (a poorer) life!

As another example of the differences between the two types of error, consider again the criminal trial setting. Since here the null hypothesis corresponds to a defendant being innocent, a Type I error occurs when a jury reaches a verdict of guilty for an innocent defendant. On the other hand, a Type II error occurs when a guilty defendant is found to be innocent by the jury. Thus, a Type I error can occur only with a jury finding of guilty and a Type II error can occur only for an innocent verdict by a jury. What are the practical implications of the two types of error for this setting? Which do you feel is the more serious?

Error Probabilities – Significance Level and Power Since the decisions we make using hypothesis tests are based on samples, it is impossible to be certain whether or not we have reached a correct decision for a given set of data. This is similar to the situation we face with interval estimation when we cannot be certain whether or not an observed confidence interval actually contains the unknown value of the parameter it is estimating. In the case of confidence intervals, we are able to rely on the structure of the sampling process to provide us with some measure of comfort (i. e., the prescribed confidence) that our observed interval does, in fact, contain the unknown parameter value. It is this randomness of the sampling process that saves us once again in evaluating the effectiveness of hypothesis tests.

Definition 6.6 The significance level, denoted by the symbol α, for a statistical test of the null hypothesis H_0 versus the alternative hypothesis H_A is the probability of making a Type I error; that is,

$$\begin{aligned}\alpha &= \text{significance level} = P\{\text{Type I error}\}\\ &= P\{\text{rejecting } H_0 \text{ when it is true}\}.\end{aligned}$$

A hypothesis test with significance level α is often called a **level α test.**

To illustrate the concept of a significance level, we return to the experiment with your friend's coin and determine the significance level α for the test that rejects the fairness of the coin (i. e., H_0: $P(\text{Heads}) = .5$) if $B \geq 6$, where B is the number of heads obtained in eight flips of the coin. When H_0 is true, the sampling distribution for B is binomial with parameters $n = 8$ and $p = .5$. It follows that the significance level for the stipulated test is given by

$$\begin{aligned}\alpha &= P(B \geq 6 \text{ when } p = .5)\\ &= P(B = 6 \text{ when } p = .5) + P(B = 7 \text{ when } p = .5) + P(B = 8 \text{ when } p = .5)\\ &= \sum_{t=6}^{8} \frac{8!}{t!(8-t)!}(.5)^8 = .1094 + .0312 + .0039 = .1445.\end{aligned}$$

Note that this probability can also be found directly using the **R** function *pbinom( )* .

```
> pbinom(q = 5, size = 8, prob = 0.5, lower.tail = FALSE)
[1] 0.1445313
```

Thus the probability is .1445 (not so small) that you will refuse to play even if she is not using the biased coin!

What about the performance of the test if your friend is using a biased coin? How likely is it that this test will detect that event? This leads us to consider properties of a test when the alternative is true. Usually the alternative hypothesis H_A includes an entire range of values for the parameter of interest. For your coin test, H_0 is $P(\text{Heads}) = 0.5$ and H_A is $P(\text{Heads}) > 0.5$. To compute how likely it is that your test will detect an unfair (weighted) coin you need to choose a particular alternative for $P(\text{Heads})$. Since you suspect that if the coin is weighted at all it is likely that $P(\text{Heads}) = 0.9$, you might use this value in your computation. We refer to this choice, $P(\text{Heads}) = 0.9$, as a *member of the set of alternatives* H_A corresponding to $P(\text{Heads}) > 0.5$.

Definition 6.7 The power of a statistical test of the null hypothesis H_0 against a particular member of the set of alternatives H_A is the probability that the test will correctly reject the null H_0 when that particular alternative hypothesis is true; that is,
Power against an alternative a^* in $H_A = P(\text{rejecting } H_0 \text{ when } a^* \text{ in } H_A \text{ is true})$.

It is common statistical language to refer to the measured power of a test for a particular member of the set of alternatives H_A as the power of the test "against that alternative". Since a Type II error occurs if we fail to reject H_0 when $a^* \in H_A$ is true, the power of a test against the alternative $a^* \in H_A$ is also related directly to the probability of committing a Type II error when $a^* \in H_A$ is true. Thus, we have

$$\begin{aligned} &\text{Power at particular alternative } \alpha^* \in H_A \\ &\quad = 1 - P(\text{failing to reject } H_0 \text{ when } \alpha^* \in H_A \text{ is true}) \\ &\quad = 1 - P(\text{Type II error when } \alpha^* \in H_A \text{ is true}). \end{aligned}$$

What about the power of the test against your friend's biased coin with $p = .9$? When this alternative is true, the sampling distribution for B is still binomial, but now with parameters $n = 8$ and $p = .9$. Therefore,

$$\{\text{Power against } p = .9\} = P(\text{Reject } H_0 \text{ when } p = .9)$$
$$= P(B = 6 \text{ when } p = .9) + P(B = 7 \text{ when } p = .9) + P(B = 8 \text{ when } p = .9)$$
$$= \sum_{t=6}^{8} \frac{8!}{t!(8-t)!} (.9)^t (.1)^{8-t} = .1488 + .3826 + .4305 = .9619.$$

Thus with this test you can be quite confident (with probability .9619) that you will be able to detect the biased coin if your friend is using it. Stated another way, this says that the probability of failing to detect your friend's biased coin if she is using it (i. e., a Type II error) is just 1 - .9619 = .0381.

Armed with this new information about its power, it is clear that the test for fairness of your friend's coin was designed to be more sensitive toward detecting the use of a biased coin than toward concern about accusing your friend of using the biased coin when she is not. Perhaps you have gone too far in that direction. Can we modify the test procedure to make it less likely that you will falsely accuse your friend of attempting to use a biased coin when she is not? Such a modification would require shrinking the critical region, since it contains the sample outcomes that lead to rejection of H_0.

Our original critical region is $C_1 = \{$ 6, 7 or 8 heads on the eight flips of the coin$\}$. Thus there are two natural modifications, namely, shrinking C_1 to $C_2 =$ {7 or 8 heads on the eight flips} or to $C_3 =$ {all eight flips are heads}. Suppose you decide to give your friendship a bit more credit and relax the stipulations for rejecting H_0 to C_2. The significance level for the test procedure using critical region C_2 is then

$$\alpha_2 = P(B \geq 7\,|p = .5) = \sum_{t=7}^{8} \frac{8!}{t!(8-t)!} (.5)^8 = .0312 + .0039 = .0351,$$

which is certainly more appealing than the significance level $\alpha = .1445$ for the original test. However, we must also see what you have given up in the way of power to obtain this reduced probability of incorrectly accusing your friend of using the biased coin when she is not. The power of the new test with critical region C_2 is

$$\begin{aligned} &\text{Power against } p = .9 \text{ using critical region } C_2 \\ &\quad = P(B \geq 7\,|p = .9) \\ &\quad = \sum_{t=7}^{8} \frac{8!}{t!(8-t)!} (.9)^t (.1)^{8-t} = .3826 + .4305 = .8131. \end{aligned}$$

Thus for this modified test with critical region C_2, the probability of failing to detect your friend's biased coin if she is using it (i. e., a Type II error) is increased to 1 - .8131 = .1869. Depending on your relative valuation of this friendship and money, C_2 might be a better critical region than C_1 to use to assess the bias of the coin. What about the third possible critical region C_3 = {exactly 8 heads on the eight flips}? You are asked in Exercise 6.B.12 to verify that the significance level for the critical region C_3 is $\alpha_3 = .0039$ and that [Power of the test with critical region C_3 against the alternative $p = .9$] = .4305.

We note that in changing from critical region C_1 to critical region C_2 and then to critical region C_3 we were able to steadily reduce the probability of a Type I error (i. e., the significance level) for the associated tests. However, this was accomplished only at the expense of corresponding increases in the probability that a Type II error (i. e., reduction in power) occurs. This is not an accident; for any reasonable test procedure there is a natural and inevitable tradeoff between these two error probabilities. For a given setting and a fixed sample size, any change in a test procedure that will result in lowering the probability of one of these two types of error will always lead to an increase in the probability that the other type of error will occur. The only way to decrease both error probabilities simultaneously for a given hypothesis test procedure is to increase the sample size and collect more data. A more common approach to hypothesis testing is to hold the Type I error probability fixed and increase the number of observations in our data collection enough to reduce the Type II error probability to an acceptable value. (For an example of this approach, see Exercise 6.A.12.)

Another unusual feature of the language of statistics should be noted here. Reducing the significance level α for a hypothesis test corresponds to reducing the Type I error probability for the test. Yet if the sample data leads to rejection of the null hypothesis H_0 at this reduced level, we would conclude that there is an *increased significance of the result*! That is, the lower the significance level at which we can reject H_0, the greater is the significance of the conclusion! While such a statement likely seems a bit contradictory at first exposure, as you increase your understanding of the role of the significance level, you will see this as being a very logical relationship between significance level and the implied significance from the rejection of the null hypothesis.

Choosing a Significance Level In the previous discussion related to evaluating your friend's coin, we took the approach of first specifying the critical region and then computing the significance level for the associated hypothesis test. While this is perfectly acceptable, it is not the standard way in which hypothesis tests are formulated or conducted. In practice, once the null and alternative hypotheses of interest are stated, an appropriate significance level α is commonly stipulated first and then the proper critical region for the test is constructed. This, of course, leaves the choice of "appropriate significance level" for a problem rather vague. It is quite natural to want our significance level to be small, since it is the probability of making a Type I error. However, we must also remember that the probability of a Type II error occurring is inversely related to the significance level. Hence, the choice of significance level depends on the relative seriousness of the consequences of making the two types of errors, something that is not always easy to evaluate and is often quite subjective. For example, suppose we are conducting a study to evaluate the effectiveness of a new medication. A Type I error for such a study corresponds to declaring the new medication to be effective, when in fact it is not. On the other hand, a Type II error occurs when the new medication is effective and we fail to detect that fact with our hypothesis test. The relative importance of the two types of errors for this setting depends heavily on the particular medical problem for which the new medication has been developed. If the medication is intended to treat a serious illness, such as lung cancer, what are the consequences of the two types of error? Which is more serious? Would your answers to these two questions change if the new medication is intended instead to be a treatment for mild cases of acne?

As statisticians have dealt with this question over the years, it has become rather standard practice to simply set the significance level to be an acceptably small number, typically ranging from .001 to .10, but most commonly taken to be .01 or .05. If the power of the test associated with such a significance level is not as high as one would like, collecting additional sample observations is the option to consider. In fact, in a number of settings statisticians have been able to predetermine the necessary sample size(s) so that both of the error probabilities for a particular hypothesis test are acceptably small. Although this development is beyond the formal scope of our text, the interested reader is encouraged to learn more about this approach in more advanced statistics texts.

***P*-values** The purpose of a hypothesis test is to assess the reasonableness of the null and alternative hypotheses using the evidence presented by the sample data. When we take the approach of pre-selecting a significance level α for our test, the eventual outcome of the test procedure will be to state whether or not the data provided adequate evidence to support the alternative hypothesis. In such cases, the interpretation of "adequate evidence" is provided by a small probability (α) of incorrectly rejecting a true null hypothesis. When one's reason for conducting a given hypothesis test is to make a decision and take some specific action based on the outcome of the test at a prescribed significance level α, then that decision is all that is important to glean from the data. However, simply stating that a particular hypothesis test rejects at significance level $\alpha = .05$, for example, does not provide the user with all of the information on this issue that is available in the sample. Knowing that a set of sample data leads to the rejection of H_0 at significance level $\alpha = .05$ does, in fact, tell us that we would also reject H_0 for any significance level greater than .05. (See Exercise 6.A.11 for more discussion.) However, it does not tell us whether or not the evidence in the sample data is also sufficient to reject H_0 for a smaller significance level such as $\alpha =$ .01. Thus the complete information about H_0 contained in the sample evidence is not totally captured in the simple statement that we reject H_0 at the specific significance level .05.

We illustrate this idea in Figs. 6.6 and 6.7, where we depict two critical regions for a test that rejects the null hypothesis H_0 for large values of a statistic T. The critical region in Fig. 6.6 consists of those possible values of T greater than or equal to 1.96 and the associated significance level is $\alpha = .025$.

Fig. 6.6 Upper-tail critical region for a test that rejects the null hypothesis H_0 for values of the test statistic T greater than or equal to 1.96

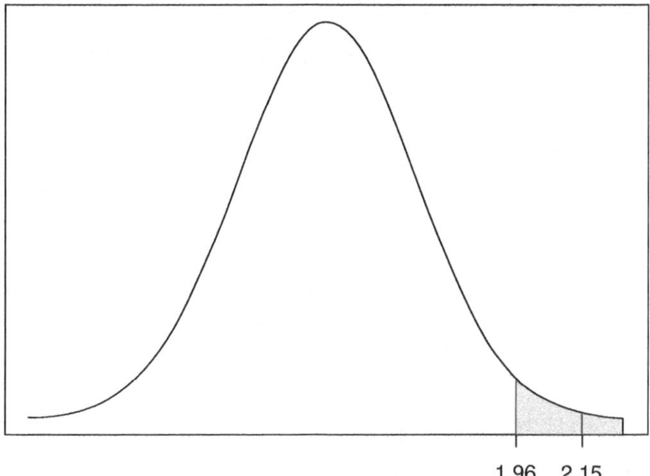

Fig. 6.7 Upper-tail critical region for a test that rejects the null hypothesis H_0 for values of the test statistic T greater than or equal to 2.33

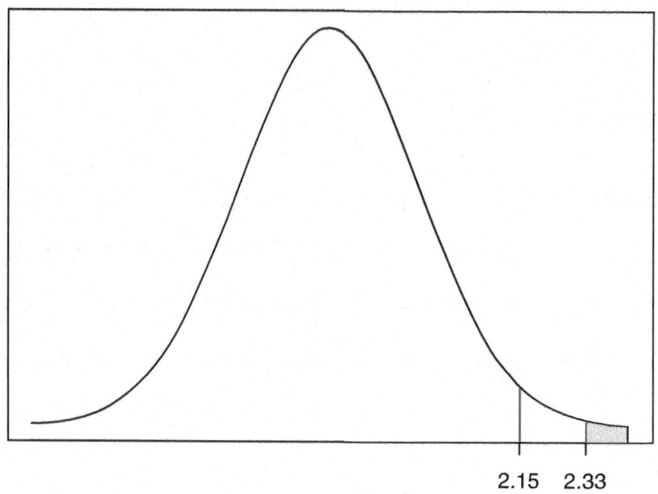

The second critical region in Fig. 6.7 contains possible values of T greater than or equal to 2.33, with corresponding significance level $\alpha = .01$. If the observed value of T is $t_{obs} = 2.15$, we see that it falls in the critical region depicted in Fig. 6.6, but not in the smaller critical region in Fig. 6.7. Thus with $t_{obs} = 2.15$, we would reject H_0 at significance level .025, but not at significance level .01.

What would our decision for this outcome be at significance level $\alpha = .015$ for $t_{obs} = 2.15$? at significance level $\alpha = .020$? Do we have to construct a separate critical region for each specific significance level in order to make the corresponding decision about whether or not to reject H_0 at that level? Fortunately, the answer is no! There is a way to impart a more complete picture of the 'sample evidence' against the null hypothesis without constructing a large number of separate critical regions for different possible significance levels. One such way is to assess just how unlikely it would have been to collect a random sample with the features in our observed data if the null hypothesis H_0 were true. This 'sample evidence' is quantified by finding the probability under the null hypothesis H_0 of getting a value of our test statistic T (B in our biased coin example) at least as unusual or extreme as the

Fig. 6.8 Area depicting the *P*-value for a test that rejects for large values of the test statistic *T* when the observed sample value of *T* is $t_{obs} = 2.15$

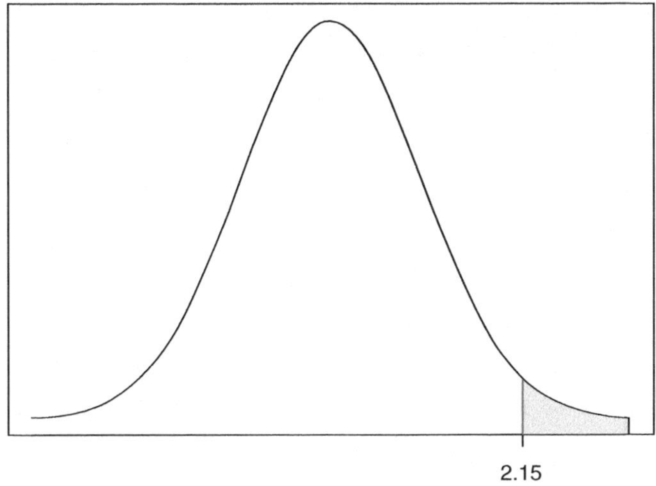

one we compute for our observed sample data. For the setting described in Figs. 6.6 and 6.7, this corresponds to finding the probability that T is greater than or equal to $t_{obs} = 2.15$ when the null hypothesis is true, as depicted in Fig. 6.8. The smaller this probability, called the ***P*-value,** is (i.e., the more unusual or extreme is our observed value of T), the more evidence we have in favor of rejecting H_0.

Definition 6.8 Let t_{obs} be the observed sample value of a statistic T used in testing the null hypothesis H_0 against the alternative hypothesis H_A. The probability when the null hypothesis is true of obtaining a random sample of size n that leads to a T value as extreme or more extreme than the observed t_{obs} is called the ***P*-value of the test.**

Thus, calculation of a *P*-value depends on the sampling distribution of the test statistic T when the null hypothesis H_0 is true. If the hypothesis test is designed to reject H_0 in favor of the alternative H_A for large values of T, the associated *P*-value is $P_0(T \geq t_{obs})$, where the subscript on P_0 indicates that the probability is being computed under the assumption that the null hypothesis

H_0 is true. For a test that rejects H_0 in favor of H_A for small values of T, the P-value is $P_0(T \leq t_{obs})$. Finally, for a test that rejects H_0 in favor of H_A for either small or large values of T, we must first ascertain whether the observed value, t_{obs}, is 'large' or 'small'; that is, we first compute the smaller of the two probabilities $P_0(T \leq t_{obs})$ and $P_0(T \geq t_{obs})$, providing the observed 'direction' of t_{obs}. The P-value for the two-sided test is then $2 \times$ minimum $\{P_0(T \leq t_{obs}), P_0(T \geq t_{obs})\}$. For any of these alternatives, the smaller the P-value, the greater the justification provided by the data for rejection of H_0 in favor of the alternative H_A.

To illustrate this P-value concept, we return one last time to your friend's coin. Suppose that when you actually flip the coin eight times, you obtain seven heads and one tail; that is, the observed value of the test statistic $B =$ [number of heads] is $b_{obs} = 7$. Since the test procedure we discussed rejects H_0: $p = .5$ in favor of H_1: $p > .5$ for large values of B, it follows that the P-value for these data is given by

$$\begin{aligned} P\text{–value} &= P\big(\text{observe an outcome as extreme or more extreme} \\ &\qquad \text{than 7 heads in 8 flips if } H_0 \text{ is true}\big) \\ &= P(B \geq 7 \,|\, p = .5) = \sum_{t=7}^{8} \frac{8!}{t!(8-t)!} (.5)^8 = .0351. \end{aligned}$$

Thus obtaining a value of B at least as large as the observed value $b_{obs} = 7$ is rather unusual if, in fact, the null hypothesis is true and the coin is fair. Hence the evidence provided by the P-value is supportive of rejecting H_0.

This P-value approach to evaluating sample evidence for testing purposes provides more information than is associated with simply rejecting H_0 at a prescribed significance level α. In fact, providing the P-value for a hypothesis test enables us to reach a conclusion as to whether or not to reject H_0 at any specific significance level. To illustrate, given a P-value for a hypothesis test of the null hypothesis H_0 versus the alternative H_A, we would reject H_0 for all significance levels α that are at least as large as the P-value, but not for any that are smaller than it. That is, **the P-value is the smallest significance level at which we would reject H_0 based on the observed sample data.** Thus, if we had, in fact, obtained seven heads and one tail in the eight flips of our friend's coin, we would reject H_0 for any significance level $\alpha \geq P\text{-value} = P(B \geq 7 \mid p = .5) = .0351$ and fail to reject H_0 for significance levels less than .0351.

Using the P-value concept in this way, a test of any null hypothesis H_0 versus an alternative H_A at significance level α can be written as

$$\text{Reject } H_0 \text{ in favor of } H_A \text{ at significance level } \alpha \text{ if and only if } P\text{–value} \leq \alpha.$$

This general format will be consistent throughout our study in this book. What will vary as we move among different statistical settings are the test statistics and null sampling distributions associated with the computation of the P-value.

What are the advantages and disadvantages of these two approaches to hypothesis testing? The fixed significance level approach provides a *preset* standard to which we must adhere once the data are collected. Observing the data cannot change our mind in this regard. If we desire our probability of a Type I Error to be less than or equal to .05, for example, then the use of a preset significance level .05 guarantees this to be the case, no matter what we find in the data. Decisions associated with a study can then be made with the assuredness of a guaranteed maximum Type I Error probability.

On the other hand, the P-value approach enables us to more accurately report the extremeness of the observed value of our test statistic T without losing the ability to make a decision at any preset significance level. This clearly provides consumers of our findings with more information than is contained in the statement that we did or did not reject H_0 at any particular prescribed significance level.

Tests for the Probability of an Event When the Sample Size n is Large In our previous coin-tossing example we discussed how to test hypotheses about p, the probability of an event A, when the sample size n is small. When the number of sample observations becomes large, use of the exact binomial distribution for B to set up appropriate critical regions or to compute P-values for an observed set of data becomes a bit unwieldy by hand. For such large sample settings, we can, of course, use **R** to obtain P-values or set up critical regions. A second alternative is to turn to the Central Limit Theorem discussed in Sect. 5.3 for help. Once again we rely on the information contained in the observed percentage of times, $\hat{p}$, that the event A occurs in a sample of n items from the underlying population. As an application of the Central Limit Theorem, we saw in Sect. 5.3 that the standardized variable

$$Z = \frac{\hat{p} - p}{\sqrt{\frac{p(1-p)}{n}}} \tag{6.9}$$

has an approximate N(0, 1) distribution when the sample size n is sufficiently large (at least 50 for a reasonably good approximation).

Now, suppose that the null hypothesis of interest corresponds to the probability of the event A being equal to a specified value p_0; that is, the null hypothesis is given by H_0: $p = p_0$. It follows from expression (6.9) that the statistic

$$Z = \frac{\hat{p} - p_0}{\sqrt{\frac{p_0(1-p_0)}{n}}} \tag{6.10}$$

has an approximate N(0, 1) distribution when the sample size n is large and the null hypothesis is true. Thus the statistic Z in (6.10) provides us with the first part of our hypothesis test, namely, an appropriate test statistic. Moreover, knowing that the sampling distribution of Z is approximately N(0, 1) when H_0 is true will enable us to determine critical values that correspond to significance levels of interest, as well as to compute P-values once the data have been collected. All that remains, therefore, is to decide on the direction of our rejection region--a decision that depends explicitly on which of the three possible alternative hypotheses H_A is of interest for the problem at hand. If the conjectured alternative is H_A: $p > p_0$, it should be intuitively clear that unusually large values of Z provide the most evidence in favor of H_A and therefore we would want to reject H_0 for such large values of Z. (This was the case in your fair-coin test, since you wanted to know whether P(Heads) = 0.5 was true or whether P(Heads) was greater than 0.5.) For an alternative H_A: $p < p_0$, the appropriate rejection region would consist of unusually small values of Z. For the most general, two-sided, alternative H_A: $p \neq p_0$ either unusually small or unusually large values of Z (i. e., large values of $|Z|$) would support rejection of H_0.

Combining these three components of a hypothesis test, we have the following formulation for testing hypotheses about the probability of an event, p, when the sample size n is large.

Hypothesis Tests for the Probability of an Event, *p*

Let p be the probability of an event A relative to an underlying population of interest and let $\hat{p}$ be the percentage of times the event A occurs in a random sample of size n from this population. A test of the null hypothesis H_0: $p = p_0$ against an alternative H_A with approximate significance level α (i. e., a level α test) has a decision rule given by one of the three forms:

$$\begin{aligned}
&\text{Reject } H_0 \text{ in favor of } H_A: p > p_0 \text{ if and only if } Z_{\text{test}} \geq z_\alpha, \\
&\text{Reject } H_0 \text{ in favor of } H_A: p < p_0 \text{ if and only if } Z_{test} \leq -z_\alpha, \\
&\text{Reject } H_0 \text{ in favor of } H_A: p \neq p_0 \text{ if and only if } |Z_{test}| \geq z_{\alpha/2},
\end{aligned} \tag{6.11}$$

where Z_{test} is given by the right-hand side of expression (6.10) and z_α is the upper α^{th} percentile for the $N(0, 1)$ distribution. These critical regions are portrayed in Figs. 6.9, 6.10 and 6.11. Note that the critical values z_α, $-z_\alpha$, and $z_{\alpha/2}$ are chosen to satisfy $P(Z_{\text{test}} \geq z_\alpha$ when H_0 is true) $\approx \alpha$, $P(Z_{\text{test}} \leq -z_\alpha$ when

Fig. 6.9 Approximate level α critical region that rejects the null hypothesis if the test statistic Z_{test} is too large

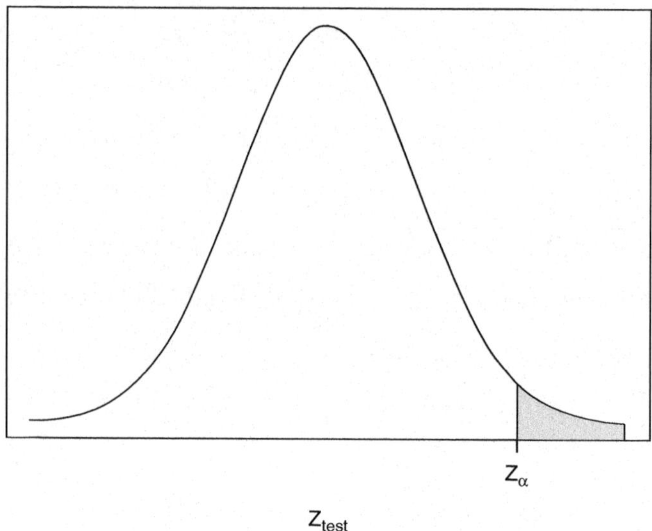

Fig. 6.10 Approximate level α critical region that rejects the null hypothesis if the test statistic Z_{test} is too small

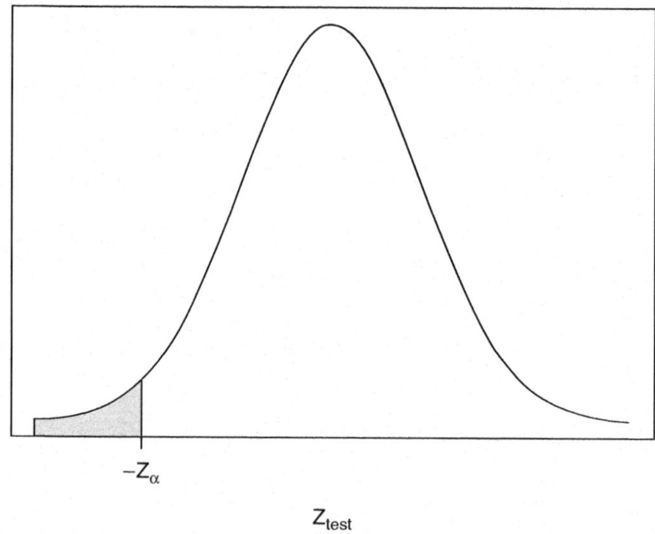

H_0 is true) $\approx \alpha$, and $P(\ |Z_{test}|\ \geq z_{\alpha/2}$ when H_0 is true) $\approx \alpha$, so that the respective tests all have approximate significance levels equal to α.

The associated approximate *P*-values for these hypothesis tests are given by:

Fig. 6.11 Approximate level α critical region that rejects the null hypothesis if the test statistic Z_{test} is either too large or too small

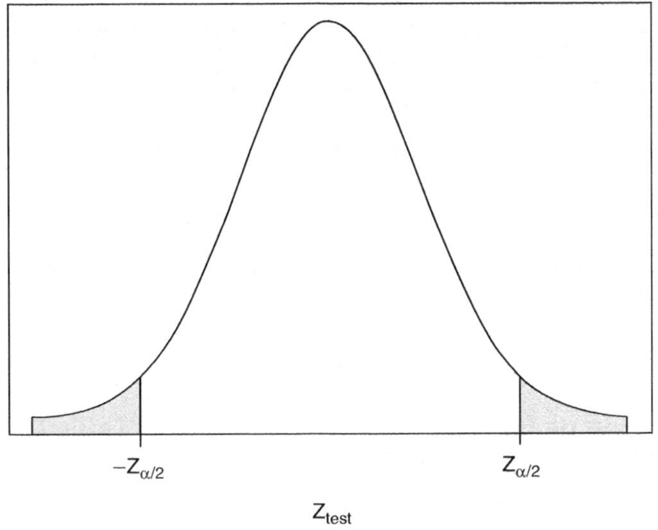

$$\begin{aligned}
&P\text{–value} \approx P(Z \geq z_{\text{test}}) \text{ for the alternative } H_A : p > p_0,\\
&P\text{–value} \approx P(Z \leq z_{\text{test}}) \text{ for the alternative } H_A : p < p_0, \qquad (6.12)\\
&P\text{–value} \approx 2P(Z \geq |z_{\text{test}}|) \text{ for the alternative } H_A : p \neq p_0,
\end{aligned}$$

where Z has a $N(0, 1)$ distribution and z_{test} is the observed value of Z_{test} (6.10) for the sample data.

Note that in order to write our hypothesis test procedures precisely we have introduced some rather complex notation. Here is a summary to help you keep things straight. We will use capital letters to denote random variables and lower case letters to denote single observations from a distribution or for the value of a statistic that we have computed from a particular sample. In our discussion here, we have:

Z_{test} is the normalized test statistic $\dfrac{\hat{p} - p_0}{\sqrt{\frac{p_0(1-p_0)}{n}}}$.

z_{test} is a particular value of this test statistic that we obtain by taking a sample and carrying out the arithmetic in the formula for Z_{test}.

Z represents the generic $N(0,1)$ variable

z_α is the numerical value such that $P(Z \geq z_\alpha) = \alpha$.

Using the appropriate P-value expression in (6.12), we note that the three hypothesis tests stipulated in (6.11) can all be expressed by the single statement:

$$\text{Reject } H_0 \text{ in favor of } H_A \text{ at approximate significance level } \alpha$$
$$\text{if and only if } P\text{–value} \leq \alpha.$$

For this initial exposure to the statistical practice of testing hypotheses, we have chosen to present the hypothesis tests for p in both the critical region format (6.11) and the P-value format (6.12). We feel it is important that the reader see both of these approaches and understand how they relate to one another. In our next example we will demonstrate how both approaches are applied to an actual data collection. However, in most of the later examples in the text we will emphasize the P-value formulation, as it is generally simpler and provides more information to the statistical consumer than does knowledge of a decision only at a single, pre-specified significance level α.

Example 6.7. Dual-career Couples Many domestic relationships involve couples where both partners work outside the home. In a report published in 1998, the research organization Catalyst presented the results of a study on working couples' opinions about such issues as whether or not the fact that both partners were working enabled each of them to have more flexibility in shaping their separate career options. They conducted interviews with 802 people involved in dual-earner relationships, seeking responses to a variety of questions on their attitudes regarding advantages and disadvantages of such work relationships. Catalyst found that 441 of the 802 people interviewed thought that they, in fact, did have more control over shaping their own careers because their partner was also working.

Does this provide sufficient evidence to conclude that the majority of all working couples feel this way? To address this question, let p denote the proportion of all individuals involved in dual-earner relationships who think that they have more control over shaping their own careers because their partner is working. Then it might be of interest to test the null hypothesis H_0: $p = .5$ versus the one-sided alternative H_A: $p > .5$. In this example we will illustrate both the critical region and P-value approaches to hypothesis testing. However, use of the P-value is both simpler and more informative than the formal critical region approach to testing. As a result, in most future examples we will test the hypotheses of interest by simply finding the appropriate P-value for the data.

Using a significance level of $\alpha = .05$, we find $z_{.05} = 1.645$ and the associated test procedure is obtained from (6.11) to be

$$\text{Reject } H_0 \text{ in favor of } H_A : p > .5 \text{ if and only if } Z_{\text{test}} \geq 1.645,$$

where

$$Z_{test} = \frac{\hat{p} - .5}{\sqrt{\frac{.5(1-.5)}{802}}}.$$

Using the data to compute the sample percentage estimator $\hat{p} = 441/802 = .5499$, we see from (6.10) that the observed value of our test statistic Z_{test} is given by

$$z_{test} = \frac{.5499 - .5}{\sqrt{\frac{.5(1-.5)}{802}}} = 2.83.$$

Since $z_{\text{test}} > 1.645$, we reject H_0: $p = .5$ at level $\alpha = .05$.

From the appropriate expression in (6.12), we also find that P-value $= P(Z \geq 2.83) \approx .0023$ for these data, so that the probability of obtaining a sample for which the value of $\hat{p}$ is at least as large as the observed .5499 is approximately .0023 if the null hypothesis $p = .5$ were true. This leads to the conclusion that, in fact, we would have rejected H_0: $p = .5$ in favor of the alternative H_A: $p > .5$ if our preset significance level α had been any value at least as large as .0023 (including, as we already knew, $\alpha = .05$). Thus the Catalyst study provides rather strong evidence that a majority of all individuals (i. e., the population) involved in dual-earner relationships do, in fact, feel that they have more control over shaping their own careers because their partner is also working.

Example 6.8. Are Good Samaritans Hard to Find? A survey by the National Highway Traffic Safety Administration (1998) addressed a number of questions about drivers' attitudes toward helping victims at traffic accidents. Among the questions asked in the survey was the following: "You come upon a traffic accident on a lonely stretch of road. No other cars are in sight. Would you stop or drive on and call for help down the road?" The study was based on interviews with 8200 drivers at least 16 years old during the winter of 1996-97. Of those interviewed, 4264 respondents indicated that they would stop their vehicle to check on the accident victims. Is this sufficient evidence to indicate that less than, say 60%, of all eligible drivers would stop to aid traffic accident victims in such circumstances? Letting p denote the proportion of all eligible drivers who would stop to assist, we might be interested in testing the null hypothesis H_0: $p = .6$ versus the one-sided alternative H_A: $p < .6$. The sample percentage of eligible drivers who said they would stop and help is $\hat{p} = 4264/8200 = .52$. Hence the value of our test statistic Z_{test} (6.10) is

$$z_{test} = \frac{.52 - .6}{\sqrt{\frac{.6(1-.6)}{8200}}} = -14.79.$$

From (6.12), the P-value for these data is P-value $= P(\ Z \leq \text{-}14.79) \approx 0$, indicating that there is virtually no chance that this observed value of $\hat{p} =$

.52 could have resulted if the null hypothesis condition $p = .6$ were true, and we would reject H_0: $p = .6$ in favor of the alternative H_A: $p < .6$ for any reasonable significance level α. Thus there is overwhelming sample evidence to support our conclusion that less than 60% of all eligible drivers would stop to aid traffic accident victims on a lonely stretch of road with no other cars in sight. The strength of this support for the alternative $p < .6$ is, of course, at least partly attributable to the very large sample size ($n = 8200$) for this survey.

We should also keep in mind that these data are the result of a self-reported survey, which, as discussed in Chap. 3, raises the issue of reliability of the responses. For example, some people who say they would not stop might actually do so in a real situation, while there are likely to be others who claim that they would stop, even though they might not in a real situation. All of this simply emphasizes how important it is to do everything possible to eliminate potential biases and collect reliable data when conducting sample surveys that involve self-reporting.

Exact *P*-values for One-sided Alternatives and Small Sample Sizes We have previously seen in the coin tossing example that the binomial distribution can be used to obtain exact *P*-values for tests of the null hypothesis H_0: $p = p_0$ versus either of the one-sided alternatives H_A: $p > p_0$ or H_A: $p < p_0$. We further illustrate this process on the cognitive subscale score data from Example 6.3, where the sample size is $n = 10$.

Example 6.9. Medical Improvements for Patients with Alzheimer's Disease - Exact *P*-values In Example 6.3 we discussed the investigation by Schneider et al. (1993) of the potential for treating Alzheimer's disease with *tacrine* or *physostigmine salicylate* plus *l-deprenyl*. They found that seven of the ten subjects (all currently being treated with either *tacrine* or sustained-release *physostigmine salicylate*) in the study showed an improvement (i.e., lowering) in the cognate subscale score on the Alzheimer's Disease Assessment Scale after a 4-week period of added treatment with *l-deprenyl* relative to a similar 4-week period with placebo as the only addition to the *tacrine* or *physostigmine salicylate* medication. Let p be the probability that a randomly chosen subject under treatment for Alzheimer's disease with either *tacrine* or *physostigmine salicylate* will show an improvement in the cognate subscale score when *l-deprenyl* is added to their treatment regime. Even if there is no effect whatsoever from the addition of *l-deprenyl*, we would expect, simply by chance alone, that roughly 50% of the population of all such subjects would exhibit improvements in their cognate subscale scores following the *l-deprenyl* treatment. Hence, it is of natural interest to test the null hypothesis H_0: $p = .5$, which corresponds to no discernible difference between the placebo effect and that of *l-deprenyl*, versus the alternative H_A: $p > .5$, corresponding to an improvement (over placebo) from the additional use of *l-deprenyl* for more than 50% of the population. Setting B_{test} = [number of subjects in the study who exhibited improvement from the use of *l-deprenyl*], we would naturally reject H_0 for large values of B_{test}. Thus the P-value associated with the observed value of B_{test}, namely, $b_{test} = 7$, is

$$P\text{–value} = P(B_{test} \geq b_{test} \text{ when } p = .5) = P(B_{test} \geq 7 \text{ when } p = .5).$$

This upper-tail probability is graphically depicted in Fig. 6.12.

When H_0: $p = .5$ is true, the test statistic B_{test} has a binomial distribution with parameters $n = 10$ and $p = .5$, and the resulting P-value for these data is

Fig. 6.12 ***P*-value for the upper-tail binomial test in Example 6.9**

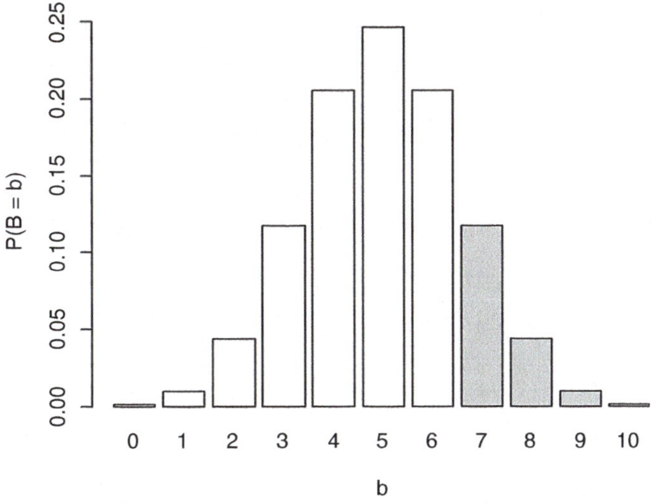

$$P\text{–value} = P(B_{\text{test}} = 7\,\text{when } p = .5) + P(B_{\text{test}} = 8 \text{ when } p = .5)$$
$$+ P(B_{\text{test}} = 9\,\text{when } p = .5) + P(B_{\text{test}} = 10 \text{ when } p = .5)]$$
$$= \sum_{t=7}^{10} \frac{10!}{t!(10-t)!}(.5)^8 = .1172 + .0439 + .0098 + .0010 = .1719.$$

Again we can use the **R** function *pbinom*( ) to calculate the .1719 *P*-value for the Alzheimer data.

```
> pbinom(6,10,.5,lower.tail=FALSE)
[1] 0.171875
```

Thus, although our point estimate of the percentage of Alzheimer patients who will show an improvement in cognitive subscale score values if *l-deprenyl* is added to treatment with a *cholinesterase* inhibitor is 70% (i. e., $\hat{p} = .7$), the sample results for the ten patients in the study do not provide compelling evidence for rejection of H_0: $p = .5$ in favor of the claimed improvement for the majority of the population. In fact, to conclude that there is such an improvement (i. e., that $p > .5$) we would have to be willing to use a significance level at least as large as the *P*-value = .1719. Would this be a reasonable Type I error rate to accept for this setting? That is, would you be willing to claim that there is a positive effect when *l-deprenyl* is added to treatment with a *cholinesterase* inhibitor if there is a 17% chance that your observed sample result could have been obtained even when addition of the *l-deprenyl* had no positive effect whatsoever? What are the consequences of such a claim? The expense of the treatment would have to be borne by someone (patient's family, hospital, insurance company), patients and their relatives would be given false hope that the treatment is helpful, and, even more seriously, this false hope might curtail further research on other potentially more effective treatments of Alzheimer's. Under such circumstances, .1719 would probably be an unacceptably high risk of a Type I error for most people.

In Example 6.8 we noted that the overwhelming sample evidence in support of the conclusion that less than 60% of all eligible drivers would stop to aid traffic accident victims on a lonely stretch of road with no other cars in sight was at least partly attributable to the very large sample size ($n = 8200$) for that survey. The opposite is true for the *l-deprenyl* study in Example 6.9. It is very difficult to obtain statistically significant results with as few as $n = 10$ Alzheimer's patients. Even if the addition of *l-deprenyl* to treatment with a *cholinesterase* inhibitor does benefit Alzheimer's patients, it would require extreme evidence in the ten trials to lead to this conclusion (i.e., the rejection of H_0). This is because the probability of a Type II error becomes the overriding constraint for such a small sample size. That is, for small samples it is not unlikely that we would fail to reject H_0 when, in fact, H_A is true. In the case of the Alzheimer study, this means that there is a rather high probability that we would not observe a positive effect in enough of the ten subjects to lead to rejection of H_0 even if the addition of *l-deprenyl* can, in fact, lead to an improved treatment of the disease.

The general expressions for obtaining exact P-values for tests of $\mathrm{H_0}$: $\mathrm{p} = \mathrm{p_0}$ versus either of the one-sided alternatives $\mathrm{H_A}$: $\mathrm{p} > \mathrm{p_0}$ or $\mathrm{H_A}$: $\mathrm{p} < \mathrm{p_0}$ are then as follows.

EXACT *P*-VALUES FOR TESTS OF H_0: $p = p_0$ VERSUS EITHER OF THE ONE-SIDED ALTERNATIVES H_A: $p > p_0$ OR H_A: $p < p_0$

Let B_{test} denote the number of times the event A occurs in a random sample of size n from a population for which $p = P(\text{event } A)$. Then the exact P-values for a test of H_0: $p = p_0$ versus a one-sided alternative are given by

$$P\text{–value} = P(B_{test} \geq b_{test} \text{ when } p = p_0) \text{ for the alternative } H_A : p > p_0$$

and

$$P\text{–value} = P(B_{test} \leq b_{test} \text{ when } p = p_0) \text{ for the alternative } H_A : p < p_0,$$

where b_{test} is the observed number of times the event A occurs in the sample of size n and B_{test} has a binomial distribution with parameters n and p_0. As we have seen previously in this chapter, the **R** function *pbinom*( ) can be used to compute such exact one-sided P-values.

Confidence Interval and Two-sided Hypothesis Test for p_A In Example 6.4 we used Buckeye State Poll sample data from the week of February 17-26, 1998 to obtain the approximate 95% confidence interval (.428, .492) for the proportion, p, of all central Ohio women who consider the television industry to be a greater danger to society than government restrictions on what can appear on television. Thus we are 95% confident that the true value of p is between .429 and .492. However, the flip side of this statement is that the sample data also provide evidence of being incompatible with other values of p either above .492 or below .428. Viewed from this perspective, it sounds very much like a statement that we would be willing to reject such proportions as possible values for p. This, in turn, is really just a conclusion about potential null hypotheses for p; that is, based on the 95% confidence interval (.428, .492), it would seem reasonable to reject the null hypothesis H_0: $p = p_0$ for all p_0 that are at least as small as .428 or at least as large as .492 and fail to reject H_0: $p = p_0$ if $.428 < p_0 < .492$.

Such an interpretation of the relationship between this particular confidence interval and hypothesis tests about p is, in fact, quite legitimate. Moreover, it is much more generally valid than just for this particular example.

Relationship Between Two-Sided Level α Hypothesis Tests and 100(1−α)% Confidence Intervals for p_A We collect a sample from a population of interest and compute $\hat{p}_A$, the proportion of times an event A occurs among these sample items. In this chapter we have learned how to use this sample statistic to make inferences about the relative frequency, p_A, of the event A in the entire population. In particular, we have discussed ways to use $\hat{p}_A$ to test hypotheses about p_A at a prescribed significance level, say .05, and how to construct confidence intervals for p_A at a desired confidence such as 95%.

(1) An approximate 95% confidence interval for p_A, for example, looks like this:

$$\left(\hat{p}_A - 1.96\left[\frac{\hat{p}_A(1-\hat{p}_A)}{n}\right]^{1/2}, \hat{p}_A + 1.96\left[\frac{\hat{p}_A(1-\hat{p}_A)}{n}\right]^{1/2}\right).$$

This interval is centered at the point estimate $\hat{p}_A$ and both the 95% confidence level and the sampling variability of $\hat{p}_A$ determine the length of the interval.

(2) For some event probability p_0 of interest, we test the null hypothesis H_0: $p_A = p_0$ versus the alternative hypothesis H_A: $p_A \neq p_0$ at significance level $\alpha = .05$, for example, by rejecting H_0 if the point estimate $\hat{p}_A$ is too far away from the hypothesized value p_0. Thus we would not reject those null hypothesis values p_0 that are sufficiently close to the sample point estimate $\hat{p}_A$. Not surprisingly, both the $\alpha = .05$ significance level

and the sampling variability of $\hat{p}_A$ also determine what constitutes "sufficiently close" in this hypothesis testing setting.

It is not just a coincidence that both the length of a 95% confidence interval for p and the decision criterion for a two-sided hypothesis test about p at the $\alpha = .05$ significance level depend explicitly on the sampling variability of the point estimate $\hat{p}_A$. In fact, there is a direct relationship between these two ways of using the sampling variability of $\hat{p}_A$ to make inferences about p. If a p_0 of interest in our null hypothesis H_0 happens to be in the 95% confidence interval described in (1), then the hypothesis test of H_0: $p_A = p_0$ versus the alternative hypothesis H_A: $p_A \neq p_0$ described in (2) will fail to reject H_0 at significance level $\alpha = .05$. Conversely, none of the values of $p = p_0$ belonging to the 95% confidence interval described in (1) would be rejected by the hypothesis test of H_0: $p_A = p_0$ versus the alternative hypothesis H_A: $p_A \neq p_0$ at the $\alpha = .05$ significance level. However, all values of p_0 that are outside (either too large or too small) the 95% confidence interval would be rejected by the $\alpha = .05$ significance level test. Thus, the 95% confidence interval for p is precisely the set of "acceptable" values of p for the corresponding two-sided hypothesis test at the $\alpha = .05$ significance level.

Thus, for example, the approximate 90% confidence interval (.6269, 1) for p_{Closed} discussed in Example 6.5 is also the set of p_{Closed} values that would not be rejected as null hypothesis values with an approximate significance level $\alpha = .10$ two-sided hypothesis test; that is, at approximate level $\alpha = .10$, $p_{\text{Closed}} = .75$ is compatible with the observed data, while $p_{\text{Closed}} = .50$ is not.

It is clear from the relationship between the level α two-sided hypothesis test in (6.11) and the corresponding 100(1- α)% confidence interval in (6.7) that the confidence interval formulation is, in some sense, a good deal more informative than is a specific hypothesis test. The confidence interval provides immediately the entire range of parameter values p_A that are compatible with the sample data, while a given hypothesis test simply certifies that one specific value of p_A is consistent with what has been observed in the sample. Of course, the P-value for a hypothesis test does provide us with a measure of the *degree of incompatibility* between an observed set of data and a specific null hypothesis, a piece of information that is not readily available from a single confidence interval.

Similar comments about the relationship between confidence intervals for a parameter and corresponding hypothesis tests about the parameter apply to many of the other inference settings discussed in this text, and we will note this when it happens.

Section 6.3 Practice Exercises

6.3.1. *V—chip and Children.* The Annenberg Public Policy Center of the University of Pennsylvania conducts annual surveys about issues related to media in the home. An issue in one of their 2010 surveys was whether or not a majority of parents would use the V-chip, if available to them, to control the programs that could be watched by their children. Interviews with parents provided data to assess the public opinion about this option. State the null hypothesis H_0 and the appropriate alternative hypothesis H_A of interest in this survey question.

6.3.2. *Participation in Team Sports.* The National Sporting Goods Association conducts an annual survey to elicit information about participation in team sports among 7–12 year old children. Suppose that last year's survey showed that 18% of all youth between 7 and 12 years old participated in team sports. If we want to use the results of this year's survey to test whether there is an increasing trend in team sports participation by this age group, what would be the appropriate null H_0 and alternative hypotheses H_A of interest?

6.3.3. *Hot Lottery Numbers.* Newspapers in states with lotteries often report numbers that are "hot" and numbers that are "not". (As you are learning statistics, you should become aware that there are no "hot" or "cold" numbers in a fair lottery. In one such article, a newspaper reported that, among the numbers 1 to 40, inclusive, "35" was a hot number and "2" was a cold number. Suppose you wish to use a sample of randomly selected integers between 1 and 40, inclusive, to test these conjectures.

(a) What are the appropriate null H_0 and alternative H_A hypotheses for testing whether "35" is a "hot" number?

(b) What are the appropriate null H_0 and alternative H_A hypotheses for testing whether "2" is a "cold" number?

6.3.4. *Doctors' Visits.* Have you ever been to a doctor with a physical complaint for which she can not find an organic cause? It has been conjectured that this happens in more than 20% of the visits to doctors' offices. Discuss how you might collect information to test this conjecture. State specifically the null H_0 and alternative H_A hypotheses that would be of interest in your study.

6.3.5. *Who Are Better Drivers?* Do you consider yourself a "better" driver than most other drivers? Do you believe that men and women might answer this question differently? Discuss how you might collect information to test the following conjectures. For each conjecture, state specifically the null H_0 and alternative H_A hypotheses that would be of interest.

(a) Conjecture: More than 70% of male drivers think they are "better" drivers than most other drivers.

(b) Conjecture: Fewer than 70% of women drivers think they are "better" drivers than most other drivers.

(c) Conjecture: The percentage of male drivers who think they are "better" drivers than most other drivers is higher than the corresponding percentage for women drivers.

6.3.6. Let p denote the probability of the event A in a population and let B equal the number of times A occurs in a random sample of size $n = 20$ from this population. Consider the test of the null hypothesis H_0: $p = .6$ versus the alternative H_A: $p > .6$ given by:

$$\text{Reject } H_0 \text{ if and only if } B \geq 15.$$

(a) Identify the test statistic, critical region, and critical value for this test.
(b) Find the significance level for the test.
(c) If the observed value of B is $b_{obs} = 17$, what conclusion do you reach with the test?
(d) If the observed value of B is $b_{obs} = 16$, what is the P-value for the test?

6.3.7. Let p denote the probability of the event A in a population and let B equal the number of times A occurs in a random sample of size $n = 15$ from this population. Consider the test of the null hypothesis H_0: $p = .4$ versus the alternative H_A: $p < .4$ given by:

$$\text{Reject } H_0 \text{ if and only if } B \leq 3.$$

(a) Identify the test statistic, critical region, and critical value for this test.
(b) Find the significance level for the test.
(c) If the observed value of B is $b_{obs} = 4$, what conclusion do you reach with the test?
(d) If the observed value of B is $b_{obs} = 2$, what is the P-value for the test?

6.3.8. Let p denote the probability of the event A in a population and let B equal the number of times A occurs in a random sample of size $n = 50$ from this population. Consider the test of the null hypothesis H_0: $p = .5$ versus the alternative H_A: $p \neq .5$ given by:

$$\text{Reject } H_0 \text{ if and only if } |Z_{test}| \geq 1.85,$$

where Z_{test} is given by the right-hand side of expression (6.10) with $p_0 = .5$.

(a) Identify the test statistic, critical region, and critical value for this test.
(b) Find the approximate significance level for the test.

(c) If the observed value of B is $b_{\text{obs}} = 17$, what conclusion do you reach with the test?

(d) If the observed value of B is $b_{\text{obs}} = 18$, what is the approximate P-value for the test?

6.3.9. Let p denote the probability of the event A in a population and let B equal the number of times A occurs in a random sample of size $n = 10$ from this population. Consider the test of the null hypothesis H_0: $p = .2$ versus the alternative H_A: $p > .2$ given by:

$$\text{Reject } H_0 \text{ if and only if } B \geq 3.$$

(a) What is the probability of a Type I Error for this test?

(b) What is the probability of a Type II Error for this test if the null hypothesis H_0 is true?

(c) What is the probability of a Type II Error for this test if the true value of p is .3? $p = .4$? What do you think happens to the probability of a Type II Error when the true value of p is even larger?

6.3.10. Let p denote the probability of the event A in a population and let B equal the number of times A occurs in a random sample of size $n = 100$ from this population. Consider the test of the null hypothesis H_0: $p = .7$ versus the alternative H_A: $p < .7$ given by:

$$\text{Reject } H_0 \text{ if and only if } Z_{\text{test}} \leq -1.75,$$

where Z_{test} is given by the right-hand side of expression (6.10) with $p_0 = .7$.

(a) What is the approximate probability of a Type I Error for this test?

(b) What is the approximate probability of a Type I Error for this test if the true value of p is .8?

(c) What is the approximate power of this test if the true value of p is .6? .5? What do you think happens to the approximate power of the test when the true value of p is even smaller?

6.3.11. Let p denote the probability of the event A in a population and let B equal the number of times A occurs in a random sample of size n from this population. Consider the approximate level α test of the null hypothesis H_0: $p = .45$ versus the alternative H_A: $p > .45$ given by expression (6.11) with $p_0 = .45$.

(a) What effect does varying the value of the approximate significance level α have on the test procedure?

(b) For approximate significance level $\alpha = .05$ and **fixed** value of the percentage, $\hat{p} = \frac{B}{n}$, of times that A occurs in the sample, what affect does varying the sample size n have on the conclusion reached by the test?

6.3.12. If the P-value for a test procedure is .073, what conclusion would you reach with this test for significance level $\alpha = .075$? $\alpha = .10$? $\alpha = .05$? $\alpha = .073$?

6.3.13. Consider a test procedure with significance level $\alpha = .01$. What conclusion do you reach with this test if the data yield a P-value $= .04$? P-value $= .01$? P-value $= .001$? P-value $= .10$?

6.3.14. Let p be the probability that an event A occurs. If the interval (.375, .698) is an approximate 95% confidence interval for p, what conclusion would you reach for the approximate level $\alpha = .05$ two-sided test of H_0: $p = .6$ versus H_A: $p \neq .6$? for the approximate level $\alpha = .05$ test of H_0: $p = .8$ versus H_A: $p \neq .8$?

6.3.15. Let p be the probability that an event A occurs. For sample size $n = 125$, consider the approximate level $\alpha = .025$ test of H_0: $p = .25$ versus the alternative H_A: $p > .25$, as given in expression (6.12). If the observed value of B is 35, what conclusion do you reach with the test?

6.3.16. Let p be the probability that an event A occurs. For sample size $n = 50$, consider testing the null hypothesis H_0: $p = .33$ versus the alternative H_A: $p < .33$ using the appropriate one-sided procedure in expression (6.12). If the observed value of B is 13, compute the value of the test statistic Z_{test} and find the associated P-value. What conclusion is reached by the approximate level $\alpha = .05$ test?

6.3.17. *Height of Students.* Let p denote the proportion of male students in your university who are at least six feet tall. You obtain the following heights (in inches) of a random sample of $n = 10$ male students who are currently enrolled in the university:

$$74, 66, 63, 69, 72, 77, 67, 68, 79, 71.$$

Find the exact P-value for a test of the null hypothesis H_0: $p = .3$ versus the alternative H_A: $p > .3$.

6.3.18. *Free Throws.* The star of your men's basketball team makes 80% of his free throws. A good friend of yours insists that she is better at shooting free throws than he is and agrees to shoot 100 free throws to convince you.

(a) State the appropriate null H_0 and alternative H_A hypotheses of interest in this setting.

(b) If your friend makes 90 of her 100 free throws, find the P-value for the test of H_0 versus H_A.

(c) What conclusion do you reach with the P-value in part (b) for approximate significance level $\alpha = .05$?

6.3.19. *Men and Candy.* Consider the approximate 96% confidence interval obtained in Exercise 6.2.25 for the percentage, p, of all men who almost never eat candy. Using only this observed 96% confidence interval, what is the conclusion of the approximate level $\alpha = .04$ test of the null hypothesis H_0: $p = .4$ versus the two-sided alternative H_A: $p \neq .4$?

6.3.20. *Women's Shoe Preferences.* Consider the approximate 95% confidence interval obtained in Exercise 6.2.8 for the percentage, p, of those women who work outside the home who typically wear "flats" to work. Using only this observed 95% confidence interval, what is the conclusion of the approximate level $\alpha = .05$ test of the null hypothesis H_0: $p = .5$ versus the two-sided alternative H_A: $p \neq .5$?

6.3.21. *Lost While Driving.* Consider the study discussed in Exercise 6.1.11 regarding a number of issues about being lost while driving. Find the approximate P-value for a test of the conjecture that more 75% of all male drivers would stop and ask for directions or consult a map if they become lost while driving.

6.3.22. *Physical Health and Psychological Needs.* Consider the study discussed in Exercise 6.1.12 regarding possible benefits to a patient's physical health from simply addressing his psychological needs. Find the approximate P-value for a test of the conjecture that addressing the psychological needs of asthma and rheumatoid arthritis patients leads to improved medical status for a majority of such patients.

6.3.23. *American Tastes in Art.* Consider the poll discussed in Exercise 6.1.13 regarding American tastes in art. Using an approximate significance level $\alpha = .01$, test whether there is sufficient evidence in the survey results to support the conclusion that more than 60% of Americans prefer the colors in a painting to be blended into each other, rather than kept separate.

6.3.24. *Intimate Partner Abuse.* How prevalent is intimate partner abuse among women? One way to obtain information about this issue is to determine the prevalence of such intimate partner abuse among female patients entering Emergency Departments for treatment. Dearwater et al. (1998) conducted a retrospective study of this type for community hospitals. For the period from 1995 to 1997 an anonymous survey was sent to 4641 women who had been treated in community hospital Emergency Departments in either Pennsylvania or California. Of the 3455 women who completed this survey, 489 reported that they had experienced physical and/or sexual abuse by an intimate partner during the past year.

(a) Find the approximate P-value for a test of the conjecture that more than 10% of all women patients treated in hospital emergency rooms have endured physical and/or sexual abuse by an intimate partner during the prior year.

(b) What do you think these findings imply about women, in general, who are involved in intimate relationships? Do you feel the percentage of all women who are abused while in an intimate relationship is likely to be higher or lower than the estimate obtained by the Dearwater et al. study of community hospital Emergency Departments? Why?

6.3.25. *Euthanasia.* Is euthanasia (i. e., physician-assisted suicide) a potential problem commonly faced by practicing physicians in the United States? Meier et al. (1998) mailed surveys to 3102 physicians whose specialities are most likely to receive requests from patients for assistance with euthanasia. The authors received completed questionnaires from 1902 physicians, including 320 who indicated that they had received at least one request for such assistance since they began practicing.

(a) Find the approximate P-value for a test of the conjecture that more than 15% of all physicians working in the designated specialities have received at least one request for euthanasia since they began practicing.

(b) How well do you feel the larger group of 3102 physicians who were sent the survey are represented by the 1902 physicians who actually responded? That is, do you believe there is a possibility of non-response bias in this study? Why or why not?

Chapter 6 Comprehensive Exercises

6.A. Conceptual

6.A.1. Explain how to properly interpret the following statement.

"(.47, .87) is a 90% confidence interval for the population percentage p."

6.A.2. Will an approximate 90% confidence interval for a population percentage be shorter or longer than an approximate 95% confidence for the percentage based on the same data collection? Justify your answer.

6.A.3. Is the following statement true or false? Justify your answer.

"If the P-value for rejecting a null hypothesis H_0 in favor of an alternative H_A is .17, then the probability is .17 that H_0 is true."

6.A.4. Is the following statement true or false? Justify your answer.

"Suppose that (.37, .64) is a 94% confidence interval for a population percentage p. Then we would reject the null hypothesis H_0: $p = .75$ in favor of an alternative H_A: $p \neq .75$ at significance level $\alpha = .06$."

6.A.5. Is the following statement true or false? Justify your answer.

"The length of an approximate 93% confidence interval for a population percentage p is a decreasing function of the sample size n."

6.A.6. Is the following statement true or false? Justify your answer.

"If we reject a null hypothesis H_0 in favor of an alternative H_A at significance level α, then we would also reject H_0 in favor of the alternative H_A for any significance level greater than α."

6.A.7. Let p be the probability that an event A occurs. If the approximate level $\alpha = .08$ test of H_0: $p = .3$ versus H_A: $p \neq .3$ leads to rejection of H_0, which of the following are possible approximate 92% confidence intervals for p?

(i) (.459, .683)
(ii) (.663, .994)
(iii) (.104, .898)
(iv) (.255, .388)
(v) (.334, .665).

6.A.8. Explain how to properly interpret the following statement.

"The P-value for rejecting a null hypothesis H_0 in favor of an alternative H_A is .07."

6.A.9. Is the following statement true or false? Justify your answer.

"If we fail to reject a null hypothesis H_0 in favor of an alternative H_A at significance level α, then we know that H_0 is true."

6.A.10. Is the following statement true or false? Justify your answer.

"If we reject a null hypothesis H_0 in favor of an alternative H_A at significance level α, then we know that H_0 is not true."

6.A.11. Is the following statement true or false? Justify your answer.

"If we fail to reject a null hypothesis H_0 in favor of an alternative H_A at significance level α, then we cannot make a Type I Error."

6.A.12. Is the following statement true or false? Justify your answer.

"If we reject a null hypothesis H_0 in favor of an alternative H_A at significance level α, then we cannot make either a Type I or a Type II error."

6.A.13. *Flowers—Colored or White?* In Example 6.5 we used the Ostler-Harper (1978) data to obtain approximate confidence intervals for the proportions of all open and closed communities that have a majority white flowers. The construction and interpretation of these confidence intervals relies heavily on the particular sample open and closed communities studied by Ostler and Harper. What must be true about the 14 open and 11 closed communities in their study in order for these two confidence intervals to provide reasonable inferences?

6.A.14. *Statistics, Earwax, and the Bering Strait.* Consider the Petrakis et al. (1967) study discussed in Exercise 6.B.12. The sample data for the study were collected from patients and non-patient visitors in wards and outpatient clinics and from family studies at well-baby clinics. Discuss any reservations you might have about this sampling technique in view of the fact that the type of earwax a person has is hereditary.

6.A.15. *Statistics, Earwax, and the Bering Strait.* Consider the Petrakis et al. (1967) study discussed in Exercise 6.B.12. The authors also collected earwax data on a sample of Sioux Indians living primarily in the upper midwest and western portions of the United States. In contradiction with their data from the Navaho Indians, the authors found a much lower percentage of dry earwax for this group (only 54 out of 147 in the sample had dry earwax). Discuss these findings in conjunction with the fact that the Sioux (and other Plains Indians) experienced substantial and sustained contact with French trappers and traders, including extensive intermarriage.

6.A.16. *Small Sample Confidence Intervals for a Percentage.* In Sect. 2 we discussed the selection of a sample size n sufficiently large to guarantee a margin of error no greater than some pre-specified level $\pm d$ in the approximate 100CL% confidence interval for the probability, p, of an event A. There we used the worst case (i.e., most variability) setting, corresponding to $p = .5$, to provide the conservative sample size stipulation given in (6.8). However, if we have reasonable prior information (either from basic knowledge about the problem or from a preliminary sample estimate of p) that the true value of p is greater than some bound p^* (or, conversely, less than p^*), we can improve on (i.e., decrease the size of the required sample) the stipulation given in (6.8). In fact, if we are reasonably certain that $p \geq p^* > .5$ (or $p \leq p^* < .5$), then the required sample size can be reduced to satisfy

$$n \geq z^2_{(1-CL)/2} p^*(1-p^*)/d^2. \tag{6.13}$$

(a) Provide an intuitive argument why the less stringent lower bound for n in (6.13) can be used to provide a rough guarantee that the margin of error in the approximate $100CL\%$ confidence interval for p will be no greater than $\pm\, d$ when it is known that either $p \geq p^* > .5$ or $p \leq p^* < .5$? (Hint: Consider the term $p(1\text{-}p)$ as a function of p. Draw a picture of this function and note that it achieves its maximum value of .25 at $p =$ .5. Also observe that the function is strictly decreasing as either p increases from .5 to 1 or p decreases from .5 to 0.)

(b) Using Eqs. (6.8) and (6.13), compare the sample size requirement necessary to provide a rough guarantee that the margin of error in a 95% confidence interval for p will be no greater than $\pm\, d$ when it is known that either $p \geq p^*$ or $p \leq p^*$ for

  (i) $p^* = .5$ (the conservative setting)

  (ii) $p^* = .6$

  (iii) $p^* = .7$

  (iv) $p^* = .9$.

6.A.17. Eighteen independent Bernoulli trials, each with probability of success p, are to be conducted.

(a) Use the **R** function *qbinom*( ) to obtain the form of the level $\alpha = .0210$ test of the null hypothesis H_0: $p = .3$ versus the alternative H_1: $p > .3$.

(b) What is the power of the test in (a) against the alternative $p = .4$? $p = .6$?

(c) What is your conclusion for the test in (a) if nine of the Bernoulli trials resulted in successes?

(d) What is the P-value for the test in (a) for the sample outcome in (c)?

6.A.18. Eighteen independent Bernoulli trials, each with probability of success p, are to be conducted.

(a) Use the **R** function *qbinom*( ) to obtain the form of the level $\alpha = .0596$ test of the null hypothesis H_0: $p = .3$ versus the alternative H_1: $p > .3$.

(b) What is the power of the test in (a) against the alternative $p = .4$? $p = .6$?

(c) What is your conclusion for the test in (a) if nine of the Bernoulli trials resulted in successes?

(d) What is the P-value for the test in (a) for the sample outcome in (c)?

(e) Compare and contrast the results of parts (a)-(d) of this exercise with the similar results for parts (a)-(d) of Exercise A.17.

6.A.19. In Example 6.5 we illustrated one of the problems that can result from the application of the formula given in (6.7) for an approximate confidence interval for the probability parameter p when the sample size n is not sufficiently large. For given confidence level CL and an observed value for B = [number of times the event A occurs in a sample of n Bernoulli trials], we can construct a confidence interval ($p_1(B)$, $p_2(B)$) for p that provides confidence of at least the desired level CL for any sample size n, small or large.

While the technical details behind the construction of this interval are beyond the assumed level of this text, it can be shown that the endpoints of this 100CL % confidence interval for p are given by

$$p_1(B) = \frac{B}{B + (n - B + 1)f_{\frac{1-CL}{2}}} \quad \text{and} \quad p_2(B) = 1 - \frac{n - B}{n - B + (B + 1)f^*_{\frac{1-CL}{2}}},$$

where $f_{\frac{1-CL}{2}}$ is the upper $(\frac{1-CL}{2})^{\text{th}}$ percentile for an F distribution with degrees of freedom $\{d_1, d_2\} = \{2(B{+}1), 2(n{-}B)\}$ and $f^*_{\frac{1-CL}{2}}$ is the upper $(\frac{1-CL}{2})^{\text{th}}$ percentile for an F distribution with degrees of freedom $\{d_1, d_2\} = \{2(n{-}B{+}1), 2B\}$. For given values of CL, n, and B, the **R** function *qf*( ) can be used to obtain the required F distribution upper percentiles and, subsequently, the 100CL% confidence interval $(p_1(B)\ ,\ p_2(B))$ for p.

(a) For $n = 15$ and an observed value of $B = 11$, find a 95% confidence interval for the probability p.

(b) For $n = 30$ and an observed value of $B = 22$, find a 95% confidence interval for the probability p.

(c) Compare your intervals obtained in parts (a) and (b). Are they centered at the same value? How do their lengths compare? Comment on these findings.

6.A.20. Consider testing $H_0 : p = .5$ versus the alternative $H_A : p > .5$. If 12 out of 15 independent Bernoulli trials with probability of success p resulted in the occurrence of the event A, use the **R** function *qbinom*( ) to obtain the exact P-value for a test of H_0 versus H_A? What is your decision with this test for significance level $\alpha = .0592$?

6.A.21. Consider testing H_0: $p = .5$ versus the alternative $H_A : p > .5$. If 12 out of 15 independent Bernoulli trials with probability of success p resulted in the occurrence of the event A, use the standard normal distribution to find an approximate P-value for a test of H_0 versus H_A. Compare this approximate P-value with the exact P-value obtained in Exercise 6.A.20.

6.A.22. Consider testing $H_0 : p = .4$ versus the alternative $H_A : p < .4$. Find the P-value under each of the following sample outcomes:

(a) Two out of ten independent Bernoulli trials with probability of success p resulted in occurrence of the event A.

(b) Four out of twenty independent Bernoulli trials with probability of success p resulted in occurrence of the event A.

(c) Forty out of two hundred independent Bernoulli trials with probability of success p resulted in occurrence of the event A.

(d) Four hundred out of two thousand independent Bernoulli trials with probability of success p resulted in occurrence of the event A.

(e) Compare the answers you obtained in parts (a) through (d). Discuss the significance of the comparison.

6.A.23. Let p be the probability of an event A relative to an underlying population. Consider the approximate level α test of H_0: $p = .4$ versus H_A: $p > .4$ given in expression (6.11).

(a) Find the form of the critical region for approximate level $\alpha = .025$.

(b) Suppose the observed value of the test statistic Z_{test} for a sample of size $n = 100$ from the population is $z_{\text{test}} = 2.04$. What decision is reached with the level $\alpha = .025$ test in part (a)?

(c) **Based only on the information given in part (b)**, can you conclude what decision would be reached with these data if we had used an approximate level of $\alpha = .05$ (instead of .025)? $\alpha = .035$? $\alpha = .03$? Find the forms of the three critical regions for the approximate levels .05, .035, and .03, respectively. Describe how they relate to each other and to the level $\alpha = .025$ critical region in part (a). Can this discussion be generalized to other levels? How?

(d) **Based only on the information given in part (b)**, can you conclude what decision would be reached with these data if we had used an approximate level of $\alpha = .01$ (instead of .025)? $\alpha = .015$? $\alpha = .02$? Find the forms of the three critical regions for the approximate levels .01, .015, and .02, respectively. Describe how they relate to each other and to the observed value of $z_{\text{test}} = 2.04$. What would your conclusions have been with each of these other levels?

(e) Find the approximate P-value for the test of H_0: $p = .4$ versus H_A: $p > .4$ with the observed value of $z_{\text{test}} = 2.04$. Discuss how this P-value can be used to consolidate the discussion in parts (c) and (d).

6.A.24. Let p be the probability of an event A relative to an underlying population and let B denote the number of times A occurs in a random sample of size n from the population. Consider the level $\alpha \approx .05$ test of H_0: $p = .5$ versus the one-sided alternative H_A: $p > .5$ that rejects H_0 if and only if $B \geq c_n$, where c_n is a constant that depends on the sample size n.

(a) Find the appropriate constants c_n for each of the sample sizes $n = 10, 15, 20, 25, 50$, and 100.

(b) What are the powers against the alternative $p = .6$ of the level $\alpha \approx .05$ tests for these six sample sizes?

(c) What are the Type II Error probabilities of the level $\alpha \approx .05$ tests for these six sample sizes when the alternative $p = .7$ is true?

(d) Discuss the implications of your calculations in parts (b) and (c).

6.A.25. *Flowers—Colored or White?* In Example 6.5 we used the Ostler-Harper (1978) data to obtain approximate confidence intervals for the proportions of open and closed communities that have a majority white flowers. The particular open community in the Ostler-Harper sample that had 52.6% white flowers was an Alpine wet meadow. Could this information alone be used to provide a confidence interval for $p_{\text{Alpine wet meadow}} =$ [proportion of white flowers in a typical Alpine wet meadow]? Why or why not?

6.B. Data Analysis/Computational

6.B.1. *How "Golden" Are the Golden Years?* Who has responsibility for aging parents? In a national survey of 1118 individuals aged 40 and older with both living parents, Visiting Angels (2013), one of the largest in-home senior care companies in the United States, found that 783 of the respondents say they do not have a plan for taking care of their aging parents.

(a) What are the relevant population and event of interest in this survey?

(b) Find the approximate P-value for a test of the conjecture that more than 50% of all individuals aged 40 and older with both living parents do not have a plan for taking care of their aging parents.

(c) Find an approximate 95% confidence interval for the percentage of all individuals aged 40 and older with both living parents who do not have a plan for taking care of their aging parents.

6.B.2. *Firearms in the Home.* The proper storage of firearms is extremely important in households with children. Baxley and Miller (2006) collected information from 314 households with children in rural Alabama about the presence of firearms and, if present, how they were stored.

(a) Of the 314 participating households, 201 were determined to contain guns. Let p denote the percentage of all households with children in rural Alabama that also contain guns. Find a point estimate of p and an approximate 95% confidence interval for p.

(b) Among the 196 parents who reported household guns **and** provided complete information about how their guns were stored, 110 of them stored all of the household guns unloaded and locked away, 61 stored at least one firearm unloaded and unlocked, and 25 stored at least one gun loaded. What is the population under consideration here and what three events are of interest? Find approximate 90% confidence intervals for the percentages of the population that belong to each of these events.

(c) Can these results be extended to other populations of households with children and firearms? Why or why not?

6.B.3. *What is the Status of Vegetarianism in America?* The *Vegetarian Times* (2008) commissioned the Harris Interactive Service Bureau to conduct a nationwide poll designed to address this question. The poll collected information on 5050 respondents to the survey.

(a) One hundred sixty two of the respondents indicated that they follow a vegetarian diet. What are the population and event of interest here? Find an estimate of and an approximate 91% confidence interval for the percentage, p, of the population that follow a vegetarian diet.

(b) Of the 162 respondents who indicated that they follow a vegetarian diet, 96 are female and 66 are male. Find the approximate P-value for a test of H_0: $p = .50$ versus H_A: $p > .50$, where p is the percentage of vegetarians who are female. At approximate significance level $\alpha = .05$, do you reject H_0?

(c) Of the 162 respondents who indicated that they follow a vegetarian diet, 68 were younger than 35 years of age. Find an estimate of and an approximate 98% confidence interval for the percentage, p, of all vegetarians who are younger than 35 years of age.

6.B.4. *What are Americans' Attitudes Toward the Environment?* In a recent survey, GfK(2015b) asked Americans the following three questions:

(i) Do you believe that brands and companies "have to be" environmentally responsible?
(ii) Do you feel guilty when not being environmentally friendly?
(iii) Do you "only buy" products and services that appeal to your beliefs, values or ideals?

GfK found that 990 out of 1500 respondents answered yes to question (i), 795 answered yes to question (ii), and 810 answered yes to question (iii).

(a) Define the population of interest in this study and identify the three percentages corresponding to questions (i), (ii), and (iii), respectively.

(b) For each of these three questions, use the survey data to obtain an estimate of and an approximate 95% confidence interval for the relevant percentage.

(c) Find the approximate *P*-value for a test of the conjecture that more than 50% of all Americans feel guilty when they are not being environmentally friendly.

(d) Find the approximate P-value for a test of the conjecture that more than 50% of all Americans believe that brands and companies "have to be" environmentally responsible. At approximate significance level $\alpha = .01$, do you reject the associated null hypothesis?

6.B.5. *Employment Satisfaction Factors.* The Society for Human Resource Management (SHRM) (2014) conducted an Employee Job Satisfaction and Engagement Survey in which participants were asked to respond to a variety of statements about their current employment, including the following two:

(i) How important is it to you that you have opportunities to use your skills and abilities in your work?

(ii) In my organization, employees are encouraged to take action when they see a problem or opportunity.

The survey involved a random sample of 600 U. S. employees. SHRM reported that 354 respondents answered "Very Important" to question (i) and 132 answered "Strongly Agree" to question (ii).

(a) Define the population of interest in this study and identify the two percentages corresponding to questions (i) and (ii), respectively.

(b) Find the approximate *P*-value for a test of H_0: $p = .50$ versus H_A: $p > .50$, where p is the percentage of all U. S. employees who would respond "Very Important" to statement (i). At approximate significance level $\alpha = .02$, do you reject H_0?

(c) Find an estimate of and an approximate 98% confidence interval for the percentage, p, of all U. S. employees who would respond "Strongly Agree" to question (ii).

6.B.6. *Do Children Know Where Household Firearms Are Stored?* Depends on who you ask!! In a survey of parents and children in households with firearms, Baxley and Miller (2006) collected information from 201 such households in rural Alabama.

(a) Parents in 60 of these households reported that their children did not know the storage location of the firearms. What are the population and event of interest here? Find an estimate of and an approximate 95% confidence interval for the percentage of the population belonging to this event.

(b) Parents in 140 of these households reported that their children had never handled a firearm in the home. What are the population and event of interest here? Find an estimate of and an approximate 90% confidence interval for the percentage of the population belonging to this event.

(c) Consider the 60 parents in part (a) who reported that their children did not know the storage location of the firearms. Children from 23 of these households reported that they did, in fact, know the storage location of the firearms in their households. What are the population and event of interest here? Find an estimate of and an approximate 90% confidence interval for the percentage of the population belonging to this event.

(d) Consider the 140 parents in part (b) who reported that their children had never handled a firearm in the home. Children from 31 of these households reported that they had, in fact, handled a firearm in their homes. What are the population and event of interest here? Find an estimate of and an approximate 92% confidence interval for the percentage of the population belonging to this event.

(e) Comment on the implications of the results in parts (c) and (d).

6.B.7. *Moving Aging Parents in to Live with You—Mom or Dad?* In a national survey of 1118 individuals aged 40 and older with both living parents, Visiting Angels (2013), one of the largest in-home senior care companies in the United States, asked respondents the following question: If you had to choose only one of your aging parents to move in and live with you, would you choose your mom or your dad? 745 of the respondents said they would choose mom over dad!

(a) What are the relevant population and event of interest in this survey?

(b) Find the approximate *P*-value for a test of the conjecture that more than half of all individuals aged 40 and older with both living parents would prefer to move mom in over dad if they could only choose one parent.

(c) Find an approximate 95% confidence interval for the percentage of all individuals aged 40 and older with both living parents who would prefer to move mom in over dad if they could only choose one parent.

(d) What do you think respondents might have given as some of the reasons for preferring mom over dad? Check out what the respondents said at www.visitingangels.com.

6.B.8. *Families—Who and How.* The Pew Research Center (PRC) (2012) conducted a wide-ranging survey among residents of the United States on a number of important issues. In particular, they asked participants whether they agreed with the following two statements:

(i) One parent can bring up a child as well as two parents together.

(ii) Favor allowing gay and lesbian couples to marry legally.

(a) PRC reported that 65% of the respondents between 18 and 29 years old (inclusive) agreed with statement (i). What is the relevant population and event of interest for this setting? Suppose you want to find an approximate 90% confidence interval for the percentage of all 18–29 year olds (inclusive) who would agree with statement (i). What additional information do you need to accomplish this? Consider three possible values for this additional information and construct the associated approximate 90% confidence intervals. Discuss your results.

(b) PRC reported that only 36% of the respondents 65 years of age or older agreed with statement (i). What is the relevant population and event of interest for this setting? Using the same possible values for the additional information from part (a), construct approximate 90% confidence intervals for the percentage of all individuals 65 years of age or older who would agree with statement (i). Discuss your results in conjunction with your findings from part (a).

(c) PRC reported that 51% of female respondents agreed with statement (ii), while only 42% of male respondents agreed. Let p_F and p_M denote the percentages of all females and males, respectively, in the United States who would agree with statement (ii). Using the largest of the three possible values for the needed additional information you considered in part (a), find the approximate P-value for a test of the conjecture that more than half of U. S. women would agree with statement (ii). Do the same for the conjecture that less than half of U. S. men would agree with statement (ii). Discuss your findings.

6.B.9. *Time Heals.* Nearly 40 years after the fall of Saigon brought an end to the Vietnam War, the Pew Research Center (PRC) (2014) conducted a survey of 1000 residents of Vietnam on a number of important issues, including their views of the United States.

(a) PRC reported that 76% of the individuals surveyed expressed a favorable opinion of the United States. Find an approximate 95% confidence interval for the percentage, p, of all the Vietnamese who had a favorable opinion of the United States at the time of the survey.

(b) PRC also reported that 56% of the respondents viewed the United States as the world's leading economic power. Find the approximate

P-value for the conjecture that more than half of all the Vietnamese had this view as the time of the survey. What decision do you make at the approximate significance level $\alpha = .03$?

6.B.10. *Online Dating Sites—Who Visits and Who Posts.* Online dating services have become increasing popular as a strategy to find a romantic and possibly lifelong partner. Valkenburg and Peter (2007) reported on some of the demographic and social attributes of individuals who use these services. In May 2005, Valkenburg and Peter conducted an online survey of 367 Dutch singles between 18 and 60 years old. They found that 157 of the respondents had visited at least one dating site and 122 of the respondents had posted a profile on at least one dating site.

(a) What two populations and events are of interest here?

(b) Find an estimate of and approximate 94% confidence interval for the percentage of all Dutch individuals between the ages of 18 and 60 who have visited at least one online dating site.

(c) Find an estimate of the percentage of all Dutch individuals between the ages of 18 and 60 who have posted a profile on at least one dating site. Find the approximate P-value for the conjecture that less than 40% of all Dutch individuals between the ages of 18 and 60 have posted a profile on at least one dating site. What decision do you make at the approximate significance level $\alpha = .07$?

6.B.11. *Treating Alzheimer's Disease.* Consider the Schneider et al. (1993) data in Table 6.2 relating to the treatment of Alzheimer's disease with a cholinesterase inhibitor (*tacrine* or *physostigmine salicylate*) plus *l-deprenyl.* Find an approximate 99% confidence interval for the probability that a randomly selected Alzheimer patient would show improvement in cognate subscale score from the combined use of *l-deprenyl* and a cholinesterase inhibitor relative to the latter in combination only with a placebo.

6.B.12. *Statistics, Earwax, and the Bering Strait.* Where did the first Americans come from? That question has intrigued scientists for many years and is far from completely answered even today. While it is pretty well agreed that one source of settlement of the Americas was a major migration across the Bering Strait roughly 10,000–12,000 years ago, there has also been considerable scientific discussion about the possibility that the Americas were initially settled much earlier than that. Evidence uncovered at Monte Verde, in southern Chile, supports the claim that the earliest confirmed settlement of the new world was at least 12,500 years ago, and there is additional support for the conjecture that the peopling of America could have begun as early as 20,000–30,000 years ago. Moreover, it is likely that there were a number of different waves of migration into the Americas, primarily across the Bering Strait, but also possibly across the Atlantic and Pacific Oceans.

One thing that is generally agreed upon, however, is that the migrations to the new world were primarily from northern Asia. An interesting article that supports this claim that many of the first Americans came from northern Asia was published a number of years ago by Petrakis et al. (1967) and provides an unusual link between this theory and the makeup of human cerumen, better known as 'earwax'.

Matsunaga (1962) had earlier documented that human cerumen occurs in two phenotypic forms, wet (sticky) and dry (hard). Moreover, you can blame your parents for the type of earwax that you have, as it is controlled by a single pair of genes in which the allele for the sticky trait is dominant. Matsunaga also found substantial differences in the frequencies of these two forms of earwax among different ethnic groups, ranging from a very high frequency of dry earwax in the peoples of Northern Asia through intermediate percentages of dry earwax in Southeast Asia to very low frequencies of dry earwax across the rest of the world.

Petrakis et al. (1967) reasoned that if the ancestors of present day Native Americans had, indeed, migrated from Northern Asia (either via the Bering Strait or by other routes), the makeup of the cerumen for such Native Americans should reflect that linkage and be predominantly dry in form (in variance with the makeup of other present day inhabitants of the Americas with ancestral links to Europe or Africa). To investigate this theory, the authors undertook a number of studies relating to the nature of the cerumen of present day 'Native Americans'. In this example, we concentrate on one aspect of their studies associated with data obtained from Navaho Indians in Arizona and California. In that portion of their study, Petrakis et al. determined (using otoscopic examination of both ears as previously employed by Matsunaga) the type of cerumen for 162 Navaho subjects (satisfactorily determined to be full-blooded, with no known interracial marriages in their ancestry) and found that 113 of them had dry earwax.

(a) Letting $p_{\mathrm{DE:N}}$ denote the percentage of all full-blooded Navaho Indians who have dry earwax, find an approximate 92% confidence interval for $p_{\mathrm{DE:N}}$.
(b) The percentages of dry earwax observed by Matsunaga (1962) among various Northern Asian ethnic groups ranged from 62.41% to 95.65%. Discuss the relevance of this fact in conjunction with the confidence interval for $p_{\mathrm{DE:N}}$ obtained in (a) and the hypothesis that the ancestors of the Navahos came to the new world from Northern Asia.

6.B.13. *Statistics, Earwax, and the Bering Strait.* Consider the Petrakis et al. (1967) study discussed in Exercise 6.B.12. The authors also gathered similar data on a sample of 45 Caucasian Americans and found that 2 of them had dry earwax. Use these data to find an approximate 92% confidence interval for $p_{\mathrm{DE:C}}$, the percentage of Caucasian Americans with dry earwax. Discuss the implication of this result in conjunction with the confidence interval for $p_{\mathrm{DE:N}}$ obtained in part (a) of Exercise 6.B.12.

6.B.14. *Statistics, Earwax, and the Bering Strait.* Consider the Petrakis et al. (1967) study discussed in Exercise 6.B.12. In part (a) of that exercise, you were asked to find an approximate 92% confidence interval for $p_{\text{DE:N}}$ utilizing the data for the $n = 162$ Navaho subjects included in the authors' sample. The margin of error associated with the confidence interval for their data is $\pm$ .0632. How many Navaho subjects should the authors have included in their sample in order to be certain that the observed margin of error for their approximate 92% confidence interval for $p_{\text{DE:N}}$ would have been no greater than $\pm$ .05?

6.B.15. *Flowers—Colored or White?* Consider the Ostler and Harper (1978) data discussed in Examples 6.2 and 6.5 on the relative proportions of white and colored flowers in open and closed communities.

(a) How many closed communities would they have had to examine if they wanted an approximate 90% confidence interval for p_{Closed} with a margin of error no greater than $\pm$ 3%?

(b) How would your answer in (a) change if you knew that at least 3/4 of all closed communities have a majority of white flowers? (Hint: See Exercise 6.A.16.)

6.B.16. *Casino Gambling in Tennessee?* Consider designing the sample survey to assess the sentiment in support of legalized casino gambling in the state of Tennessee, as discussed in Example 6.6. How many eligible state voters would you have to interview in order to obtain an approximate 90% confidence interval for the proportion, p, of eligible voters in Tennessee who favor legalizing casino gambling in the state, with margin of error no greater than $\pm$ 2% if, in fact, you were relatively confident that p cannot possibly exceed .40? (Hint: See Exercise 6.A.16.)

6.B.17. *Statistics, Earwax, and the Bering Strait.* Consider the Petrakis et al. (1967) study discussed in Exercise 6.B.12. In part (b) of that exercise, we note that the percentages of dry earwax observed by Matsunaga (1962) for various Northern Asian ethnic groups ranged from 62.41% to 95.65%. Taking the midpoint of this range (62.42% + 95.65%)/2 = 79.04% to be 'representative' of the percentage of dry earwax for Northern Asians, find the approximate P-value for a test of H_0: $p_{DE:N} = .79$ versus the alternative H_A: $p_{DE:N} < .79$, where $p_{DE:N}$ denotes the percentage of all full-blooded Navaho Indians who have dry earwax. What is your conclusion for approximate significance level $\alpha = .05$? Discuss the implication of your analysis.

6.B.18. *Dual-Career Couples.* In Example 6.7 we discussed a study by the research group Catalyst (1998) dealing with dual-career couples' opinions regarding whether they felt they have more control over shaping their own careers because their partner is also working. Another issue addressed in that study related to how such couples viewed the relative importance of their respective jobs in the relationship. Catalyst found that 196 of the 401 women interviewed felt that their jobs were treated equally in the dual-career relationship, while 233 of the 401 men interviewed expressed this opinion.

(a) Find an approximate 95% confidence interval for the percentage of all men in dual-career relationships who feel that both partners' jobs are treated equally in their relationships.

(b) Find an approximate 95% confidence interval for the percentage of all women in dual-career relationships who feel that both partners' jobs are treated equally in their relationships.

(c) Comment, informally, on the results obtained in (a) and (b).

6.B.19. *Statistics, Earwax, and the Bering Strait.* Consider the Petrakis et al. (1967) study discussed in Exercise 6.B.12. The percentages of dry earwax observed by Matsunaga (1962) among various Caucasian ethnic groups were generally below 10%.

(a) Find the approximate P-value for a test of H_0: $p_{DE:N} = .10$ versus the alternative H_A: $p_{DE:N} > .10$, where $p_{DE:N}$ denotes the percentage of all full-blooded Navaho Indians who have dry earwax. What is your conclusion for approximate significance level $\alpha = .05$? Discuss the implication of your analysis.

(b) Compare and contrast the two approaches taken in part (a) of this exercise and in Exercise 6.B.17 to provide evidence for the Northern Asian ancestry of the Navaho Indians. In particular, discuss the different implications associated with Type I and Type II Errors for the two formulations; that is, what are the consequences of falsely rejecting a true null hypothesis or incorrectly failing to reject a false null hypothesis for each of these testing approaches?

6.B.20. *Flowers—Colored or White?* Consider the flower color data for open and closed communities, as discussed in Examples 6.2 and 6.5. Use the data from those examples and the small sample procedure discussed in Exercise 6. A.19 to obtain a 95% confidence interval for p_{Open} = [proportion of open communities that contain a majority white flowers]. Use the **R** function *qf*( ) to generate the necessary F distribution percentiles. Compare this small sample 95% confidence interval for p_{Open} with the approximate 95% confidence interval for p_{Open} previously obtained in Example 6.5.

6.B.21. *Coin Tossing with a Friend.* Consider the third critical region C_3 = {exactly 8 heads on the eight flips} for the coin tossing experiment with your friend discussed in Sect. 3. Verify that this critical region has significance level $\alpha = .0039$ for testing the null hypothesis of a fair coin (i.e., H_0: $p = .5$) and that it has power .4305 against the biased coin alternative (i.e., H_A: $p = .9$).

6.B.22. *Flowers—Colored or White?* Consider the flower color data for open and closed communities, as discussed in Examples 6.2 and 6.5. Use the data from those examples and the small sample procedure discussed in Exercise 6. A.19 to obtain a 95% confidence interval for p_{Closed} = [proportion of closed communities that contain a majority white flowers]. Use the **R** function qf( ) to generate the necessary F distribution percentiles. Compare this small sample

95% confidence interval for p_{Closed} with the approximate 95% confidence interval for p_{Closed} previously obtained in Example 6.5.

6.C. Activities

6.C.1. *How Old Are Your Pennies?* What is the percentage, p, of pennies in circulation that were minted in the 1970's? Design an experiment to obtain data for addressing this question. Carry out your experiment and find a point estimate for p and an approximate 95% confidence interval for p. At approximate significance level $\alpha = .05$, do you reject H_0: $p = .25$ in favor of H_A: $p \neq .25$? Prepare a short report that discusses your experiment/data collection process and provides details of your data analysis.

6.C.2. *Where Are Your Coins Minted?* U. S. coins are currently produced at two mints, one in Philadelphia and one in Denver. Pennies minted in Philadelphia do not have a mint mark below the date, while pennies minted in Denver have a mint mark D below the date. Let p denote the percentage of U. S. pennies in circulation that were minted in Philadelphia. Design and conduct an experiment to provide information about p. Use your sample data to obtain a point estimate of p and an approximate 92% confidence interval for p. In your data collection you might run across one or more pennies with a mint mark S below the date. What does the S stand for? Do you have to discard such S pennies from your study? Why or why not? Prepare a short report that discusses your experiment/data collection process and provides details of your data analysis.

6.C.3. *College Students and Sleep.* What percentage, p, of college students get an average of at least 6 h of sleep a night? Design and conduct an experiment to provide information about p. Use your sample data to obtain a point estimate of p and an approximate 98% confidence interval for p. Find the approximate P-value for a test of H_0: $p = .40$ versus H_A: $p < .40$. At approximate significance level $\alpha = .05$, do you reject H_0? Prepare a short report that discusses your experiment/data collection process and provides details of your data analysis.

6.C.4. *Semi Trailers on Interstates.* It seems like we are always either passing or being passed by semi trailers when we drive on interstate highways. Is it reality or simply a figment of our imagination? Design and conduct an experiment to provide information about the percentage, p, of semi trailers among the vehicles traveling on interstate highways. Prepare a short report that discusses your experiment/data collection process and provides details of an appropriate statistical analysis of your sample data.

6.C.5. *Mother's Day or Father's Day?* Suppose the powers that be decided to cancel either Mother's Day or Father's Day. What percentage, p, of college students do you think would prefer to see Father's Day cancelled in favor of Mother's Day? Design and conduct an experiment to provide information about p. Use your sample data to obtain a point estimate of p and an approximate 94% confidence interval for p. Find the approximate P-value for a test of H_0: $p = .50$ versus H_A: $p > .50$. At approximate significance level $\alpha = .10$, do you reject H_0? Prepare a short report that discusses your experiment/data collection process and provides details of your data analysis.

6.C.6. *College Students and Meat.* What percentage, p, of college students are vegetarian? Design and conduct an experiment to provide information about p. Use your sample data to obtain a point estimate of p and an approximate 94% confidence interval for p. Find the approximate P-value for a test of H_0: $p = .10$ versus H_A: $p < .10$. At approximate significance level $\alpha = .01$, do you reject H_0? Prepare a short report that discusses your experiment/data collection process and provides details of your data analysis.

6.C.7. *One-Eyed Autos.* When driving at night we often encounter cars with only a single headlight (or even occasionally NO headlights). How common an occurrence do you think this is? Design and conduct an experiment to provide information about the percentage, p, of cars that are driven at night with less than one headlight on. Use your sample data to obtain a point estimate of p and an approximate 96% confidence interval for p. Prepare a short report that discusses your experiment/data collection process and provides details of your data analysis.

6.C.8. *Special Plates for Special People.* When purchasing license plates for an auto we are usually given the opportunity to specialize the plates (for a fee, of course) by either selecting numbers/letters that mean something to us personally (such as "I M FINE" or "BM ME UP" or "ORTHODOC") or by selecting a license plate associated with a designated organization or society (such as a "university affiliation" or "Save the Seashores" or a "wildlife picture"). Design and conduct an experiment to provide information about the two percentages, p_1 = percentage of cars that have license plates with personalized numbers/letters and p_2 = percentage of cars that have license plates associated with designated organizations. Prepare a short report that discusses your experiment/data collection process and provides details of an appropriate statistical analysis of your sample data.

6.C.9. *Pine Tree Growth.* Using the data in the dataset *pines_1997*, select a random sample of 100 pine trees from the population of 1000 such trees planted by Kenyon College student and faculty volunteers in April 1990.

(a) Use your sample to estimate the percentage of the 1000 trees planted in 1990 that grew more than 5 *cm* from 1996 to 1997.

(b) Find an approximate 94% confidence interval for the probability that a randomly selected tree planted in 1990 grew less than 4 *cm* from 1995 to 1996.

6.C.10. *Home Prices in Columbus, Ohio.* Find a published list of the prices of Style 3 homes (split level) that are currently for sale in the Columbus, Ohio area. Select a random sample of 75 homes from this list. Use this sample to find an approximate 98% confidence interval for the proportion of all Style 3 homes for sale in central Ohio that have a listed selling price of more than $150,000.

6.C.11. *Home Prices in Atlanta, Georgia.* Find a published list of the prices of Style 3 homes (split level) that are currently for sale in the Atlanta area. Select a random sample of 75 homes from this list. Use this sample to find an approximate 98% confidence interval for the proportion of all Style 3 homes currently for sale in the Atlanta area that have a listed selling price of more than $150,000. Compare your confidence interval with that found in Exercise 6.C.10 for the Columbus, Ohio area.

6.D. Internet Archives

6.D.1. *Social Issues.* Search the Internet to find a site that describes a survey related to a social issue of particular interest to you. List some of the factors that were investigated in the study and write a short report discussing the most important results from the survey.

6.D.2. *Political Polls.* Search the Internet to find a site that describes a political poll involving at least three candidates for a statewide or national office. Discuss the details of the poll and write a short report outlining the most important conclusions from the poll.

6.D.3. *College Students and Participatory Sports.* Search the Internet to find a site that discusses participatory sport preferences for college students. Discuss the experimental design of the study and write a short report detailing the most important conclusions from the study.

6.D.4. *Public Opinion Poll.* Search the Internet to find a site that presents the results of a public opinion poll on a currently relevant topic of general interest. Discuss the experimental design of the study and write a short report detailing the most important conclusions from the poll.

6.D.5. *Disease Prevalence.* Search the Internet to find a site that discusses the prevalence of a specific serious disease among the U. S. population, broken down across geographical regions of the country. Write a short report detailing the most important conclusions from the study.

6.D.6. *Medical Treatment Effectiveness.* Search the Internet to find a site that discusses the effectiveness of a specific medical treatment for HIV/AIDS. Write a short report detailing the most important conclusions from the study.

6.D.7. *Presidential Election Voter Opinion Polls.* Gallup, Inc. is a private research group that conducts opinion polls on a variety of issues, including presidential elections. At its website, gallup.com, Gallup (2015a) reports the results of its final pre-election Gallup Voter Opinion Poll and the eventual election results for every presidential election from 1936 through 2012. These data are also contained in the dataset *presidential_elections_polls*. Assume for sake of illustration that each of these final Gallop Voter Opinion Polls was based on interviews with $n = 1000$ eligible and 'likely' voters.

(a) Select at least three of the presidential elections between 1936 and 2012. For each of them, use the final Gallop Voter Opinion Poll data to construct an approximate 95% confidence interval for the percentage of all eligible and 'likely' voters who would have (at the time of the final poll) voted for the eventual election winner.

(b) How many of the confidence intervals constructed in (a) contained the value of the final percentage of popular votes received by the winner of the election? Is this a surprise number? Why or why not?

(c) For each of the election years you chose, use the confidence interval obtained in (a) to test at approximate significance level $\alpha = .05$ the conjecture that at least 50% of all eligible and 'likely' voters would have voted (at the time of the final Gallup Voter Opinion Poll) for the eventual winner in that election. How many of these hypothesis tests support that conjecture? Is this a surprise? Why or why not?

(d) Repeat parts (a) – (c) if the interviews had been based on $n = 100$, instead of $n = 1000$, eligible and 'likely' voters.

(e) Repeat parts (a) – (c) if the interviews had been based on $n = 500$, instead of $n = 1000$, eligible and 'likely' voters.

(f) Compare and contrast your results for the three values of $n = 100, 500$, and 1000.

7 Statistical Inference for the Center of a Population

In this chapter we consider the commonly encountered statistical problem of using sample data for a quantitative variable along with the sampling distribution of an appropriate summary statistic to make inferences about the center of the corresponding population distribution. For example, Sciulli and Carlisle (1975) used skeletal remains to obtain sample data on the stature of a number of prehistoric Amerindian populations living in the Ohio Valley over the years from 200 BC to 1200 AD. A number of questions naturally arose in this study. What was the typical height for male and female Amerindians living in this region of the country during that period of time? As the degree of plant cultivation increased and the reliance on the availability or scarcity of fresh game decreased over the years, was there a noticeable change in the stature of the Amerindian populations? Questions such as these can be addressed only through the use of appropriate statistical techniques.

The emphasis in this chapter will be on inference about the center of the distribution of values for a single variable. There are a variety of statistical

© Springer International Publishing AG 2017
D.A. Wolfe, G. Schneider, *Intuitive Introductory Statistics*, Springer Texts in Statistics,
DOI 10.1007/978-3-319-56072-4_7

approaches to making such inferences, depending on one's knowledge about the population under study. We begin with those procedures that are broadest in their applicability (i. e., require the least conditions on the population distribution). Then we move progressively to procedures that are more effective when it is reasonable to make additional assumptions about the population distribution, for example, that it is symmetric or perhaps even normal. The first three sections of the chapter are devoted to exact procedures that are appropriate under three different assumptions about the underlying population. In the fourth section we discuss comparative features of these three approaches to inference. Approximate procedures that can be used when we are fortunate enough to have a larger number of sample observations are presented in Sect. 5. Finally, in Sect. 6 we illustrate how the technique of bootstrapping can be used to provide approximate inferences for the median of an arbitrary distribution.

General Setting and Notation Let X_1, ..., X_n denote the items of a random sample of size n from the population of a quantitative variable and consider the center of the distribution for this variable. As we shall see, this center can correspond to either the mean, μ, or the median, η, of the distribution, depending on both the information that is available to us about the population and our primary reasons for conducting the study. Since the order in which we collect our sample observations is not important in this context, the available information about the center of the distribution is contained solely in the list of sample values sorted from smallest to largest (i. e., the order statistics for the sample). We write the sorted observations as $X_{(1)}$, $X_{(2)}$, . . ., $X_{(n)}$, so that $X_{(1)}$ is the smallest sample value and $X_{(n)}$ is the largest.

The most appropriate choice of a summary statistic and associated statistical methodology for making inferences about the center of a population depends directly on what information is available about the form of the underlying distribution for the variable of interest. For instance, is the underlying distribution symmetric? What is the probability that a sample from this distribution will contain outliers? The more we know about the form of this underlying distribution, the more focused our inferences can be, resulting in procedures with better properties. Later in this chapter we will discuss graphical techniques that can be used to evaluate the reasonableness of certain assumptions about an underlying population.

For a particular problem of interest, you should give careful consideration to what you know about the underlying distribution and make as many assumptions as can be supported safely by this information and appropriate graphical assessments of the observed sample data. The statistical procedures associated with this set of assumptions should then be applied to the data to provide the desired inferences about the center of the distribution.

7.1 Exact Inference for the Center of a Population under a Minimal Assumption

When little is known about the distribution of the variable of interest or when what is known about the distribution indicates that it is probably asymmetric (skewed in one direction) and/or heavy-tailed (likely to produce samples with outliers), then the center of the distribution that most appropriately represents the typical value for the distribution is the population median η. In this section we present statistical procedures which are appropriate for making inferences about a population median η under the single assumption that $P(X = \eta) = 0$; that is, the variable X never takes on its median value. Throughout this chapter we will refer to this condition on the underlying population as the *minimal assumption*. At first this may seem like a strange assumption, but we point out that it is automatically satisfied for every continuous variable X. We note that the assumption does place a very mild limitation on settings where these procedures can be used for discrete variables. For a discrete population, for example, for which the median value η has high probability of occurrence, even these procedures developed for this minimal assumption must be used with caution, if at all. With only this minimal requirement on the underlying population, both point and interval estimation of the population median η are based directly on the order statistics. The natural point estimator for η is the corresponding sample median, $\tilde{X}$, as defined in Eqs. (1.2) and (1.3) for odd and even sample sizes, respectively. For confidence intervals and bounds for η we need to also account for the variability that is associated with our measurements. To accomplish this, we utilize the observed spread among the order statistics and the probability distribution of B = [number of sample observations greater than the population median η].

Under the minimal assumption the population median η is not a possible value for the variable X. Hence, each individual observation in our random sample $X_1, \ldots, X_n$ has probability 1/2 of being larger than η and probability 1/2 of being less than η. Moreover, we know that the sample observations are independent. Hence, if we let $Y_i = 1$ or 0 depending on whether the ith sample observation, X_i, is greater than or less than η, respectively, then the probability distribution of $B = \sum_{i=1}^{n} Y_i = [\text{number of sample observations greater than } \eta]$ is Binomial with parameters n and $p = 1/2$; that is, $B \sim Binom(n, 1/2)$.

We now show how to use this $Binom(n, 1/2)$ probability distribution to construct confidence intervals for η that are valid under the minimal assumption. Since all that we have to work with in this setting are the sample values themselves, we will use appropriate order statistics $X_{(1)}, \ldots, X_{(n)}$ as the end points of our confidence intervals for η, corresponding to the same number of observations from the top as from the bottom of the ordered sample values. Thus, for example, if our sample size is $n = 7$, there are only three possible confidence intervals using this approach, namely, $X_{(1)} < \eta < X_{(7)}$, $X_{(2)} < \eta < X_{(6)}$, and $X_{(3)} < \eta < X_{(5)}$. In general, for arbitrary sample size n the possible confidence intervals are $X_{(j)} < \eta < X_{(n-j+1)}$, where j can be any positive integer less than $\frac{n+1}{2}$. For $n = 7$, we have $\frac{n+1}{2} = 4$, and, as previously noted, j can be any of the integers 1, 2, or 3.

Now, to evaluate the confidence level associated with a confidence interval of the form $X_{(j)} < \eta < X_{(n-j+1)}$, we need to compute $P(X_{(j)} < \eta < X_{(n-j+1)})$. First, suppose that η is very close to the lower end point (i. e., bottom) of our confidence interval; that is,

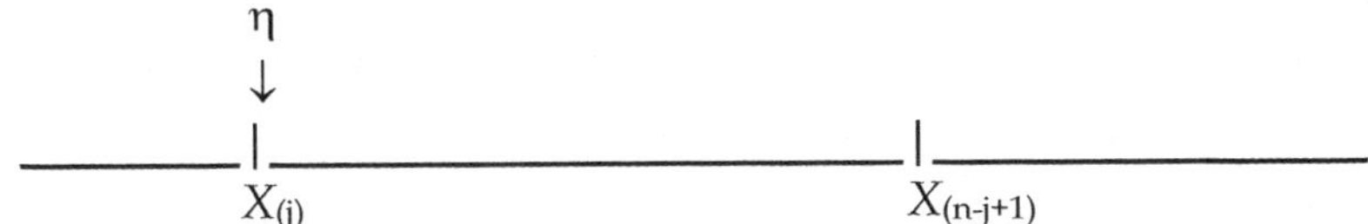

Then there are exactly j observations less than η, namely, $X_{(1)}$ through $X_{(j)}$, and $(n-j)$ observations greater than η. On the other hand, if η is very close to the upper end point (i. e., top) of our confidence interval,

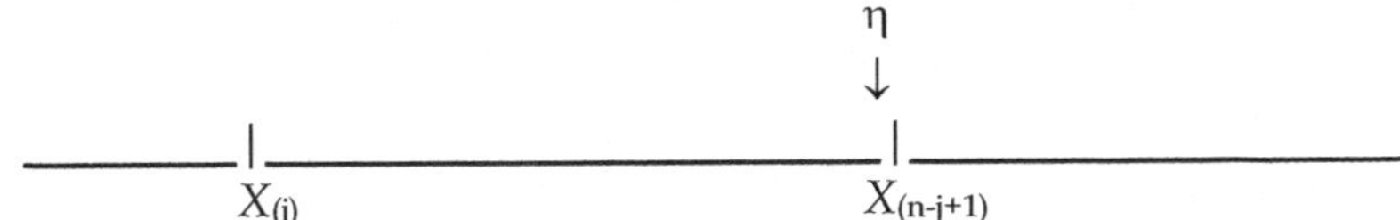

then exactly $n-j$ of the order statistics (up to, but not including, $X_{(n-j+1)}$) are less than η and there are j observations greater than η. Thus

$$\begin{aligned} P(X_{(j)} < \eta < X_{(n-j+1)}) &= P(\text{ the number of sample observations greater than } \eta \\ &\qquad \text{is between } j \text{ and } n-j, \text{ including these values}) \\ &= P\{j \le B \le (n-j)\}, \end{aligned}$$

since B counts the number of sample observations greater than η. Using the fact that the probability distribution of B is $Binom(n, 1/2)$, the confidence level CL associated with a confidence interval of the form $X_{(j)} < \eta < X_{(n-j+1)}$ is then given by

$$CL = P\left(X_{(j)} < \eta < X_{(n-j+1)}\right) = \sum_{t=j}^{n-j} \binom{n}{t} (.5)^n. \tag{7.1}$$

Thus the probability is $\sum_{t=j}^{n-j} \binom{n}{t}(.5)^n$ that we will collect a random sample for which the observed interval $(X_{(j)}, X_{(n-j+1)})$ contains the unknown value of η. Selecting the integer j so that $\sum_{t=j}^{n-j} \binom{n}{t}(.5)^n = CL$ then leads to a $100CL\%$ confidence interval for the population median η.

Point and Interval Estimation of the Population Median η under Minimal Assumptions Let $X_{(1)} \leq \ldots \leq X_{(n)}$ be the order statistics for a random sample of size n from a population with median η for which it is reasonable to assume the minimal condition that $P(X = \eta) = 0$. Then the point estimator for the population median η is the sample median, $\tilde{X}$. Moreover, for any positive integer $j < (n+1)/2$, the interval $(X_{(j)}, X_{(n-j+1)})$ provides a confidence interval for the population median η with confidence level given by

$$CL = \sum_{t=j}^{n-j} \binom{n}{t} (.5)^n. \tag{7.2}$$

Corresponding lower and upper confidence bounds for the population median η can also be obtained from the order statistics. If $\eta < X_{(n-j)}$, then the number of sample observations greater than η must be somewhere between j and n, inclusive, so that the upper confidence bound $X_{(n-j)}$ has confidence level

$$CL = \sum_{t=j}^{n} \binom{n}{t} (.5)^n. \tag{7.3}$$

By similar reasoning, the order statistic $X_{(j)}$ provides a lower confidence bound with confidence level also given by (7.3).

Since the probability expressions in Eqs. (7.2) and (7.3) are valid for **any** underlying population which satisfies the very mild condition $P(X = \eta) = 0$, the associated exact $100CL\%$ confidence interval and bounds are valid over this large class of populations.

Example 7.1. Does Regular Aerobic Activity Affect High-Density Lipoprotein (HDL) Cholesterol Levels? It has been established that higher levels of high-density lipoprotein (HDL) cholesterol (i. e., the 'good' cholesterol) are associated with lessened risk of coronary heart disease. It is known that routine physical activity can reduce this risk as well. That raises the natural question of linkage between a person's HDL level and their normal physical activity pattern. Kerr (1983) examined that issue with data collected from 12 women, ranging in age from 25 to 32, who participated in a spa program (including aerobic dancing, cycling, and jogging in place) 3–4 times per week for 45 min – 1 h per session. HDL levels (in mg/dl) were measured for these women at the Ohio State University Hospital from two blood samples taken 24 h apart, each following 12 h of fasting. The ordered values of the averages of the two HDL measurements for each of these twelve women are reported in Table 7.1.

With an even sample size, $n = 12$, we estimate the median HDL level for active women in the age range studied to be $\tilde{x} = [x_{(6)} + x_{(7)}]/2 = [52 + 53]/2 = 52.5\ mg/dl$. To obtain a confidence interval for the median HDL level, we use expression (7.2) to provide an appropriate confidence level. However, since the $B(12, 1/2)$ distribution is discrete, we have only a limited number of such confidence levels which can be used. In fact, for $n = 12$ the available confidence levels are given by the expression $\sum_{b=j}^{12-j} \binom{12}{b}(.5)^{12}$, with $j = 1, \ldots, 6$. Thus, the six available confidence levels for $n = 12$ are $CL = .9996, .9936, .9614, .8540, .6124$, and $.2256$, corresponding to $j = 1, 2, 3, 4, 5$, and 6, respectively. Taking

$j = 3$ in expression (7.2), it follows that $(x_{(3)}, x_{(10)}) = (41, 54)$ mg/dl is an exact 96.14% confidence interval for the median HDL level, η, for active women in the age range 25-32. Thus, with only the minimal assumption about the underlying population, we are 96.14% confident that the median HDL level for active women in the studied age range will be somewhere between 41 and 54 mg/dl.

We note we can also use **R** to provide these confidence intervals for the population median under the minimal assumption. For example, for the average HDL level data in Table 7.1 (available as the dataset *average_HDL_levels*), we can use the **R** function *SIGN.test*( ) from the BSDA package to provide the following output, which includes the point estimator and the 96.14% confidence interval obtained above.

Table 7.1 Average HDL levels (*mg/dl*) for women involved in regular active exercise

Subject number	Average HDL level
1	28
2	41
3	41
4	45
5	51
6	52
7	53
8	53
9	53
10	54
11	59
12	66

Source: Kerr (1983)

```
> SIGN.test(average_HDL_levels, conf.level = 0.9614)

        One-sample Sign-Test

data:  average_HDL_levels
s = 12, p-value = 0.0004883
alternative hypothesis: true median is not equal to 0
96.14 percent confidence interval:
 41.00096 53.99976
sample estimates:
median of x
       52.5

                    Conf.Level L.E.pt  U.E.pt
Lower Achieved CI       0.8540 45.000 53.0000
Interpolated CI         0.9614 41.001 53.9998
Upper Achieved CI       0.9614 41.000 54.0000
```

As we previously noted, there are only six available exact confidence levels for $n = 12$. The *SIGN.test*( ) function reports the CI at the confidence levels immediately below and immediately above the level given by the *conf.level* argument. It then uses these two intervals to interpolate or "fill in" the approximate (in contrast to the "Achieved CI"s, which are exact) confidence interval at the specified *conf.level*.

To test hypotheses about the population median under the minimal population assumption, the individual sample observations once again play the central role. However, for testing purposes we also have a hypothesized value for the population median, say, η_0, against which the observed data are to be compared. Hence, the information in our sample data that is relevant for testing the null hypothesis H_0: $[\eta = \eta_0]$ is contained in the n differences, $D_1 = X_1\text{-}\eta_0, \ldots, D_n = X_n\text{-}\eta_0$, between the observed values and the hypothesized median. Under the minimal assumption, we utilize only the information about η that is contained in the signs of these differences. In particular, the *sign statistic*, B, is simply equal to the number of these signs which are positive; that is,

$$B = \left[\text{the number of sample observations which exceed the hypothesized median value } \eta_0\right]. \tag{7.4}$$

Moreover, when H_0: $[\eta = \eta_0]$ is true, the null sampling distribution of B is binomial with parameters n and $p = P(X > \eta_0) = 1/2$; that is, $B \sim Binom(n, 1/2)$ when H_0 is true.

Hypothesis Tests About the Population Median Under the Minimal Assumption–Sign Test To test the null hypothesis H_0: $[\eta = \eta_0]$ under the minimal assumption about the underlying population, compute the sign statistic B (7.4) and let b_{obs} be the attained value of B. The exact P-values for a test of H_0 against the possible alternatives H_A are then:

H_A	P-value	
$\eta > \eta_0$	$P(B \geq b_{\text{obs}})$	
$\eta < \eta_0$	$P(B \leq b_{\text{obs}})$	
$\eta \neq \eta_0$	$2P(B \geq b_{\text{obs}})$,	if $b_{\text{obs}} > \frac{n}{2}$
	$2P(B \leq b_{\text{obs}})$,	if $b_{\text{obs}} < \frac{n}{2}$
	1,	if $b_{\text{obs}} = \frac{n}{2}$.

We compute all of these probabilities under the assumption that the null hypothesis is true, so that B has the $Binom(n, .5)$ distribution. Thus we have:

H_A **P-value**

$$\eta > \eta_0 \qquad \sum_{t=b_{obs}}^{n} \binom{n}{t} (.5)^n \tag{7.5}$$

$$\eta < \eta_0 \qquad \sum_{t=0}^{b_{obs}} \binom{n}{t} (.5)^n \tag{7.6}$$

$$\eta \neq \eta_0 \qquad \begin{cases} 2\sum_{t=b_{obs}}^{n} \binom{n}{t} (.5)^n, & \text{if } b_{\text{obs}} > \frac{n}{2} \\ 2\sum_{t=0}^{b_{obs}} \binom{n}{t} (.5)^n, & \text{if } b_{\text{obs}} < \frac{n}{2} \\ 1, & \text{if } b_{\text{obs}} = \frac{n}{2} \end{cases} \tag{7.7}$$

We note that these sign test procedures are really nothing more than special cases of the corresponding test procedures for the probability of the event $A = \{X > \eta_0\}$, as discussed in Sect. 6.3. Following the discussion in that section, the corresponding probability $p_A = P(A) = P(X > \eta_0)$ is naturally estimated by $\hat{p}_A = B/n =$ [percentage of the sample observations which exceed η_0] and approximate confidence intervals for the probability p_A can be obtained from Eq. (6.7).

Example 7.2. High-Density Lipoprotein (HDL) Cholesterol Levels In his study of the effect that regular aerobic activity has on HDL cholesterol levels, Kerr (1983) also obtained HDL levels for a number of women who were not active participants in any routine vigorous activity. For this inactive group, he found that the median HDL level was 33 *mg/dl*. Is there evidence in the collected data to suggest that regular aerobic activity leads to a median HDL level greater than this baseline of 33 *mg/dl*? Letting η correspond to the median HDL level for women of this age group who pursue regular aerobic activities, we are interested in testing the null hypothesis of no difference, corresponding to H_0: $\eta = 33$, against the conjectured one-sided alternative H_A: $\eta > 33$ associated with a larger median HDL level for the exercising group. Here $\eta_0 = 33$ and the observed value of the sign statistic $B =$ [number of sample observations greater than 33] is $b_{obs} = 11$, since only subject number 2 had an HDL level that did not exceed 33. Hence, from (7.5), the P-value for our test of H_0: $\eta = 33$ against H_A: $\eta > 33$ is

$$P\text{–value} = \sum_{t=11}^{12} \binom{12}{t} (.5)^{12} = .0030 + .0002 = .0032.$$

We would reject H_0 for any significance level greater than or equal to this P-value $= .0032$, providing strong sample evidence that the median HDL level is greater than 33 *mg/dl* for women in the age group 25–32 who pursue regular aerobic activity.

We note that the sample estimate of $p_A = P(X > 33)$ is $\hat{p}_A = 11/12 = .9167$. Thus we estimate that 91.67% of all women in the age group 25–32 who pursue regular aerobic activity will have an HDL level greater than 33 mg/dl.

The **R** function *SIGN.test*( ) can also be used (with the *md* and *alternative* arguments specified to be 33 and 'greater', respectively) on the *average_HDL_levels* dataset to obtain the following output, which includes the observed value of B (denoted as s in the **R** output) and the associated P-value for this HDL cholesterol level hypothesis test.

```
> SIGN.test(average_HDL_levels, md = 33, alternative = 'greater')

        One-sample Sign-Test

data:  average_HDL_levels
s = 11, p-value = 0.003174
alternative hypothesis: true median is greater than 33
95 percent confidence interval:
 43.28727      Inf
sample estimates:
median of x
       52.5

                    Conf.Level  L.E.pt U.E.pt
Lower Achieved CI       0.9270 45.0000    Inf
Interpolated CI         0.9500 43.2873    Inf
Upper Achieved CI       0.9807 41.0000    Inf
```

Section 7.1 Practice Exercises

7.1.1. Consider the discrete random variable X with probability distribution as follows:

$$P(X = -2) = .25 \quad P(X = -1) = .1 \quad P(X = -.5) = .15$$
$$P(X = .5\,) = .15 \quad P(X = 1) = .1 \quad P(X = 2) = .25.$$

(a) What is the median for this probability distribution?

(b) Show that the probability distribution is symmetric.

(c) Does this probability distribution satisfy the minimal assumption? Justify your answer.

7.1.2. Consider the discrete random variable X with probability distribution as follows:

$$P(X = -6) = .15 \quad P(X = -3) = .2 \quad P(X = 0) = .3$$
$$P(X = 3) = .2 \quad P(X = 6) = .15$$

(a) What is the median for this probability distribution?
(b) Show that the probability distribution is symmetric.
(c) Does this probability distribution satisfy the minimal assumption? Justify your answer.

7.1.3. Let $X_1, \ldots, X_{15}$ be a random sample of size $n = 15$ from a distribution that satisfies the minimal assumption with median η. Let $B =$ [number of sample observations greater than η].

(a) What are the possible values for B?
(b) Find $P(B \geq 11)$.
(c) Find $P(B \leq 13)$.
(d) Use the symmetry of the distribution of B to find $P(B \leq 4)$.

7.1.4. Let $X_1, \ldots, X_{20}$ be a random sample of size $n = 20$ from a distribution that satisfies the minimal assumption with median η. Let $B =$ [number of sample observations greater than η].

(a) What are the possible values for B?
(b) Find $P(B \geq 16)$.
(c) Find $P(B \leq 12)$.
(d) Use the symmetry of the distribution of B to find $P(B \leq 8)$.

7.1.5. Let B have a binomial distribution with parameters n and $p = .5$. Show that the probability distribution is symmetric about its mean $np = .5n$; that is, show that $P(B = b) = P(B = n - b)$ for all $b = 0, 1, \ldots, n$.

7.1.6. Let $X_1, \ldots, X_{15}$ be a random sample of size $n = 15$ from a population that satisfies the minimal assumption with median η and let $X_{(1)} \leq \ldots \leq X_{(15)}$ be the order statistics for the sample.

(a) List all possible confidence intervals for η that could be constructed from these order statistics.

(b) What are the exact confidence levels associated with these confidence intervals?

7.1.7. Let $X_1, \ldots, X_{20}$ be a random sample of size $n = 20$ from a population that satisfies the minimal assumption with median η and let $X_{(1)} \leq \ldots \leq X_{(20)}$ be the order statistics for the sample.

(a) List all possible confidence intervals for η that could be constructed from these order statistics.

(b) What are the exact confidence levels associated with these confidence intervals?

7.1.8. Let $X_1, \ldots, X_{20}$ be a random sample of size $n = 20$ from a population that satisfies the minimal assumption with median η and let $X_{(1)} \leq \ldots \leq X_{(20)}$ be the order statistics for the sample.

(a) List all possible upper confidence bounds for η with confidence levels CL at least .5 that could be constructed from these order statistics.

(b) What are the exact confidence levels associated with these upper confidence bounds?

(c) Compare these upper confidence bounds with the upper endpoints of the corresponding confidence intervals obtained in Exercise 7.1.7.

7.1.9. Let $X_1, \ldots, X_{18}$ be a random sample of size $n = 18$ from a population that satisfies the minimal assumption with median η and let $X_{(1)} \leq \ldots \leq X_{(18)}$ be the order statistics for the sample.

(a) List all possible lower confidence bounds for η with confidence levels CL at least .5 that could be constructed from these order statistics.

(b) What are the exact confidence levels associated with these lower confidence bounds?

7.1.10. Let {4.5, 6.6, 8.9, 12.3, -3.4, 18, 37.4, -23.5, -16.6, 13.3, 2, -8.4, 12.8} be a random sample of size $n = 13$ from a population that satisfies the minimal assumption with median η.

(a) Find the value, b_{obs}, of the sign statistic B (7.4) for testing H_0: $\eta = 3$.

(b) If $\eta = 3$, compute $P(B \geq b_{obs})$ and $P(B \leq b_{obs})$.

7.1.11. Let {33.7, 16.9, -11, 6.7, 12.5, 19.2, - 6.6, 18, -4, 6.9, -4.4, 44} be a random sample of size $n = 12$ from a population that satisfies the minimal assumption with median η.

(a) Compute the value, b_{obs}, of the sign statistic B (7.4) for testing H_0: $\eta = 0$.

(b) If $\eta = 0$, compute $P(B \geq b_{obs})$ and $P(B \leq b_{obs})$.

7.1.12. Let {4.5, 6.6, 8.9, 12.3, -3.4, 18, 37.4, -23.5, -16.6, 13.3, 2, -8.4, 12.8} be a random sample of size $n = 13$ from a population that satisfies the minimal assumption with median η.

(a) Find a 97.76% confidence interval for η.

(b) Find a 98.88% lower confidence bound for η.

7.1.13. Let {33.7, 16.9, -11, 6.7, 12.5, 19.2, - 6.6, 18, -4, 6.9, -4.4, 44} be a random sample of size $n = 12$ from a population that satisfies the minimal assumption with median η.

(a) Find a 99.36% confidence interval for η.

(b) Find a 92.7% upper confidence bound for η.

7.1.14. *Bird Variety.* Groom (1999) recorded breeding-bird counts for riparian habitat along the Big and Little Darby Creeks in central Ohio. The data for thirty-nine 5-min periods in June 1998 are presented in Table 1.7. Viewing this data collection as a random sample of size $n = 39$ from the population of all such 5-min periods in riparian habitat along the Big and Little Darby Creeks, complete the following statistical analyses. Assume only the minimal assumption for the underlying population with unknown median η.

(a) Obtain a point estimate of η and find a 94.7% confidence interval for η.

(b) Find the P-value for a test of the null hypothesis H_0: $\eta = 13.5$ versus the two-sided alternative H_A: $\eta \neq 13.5$.

(c) Comment on the assumption that these data represent a random sample of size $n = 39$ from the population of breeding-bird counts for all 5-min periods in riparian habitat along the Big and Little Darby Creeks.

7.1.15. *Ultrasound Probes and Bacterial Infections.* One of the major sources for spreading nosocomial (hospital-acquired) infections from patient to patient is through the use of ultrasound probes at tertiary care facilities. Ali et al. (2015) presented data for 25 culture swabs from ultrasound probes conducted at the Radiology Department of the Aga Khan University Hospital in Karachi, Pakistan. The Colony Forming Unit (CFU) of bacterial counts using a standard agar plate for the 75 probes are presented in Table 7.2.

Viewing this data collection as a random sample of size $n = 25$ from the population of CFU bacterial counts on all ultrasound probes, complete the following statistical analyses. Assume only the minimal assumption for the underlying population with unknown median η.

(a) Obtain a point estimate of η and find a 95.7% confidence interval for η.

(b) Find the P-value for a test of the null hypothesis H_0: $\eta = 150$ colonies versus the one-sided alternative H_A: $\eta > 150$ colonies.

7.1.16. *Vertical Cliffs and Ages of Trees.* Vertical cliffs are known to support populations of trees possessing characteristics quite different from other tree populations. For example, they often contain unusual numbers of old and deformed trees. In addition, cliff trees are commonly slow growing and

Table 7.2 Number of Colony Forming Units (CFU) of bacterial counts for ultrasound probes

Probe number	Number of Colony Forming Units
1	350
2	142
3	190
4	300
5	409
6	390
7	159
8	198
9	302
10	296
11	322
12	172
13	104
14	151
15	133
16	202
17	102
18	109
19	167
20	79
21	107
22	89
23	202
24	197
25	106

Source: Ali et al. (2015)

widely dispersed over the terrain. Larson et al. (2000) studied the ages and radial growth rates for trees on selected cliffs in the United States and in Western Europe. The estimated ages for a sample of 30 trees from cliffs in eastern and southern Germany, eastern and southern France, central and northern England and Wales are presented in Table 7.3.

Table 7.3 Age estimates (in years) for trees growing on selected cliffs in Western Europe

40	44	60	80	100	110	140	145	152	155
157	159	162	165	170	220	245	280	295	310
325	345	360	375	440	640	800	920	980	1200

Source: Larson (1999)

Viewing this data collection as a random sample of size $n = 30$ from the population of all trees growing on cliffs in Western Europe, complete the following statistical analyses. Assume only the minimal assumption for the underlying population with unknown median η.

(a) Obtain a point estimate of η and find a 97.9% upper confidence bound for η.

(b) Find the P-value for a test of the null hypothesis H_0: $\eta = 300$ years versus the one-sided alternative H_A: $\eta < 300$ years.

7.2 Exact Inference for the Center of a Continuous Population Under the Assumption of Population Symmetry

When we have reliable information that a continuous measurement has a distribution that is approximately symmetric about its population median η, we can use inference procedures that take advantage of this symmetry condition. As we might expect, these specialized procedures are generally better than the corresponding sign statistic procedures (developed under the minimal assumption and discussed in Sect. 1) when the symmetry assumption is justified.

Let X_1, ..., X_n be a random sample of size n from a population that is symmetric about its median η. We know from the symmetry condition that the observed value for any of the X_i's is equally likely to be a given amount above or below η; that is, the probability that an observation will be d units above η is the same as the probability that it will be d units below η for an arbitrary value d. Thus the symmetry of the population suggests that the **relative** values of pairs of sample observations contain additional information about the median η. One particular consequence of this fact is that we would expect a **typical** mini-average of a pair of observations, say, $(X_i + X_j)/2$, to be close to the population median η. Since these mini-averages can be computed for every pair of sample observations (including an observation with itself), we have available to us many more individual pieces of information about η than what is provided just by the n sample observations themselves. It is this fact that drives the development of point and interval estimation of the median η for a symmetric population based on these mini-averages, $(X_i + X_j)/2$.

Definition 7.1. Let $X_1, \ldots, X_n$ be a random sample of size n from a population that is symmetric about its median. Every mini-average of the form $W_{ij} = (X_i + X_j)/2$, for $1 \leq i \leq j \leq n$, is called a **Walsh average**. Since we include averages of each observation with itself (corresponding to $i = j$), there are a total of $M = \binom{n}{2} + n = \frac{n(n+1)}{2}$ individual Walsh averages associated with the random sample $X_1, \ldots, X_n$. We let $W_{(1)} \leq \ldots \leq W_{(M)}$ denote the M ordered Walsh averages associated with the random sample $X_1, \ldots, X_n$.

Under the assumption of an underlying symmetric distribution we would expect about half of the $M = n(n+1)/2$ Walsh averages to be greater than the population median η and the other half to be smaller than η. It follows that the median of the Walsh averages, namely, $\tilde{W} = median\{W_{ij}, 1 \leq i \leq j \leq n\}$, is a natural point estimator of η under this symmetry assumption.

Confidence intervals and bounds for η under the assumption of population symmetry are also naturally based on using the additional symmetry information contained in the Walsh averages. For this, however, we must incorporate the sampling variability associated with the Walsh averages. Do you think the Walsh averages will be more or less variable than the n original sample observations? Clearly each of the n Walsh averages that is an average of a single sample observation with itself has the same variability as that observation. However, the vast majority of the Walsh averages ($\frac{n(n-1)}{2}$ of them) are averages of two different sample observations. From our work in Chap. 4 we know that the variance for the average of two sample observations is only one-half as large as the variance of each of the individual observations. Thus the Walsh averages are, in general, less variable that the original observations. (See Exercises 7.B.14 and 7.B.15 for more about this difference in variability between the Walsh averages and the sample observations.) This reduced variability for the Walsh averages permits the construction of confidence intervals for the median η for a symmetric population that are generally shorter than the confidence intervals that are constructed under the minimal assumption using only the sample observations themselves (as discussed in Sect. 1).

To construct confidence intervals for η from the Walsh averages, we consider the probability distribution of the variable W = [number of Walsh averages greater than the population median η]. From the fact that the underlying distribution is symmetric about η, it follows that each of our M Walsh averages has probability 1/2 of exceeding η. Since W counts the number of times a Walsh average exceeds the median η, it would appear, at first glance, that W might have a binomial distribution. However, unlike the original sample observations $X_1, \ldots, X_n$, the M Walsh averages are not mutually independent, as there are n individual Walsh averages involving each of the sample observations. Thus, it is not appropriate to use the Binomial distribution to account for the variability among the Walsh averages. Fortunately, however, the probability distribution of W has been developed and can be obtained using the **R** function *psignrank*( ). (For more on the construction of the probability distribution of W, see Exercises 7.B.16 and 7. B.17.) For example, if $n = 10$, we can use the following **R** commands to find that $P(W \geq 45) = .042$ and $P(W \geq 40) = .116$.

```
> psignrank(q = 44, n = 10, lower.tail = FALSE)
[1] 0.04199219

> psignrank(q = 39, n = 10, lower.tail = FALSE)
[1] 0.1162109
```

Note that *psignrank*( ) will return $P(W \leq q)$ by default (where q is specified by the q argument). We specify that we would like the upper-tail probability by specifying *lower.tail* = *FALSE*. This will then return $P(W > q)$, but since we want $P(W \geq w) = P(W > w\text{-}1)$, we set the q argument to be w-1. In the examples above, this corresponds to 44 and 39, respectively.

Under the assumption of population symmetry, the probability distribution of W is symmetric about the point $n(n+1)/4$, corresponding to half of the total number of Walsh averages. This implies that $P\left(W \leq \left[\frac{n(n+1)}{2}\right] - w\right) = P(W \geq w)$ for sample size n and every possible value w. (See Exercises 7.2.5 and 7.2.6.) Again, for $n = 10$, this symmetry for the distribution of W implies that $.042 = P(W \geq 45) = P\left(W \leq \left[\frac{10(10+1)}{2}\right] - 45\right) = P(W \leq 10)$ and $.116 = P(W \geq 40) = P\left(W \leq \left[\frac{10(10+1)}{2}\right] - 40\right) = P(W \leq 15)$. As you might expect, we can also obtain these probabilities using the **R** function *psignrank*( ) with the argument *lower.tail* = *TRUE* (which is the default, so we do not need to explicitly specify this) as follows.

```
> psignrank(q = 15, n = 10)
[1] 0.1162109

> psignrank(q = 10, n = 10)
[1] 0.04199219
```

We now show how to use the probability distribution for W and the relationship between W and η to construct confidence intervals for η that are valid under the condition of population symmetry. We employ a counting and interval construction scheme similar to that used in Sect. 1 under the minimal condition, except here our basic sample pieces of information are the $M = \frac{n(n+1)}{2}$ Walsh averages, rather than the n sample observations themselves. Letting $W_{(1)} \leq W_{(2)} \leq \ldots \leq W_{(M)}$ denote the M ordered Walsh averages, we construct confidence intervals for η that have endpoints which are the same number of ordered Walsh averages away from the largest and the smallest Walsh average, respectively. Thus, for example, if our sample size is $n = 5$, we have $M = \frac{5(6)}{2} = 15$ Walsh averages and there are seven possible confidence intervals using this approach, namely, $W_{(1)} < \eta < W_{(15)}$, $W_{(2)} < \eta < W_{(14)}$, $W_{(3)} < \eta < W_{(13)}$, $W_{(4)} < \eta < W_{(12)}$, $W_{(5)} < \eta < W_{(11)}$, $W_{(6)} < \eta < W_{(10)}$, and $W_{(7)} < \eta < W_{(9)}$. In general, for arbitrary sample size n the possible confidence intervals are $W_{(q)} < \eta < W_{(M-q+1)}$, where q can be any positive integer less than $\frac{M+1}{2}$. For $n = 5$, we have $\frac{M+1}{2} = 8$, and, as previously noted, q can be any of the integers from 1 through 7.

Now, to evaluate the confidence level associated with a confidence interval of the form $W_{(q)} < \eta < W_{(M-q+1)}$ we need to compute $P(W_{(q)} < \eta < W_{(M-q+1)})$. First, suppose that η is very close to (but greater than) the lower end point (i.e., bottom) of our confidence interval; that is,

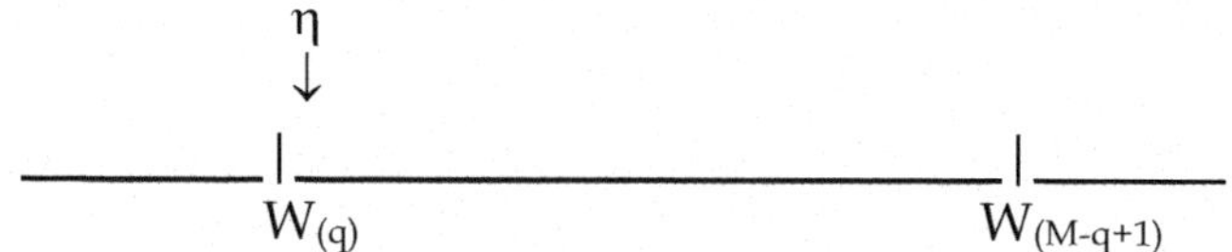

Then there are exactly q Walsh averages less than η, namely, $W_{(1)}$ through $W_{(q)}$, and $(M - q)$ Walsh averages greater than η. On the other hand, if η is very close to (but less than) the upper end point (i. e., top) of our confidence interval,

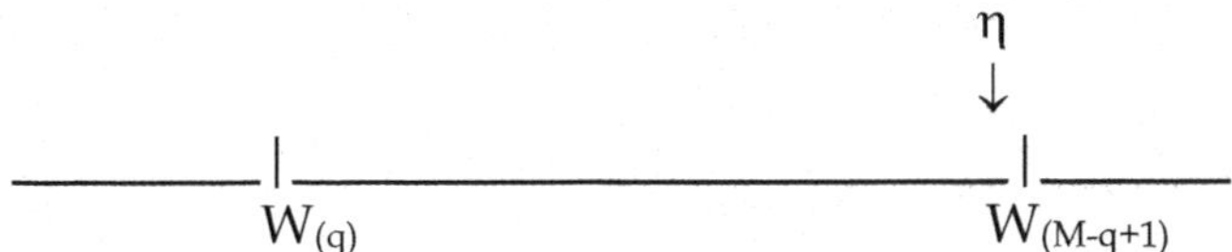

then exactly $M - q$ of the Walsh averages (up to, but not including, $W_{(M-q+1)}$) are less than η and there are q Walsh averages greater than η. Thus

$$\begin{aligned} P\left(W_{(q)} < \eta < W_{(M-q+1)}\right) &= P(\text{the number of Walsh averages greater than } \eta \text{ is} \\ &\quad \text{between } q \text{ and } M-q, \text{ including these values}) \\ &= P\{q \leq W \leq (M-q)\}, \end{aligned} \tag{7.8}$$

since W counts the number of Walsh averages greater than η.

Now, if we choose q so that $P\{W \geq (M-q+1)\} = P\{W \leq (q-1)\} = \frac{(1-CL)}{2}$, then we have

$$CL = P\{q \leq W \leq (M-q)\}. \tag{7.9}$$

From (7.8) and (7.9) it follows that CL is the probability that we will collect a random sample for which the observed interval $\left(W_{(q)}, W_{(M-q+1)}\right)$ contains the unknown value of η. Selecting the integer q so that $P\{q \leq W \leq (M - q)\} = CL$ then leads to a $100CL\%$ confidence interval for the population median η.

Point and Interval Estimation of the Population Median η Under the Assumption of Population Symmetry Let $W_{(1)} \leq \ldots \leq W_{(M)}$, with $M = n(n+1)/2$, be the ordered Walsh averages for a random sample of size n from a population that is symmetric about its median η. Then the point estimator for the population point of symmetry (median) η is the median of the $M = n(n+1)/2$ Walsh averages, . Moreover, for any positive integer $q < (M+1)/2$, the interval $\left(W_{(q)}, W_{(M-q+1)}\right)$ provides a confidence interval for the population point of symmetry η with confidence level CL given by

$$CL = P\{q \leq W \leq (M - q)\}. \tag{7.10}$$

Since the distribution of W is symmetric, Eq. (7.10) is equivalent to using the **R** function *qsignrank*( ) to find the integer q such that $P\{ W \leq q\text{-}1 \} = (1\text{-}CL)/2$. Once this value of q is obtained, the lower and upper endpoints of the 100CL% confidence interval $\left(W_{(q)}, W_{(M-q+1)}\right)$ for η are simply the qth smallest (up from the bottom) and qth largest (down from the top) ordered Walsh averages, respectively. Separate lower and upper confidence bounds for the population point of symmetry η with confidence level

$$(1 - CL) \quad = \quad P\{W \leq q - 1\} \tag{7.11}$$

are given by the qth smallest, $W_{(q)}$, and qth largest, $W_{(M - q + 1)}$, ordered Walsh averages, respectively.

Since the probability expressions in Eqs. (7.10) and (7.11) are valid for any underlying continuous population that is symmetric about its median η, the associated 100CL% confidence interval and bounds are also valid over this extensive class of populations.

Example 7.3. Peptides–to Direct or Not to Direct, That Is the Question Biologically active peptides are chemical substances that are present in all of us to maintain normal host defenses against unwanted microorganisms and to help the body respond to immunologically induced inflammation and tissue damage. In times of good health, the polymorphonuclear leukocytes (PMN) in our blood exhibit random movements. However, C5-derived peptides have the ability to interact with human PMN, causing it to migrate in a directed (non-random) fashion. Such activity has the potential to lead to worsened conditions for people already dealing with other serious diseases. Perez et al. (1983) were interested in whether this C5-directed migration of PMN plays an important role in the origination and development of a potentially lethal, respiratory insufficiency known as

Table 7.4 PMN migration in sera from patients with acute pancreatitis ($\mu m/35\ min$)

Patient number	Migration of PMN
1	75.2
2	109.9
3	119.2
4	80.5
5	114.7
6	103.6
7	75.5
8	108.1

Source: Perez et al. (1983)

'shock lung' in patients with acute pancreatitis. For normal human serum, the migration of PMN, measured in units of $\mu m/35\ min$, is 76.8. Perez *et al* measured the PMN migration in the sera of eight patients with acute pancreatitis and these values are presented in Table 7.4.

PMN migration is clearly a continuous variable. Moreover, from either a stemplot or histogram of the data in Table 7.3 (check this out using the *stem*( ) and *hist*( ) functions in **R**!), it does not seem unreasonable to assume symmetry for the underlying distribution of PMN migration for subjects with acute pancreatitis. Here, the sample size is $n = 8$ and the required number of Walsh averages is $M = 8(9)/2 = 36$. The **R** function *walsh*( ) from the Rfit package can be used with the *pmn_migration* dataset to obtain these Walsh averages. The 36 ordered Walsh averages for the PMN migration data are obtained using the following command and are presented in Table 7.5.

```
> sort(walsh(pmn_migration))
```

The estimate of the median PMN migration rate is $\tilde{w} = \left[w_{(18)} + w_{(19)}\right]/2 = [97.20 + 97.35]/2 = 97.275\mu m/35min.$ Moreover, since increased PMN migration activity has the potential to lead to worsened conditions for people already dealing with other serious diseases, it is natural to be interested in a

Table 7.5 Ordered Walsh averages for the PMN migration data in Table 7.4

75.20	89.40	92.70	97.35	108.10	112.30
75.35	89.55	94.30	97.60	109.00	113.65
75.50	91.65	94.95	99.85	109.15	114.55
77.85	91.80	95.10	103.60	109.90	114.70
78.00	92.05	95.20	105.85	111.40	116.95
80.50	92.55	97.20	106.75	111.40	119.20

lower confidence bound for η, the median PMN migration activity level. To obtain a 98.05% lower confidence bound for η, for example, we first use the **R** function *qsignrank*( ) with arguments $p = 0.0195$ and $n = 8$ to find that $P\{ W \leq 3 \} = .0195$.

```
> qsignrank(p = 0.0195, n = 8)
[1] 3
```

Setting q^*-1 = 3, it follows that $q^* = 4$. Hence, the exact 98.05% lower confidence bound for η corresponds to the 4th smallest ordered Walsh average $w_{(4)} = 77.85\ \mu m/35\ min$. Thus we are 98.05% confident that the median PMN migration rate for patients with acute pancreatitis is at least $77.85\ \mu m/35\ min$.

The **R** function *wilcox.test*( ) can also be used with the *conf.int* argument specified to be *TRUE* and the *conf.level* argument specified to be *0.*9610 in order to obtain the following output, which includes the point estimator and the lower confidence bound for η.

```
> wilcox.test(pmn_migration, conf.int = TRUE, conf.level = 0.9610)

        Wilcoxon signed rank test

data:  pmn_migration
V = 36, p-value = 0.007813
alternative hypothesis: true location is not equal to 0
96.1 percent confidence interval:
  77.85 114.55
sample estimates:
(pseudo)median
        97.275
```

Notice that the 98.05% lower confidence bound for η corresponds to the lower endpoint of a 96.10% confidence interval for η in this **R** output. (Do you see why the symmetry of the distribution of W causes this to be the case?)

Hypothesis Tests About the Population Median η Under the Assumption of Population Symmetry—Signed Rank Test Up to this point in Sect. 2 we have been working entirely with the Walsh averages to obtain point and interval estimates of the median η under the more informative assumption that the underlying distribution is symmetric. We now turn our attention to the problem of hypothesis testing for the median under this additional assumption. For this purpose, we return to consideration of the sample values $X_1, \ldots, X_n$ themselves, rather than the Walsh averages.

In Sect. 1 under the minimal assumption about the underlying distribution, we used the signs of the differences $D_1 = X_1 - \eta_0, \ldots, D_n = X_n - \eta_0$ to test the null hypothesis H_0: $\eta = \eta_0$. However, when we also know that the underlying distribution is symmetric about its median η, we can use more than just the signs of the D's. In fact, the relative distances given by the absolute values of the D's, namely, $|D_1|, \ldots, |D_n|$, provide us with additional information about the plausibility of the null hypothesis H_0. The relative magnitudes of these distances are captured in their ordered ranks. Let R_i denote the rank, from least to greatest, of $|D_i|$ among $|D_1|, \ldots, |D_n|$, for $i = 1, \ldots, n$.[1] Thus the greatest weight (n) is assigned to the observation that is furthest from the hypothesized median η_0 (providing the most evidence against the null hypothesis) and the least weight (1) is assigned to the observation that is closest to η_0 (providing the least evidence against the null

[1] (If there are ties among the $|D|$'s, we assign average ranks to the tied values. For example, if $n = 5$ and $D_1 = -4$, $D_2 = 9$, $D_3 = -3$, $D_4 = 4$, and $D_5 = 6$, the ordered $|D|$ values are 3, 4, 4, 6, 9. The average rank of $\frac{2+3}{2} = 2.5$ is assigned to each of the tied absolute values $|D_1| = |D_4| = 4$. The complete set of ranks for $(|D_1|, \ldots, |D_5|)$ is then $(R_1, \ldots, R_5) = (2.5, 5, 1, 2.5, 4)$.)

hypothesis). The signed rank test statistic, W^+, then adds these absolute value ranks for those observations that are greater than η_0.

Definition 7.2. The *signed rank statistic,* W^+, is the sum of the absolute value ranks for those observations that are greater than the hypothesized median η_0; that is,

$$W^+ = \sum_{\{X_i's>\eta_0\}} R_i = \sum_{\{d_i>0\}} R_i. \tag{7.12}$$

Example 7.4. Computation of the Signed Rank Statistic, W^+ We illustrate the computation of the signed rank statistic W^+ (7.12) for the PMN migration data in Table 7.4. Here $n = 8$ and we take $\eta_0 = 76.8\ \mu m/35\ min$ to be the migration rate for PMN for healthy human serum. The sample $|d| = |x - 76.8|$ values, their signs, and their absolute value ranks are presented in Table 7.6. Summing the $|d|$ ranks for those x's that are greater than 76.8 (i. e., for those x's with positive d's), we obtain the observed value of the signed rank statistic to be $w^+ = [6 + 8 + 3 + 7 + 4 + 5] = 33$ for these PMN migration data. (Once again, we use the upper case W^+ to denote the computational form of the

Table 7.6 Computation of the signed rank statistic, W^+, for the PMN migration data in Table 7.4

Patient i	x_i	$d_i = x_i - 76.8$	$\lvert d_i \rvert$	sign of d_i	R_i, rank of $\lvert d_i \rvert$
1	75.2	−1.6	1.6	−	2
2	109.9	33.1	33.1	+	6
3	119.2	42.4	42.4	+	8
4	80.5	3.7	3.7	+	3
5	114.7	37.9	37.9	+	7
6	103.6	26.8	26.8	+	4
7	75.5	−1.3	1.3	−	1
8	108.1	31.3	31.3	+	5

signed rank statistic and the lower case w^+ to represent the observed value of the statistic for a particular set of sample data.)

Now we must address the issue of how to use the signed rank statistic to construct appropriate tests of the null hypothesis H_0: $[\eta = \eta_0]$. If H_0 is true, we would expect about one-half of the sample observations to be greater than η_0 and the other one-half to be less than η_0. Moreover, we would expect the absolute value ranks 1, ..., n to be roughly evenly distributed across the positively and negatively signed differences. However, when H_0 is not true and η is greater than the hypothesized median η_0, we would expect more than one-half of the observations to be greater than η_0 and these observations will be more likely to have the larger absolute value ranks. When η is less than η_0 the opposite is true, as we would expect more than one-half of the observations to be less than η_0 and to have the larger absolute value ranks. These intuitive features of the test statistic W^+ lead to the signed rank procedures for testing H_0: $[\eta = \eta_0]$ under the assumption of population symmetry.

Hypothesis Tests About the Population Median η Under the Assumption of Population Symmetry To test the null hypothesis H_0: $[\eta = \eta_0]$ under the additional information that the underlying population is symmetric, compute the signed rank statistic W^+ (7.12) and let w^+ be the attained value of W^+. The exact P-values for a test of H_0 against the possible alternatives H_A are then:

H_A	P-value	
$\eta > \eta_0$	$P(W^+ \geq w^+)$	(7.13)
$\eta < \eta_0$	$P(W^+ \leq w^+)$	(7.14)
$\eta \neq \eta_0$	$2P(W^+ \geq w^+)$, if $w^+ > \frac{n(n+1)}{4}$	
	$2P(W^+ \leq w^+)$, if $w^+ < \frac{n(n+1)}{4}$	(7.15)
	1, if $w^+ = \frac{n(n+1)}{4}$.	

To compute any of these P-values for a given sample size n and observed value w^+, we need the null sampling distribution of the statistic W^+. When H_0 is true, the symmetry of the underlying population dictates that the individual signs are independent of the corresponding absolute value ranks $R = (R_1, \ldots, R_n)$. Moreover, under H_0, the signs are independent Bernoulli variables with probability 1/2 that any individual observation exceeds η_0. These two facts lead directly to the null sampling distribution of the signed rank statistic W^+ (7.12). (See Exercise 7.A.7 for more details.) To compute the probability $P(W^+ \leq w^+ \mid \eta = \eta_0)$, we use the **R** function *wilcox.test*( ).

Example 7.5. PMN Migration Rates In their study of the PMN migration rate for patients with acute pancreatitis, Perez et al. noted that the migration rate for PMN for healthy human serum is 76.8 $\mu m/35\ min$. That naturally raises the question of whether the evidence collected in their sample from patients with acute pancreatitis warrants describing the median PMN migration rate for such patients as being higher than this norm. If η is the median PMN migration rate for patients with acute pancreatitis, then H_0: $\eta = 76.8$ is the appropriate null hypothesis and H_A: $\eta > 76.8$ is the conjectured one-sided alternative. We can obtain the value of the signed rank statistic for the data in this study, which is $w^+ = 33$ (as also found in Example 7.4), and the P-value for this test, which is $P(W^+ \geq 33 \mid \eta = \eta_0) = .01953$, using the following **R** command.

```
> wilcox.test(pmn_migration, mu = 76.8, alternative = "greater")

        Wilcoxon signed rank test

data:  pmn_migration
V = 33, p-value = 0.01953
alternative hypothesis: true location is greater than 76.8
```

Summarizing the **R** output, we have:

$$\begin{array}{ll} \text{Null Hypothesis:} & H_0\text{: } \eta = 76.8 \\ \text{Alternative Hypothesis:} & H_A\text{: } \eta > 76.8 \\ \text{Observed Value of } W^+\text{:} & w^+ = 33 \\ P\text{-value:} & .020 \end{array} \tag{7.16}$$

Thus we would reject H_0 for any significance level greater than or equal to .020, suggesting that the median PMN migration rate for patients with acute pancreatitis is, indeed, greater than the healthy norm of 76.8 $\mu m/35\ min$. In addition, since 6 of the 8 signs in Table 7.5 are +, our sample estimate is that 6/8, or 75%, of all patients with acute pancreatitis will have a PMN migration rate greater than 76.8 $\mu m/35\ min$.

Section 7.2 Practice Exercises

7.2.1. Let { 2, 3.2, 6.8, 10.6, 12 } be a random sample of size $n = 5$ from a symmetric population.

(a) How many Walsh averages are there for this data collection?

(b) Compute the Walsh averages for this data collection and order them from least to greatest.

(c) What is the median of the set of Walsh averages computed in part (b)?

7.2.2. Let { -1.3, 0, –4.4, 5.7, 6.4, 10 } be a random sample of size $n = 6$ from a symmetric population.

(a) How many Walsh averages are there for this data collection?

(b) Compute the Walsh averages for this data collection and order them from least to greatest.

(c) What is the median of the set of Walsh averages computed in part (b)?

7.2.3. Let $X_1, \ldots, X_8$ be a random sample of size $n = 8$ from a continuous population that is symmetric about its median η. Let W = [number of Walsh averages for this sample that are greater than η]. Use the **R** functions *psignrank*

( ) and *qsignrank*( ) to answer the following questions about the probability distribution of W.

(a) What are the possible values for W?

(b) Find $P(W \geq 28)$.

(c) Find $P(W \leq 25)$.

(d) Use the symmetry of the distribution of W to find $P(W \leq 8)$.

7.2.4. Let $X_1, \ldots, X_{11}$ be a random sample of size $n = 11$ from a continuous population that is symmetric about its median η. Let W = [number of Walsh averages for this sample that are greater than η]. Use the **R** functions *psignrank* ( ) and *qsignrank*( ) to answer the following questions about the probability distribution of W.

(a) What are the possible values for W?

(b) Find $P(W \geq 50)$.

(c) Find $P(W \leq 41)$.

(d) Use the symmetry of the distribution of W to find $P(W \leq 21)$.

7.2.5. Let $X_1, \ldots, X_6$ be a random sample of size $n = 6$ from a continuous population that is symmetric about its median η. Let W = [number of Walsh averages for this sample that are greater than η]. For any such continuous population, the probability distribution for W is given by:

Value of w	**$P(W = w)$**	**Value of w**	**$P(W = w)$**
0	.016	11	.078
1	.015	12	.078
2	.016	13	.063
3	.031	14	.062
4	.031	15	.063
5	.047	16	.047
6	.063	17	.031
7	.062	18	.031
8	.063	19	.016
9	.078	20	.015
10	.078	21	.016

(a) Use an appropriate graphical technique of your choice to show that the probability distribution of W is symmetric.

(b) Show numerically that the probability distribution of W is symmetric by matching up pairs of possible values of W.

(c) What is the point of symmetry (i.e. median) for the probability distribution of W?

(d) Find $P(W \geq 14)$ directly from the probabilities given above.

(e) Find $P(W \leq 7)$ directly from the probabilities given above.

(f) Use the result in part (d) and the symmetry of the distribution of W to verify the value for $P(W \leq 7)$ obtained directly in part (e).

7.2.6. Let $X_1, \ldots, X_4$ be a random sample of size $n = 4$ from a continuous population that is symmetric about its median η. Let W = [number of Walsh averages for this sample that are greater than η]. For any such continuous population, the probability distribution for W is given by:

Value of w	$P(W = w)$	Value of w	$P(W = w)$
0	.062	6	.126
1	.063	7	.124
2	.063	8	.063
3	.124	9	.063
4	.126	10	.062
5	.124		

(a) Use an appropriate graphical technique of your choice to show that the probability distribution of W is symmetric.

(b) Show numerically that the probability distribution of W is symmetric by matching up pairs of possible values of W.

(c) What is the point of symmetry (i.e., median) for the probability distribution of W?

(d) Find $P(W \geq 8)$ directly from the probabilities given above.

(e) Find $P(W \leq 2)$ directly from the probabilities given above.

(f) Use the result in part (d) and the symmetry of the distribution of W to verify the value for $P(W \leq 2)$ obtained directly in part (e).

7.2.7. Let $X_1, \ldots, X_4$ be a random sample of size $n = 4$ from a continuous population that is symmetric about its median η and let $W_{(1)} \leq \ldots \leq W_{(10)}$ be the ordered Walsh averages for this sample.

(a) List all possible confidence intervals for η that could be constructed from these Walsh averages.

(b) What are the exact confidence levels associated with these confidence intervals?

7.2.8. Let $X_1, \ldots, X_7$ be a random sample of size $n = 7$ from a continuous population that is symmetric about its median η and let $W_{(1)} \leq \ldots \leq W_{(28)}$ be the ordered Walsh averages for this sample.

(a) List all possible confidence intervals for η that could be constructed from these Walsh averages.

(b) What are the exact confidence levels associated with these confidence intervals?

7.2.9. Let $X_1, \ldots, X_6$ be a random sample of size $n = 6$ from a continuous population that is symmetric about its median η and let $W_{(1)} \leq \ldots \leq W_{(21)}$ be the ordered Walsh averages for this sample.

(a) List all possible upper confidence bounds for η with confidence levels CL at least .50 that could be constructed from these Walsh averages.

(b) What are the exact confidence levels associated with these upper confidence bounds?

7.2.10. Let $X_1, \ldots, X_7$ be a random sample of size $n = 7$ from a continuous population that is symmetric about its median η and let $W_{(1)} \leq \ldots \leq W_{(28)}$ be the ordered Walsh averages for this sample.

(a) List all possible lower confidence bounds for η with confidence levels CL at least .50 that could be constructed from these Walsh averages.

(b) What are the exact confidence levels associated with these lower confidence bounds?

(c) Compare these lower confidence bounds with the lower endpoints of the corresponding confidence intervals obtained in Exercise 7.2.8.

7.2.11. Let {-4.5, -3.6, 0.3, 2, 3.2, 6.8, -5.5, 10.6, 12, 2.2} be a random sample of size $n = 10$ from a population that is symmetric about η.

(a) Find the value, w^+, of the signed rank statistic W^+ (7.12) for testing H_0: $\eta = 0$.

(b) If $\eta = 0$, use the **R** function *wilcox.test*( ) to compute $P(W^+ \geq w^+)$ and $P(W^+ \leq w^+)$.

7.2.12. Let {4.4, 1.3, 9.9, 18, 6.7, 12, 37.8, -2.0, 0.4} be a random sample of size $n = 9$ from a population that is symmetric about η.

(a) Find the value, w^+, of the signed rank statistic W^+(7.12) for testing H_0: $\eta = 7$.

(b) If $\eta = 7$, use the **R** function *wilcox.test*( ) to compute $P(W^+ \geq w^+)$ and $P(W^+ \leq w^+)$.

7.2.13. Let {4.4, 1.3, 9.9, 18, 6.7, 12, 37.8, -2.0, 0.4} be a random sample of size $n = 9$ from a population that is symmetric about η.

(a) Find a 92.6% confidence interval for η.

(b) Find a 98% lower confidence bound for η.

7.2.14. Let {-4.5, -3.6, 0.3, 2, 3.2, 6.8, -5.5, 10.6, 12, 2.2} be a random sample of size $n = 10$ from a population that is symmetric about η.

(a) Find a 95.2% confidence interval for η.

(b) Find a 93.5% upper confidence bound for η.

7.2.15. *Fir Trees and Fish Habitat.* Kayle (1984) studied the mean interstitial length for a variety of fir trees, with an eye on their potential use in habitat modification for fish populations. The mean interstitial lengths for 12 blue spruce trees are presented in Table 1.24. Viewing these data as a random sample of size $n = 12$ from the population of all blue spruce trees, complete the following statistical analyses under the assumption that this population is symmetrically distributed about its unknown median η.

(a) Obtain a point estimate of η and find a 97.4% upper confidence bound for η.

(b) Find the P-value for a test of the null hypothesis H_0: $\eta = 75\ mm$ versus the one-sided alternative H_A: $\eta < 75\ mm$.

(c) Are you comfortable with the assumption that mean interstitial length for the population of blue spruce trees is symmetrically distributed about its median η? Why or why not?

7.2.16. *Lead-poisoned Geese.* March et al. (1976) studied the plasma glucose levels for healthy Canadian geese. The plasma glucose levels for eight healthy Canadian geese are presented in Table 1.30. Viewing these data as a random sample of size $n = 8$ from the population of all healthy Canadian geese, complete the following statistical analyses under the assumption that this population is symmetrically distributed about its unknown median η.

(a) Obtain a point estimate of η and find a 97.4% upper confidence bound for η.

(b) Find the P-value for a test of the null hypothesis H_0: $\eta = 275\ mg/100\ ml$ plasma versus the one-sided alternative H_A: $\eta < 275\ mg/100\ ml$ plasma.

(c) Are you comfortable with the assumption that the plasma glucose level for the population of healthy Canadian geese is symmetrically distributed about its median η? Why or why not?

7.2.17. *Zooplankton on South Bass Island, Ohio.* Terwilliger's Pond is a shallow embayment of Lake Erie (not a true pond) on South Bass Island, Ohio,

Table 7.7 Numbers per liter of the rotifer *Keratella cochlearis* in twenty-one samples from Terwilliger's Pond

13	56	14	7	116	5	28	30	18	69	263
34	199	12	150	15	30	313	42	199	12	

Source: Hines (1999)

connecting to the open lake through a channel that is approximately twelve feet wide. Hines (1999) was interested in the species composition of zooplankton during the summer months (July and August) in Terwilliger's Pond. Using plankton traps, Hines collected seven samples from each of three different locations in the embayment, and recorded the average numbers per liter of water of different plankton species for each of the twenty-one samples. The specific counts for the rotifer *Keratella cochlearis* for the twenty-one samples are given in Table 7.7. Viewing these data as a random sample of size $n = 21$ from Terwilliger's Pond, complete the following statistical analyses under the assumption that the population of *Keratella cochlearis* densities in Terwilliger's Pond in the summer months is symmetrically distributed about its unknown median η.

(a) Obtain a point estimate of η and find a 95.4% confidence interval for η.

(b) Find the P-value for a test of the null hypothesis H_0: $\eta = 50$ organisms per liter versus the two-sided alternative H_A: $\eta \neq 50$ organisms per liter.

(c) Are you comfortable with the assumption that the population of *Keratella cochlearis* densities in Terwilliger's Pond in the summer months is symmetrically distributed about its unknown median η. Why or why not?

7.3 Inference for the Center of a Normal Distribution–Procedures Associated with the Sample Mean and Sample Standard Deviation

If conditions are such that a continuous measurement of interest is known to have an underlying normal distribution, then we can construct more effective inference procedures that utilize this additional information. In such settings, the center of the population is represented equally well by either the median, η, or the mean, μ, since these two features coincide for all symmetric distributions, which includes normal distributions. However, our previous discussion in Chaps. 4 and 5 relied exclusively on the common statistical practice of taking the mean μ to be the descriptor to represent the center of a normal distribution. We follow the same approach in this section to describe inferential procedures for the center of a normal population.

***t*-Distribution** When we have a simple random sample $X_1, \ldots, X_n$ from a $N(\mu, \sigma)$ population where both μ and σ are unknown, the sample information about the center and variability for the distribution is best summarized by the sample mean $\bar{X}$ and sample standard deviation S, respectively. In such settings, it is both natural and optimal to base our inferences about the center of the distribution on these two statistics.

For such normally distributed data, the optimal point estimator for μ is quite naturally the sample mean, $\hat{\mu} = \bar{X}$. To construct confidence intervals and bounds for μ, however, we also need to incorporate the variability associated with this point estimator through its sampling distribution. We have previously seen in Sect. 5.1 that for a normally distributed random sample the probability distribution of $\bar{X}$ is $N(\mu, \sigma/\sqrt{n})$, which depends on the unknown σ. The standardized sample mean

$$Z = \frac{\bar{X} - \mu}{\sigma/\sqrt{n}}$$

has a standard $N(0,1)$ distribution. However, using the sample standard deviation S to estimate σ in this expression adds additional variability to the process and the resulting variable no longer has a distribution that is $N(0, 1)$. Fortunately, the impact on the sampling distribution of the standardized $\bar{X}$ that results from estimating σ by S is easily quantified, leading to another well-known distribution, called a *t-distribution.*

Definition 7.3. The class of *t*-distributions. Suppose that $X_1, ..., X_n$ is a random sample from a $N(\mu, \sigma)$ population and let $\bar{X}$ and S denote the sample mean and standard deviation, respectively. Then the variable

$$T = \frac{\bar{X} - \mu}{S/\sqrt{n}} \tag{7.17}$$

has a ***t*-distribution with *n* - 1 degrees of freedom** and we write $T \sim t(n\text{-}1)$.

Like the class of normal distributions, which are indexed by values of the mean μ and standard deviation σ, there is an entire collection of *t*-distributions indexed by a single parameter, its *degrees of freedom.* This name for this indexing parameter for the collection of *t*-distributions comes quite naturally from applied situations where such distributions often arise as the appropriate sampling distributions. In the construction of the variable T, for example, we compute the sample mean $\bar{X}$ from the n independent pieces of information initially provided by our random sample $X_1, ..., X_n$. Thus we begin with n 'degrees of freedom', corresponding to these n independent pieces of information. However, we spend one of these degrees of freedom to compensate for the fact that we do not know the population standard deviation σ and must use the sample standard deviation S to estimate it, resulting in n-1 remaining degrees of freedom from our data in the computation of the T (7.17) variable. Thus, while we are able to use n independent sample observations $X_1, \ldots, X_n$ to compute the sample mean, the loss of one degree of freedom is the price we have to pay for the need to estimate the unknown standard deviation σ by the sample standard deviation S.

The general shapes of the density curves for the t-distributions are similar in many ways to that of the standard normal distribution. For example, all of the t-distributions are symmetric about 0 and bell-shaped. However, the t-distributions have variances that are greater than one (the variance of the standard normal distribution) and slightly greater probabilities associated with extreme values (i. e., heavier tails) than does the standard normal distribution. In the case of T (7.17), this increased probability in the tails of the $t(n\text{-}1)$ distribution is a direct result of the extra variability introduced through the use of S to estimate the unknown σ. As we encounter other variables with t-distributions later in this course, there will be similar reasons for the increased tail probabilities associated with their t-distributions.

The class of t-distributions also has another very important link to the standard normal distribution. As the degrees of freedom, n-1, for a t-distribution increases, the $t(n\text{-}1)$ density curve approaches that of the $N(0, 1)$ distribution! This is actually not surprising, since an increase in the degrees of freedom is directly associated with an increase in the sample size n and, as the sample size increases, S will provide a more accurate estimate for σ. Thus, as the sample size increases the variable T will behave more and more like the variable $Z = (\bar{X} - \mu)/(\sigma/\sqrt{n})$, which we know has a $N(0, 1)$ distribution.

In Fig. 7.1 we present the density curves for t-distributions with 4 and 10 degrees of freedom and the $N(0, 1)$ density curve. While all three are clearly similar in shape (i.e., bell-shaped and symmetric about the common center 0),

Fig. 7.1 Density curves for the *t*-distribution with 4 degrees of freedom, the *t*-distribution with 10 degrees of freedom, and the N(0,1) Distribution

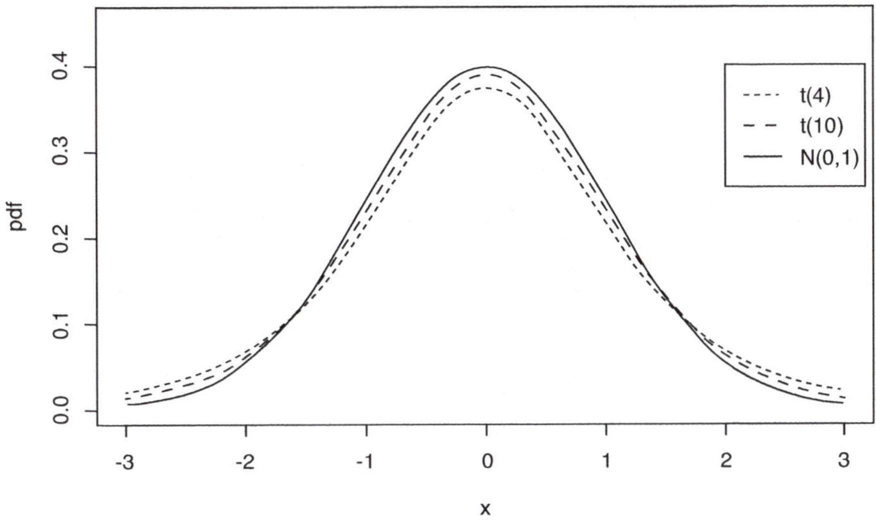

the additional probability in the tails of both of the *t*-distributions is apparent. It is also clear from the figures that this difference narrows as the degrees of freedom for the *t*-distribution increases.

Upper tail probabilities for a specific *t*-distribution can be obtained using the **R** function *pt*( ). For example, we can get the probability that an observation from the *t*-distribution with 12 degrees of freedom will exceed 1.782 as follows.

```
> pt(q = 1.782, df = 12, lower.tail = FALSE)
[1] 0.05002442
```

Also, we can find that 1.782 is the upper 5th percentile for this distribution using the **R** function *qt*( ). We can write this compactly as $t_{12,.05} = 1.782$.

```
> qt(p = 0.05, df = 12, lower.tail = FALSE)
[1] 1.782288
```

Armed with the fact that T (7.17) has a $t(n\text{-}1)$ probability distribution and our new knowledge about this class of *t*-distributions, we now proceed with the discussion concerning statistical inference about the center (mean) of a normal population. Interval estimates for μ are centered at the point estimator $\bar{X}$ and the width of such an interval is determined by both the estimated standard error $\frac{S}{\sqrt{n}}$ for $\bar{X}$ and the confidence level CL through the appropriate percentiles of the $t(n\text{-}1)$ distribution. To construct a 100CL% confidence interval for μ, we use the upper $\frac{(1-CL)}{2}th$ percentile for the t-distribution with n-1 degrees of freedom, namely, $t_{n-1,\frac{(1-CL)}{2}}$. For example, for confidence level $CL = .90$, we have $\frac{(1-CL)}{2} = \frac{.10}{2} = .05$ and we would use the percentile $t_{n-1,.05}$, corresponding to n-1 degrees of freedom.

We now describe the procedure for building a confidence interval for a normal population mean μ with a given confidence level CL. For sake of illustration, suppose we have sample size $n = 13$ and that our desired confidence level is $CL = .90$. There are then $d = (n-1) = 12$ degrees of freedom and from the $qt(\,)$ command on the previous page, we have $t_{12,.05} = 1.782$.

From the fact that the $t(12)$ distribution is symmetric about 0 (see Fig. 7.2), it follows that

$$P\left(-1.782 < \frac{\bar{X}-\mu}{S/\sqrt{n}} < 1.782\right) = 1 - (.05 + .05) = .90. \tag{7.18}$$

This leads directly to the probability statement

$$P\left(\bar{X} - 1.782\frac{S}{\sqrt{n}} < \mu < \bar{X} + 1.782\frac{S}{\sqrt{n}}\right) = .90 \tag{7.19}$$

and the desired 90% confidence interval ($\bar{X} - 1.782\frac{S}{\sqrt{n}}$, $\bar{X} + 1.782\frac{S}{\sqrt{n}}$) for the normal population mean μ.

This same approach can be used to obtain the general form for a 100CL% confidence interval for the mean μ of a normal population for an arbitrary sample size n and confidence level CL.

Fig. 7.2. Central $CL = .90$ probability for the $t(12)$ distribution

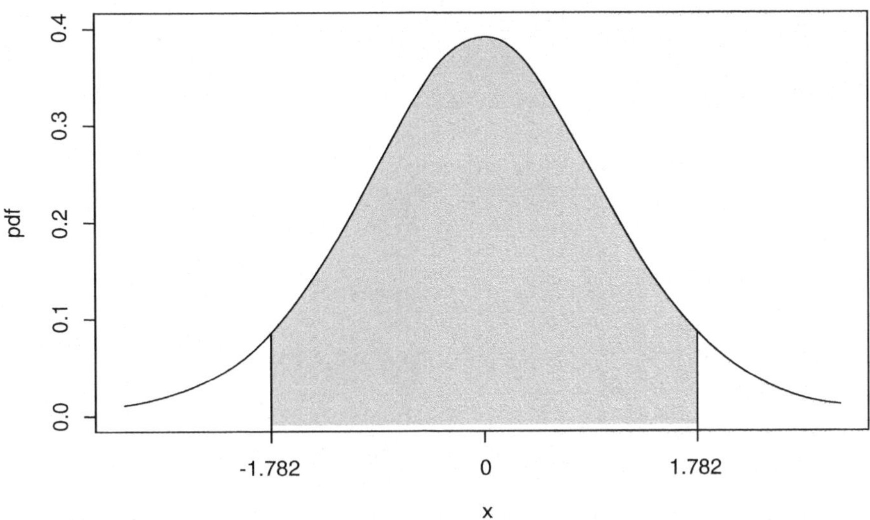

Point and Interval Estimation of the Mean μ of a Normal Population Let $\bar{X}$ and S be the sample mean and sample standard deviation, respectively, for a random sample $X_1, \ldots, X_n$ from a $N(\mu, \sigma)$ population. Then the point estimator for the mean μ is $\bar{X}$. Moreover, an exact 100CL% confidence interval for μ is provided by the interval

$$\left(\bar{X} - t_{n-1,\frac{(1-CL)}{2}}\frac{S}{\sqrt{n}}, \bar{X} + t_{n-1,\frac{(1-CL)}{2}}\frac{S}{\sqrt{n}}\right), \tag{7.20}$$

where $t_{n-1,\frac{(1-CL)}{2}}$ is the upper $\left(\frac{1-CL}{2}\right)^{\text{th}}$ percentile for the t-distribution with n-1 degrees of freedom. The corresponding 100CL% lower and upper confidence bounds for μ are given by $\bar{X} - t_{n-1,(1-CL)}\frac{S}{\sqrt{n}}$ and $\bar{X} + t_{n-1,(1-CL)}\frac{S}{\sqrt{n}}$, respectively.

For example, suppose a sample of size $n = 17$ from a normal population yields a sample mean $\bar{x} = 8.2$ and sample standard deviation $s = 1.5$. With confidence level $CL = .95$ and degrees of freedom $n - 1 = 16$, we use the following **R** command to find that $t_{n-1,\frac{(1-CL)}{2}} = t_{16,.025} = 2.120$.

```
> qt(0.025, 16, lower.tail = FALSE)
[1] 2.119905
```

From (7.20) it follows that the 95% confidence interval for the mean μ is given by

$$\left(8.2 - 2.12\left(\frac{1.5}{\sqrt{17}}\right), 8.2 + 2.12\left(\frac{1.5}{\sqrt{17}}\right)\right) = (7.429, 8.971).$$

On the other hand, the 95% upper confidence bound for μ for these data would be $8.2 + 1.746\left(\frac{1.5}{\sqrt{17}}\right) = 8.835$. Notice that the upper confidence bound provides a sharper statement about the value of μ than does the confidence interval with the same confidence level $CL = .95$. This is always the case for any data collection and any confidence level. (See Exercise 7.A.9.)

Example 7.6. Stature of Prehistoric Ohio Hopewell Amerindians Skeletal analyses are important tools for the study of prehistoric populations. The stature of an individual is commonly understood to be a result of many things, including diet, diseases, environmental factors, and genetic makeup. In prehistoric times a number of different stresses would surely also have had their impact on the stature of the existing populations. Reliance on animal proteins must have led to vitamin deficiencies and even those populations that had developed agricultural features had to deal with droughts, famines, and crop diseases.

Sciulli and Carlisle (1975) studied a number of prehistoric Amerindian populations living in the area of the Ohio Valley over the years from 200 BC to 1200 AD (known as the Woodland period). In particular, the Ohio Hopewell period (approximately 200-400 AD) is known to have involved a number of changes in both settlement and living patterns, including a substantial increase in the use of cultivated plants as sources of food. Hence, the stature of individuals living during this period of time in the Ohio Valley relative to

Table 7.8. Stature (centimeters) of twenty female Amerindians from the Turner site in Cincinnati, Ohio, representing the Ohio Hopewell period (200-400 AD)

150.8	150.5	149.3	150.2
154.7	153.2	154.8	153.9
137.5	152.2	155.3	154.8
149.0	147.3	152.0	155.4
149.8	144.2	155.3	150.3

Source: Sciulli and Carlisle (1975)

similar groups living in the area in previous time periods should provide some idea of the effect that this increased plant cultivation had on the populations. Scuilli and Carlisle measured the stature (in centimeters) of 20 female Amerindians from this Ohio Hopewell period recovered from the Turner site in Cincinnati, Ohio. The heights for these 20 skeletons are reported in Table 7.8.

One of the conditions affecting the appropriateness of the application of the t-confidence interval in (7.20) to a data collection is whether the underlying population from which the sample was collected is normally distributed. There are several graphical ways to check the reasonableness of this assumption, including the previously discussed stemplots, histograms, and normal probability plots. These graphical displays can be obtained for a data collection using any of a number of different statistical software packages. While such displays do not provide a definitive verification of this condition, they can be used to point out possible areas of concern regarding normality of the underlying population.

In Fig. 7.3 we show the normal probability plot for the female Hopewell Amerindian stature data (available as the **R** dataset *female_amerindians*) in Table 7.8. There is obviously some departure from the straight line that we should expect for this plot if the underlying population of female Hopewell Amerindian heights is normally distributed. For instance, the one unusually small height in the data collection appears clearly as an outlier on the plot and

Fig. 7.3. Normal probability plot for the data in Table 7.8

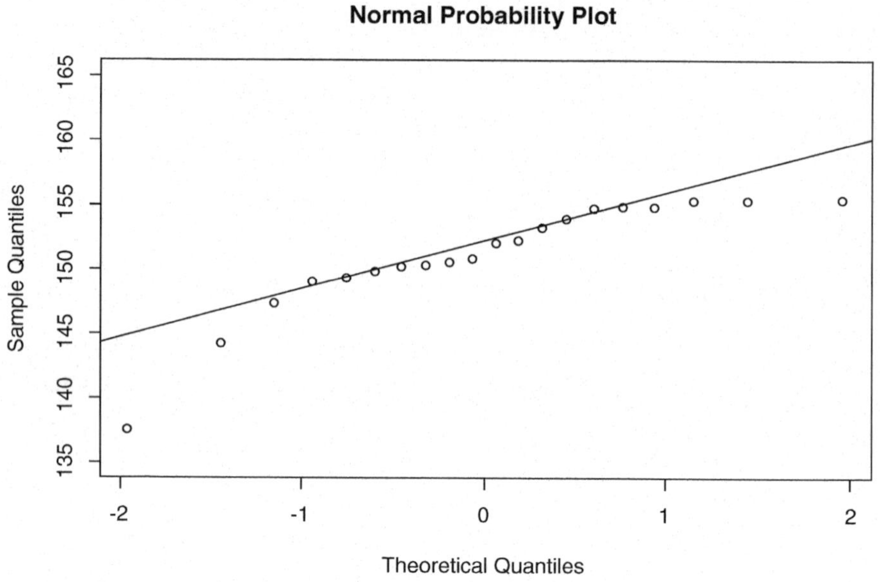

the largest data values deviate quite a bit from what should be expected of normally distributed data. Is this sufficient deviation from the straight line to question the normality assumption? Will statistical analyses based on the normality assumption lead to invalid conclusions?

For now, we assume that the observed signs of non-normality for the female Hopewell Amerindian stature data are not serious enough to invalidate the application of the procedures discussed in this section and we apply the t-confidence interval procedure in (7.20) to the data collection in Table 7.8. However, we return to the issue of normality of this data collection in Sect. 4.

Here the sample size is $n = 20$ and the sample mean and sample standard deviation are $\bar{x} = 151.025$ and $s = 4.392$, respectively. With degrees of freedom $d = 20\text{-}1 = 19$, we find, using the **R** function $qt(\ )$, that $t_{19,.025} = 2.093$. Hence we estimate the mean stature of all female Amerindians living in the Ohio Valley during the Ohio Hopewell Period to be $\bar{x} = 151.025\ cm$ and the 95% confidence interval for μ is

$$\bar{x} \pm t_{19,.025}\frac{s}{\sqrt{20}} = 151.025 \pm 2.093\left(\frac{4.392}{\sqrt{20}}\right)$$
$$= 151.025 \pm 2.055 = (148.97, 153.08)cm.$$

Thus we are 95% confident that the mean height for all female Amerindians living in the Ohio Valley during the Ohio Hopewell period is between 148.97 and 153.08 centimeters. (If you are not comfortable with heights measured in centimeters, the range is from about 4 feet $10\frac{1}{2}$ inches to 5 feet $\frac{1}{4}$ inch.)

In addition to supplying the appropriate t-percentile for these confidence intervals, **R** can also be used to obtain these confidence intervals for the mean of a normal population. For example, for the Amerindian stature data in Table 7.8, we can use the *t.test*( ) function to provide the following output, which includes the point estimator and the 95% confidence interval for μ.

```
> t.test(female_amerindians)

        One Sample t-test

data:  female_amerindians
t = 153.78, df = 19, p-value < 2.2e-16
alternative hypothesis: true mean is not equal to 0
95 percent confidence interval:
 148.9694 153.0806
sample estimates:
mean of x
  151.025
```

The approach to conducting hypothesis tests about the mean μ for normal populations is very similar to that used in constructing confidence intervals and bounds for μ under normality. The variable T (7.17) again plays the key role in the development.

Hypothesis Tests About the Mean of a Normal Population To test the null hypothesis H_0: $[\mu = \mu_0]$ for a $N(\mu, \sigma)$ population, compute the statistic T(7.17) under the null hypothesis condition that $\mu = \mu_0$, namely,

$$T = \frac{\bar{X} - \mu_0}{S/\sqrt{n}}, \tag{7.21}$$

and let t_{obs} be the attained value of T. The exact P-values for normal populations for a test of H_0 against the possible alternatives H_A are then:

H_A	P-value	
$\mu > \mu_0$	$P(T \geq t_{obs})$	(7.22)
$\mu < \mu_0$	$P(T \leq t_{obs})$	(7.23)
$\mu \neq \mu_0$	$2P(T \geq t_{obs})$, if $t_{obs} \geq 0$ $2P(T \leq t_{obs})$, if $t_{obs} < 0$,	(7.24)

where $T \sim t(n\text{-}1)$.

Example 7.7. Ohio Hopewell Amerindians Sciulli and Carlisle (1975) indicate that the average height of female Amerindians residing in the Ohio Valley over the years from 200-0 BC (Early Woodland period) was 150.30 *cm*. This Early Woodland era was marked by an economy based almost exclusively on hunting, fishing, and gathering, with seasonal occupation of various sites. These differences in economy between the Early Woodland and Ohio Hopewell periods leads to speculation that the development of at least a partially agriculture-based economy during the Ohio Hopewell period would also have led to a more stabilized diet and, therefore, to an increase in average stature for the Amerindians of that period. Do the data presented in Table 7.8 for 20 female Amerindians residing in the Ohio Valley during the Ohio Hopewell period (200-400 AD) provide evidence to support this conjecture?

Letting μ denote the average height for female Amerindians residing in the Ohio Valley during the Ohio Hopewell period, we are interested in testing the null hypothesis H_0: [$\mu = 150.3$] versus the one-sided alternative H_A: [$\mu > 150.3$]. In Example 7.5 we checked that the underlying normality assumption is not unreasonable for the stature data collection in Table 7.8 and we found the sample average and standard deviation for these data to be $\bar{x} = 151.025$ and $s = 4.392$, respectively. Computing the test statistic T (7.21), we see that $t_{obs} = (151.025 - 150.3)/\left(4.392/\sqrt{20}\right) = .738$. Hence from (7.22) the P-value for our test of H_0: [$\mu = 150.3$] against H_A: [$\mu > 150.3$] is found using the *pt*( ) function with n-1 = 19 degrees of freedom to be P-value = $P(T \geq .738) = P(t(19) \geq .738) = .235$.

```
> pt(0.738, df = 19, lower.tail = FALSE)
[1] 0.2347661
```

Thus there is virtually no evidence in the stature data from the Turner site in Cincinnati, Ohio to support the conjecture that the average height for female Amerindians increased from the Early Woodland period (200-0 BC) to the Ohio Hopewell period (200-400 AD).

The **R** function *t.test*( ) can also be used with the *mu* and *alternative* arguments specified to be 150.3 and "greater", respectively, to obtain the following output, which includes the observed value of T and the associated P-value for this Amerindian stature hypothesis test.

```
> t.test(female_amerindians, mu = 150.3, alternative =
"greater")

        One Sample t-test

data:  female_amerindians
t = 0.73821, df = 19, p-value = 0.2347
alternative hypothesis: true mean is greater than 150.3
95 percent confidence interval:
 149.3268      Inf
sample estimates:
mean of x
  151.025
```

We note in passing that the average height of 150.30 *cm* for female Amerindians residing in the Ohio Valley in the Early Woodlands period is itself based on sample information obtained in other previous studies. Thus, there is variability associated with that number as well. In Chap. 9 we will discuss methods for making inferences about two population means (in this example, average heights for female Amerindians from the Early Woodland

and Ohio Hopewell periods) based on random samples from each of the two populations.

Section 7.3 Practice Exercises

7.3.1. Find the upper $\frac{(1-CL)}{2}th$ percentile for the t-distribution with n-1 degrees of freedom, namely, $t_{n-1,\frac{(1-CL)}{2}}$, for the following values of n and CL.

(a) $n = 9, CL = .95$

(b) $n = 20, CL = .975$

(c) $n = 15, CL = .90$

(d) $n = 18, CL = .99$.

7.3.2. Find the designated upper-tail probabilities for the following t-distributions.

(a) $T \sim t(14), P(T \geq 1.761)$

(b) $T \sim t(20), P(T \geq 2.528)$

(c) $T \sim t(4), P(T \geq 2.776)$

(d) $T \sim t(9), P(T \geq 3.250)$.

7.3.3. Use the **R** function *rt*( ) to simulate a random sample of 1000 observations from each of the following t-distributions and graphically display each of the sets of sample observations in a normal probability plot.

(a) $t(4)$

(b) $t(8)$

(c) $t(12)$

(d) $t(18)$

(e) $t(25)$

(f) $t(30)$

(g) $t(40)$

Comment on these normal probability plots in view of the fact that the t-distribution with d degrees of freedom approaches the $N(0, 1)$ distribution as d increases. Based on these plots, how large do you feel the degrees of freedom d needs to be for the $N(0, 1)$ distribution to provide a good approximation for the t-distribution?

7.3.4. Using the **R** function *rnorm*( ), simulate 1000 independent random samples of size $n = 10$ each from the $N(5, 10)$ distribution and compute the value of the T (7.17) variable for each of these random samples. Let $T_{(1)} \leq T_{(2)} \leq \ldots \leq T_{(1000)}$ denote the ordered values of T for the 1000 random samples. Let $t_{9,.90}$, $t_{9,.80}$, $t_{9,.70}$, $t_{9,.60}$, $t_{9,.50}$, $t_{9,.40}$, $t_{9,.30}$, $t_{9,.20}$, and $t_{9,.10}$ be the nine deciles for the $t(9)$ distribution.

(a) Use these deciles as fixed fenceposts to summarize the observed values of T (7.17) for your 1000 samples.

(b) Compute $|T_{(10j)} - t_{9,\,(1\text{-}.1j)}|$, for $j = 1, 2, \ldots, 9$. Comment on the values of the differences.

(c) What do your findings in parts (a) and (b) tell you about the probability distribution of the variable T (7.17) for data arising from the $N(5, 10)$ distribution?

7.3.5. Using the **R** function *rnorm*( ), simulate 1000 independent random samples of size $n = 10$ each from the $N(5, 100)$ distribution and compute the value of the T (7.17) variable for each of these random samples. Let $T_{(1)} \leq T_{(2)} \leq \ldots \leq T_{(1000)}$ denote the ordered values of T for the 1000 random samples. Let $t_{9,.90}$, $t_{9,.80}$, $t_{9,.70}$, $t_{9,.60}$, $t_{9,.50}$, $t_{9,.40}$, $t_{9,.30}$, $t_{9,.20}$, and $t_{9,.10}$ be the nine deciles for the $t(9)$ distribution.

(a) Use these deciles as fixed fenceposts to summarize the observed values of T (7.17) for your 1000 samples.

(b) Compute $|T_{(10j)} - t_{9,\,(1\text{-}.1j)}|$, for $j = 1, 2, \ldots, 9$. Comment on the values of the differences.

(c) What do your findings in parts (a) and (b) tell you about the probability distribution of the variable T (7.17) for data arising from the $N(5, 100)$ distribution?

(d) Compare your results with those obtained in Exercise 7.3.4.

7.3.6. Using the **R** function *rnorm*( ), simulate 1000 independent random samples of size $n = 10$ each from the $N(25, 10)$ distribution and compute the value of the T (7.17) variable for each of these random samples. Let $T_{(1)} \le T_{(2)} \le \ldots \le T_{(1000)}$ denote the ordered values of T for the 1000 random samples. Let $t_{9,.90}$, $t_{9,.80}$, $t_{9,.70}$, $t_{9,.60}$, $t_{9,.50}$, $t_{9,.40}$, $t_{9,.30}$, $t_{9,.20}$, and $t_{9,.10}$ be the nine deciles for the $t(9)$ distribution.

(a) Use these deciles as fixed fenceposts to summarize the observed values of T (7.17) for your 1000 samples.

(b) Compute $| T_{(10j)} - t_{9,\,(1-.1j)} |$, for $j = 1, 2, \ldots, 9$. Comment on the values of the differences.

(c) What do your findings in parts (a) and (b) tell you about the probability distribution of the variable T (7.17) for data arising from the $N(25, 10)$ distribution?

(d) Compare your results with those obtained in Exercise 7.3.4.

7.3.7. Using the **R** function *rnorm*( ), simulate 1000 independent random samples of size $n = 10$ each from the uniform distribution on the interval (-1, 1) and compute the value of the T (7.17) variable for each of these random samples. Let $T_{(1)} \le T_{(2)} \le \ldots \le T_{(1000)}$ denote the ordered values of T for the 1000 random samples. Let $t_{9,.90}$, $t_{9,.80}$, $t_{9,.70}$, $t_{9,.60}$, $t_{9,.50}$, $t_{9,.40}$, $t_{9,.30}$, $t_{9,.20}$, and $t_{9,.10}$ be the nine deciles for the $t(9)$ distribution.

(a) Use these deciles as fixed fenceposts to summarize the observed values of T (7.17) for your 1000 samples.

(b) Compute $|T_{(10j)} - t_{9,\,(1-.1j)}|$, for $j = 1, 2, \ldots, 9$. Comment on the values of the differences.

(c) What do your findings in parts (a) and (b) tell you about the probability distribution of the variable T (7.17) for data arising from this uniform distribution on the interval (-1, 1)?

7.3.8. The following data collection is a random sample of size $n = 50$ from a continuous population:

{14.50, 9.77, 8.18, 11.94, 6.55, 9.62, 7.54, 13.05, 9.27, 8.75, 11.12, 7.84, 8.09, 11.63, 7.91, 5.02, 11.80, 9.84, 9.77, 7.23, 10.17, 14.21, 10.68, 6.23, 14.18, 10.26, 8.65, 10.63, 8.25, 10.78, 8.84, 10.29, 10.37, 11.84, 11.39, 11.77, 13.03, 9.34, 8.80, 13.43, 10.62, 7.25, 8.47, 10.61, 13.21, 11.07, 7.79, 10.57, 10.56, 12.55}.

Use a variety of graphical techniques (e. g., stemplot, histogram, normal probability plot, etc.) to assess whether it is likely that the underlying population from which this sample was collected has a normal distribution.

7.3.9. The following data collection is a random sample of size $n = 100$ from a continuous population:

{4.78, -4.25, 5.92, 5.00, 6.02, 7.08, -3.13, 1.11, 7.52, 2.55, 8.28, 2.14, -0.92, 2.71, 2.99, 1.58, 8.14, 2.10, 5.64, 3.27, -7.69, 11.47, -1.98, 5.27, 2.43, 0.23, -1.24, -4.23, 5.83, 1.87, 6.40, 7.80, -1.14, 0.43, 1.49, 6.64, 4.04, 2.85, 4.71, 4.99, 3.32, 7.15, 4.27, 8.65, -0.09, -2.35, 1.07, 2.45, 3.12, 6.68, 4.74, 6.16, -3.37, 4.15, 5.08, 6.54, 1.02, 5.94, 5.21, 2.81, 3.87, 13.94, 10.56, 2.85, 8.82, -6.50, 2.41, 8.71, 5.82, -0.45, 2.60, 4.12, 1.05, -1.37, 9.54, 7.84, 10.27, 6.04, 2.10, 4.57, 0.08, 4.67, 3.14, 10.94, 3.94, 14.50, 5.66, 1.36, 6.56, 5.60, -1.22, -1.62, 11.92, 10.06, -1.42, 0.81, -2.66, 4.66, 10.32, 9.79}.

Use a variety of graphical techniques (e. g., stemplot, histogram, normal probability plot, etc.) to assess whether it is likely that the underlying population from which this sample was collected has a normal distribution.

7.3.10. The following data collection is a random sample of size $n = 50$ from a continuous population:

{3.10, -1.80, 7.85, 6.85, 1.02, 4.70, 5.71, 0.95, 4.72, -5.28, 2.06, -6.52, -19.07, 4.47, 3.39, 1.50, 1.21, 1.26, -2.01, 1.07, 7.40, 5.31, 1.86, 9.54, 153.97, 2.91, 0.78, 3.49,

2.99, 9.53, 1.64, -0.08, 4.50, 5.35, 3.16, -20.71, -341.50, 4.27, -37.20, 18.91, -15.36, -7.24, 2.67, 146.75, 15.70, -26.88, 23.34, 7.34, 11.24, 2.85}.

Use a variety of graphical techniques (e. g., stemplot, histogram, normal probability plot, etc.) to assess whether it is likely that the underlying population from which this sample was collected has a normal distribution.

7.3.11. The following data collection is a random sample of size $n = 100$ from a continuous population:

{3.13, 12.80, 3.79, 0.77, 6.91, 11.97, 13.02, 1.87, 13.52, 0.54, 6.73, 3.13, 1.12, 8.67, 2.65, 1.06, 10.87, 7.63, 6.18, 2.95, 1.44, 3.64, 0.25, 0.20, 0.17, 1.65, 6.69, 10.51, 2.37, 3.40, 19.35, 0.64, 10.83, 2.38, 1.64, 33.34, 3.55, 2.28, 2.38, 3.84, 6.67, 3.23, 5.48, 4.11, 9.14, 8.78, 17.82, 14.98, 1.38, 5.92, 6.93, 5.15, 0.53, 0.46, 1.41, 17.08, 1.36, 2.27, 3.01, 0.26, 0.99, 2.18, 1.22, 12.68, 8.49, 5.17, 2.15, 3.82, 3.51, 3.16, 4.56, 21.67, 2.75, 2.14, 0.30, 2.09, 0.52, 7.00, 10.58, 7.30, 8.71, 2.84, 15.01, 2.74, 5.76, 3.20, 8.93, 6.00, 1.60, 0.55, 41.49, 1.67, 3.61, 2.36, 1.73, 3.15, 0.39, 0.55, 1.67, 1.64}.

Use a variety of graphical techniques (e. g., stemplot, histogram, normal probability plot, etc.) to assess whether it is likely that the underlying population from which this sample was collected has a normal distribution.

7.3.12. *Asbestos Workers and Lung Function.* Al Jarad et al. (1993) studied the percentage decrease in lung function over roughly a 4-year period for asbestos workers who do not yet have asbestosis. Their results are presented in Table 1.18. Viewing these data as a random sample of size $n = 20$ from the population of all such asbestos workers, complete the following statistical analyses under the assumption that the underlying population is normal with unknown mean μ.

(a) Obtain a point estimate of μ and find a 95% confidence interval for μ.

(b) Find the P-value for a test of the null hypothesis H_0: $\mu = 7\%$ versus the two-sided alternative H_A: $\mu \neq 7\%$.

(c) Are you comfortable with the assumption that the 4-year percentage decrease in lung function for the population of all asbestos workers who do not have asbestosis follows a normal distribution? Why or why not?

7.3.13. *Ultrasound Probes and Bacterial Infections.* Consider the CFU bacterial counts data for ultrasound probes as presented in Table 7.2. Viewing this data collection as a random sample of size $n = 25$ from the population of CFU bacterial counts on all ultrasound probes, complete the following statistical analyses under the assumption that the underlying population is normal with unknown mean μ.

(a) Obtain a point estimate of μ and find a 95.7% confidence interval for μ.

(b) Find the P-value for a test of the null hypothesis H_0: $\mu = 150$ colonies versus the one-sided alternative H_A: $\mu > 150$ colonies.

(c) Compare your findings with those obtained in Exercise 7.1.15 with only the minimal assumption about the underlying population.

(d) Are you comfortable with the assumption that the CFU bacterial counts for all ultrasound probes follows a normal distribution? Why or why not?

7.3.14. *Anesthetics and Treesnakes.* Determination of an appropriate anesthetic for use in short-term medical procedures is an important aspect of the proper care of animals. Anderson (1999) studied the use of *propofol* (a non-barbiturate, substituted isopropyl phenol) in anesthetic induction for brown treesnakes (*Boiga irregularis*). She was interested in the effect that *propofol* has on various cardiac blood gas values, including partial carbon dioxide pressure, partial oxygen pressure, bicarbonate level, and percent hemoglobin saturation. Pre-anesthetic blood gas values were obtained using cardiocentesis for nine brown treesnakes. The snakes were then administered a 5.0 *mg/kg* weight dose of *propofol* by the intracardiac route using an 0.5 *ml* insulin syringe and a 27.5 gauge needle. A second blood sample for blood gas analysis was taken from

Table 7.9 Partial carbon dioxide pressure levels (*mm* Hg) for brown treesnakes before and after administration of 5 *mg/kg Propofol*

Treesnake number	Before *Propofol*	After *Propofol*
1	50.2	39.8
2	43.7	37.7
3	51.5	51.3
4	42.3	46.2
5	38.7	25.3
6	35.2	46.8
7	66.8	58.8
8	35.3	48.5
9	49.5	34.1

Source: N. L. Anderson (1999)

each of the snakes 15 min after the administration of the *propofol*. The pre- and post-anesthetic partial carbon dioxide pressure values for the nine brown treesnakes are given in Table 7.9. Viewing these data as a random sample of size $n = 9$ from the population of all brown treesnakes, complete the following statistical analyses under the assumption that the change in partial carbon dioxide pressure for brown treesnakes due to treatment with a 5.0 mg/kg weight dose of *propofol* follows a normal distribution with unknown mean μ.

(a) Obtain a point estimate of μ and find a 90% confidence interval for μ.

(b) Find the *P*-value for a test of the null hypothesis H_0: $\mu = 0$ *mm* Hg versus the two-sided alternative H_A: $\mu \neq 0$ *mm* Hg. What does the null hypothesis $\mu = 0$ correspond to in relation to the use of this dose of *propofol* as an anesthetic for brown treesnakes?

(c) Are you comfortable with the assumption that the change in partial carbon dioxide pressure for brown treesnakes due to treatment with a 5.0 mg/kg weight dose of *propofol* follows a normal distribution with unknown mean μ? Why or why not?

7.4 Discussion of Methods of Inference for the Center of a Population

In Sects. 1 through 3 we discussed three complete sets of procedures for making statistical inferences about the center of a population. These procedures are differentiated by the assumptions they place on the underlying population from which the sample data arise, ranging from virtually no assumptions about the population for the sign procedures of Sect. 1, an additional assumption of symmetry for the signed rank procedures of Sect. 2, to the very specific normality assumptions for the procedures based on the sample mean and standard deviation discussed in Sect. 3. Clearly, the sign procedures can be applied to virtually any univariate random sample, while judgments must be made before accepting the more restrictive conditions of the signed rank and t procedures. However, if our decision to use a more restrictive procedure is correct (i. e., the underlying population is, as assumed, symmetric or normal), then the appropriate procedures for that set of assumptions will be more efficient and provide better inference for the data (e. g. , estimators with smaller sampling variability, shorter confidence intervals, and more powerful tests that are better at detecting correct alternative hypotheses).

This raises the natural question of how to decide which set of procedures should be used to analyze a specific set of data. While there is no definitive answer to this question, some preliminary evaluation of the sample data can prove quite useful in the selection process. The data themselves should be examined for signs of asymmetry or other non-normalities. One or more of the exploratory data analysis techniques (e. g., dotplots, stemplots, histograms, boxplots, fixed or variable fenceposts) discussed in Chap. 1 as well as the normal probability plot discussed in Chap. 4 would certainly be useful in this regard. If the data are strongly asymmetric, it is always safest to use the sign procedures of Sect. 1. If the asymmetry is only slight but there are unusually large and or small observations (outliers), the signed rank procedures of Sect. 2 are a better choice. Finally, for data that exhibit a roughly bell-shaped appearance with no appreciable evidence of outliers, the t-procedures of Sect. 3 generally provide more efficient inference.

We illustrate this screening process for determining which set of procedures to use with two very specific data sets and then illustrate the consequences of applying each of the three sets of procedures to the data.

First, consider the Amerindian stature data presented in Table 7.8. In Sect. 3 we noted some evidences of non-normality for this data collection in the normal probability plot presented in Fig. 7.3. However, in Examples 7.6 and 7.7 we chose to analyze these data using the t-procedures of Sect. 3. Was this a good choice? To shed additional light on this question of normality, a stemplot and boxplot for these data are displayed in Figs. 7.4 and 7.5, respectively.

It is clear from these two displays that the Amerindian stature data are, indeed, somewhat skewed to the left (toward smaller values) and that the one individual with height 137.5 *cm* might very well have the influence of an outlier. (Similar insights are obtained from looking at a histogram.) This agrees with our earlier observations about the normal probability plot for these data. Are these serious enough to cause concern with our analyses in Examples 7.6 and 7.7 under the assumption of normality? One way to evaluate this is to analyze the same data under both of the other less restrictive sets of assumptions. Using the sign procedures (and the **R** function *SIGN.test*( ) from the BSDA package) of Sect. 1, we obtain the following analogues to the normality-based results in Examples 7.6 and 7.7:

Fig. 7.4 Stemplot of female Ohio Hopewell Amerindian stature data from Table 7.8

137	5
138	
139	
140	
141	
142	
143	
144	2
145	
146	
147	3
148	
149	038
150	2358
151	
152	02
153	29
154	788
155	334

Fig. 7.5 Boxplot of female Ohio Hopewell Amerindian stature data from Table 7.8

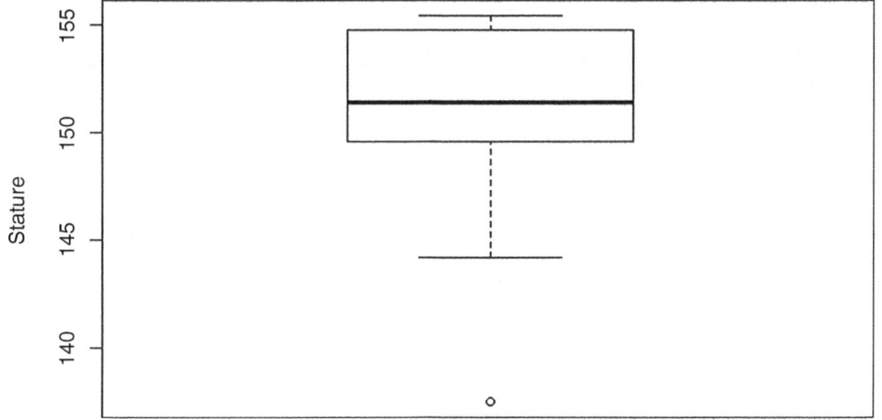

```
> SIGN.test(female_amerindians, md = 150.3, alternative = "greater")

        One-sample Sign-Test

data:  female_amerindians
s = 12, p-value = 0.1796
alternative hypothesis: true median is greater than 150.3
95 percent confidence interval:
 150.1171      Inf
sample estimates:
median of x
      151.4

                    Conf.Level   L.E.pt U.E.pt
Lower Achieved CI       0.9423 150.2000    Inf
Interpolated CI         0.9500 150.1171    Inf
Upper Achieved CI       0.9793 149.8000    Inf
```

Comparison with the results obtained in Examples 7.6 and 7.7 under the normality assumption shows clearly that there is little difference in the statistical inferences from applying either the sign procedures of Sect. 1 or the analogous t-procedures of Sect. 3 to the female Ohio Hopewell Amerindian stature data in Table 7.8. (You are asked in Exercise 7.B.13 to show that this is also the case for the corresponding signed rank procedures of Sect. 2.)

Thus the presence of some skewness and an apparent outlier, as clearly indicated by both the stemplot (Fig. 7.4) and boxplot (Fig. 7.5), in the Ohio Hopewell Amerindian stature data are not serious enough to have a negative effect on the statistical inferences that are obtained from operating under the normality assumption. The important thing to note here is that using either the sign or signed rank procedure for these data provides the same conclusions as those that result from imposing the normality assumption. Thus, in situations where there is some doubt about the normality assumption (as there was with the Ohio Hopewell Amerindian stature data), it is often wise to play it safe by using either the sign or the signed rank procedure. If the effect from the non-normality of the data is minimal (as is apparently the case with the Ohio Hopewell Amerindian stature data), it is not unusual that the inferences will be the same for all three procedures. However, playing it safe by using the sign or signed rank procedure can make a big difference in

Table 7.10 Total engineering drawing hours for seventeen pieces of equipment in class D

16	22	24	26	230	16
24	24	34	18	24	26
34	22	24	26	104	

reaching the proper conclusions for some data sets where the non-normality is severe, as we now demonstrate in our second example.

Consider the total number of engineering drawing hours that contributed to the cost of a particular class of machinery for a major Ohio-based company. A random sample of $n = 17$ pieces of this equipment yielded the total engineering drawing hours presented in Table 7.10.

Looking at any of the evaluative criteria reveals very non-normal features for these drawing hours data. For example, the stemplot and the normal probability plot for the data are presented in Figs. 7.6 and 7.7 and provide clear evidence of non-normality for these data. What effect does such strong non-normality have on the statistical inferences for the sign, signed rank and t-based procedures?

As there is no natural null hypothesis value for these drawing hours data, we consider here only a comparison of the point and interval estimation results under the three sets of assumptions. Employing a confidence coefficient of approximately .95, the three sets of estimation results are as follows:

Minimal Assumptions–Sign Procedures:

Point estimator for the median η of the population: $\tilde{x} = 24$ hours
95.1% confidence interval for η: (22, 26) hours

Symmetry Assumption–Signed Rank Procedures:

Point estimator for the median η of the population: $\tilde{w} = 25$ hours
94.8% confidence interval for η: (22, 34) hours

Fig. 7.6 Stemplot for total engineering drawing hours data in Table 7.10

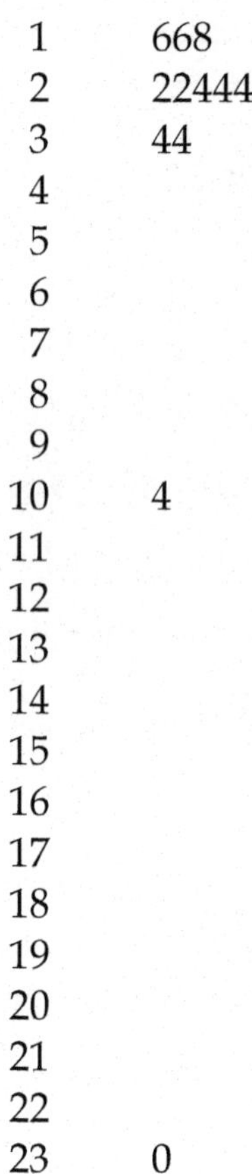

Fig. 7.7 Normal probability plot for total engineering drawing hours data in Table 7.10

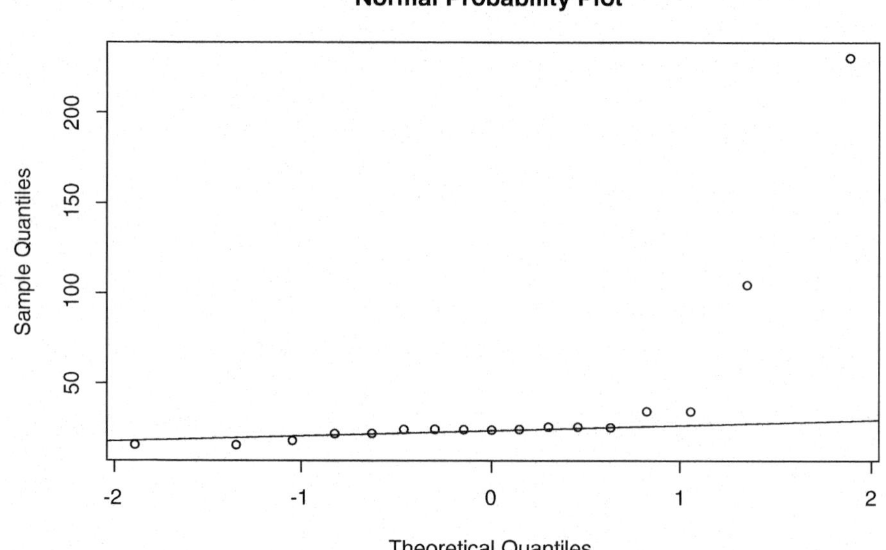

Normality Assumption–t-procedures:

Point estimator for the mean of the population: $\bar{x} = 40.8$ hours
95% confidence interval for μ: (13.7, 67.9) hours.

Since we know that the sample mean and sample median for this data collection are quite different, the symmetry assumption is probably not correct. However, the point and interval estimates obtained under the minimal assumptions and the assumption of symmetry are quite similar. Both of them do a good job of representing the true makeup of the sample data, although the upper endpoint of the confidence interval based on Walsh averages is a bit more influenced by the skewness of the data than is the sign procedure interval. Operating under either of these two sets of assumptions seems quite reasonable for these data.

However, it is also very clear from our calculations that operating under the normality assumption is not a wise idea for this setting. The data are, of course, very non-normal, being skewed right and containing two extreme outliers. However, in this case the non-normal nature of the data (primarily because of the outliers) takes its toll on the t-procedures. The t confidence interval is simply not acceptable, as it is over four times as long as the one based on Walsh averages and over THIRTEEN times as long as the sign confidence interval. Part of this is due to the fact that the t confidence interval is for the mean of the population, while both the sign and signed rank confidence intervals are for the median of the population. However, it is quite clear that the sample mean, $\bar{x} = 40.8$ hours, does a very poor job of representing the visual center (or typical value) of the sample data. All but two of the sample observations are smaller than $\bar{x}$. In addition, the estimate provided by the sample mean is not even considered among the possible values for the center of the population by either of the sign or signed rank 95% confidence intervals!

With these two examples in mind, what then should we do when we want to use a set of sample data to estimate the center of a population? Unfortunately, there is no definitive answer. However, it is clear from these examples that it is an important issue to address in any data analysis. A researcher should give serious thought to the question of which population parameter, the median or the mean, best represents the feature of most interest in the study. If the underlying distribution is thought to be heavily skewed or have a rather high probability of producing outliers, we suggest that the median provides a more realistic picture of the center of such a population. If the underlying distribution is thought to be roughly symmetric with low probability of outliers, then the mean and the median provide similar pictures of the center of the population. In such settings, there is no automatic response as to which measure is preferred. All that we recommend in these cases is that you should not jump blindly into making more assumptions about your underlying population than seems reasonable with the given collection of sample data. Use any number of the exploratory data analysis tools we discussed in Chap. 1 to get a good, clear picture of the structure of your observed data set. We particularly recommend that this snapshot of the data include a normal probability plot, as illustrated in our two examples. (See Exercises 7.4.4 through 7.4.7 for additional guidance in interpreting such normal probability plots.)

If there do not appear to be any obvious non-normalities exhibited in your data set, it is probably reasonable to proceed with inferences under the normality assumption. However, if you have concerns about non-normality with your data, we would recommend that you proceed with your inferences based on the sign or signed rank procedures, with the latter being preferred when the data do not exhibit strong asymmetries.

One good thing to keep in mind through all of this discussion is that the inference procedures developed under the minimal assumptions are valid for virtually all populations, while assuming symmetry restricts you to a smaller set of populations, and assuming normality restricts you to the smallest set. If the data arise from a population that satisfies one set of assumptions, but not the next more stringent set of assumptions, then those procedures based on the most stringent set of assumptions that are appropriate will provide more informative results (e. g., shorter confidence intervals and more powerful tests). Thus, if the data arise from a normal population, the t-based procedures are the approach of choice. However, as we saw with the total engineering drawing hours data, if the three sets of procedures are not in relatively close agreement, then the normality assumption is probably not justified and using the results from one of the other two procedures makes more sense.

Section 7.4 Practice Exercises

7.4.1. *Vertical Cliffs and Ages of Trees.* Consider the age estimate data for a sample of cliff trees in western Europe, as presented in Table 7.3 and discussed in Exercise 7.1.16.

(a) Conduct statistical analyses similar to those of Exercise 7.1.16, but now under the more stringent assumption that the age population for cliff trees in Western Europe has a distribution that is symmetric about its unknown median η.

(b) Conduct statistical analyses similar to those of Exercise 7.1.16 , but now under the assumption that the age population for cliff trees in Western Europe has a normal distribution with unknown mean μ.

(c) Compare and contrast the results from Exercise 7.1.16 with those obtained in parts (a) and (b). Which of the three sets of analyses do you prefer and why?

7.4.2. *Zooplankton on South Bass Island, Ohio.* Consider the rotifer *Keratella cochlearis* density data for Terwilliger's Pond in South Bass Island, Ohio, as presented in Table 7.7 and discussed in Exercise 7.2.17.

(a) Conduct statistical analyses similar to those of Exercise 7.2.17, but now under the less stringent minimal assumption about the population of *Keratella cochlearis* densities in Terwilliger's Pond in the summer months.

(b) Conduct statistical analyses similar to those of Exercise 7.2.17, but now under the assumption that the population of *Keratella cochlearis* densities in Terwilliger's Pond in the summer months follows a normal distribution with unknown mean μ.

(c) Compare and contrast the results from Exercise 7.2.17 with those obtained in parts (a) and (b). Which of the three sets of analyses do you prefer and why?

7.4.3. *Anesthetics and Treesnakes.* Consider the partial carbon dioxide pressure level data for brown treesnakes before and after administration of *propofol*, as presented in Table 7.9 and discussed in Exercise 7.3.14.

(a) Conduct statistical analyses similar to those of Exercise 7.3.14, but now under the least stringent minimal assumption about the distribution of the change in partial carbon dioxide pressure for brown treesnakes after treatment with a 5.0 mg/kg weight dose of *propofol*.

(b) Conduct statistical analyses similar to those of Exercise 7.3.14, but now with only the assumption that the distribution of the change in partial carbon dioxide pressure for brown treesnakes after treatment with a 5.0 mg/kg weight dose of *propofol* is symmetric about its unknown median η

(c) Compare and contrast the results from Exercise 7.3.14 with those obtained in parts (a) and (b). Which of the three sets of analyses do you prefer and why?

7.4.4. The following data collection of $n = 30$ sample observations was generated from a normal population. Construct a normal probability plot for these data and describe its features.

{0.810, 6.808, 3.230, -8.330, 1.817, 1.040, 0.232, 3.117, -3.734, -2.340, -1.808, 5.265, 3.017, -4.773, -0.699, -0.806, -6.129, -0.560, -3.447, -4.271, -0.563, 1.060, -6.508, -12.104, -0.831, -2.266, -2.329, -5.240, 5.230, -3.460}.

7.4.5. The following data collection of $n = 30$ sample observations was generated from a truncated light-tailed population. Construct a normal probability plot for these data and describe its features. Compare to the normal probability plot constructed in Exercise 7.4.4 for data from a normal population.

{-0.292, -0.678, 0.945, 0.539, 0.642, -0.717, -0.897, -0.888, 0.862, 0.592, 0.925, 0.286, -0.590, -0.906, -0.357, -0.421, 0.756, -0.017, -0.152, 0.422, 0.503, -0.215, -0.331, 0.118, -0.470, -0.954, -0.882, -0.872, -0.438, -0.607}.

7.4.6. The following data collection of $n = 30$ sample observations was generated from a population that is skewed to the right. Construct a normal probability plot for these data and describe its features. Compare to the normal probability plot constructed in Exercise 7.4.4 for data from a normal population.

{4.68, 2.41, 11.64, 16.47, 8.01, 7.03, 1.36, 11.34, 8.42, 12.00, 44.43, 28.80, 15.71, 4.76, 11.50, 0.79, 0.91, 13.26, 11.12, 9.44, 1.63, 1.54, 9.55, 18.89, 30.21, 7.94, 5.61, 1.32, 0.43, 6.63}.

7.4.7. The following data collection of $n = 30$ sample observations was generated from a heavy-tailed population that contains both unusually large and unusually small values. Construct a normal probability plot for these data and describe its features. Compare to the normal probability plot constructed in Exercise 7.4.4 for data from a normal population.

{19.79, -9.22, -2.80, 3.06, -0.37, 22.51, 40.84, 3.65, -107.92, -168.18, 4.37, 11.20, 4.42, 33.44, 7.92, 17.36, 9.10, -7.00, 376.89, -8.70, -0.04, -0.68, -2.93, 7.32, -15.65, 11.84, 4.81, -34.73, -3.37, -0.87}.

7.5 Approximate Inference for the Center of a Population when the Number of Sample Observations is Large

In Sects. 1 through 3 we considered exact statistical inference procedures developed under three different sets of assumptions about the underlying population. The values of the point estimators for these three different settings depend solely on the observed sample data collection. However, to compute confidence intervals or carry out hypothesis tests, we need to use the sampling distribution of a related statistic to assess the sampling variability of our estimate for the center of the population. Thus, for example, in the confidence interval for the population median η discussed in Sect. 1 and presented in Eq. (7.2), we require values of appropriate Binomial percentiles. Similarly for the t-test procedures for normal populations presented in Eqs. (7.22)-(7.24), the appropriate t-distribution is used to obtain the P-values for the observed data. These sampling distributions (Binomial and t-distributions, as well as those associated with the Walsh average and signed rank procedures) depend directly on the number of observations, n, in our sample.

Percentiles for these sampling distributions can be obtained using the **R** functions *qbinom*( ) and *qt*( ). However, as we learned in Chap. 4, statistics that are based on sums or averages of a large number of observations are often approximately normally distributed. We can use this fact to develop approximations for percentiles of the sampling distributions we have used in this section that are quite accurate when the sample size is large. These approximations will involve the standard normal distribution.

When a random variable X has a $N(\mu, \sigma)$ distribution, we are able to use the fact that the standardized variable $Z = (X\text{-}\mu)/\sigma$ has a $N(0,1)$ distribution to compute exact probabilities for events associated with X. However, the role of the standard normal distribution in statistics is much more extensive than simply computing probabilities for normally distributed random variables. It also can be used to provide approximations for probabilities associated with the sampling distributions of many important statistics when the number of sample observations is large. This is a consequence of the series of theoretical results called central limit theorems that we discussed in Chap. 4. These theorems tell us about the sampling distribution of sums of random variables when the number of observations is large. In particular, such results can be used to approximate probabilities associated with the null hypothesis sampling distributions of all three of the test statistics B (7.4), W^+ (7.12), and T (7.21).

Large Sample Approximations for Inference About the Population Median η Based on the Sign Statistic *B*. First we consider the use of an appropriate central limit theorem to approximate percentiles for the null sampling distribution of the sign statistic B (7.4). When the null hypothesis H_0: $[\eta = \eta_0]$ is true, the sampling distribution of the standardized variable

$$B^* = \frac{B - E_0(B)}{\sqrt{\text{var}_0(B)}} = \frac{B - \frac{n}{2}}{\sqrt{\frac{n}{4}}}$$

can be well-approximated by a standard normal distribution when the sample size n is large. In particular, if b is the attained sample value of B, then, for large sample size n, the standard normal approximations to the probabilities $P(B \geq b)$ and $P(B \leq b)$ when $\eta = \eta_0$ are given by

$$P(B \geq b) \approx P\left(Z \geq \frac{b - \frac{n}{2}}{\sqrt{\frac{n}{4}}}\right) \tag{7.25}$$

and

$$P(B \leq b) \approx P\left(Z \leq \frac{b - \frac{n}{2}}{\sqrt{\frac{n}{4}}}\right), \tag{7.26}$$

respectively, where $Z \sim N(0,1)$. Thus, for large sample size n, these standard normal probabilities provide approximations for the P-value expressions given in (7.5), (7.6), and (7.7).

To illustrate this large sample approximation, consider once again the HDL cholesterol level data discussed in Example 7.2. There we found the observed value of B was $b = 11$ and the exact P-value for sample size $n = 12$ was P-value $= P(B \geq 11) = .0032$. Using the large sample approximation in (7.25) we find

$$P(B \geq 11) \approx P\left(Z \geq \frac{11 - \frac{12}{2}}{\sqrt{\frac{12}{4}}}\right) = P\left(Z \geq \frac{11 - 6}{\sqrt{3}}\right) = P(Z \geq 2.89) = .0019.$$

While this approximation is a bit smaller than the exact P-value, it is not too far off, especially for a sample size as small as $n = 12$. It is generally the case that for sample sizes $n \geq 25$, the exact and approximate P-values will be in reasonably close agreement.

We also note that the integers j in expressions (7.2) and (7.3) that lead to confidence intervals and confidence bounds, respectively, for the population median η under the minimal assumption that $P(X = \eta) = 0$ are the integers closest to the following expressions:

$$j \approx \frac{n}{2} - z_{\frac{(1-CL)}{2}}\sqrt{\frac{n}{4}} \tag{7.27}$$

and

$$j \approx \frac{n}{2} - z_{1-CL}\sqrt{\frac{n}{4}}, \tag{7.28}$$

respectively, where $z_{(1\text{-}CL)/2}$ and $z_{(1\text{-}CL)}$ are the ({1-CL}/2)th and (1-CL)th percentiles for the standard normal distribution.

To illustrate this approximation, consider the confidence interval for the median HDL level for active women in the age range 25-32, as discussed in Example 7.1. There we found that $(x_{(3)}, x_{(10)}) = (41, 54)$ mg/dl is an exact 96.14% confidence interval for the median HDL level. For this example, we have $n = 12$ and $CL = .9614$. The approximate value for j given in (7.27) is

$$j \approx \frac{12}{2} - z_{\frac{(1-.9614)}{2}} \sqrt{\frac{12}{4}} = 6 - 2.07\sqrt{3} = 2.415.$$

Since 2 is the integer closest to 2.415, the approximate 96.14% confidence interval for the median HDL level corresponding to expression (7.27) would be $(x_{(2)}, x_{(11)}) = (41, 59)$ mg/dl. While this differs slightly from the exact 96.14% confidence interval, the sample size $n = 12$ is not particularly large. It is generally the case that the exact and approximate confidence intervals and bounds will be in relative agreement for sample sizes $n \geq 25$.

Large Sample Approximations for Inference About the Population Median η Based on the Signed Rank Statistic W^+. An appropriate central limit theorem is also available to provide normal approximations to the percentiles of the null sampling distribution of the standardized signed rank statistic W^+ (7.12). When the null hypothesis H_0: $[\eta = \eta_0]$ is true, the sampling distribution of the standardized variable

$$W^{+*} = \frac{W^+ - E_0(W^+)}{\sqrt{\mathrm{var}_0(W^+)}} = \frac{W^+ - \frac{n(n+1)}{4}}{\sqrt{\frac{n(n+1)(2n+1)}{24}}}$$

can be well-approximated by a standard normal distribution when the sample size n is large. In particular, if w^+ is the attained sample value of W^+, then, for large sample size n, the standard normal approximations to the probabilities $P(W^+ \geq w^+)$ and $P(W^+ \leq w^+)$when $\eta = \eta_0$ are given by

$$P(W^+ \geq w^+) \approx P\left(Z \geq \frac{w^+ - \frac{n(n+1)}{4}}{\sqrt{\frac{n(n+1)(2n+1)}{24}}}\right) \tag{7.29}$$

and

$$P(W^+ \leq w^+) \approx P\left(Z \leq \frac{w^+ - \frac{n(n+1)}{4}}{\sqrt{\frac{n(n+1)(2n+1)}{24}}}\right), \tag{7.30}$$

respectively, where $Z \sim N(0,1)$. Thus, for large sample size n, these standard normal probabilities provide approximations for the P-value expressions given in (7.13), (7.14), and (7.15). While these expressions are a bit complicated, the argument *exact* can be specified to be *FALSE* in the **R** function *wilcox.test*( ) to compute the approximation directly from the sample data and avoid doing the calculations by hand. (Note that the documentation of *wilcox.test*( ) indicates that if the *exact* argument is not specified and the sample size is larger than $n = 50$, the approximation will be used by default.)

We illustrate this large sample approximation for the PMN migration data discussed in Example 7.5. There we found the observed value of W^+ was $w^+ = 33$ and the exact P-value for sample size $n = 8$ was P-value $= P(W^+ \geq 33) = .020$. Using the large sample approximation in (7.29) we find

$$P(W^+ \geq 33) \approx P\left(Z \geq \frac{33 - \frac{8(9)}{4}}{\sqrt{\frac{8(9)(17)}{24}}}\right) = P\left(Z \geq \frac{33 - 18}{\sqrt{51}}\right) = P(Z \geq 2.10) = .0179.$$

While this approximation is slightly smaller than the exact P-value, it is remarkably close in view of the fact that the sample size of $n = 8$ would hardly be considered large. For sample sizes $n \geq 25$, the exact and approximate P-values will generally be in very close agreement.

In addition, the integers q in expressions (7.10) and (7.11) that lead to confidence intervals and confidence bounds, respectively, for the population median η associated with the signed rank statistic W^+ are, for large n, approximated by the integers closest to the following expressions:

$$q \approx \frac{n(n+1)}{4} - z_{\frac{(1-CL)}{2}} \sqrt{\frac{n(n+1)(2n+1)}{24}} \tag{7.31}$$

and

$$q \approx \frac{n(n+1)}{4} - z_{(1-CL)} \sqrt{\frac{n(n+1)(2n+1)}{24}}, \tag{7.32}$$

respectively, where $z_{(1\text{-CL})/2}$ and $z_{(1\text{-CL})}$ are the ({1-*CL*}/2)th and (1-*CL*)th percentiles for the standard normal distribution.

To illustrate this approximation, consider the lower confidence bound for the median PMN migration rate for patients with acute pancreatitis, as discussed in Example 7.3. There we found that $w_{(4)} = 77.85\ \mu m/35\ min$ is an exact 98% lower confidence bound for the median PMN migration rate. For this example, we have $n = 8$ and $CL = .98$. The approximate value for q given in (7.32) is

$$q \approx \frac{8(9)}{4} - z_{(1-.98)} \sqrt{\frac{8(9)(17)}{24}} = 18 - 2.055\sqrt{51} = 3.324.$$

Since 3 is the integer closest to 3.324, the approximate 98% lower confidence bound for the median PMN migration rate corresponding to expression (7.32) would be $w_{(3)} = 75.50\ \mu m/35\ min$. In view of the small sample size $n = 8$, this slight difference from the exact 98% lower confidence bound is not too bad. It is generally the case that the exact and approximate confidence intervals and bounds will be in relatively good agreement for sample sizes $n \geq 25$.

Large Sample Approximations for Inference about the Population Mean μ Based on the *t*-statistic *T*. In Sect. 7.3 we discussed the fact that the variable T (7.17) has a *t*-distribution with n-1 degrees of freedom, provided that the random sample $X_1, \ldots, X_n$ is obtained from a $N(\mu, \sigma)$ population. We also pointed out that as the degrees of freedom, n-1, increases, the $t(n\text{-}1)$ density curve approaches that of the $N(0, 1)$ distribution. This fact is a direct consequence of yet another central limit theorem associated with the sample average, $\bar{X}$, and the fact that the population standard deviation, σ, is well approximated by the sample standard deviation, S, for n large. However, the central limit theorem for the sample average provides an additional bonus that permits far wider applicability of the inference procedures associated with the T statistic. Although the underlying population for the sample data is required to be normal in order to use the $t(n\text{-}1)$ density curve for the sampling distribution of T when the sample size n is small, this is no longer the case for large sample sizes; that is, the sampling distribution of T can be well approximated by the standard normal distribution for ANY underlying distribution, as long as it satisfies the very mild condition that its standard deviation, σ, is finite. This is, in fact, the case for most populations you are likely to encounter in practical settings.

In particular, if t is the attained value of the test statistic T (7.21), then when $\mu = \mu_0$ and the sample size n is large, central limit theorems for sample averages provide the approximations

$$P(T \geq t) \approx P(Z \geq t) \tag{7.33}$$

and

$$P(T \leq t) \approx P(Z \leq t), \tag{7.34}$$

where $Z \sim N(0,1)$. Thus, for large sample size n, standard normal probabilities can be used to provide approximations for the P-value expressions given in (7.22), (7.23), and *(7.24) even when the underlying population is not necessarily normal.* The accuracy of these approximations to the sampling distribution of T depends on the nature of the underlying distribution from which the sample data arise, particularly how skewed it is and its probability of producing outliers. However, for most practical settings the standard normal approximations in (7.33) and (7.34) are generally considered adequate for sample sizes $n \geq 30$, unless the data collection is heavily skewed in one direction or contains a number of unusually large and/or small observations. In the latter settings, sample sizes of at least 50 may be necessary to assure acceptable standard normal approximations.

Once again using the central limit theorem for a sample average, an approximate 100*CL*% confidence interval for the population mean μ is given by

$$\left(\bar{X} - z_{\frac{(1-CL)}{2}}\frac{S}{\sqrt{n}}, \bar{X} + z_{\frac{(1-CL)}{2}}\frac{S}{\sqrt{n}}\right) \tag{7.35}$$

and the corresponding approximate 100*CL*% lower and upper confidence bounds for μ are $\bar{X} - z_{1-CL}\frac{S}{\sqrt{n}}$ and $\bar{X} + z_{1-CL}\frac{S}{\sqrt{n}}$, respectively. Moreover, these large sample approximate confidence intervals and bounds no longer require that the underlying population be normal.

We must keep in mind that all of these large sample approximate inference procedures associated with T are based directly on the sample mean, $\bar{X}$, and sample standard deviation, S. As these sample measures are not resistant to outliers or asymmetry, caution is still the rule when applying these approximate procedures based on T when the data are rather asymmetric or involve a number of outliers. In other words, even with large sample sizes the primary issue remains to select a set of inference procedures which best address the population measure of center (mean or median) that is of most interest in a study. In other words, even though central limit theorems for sample averages guarantee that these large sample approximate procedures based on T CAN be used to analyze such data, there is no corresponding guarantee that they will provide effective inferences (i. e., short confidence intervals, small P-values, etc.). Thus, even in the case of a large sample size, inferences associated with either the sign statistic B or the signed rank statistic W^+ are generally preferred to those based on T when a data collection is strongly asymmetric or contains a considerable number of outliers. For more on such comparisons between procedures based on B, W^+, and T, see Hollander, Wolfe, and Chicken (2014), for example.

Section 7.5. Practice Exercises

7.5.1. Let B (7.4) be the sign statistic. When the null hypothesis H_0: $[\eta = \eta_0]$ is true, the exact sampling distribution of B is $Binom(n, .5)$. We have also seen in this section that the sampling distribution of the standardized statistic

$$B^* = \frac{B - \frac{n}{2}}{\sqrt{\frac{n}{4}}}$$

can be approximated by the standard normal distribution when the number of observations n is large. To evaluate the accuracy of this normal approximation as a function of the sample size n, compute the following null hypothesis probabilities using both the exact binomial sampling distribution and the prescribed normal approximation:

(a) $n = 10$, $P(B \geq 8)$, $P(B \leq 8)$, $P(B \leq 2)$
(b) $n = 15$, $P(B \geq 11)$, $P(B \leq 4)$, $P(B \geq 2)$
(c) $n = 20$, $P(B \geq 14)$, $P(B \leq 1)$, $P(B \leq 7)$
(d) $n = 25$, $P(B \geq 13)$, $P(B \leq 5)$, $P(B \geq 20)$
(e) $n = 30$, $P(B \geq 28)$, $P(B \leq 2)$, $P(B \leq 17)$.

7.5.2. Let W^+ (7.12) be the signed rank statistic. When the null hypothesis H_0: $[\eta = \eta_0]$ is true, the exact sampling distribution of W^+ is provided by the **R** function *wilcox.test*( ). We have also seen in this section that the sampling distribution of the standardized statistic

$$W^{+*} = \frac{W^{+} - \frac{n(n+1)}{4}}{\sqrt{\frac{n(n+1)(2n+1)}{24}}}$$

can be approximated by the standard normal distribution when the number of observations n is large. To evaluate the accuracy of this normal approximation as a function of the sample size n, compute the following null hypothesis probabilities using both the exact sampling distribution for W^+ and the prescribed normal approximation:

(a) $n = 10$, $P(W^+ \geq 41)$, $P(W^+ \leq 8)$
(b) $n = 15$, $P(W^+ \geq 82)$, $P(W^+ \leq 35)$
(c) $n = 20$, $P(W^+ \geq 150)$, $P(W^+ \leq 70)$
(d) $n = 25$, $P(W^+ \geq 235)$, $P(W^+ \leq 109)$
(e) $n = 30$, $P(W^+ \geq 304)$, $P(W^+ \leq 151)$.

7.5.3. Let T be the test statistic given in (7.21). When the null hypothesis H_0: $[\mu = \mu_0]$ is true, the exact sampling distribution of T for data arising from a normal population is the $t(n\text{-}1)$ distribution. We have also seen in this section that the $t(n\text{-}1)$ distribution can be approximated by the standard normal distribution when the number of observations n is large. To evaluate the accuracy of this normal approximation as a function of the sample size n, compute the following null hypothesis probabilities using both the exact $t(n\text{-}1)$ distribution and the normal approximation:

(a) $n = 10$, $P(T \geq 2.262)$, $P(T \leq \text{-}2.821)$
(b) $n = 15$, $P(T \geq 1.761)$, $P(T \leq \text{-}1.076)$
(c) $n = 20$, $P(T \geq 2.539)$, $P(T \leq \text{-}1.328)$
(d) $n = 25$, $P(T \geq 2.172)$, $P(T \leq \text{-}2.797)$
(e) $n = 30$, $P(T \geq 2.045)$, $P(T \leq \text{-}3.396)$.

7.5.4. Let $X_{(1)} \leq X_{(2)} \leq \ldots \leq X_{(15)}$ be the order statistics for a random sample of $n = 15$ from a population that satisfies the minimal assumption with median η.

(a) Find the integer j so that $(X_{(j)}, X_{(n\text{-}j+1)})$ is an exact 96.48% confidence interval for η.

(b) Evaluate the accuracy of the normal approximation for j given in (7.27).

7.5.5. Let $W_{(1)} \leq W_{(2)} \leq \ldots \leq W_{(210)}$ be the $M = \frac{20(20+1)}{2} = 210$ ordered Walsh averages for a random sample of size $n = 20$ from a population that is symmetric about its median η.

(a) Find the integer q so that $(W_{(q)}, W_{(211\text{-}q)})$ is an exact 94.68% confidence interval for η.

(b) Evaluate the accuracy of the normal approximation for q given in (7.31).

7.5.6. Consider the random sample of $n = 50$ observations given in Exercise 7.3.8. Use the normal approximations discussed in this section to conduct the following statistical analyses of these data under the assumption that the underlying population from which these data were collected is normal with mean μ.

(a) Find the approximate P-value for the appropriate statistical test of the null hypothesis H_0: $\mu = 9$ versus the one-sided alternative H_A: $\mu > 9$.

(b) Find an approximate 97.5% lower confidence bound for μ.

7.5.7. Consider the random sample of $n = 50$ observations given in Exercise 7.3.10. Use the normal approximations discussed in this section to conduct the following statistical analyses of these data under the assumption that the underlying population from which these data were collected is symmetric about its median η.

(a) Find the approximate P-value for the appropriate statistical test of the null hypothesis H_0: $\eta = 3$ versus the two-sided alternative H_A: $\eta \neq 3$.

(b) Find an approximate 95% confidence interval for η.

7.5.8. Consider the random sample of $n = 100$ observations given in Exercise 7.3.11. Use the normal approximations discussed in this section to conduct the following statistical analyses of these data with only the minimal assumption about the underlying population with median η from which these data were collected.

(a) Find the approximate P-value for the appropriate statistical test of the null hypothesis H_0: $\eta = 6$ versus the one-sided alternative H_A: $\eta < 6$.

(b) Find an approximate 90% upper confidence bound for η.

7.5.9. Consider the random sample of $n = 100$ observations given in Exercise 7.3.9. Use the normal approximations discussed in this section to conduct the following statistical analyses of these data under the assumption that the underlying population from which these data were collected is normal with mean μ.

(a) Find the approximate P-value for the appropriate statistical test of the null hypothesis H_0: $\mu = 9$ versus the one-sided alternative H_A: $\mu > 9$.

(b) Find an approximate 97.5% upper confidence bound for μ.

7.5.10. Repeat Exercise 7.5.6 under the assumption that the underlying population from which the data were collected is symmetric about its median η, but that it is not necessarily normal. Compare your answers with those obtained in Exercise 7.5.6.

7.5.11. Repeat Exercise 7.5.7 under the assumption that the underlying population from which the data were collected is normal with mean (and median) μ. Compare your answers with those obtained in Exercise 7.5.7.

7.5.12. Repeat Exercise 7.5.8 under the stronger assumption that the underlying population from which the data were collected is symmetric about its median η. Compare your answers with those obtained in Exercise 7.5.8.

7.5.13. Repeat Exercise 7.5.8 under the most stringent assumption that the underlying population from which the data were collected is normal with mean (and median) μ. Compare your answers with those obtained in Exercises 7.5.8 and 7.5.12.

7.5.14. Repeat Exercise 7.5.9 with only the minimal assumption about the underlying population with median η from which the data were collected. Compare your answers with those obtained in Exercise 7.5.9.

7.6 Approximate Inference for the Median of an Arbitrary Distribution – Bootstrapping the Sample Median

As we saw in Sect. 5.4, bootstrapping is a powerful tool that allows us to approximate the distribution of an arbitrary statistic using only a representative sample from a population. Using the **R** dataset *pennies_age* and various sample sizes, we investigated the approximate sampling distribution of the mean and median age of pennies. Now, we extend these ideas to demonstrate an alternative (and often more intuitive than those discussed previously in this chapter) approach to inference for the population median η.

To conduct hypothesis tests or construct confidence intervals, we need some way to obtain the standard error of our estimate. Bootstrapping allows us to approximate this standard error by simulation (as opposed to repeating the original sampling process many more times, which would typically be very costly!).

Given the original sample of size n from a population, we draw (with replacement) a subsample of size n from this sample. This procedure is then repeated r times to produce the bootstrap replicates. For each of these replicates, we compute the median $\tilde{X}_i$ (or any other arbitrary statistic of interest) based on the subsampled values. From these r subsample medians $\tilde{X}_1, \tilde{X}_2, \ldots, \tilde{X}_r$, we estimate the standard error of the replicate values. This estimate can then be combined with the average of the subsampled medians to provide inference about the population median.

The bootstrap estimate of this standard error (or standard deviation of the *estimate*) is given by

$$SE_r = \sqrt{\frac{\sum_{i=1}^{r} \left(\tilde{X}_i - \overline{\tilde{X}}\right)^2}{r-1}}, \tag{7.36}$$

where $\overline{\tilde{X}}$ is simply the average of the r subsample medians $\tilde{X}_1, \tilde{X}_2, \ldots, \tilde{X}_r$.

When the null hypothesis H_0: [$\eta = \eta_0$] is true, the sampling distribution of the standardized variable $\overline{\tilde{X}}^* = \dfrac{\overline{\tilde{X}} - \eta_0}{SE_r}$ can be approximated by a t-distribution with r - 1 degrees of freedom. In particular, if $\overline{\tilde{x}}_{obs}$ is the attained sample value of $\overline{\tilde{X}}$, then the t-distribution can be used to approximate the probabilities $P\left(\overline{\tilde{X}} \geq \overline{\tilde{x}}_{obs}\right)$ and $P\left(\overline{\tilde{X}} \leq \overline{\tilde{x}}_{obs}\right)$ when $\eta = \eta_0$. These approximate probabilities are given by

$$P\left(\overline{\tilde{X}} \geq \overline{\tilde{x}}_{obs}\right) \approx P\left(T \geq \frac{\overline{\tilde{x}}_{obs} - \eta_0}{SE_r}\right) \tag{7.37}$$

and

$$P\left(\overline{\tilde{X}} \leq \overline{\tilde{x}}_{obs}\right) \approx P\left(T \leq \frac{\overline{\tilde{x}}_{obs} - \eta_0}{SE_r}\right), \tag{7.38}$$

where T has a t-distribution with $r - 1$ degrees of freedom.

Similarly, approximate bootstrap confidence intervals for η can be constructed at level CL using the upper (1-CL)/2 percentile of the relevant t-distribution, similar to the construction of the interval for μ in (7.20). That is, an approximate 100CL% confidence interval for η is provided by the interval

$$\left(\overline{\tilde{X}} - t_{r-1,\frac{(1-CL)}{2}} SE_r, \overline{\tilde{X}} + t_{r-1,\frac{(1-CL)}{2}} SE_r\right), \tag{7.39}$$

where $t_{r-1,\frac{(1-CL)}{2}}$ is the upper $\left(\frac{1-CL}{2}\right)^{\text{th}}$ percentile for the t-distribution with $r-1$ degrees of freedom.

Example 7.8. We return to the **R** dataset *pennies_age* examined in Example 5.12 to create an approximate 90% confidence interval for η. We'll use the **R** functions *replicate*( ) and *sample*( ) to generate the bootstrap samples for the *pennies_age* data using $r = 10{,}000$ replicates and store the subsample medians as the variable *pennies_bootstrap_replicates*.

```
> pennies_bootstrap_replicates <-
    replicate(median(sample(x = pennies_age,
                            size = length(pennies_age),
                            replace = TRUE)),
              n = 10000)
```

From these subsample medians, we compute the average value and the standard error, which we save as the variables *pennies_bootstrap_average* and *pennies_bootstrap_standard_error*, respectively.

```
> pennies_bootstrap_average <- mean(pennies_bootstrap_replicates)
> pennies_bootstrap_standard_error <-
    sqrt(
      sum((pennies_bootstrap_replicates -
             pennies_bootstrap_average)^2) /
        (length(pennies_bootstrap_replicates) - 1)
      )
```

Finally, we define CL to be 0.90 and calculate the lower and upper endpoints of the approximate 90% confidence interval to be 5.30 and 7.24.

```
> CL <- 0.90
> pennies_bootstrap_average -
    qt(p = (1-CL)/2,
       df = length(pennies_age) - 1,
       lower.tail = FALSE) *
    pennies_bootstrap_standard_error
[1] 5.300624
> pennies_bootstrap_average +
    qt(p = (1-CL)/2,
       df = length(pennies_age) - 1,
       lower.tail = FALSE) *
    pennies_bootstrap_standard_error
[1] 7.239476
```

Note that a crucial assumption for the validity of the approximation when we use bootstrapping is that the original sample used to construct the bootstrap replicates is *representative* of the population. Implicitly, this means that this base sample needs to be large enough to capture all of the important features of the population. What is "large enough" will depend on how well behaved the underlying population is.

Also, since this approximate inference assumes normality of the statistics calculated from the subsamples, it is generally a good idea to check that the distribution of these r sample values is roughly normal by using a histogram or any of the other methods for assessing normality that were previously discussed.

When there is reason to question the normality assumption, we can alternatively use the bootstrap samples to construct a nonparametric confidence interval by selecting the $r(1\text{-}CL)/2$ smallest and $r(1\text{-}CL)/2$ largest values of the subsample medians. The following commands will construct such an interval for $CL = 0.90$ based on the *pennies_bootstrap_replicates* vector previously obtained.

```
> index_of_value_to_select <-
    (1 - CL) / 2 *
    length(pennies_bootstrap_replicates)
> sort(pennies_bootstrap_replicates)[index_of_value_to_select]
[1] 5
> sort(pennies_bootstrap_replicates,
+      decreasing = TRUE)[index_of_value_to_select]
[1] 7
```

Does it make sense that this interval is similar to but slightly wider than the one constructed with the additional assumption of normality?

Section 7.6 Practice Exercises

7.6.1. Out of the three values {12, 17, 86} how many possible bootstrap samples are there? List the possible replicates and the median value for each of the possible replicates. In light of this, do you think that confidence intervals constructed based on $r=100$, $r=1000$, and $r=10{,}000$ replicates will be very different? Why or why not?

7.6.2. Use the **R** functions *replicate*( ) and *sample*( ) to construct a 95% confidence interval for the median assessed value of homes in the **R** dataset *homes_prices*. Using a number of replicates that you feel is appropriate, report your individual sample median values and your confidence interval.

7.6.3. Repeat the construction of the 90% confidence interval from Example 7.8 using 100 bootstrap samples instead of 10,000. Does this affect your interval? Why or why not?

7.6.4. Use the **R** dataset *body_temperature_and_heart_rate* (see Exercise 5.B.14) to simulate 500 bootstrap samples and construct a 95% confidence interval for the median heart rate of the male subjects. Repeat this procedure for the female subjects. Which interval is wider? Does there appear to be a difference in the median heart rates between men and women?

7.6.5. Compare and contrast the 90% confidence intervals based on $r=1000$ bootstrap samples using the normal method and the nonparametric method for the median heart rate of the female subjects in the **R** dataset *body_temperature_and_heart_rate*. Which interval do you prefer?

Chapter 7 Comprehensive Exercises

7.A. Conceptual

7.A.1. Consider a random sample of size $n = 5$ from a population with median $\eta = 0$. If the value of the signed rank statistic for these data is $W^+ = 13$, what are the possible values for the sign statistic B?

7.A.2. Consider a random sample of size $n = 8$ from a population with median $\eta = 0$. If the value of the sign statistic for these data is $B = 3$, what are the possible values for the signed rank statistic W^+?

7.A.3. Consider a random sample of size $n = 10$ from a population with median $\eta = 0$. If the value of the signed rank statistic for these data is $W^+ = 9$, what are the possible values for the sign statistic B?

7.A.4. Consider a random sample of size $n = 15$ from a population with median $\eta = 0$. If the value of the sign statistic for these data is $B = 10$, what are the possible values for the signed rank statistic W^+?

7.A.5. Construct a set of data for which the sign test rejects H_0: $\eta = 0$ in favor of H_A: $\eta > 0$ at significance level $\alpha = .05$ but the t-test does not reject H_0 in favor of H_A.

7.A.6. Construct a set of data for which the signed rank test rejects H_0: $\eta = 0$ in favor of H_A: $\eta > 0$ at significance level $\alpha = .05$ but the t-test does not reject H_0 in favor of H_A.

7.A.7. Construct a set of data for which the t-test rejects H_0: $\eta = 0$ in favor of H_A: $\eta > 0$ at significance level $\alpha = .05$ but the sign test does not reject H_0 in favor of H_A.

7.A.8. Construct a set of data for which the t-test rejects H_0: $\eta = 0$ in favor of H_A: $\eta > 0$ at significance level $\alpha = .05$ but the signed rank test does not reject H_0 in favor of H_A.

7.A.9. Construct a set of data for which the signed rank test rejects H_0: $\eta = 0$ in favor of H_A: $\eta > 0$ at significance level $\alpha = .05$ but the sign test does not reject H_0 in favor of H_A.

7.A.10. Construct a set of data for which the sign test rejects H_0: $\eta = 0$ in favor of H_A: $\eta > 0$ at significance level $\alpha = .05$ but the signed rank test does not reject H_0 in favor of H_A.

7.A.11. Construct a set of data for which the t-test rejects H_0: $\eta = 0$ in favor of H_A: $\eta > 0$ at significance level $\alpha = .05$ but neither the sign test nor the signed rank test rejects H_0 in favor of H_A.

7.A.12. Construct a set of data for which the sign test rejects H_0: $\eta = 0$ in favor of H_A: $\eta > 0$ at significance level $\alpha = .05$ but neither the t-test nor the signed rank test rejects H_0 in favor of H_A.

7.A.13. Construct a set of data for which the signed rank test rejects H_0: $\eta = 0$ in favor of H_A: $\eta > 0$ at significance level $\alpha = .05$ but neither the sign test nor the t-test rejects H_0 in favor of H_A.

7.A.14. How many possible values are there for the sign statistic B and signed rank statistic W^+ for a sample of 15 observations?

7.A.15. How many possible values are there for the sign statistic B for a sample of 10 observations? How large would you have to increase the sample size to double the number of possible values for B?

7.A.16. How many possible values are there for the signed rank statistic W^+ for a sample of 10 observations? How large would you have to increase the sample size to double the number of possible values for W^+?

7.A.17. Suppose we are interested in a $100CL\%$ confidence interval for the median of a population η based on a random sample of $n = 15$ sample observations. List the exact levels CL that are available if we base our inferences on the sign statistic.

7.A.18. Suppose we are interested in a $100CL\%$ confidence interval for the median of a population η based on a random sample of $n = 15$ sample observations. List the exact levels CL that are available if we base our inferences on the signed rank statistic.

7.A.19. Construct a set of sample observations for which the sample median, $\tilde{X}$, is positive, but the median of the Walsh averages, $\tilde{W}$, is negative.

7.A.20. Construct a set of sample observations for which the sample median, $\tilde{X}$, is negative, but the median of the Walsh averages, $\tilde{W}$, is positive.

7.A.21. Consider a sample of $n = 20$ observations for which the observation with the largest absolute value is positive. What is the minimum number of Walsh averages that could be positive?

7.A.22. Consider a sample of $n = 15$ observations for which the observations with the largest and second largest absolute values are both negative. What is the maximum number of Walsh averages that could be positive?

7.A.23. Consider a sample of $n = 10$ observations for which the observation with the largest absolute value is positive and the observation with the second largest absolute value is negative. What are the minimum and maximum number of Walsh averages that could be positive?

7.B. Data Analysis/Computational

7.B.1. *Diamonds.* In the February 18, 2000 edition of Singapore's *Business Times*, an advertisement (as discussed in Chu, 2001) listed data (weight in carats, color purity, grade of clarity, certification body, and value in Singapore dollars) for 308 round diamond stones. These data are provided in the dataset *diamonds_carats_color_cost*. Viewing these data as a random sample of size $n = 308$ from the population of all round diamond stones, complete the following statistical analyses. Assume only the minimal assumption for the underlying population with unknown median diamond size (in carats) η.

(a) Obtain a point estimate of η and find an approximate 94% confidence interval for η.

(b) Find the *P*-value for a test of the null hypothesis H_0: $\eta = 0.5$ carats versus the one-sided alternative H_A: $\eta > 0.5$ carats.

7.B.2. *Diamonds Round Two.* Carry out the same statistical analyses prescribed in Exercise 7.B.1, but now under the more stringent assumption that the population of round diamond stone sizes is symmetrically distributed about its unknown median size (in carats) η. Compare your findings with those obtained under the minimal assumption of Exercise 7.B.1. Are you comfortable with the assumption of distributional symmetry? Why or why not?

7.B.3. *Diamonds Round Three.* Carry out the same statistical analyses prescribed in Exercise 7.B.1, but now under the even more stringent assumption that the population of round diamond stone sizes (in carats) is normal with mean (and median) μ. Compare your findings with those obtained in Exercises 7.B.1 and 7.B.2 under minimal and symmetry assumptions, respectively. Are you comfortable with the assumption that the population of sizes (in carats) for round diamond stones is normally distributed? Why or why not?

7.B.4. *How Much Do Euros Weigh?* The Euro is the common currency coin for the twenty-eight countries comprising the European Union. According to information from the "National Bank of Belgium", the 1 Euro coin is stipulated to weigh 7.5 grams. Shkedy et al. (2006) obtained eight separate packages of 250 Euros each from a Belgian bank and their assistants Sofie Bogaerts and Saskia Litière individually weighed each of these 2000 coins using a weighing scale of the type Sartorius BP310, which provides an accurate reading up to one thousandth of a gram. These two thousand weights, indexed by package number, are provided in the dataset *weight_of_Euros*. Using only the 250 coins from package number 1, conduct the following analyses under the assumption that the population of Euro weights is normally distributed with mean μ.

(a) Obtain a point estimate of μ and find an approximate 96% confidence interval for μ.

(b) Find the P-value for a test of the null hypothesis H_0: $\mu = 7.5$ grams versus the one-sided alternative H_A: $\mu > 7.5$ grams.

7.B.5. *How Much Do Euros Weigh—Again?* Repeat the statistical analyses from Exercise 7.B.4 using the 500 Euros obtained from combining packages 1 and 2. Compare and contrast the outcome of these two sets of statistical analyses.

7.B.6. *How Much Do Euros Weigh—Once More?* Repeat the statistical analyses from Exercise 7.B.4 using only the 250 Euros from package number 8. Compare and contrast the results for package 1 versus the results for package 8.

7.B.7. *Is 98.6 Degrees Fahrenheit Truly the Mean Body Temperature?* It is a widely held belief that the normal body temperature for humans is 98.6°F. Mackowiak et al. (1992) provide a critical evaluation of this statement through the collection of data from 148 individuals aged 18 through 40 years. The dataset *body_temperature_and_heart_rate* contains body temperature and heart rate values (artificially generated by Shoemaker, 1996, to closely recreate the original data considered by Mackowiak et al.) for 65 male and 65 female subjects. Conduct the following analyses under the assumption that the population of human body temperatures for healthy individuals is normally distributed with mean μ.

(a) Obtain a point estimate of μ and find an approximate 94% upper confidence interval for μ.

(b) Find the *P*-value for a test of the null hypothesis H_0: $\mu = 98.6°F$ versus the one-sided alternative H_A: $\mu < 98.6°F$.

(c) Carry out similar analyses separately for the 65 male and 65 female subjects. Discuss the results.

7.B.8. *House Sizes in North Carolina.* The dataset *house_lot_sizes* contains the information about house and lot sizes for a random sample of 100 properties in Wake County, North Carolina, as collected by Woodard and Leone (2008). Assume only the minimal assumption for the underlying population of house sizes in Wake County, North Carolina with unknown median η.

(a) Obtain a point estimate of η.

(b) Find an approximate 96% confidence interval for η.

7.B.9. *House Sizes in North Carolina Round Two.* Carry out the same statistical analyses prescribed in Exercise 7.B.8, but now under the more stringent assumption that the population of house sizes in Wake County, North Carolina is symmetrically distributed about its unknown median η. Compare your findings with those obtained under the minimal assumption of Exercise 7.B.8. Are you comfortable with the assumption of distributional symmetry? Why or why not?

7.B.10. *Lot Sizes in North Carolina.* The dataset *house_lot_sizes* contains the information about house and lot sizes for a random sample of 100 properties in Wake County, North Carolina, as collected by Woodard and Leone (2008). Assume only the minimal assumption for the population of lot sizes in Wake County, North Carolina with unknown median η.

(a) Obtain a point estimate of η.

(b) Find an approximate 92% confidence interval for η.

7.B.11. *Lot Sizes in North Carolina Round Two.* Carry out the same statistical analyses prescribed in Exercise 7.B.10, but now under the more stringent assumption that the population of lot sizes in Wake County, North Carolina is normal with mean (and median) μ. Compare your findings with those obtained in Exercises 7.B.10 under only the minimal assumption about the population. Are you comfortable with the assumption that the population of lot sizes is normally distributed? Why or why not?

7.B.12. *Healthy Heart Rate.* In Exercise 7.B.7 we discussed the dataset *body_temperature_and_heart_rate* generated by Shoemaker (1996) for 65 healthy female and 65 healthy male subjects. Conduct the following analyses under the assumption that the population of human heart rates for healthy individuals is normally distributed with mean μ.

(a) Obtain a point estimate of μ and find an approximate 95% confidence interval for μ.

(b) Repeat the calculations in part (a) separately for the 65 male and 65 female subjects. Discuss the results.

7.B.13. *Movie Ratings.* The *Movie and Video Guide* is a ratings and information guide to movies prepared annually by Leonard Maltin. Moore (2006) selected a random sample of 100 movies from the 1996 edition of the *Guide*. He compiled the dataset *movie_facts* containing relevant information about the selected movies. One of the pieces of information provided is the rating that Maltin gave to each of the movies on a rising (worst to best) scale of 1, 1.5, 2, 2.5, 3, 3.5, 4. Assume only the minimal assumption for the population of moving ratings with unknown median η.

(a) Obtain a point estimate of η and find an approximate 90% confidence interval for η.

(b) Find the P-value for a test of the null hypothesis H_0: $\eta = 2.5$ versus the one-sided alternative H_A: $\eta < 2.5$.

7.B.14. *How Long Are Movies?* The *Movie and Video Guide* is a ratings and information guide to movies prepared annually by Leonard Maltin. Moore (2006) selected a random sample of 100 movies from the 1996 edition of the *Guide*. He compiled the dataset *movie_facts* containing relevant information about the selected movies. One of the pieces of information provided is the running length of the movies, in minutes. Conduct the following analyses under the assumption that the running length of movies is normally distributed with mean μ.

(a) Obtain a point estimate of μ and find an approximate 93% confidence interval for μ.

(b) Find the P-value for a test of the null hypothesis H_0: $\mu = 90$ min versus the one-sided alternative H_A: $\mu > 90$ min.

7.B.15. *How Big Are Movies?* The *Movie and Video Guide* is a ratings and information guide to movies prepared annually by Leonard Maltin. Moore (2006) selected a random sample of 100 movies from the 1996 edition of the *Guide*. He compiled the dataset *movie_facts* containing relevant information about the selected movies. One of the pieces of information provided is the number of listed cast members in each movie. Conduct the following analyses under the assumption that the population of number of cast members in a movie is symmetrically distributed about its median η

(a) Obtain a point estimate of η and find an approximate 97% confidence interval for η.

(b) Find the P-value for a test of the null hypothesis H_0: $\eta = 6.5$ versus the one-sided alternative H_A: $\eta > 6.5$.

7.C. Activities

7.C.1. *Coffee, Coffee, Coffee.* How many cups of coffee does a typical college student drink in a given day? Design a study to collect data to address this question. Carry out an appropriate statistical analysis of your collected data.

7.C.2. *Beer, Beer, Beer.* How many bottles of beer does a typical college student drink in a given week? Design a study to collect data to address this question. Carry out an appropriate statistical analysis of your collected data.

7.C.3. *Sleep, Sleep, Sleep.* How many hours of sleep does a typical college student get during the "school nights" of Sunday through Thursday? Design a study to collect data to address this question. Carry out an appropriate statistical analysis of your collected data.

7.C.4. *Sleep, Sleep, Sleep?* How many hours of sleep does a typical college student get during the weekend nights of Friday and Saturday? Design a study to collect data to address this question. Carry out an appropriate statistical analysis of your collected data. Compare your findings with those you obtained for "school nights" in Exercise 7.C.3.

7.C.5. *U.S. Pennies.* How long do U.S. pennies remain in circulation? Design a study to collect data to address this question. Carry out an appropriate statistical analysis of your collected data. Does your result depend on where the coins were minted (Denver or Philadelphia)?

7.C.6. *U.S. Nickels.* How long do U.S. nickels remain in circulation? Design a study to collect data to address this question. Carry out an appropriate statistical analysis of your collected data. Does your result depend on where the coins were minted (Denver or Philadelphia)? Compare your findings with those you obtained for U.S. pennies in Exercise 7.C.5.

7.C.7. *Smart Phones.* How much daily time do college students spend on their smart phones? Design a study to collect data to address this question. Carry out an appropriate statistical analysis of your collected data.

7.C.8. *Exercise.* How many hours per week do college students spend exercising? Design a study to collect data to address this question. Carry out an appropriate statistical analysis of your collected data.

7.C.9. *Coursework/Studying.* How many hours per week (outside of the classroom) do college students spend on coursework, including studying for examinations? Design a study to collect data to address this question. Carry out an appropriate statistical analysis of your collected data.

7.C.10. *Solitary Time.* How many hours per week (other than sleeping) do college students spend without conversing with another person, either face to face or through an electronic device? Design a study to collect data to address this question. Carry out an appropriate statistical analysis of your collected data.

7.D. Internet Archives

7.D.1. *Social Issues.* Search the Internet to find a published research article that uses data from a random sample to address a social issue of particular interest to you. Discuss their statistical findings in the context of the one-sample setting of Chap. 7.

7.D.2. *Treating Disease.* Search the Internet to find a published research article that uses data from a random sample to address ways to treat a potentially fatal disease. Discuss their statistical findings in the context of the one-sample setting of Chap. 7.

7.D.3. *Income Inequality.* Search the Internet to find a published research article that uses data from a random sample to address the issue of income inequality in the United States. Discuss their statistical findings in the context of the one-sample setting of Chap. 7.

7.D.4. *Sports Injuries.* Search the Internet to find a published research article that uses data from a random sample to address the prevalence of youth sports injuries. Discuss their statistical findings in the context of the one-sample setting of Chap. 7.

7.D.5. *Student Loan Debt at Public Universities.* Search the Internet to find a published research article that uses data from a random sample to address the issue of student loan debt for graduates of public universities. Discuss their statistical findings in the context of the one-sample setting of Chap. 7.

7.D.6. *Student Loan Debt at Private Universities.* Search the Internet to find a published research article that uses data from a random sample to address the issue of student loan debt for graduates of private universities. Discuss their statistical findings in the context of the one-sample setting of Chap. 7 and compare the findings with those in Exercise 7.D.5.

7.D.7. *Global Warming.* Search the Internet to find a published article that uses data to address an issue related to global warming. Discuss their statistical findings in the context of the one-sample setting of Chap. 7.

7.D.8. *Trans Fat.* Search the Internet to find a published article that uses data to address the health impact of trans fat in our diets. Discuss their statistical findings in the context of the one-sample setting of Chap. 7.

7.D.9. *Fracking.* Search the Internet to find a published article that uses sample data to address the impact of fracking on society. Discuss their statistical findings in the context of the one-sample setting of Chap. 7.

7.D.10. *Toxic Algae Bloom.* Search the Internet to find a published article that uses sample data to address the causes of toxic algae blooms in a water system. Discuss their statistical findings in the context of the one-sample setting of Chap. 7.

8 Statistical Inference for Matched Pairs or Paired Replicates Data

In many experimental situations it is of interest to assess how a set of circumstances or a treatment affects a population of subjects. In such studies it is important that we take care to control as much as possible for any additional circumstances or characteristics other than those under investigation that might also affect the outcome of the measurements to be collected. For example, gender and age would be important factors to take into account when evaluating the effect of a medication to treat high blood pressure. Similarly, environmental living conditions would be an important factor to consider when comparing subjects with a respiratory disease. Two commonly used approaches to facilitate experimental control of possible confounding factors are provided by the experimental research designs known as *matched pairs* and *paired replicates*.

In a matched pairs design, pairs of subjects from the population of interest are formed so that members of a given pair are similar with respect to one or more characteristics that might potentially affect the measurement to be

© Springer International Publishing AG 2017

D.A. Wolfe, G. Schneider, *Intuitive Introductory Statistics*, Springer Texts in Statistics,
DOI 10.1007/978-3-319-56072-4_8

investigated in the study. Such pairing characteristics could be routine physiological traits such as weight, age, or extent of regular exercise or they could be directly related to the treatment of interest, such as smoking habits, the severity of a medical condition, or genetic linkage (e.g., siblings). Comparisons of the measurement of interest are then made within each pair of subjects.

For a paired replicates study, subjects serve as their own controls in providing baseline measurements prior to being exposed to the experimental set of circumstances or treatment and then they are measured once again after such exposure. Differences between the pre- and post-observations are then used to assess the effect of the treatment on the sample subjects.

Matched pairs or paired replicate designs are certainly among the most commonly used experimental designs. Such designs would be appropriate, for example, for answering such questions as:

1. Is glaucoma (or the tendency for it) a genetically inherited disease? Here we would use a matched pairs design with genetic linkage (e. g., twins or at least siblings) being the factor that matches the pairs.
2. Does aspirin increase the time it takes for a person's blood to clot? Here it is natural to use a paired replicates design with pre-aspirin and post-aspirin blood-clotting times as the sample measurements.
3. Will an experimental medication for high blood pressure have the desired effect to lower the blood pressure? Here it might be advisable to use aspects of both the matched pairs and paired replicates designs. Study subjects would first be matched with respect to the severity of their high blood pressure and then within the sets of matched pairs (i.e., common severity of high blood pressure), the subjects would serve as their own controls for the purpose of baseline blood pressure measurements.

As is the case for the one-sample data discussed in Chap. 7, the particular inference techniques that are best for analyzing matched pairs or paired replicates data can depend on what we know (or can reasonably assume) about the population distribution of matched pairs or paired replicate measurements.

General Setting and Notation Let $(X_1, Y_1), \ldots, (X_n, Y_n)$ denote the items of a random sample of n matched pairs or paired replicates. The information about the additive effect of the experimental conditions or treatment is then contained in the differences $D_i = Y_i - X_i$, $i = 1, \ldots, n$, and we are interested in making inferences about the center of the distribution for the population of these differences. As is the case for the one-sample setting discussed in Chap. 7, the information about the center for the distribution of the differences can sometimes be represented best by its mean, μ_D, and sometimes best by its median, η_D, depending on the form of the probability distribution for the differences.

In Chap. 7 we discussed statistical procedures that are appropriate under a variety of circumstances for making inferences about the distribution of a single quantitative variable. Since the difference $D = Y - X$ is itself such a quantitative variable, it should be no surprise that any of the statistical procedures discussed in Chap. 7 can be applied under the proper conditions to the sample differences $D_1, \ldots, D_n$ to make inferences about the center of the difference distribution. Thus, for example, if X represents the radioactivity level in a toxic waste dump before treatment with a biological agent designed to reduce the radioactivity and Y represents the level after treatment with the agent, we can assess the effectiveness of the treatment by analyzing a collection of differences $D_i = Y_i - X_i$, $i = 1, \ldots, n$, for samples taken before and after administering the agent.

8.1 Inference for Continuous Paired Replicates or Matched Pairs Data

In Chap. 7 we considered three different sets of distributional assumptions, namely, the minimal assumption that the population median is not a possible value for the variable of interest, the more restrictive assumption that the variable is continuous and that its distribution is symmetric about its median, and the most restrictive assumption that the variable is normally distributed.

When we have data that are differences of continuous paired replicates or matched paired variables, the symmetry of the underlying distribution for these differences is inherently satisfied in most situations. In particular, if a pair of variables (X, Y) are such that X and Y arise from populations that differ only in their centers, then the difference $D = Y - X$ will have a probability distribution that is symmetric about the median of the difference, η_D. Since this condition is quite natural for most paired replicates or matched pairs data, we can restrict our consideration to the inference procedures associated with the signed rank statistic or t-statistic discussed in Chap. 7 and applied here to the differences $D_1, \ldots, D_n$. As is the case for the one-sample setting, the choice between the use of signed rank or t procedures will depend on the reasonableness of the assumption of normality for the distribution of the differences and the sample size n. If the distribution of the differences can be appropriately represented by a normal distribution, inferences associated with the appropriate t-procedures are preferred, no matter what the size of the sample. If the sample size is small and the distribution of the D differences is not reasonably represented by a normal distribution, inferences associated with the signed rank statistic are the only appropriate choices.

If the sample size is large, the choice is a bit less clear. In such a setting, the approximate procedures associated with either the signed rank statistic or the t-statistic (discussed in Sect. 7.5 for the one-sample setting) can be applied to make inferences about the center of the distribution of the D differences. Which of these approaches is preferred for a given set of differences depends primarily on the extent of the non-normality (e. g., presence of outliers or skewness) exhibited by the collected sample data. The greater this non-normality for the difference distribution, the more appropriate it is to use the signed rank procedures rather than the t-procedures to make inferences about the center of the distribution even if the sample size is large.

Since computational details of the signed rank and t-procedures are provided in Chap. 7, we simply illustrate the application of these procedures to differences obtained from paired replicates and matched pairs experiments via a pair of examples.

Example 8.1. What Effect Does Nest-Sharing Have on the Hatching of Wood Duck Eggs? Dump nesting is an approach often utilized by wild ducks to accommodate for a scarcity of suitable nesting sites in a habitat. Dump nesting is where two or more duck hens lay their eggs in a common nest or nest box, but where the incubation of the eggs is generally the assumed responsibility of a single female duck. In an effort to assess the effect that dump nesting might have on the hatching success of the eggs, Clawson et al. (1979) collected data for the wood duck population on the Duck Creek Wildlife Management Area in southeastern Missouri. The primary observation site was a man-made reservoir of 718 *ha* (hectares) where 71–118 nest boxes were available during the 9 years of the study from 1966–1974. Based on periodic checks of the nest boxes, the authors' research team gathered data on the numbers of eggs in both the dump nests and normal nests (i.e., nests with eggs from a single female wood duck) and the percentages of eggs that eventually hatched for those nests that were successful (i.e., nests in which at least one egg hatched). The data in Table 8.1 represent these hatching percentages for successful nests over the 9-year period of the study.

Table 8.1 Percentages of hatched eggs from successful dump and normal wood duck nests on a portion of the Duck Creek Wildlife Management Area in Missouri for the period 1966–1974

	Percentage hatched = (Number hatched eggs/Number eggs)	
Year	**Normal nests (Y)**	**Dump nests (X)**
1966	267/323 = 82.66%	76/94 = 80.85%
1967	553/683 = 80.96%	135/163 = 82.82%
1968	268/365 = 73.42%	556/957 = 58.10%
1969	58/82 = 70.73%	607/1071 = 56.68%
1970	87/137 = 63.50%	596/1078 = 55.29%
1971	228/307 = 74.27%	365/624 = 58.49%
1972	190/221 = 85.97%	358/506 = 70.75%
1973	329/403 = 81.64%	656/957 = 68.55%
1974	141/202 = 69.80%	847/1222 = 69.31%

Source: Clawson et al. (1979)

Since environmental conditions could have a significant effect on both the usage and effectiveness of dump nesting, it is important to compare the outcomes for normal and dump nests under the same set of environmental conditions. As a result, we will compare the percentage-hatched figure for dump nests in any given year only with the percentage-hatched figure for normal nests in the same year (i.e., under the same environmental conditions). Thus the data in Table 8.1 represent a sample of size $n = 9$ matched pairs data $(X_1, Y_1) = (80.85, 82.66), \ldots, (X_9, Y_9) = (69.31, 69.80)$, where we have arbitrarily chosen to label the dump nest percentages as X's and the normal nest percentages as Y's.

With these labels, the differences $D_i = Y_i - X_i$, $i = 1, \ldots, 9$, for the nine years are found to be:

$D_1 = 82.66 - 80.85 = 1.81,$	$D_2 = 80.96 - 82.82 = -1.86,$
$D_3 = 73.42 - 58.10 = 15.32,$	$D_4 = 70.73 - 56.68 = 14.05,$
$D_5 = 63.50 - 55.29 = 8.21,$	$D_6 = 74.27 - 58.49 = 15.78,$
$D_7 = 85.97 - 70.75 = 15.22,$	$D_8 = 81.64 - 68.55 = 13.09,$

$$D_9 = 69.80 - 69.31 = 0.49.$$

Table 8.2 Ordered Walsh averages of the differences for the percentage hatched data in Table 8.1

−1.860	4.350	6.960	8.210	11.765	14.435	15.500
−0.685	5.010	7.270	8.515	11.995	14.635	15.550
−0.025	5.615	7.450	8.565	13.090	14.685	15.780
0.490	6.095	7.855	8.795	13.570	14.915	
1.150	6.680	7.905	10.650	14.050	15.220	
1.810	6.730	7.930	11.130	14.155	15.270	
3.175	6.790	8.135	11.715	14.205	15.320	

Since the difference −1.86 gives the appearance of a possible outlier, we choose to apply the signed rank procedures to analyze these differences.

Using the **R** function *walsh*( ) on the dataset *percentage_hatched_eggs*, we obtain the $M = 9(10)/2 = 45$ ordered Walsh averages for these $n = 9$ differences presented in Table 8.2 as follows.

```
> #Obtain the differences
> differences = percentage_hatched_eggs$normal_nests-
percentage_hatched_eggs$dump_nests

> #Get the Walsh averages for the differences
> walsh_averages = walsh(differences)

> #Sort the Walsh averages, as in Table 8.2
> sort(walsh_averages)
```

The point estimate of the median, η_D, for the D differences is the median of the Walsh averages, corresponding to the unique middle ordered Walsh average,$\tilde{w} = w_{(23)} = 8.515\%$. You can obtain this value using the **R** function *median*( ) with the following command.

```
> median(walsh_averages)
[1] 8.515
```

We now show how to find a 90.2% confidence interval for η_D. First, why not choose a "nicer" number such as 90% or 91%? Using the **R** function *psignrank*( ), we see that these confidence levels are not available for the Wilcoxon signed rank statistic based on $n = 9$ matched pairs.

```
> psignrank(7:10 , 9)
[1] 0.03710938 0.04882812 0.06445312 0.08203125
```

The **R** output here tells us that, rounded to two decimal places, $P(W \leq 7) = 0.037$, $P(W \leq 8) = 0.049$, $P(W \leq 9) = 0.064$, and $P(W \leq 10) = 0.082$. That is, we could choose (among many other possibilities) to construct $1 - 2 \times 0.037 = 92.6\%$, $1 - 2 \times 0.049 = 90.2\%$, $1 - 2 \times 0.064 = 87.2\%$, or $1 - 2 \times 0.082 = 83.6\%$ confidence intervals. Alternatively, we can used the **R** function *qsignrank*( ) to find q.

```
> qsignrank(0.049 , 9)
[1] 9
```

Hence the endpoints of the exact 90.2% confidence interval for η_D correspond to the 9th smallest and 9th largest ordered Walsh average, namely, $(w_{(9)}, w_{(37)}) = (5.010, 14.635)$. Thus we are 90.2% confident that the median hatching percentage from normal nests is somewhere between 5.010 and 14.635% higher than that for dump nests under similar environmental conditions, as accounted for by the use of matched pairs within years. We point out that you can obtain these estimation results using the **R** function *wilcox.test*( ) as well. Note that we specify the *conf.int* and *conf.level* arguments to indicate that we would like the 90.2% confidence interval included in the **R** output.

```
> wilcox.test(differences, conf.int = TRUE, conf.level = 0.902)

        Wilcoxon signed rank test

data:  differences
V = 42, p-value = 0.01953
alternative hypothesis: true location is not equal to 0
90.2 percent confidence interval:
  5.010 14.635
sample estimates:
(pseudo)median
         8.515
```

The natural null hypothesis of interest for the data in Table 8.1 is that of no difference in percentage hatched between normal and dump nests, corresponding to $\eta_D = 0$. Applying the computational approach illustrated in Table 7.4 with $\eta_0 = 0$ to the percentage-hatched differences $D_1, \ldots, D_9$, we obtain the value of the signed rank statistic for these differences to be w^+

$= [2 + 8 + 6 + 4 + 9 + 7 + 5 + 1] = 42$. Hence, from (7.15) the P-value for a test of H_0: $\eta_D = 0$ versus the two-sided alternative H_A: $\eta_D \neq 0$ is given by

$$\text{P-value} = 2 \times \text{minimum}\{P(W^+ \geq 42|\eta = 0), P(W^+ \leq 42|\eta = 0)\},$$

which we can obtain via the **R** function *psignrank*( ). Since $P(W^+ \geq 42 \mid \eta = 0) = P(W^+ > 41 \mid \eta = 0)$, we use the following command (with the *lower.tail* argument specified to be FALSE since we want an upper tail probability) to obtain $P(W^+ > 41 \mid \eta = 0)$.

```
> psignrank(41, 9, lower.tail = FALSE)
[1] 0.009765625
```

Multiplying this number by 2 gives us the P-value of 0.195 (rounded to 3 places).

Thus we would reject H_0: $\eta_D = 0$ in favor of the two-sided alternative H_A: $\eta_D \neq 0$ for any significance level greater than or equal to .0195, suggesting that there is rather strong sample evidence that there is a difference between normal and dump nesting in the median percentages of eggs hatched. As noted in Chap. 7, we can also use the **R** function *wilcox.test*( ) to obtain the *approximate* P-value for this hypothesis test by specifying the *exact* argument to be FALSE.

```
> wilcox.test(percentage_hatched_eggs$normal_nests -
percentage_hatched_eggs$dump_nests,
alternative = 'two.sided',exact=FALSE)

        Wilcoxon signed rank test with continuity correction

data:  percentage_hatched_eggs$normal_nests -
percentage_hatched_eggs$dump_nests
V = 42, p-value = 0.02439
alternative hypothesis: true location is not equal to 0
```

Example 8.2. Will Treatment with Desimipramine, in Conjunction with Psychotherapy, Help with Weight Gain for Subjects with Anorexia Nervosa? Anorexia nervosa is a psychosomatic disorder typified by self-starvation, with substantial weight loss involving more than twenty-five percent of body weight being its chief physical symptom. It occurs most often among adolescent females from middle and upper-middle class families. Patients with anorexia nervosa are generally treated with a combination of psychotherapy and pharmacotherapy (i.e., medication). Brambilla et al. (1985) studied the effect of treatment with the medication desimipramine in conjunction with psychotherapy on the weight gain of subjects with anorexia nervosa.

Twelve female patients, ranging in age from 15 through 40 years, participated in the study. In conjunction with psychotherapy, desimipramine was given orally, three times a day for 30 days. For the first 15 days, the dosage for a patient was 1.5 *mg* per *kg* of body weight. For the remaining 15 days, the dosage was 2.0 *mg* per *kg* of body weight. Blood levels of desimipramine were monitored periodically to ensure that the drug had been administered properly.

The body weight (in *kg*) for each patient in the study was recorded prior to the beginning of the desimipramine administration (i.e., pre-treatment) and then again after the last dose of desimipramine had been given (i.e., post-

Table 8.3 Body weight (in *kg*) for patients with anorexia nervosa prior to and after desimipramine treatment in conjunction with psychotherapy

Patient	Pre-desimipramine	Post-desimipramine
1	37.0	39.0
2	46.0	47.0
3	35.0	34.6
4	39.8	42.0
5	41.6	43.9
6	30.0	30.5
7	39.0	40.0
8	47.5	51.0
9	40.0	40.5
10	47.5	49.0
11	42.0	42.5
12	35.0	34.0

Source: Brambilla et al. (1985)

treatment). The data in Table 8.3 represent these pre-treatment and post-treatment body weights for the twelve participants in the study.

Since both the pre-desimipramine and post-desimipramine observations are collected on the same subjects, the data in Table 8.3 represent paired replicates data. The sample size is $n = 12$ and the data pairs are $(X_1, Y_1) = (37.0, 39.0), \ldots, (X_{12}, Y_{12}) = (35.0, 34.0)$, where we have chosen to label the pre-desimipramine weights as X's and the post-desimipramine weights as Y's. With these labels, the twelve differences $D_i = Y_i - X_i$, $i = 1, \ldots, 12$, are given by:

$D_1 = 39.0 - 37.0 = 2.0,$	$D_2 = 47.0 - 46.0 = 1.0,$
$D_3 = 34.6 - 35.0 = -0.4,$	$D_4 = 42.0 - 39.8 = 2.2,$
$D_5 = 43.9 - 41.6 = 2.3,$	$D_6 = 30.5 - 30.0 = 0.5,$
$D_7 = 40.0 - 39.0 = 1.0,$	$D_8 = 51.0 - 47.5 = 3.5,$
$D_9 = 40.5 - 40.0 = 0.5,$	$D_{10} = 49.0 - 47.5 = 1.5,$
$D_{11} = 42.5 - 42.0 = 0.5,$	$D_{12} = 34.0 - 35.0 = -1.0.$

As there are no apparent indications of non-normality in these differences, we will use the appropriate t-procedures to analyze the data. The mean and standard deviation of the observed sample differences $d_1, \ldots, d_{12}$ are $\bar{d} = 1.133$ and $s_d = 1.248$, respectively. We estimate the mean weight change effect, μ_D, of the combination desimipramine with psychotherapy treatment to be $\bar{d} = 1.133$ kilograms. To construct a 95% confidence interval for μ_D, we first use the **R** function *qt*( ) to obtain the upper .025 percentile for the t-distribution with $n - 1 = 11$ degrees of freedom, namely, $t_{11,\ .025} = 2.201$.

```
> qt(.025, 11, lower.tail = FALSE)
[1] 2.200985
```

Then the 95% confidence interval for μ_D is given by

$$\bar{d} \pm t_{11,.025} \frac{s_d}{\sqrt{12}} = 1.133 \pm 2.201 \left(\frac{1.248}{\sqrt{12}} \right)$$
$$= 1.133 \pm 0.793 = (0.34, 1.926) \text{ kilograms.}$$

(Note that if we had decided instead to take the pre-desimipramine weights to be the Y's and the post-desimipramine weights to be the X's, then the resulting 95% confidence interval for the associated μ_D based on the pre- minus post- desimipramine differences would be (−1.926, −0.34) kilograms. Does this make intuitive sense?)

The natural null hypothesis here is that of no effect on body weight from the combination desimipramine with psychotherapy treatment, corresponding to $\mu_D = 0$, and the alternative of interest is $\mu_D > 0$, corresponding to a positive (i.e., increased body weight) effect from the regime. We compute

$$t^*_{obs} = \frac{\bar{d} - 0}{s_d/\sqrt{12}} = \frac{1.133}{(1.248/\sqrt{12})} = 3.14.$$

You can use the **R** function *t.test*( ) to obtain the *P*-value of 0.0047 for the dataset *desimipramine* using the following command and output.

```
> t.test(desimipramine$'Post-Desimipramine',
+        desimipramine$'Pre-Desimipramine',
+        paired = TRUE,
+        alternative = 'greater')

	Paired t-test

data:  desimipramine$"Post-Desimipramine" and desimipramine$"Pre-
Desimipramine"
t = 3.1464, df = 11, p-value = 0.004651
alternative hypothesis: true difference in means is greater than 0
95 percent confidence interval:
 0.4864471       Inf
sample estimates:
mean of the differences
               1.133333
```

Thus there is clearly sample evidence to indicate that the combination desimipramine with psychotherapy treatment does have the positive effect of increasing the average body weight for patients with anorexia nervosa. (As illustrated in Chap. 7, **R** can also be used to obtain both the 95%

confidence interval for μ_D and the P-value for this hypothesis test from the observed differences $d_1, \ldots, d_{12}$.)

Notice that in both Examples 8.1 and 8.2 the null hypothesis is H_0: $\mu_D = 0$, corresponding to no effect from the experimental sets of circumstances or treatments. While this is by far the most commonly tested null hypothesis for these matched pairs or paired replicates settings, we can just as easily use the signed rank or t-test procedures of Chap. 7 to test for nonzero effects of a specific magnitude for the differences $D_1, \ldots, D_n$, if that is more appropriate for the problem of interest. (See also Exercises 8.A.6 and 8.B.3.)

Section 8.1 Practice Exercises

8.1.1. Let $(X_1, Y_1) = (2.3, 4.5)$, $(X_2, Y_2) = (6.5, 9.7)$, $(X_3, Y_3) = (-3.4, -2.2)$, $(X_4, Y_4) = (5.0, 5.5)$, $(X_5, Y_5) = (4.8, 5.2)$, $(X_6, Y_6) = (-1.9, 1.3)$, $(X_7, Y_7) = (6.7, 5.6)$, $(X_8, Y_8) = (10.5, 8.9)$, $(X_9, Y_9) = (2.2, 3.3)$, and $(X_{10}, Y_{10}) = (9.9, 11.2)$ be the observed values for a sample of $n = 10$ paired replicates data.

(a) Compute the sample averages, $\bar{X}$ and $\bar{Y}$, for the X and Y values.

(b) Compute the sample variances, S_x^2 and S_y^2, for the X and Y values.

(c) Compute the ten differences $D_i = Y_i - X_i$, $i = 1, \ldots, 10$.

(d) Compute the average, $\bar{D}$, and variance, S_d^2, for the differences in part (c).

(e) Compare $\bar{D}$ with $\bar{Y} - \bar{X}$ and S_d^2 with $S_x^2 + S_y^2$. In order to apply any of the t-procedures to these paired replicates data, is it actually necessary for us to compute the ten differences $D_1, \ldots, D_{10}$? Why or why not?

8.1.2. *Bison and Species Diversity*. Land-use change, among other factors, has led to a serious decline in species diversity in the planet's ecosystems. This has led to an urgent need to provide ecological mechanisms to restore or at least maintain biodiversity. Collins et al. (1998) conducted two long-term field

Table 8.4 Total number of plant species on sixteen pairs of 50-m^2 matched plots of grassland on the Konza Prairie

Plot pair	Burned only	Burned with Bison Grazing
1	31	41
2	30	53
3	36	59
4	48	69
5	35	36
6	37	54
7	40	66
8	49	76
9	30	54
10	32	60
11	33	71
12	36	81
13	32	49
14	33	52
15	43	64
16	47	89

Source: Collins et al. (1998)

experiments in native grassland to assess the affects of a number of factors, including fire, nitrogen addition, and grazing or mowing, on plant species diversity. The data in Table 8.4 are the total numbers of plant species observed on 16 pairs of 50-m^2 matched plots of grassland on the Konza Prairie in Kansas. For each matched pair of land plots, one of the plots was subjected to some form of burning only. The second plot in that pair was subjected to the same burning scheme, but grazing by bison (*Bos bison*) was permitted following the burning.

(a) Find the P-value for a test of the conjecture that permitting bison to graze after burning grassland leads to an increase in the number of plant species over simply burning the grassland. What is your conclusion at significance level .025?

(b) Find a lower confidence bound for the median change in the number of plant species as a result of permitting bison to graze after burning grassland. Select your own reasonable confidence coefficient.

8.1.3. *Dilemma Zones and High Speed Accidents.* A "dilemma zone" is defined to be a section of roadway immediately in advance of a signalized intersection where a driver upon receiving a yellow light must decide whether to brake to a stop or accelerate to clear the intersection before a red light is displayed. Such dilemma zones are particularly critical for truck drivers traveling at high speeds, since semi-trucks take longer to come to a complete stop than the typical family car. "Prepare to Stop When Flashing" signs are erected on freeways and expressways in advance of signalized intersections where standard traffic control devices have failed to solve critical accident problems associated with a "dilemma zone". In order to determine the effectiveness of these signs, the Ohio Department of Transportation selected four dilemma zones involving high speed approaches (55 *mph* zones) for study. The number of accidents at each of these four dilemma zones was recorded at four different times of the year prior to erection of "Prepare to Stop When Flashing" signs and then during the same time periods during the following year after erection of the signs. (Since the signs were only erected for one or two of the approaches at these locations, the only accidents used in the evaluation were those that involved vehicles on one of the approaches with such a sign.) The numbers of accidents per million vehicles for these dilemma zones with and without the warning signs for the different times of year are reported in Table 8.5.

(a) Find the P-value for a test of the conjecture that erection of the "Prepare to Stop When Flashing" signs has a positive effect (i.e., leads to a decrease in traffic accidents per vehicle miles) at high speed dilemma zones. What is your conclusion at significance level .075?

(b) Find a confidence interval for the median change in traffic accidents per vehicle miles at high speed dilemma zones that results from

Table 8.5 Number of traffic accidents per million vehicles for high speed approach dilemma zones with and without "Prepare to Stop When Flashing" signs at four different times of the year

	Number of accidents per million vehicle miles	
Site-time	**Without signs**	**With signs**
Site 1 – Mar–May	2.144	0
Site 1 – June–Aug	3.574	0.715
Site 1 – Sept–Nov	2.144	2.144
Site 1 – Dec–Feb	1.430	0.715
Site 2 – Mar–May	2.144	0
Site 2 – June–Aug	1.430	0
Site 2 – Sept–Nov	0.715	0
Site 2 – Dec–Feb	0	0
Site 3 – Mar–May	0.858	0
Site 3 – June–Aug	0	0
Site 3 – Sept–Nov	0	0.858
Site 3 – Dec–Feb	0	0
Site 4 – Mar–May	0.858	0.858
Site 4 – June–Aug	0.858	1.716
Site 4 – Sept–Nov	0	0.858
Site 4 – Dec–Feb	0	0.858

Source: Ohio Department of Transportation (1980)

erecting the "Prepare to Stop When Flashing" signs. Select your own reasonable confidence coefficient.

8.1.4. *Percentage Hatched Eggs.* Consider the percentage hatched eggs data in Table 8.1. In Example 8.1 we used signed rank procedures to analyze this data collection. Carry out similar analyses using the *t*-procedures under the assumption of normality. Compare and contrast the two sets of results.

8.1.5. *Weight Gain for Anorexia Nervosa Subjects.* Consider the desimipramine treatment data in Table 8.2. In Example 8.2 we assumed normality and used *t*-procedures to analyze this data collection. Carry out similar analyses using the signed rank procedures without the assumption of normality. Compare and contrast the two sets of results.

8.2 Inference for Qualitative Differences—Data from Paired Replicates or Matched Pairs Experiments

The discussion in Sect. 8.1 centered on continuous paired replicates or matched pairs data. While this is certainly the most common form for such data, there are also situations where the information collected from a paired replicates/matched pairs experiment is purely qualitative in nature. For example, a random sample of subjects might assess which of two competing products they prefer based on data collected from trial uses of the products. In a similar vein, a participant in a medical study might simply report whether they considered that a prescribed treatment had improved their medical condition (or an impartial medical observer might provide her/his professional opinion about such improvement), without a numerical quantity being assigned to the degree of improvement. While there is clearly no information provided by such qualitative data for estimation (either point or interval) of a quantitative median effect, it is still natural to wish to test appropriate hypotheses about such attributes of the products or the treatment. Since both the signed rank and the t procedures illustrated in Sect. 8.1 require quantitative data, they cannot be used to analyze the qualitative difference-data collected from such studies. However, procedures similar to those discussed in Sect. 6.3 can be applied to conduct appropriate hypothesis tests for such qualitative difference-data. We illustrate this approach in the following example.

Example 8.3. Can Infants Less Than Six Months Old Distinguish Patterns of Visual Stimuli? One item of interest in the eye development of young infants is the age at which they can begin to effectively differentiate between a solid visual pattern and a more elaborate visual pattern, such as stripes or polka dots. In a study designed to address this question, Frantz et al. (1962) worked with infants varying in age from 2 to 6 months. Each infant in the study was placed in a chamber that was illuminated by four 60-watt incandescent lamps with reflectors in each of the bottom four corners of the chamber, below the infant's eyes. To reduce glare on the chamber walls above the lights, they were covered with finely knit, medium blue jersey cloth. The chamber ceiling was also light blue to enhance the infant's acute vision.

Two visual patterns, one a solid gray pattern and the other consisting of a striped pattern of white and gray, were placed directly on top of the chamber. These patterns were visible through two separate holes in the ceiling, each 7 by 5 inches wide, and 12 inches apart, center to center, and 15 1/2 inches from the infant's eyes. The ceiling was sufficiently darker than the patterned surfaces to make reflections of the pattern being looked at in the infant's eyes. Each baby was placed in the chamber for a period of 40 sec and observed by two separate investigators looking through two ¼-inch holes near the infant's feet. One observer recorded the percentage of time that the infant's eyes were focused on the solid gray pattern (X), while the second observer recorded the percentage of time that the infant's eyes were focused on the white/gray striped pattern (Y). (The infant's eyes were not focused on either pattern for the remaining percentage of the 40-s period.)

Frantz et al. (1962) chose to report only which of the two patterns was observed for the greater percentage of time for each of the infants in their study, rather than the actual percentages themselves. Thus, for the paired replicates data (X, Y), they simply indicated which of the two patterns was qualitatively preferred by each of the infants, with $Y>X$ corresponding to preference for the gray/white striped pattern and $Y<X$ corresponding to preference for the solid gray pattern.

To evaluate an additional possible effect due to stripe width size, the authors studied four different stripe widths (1/64, 1/32, 1/16, and 1/8 inch). Each of forty infants aged from 2 to 4 months was placed in the chamber and the investigators recorded how many of the infants preferred the striped pattern over the plain gray pattern (i.e., $Y > X$) separately for each of the four stripe widths. The results for the forty infants and the four stripe widths are presented in Table 8.6.

Table 8.6 Visual preferences by infants aged 2–4 months for striped patterns of various sizes as opposed to a plain gray pattern

Stripe width	Number preferring the striped pattern ($Y > X$)	Number preferring the plain gray pattern ($Y < X$)
1/64 in	22	18
1/32 in	30	10
1/16 in	32[a]	7[a]
1/8 in	34	6

Source: Frantz et al. (1962)
[a]The observation for one infant is missing for the 1/16 inch striped pattern

Clearly these data are, in fact, paired replicates since a given infant was simultaneously exposed to a particular width striped pattern and the plain gray pattern and the percentages of observed times for each were noted. However, Frantz et al. reported only the qualitative preference shown by each infant for one or the other of the patterns. As a result, the signed rank and *t*-procedures for paired data cannot be used to analyze the data in Table 8.6, since the necessary $D = Y - X$ differences required by both the signed rank and *t*-procedures for paired replicates are simply not reported for these infant preferences.

Fortunately an adaptation of the sign statistic discussed in Sect. 7.1 can be applied to conduct an appropriate hypothesis test for these preferences. Let $p_{\frac{1}{32}}$ $= P$(randomly chosen infant aged 2–4 months prefers the 1/32 inch striped pattern over the plain gray pattern). We are interested in testing the null hypothesis H_0: $[p_{\frac{1}{32}} = \frac{1}{2}]$, corresponding to no preference between the striped and plain gray patterns, versus the alternative H_A: $[p_{\frac{1}{32}} > \frac{1}{2}]$, corresponding to a preference for the striped pattern over the plain gray. Letting $\hat{p}_{\frac{1}{32}}$ denote the percentage of infants aged 2–4 months in the study who preferred the 1/32 inch striped pattern over the plain gray pattern, we see that the sample point estimate of $p_{\frac{1}{32}}$ is $\hat{p}_{\frac{1}{32}} = \frac{30}{40}$. The corresponding value of the standardized statistic $Z_{\hat{p},0}$ (6.10) is then

$$z_{\hat{p},0(observed)} = \frac{\frac{30}{40} - \frac{1}{2}}{\sqrt{\frac{\frac{1}{2}\left(1-\frac{1}{2}\right)}{40}}} = 3.16,$$

and it follows from (6.12) that the associated approximate P-value for the test of H_0: $[p_{\frac{1}{32}} = \frac{1}{2}]$ versus H_A: $[p_{\frac{1}{32}} > \frac{1}{2}]$ is $P(\,Z \geq 3.16) = .0008$. Thus there is strong evidence that infants aged 2–4 months can, indeed, differentiate between the 1/32 inch striped pattern (which they prefer) and the plain gray pattern.

We note that approximate confidence intervals for $p_{\frac{1}{32}}$ can also be obtained using expression (6.7). Thus, an approximate 90% confidence interval for $p_{\frac{1}{32}}$ is given by

$$\hat{p}_{\frac{1}{32}} \pm z_{.05}\sqrt{\hat{p}_{\frac{1}{32}}\left(1-\hat{p}_{\frac{1}{32}}\right)/40} = \frac{3}{4} \pm 1.645\sqrt{\frac{3}{4}\left(1-\frac{3}{4}\right)/40}$$
$$= .75 \pm .113 = (.637, .863).$$

We can use the **R** function *binconf*( ) to obtain this confidence interval as well. We specify the *alpha* argument to be 0.10 and the *method* argument to be "asymptotic" (as opposed to "exact", which would give an exact, rather than approximate, confidence interval).

```
> binconf(30, 40, alpha = 0.10, method='asymptotic')
 PointEst     Lower     Upper
     0.75 0.6373846 0.8626154
```

Section 8.2 Practice Exercises

8.2.1. *Psychological Needs and Physical Health.* Can supporting a patient's psychological needs lead to improvement in the condition of the patient's physical health? Smyth et al. (1999) addressed this question by studying the effect that writing about stressful life experiences might have on the medical conditions of rheumatoid arthritis patients. Thirty-one rheumatoid arthritis patients participated in the study. Rheumatologists used a five-category rating scheme (asymptomatic, mild, moderate, severe, very severe) to clinically evaluate each of the subjects prior to beginning the study protocol. Participants were then asked to write for 20 min on each of three consecutive days about the most stressful life experiences that they had undergone. The patients were clinically evaluated again 4 months after they had completed their writing projects. Smyth et al. report that 15 of the participating arthritis patients were diagnosed as being medically improved (a shift of one or more rating categories toward asymptomatic) 4 months after the 3-day writing periods. Let p = [proportion of all rheumatoid arthritis patients who would exhibit such medical improvement 4 months after writing about their most stressful life experiences].

(a) Find an approximate 98% confidence interval for p.

(b) Using the confidence interval obtained in part (a), what decision would you reach in a test of H_0: $p = .5$ versus H_A: $p \neq .5$ at approximate significance level $\alpha = .02$?

8.2.2. *Can Stressful Writing Help Subjects With Rheumatoid Arthritis?* Consider the stressful writing study discussed in Exercise 8.2.1. For this study, the value $p = .5$ tested in part (b) of Exercise 8.2.1 does not have much relevance. What is more important is a comparison of those rheumatoid arthritis patients who participated in the stressful writing protocol with another control group of rheumatoid arthritis patients who did not write about their most stressful lifetime experiences. Smyth et al. (1999) also studied such a control group, members of which were treated exactly the same as the treatment group, except for the absence of the stressful writing experiences. Smyth et al. found that only 4 of the 17 subjects in their control group of arthritis patients were diagnosed as being medically improved (a shift of one or more rating categories toward asymptomatic) 4 months after the beginning of the study. Let p_C = [proportion of all rheumatoid arthritis patients who would exhibit such medical improvement 4 months after onset of the study without writing about their most stressful lifetime experiences].

(a) Find an approximate 98% confidence interval for p_C.

(b) Compare the confidence interval for p (treatment group) obtained in part (a) of Exercise 8.2.1 with the confidence interval for p_C in part (a) of this exercise. Discuss the implications of your comparison. What hypothesis would be of most interest to test in this setting? (We will return to this study in Chap. 9 and address this issue once again.)

8.2.3. *Wisk Versus Tide.* A laboratory scientist purchases ten identical new white towels to be used in a test of the relative effectiveness of the two detergents Wisk and Tide. Each of the towels is stained with a combination of grass, strawberry jam, and spaghetti sauce. Five of the towels are washed in Wisk and the other five towels are washed in Tide. The ten washed towels are then presented to a panel of eighteen consumers. Each member of the panel is asked to select the five washed towels they believe to be the whitest. Fourteen of the panelists select a majority of towels washed in Tide. Let p denote the proportion of all consumers who believe that Tide does a better job of removing stains than does Wisk.

(a) Find an approximate 99% confidence interval for p.

(b) Find the approximate P-value for a test of the null hypothesis H_0: $p = .5$ versus the one-sided alternative H_A: $p > .5$.

8.2.4. *Do Infants Prefer Striped or Plain Patterns?* Consider the Frantz et al. (1962) infant visual response data given in Table 8.6.

(a) Find an approximate 95% confidence interval for $p_{\frac{1}{16}} = P$(randomly chosen infant aged 2–4 months will prefer the 1/16 inch striped pattern over the plain gray pattern).

(b) Use the result in (a) to test the null hypothesis H_0: $[p_{\frac{1}{16}} = .5]$ versus the two-sided alternative hypothesis H_A: $[p_{\frac{1}{16}} \neq .5]$ at approximate significance level .05.

Chapter 8 Comprehensive Exercises

8.A. Conceptual

8.A.1. Compare and contrast the differences between matched pairs and paired replicates designs and associated data. Explain how each of these designs controls for circumstances or characteristics other than those under investigation that might also affect the outcome of the measurements of interest.

8.A.2. Consider the following experimental design for data collection. Is it a matched pairs design or a paired replicates design?

Experimental Design A laboratory scientist purchases eighteen identical new white towels to be used in a test of the relative effectiveness of the two detergents Wisk and Tide. Each of the following nine stains are applied identically to a different pair of these towels: grass, strawberry jam, mud, spaghetti sauce, orange juice, rust, wine, oil, and beef gravy. One towel of each pair is washed in Wisk and the other in Tide. A consumer panel is then asked to evaluate each such pair and come to a decision whether Tide or Wisk did a better job of removing that particular stain.

8.A.3. Consider the following experimental design for data collection. Is it a matched pairs design or a paired replicates design?

Experimental Design Twenty subjects with mildly high blood pressure volunteer to test a new diet regimen designed to treat such a condition. Each subject has his/her blood pressure measured each of two consecutive days prior to beginning the diet regimen and then every other day for 2 months on the diet.

8.A.4. Consider the following experimental design for data collection. Is it a matched pairs design or a paired replicates design?

Experimental Design A potential advertiser for a new television program wishes to ascertain if it is more likely to be watched by men or women. Each of thirty husband and wife couples is asked to watch five episodes of the show and then separately rank the show on a scale of 1 (worst) to 10 (best).

8.A.5. Discuss, in general, why the interpretation of an analysis of matched pairs or paired replicates data is not dependent on which of the pairs is labeled Y and which is labeled X.

8.A.6. Let $D_i = Y_i - X_i$, for $i = 1, \ldots, n$, be a set of n differences for paired replicate or matched pairs data. In Example 8.1 we illustrated how to use $D_1, \ldots, D_n$ to test the null hypothesis H_0: $\eta_D = 0$, where η_D is the median of the distribution for the $Y - X$ differences.

(a) Let $\eta^* \neq 0$ be some fixed, known value. Discuss how to modify the D_i values in order to provide a test of the more general null hypothesis H_0: $\eta_D = \eta^*$.

(b) Describe how an approach similar to that described in part (a) could also be used in conjunction with Example 8.2 to provide a test of a more general null hypothesis about the mean, μ_D, of the distribution for the $Y - X$ differences.

8.A.7. Let X and Y be random variables measured in the same units with variances σ_x^2 and σ_y^2, respectively, and let $D = Y - X$. In Chap. 4 we learned that if X and Y are independent random variables, then the variance of the difference D is the sum of the X and Y variances; that is, $\sigma_d^2 = \sigma_x^2 + \sigma_y^2$. However, when the pair (X, Y) are paired replicates or matched pairs variables, it is most likely that X and Y are dependent random variables. When this is the case, the variance of the difference, σ_d^2, can be either larger or smaller than the sum, $\sigma_x^2 + \sigma_y^2$, of the individual variances, depending on how X and Y are related.

(a) Consider the pair $(X, 2X)$; that is, let $Y = 2X$. How does σ_d^2 compare to $\sigma_x^2 + \sigma_y^2$ for this choice of Y?

(b) Consider the pair $(X, -2X)$; that is, let $Y = -2X$. How does σ_d^2 compare to $\sigma_x^2 + \sigma_y^2$ for this choice of Y?

(c) Based on your findings in parts (a) and (b), what type of relationship between X and Y do you think will lead to a value of σ_d^2 that is greater than $\sigma_x^2 + \sigma_y^2$? Smaller than $\sigma_x^2 + \sigma_y^2$?

(d) Let S_x^2 and S_y^2 be the X and Y sample variances for the paired data $(X_1, Y_1), \ldots, (X_n, Y_n)$, and let S_d^2 be the sample variance for the differences $D_1 = Y_1 - X_1, \ldots, D_n = Y_n - X_n$. Choose a small sample size n and construct a paired replicates data collection for which $S_d^2 > S_x^2 + S_y^2$. For the same sample size, construct a second data collection for which $S_d^2 < S_x^2 + S_y^2$.

8.B. Data Analysis/Computational

8.B.1. *Percentage Hatched Eggs.* Consider the percentage hatched values provided in Table 8.1 (and the dataset *percentage_hatched_eggs*) for normal and dump wood duck nests. It seems quite natural to think that dump nesting might actually be *less* efficient than normal nesting in regard to leading to hatched eggs.

(a) What is the one-sided alternative hypothesis that corresponds to this conjecture?

(b) Use the data in Table 8.1 to find the P-value for an appropriate test of the null hypothesis H_0: $\eta_D = 0$ against the one-sided alternative hypothesis in (a).

(c) Which confidence bound corresponds to the one-sided alternative in (a)? Select a reasonable confidence level and find the appropriate confidence bound for the percentage-hatched data In Table 8.1.

8.B.2. *Percentage Hatched Eggs.* Consider the wood duck nesting data in Table 8.1. Let η_D denote the median for the $D = Y - X$ differences, where X and Y represent the dump and normal nest hatching percentages, respectively. Find the P-value for a test of the conjecture that the hatching percentage for normal nests is at least 8% higher than that for dump nests. (See Exercise A.6.)

8.B.3. *Percentage Hatched Eggs.* In Example 8.1 we arbitrarily chose to let X and Y correspond to the dump nest and normal nest percentage hatched, respectively. With those labels, we found that (5.01%, 14.635%) is a 90.3% confidence interval for the median, η_D, of the distribution for the variable Y - X. Suppose that we had chosen instead to reverse these arbitrary labels and let X and Y correspond to the normal nest and dump nest percentage hatched, respectively. Describe the resulting confidence interval and provide an interpretation similar to that given in Example 8.1.

8.B.4. *Right Versus Left Side of Your Brain.* What roles do the right and left sides of your brain play in distinguishing similarity of objects? This was among a number of questions addressed by Atkinson and Egeth (1973). Fifteen right-handed students at The Johns Hopkins University served as paid volunteers for the study. On each trial, two lines, one directly above the other, were presented simultaneously for 150 *msec*, randomly in either the left visual field (LVF) or in the right visual field (RVF). The subjects were told to indicate whether or not the two lines were parallel. The authors employed six different orientations for the pairs of lines: $\pm 15°$ from vertical, $\pm 45°$ from vertical, and $\pm 75°$ from vertical. On half of the trials the lines were parallel. On the other half of the trials the lines were either both tilted left or both tilted right, but were not parallel. The pairs of lines used in the right visual field were identical with the pairs of lines used in the left visual field. The mean reaction times (*msec*) for the subjects to reach their conclusions about same

Table 8.7 Mean reaction times (*msec*) for subjects to reach conclusions about same (parallel) or different line orientations for twelve LVF and twelve RVF pairs of lines

Subject	LVF	RVF
1	474.5	487
2	489	518
3	524	412.5
4	452	471
5	428.5	426
6	490.5	485
7	432	433
8	401	438.5
9	365	387.5
10	383	407.5
11	574.5	598.5
12	427	457.5
13	447.5	483.5
14	419	431.5
15	408.5	427

Source: Atkinson and Egeth (1973)

(parallel) or different orientations for the twelve LVF and twelve RVF pairs of lines are given in Table 8.7.

(a) Find the P-value for a test to evaluate the conjecture that right-handed subjects differentially use the right and left sides of their brain to distinguish whether objects are similar or not.

(b) Find a confidence interval for the difference between LVF and RVF mean reaction times for right handed people to distinguish between the same (parallel) or different pairs of lines. Choose your own reasonable confidence coefficient.

8.B.5. *Dump Nests Versus Normal Nests.* In Example 8.1 we discussed the difference in percentage-hatched eggs = (total number of hatched eggs/total number of successful eggs) for normal and dump wood duck nests at the

Table 8.8 Number of hatched eggs and number of dump and normal wood duck nests on a portion of the Duck Creek Wildlife Management Area in Missouri for the period 1966–1974

Year	Normal nests		Dump nests	
	# Nests	#Hatched eggs	#Nests	#Hatched eggs
1966	42	267	5	76
1967	89	553	9	135
1968	57	268	47	556
1969	38	58	91	607
1970	35	87	79	596
1971	68	228	53	365
1972	69	190	40	358
1973	64	329	91	656
1974	72	141	109	847

Source: Clawson et al. (1979)

Duck Creek Wildlife Management Area in Missouri over the years 1966–1974. Clawson et al. (1979) also obtained the data in Table 8.8 on the number of eggs hatched and number of nests (dump and normal) over that same time period.

(a) Consider the variables corresponding to the numbers of hatched eggs per nest for normal and dump nests. Find the *P*-value for a test of the conjecture that the hatching rate per nest is greater for dump nests than for normal nests over the period 1966–1974. What is your conclusion at significance level .05?

(b) Find a confidence interval for the difference in hatching rate per nests for dump nests and normal nests. Select your own reasonable confidence coefficient.

(c) Compare and contrast the results obtained in (a) and (b) with those found in Example 8.1 and Exercise 8.B.1.

8.B.6. *Do Infants Prefer Striped or Plain Patterns?* Consider the Frantz et al. (1962) visual response data given in Table 8.6 for infants aged 2–4 months. Let

Table 8.9 Visual preferences by infants aged 4–6 months to striped patterns of various sizes as opposed to a plain gray pattern

Stripe width (in)	Number preferring the striped pattern (Y>X)	Number preferring the plain gray pattern (Y<X)
1/64	16[a]	7[a]
1/32	20	4
1/16	21	3
1/8	24	0

[a]The observation for one infant is missing for the 1/64 inch striped pattern
Source: Frantz et al. (1962)

$p_{\frac{1}{8}} = P$(randomly chosen infant aged 2–4 months prefers the 1/8 inch striped pattern over the plain gray pattern). Find the approximate P-value for a test of the null hypothesis H_0: $[p_{\frac{1}{8}} = .2]$ versus the one-sided alternative hypothesis H_A: $[p_{\frac{1}{8}} < .2]$.

8.B.7. *Do Infants Prefer Striped or Plain Patterns?* In addition to the visual response data for infants aged 2–4 months, Frantz et al. (1962) also collected similar data for infants aged 4–6 months. Their findings for this age group are reported in Table 8.9. Carry out the same analyses discussed in Example 8.3 for the data in Table 8.9. Comment on the differences between these results for infants aged 4–6 months and those obtained in Example 8.3 for infants aged 2–4 months.

8.B.8. *Do Infants Prefer Striped or Plain Patterns?* Consider the Frantz et al. (1962) visual response data for infants aged 4–6 months given in Table 8.9. Let $p_{\frac{1}{64}} = P$(randomly chosen infant aged 4–6 months prefers the 1/64 inch striped pattern over the plain gray pattern). Find an approximate 92% lower confidence bound for $p_{\frac{1}{64}}$.

8.B.9. *Percentage Eggs Hatched.* Consider the wood duck nesting data in Table 8.1 (and the dataset *percentage_hatched_eggs*) and let π denote the probability that the percentage hatched from normal nests will be higher than the percentage hatched from dump nests in any given year in the Duck Creek Wildlife Management Area in Missouri. Find an approximate 90% confidence interval for π.

8.B.10. *Weight Gain for Anorexia Nervosa Subjects.* Consider the desimipramine/psychotherapy treatment data in Table 8.3 (and the dataset *desimipramine*). Let μ_D denote the median for the $D = Y - X$ differences, where X and Y represent the pre- and post-desimipramine body weights, respectively. Use the results obtained in Example 8.2 to test the null hypothesis H_0: $[\mu_D = 2]$ against the two-sided alternative H_A: $[\mu_D \neq 2]$ at significance level .05.

8.B.11. *Bison and Species Diversity.* In their study of ways to restore or maintain biodiversity (see Exercise B.1), Collins et al. (1998) obtained data on the numbers of plant species for plots of grassland that were grazed by bison without burning first and on control plots (no grazing or burning). The data in Table 8.10 are the total numbers of plant species observed on 16 pairs of 50-m^2 matched plots of grassland on the Konza Prairie. For each matched pair of land plots, one of the plots served as the control where nothing was done to the grassland. Bison were permitted to graze on the second plot in that pair, but there was no burning of the grassland first.

(a) Find the P-value for a test of the conjecture that permitting bison to graze on grassland leads to an increase in the number of plant species. What is your conclusion at significance level .075?

(b) Find a confidence interval for the median change in the number of plant species as a result of permitting bison to graze on grassland. Select your own reasonable confidence coefficient.

8.B.12. *Dilemma Zones and Low Speed Accidents.* In their study of the effectiveness of "Prepare to Stop When Flashing" signs (see Exercise 8.1.3), the Ohio Department of Transportation also evaluated these signs at dilemma

Table 8.10 Total number of plant species on sixteen pairs of 50-m^2 matched plots of grassland on the Konza Prairie

Plot pair	Control	Bison Grazing
1	34	38
2	46	55
3	51	59
4	55	76
5	34	55
6	40	63
7	46	65
8	55	82
9	34	63
10	39	69
11	49	71
12	62	88
13	36	46
14	41	51
15	54	55
16	63	90

Source: Collins et al. (1998)

zones with low speed (35 *mph* zones) approaches. The number of accidents at each of four low speed approaches dilemma zones was recorded at four different times of the year prior to erection of "Prepare to Stop When Flashing" signs and then during the same time periods during the following year after erection of the signs. (Since the signs were only erected for one or two of the approaches at these locations, the only accidents used in the evaluation were those that involved vehicles on one of the approaches with such a sign.) The numbers of accidents per million vehicles for these four low speed dilemma zones with and without the warning signs for the different times of year are reported in Table 8.11.

(a) Find the P-value for a test of the conjecture that erection of the "Prepare to Stop When Flashing" signs has a positive effect (i.e., leads to a

Table 8.11 Number of traffic accidents per million vehicles for low speed approach dilemma zones with and without "Prepare to Stop When Flashing" signs at four different times of the year

	Number of accidents per million vehicle miles	
Site-time	**Without signs**	**With signs**
Site 1 – Mar–May	0.981	2.452
Site 1 – June–Aug	0.490	0
Site 1 – Sept–Nov	0.981	0.981
Site 1 – Dec–Feb	1.471	0
Site 2 – Mar–May	0.638	1.276
Site 2 – June–Aug	2.553	2.553
Site 2 – Sept–Nov	0.638	1.915
Site 2 – Dec–Feb	0.638	2.553

Source: Ohio Department of Transportation (1980)

decrease in traffic accidents per vehicle miles) at low speed dilemma zones. What is your conclusion at significance level .075?

(b) Find a confidence interval for the median change in traffic accidents per vehicle miles at low speed dilemma zones that results from erecting the "Prepare to Stop When Flashing" signs. Select your own reasonable confidence coefficient.

8.B.13. *Right Versus Left Side of Your Brain.* In Exercise B.4 we considered whether the right and left sides of one's brain play different roles in distinguishing similarity of objects. Atkinson and Egeth (1973) also considered the question of whether one side of the brain might be better attuned to identifying similar objects, while the other side of the brain might do better at distinguishing when objects are different. . The mean reaction times (*msec*) for the subjects to reach their conclusions about six sets of differently oriented parallel lines (i.e., same lines) in the left visual field (LVF) and the same six sets of parallel lines in the right visual field (RVF) are given in Table 8.12. The mean reaction times (*msec)* for the subjects to reach their conclusions about six sets of differently oriented non-parallel lines (i.e., different lines) in LVF and

Table 8.12 Mean reaction times (*msec*) for subjects to reach conclusions about line orientations for six LVF and six RVF parallel pairs of lines

Subject	LVF	RVF
1	471	493
2	477	524
3	303	345
4	428	444
5	418	423
6	517	499
7	412	385
8	381	444
9	343	363
10	390	390
11	501	514
12	429	467
13	439	473
14	377	390
15	401	429

Source: Atkinson and Egeth (1973)

the same six sets of non-parallel lines in RVF are given in Table 8.13. (We note that the mean reaction times in Table 8.7 that are used for the overall assessment of right and left brain activity in Exercise B.4 are simply the averages of the mean reaction times for the parallel lines given in Table 8.12 and those for the non-parallel lines in Table 8.13).

(a) Find the *P*-value for a test to evaluate the conjecture that right-handed subjects differentially use the right and left sides of their brain to identify similar objects.

(b) Find a confidence interval for the difference between LVF and RVF mean reaction times for right-handed people to identify parallel pairs of lines. Choose your own reasonable confidence coefficient.

Table 8.13 Mean reaction times (*msec*) for subjects to reach conclusions about line orientations for six LVF and six RVF non-parallel pairs of lines

Subject	LVF	RVF
1	478	481
2	501	512
3	745	480
4	476	498
5	439	429
6	464	471
7	452	481
8	421	433
9	387	412
10	376	425
11	648	683
12	425	448
13	456	494
14	461	473
15	416	425

Source: Atkinson and Egeth (1973)

(c) Find the P-value for a test to evaluate the conjecture that right-handed subjects differentially use the right and left sides of their brain to distinguish between non-parallel lines.

(d) Find a confidence interval for the difference between LVF and RVF mean reaction times for right-handed people to distinguish between non-parallel pairs of lines. Choose your own reasonable confidence coefficient.

8.C. Activities

8.C.1. *Dominant Hand.* Is your dominant hand (i.e., the one you write with) physically stronger than your other hand? Design an experiment (including the appropriate data to collect and how to collect it) that will enable you to statistically address this conjecture. Collect the relevant data for a sample of 15 of your friends, conduct the appropriate statistical analyses, and write a two-page report describing your experiment and statistical conclusions.

8.C.2. *Exercise and Pulse Rate.* How much does exercise increase one's pulse rate? Design an experiment (including the appropriate data to collect and how to collect it) that will enable you to statistically address this question. Collect the relevant data for a sample of 15 of your friends, conduct the appropriate statistical analyses, and write a two-page report describing your experiment and statistical conclusions.

8.C.3. *Paired Replicates.* Find a journal article in a field of your interest that presents the results of a study that utilized paired replicates data. Prepare a short (2–3 pages) summary report of the statistical findings in the article and attach a copy of the original paper with your summary.

8.C.4. *Matched Pairs.* Find a journal article in a field of your interest that presents the results of a study that utilized matched pairs data. Prepare a short (2–3 pages) summary report of the statistical findings in the article and attach a copy of the original paper with your summary.

8.D. Internet Archives

8.D.1. *Precipitation for U.S. Cities.* Search the Internet to find a site that provides precipitation amounts for U. S. cities for each of the calendar years 2013 and 2014. Do these precipitation amounts across these U. S. cities represent paired replicates data or matched pairs data? How would you use these data to compare the precipitation pattern in the United States in these two calendar years?

8.D.2. *Statewise Average Family Income.* Search the Internet to find a site that provides average family income for each of the fifty states in the United States for calendar years 2013 and 2014. Do these average family incomes across the fifty states represent paired replicates data or matched pairs data? How would you use these data to compare the economic status of families across the United States during this period of time?

9 Statistical Inference for Two Populations–Independent Samples

William Shakespeare penned the famous quote "A rose by any other name would smell as sweet". Does this sentiment carry over to names given you by your parents? Christenfeld et al. (1999) were not so sure after studying the possible effect that the initials of your name may have on your life expectancy! Dividing names into those with "bad initials", such as DED, SIC, UGH, ROT, etc., and those with "good initials", such as GOD, HUG, VIP, WIN, etc., they studied California death certificates to see if there appeared to be a difference in age at death for people with initials in these two general categories.

Problems such as this are called *two-sample problems* because they involve the comparison of data collected from two distinct populations. They are among the most common statistical problems encountered in practical applications. The collected data could be the results of observational studies, as is the case with the initials data discussed above, or they could be obtained through a designed experiment where the data are a result of independent random samples collected separately from each of the involved populations.

© Springer International Publishing AG 2017

D.A. Wolfe, G. Schneider, *Intuitive Introductory Statistics*, Springer Texts in Statistics,
DOI 10.1007/978-3-319-56072-4_9

In Sect. 1.3 we discussed a variety of data analysis techniques for comparing such two-sample data. In this chapter we present appropriate procedures to make formal statistical inferences from these data. As is the case with the one-sample and paired replicates/matched pairs settings in Chaps. 7 and 8, respectively, a proper choice of two-sample statistical inference procedures will depend on both the type of data collected (count or numerical) and the particular assumptions which are reasonable to make about the two underlying populations.

General Setting and Notation Let $X_1, \ldots, X_m$ and $Y_1, \ldots, Y_n$ denote the items of independent random samples from two distinct populations. Choice of the most appropriate statistical techniques for comparing typical values from these populations depends primarily on two issues: (1) Are the data counts (i. e., tallies of how often a particular event happens) or quantitative in nature? and (2) What information is available or can reasonably be assumed about the shapes of the distributions for the two underlying populations?

In Sect. 1 we deal exclusively with procedures designed to produce appropriate inferences for two-sample count data. On the other hand, when the sample data from our two populations are numerical in nature, we are naturally interested in possible differences in the centers of the two populations. If we had just a single X and a single Y observation, we would estimate the difference between the centers of the Y and X populations by the value of the single sample difference $Y - X$. The information about the respective centers contained in this difference consists of two components, its sign and its magnitude. The sign provides evidence as to whether a typical observation from the Y population is larger or smaller than a typical observation from the X population. The magnitude $|Y - X|$, on the other hand, helps assess the size of the difference between typical observations from the two populations.

Section 2 presents procedures designed to produce appropriate statistical inferences about possible differences in the two population medians, η_X and η_Y, for arbitrary continuous populations. The final two Sects. 3 and 4 are devoted to statistical inference procedures for comparing two population means, μ_X and μ_Y, when it is reasonable to assume that both underlying distributions are normal (Sect. 3) or for non-normal settings where both sample sizes, m and n, are relatively large (Sect. 4). In Sect. 5 we discuss and compare these competing approaches to making statistical inferences about the difference in the centers of two distributions.

9.1 Approximate Inference for the Difference in Proportions for Two Populations

In Chap. 6 we discussed statistical procedures for making inferences about the probability that an event, say A, occurs when we sample an observation from a single population; that is, to provide information about the proportion of the population that possesses the attribute described by the event A. One of the primary questions of interest in many practical statistical applications is how the probabilities for such an event A might differ for two distinct populations; that is, do the proportions in the two populations possessing the attribute described by the event A differ? For example, we might be interested in whether men and women differ in regard to their attitudes toward saving for retirement or whether there is a difference in hearing loss between smokers and non-smokers or whether athletes and non-athletes differ relative to the extent of alcohol-related problems on college campuses.

For such questions, the relevant parameters are the proportions, p_X and p_Y, of the X and Y populations, respectively, which possess the attribute described by the event A and our interest is in making inferences about the difference p_Y - p_X. The sample data X_1, ..., X_m and Y_1, ..., Y_n for these settings are simply indicators of whether or not the sampled units possess the

A-attribute and statistical inferences about p_Y - p_X are naturally based on the proportions of sample observations from each of the two populations with the A-attribute, namely,

$$\hat{p}_X = [\text{number of } X\text{–sample units with attribute } A]/m \tag{9.1}$$

and

$$\hat{p}_Y = [\text{number of } Y\text{–sample units with attribute } A]/n. \tag{9.2}$$

The natural estimate for the difference p_Y - p_X is the analogous observed difference in the sample proportions $V = \hat{p}_Y - \hat{p}_X$. It follows from the standard rule for the mean of a difference between two variables (see Sect. 4.5) that the mean for V is $\mu_V = p_Y - p_X$, since the sample proportions are unbiased estimators of the corresponding population proportions. Moreover, since the sample proportions $\hat{p}_X$ and $\hat{p}_Y$ are based on independent random samples from the X and Y populations, respectively, the variance for V corresponds to the sum of the variances for the separate sample proportions (again, see Sect. 4.5); that is,

$$\sigma_V^2 = \frac{p_Y(1-p_Y)}{n} + \frac{p_X(1-p_X)}{m},$$

and the corresponding standard deviation for V is.

$$\sigma_V = \sqrt{\frac{p_Y(1-p_Y)}{n} + \frac{p_X(1-p_X)}{m}}.$$

To construct confidence intervals and conduct hypothesis tests about p_Y - p_X we make use of these expressions for the mean and standard deviation for V and a central limit theorem (see Sect. 5.3) to infer that the sampling distribution of the statistic V can be well-approximated for large sample sizes m and n by the normal distribution with mean μ_V and standard deviation σ_V; that is, when m and n are large, inferences about p_Y - p_X can be based on the fact that V is approximately normally distributed with mean μ_V and standard deviation σ_V.

Interval Estimation of the Difference in Two Population Proportions Since the point estimator $V = \hat{p}_Y - \hat{p}_X$ for the difference in proportions p_Y - p_X is approximately normally distributed it is natural to expect an approximate confidence interval for p_Y - p_X to be of the form $V \pm z_{\frac{1-CL}{2}} \sigma_V$, where $\frac{z_{1-CL}}{2}$ is the upper ((1-CL)/2)th percentile for the standard normal distribution. However, the standard deviation σ_V for V depends on the unknown population proportions p_Y and p_X. Hence, we must first estimate σ_V by its sample analogue

$$\widehat{\sigma}_v = \sqrt{\frac{\hat{p}_Y(1-\hat{p}_Y)}{n} + \frac{\hat{p}_X(1-\hat{p}_X)}{m}}. \tag{9.3}$$

The approximate 100CL% confidence interval for p_Y - p_X is then provided by the interval

$$V \pm z_{\frac{1-CL}{2}} \widehat{\sigma}_V = \left(\hat{p}_Y - \hat{p}_X - z_{\frac{1-CL}{2}} \widehat{\sigma}_V, \hat{p}_Y - \hat{p}_X + z_{\frac{1-CL}{2}} \widehat{\sigma}_V\right), \tag{9.4}$$

where $\widehat{\sigma}_v$ is given by the expression in (9.3). The corresponding approximate 100CL% lower and upper confidence bounds for p_Y - p_X are given by $\hat{p}_Y - \hat{p}_X - z_{1-CL}\widehat{\sigma}_v$ and $\hat{p}_Y - \hat{p}_X + z_{1-CL}\widehat{\sigma}_v$, respectively.

Example 9.1. College Athletes—Are They Health-Conscious Role Models? Alcoholic consumption by college students has been the topic of a substantial amount of research over the past few decades. The results of this research show clearly that alcohol is the most commonly used drug on American college campuses and that the majority of college students drink on a regular basis. These findings naturally led to questions regarding possible differences in the demographic characteristics of students who are involved in heavy drinking and those who are not. In particular, a number of researchers became interested in a possible connection between participation in intercollegiate athletics and drinking. Some early investigators conjectured that athletes would, of necessity, be **less likely** to drink excessively than non-athlete college students in order to maintain the peak physical fitness required by their sports. However, the findings from a number of studies simply did not support this conjecture. In fact, it was often found that athletes are actually **more likely** to participate in excessive and potentially harmful alcohol consumption than the typical non-athlete student.

In one such definitive study, Leichliter et al. (1998) studied sample data collected from 51,483 students who participated in Core and Alcohol Surveys between October 1994 and May 1996 at 125 institutions of higher education across the United States. Of these sampled students, 8749 were involved in at least one intercollegiate sport, while the remaining 42,734 were nonparticipants is such activities. One piece of information gathered from the survey was whether or not a student had engaged in binge drinking in the 2 weeks previous to completing their survey. The data in Table 9.1 represent the findings from the study.

Letting X and Y denote the nonparticipant and participant populations, respectively, and taking A to be the event that a student engaged in binge drinking at least once in the past 2 weeks, the parameters of interest in this study are p_X = [proportion of students who do not participate in

Table 9.1 Number of students reporting involvement in binge drinking in the 2 weeks prior to completing the core and alcohol survey

	Involvement in intercollegiate sports	
	Participants	**Nonparticipants**
Number completing survey	8749	42,734
Number engaging in binge drinking in the previous two weeks	4835	15,513

Source: Leichliter et al. (1998)

intercollegiate sports and engage in binge drinking at least once every 2 weeks] and p_Y = [proportion of student participants in intercollegiate sports who engage in binge drinking at least once every 2 weeks]. Our point estimate for the difference p_Y - p_X is then the observed difference in the sample proportions $V = \hat{p}_Y - \hat{p}_X = \frac{4835}{8749} - \frac{15,513}{42,734} = .553 - .363 = .19$. For an approximate 94% confidence interval for p_Y - p_X, we first find the estimated standard deviation for $V = \hat{p}_Y - \hat{p}_X$ from Eq. (9.3) to be

$$\widehat{\sigma}_V = \sqrt{\frac{.533(1 - .533)}{8749} + \frac{.363(1 - .363)}{42,734}} = .00582.$$

With $\frac{z_{(1-.94)}}{2} = z_{.03} = 1.88$, the approximate 94% confidence interval for p_Y - p_X is then given by expression (9.4) to be

$$(.19 - 1.88(.00582), .19 + 1.88(.00582)) = (.179, .201).$$

Thus we are approximately 94% confident that the proportion of student participants in intercollegiate sports who engage in binge drinking at least once every 2 weeks is somewhere between .179 and .201 higher than the proportion of nonparticipating students who engage in binge drinking that often.

Hypothesis Tests for the Difference between Two Population Proportions To test the null hypothesis H_0: $p_Y = p_X$, corresponding to no differences in the proportions of the X and Y populations that possess the attribute described by the event A, against an appropriate alternative of interest, we once again rely on the approximate normality of the point estimator $V = \hat{p}_Y - \hat{p}_X$. Under the null hypothesis that $p_Y = p_X = p$, the statistic V has an approximate normal distribution with mean 0 and standard deviation.

$$\sigma_V = \sqrt{\frac{p(1-p)}{n} + \frac{p(1-p)}{m}} = \sqrt{p(1-p)\left(\frac{1}{n} + \frac{1}{m}\right)}. \tag{9.5}$$

When H_0 is true, it is natural to estimate the common proportion p by the overall proportion of sample items possessing the attribute described by the event A; that is, we combine or *pool* the counts from both the X and Y samples to produce the *pooled estimate* of p given by.

$$\hat{p} = \frac{[\textit{number of X sample units with attribute A}] + [\textit{number of Y sample units with attribute A}]}{m+n}. \tag{9.6}$$

Combining (9.5) and (9.6), our pooled point estimate for the standard deviation of V under the null hypothesis that $p_Y = p_X = p$ is

$$\widehat{\sigma}_V = \sqrt{\hat{p}(1-\hat{p})\left(\frac{1}{n} + \frac{1}{m}\right)}. \tag{9.7}$$

Statistical procedures for testing the null hypothesis H_0: $[p_Y = p_X]$ against appropriate alternatives are then based on the fact that the standardized statistic $V^* = \frac{V}{\widehat{\sigma}_V}$ has a null (H_0) sampling distribution that can be well-approximated by the standard normal distribution when the sample sizes m and n are both large. To test H_0: $[p_Y = p_X\}$ compute the standardized statistic

$$V^* = \frac{\hat{p}_Y - \hat{p}_X}{\sqrt{\hat{p}(1-\hat{p})\left(\frac{1}{n} + \frac{1}{m}\right)}}, \tag{9.8}$$

with $\hat{p}_X$, $\hat{p}_Y$, and $\hat{p}$ given by (9.1), (9.2), and (9.6), respectively. Letting v^* denote the observed value of the test statistic V^*, the approximate P-value for a test of H_0: $p_Y = p_X$ against the alternatives H_A are:

H_A	Approximate P-value	
$p_Y > p_X$	$\approx P(Z \geq v^*)$	(9.9)
$p_Y < p_X$	$\approx P(Z \leq v^*)$	(9.10)
$p_Y \neq p_X$	$\approx 2P(Z \geq \lvert v^* \rvert)$,	(9.11)

where $Z \sim N(0, 1)$.

Example 9.2. Is there Gender Bias in Children's Preferences for Musical Instruments? A number of studies have clearly demonstrated that boys and girls at young ages generally do not exhibit preferences for the same musical instruments. Moreover, children's choices of musical instruments tend to be in agreement with the way such instruments are gender stereotyped by adults. One study by Abeles and Porter (1978) found that adults were more likely to select a 'feminine' instrument (i. e., clarinet, flute or violin) for a daughter and a 'masculine' instrument (i. e., drum, trombone or trumpet) for a son.

In a study designed to assess whether similar gender biases are already present in young boys and girls, O'Neill and Boulton (1996) conducted interviews with 72 female and 81 male children between the ages of 9 and 11 residing in north-west England. Among the questions asked of these youngsters was a question regarding whether or not it was inappropriate for boys or girls to play specific instruments. The data from this portion of the interviews by O'Neill and Boulton for six different instruments are presented in Table 9.2.

Your initial impression from these data is probably that there is striking agreement between the male and female participants in this study with respect to which instruments are inappropriate for boys and girls. But notice that only 10 of 72 girls say the violin is not appropriate for boys, while 29 of

Table 9.2 Numbers of survey participants who indicated that girls and boys should not play each of six different instruments

	Female participants		Male participants	
	Indicated that the instrument should not be played by			
	Boys	Girls	Boys	Girls
Flute	35	0	36	1
Piano	16	0	21	1
Violin	10	0	29	4
Trumpet	2	9	3	19
Guitar	0	32	1	36
Drums	0	44	0	54

Source: O'Neill and Boulton (1996)

81 boys hold this opinion. Using these sample percentages, we would like to test whether the corresponding proportions for the populations of all 9–11 year old boys and girls differ. Letting A correspond to the opinion that the violin should not be played by boys and denoting the population of 9-11 year old boys by X and that of 9–11 year old girls by Y, we have p_X = [proportion of 9–11 year old boys who believe that boys should not play the violin] and p_Y = [proportion of 9–11 year old girls who believe that boys should not play the violin]. We are interested in testing the null hypothesis H_0: $p_Y = p_X$, corresponding to no gender difference in opinions about whether boys should play the violin, versus the two-sided alternative H_A: $p_Y \neq p_X$ that 9-11 year old boys and girls do, in fact, differ in their opinions about whether boys should play the violin.

From the data in Table 9.2, we see that $\hat{p}_X = \frac{29}{81} = .358$ and $\hat{p}_Y = \frac{10}{72} = .139$. From (9.6) the pooled estimate of the common p under H_0 is $\hat{p} = \frac{29+10}{81+72} = .255$. It follows from (9.8) that the observed value of V^* is

$$v^* = \frac{.139 - .358}{\sqrt{.255(1 - .255)\left(\frac{1}{72} + \frac{1}{81}\right)}} = -3.10.$$

Hence, from (9.11) the approximate P-value for our test of H_0: $p_Y = p_X$ against the two-sided alternative H_A: $p_Y \neq p_X$ is found using the **R** function *pnorm*( ) to be $2 \times P(Z \geq |-3.10|) = 2 \times (.00096) = .0019$.

```
> 2 * pnorm(3.1, lower.tail = FALSE)
[1] 0.001935206
```

Thus there is considerable evidence in the survey results to indicate that boys and girls aged 9–11 do, indeed, differ in their views as to whether boys should play the violin. (What about some of the other instruments for which survey data are given in Table 9.2? See Exercises 9.B.3 and 9.B.4, for example.)

Section 9.1 Practice Exercises

9.1.1. We are interested in the proportions, p_X and p_Y, of two populations X and Y, respectively, that possess the attribute described by an event A. Consider independent random samples X_1, X_2, X_3, X_4, X_5 and Y_1, Y_2, Y_3, Y_4 from the X and Y populations and let $\hat{p}_X$ (9.1) and $\hat{p}_Y$ (9.2) denote the proportions of these sample observations with the A-attribute.

(a) List the possible sample values for the sample proportions $\hat{p}_X$ and $\hat{p}_Y$?

(b) List the possible sample values for the difference in sample proportions $V = \hat{p}_Y - \hat{p}_X$?

9.1.2. The expression for an approximate $100CL\%$ confidence interval for p_Y - p_X is given in (9.4). We know that this confidence interval is centered at the observed difference in sample proportions $\hat{p}_Y - \hat{p}_X$. What factors affect the length of the confidence interval?

9.1.3. Let p_X and p_Y denote the proportions of the X and Y populations, respectively, that possess an attribute described by the event A. Based on independent random samples $X_1, \ldots, X_{100}$ and $Y_1, \ldots, Y_{100}$ from the X and Y populations, let $\hat{p}_X$ (9.1) and $\hat{p}_Y$ (9.2) denote the proportions of these sample observations with the attribute A. Evaluate the approximate 95% confidence interval for $p_Y - p_X$ given in expression (9.4) for the following five possible outcomes for the pair $(\hat{p}_X, \hat{p}_Y)$:

$$(\hat{p}_X, \hat{p}_Y) = (.4, .5), (.3, .4), (.2, .3), (.1, .2), (0, .1).$$

(a) Where are the five 95% confidence intervals centered?
(b) What are the lengths of the five 95% confidence intervals?
(c) Comment on your findings in (a) and (b).

9.1.4. Let p_X and p_Y denote the proportions of the X and Y populations, respectively, that possess an attribute described by the event A. Based on equal size ($m = n$) independent random samples $X_1, \ldots, X_m$ and $Y_1, \ldots, Y_m$ from the X and Y populations, let $\hat{p}_X = .4$ and $\hat{p}_Y = .5$ denote the proportions of these sample observations with the attribute A. Evaluate the approximate 95% confidence interval for $p_Y - p_X$ given in expression (9.4) if the common sample size $m = n$ is 10, 20, 40, 80, 160, or 500.

(a) Where are the six 95% confidence intervals centered?
(b) What are the lengths of the six 95% confidence intervals?
(c) Comment on your findings in (a) and (b).

9.1.5. *Taking Care of One Another—Conservatives Versus Liberals.* Consider the data on American values with respect to taking care of each other, as collected by Princeton Survey Research Associates of Princeton, New Jersey (1998) and discussed in Exercise 1.B.3. Let p_C and p_L denote the proportions of all Conservatives and Liberals, respectively, who feel that people should expect help from the government to take care of their parents if they become sick or disabled. Using the sample data, find an approximate 97% confidence interval for the difference in proportions, $p_C - p_L$.

9.1.6. *Taking Care of Each Other—Conservatives Versus Moderates.* Consider the data on American values with respect to taking care of each other, as collected by Princeton Survey Research Associates of Princeton, New Jersey (1998) and discussed in Exercise 1.B.3. Let p_C and p_M denote the proportions of all Conservatives and Moderates, respectively, who feel that people should expect help from the government to take care of their parents if they become sick or disabled.

(a) Compute the values of the sample estimates, $\hat{p}_C$ and $\hat{p}_M$, for p_C and p_M.

(b) Compute the value of the pooled point estimate for the standard deviation of $V = \hat{p}_C - \hat{p}_M$ under the null hypothesis that $p_C = p_M$.

(c) Compute the value of the standardized statistic V^* (9.8).

9.1.7. Let p_X and p_Y denote the proportions of the X and Y populations, respectively, that possess an attribute described by the event A. Based on independent random samples $X_1, \ldots, X_{100}$ and $Y_1, \ldots, Y_{100}$ from the X and Y populations, let $\hat{p}_X$ (9.1) and $\hat{p}_Y$ (9.2) denote the proportions of these sample observations with the attribute A. Find the approximate P-value for testing the null hypothesis H_0: $p_Y = p_X$ against the one-sided alternative H_A: $p_Y > p_X$ for each of the following five possible outcomes for the pair $(\hat{p}_X, \hat{p}_Y)$:

$$(\hat{p}_X, \hat{p}_Y) = (.4, .5), (.3, .4), (.2, .3), (.1, .2), (0, .1).$$

Comment on these findings.

9.1.8. Let p_X and p_Y denote the proportions of the X and Y populations, respectively, that possess an attribute described by the event A. Based on equal size ($m = n$) independent random samples $X_1, \ldots, X_m$ and $Y_1, \ldots, Y_m$ from the X and Y populations, let $\hat{p}_X = .4$ and $\hat{p}_Y = .5$ denote the proportions of these sample observations with the attribute A. Find the approximate P-value for testing the null hypothesis H_0: $p_Y = p_X$ against the one-sided alternative H_A: $p_Y > p_X$ if the common sample size $m = n$ is 10, 20, 40, 80, 160, or 500. Comment on these findings.

9.1.9. *Taking Care of One Other—Moderates Versus Liberals.* Consider the data on American values with respect to taking care of each other, as collected by Princeton Survey Research Associates of Princeton, New Jersey (1998) and discussed in Exercise 1.B.3. Let p_M and p_L denote the proportions of all Moderates and Liberals, respectively, who feel that people should expect help from the government to take care of their parents if they become sick or disabled. Find the approximate *P*-value for a test of the conjecture that Liberals are more inclined than Moderates to feel that people should expect help from the government to take care of their parents if they become sick or disabled.

9.1.10. *Young of Year Gizzard Shad.* Consider the length of YOY gizzard shad data displayed in Table 1.38 in Exercise 1.B.13. Pool the ten observations from Sites 1 and 2 to constitute a single random sample of 20 observations from Site "C" and do the same for the ten observations from Sites 3 and 4 to constitute a single random sample of 20 observations from Site "D". Let p_C and p_D denote the proportion of all the YOY gizzard shad from Sites "C" and "D", respectively, that were at least 30 *mm* in length at the time that Johnson (1984) conducted his study.

(a) Compute the values of the sample estimates, $\hat{p}_C$ and $\hat{p}_D$, for p_C and p_D.
(b) Find an approximate 95% confidence interval for the difference in proportions, $p_C - p_D$.

9.1.11. *Art and the Color Purple.* Wypijewski (1997) reported on the results of a comprehensive scientific poll of American tastes in art, as commissioned by Vitaly Komar and Alexander Melamid in conjunction with The Nation Institute, a nonprofit offshoot of *The Nation* magazine. Random samples of 475 males and 526 females were asked to name their favorite color. Thirty-seven females and six males in the samples named purple as their favorite color.

(a) Estimate the proportions of all American females and all American males for whom purple is their favorite color.
(b) Find the approximate *P*-value for a test of the conjecture that the proportion of American females for whom purple is their favorite color is greater than the proportion of American males who prefer purple.

9.1.12. *Intimate Partner Abuse.* Often a hospital Emergency Department is the initial (and sometimes only) contact that abused women have with health care clinicians. The identification and treatment of abused women by Emergency Department personnel has become increasingly important as a potential means for preventing future abuse. In a study to determine the prevalence of intimate partner abuse among female patients seeking treatment in community hospital Emergency Departments, Dearwater et al. (1998) conducted an anonymous survey during the period 1995 through 1997. They inquired about physical, sexual, and emotional abuse among women aged 18 years or older who sought treatment in Emergency Departments during selected shifts in eleven community Emergency Departments in California and Pennsylvania. A survey respondent was classified as having suffered "past-year prevalence of physical or sexual abuse"if she answered yes to either or both of the following questions:

(i) "Within the past year, have you been pushed, shoved, hit, slapped, kicked, or otherwise physically hurt by your husband, boyfriend, or partner (or ex-husband, ex-boyfriend, or ex-partner)?
(ii) "Within the past year, has your husband, boyfriend, or partner forced you to have sexual activities (or ex-husband, ex-boyfriend, or ex-partner)?

Of the 1538 California respondents to the survey, 268 indicated that they had suffered "past-year prevalence of physical or sexual abuse". Of the 1917

Pennsylvania respondents to the survey, 230 indicated that they had suffered "past-year prevalence of physical or sexual abuse". Among all women aged 18 years or older who might seek treatment in an Emergency Department in California or Pennsylvania, let p_C and p_P denote the percentages in California and Pennsylvania, respectively, who have suffered "past-year prevalence of physical or sexual abuse".

(a) Estimate p_C and p_P.

(b) Find the approximate *P*-value for a test of the conjecture that p_C and p_P are not the same.

(c) Find an approximate 94% confidence interval for p_C - p_P .

9.1.13. *Underwear—Importance of Design and Appearance.* Wypijewski (1997) reported on the results of a comprehensive scientific poll of American tastes in art, as commissioned by Vitaly Komar and Alexander Melamid in conjunction with The Nation Institute, a nonprofit offshoot of *The Nation* magazine. Random samples of 779 white, 100 African American, and 66 Hispanic subjects were asked the question: How important would the appearance or design of underwear be in your decision about whether or not to buy it? One hundred forty white, thirty-two African American, and twelve Hispanic respondents indicated that appearance or design was very important in their decisions.

(a) Estimate the proportions of all white Americans, all African Americans, and all Hispanic Americans for whom appearance or design is very important in their decisions about whether or not to buy underwear.

(b) Find an approximate 96% confidence interval for the difference between the proportion of all white Americans and the proportion of all African Americans for whom appearance or design is very important in their decisions about whether or not to buy underwear.

(c) Find an approximate 96% confidence interval for the difference between the proportion of all white Americans and the proportion of all Hispanic Americans for whom appearance or design is very important in their decisions about whether or not to buy underwear.

(d) Find an approximate 96% confidence interval for the difference between the proportion of all African Americans and the proportion of all Hispanic Americans for whom appearance or design is very important in their decisions about whether or not to buy underwear.

(e) Discuss your findings in parts (b) – (d).

9.1.14. *Intimate Partner Abuse—Age Factor.* Consider the study by Dearwater et al. (1998) to determine the prevalence of intimate partner abuse among female patients seeking treatment in community hospital Emergency Departments, as discussed in Exercise 9.1.12. One of the questions of interest to the investigators was whether or not age was a factor in the prevalence of physical or sexual abuse. Of the 3455 respondents to the survey (California and Pennsylvania combined), 1693 were in the age group 18-39 and 400 of these women were classified as having suffered "past year prevalence of physical or sexual abuse". One hundred of the 1762 respondents who were at least 40 years old were classified as having suffered "past year prevalence of physical or sexual abuse". Among the population of all women who might seek treatment in an Emergency Department in either California or Pennsylvania, let $p_{18\text{-}39}$ and $p_{\geq 40}$ denote the percentages of women in the age groups 18-39 and $\geq$40, respectively, who have suffered "past-year prevalence of physical or sexual abuse".

(a) Estimate $p_{18\text{-}39}$ and $p_{\geq 40}$.

(b) Find the approximate P-value for a test of the conjecture that among women who might seek treatment in an Emergency Department in either California or Pennsylvania, those women in the age group

18-39 are more likely to have suffered "past-year prevalence of physical or sexual abuse" than are women who are at least 40 years old.

(c) Find an approximate 98% lower confidence bound for $p_{18\text{-}39} - p_{\geq 40}$.

9.1.15. *Intimate Partner Abuse—Survey Respondents Versus Non-Respondents.* Consider the study by Dearwater et al. (1998) to determine the prevalence of intimate partner abuse among female patients seeking treatment in community hospital Emergency Departments, as discussed in Exercises 9.1.12 and 9.1.14. A total of 4641 patients were seen at the eleven Emergency Departments in the study during the periods of data collection. As noted previously, 3455 of these patients voluntarily agreed to respond to the survey and be included (anonymously) in the study.

(a) What percentage of the patients voluntarily agreed to participate in the study?

(b) Discuss possible areas of concern over potential differences between the groups of respondents and non-respondents to this survey. How might these potential differences impact the analyses discussed in Exercises 9.1.12 and 9.1.14?

(c) The authors noted that the respondents to the survey were significantly younger than were the non-respondents. Discuss the potential impact this might have on the statistical conclusions reached in Exercises 9.1.12 and 9.1.14.

9.1.16. *Intimate Partner Abuse—Extrapolation?* Consider the study by Dearwater et al. (1998) to determine the prevalence of intimate partner abuse among female patients seeking treatment in community hospital Emergency Departments, as discussed in Exercises 9.1.12 and 9.1.14. In view of the results obtained in those two exercises, do you feel that it would be reasonable to infer that the results obtained from these data are also applicable to:

(a) all women aged 18 years or older who might seek treatment in an Emergency Department anywhere in the United States?
(b) all women aged 18 years or older in Pennsylvania (using only the Pennsylvania data) or all women aged 18 years or older in California (using only the California data), regardless of whether or not they might seek treatment in an Emergency Department in that state?
(c) Provide reasons for your answers.

9.2 Inference for the Difference in Medians for Any Two Continuous Populations

When sample data $X_1, \ldots, X_m$ and $Y_1, \ldots, Y_n$ from two populations are numerical in nature, we are naturally interested in possible differences in the centers of the two populations. If we had just a single X and a single Y observation, we would estimate the difference between the centers of the Y and X populations by the value of the single sample difference $Y - X$. The information about the respective centers contained in this difference consists of two components, its sign and its magnitude. The sign provides evidence as to whether a typical observation from the Y population is larger or smaller than a typical observation from the X population. The magnitude $|Y - X|$, on the other hand, helps us assess the size of the difference between typical observations from the two populations.

In order to use all of the data when there are m sample observations $X_1, \ldots,$ X_m from the X population and n sample observations $Y_1, \ldots, Y_n$ from the Y population, it is, therefore, quite natural to base any statistical assessment of possible differences in the typical values and/or centers of the two populations solely on the mn sample differences $D_{ij} = Y_j - X_i$, for $i = 1, \ldots, m$ and $j = 1, \ldots, n$. Any reasonable statistical procedure will rely on these D_{ij} differences to summarize the sample information in order to make inferences about the typical values and/or centers for the X and Y populations.

Where the various statistical inference procedures differ is in how they choose to utilize this information.

If both the X and Y populations correspond to continuous measurements, but we do not have any additional information about the shapes of the populations (e. g., symmetry, outliers, gaps, etc.), we would typically use the difference between the two population medians η_X and η_Y, say $\Delta = \eta_Y - \eta_X$, to assess the difference between the centers of the populations. An additional indicator of the difference between typical values from the two populations for this setting is provided by the probability statement $P(X < Y)$. Note that $P(X < Y)$ represents the likelihood that a randomly chosen member from the Y population will have a larger value than a randomly chosen member from the X population, while $\Delta = \eta_Y - \eta_X$ measures the typical size of the difference between such randomly chosen X and Y values.

Point Estimation of $P(X < Y)$ and $\Delta = \eta_Y - \eta_X$ Each of the mn sample differences $D_{ij} = Y_j - X_i$ contains information about both $P(X < Y)$ and Δ. For estimation of $P(X < Y)$, we require only the signs of the D_{ij} differences, and the associated point estimator for $P(X < Y)$ is

$$\hat{P}(X < Y) = [\text{number of positive } D's + \text{half the number of zero } D's]/mn.$$

To calculate this estimator, we can define an indicator function $I(\cdot)$ as follows:

$$I(D_{ij}) = \begin{matrix} 1 & \text{if } D_{ij} > 0 \\ .5 & \text{if } D_{ij} = 0 \\ 0 & \text{if } D_{ij} < 0 \end{matrix} \tag{9.12}$$

for $i = 1, \ldots, m$ and $j = 1, \ldots, n$. Using $I(\cdot)$, we can write the point estimator for $P(X < Y)$ as

$$\hat{P}(X < Y) = \sum_{i=1}^{m} \sum_{j=1}^{n} \frac{I(D_{ij})}{mn}. \tag{9.13}$$

Note that in the case of no zero differences, the estimator for $P(X < Y)$ is simply the proportion of sample (X_i, Y_j) pairs for which $X_i < Y_j$.

To estimate $\Delta = \eta_Y - \eta_X$, the signs of the D_{ij}'s are no longer adequate, as we now require the actual magnitudes of these differences. The appropriate estimator for Δ without additional information about the shapes of the populations is then the median of these differences; that is, our point estimator for $\Delta = \eta_Y - \eta_X$ is

$$\hat{\Delta} = \tilde{D} = \text{median}\{D_{ij}\} = \text{median}\{Y_j - X_i\}. \quad (9.14)$$

Example 9.3. You Can Lead a Slug to Food But Will It Eat? Herbivores are animals that feed solely on plants. The interaction between herbivores and their food sources can lead to highly coevolved systems. In particular, a herbivore's particular liking for a specific plant species can lead to the evolution of new plant defenses by the species through the creation over time of internal chemical changes which are toxic (but not necessarily lethal) to the herbivore. Such changes usually lead to decreased foraging of the now unacceptable plant species by the herbivore, thereby preserving the species.

This naturally raises the question as to how quickly a herbivore is able to recognize the increased toxicity (non-lethal, of course, as lethal toxicity is recognized quite soon!) in a plant species. Suppose a herbivore from one ecological environment is suddenly exposed for the first time to a toxic plant species from another ecological environment. Will the herbivore know immediately to select alternative non-toxic food sources? If not, will it learn to do so relatively quickly?

One study designed to address such questions was conducted by Whelan (1982) and part of the study involved common slugs of the species *Arion subfuscus*. These slugs occur naturally in both waste sites (such as old vegetable plots, or dumps) and woodland sites. Ten medium-sized slugs were obtained from a relatively undisturbed patch of oak wood (woodland site) at Gadlys Farm, Llansadwrn, North Wales and ten similar-sized slugs were obtained from an old vegetable plot (waste site) on the same farm (but separated from the woodland site by a road and a row of houses). Control discs were prepared containing a gel composed of a mixture of powdered wheat germ, powdered milk, powdered bran, sodium alginate, dissolved in lettuce extract prepared by crushing fresh, young lettuce leaves in distilled water. Test discs were also prepared containing a similar gel with fresh leaves of an unacceptably toxic plant species *Allium ursinum* (wild garlic) found abundantly in the woodland site of the study, but absent from the waste site. (Thus the woodland slugs would have been previously exposed to the toxic nature of wild garlic, while it would be new to the waste site slugs.) After one evening of starvation during which the animals became accustomed to their test arenas, each of the twenty slugs was exposed to an equal number of alternating discs of control gel and test gel. The data in Table 9.3 are measured acceptability indices (AI), corresponding to the ratio of the area of test (unacceptable) gel eaten to the total area of gels eaten, for the ten woodland site and ten waste site slugs in the study.

Table 9.3 Acceptability indices (AI) for *Arion Subfuscus* from woodland and waste sites with the toxic woodland plant *Allium Ursinum* (wild garlic) as test gel

Woodland site slugs (X)	Waste site slugs (Y)
0.08	0.45
0.24	0.54
0.13	0.38
0.28	0.48
0.42	0.23
0.10	0.41
0.31	0.53
0.19	0.09
0.36	0.08
0.42	0.39

Source: Whelan (1982)

Letting the X and Y labels correspond to the woodland site and waste site slug populations, respectively, we see that we have $m = n = 10$ sample observations from each population. Thus we need to compute $mn = 100$ differences $D_{ij} = Y_j - X_i$, $i = 1, \ldots, 10$ and $j = 1, \ldots, 10$. The **R** functions *outer*( ) and *sort*( ) can be used on the columns of the *arion_subfuscus* data. frame to obtain and order these differences as follows.

```
> differences <- outer(arion_subfuscus$waste,
arion_subfuscus$woodland,"-")
> sort(differences)
```

These calls to the **R** functions return the 100 ordered differences presented in Table 9.4 for the acceptability index data in Table 9.3.

Table 9.4 Ordered differences $D_{ij} = Y_j - X_i$ for the acceptability index (AI) data in Table 9.3

−0.34	−.04	.07	.17	.29
−0.34	−.04	.08	.17	.30
−0.33	−.04	.09	.18	.30
−0.33	−.03	.10	.19	.31
−0.28	−.03	.10	.20	.31
−0.27	−.02	.10	.20	.32
−0.23	−.01	.11	.21	.33
−0.22	−.01	.11	.22	.34
−0.20	−.01	.11	.22	.35
−0.19	−.01	.12	.23	.35
−0.19	.00	.12	.24	.35
−0.19	.01	.12	.25	.37
−0.16	.02	.13	.25	.38
−0.15	.03	.13	.26	.40
−0.13	.03	.14	.26	.40
−0.11	.03	.14	.26	.41
−0.10	.04	.15	.28	.43
−0.08	.05	.15	.28	.44
−0.05	.06	.17	.29	.45
−0.05	.06	.17	.29	.46

Since 69 of these differences are positive and there is one zero difference, it follows from (9.13) that the estimate for $P(X < Y)$ is $(69 + .5)/10(10) = .695$. Moreover, from expression (9.14), we see that the estimate of $\Delta = \eta_Y - \eta_X$ is $\tilde{D} = \frac{D_{(50)} + D_{(51)}}{2}$, where $D_{(1)} \leq \ldots \leq D_{(100)}$ denote the 100 ordered D_{ij} differences. Using the ordered values in Table 9.4, we see that our estimate of Δ is $\tilde{D} = \frac{.12 + .12}{2} = .12$. Thus, we estimate that the probability is .695 that a randomly chosen *Arion Subfuscus* slug from a waste site will have a higher AI value for the unacceptable woodland plant *Allium ursinum* (i. e., eat a higher percentage of it) than a randomly chosen slug from a woodland site. In addition, we estimate that the ratio of the area of test (unacceptable) gel eaten to the total area of gels eaten for a typical *Arion Subfuscus* slug from a waste site will be about .12 higher than the ratio for a typical slug from a woodland site. Both estimates indicate that waste site slugs of this species are more likely to eat the unfamiliar toxic plant *Allium ursinum* the first time they are exposed to it than are similar slugs from woodland sites where they encounter the unacceptable plant on a regular basis. Whelan (1982) also addressed the question of whether the waste site slugs became more aware of the toxicity of *Allium ursinum* and adjusted their eating habits after initial exposure to the plant. (See Exercise 9.B.17.)

Confidence intervals and bounds for $\Delta = \eta_Y - \eta_X$ are also naturally based on the mn ordered differences $D_{(1)} \leq \ldots \leq D_{(mn)}$. Here, however, we must also take into account the sampling variability associated with these differences. To do so we consider the probability distribution of the variable

$$U_\Delta = \left[\text{number of differences } D_{ij} = Y_j - X_i, i = 1, \ldots, m \text{ and } j = 1, \ldots, n, \text{ that are greater than the difference in population medians } \Delta = \eta_Y - \eta_X\right]. \tag{9.15}$$

If the only possible difference between the X and Y distributions is in their population medians, the probability is 1/2 that an arbitrary random

Y - *X* difference will exceed Δ and we might immediately think that U_Δ (which is a sum of such counts) has a binomial distribution. However, the D_{ij} differences are not mutually independent, since, for example, there are *m* differences that all involve Y_1, and as a result the sampling distribution of U_Δ is not binomial. At first glance, we might also incorrectly think that the sampling distribution of U_Δ depends on Δ. Derivation of the probability distribution of U_Δ shows, however, that this is not the case. Lower tail probabilities for U_Δ can be obtained using the **R** function *pwilcox*( ) by specifying sample sizes *m* and *n* and the value of U_Δ (denoted by *q* in the call of the *pwilcox*( ) function). Upper tail probabilities can easily be found by specifying the *lower.tail* argument to be *FALSE*. Alternatively, the probability distribution of U_Δ is symmetric about the point $mn/2$, corresponding to exactly one-half of the total number of D_{ij} differences. Thus, if $u_{m,n,\alpha}$ denotes the *upper αth percentile* for the probability distribution of U_Δ for samples of *m* X's and *n* Y's, the symmetry implies that $2\left[\frac{mn}{2}\right] - u_{m,n,\alpha} = mn - u_{m,n,\alpha}$ is the *lower αth percentile* for the probability distribution of U_Δ for samples of *m* X's and *n* Y's. This symmetric nature of the distribution of U_Δ is illustrated in the histogram in Fig. 9.1 for the case of $m = 8$ and $n = 7$. The distribution is clearly

Fig. 9.1 Histogram for the distribution of U_Δ for samples of $m = 8$ *X*'s and $n = 7$ *Y*'s

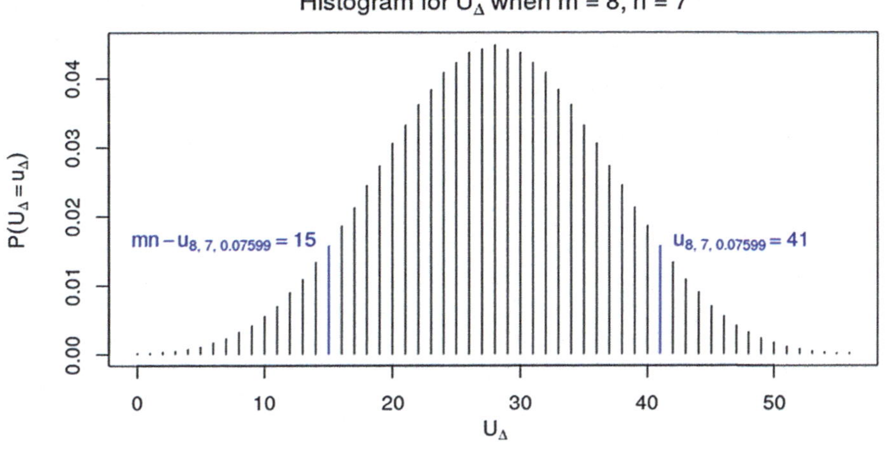

symmetric about the point $mn/2 = 8(7)/2 = 28$. The upper $\alpha = .076$ percentile for the distribution is indicated on the histogram to be $u_{8,7,.076} = 41$ and it follows from the symmetry of the distribution that the corresponding lower $\alpha = .076$ percentile for the distribution is $mn - u_{8,7,.076} = 56 - 41 = 15$, as also indicated on the histogram.

To obtain confidence intervals or bounds for the difference in medians $\Delta = \eta_Y - \eta_X$, we use the relationship between U_Δ and Δ and the distribution of U_Δ (which can be obtained using the **R** function *pwilcox*( )) . For any integer $q < (mn/2)$, we have

$$P\{\text{number of differences } D_{ij} \text{ greater than } \Delta \text{ is somewhere between } q \text{ and } (mn - q), \text{ inclusive}\} = P\{q \le U_\Delta \le (mn - q)\}. \tag{9.16}$$

But the event {number of differences D_{ij} greater than Δ is somewhere between q and $(mn-q)$, inclusive} is equivalent to the event $\{D_{(q)} < \Delta < D_{(mn-q+1)}\}$ involving the mn ordered $Y - X$ differences. It follows that.

$$P\{D_{(q)} < \Delta < D_{(mn-q+1)}\} = P\{q \le U_\Delta \le (mn - q)\}, \tag{9.17}$$

which leads directly to the desired confidence interval for the difference in population medians $\Delta = \eta_Y - \eta_X$ based on the ordered $Y - X$ differences.

Interval Estimation of $\Delta = \eta_Y - \eta_X$, the Difference in Two Population Medians Let $D_{(1)} \le ... \le D_{(mn)}$ be the ordered $D_{ij} = Y_j - X_i$ differences for random samples of sizes m and n from the X and Y populations, respectively. For any positive integer $q < mn/2$, the interval $(D_{(q)}, D_{(mn-q+1)})$ provides a confidence interval for the difference in population medians $\Delta = \eta_Y - \eta_X$ with confidence level given by.

$$CL = P\{q \le U_\Delta \le (mn - q)\}. \tag{9.18}$$

You can use the **R** function *qwilcox*( ) to choose a value of q that gives you the confidence level you want. Once the value of q is chosen, the lower and upper endpoints of the 100CL% confidence interval ($D_{(q)}$, $D_{(mn-q+1)}$) for $\Delta = \eta_Y - \eta_X$ are simply the qth smallest (up from the bottom) and qth largest (down from the top) ordered Y - X differences, respectively. Separate lower and upper confidence bounds for the difference in medians $\Delta = \eta_Y - \eta_X$ with confidence level

$$CL = 1 - P\{U_\Delta \geq (mn - q^* + 1)\} \tag{9.19}$$

are given by the q^*th smallest, $D_{(q^*)}$, and q^*th largest, $D_{(mn-q^*+1)}$, ordered differences, respectively.

Example 9.4. Interval Estimation for the Difference in Median Acceptability Indices for the Two Slug Populations Suppose we want roughly a 95% confidence interval for $\Delta = \eta_Y - \eta_X$. We use the **R** functions *qwilcox*( ) and *pwilcox*( ) to find that, with $m = 10$ and $n = 10$, $P\{U_\Delta \geq 76\} = .026 = (1 - .948)/2$.

We first find the smallest value of $(mn - q + 1)$ such that $P\{U_\Delta \geq (mn - q + 1)\} \geq 0.05/2 = 0.025$ by calling the *qwilcox*( ) function with $p = 0.025$.

```
> qwilcox(p = 0.025, m = 10, n = 10, lower.tail = FALSE)
[1] 76
```

We then obtain $P\{U_\Delta \geq 76\} = P\{U_\Delta > 75\}$ by setting $(mn - q + 1) = 75$ and calling *pwilcox*( ).

```
> pwilcox(75, m = 10, n = 10, lower.tail = FALSE)
[1] 0.02621295
```

Thus, taking $(mn - q + 1) = \{10(10) - q + 1\} = 76$ provides us with a 94.8% confidence interval for Δ, very close to our target of 95% confidence. Moreover, this tells us that $D_{(76)}$ will be the upper endpoint of our confidence interval. To find the lower endpoint we solve $76 = mn - q + 1$ for q, to obtain $q = 101 - 76 = 25$ so that our lower endpoint is $D_{(25)}$. (Note that we are really just finding how far down the 76th ordered D is from the largest D and counting the same number (25) of ordered D's up from the smallest D.)

Hence, the 94.8% confidence interval for $\Delta = \eta_Y - \eta_X$ is defined by the 25th smallest and 25th largest ordered D_{ij} differences. Using the ordered acceptability indices differences presented in Table 9.4, we are thus 94.8% confident that $\Delta = \eta_Y - \eta_X$ is somewhere in the interval $(D_{(25)}, D_{(100-25+1)}) = (D_{(25)}, D_{(76)}) = (-.03, .26)$.

Fortunately, we can avoid all of the trouble above by using the **R** function *wilcox.test*( ) to construct the confidence interval for Δ. (Note, however, that we include the discussion above because it's useful to know what's going on behind the scenes in these functions!)

The following call to **R** provides both the point estimate of Δ and the 94.8% confidence interval for Δ.

```
> wilcox.test(arion_subfuscus$waste,
          arion_subfuscus$woodland,
          conf.int = TRUE,
          conf.level = 0.948,
          correct = FALSE)

    Wilcoxon rank sum test

data:  arion_subfuscus$waste and arion_subfuscus$woodland
W = 69.5, p-value = 0.1402
alternative hypothesis: true location shift is not equal to 0
94.8 percent confidence interval:
-0.02997508  0.26005092
sample estimates:
difference in location
        0.1199828
```

To test the null hypothesis H_0: $\eta_Y = \eta_X$ ($\Delta = 0$) we use the statistic $U = mn\hat{P}(X < Y)$, where $\hat{P}(X < Y)$ is the point estimator of $P(X < Y)$ given in (9.13). Thus

$$U = \sum_{i=1}^{m} \sum_{j=1}^{n} D_{ij}. \tag{9.20}$$

Example 9.5. Computation of the Statistic *U*. We illustrate the computation of the statistic *U* (9.20) for the slug Acceptability Indices data in Table 9.3. Letting the waste site slugs correspond to the *Y* population and the woodland slugs to the X population, we found in Example 9.3 that $\hat{P}(X < Y) = .695$ for these data. Since $m = n = 10$, it follows that $U = (10)(10(.695) = 69.5$.

The next important question is how to use the value of the statistic U computed from the X and Y samples to conduct appropriate tests of the null hypothesis H_0: $\eta_Y = \eta_X$ ($\Delta = 0$). If H_0 is true and the X and Y population medians are equal, then we would expect roughly one-half of the D_{ij} differences are be positive, corresponding to U being close to $mn/2$. Large values of U (closer to mn) would be indicative of the alternative $\Delta = \eta_Y - \eta_X > 0$, and similarly, small values of U (closer to 0) would be indicative of the alternative $\Delta < 0$. To assess the implication of a particular observed value of U, we can rely on the probability distribution of U when the null hypothesis H_0: $\Delta = 0$ is true. Since U corresponds to U_Δ (9.15) under the null hypothesis $\Delta = 0$, our previous discussion about the probability distribution of U_0 can be used here as well. Lower tail probabilities for U can be obtained using the **R** function *pwilcox*( ) by specifying sample sizes m and n and the value of U (denoted by q in the call of the *pwilcox*( ) function). Upper tail probabilities can easily be found by specifying the *lower.tail* argument to be *FALSE*.

Hypothesis Tests for the Difference in Two Population Medians $\Delta = \eta_Y - \eta_X$ To test the null hypothesis H_0: $\eta_Y = \eta_X$ ($\Delta = 0$) for arbitrary continuous X and Y populations, compute the statistic U (9.20) and let u_{obs} be the attained sample value of U. The exact P-values for a test of H_0 against the alternatives H_A are then:

H_A	P-value	
$\eta_Y > \eta_X$	$P(U \geq u_{obs})$	(9.21)
$\eta_Y < \eta_X$	$P(U \leq u_{obs})$	(9.22)
$\eta_Y \neq \eta_X$	$2P(U \geq u_{obs})$, if $u_{obs} \geq \frac{mn}{2}$	
	$2P(U \leq u_{obs})$, if $u_{obs} < \frac{mn}{2}$.	(9.23)

To compute any of these P-values for given sample sizes m and n and observed value u_{obs}, we can use the **R** function *pwilcox*( ).

Example 9.6. Testing for Median Acceptability Indices of Woodland Slugs Versus Waste Site Slugs In his study, Whelan (1982) was interested in whether there would be a difference in the median feeding habits for the woodland slugs, who had prior exposure to the toxic woodland plant *Allium Ursinum* (wild garlic) and that for the waste site slugs, for whom this was their first exposures to the plant. We consider here the test of H_0: $\eta_Y = \eta_X$ versus the two-sided alternative H_A: $\eta_Y \neq \eta_X$, since the discussion in Whelan (1982) provides reasonable arguments for the possibility of either directional alternative to the null H_0. The observed value of the statistic U (9.20) was found in Example 9.5 to be $u_{obs} = 96.5$, which is greater than $\frac{mn}{2} = \frac{10(10)}{2} = 50$. Using the **R** function *wilcox.test*( ) we find that the P-value for our test of H_0: $\eta_Y = \eta_X$ versus H_A: $\eta_Y \neq \eta_X$ is $2\ P(U \geq 69.5) = 0.1506$.

```
> wilcox.test(arion_subfuscus$waste, arion_subfuscus$woodland)

        Wilcoxon rank sum test with continuity correction

data:  arion_subfuscus$waste and arion_subfuscus$woodland
W = 69.5, p-value = 0.1506
alternative hypothesis: true location shift is not equal to 0
```

(Note that the **R** function refers to U as W. Don't let this confuse you!)

Thus we would reject H_0: $\eta_Y = \eta_X$ in favor of the two-sided alternative H_A: $\eta_Y \neq \eta_X$ only for significance levels greater than or equal to .1056. This suggests that there is not sufficient evidence in the sample data to indicate that the median acceptability indices (to *Allium Ursinum*) for the two populations of slugs are different.

The result of our hypothesis test is in agreement with the 94.8% confidence interval $(D_{(25)}, D_{(76)}) = (-.03, .26)$ for $\Delta = \eta_Y - \eta_X$ previously obtained in

Example 9.4, since 0 belongs to that interval and therefore cannot be rejected as a possible value for Δ. On the other hand, recall that both of our point estimates $\hat{P}(X < Y) = .695$ and $\hat{\Delta} = .12$ from Example 9.3 had suggested that waste site slugs of this species are more likely to eat wild garlic the first time they are exposed to it than are similar slugs from woodland sites where they encounter the plant on a regular basis. However, once we take into account the innate variability associated with our sampling process (through the use of either the confidence interval or the hypothesis test) we realize that we cannot attach statistical significance to these point estimates. This illustrates an important fact: Statistical significance cannot be concluded from point estimates alone. We must take into account the natural variability associated with the sampling process through either a confidence interval or a hypothesis test before we can assert that our results are statistically significant.

Large Sample Approximations The confidence intervals/bounds and hypothesis tests for the difference in medians $\Delta = \eta_Y - \eta_X$ make use of the probability distribution of the variable U_Δ (9.15) for arbitrary Δ and under the null hypothesis setting corresponding to $\Delta = 0$. A third option is provided by a two-sample central limit theorem (similar to those used in Chap. 7 to approximate the probability distributions of several important one-sample statistics). See, for example, Chap. 4 in Hollander, Wolfe, and Chicken (2014) for details.

Rank Sum Statistic. The procedures prescribed in Eqs. (9.21) – (9.23) for testing H_0: $\eta_Y = \eta_X$ against one- or two-sided alternatives are based on the counting statistic U (9.20). These tests can also be based on the joint ranks of the m X's and n Y's in the combined sample of $N = (m + n)$ observations, with average ranks used to break tied observations. (In Chap. 1 we employed this same approach to compare typical observations from two data collections. See

Example 1.22 for a numerical illustration of how average ranks are used to break tied observations in the joint ranking.) Define the *rank sum statistic, W,* to be the sum of these joint ranks assigned to the Y observations. Letting $R_1, \ldots,$ R_n denote the joint ranks assigned to the n Y observations $Y_1, \ldots, Y_n$, respectively, the statistic W corresponds to the sum of these Y sample joint ranks; that is,

$$W = \sum_{j=1}^{n} R_j.$$

We illustrate the computation of the statistic W using the slug Acceptability Indices data in Table 9.3. The joint ranks (using average ranks to break the ranking ties for the two 0.08 and two 0.42 values) for the observations in the two samples are given in Table 9.5.

Thus the observed joint ranks for the waste site slugs (Y's) are: $r_1 = 17$, $r_2 = 20$, $r_3 = 12$, $r_4 = 18$, $r_5 = 7$, $r_6 = 14$, $r_7 = 19$, $r_8 = 3$, $r_9 = 1.5$, and $r_{10} = 13$. Summing these combined sample Y-ranks, we find the observed value of the rank sum statistic W to be $w_{obs} = [17+20+12+18+7+14+19+3+1.5+13] = 124.5$ for the slug Acceptability Indices data.

We note that the rank sum statistic W is directly related to the count statistic U (9.20) that is used in our tests of H_0: $\eta_Y = \eta_X$. In fact, if there are

Table 9.5 Joint ranks for the acceptability indices (AI) data in Table 9.3

Woodland site	Joint ranks	Waste site	Joint ranks
0.08	1.5	0.45	17
0.24	8	0.54	20
0.13	5	0.38	12
0.28	9	0.48	18
0.42	15.5	0.23	7
0.10	4	0.41	14
0.31	10	0.53	19
0.19	6	0.09	3
0.36	11	0.08	1.5
0.42	15.5	0.39	13

no ties among the combined sample observations they are linearly related by the expression

$$W = U + \frac{n(n+1)}{2}.$$

You are asked to prove this relationship in Conceptual Exercise 9.A.1.

Section 9.2 Practice Exercises

9.2.1. For random samples of $m = 3$ and $n = 2$ observations from the X and Y populations, respectively, list all possible values for the estimator $\hat{P}(X < Y)$ in expression (9.13) when there are no tied observations.

9.2.2. Let {3, 17, −4, 19, 6, 22, 76} and {4, 12, 39, 0, 15, −12} be independent random samples for the continuous random variables X and Y, respectively. Use these sample data to estimate $P(X < Y)$.

9.2.3. Let {2.6, 3.5, −6.7, 12.2, 14.8, 19.3, −26.9, 18.8, 97.9, 0.4} and {3.3, 18.9, −5.5, 22.4, 17.0, −16.8, 9.0, 5.5} be independent random samples for the continuous random variables X and Y, respectively. Use these sample data to estimate $P(X < Y)$.

9.2.4. Let {2, 5.3, 9} and {4, 6.6, 17.3, 10} be independent random samples for the continuous random variables X and Y with medians η_X and η_Y, respectively. Let $\Delta = \eta_Y - \eta_X$ and evaluate the median difference estimator $\hat{\Delta} = \tilde{D}$ (9.14) for these data.

9.2.5. Let {2.6, 3.5, −6.7, 12.2, 14.8, 19.3, −26.9, 18.8, 97.9, 0.4} and {3.3, 18.9, −5.5, 22.4, 17.0, −16.8, 9.0, 5.5} be independent random samples for the continuous random variables X and Y with medians η_X and η_Y, respectively. Let $\Delta = \eta_Y - \eta_X$ and evaluate the median estimator $\hat{\Delta} = \tilde{D}$ (9.14) for these data.

9.2.6. Let {3.5, 3.7, 4.0, 3.6, 3.7} and {3.8, 3.9, 4.3, 4.2, 4.6} be independent random samples for the continuous random variables X and Y, respectively. Use these sample data to estimate $P(X < Y)$ and evaluate the median estimator $\hat{\Delta} = \tilde{D}$ (9.14). Comment on the differences in the information about the sample data conveyed by these two estimates.

9.2.7. Let {3.5, 3.7, 4.0, 3.6, 3.7} and {27.4, 29.3, 3.0, 3.3, 20.6} be independent random samples for the continuous random variables X and Y, respectively. Use these sample data to estimate $P(X < Y)$ and evaluate the median estimator $\hat{\Delta} = \tilde{D}$ (9.14). Comment on the differences in the information about the sample data conveyed by these two estimates. Contrast this to the situation for the sample data in Exercise 9.2.6.

9.2.8. Consider the setting where we have independent random samples of $m = 8$ X's and $n = 9$ Y's. Use the **R** function *qwilcox*( ) to find the value of q so that $(D_{(q)}, D_{(73\text{-}q)})$ provides a confidence interval for the difference in population medians $\Delta = \eta_Y - \eta_Y$ with confidence level CL close to .95.

9.2.9. Consider the setting where we have independent random samples of $m = 7$ X's and $n = 6$ Y's. Use the **R** function *qwilcox*( ) to find the value of q^* so that $D_{(43\text{-}q^*)}$ provides an upper confidence bound for the difference in population medians $\Delta = \eta_Y - \eta_Y$ with confidence level CL close to .90.

9.2.10. Let {2, 5.3, 9, 6.6} and {4, 17.3, 10} be independent random samples for the continuous random variables X and Y with medians η_X and η_Y, respectively, and let $\Delta = \eta_Y - \eta_X$.

(a) Find a 94.2% confidence interval for Δ.

(b) Find a 94.3% lower confidence bound for Δ.

9.2.11. Let {2.6, 3.5, −6.7, 12.2, 14.8, 19.3, −26.9, 18.8, 97.9, 0.4} and {3.3, 18.9, −5.5, 22.4, 17.0, −16.8, 9.0, 5.5} be independent random samples for the continuous random variables X and Y with medians η_X and η_Y, respectively, and let $\Delta = \eta_Y - \eta_X$.

(a) Find a 95.6% confidence interval for Δ.

(b) Find a 98.3% upper confidence bound for Δ.

9.2.12. For random samples of $m = 3$ and $n = 2$ observations from the X and Y populations, respectively, list all possible values for the count statistic U (9.20) when there are no tied observations.

9.2.13. Let $X_1, \ldots, X_5$ and $Y_1, \ldots, Y_6$ be independent random samples of sizes $m = 6$ and $n = 5$ from a continuous distribution. Let U be the count statistic in (9.20) and assume no ties between the X and Y observations.

(a) What are the possible values for U?

(b) Find $P(U \geq 26)$.

(c) Find $P(U \leq 5)$.

(d) Use symmetry of the distribution of U to find $P(U \leq 4)$.

9.2.14. Let $X_1, \ldots, X_5$ and $Y_1, \ldots, Y_5$ be independent random samples of sizes $m = 5$ and $n = 5$ from a continuous distribution. Let U be the count statistic in (9.20) and assume no ties between the X and Y observations.

(a) What are the possible values for U?

(b) Find $P(U \geq 19)$.

(c) Find $P(U \leq 3)$.

(d) Use symmetry of the distribution of U to find $P(U \geq 22)$.

9.2.15. Let {2, 5.3, 9, 6.6} and {4, 17.3, 10} be independent random samples for the continuous random variables X and Y with medians η_X and η_Y, respectively, and let $\Delta = \eta_Y - \eta_X$.

(a) Find the value, u_{obs}, of the count statistic U (9.20) for testing H_0: $\Delta = 0$.

(b) If $\Delta = 0$, compute $P(U \geq u_{obs})$ and $P(U \leq u_{obs})$.

9.2.16. Let {2.6, 3.5, −6.7, 12.2, 14.8, 19.3, −26.9, 18.8, 97.9, 0.4} and {3.3, 18.9, −5.5, 22.4, 17.0, −16.8, 9.0, 5.5} be independent random samples for the continuous random variables X and Y with medians η_X and η_Y, respectively, and let $\Delta = \eta_Y - \eta_X$.

(a) Find the value, u_{obs}, of the count statistic U (9.20) for testing H_0: $\Delta = 0$.

(b) If $\Delta = 0$, compute $P(U \geq u_{obs})$ and $P(U \leq u_{obs})$.

9.2.17. *House Sizes as Related to Lot Sizes in North Carolina.* The dataset *house_lot_sizes* contains information about house and lot sizes for a random sample of 100 properties in Wake County, North Carolina, as collected by Woodard and Leone (2008). Consider two subsets of this dataset corresponding to the smallest 25 lot sizes (in acreage) and the largest 25 lot sizes (in acreage), respectively. Viewing these two subsets as representative samples of "*small*" and "*large*" lot sizes in Wake County, North Carolina, conduct the following analyses under the assumption that the populations of house sizes (in square feet) in Wake County, North Carolina for these two categories have medians η_{small} and η_{large}, respectively.

(a) Estimate the difference in median house sizes, $\eta_{large} - \eta_{small}$.

(b) Find a confidence interval for $\eta_{large} - \eta_{small}$. Choose your own reasonable confidence level.

(c) Find the P-value for a test of H_0: $\eta_{large} = \eta_{small}$ against the one-sided alternative H_A: $\eta_{large} > \eta_{small}$. What is your decision at significance level .02?

9.2.18. *Do Women's Hearts Beat Faster than Men's?* Mackowiak et al. (1992) collected heart rate data from 148 individuals aged 18 through 40 years. The dataset *body_temperature_and_heart_rate* contains heart rate values (artificially generated by Shoemaker 1996, to closely recreate the original data considered by Mackowiak et al.) for 65 male and 65 female subjects. Conduct the following analyses under the assumption that the median heart rate for women is η_F and the median heart rate for men is η_M.

(a) Estimate the difference in median heart rates $\eta_F - \eta_M$.

(b) Find a confidence interval for $\eta_F - \eta_M$. Choose your own reasonable confidence level.

(c) Find the *P*-value for a test of $H_0 : \eta_F = \eta_M$ against the one-sided alternative H_A: $\eta_F > \eta_M$. What is your decision at significance level .030?

9.2.19. *Diamonds—Does Color Matter?* In the February 18, 2000 edition of Singapore's *Business Times*, an advertisement (discussed in Chu, 2001) listed data (weight in carats, color purity, grade of clarity, certification body, and value in Singapore dollars) for 308 round diamond stones. These data are provided in the dataset *diamonds_carats_color_cost*. The top color purity grade is D and the rating moves down the alphabet E, F, G, ….for successively lower grades of color purity. Separate the 308 diamonds in the dataset into two groupings, those with either a D or E color purity grade (60 diamonds) and those with color purity grades of F or lower (248 diamonds). Viewing these data as random samples of sizes $m = 60$ and $n = 248$ diamonds from the populations of all round diamond stones with a color purity grade of D or E and those with a color purity grade of F or lower, respectively, complete the following statistical analyses under the assumption that the populations for these two color categories have median values (in Singapore dollars) $\eta_{D,E}$ and $\eta_{F\ or\ lower}$, respectively.

(a) Estimate the difference in median diamond value $\eta_{D,E} - \eta_{F\ or\ lower}$.

(b) Find a lower confidence bound for $\eta_{D,E} - \eta_{F\ or\ lower}$. Choose your own reasonable confidence level.

(c) Find the *P*-value for a test of H_0: $\eta_{D,E} = \eta_{F\ or\ lower}$ against the one-sided alternative H_A: $\eta_{D,E} > \eta_{F\ or\ lower}$. What is your decision at significance level .040?

9.2.20. *Movie Lengths and Ratings*. The *Movie and Video Guide* is a ratings and information guide to movies prepared annually by Leonard Maltin. Moore (2006) selected a random sample of 100 movies from the 1996 edition of the *Guide*. He compiled the dataset *movie_facts* containing relevant information about the selected movies. Two of the pieces of information provided are the rating that Maltin gave each of the movies on a rising (worst to best) scale of 1, 1.5, 2, 2.5, 3, 3.5, 4 and the running time of the movie in minutes. Divide the 100 movies in the sample into two subsets, corresponding to movies with ratings of at least 3 ($n = 31$ movies) and those with ratings of less than 3 ($m = 69$ movies). Complete the following statistical analyses under the assumption that the populations of high rated (≥ 3) and low rated (< 3) movies have median running times $\eta_{\geq 3}$ and $\eta_{<3}$, respectively.

(a) Estimate the difference in median running times $\eta_{\geq 3} - \eta_{<3}$.

(b) Find a confidence interval for $\eta_{\geq 3} - \eta_{<3}$. Choose your own reasonable confidence level.

(c) Find the *P*-value for a test of H_0: $\eta_{\geq 3} = \eta_{<3}$ against the two-sided alternative H_A: $\eta_{\geq 3} \neq \eta_{<3}$. What is your decision at significance level .070?

9.2.21. *How Well Does Your Beer Hold Its Foam?* Two features of bottled beer that are important to beer consumers are the amount of initial head formation when a beer is poured and how long the head lasts. Ault et al. (1967) measured the height of the initial head formation upon pouring, the percentage adhesion of the head, and the percentage collapse of the head 4 min after pouring for 20 bottles selected from two different production lots of the beer. The dataset *beer_head* contains the results of their study. Complete the following statistical analyses under the assumption that η_1 and η_2 are the median percentage head collapse 4 min after pouring for bottles of beer from the first and second production lots, respectively.

(a) Estimate the difference in median percentage collapse $\eta_1 - \eta_2$.

(b) Find a confidence interval for $\eta_1 - \eta_2$. Choose your own reasonable confidence level.

(c) Find the P-value for a test of H_0: $\eta_1 = \eta_2$ against the two-sided alternative H_A: $\eta_1 \neq \eta_2$. What is your decision at significance level .033?

9.2.22. *Meniscal Repair—FasT-Fix Sutures or Arrows?* Surgery is most often the only option when faced with a torn medial meniscus—but what is the best surgical method for the repair? Borden et al. (2003) studied the performance characteristics of three different meniscal repair techniques, namely, the FasT-Fix Meniscal Repair Suture System (FasT-Fix), the use of biodegradable Meniscus Arrows (MA), and the Vertical Mattress Sutures (VMS) approach. Human cadaveric knees were used in the study, with six randomly assigned to each of the three meniscal surgery techniques. Each repaired meniscus was loaded into a servohydraulic device and cycled between 5 and 50 Newtons (N) at 1 Hz for 500 cycles. After cycle testing, the meniscus was subjected to tension loading at a slow rate of 5 mm/min (similar to the type of stresses that a meniscus might have to deal with during early rehabilitation) until failure of the repair occurred. The dataset *meniscal_repairs_load_at_failure* contains the load at failure (Newtons (N)), a displacement measure (mm), and a stiffness measure (N/mm) for each of the 18 repaired menisci. Conduct the following analyses under the assumption that the median load to failure for the FasT-Fix repair technique is $\eta_{FastT\ -\ Fix}$ and the median load to failure for the Meniscus Arrows (MA) technique is η_{MA}.

(a) Estimate the difference in median load to failure $\eta_{FasT\ -\ Fix}$ - η_{MA}.

(b) Find a confidence interval for $\eta_{FasT\ -\ Fix} - \eta_{MA}$. Choose your own reasonable confidence level.

(c) Find the P-value for a test of H_0: $\eta_{FasT\ -\ Fix} = \eta_{MA}$ against the one-sided alternative H_A: $\eta_{FasT\ -\ Fix} > \eta_{MA}$. What is your decision at significance level .047?

9.2.23. *Meniscal Repair—FasT-Fix Sutures or Vertical Mattress Sutures?* Consider the study of meniscal repair techniques by Borden et al. (2003) discussed in Exercise 9.2.22. Conduct the following analyses under the assumption that the median displacement (*mm*) for the FasT-Fix repair technique is $\eta_{FasT-Fix}$ and the median displacement (*mm*) for the Vertical Mattress Sutures (VMS) technique is η_{VMS}.

(a) Estimate the difference in median displacement $\eta_{FasT-Fix} - \eta_{VMS}$.

(b) Find a confidence interval for $\eta_{FasT-Fix} - \eta_{VMS}$. Choose your own reasonable confidence level.

(c) Find the *P*-value for a test of $H_0 : \eta_{FasT-Fix} = \eta_{VMS}$ against the two-sided alternative $H_A : \eta_{FasT-Fix} \neq \eta_{VMS}$. What is your decision at significance level .078?

9.3 Approximate Inference for the Difference in Means for Two Populations–Procedures Based on the Two Sample Averages and Sample Standard Deviations

In Sect. 2 we discussed exact statistical inference procedures for the difference in medians of two arbitrary continuous distributions. In this section, we consider an approximate approach to making statistical inferences about the difference in population means that is applicable to virtually all underlying distributions.[1] Once again these approximate inferences will rely on a central limit theorem for their justification, this time one which requires that both sample sizes, *m* and *n*, are sufficiently large.

Two-Sample Central Limit Theorem for Sample Averages. Let $\bar{X}$ and $\bar{Y}$ be the sample averages and S_X and S_Y the sample standard deviations for independent random samples $X_1, \ldots, X_m$ and $Y_1, \ldots, Y_n$, from arbitrary

[1] The only requirement is that the variances exist for both populations. Neither underlying normality nor equal population variances is required for the procedures of this section.

distributions for which the population means, μ_X and μ_Y, and the population standard deviations, σ_X and σ_Y, are all finite, but unknown. The natural estimator for the difference μ_Y - μ_X is the observed difference in the sample averages $\bar{Y} - \bar{X}$. The mean for $\bar{Y} - \bar{X}$ is $\mu_{\bar{Y}-\bar{X}} = \mu_Y - \mu_X$ and the variance for $\bar{Y} - \bar{X}$ is given by the sum of the variances for the separate sample averages; that is,

$$\sigma^2_{\bar{Y}-\bar{X}} = \sigma^2_{\bar{Y}} + \sigma^2_{\bar{X}} = \frac{\sigma^2_Y}{n} + \frac{\sigma^2_X}{m}$$

and the corresponding standard deviation for $\bar{Y} - \bar{X}$ is.

$$\sigma_{\bar{Y}-\bar{X}} = \sqrt{\frac{\sigma^2_Y}{n} + \frac{\sigma^2_X}{m}}.$$

It then follows from a standard central limit theorem that for large sample sizes, m and n, the standardized variable

$$Z = \frac{(\bar{Y} - \bar{X}) - (\mu_Y - \mu_X)}{\sqrt{\frac{\sigma^2_Y}{n} + \frac{\sigma^2_X}{m}}} \qquad (9.24)$$

has an approximate standard $N(0, 1)$ distribution. Of course, Z (9.24) cannot be used directly to make inferences about μ_Y - μ_X, since the population variances σ^2_X and σ^2_Y are both unknown. However, we can use the X and Y sample variances, S^2_X and S^2_Y, to estimate σ^2_X and σ^2_Y, respectively, in expression (9.24) to yield

$$Z^* = \frac{(\bar{Y} - \bar{X}) - (\mu_Y - \mu_X)}{\sqrt{\frac{S^2_Y}{n} + \frac{S^2_X}{m}}}. \qquad (9.25)$$

Fortunately, we have available yet another central limit theorem that enables us to state that the sampling distribution of the variable Z^* in (9.25) can also be well-approximated by the standard $N(0, 1)$ distribution when the sample sizes, m and n, are both large.

This fact is exactly what is needed to establish approximate interval estimates for the difference in means, $\mu_Y - \mu_X$. These approximate interval estimates are centered at the point estimator $\bar{Y} - \bar{X}$. The length of the interval is determined in part by $\sqrt{\frac{S_Y^2}{n} + \frac{S_X^2}{m}}$, the estimated standard error of $\bar{Y} - \bar{X}$, and in part by the desired confidence. The target confidence interval endpoints are obtained from the appropriate percentiles of the $N(0, 1)$ distribution.

Approximate Interval Estimation of the Difference in Population Means, $\Delta = \mu_Y - \mu_X$, for Two Populations with Finite Variances The approximate 100*CL*% confidence interval for $\mu_Y - \mu_X$ is given by the interval.

$$\left(\bar{Y} - \bar{X} - z_{(1-CL)/2}\sqrt{\frac{S_Y^2}{n} + \frac{S_X^2}{m}}, \bar{Y} - \bar{X} + z_{(1-CL)/2}\sqrt{\frac{S_Y^2}{n} + \frac{S_X^2}{m}}\right), \qquad (9.26)$$

where $z_{(1-CL)/2}$ is the upper ((1-*CL*)/2)th percentile for the standard normal distribution. The corresponding approximate 100*CL*% lower and upper confidence bounds for $\mu_Y - \mu_X$ are given by $\bar{Y} - \bar{X} - z_{1-CL}\sqrt{\frac{S_Y^2}{n} + \frac{S_X^2}{m}}$ and $\bar{Y} - \bar{X} + z_{1-CL}\sqrt{\frac{S_Y^2}{n} + \frac{S_X^2}{m}}$, respectively.

Example 9.7. Are there Ethnic Differences in Smoking Habits? Lung cancer and chronic obstructive pulmonary disease (COPD) are diseases primarily associated with cigarette smokers. Within the population of cigarette smokers, however, the incidence and mortality of these two diseases differ between black and white smokers, indicating potentially different cigarette-smoking behaviors or patterns for these two groups of smokers. In a study to gather information about such possible differences, Pérez-Stable et al. (1998)

gathered data on 40 African American and 39 Caucasian smokers in the San Francisco area. Participating smokers were screened to be in good health, between the ages of 21 and 64, and self-identified as non-Latino Caucasian or African American. Potential subjects were excluded from the study for habitual use of any prescription medication, narcotic or sedative drug addiction, or long-term alcoholism. Pregnant females were also excluded from participation in the study. The 40 African American and 39 Caucasian smokers selected to participate in the study were matched by gender (approximately 50% female in each group) and age (within 5 years), as well as by self-reported cigarette consumption of either 1 to 9 or 10 or more cigarettes per day. Pérez-Stable et al. measured a large number of variables for these participants in an attempt to provide insight into the differential smoking habits for these two ethnic groups. Among the variables measured were the type of cigarette smoked (menthol versus non-menthol) and the associated nicotine intake per cigarette smoked. They found that 31 of the 40 African American participants smoked menthol cigarettes, while only 2 of the 39 Caucasian participants were menthol smokers. (You are asked in Exercise 9.B.19 to use these data to obtain a confidence interval for the difference between the proportions of all African American and Caucasian smokers who use menthol cigarettes.) Pérez-Stable et al. also found that the average nicotine intake per cigarette for the 40 African American smokers in the study was 1.41 *mg*, with a sample standard deviation of 0.80 *mg*. For the 39 Caucasian smokers in the study, the average nicotine intake per cigarette was 1.09 *mg*, with a sample standard deviation of 0.74 *mg*. Letting X correspond to the population of Caucasian smokers and Y to the population of black smokers, we estimate the difference, $\mu_Y - \mu_X$, in mean nicotine intake per cigarette smoked between African American and Caucasian smokers to be $\bar{y} - \bar{x} = 1.41 - 1.09 = .32\ mg$. To obtain an approximate 94% lower confidence bound for $\mu_Y - \mu_X$, we first find the appropriate standard normal percentile, $z_{.06} = 1.555$. The approximate 94% lower confidence bound for $\mu_Y - \mu_X$ is then given by

$$(\bar{y} - \bar{x}) - z_{.06}\sqrt{\frac{s_Y^2}{n} + \frac{s_X^2}{m}} = .32 - 1.555\sqrt{\frac{(.80)^2}{40} + \frac{(.74)^2}{39}}$$
$$= .32 - .2695 = .0505\,mg.$$

Thus we are approximately 94% confident that the average nicotine intake per cigarette for African American smokers is at least .0505 *mg* higher than the average nicotine intake per cigarette for Caucasian smokers.

The magnitude (.0505 *mg*) of the lower bound is not as important in this study as the simple fact that it is positive. The lower bound being positive implies that there is strong sample evidence that the per-cigarette nicotine intake for a typical African American smoker is **greater** than the per-cigarette nicotine intake for a typical Caucasian smoker. This result should not, of course, be taken as any indication that a particular brand of cigarettes affects African American and Caucasian smokers differently. The observed difference in per-cigarette nicotine intake is most definitely tied to the **type** (menthol versus non-menthol) of cigarettes preferred by African American and Caucasian smokers (see Exercise 9.B.19). In addition, nothing in our discussion here addresses possible differences in the **total** amount of nicotine intake for typical African American and Caucasian smokers. That issue would depend not only on the type but also on the typical number of cigarettes smoked per day by each group.

To find approximate *P*-values for hypothesis tests about the difference in means $\mu_Y - \mu_X$ for arbitrary populations with finite variances, we once again make use of the variable Z^* (9.25).

Approximate Hypothesis Tests about the Difference in Population Means, $\Delta = \mu_Y - \mu_X$, for Two Populations with Finite Variances To test the null hypothesis H_0: $[\mu_X = \mu_Y]$ with two-sample data from arbitrary populations with finite variances, compute the statistic Z^* (9.25) under the null hypothesis condition that $\mu_X = \mu_Y$, namely,

$$Z^* = \frac{\bar{Y} - \bar{X}}{\sqrt{\frac{S_Y^2}{n} + \frac{S_X^2}{m}}}, \tag{9.27}$$

and let z^* be the attained value of Z^*. Then, the approximate P-value for populations with finite variances for a test of H_0 against the alternatives H_A are:

H_A	P-value	
$\mu_Y > \mu_X$	$\approx P(Z^* \geq z^*)$	(9.28)
$\mu_Y < \mu_X$	$\approx P(Z^* \leq z^*)$	(9.29)
$\mu_Y \neq \mu_X$	$\approx 2P(Z^* \geq z^*)$, if $z^* \geq 0$	
	$\approx 2P(Z^* \leq z^*)$, if $z^* < 0$,	(9.30)

where $Z^* \sim N(0, 1)$.

Example 9.8. Ethnic Differences in Smoking Habits. In searching for possible reasons for the higher incidence and morbidity of lung cancer among African American smokers, Pérez-Stable et al. (1998) were interested in, among other things, testing whether cigarettes chosen by African American smokers provided more nicotine intake per cigarette on the average than do those selected by Caucasian smokers. Using the notation of Example 9.7, this corresponds to testing the null hypothesis H_0: $[\mu_Y = \mu_X]$ against the one-sided alternative H_A: $[\mu_Y > \mu_X]$. From Example 9.7, we know that the difference in sample averages is $\bar{y} - \bar{x} = .32\ mg$ and the standard deviations for the samples of Caucasian and African American smokers are $s_X = .74$ mg and $s_Y = .80$ mg, respectively. Computing the statistic Z^* (9.40), we see that.

$$z^* = \frac{\bar{y} - \bar{x}}{\sqrt{\frac{s_Y^2}{n} + \frac{s_X^2}{m}}} = \frac{.32}{\sqrt{\frac{(.80)^2}{40} + \frac{(.74)^2}{39}}} = 1.846.$$

Hence, using (9.28) and the standard normal distribution, the approximate P-value for our test of H_0: $[\mu_Y = \mu_X]$ versus H_A: $[\mu_Y > \mu_X]$ is approximately

$P(Z^* \geq 1.846) = .0322$. Thus there is relatively strong evidence in these sample data to support the Pérez-Stable et al. conjecture that the average nicotine intake per cigarette is higher for cigarettes (primarily menthol) chosen by African American smokers than for cigarettes (primarily non-menthol) of choice for Caucasian smokers.

Improved Approximations for Moderate Sample Sizes. We invoked two-sample central limit theorems to justify using the standard normal distribution as an approximation to the exact distribution of the variable Z^* (9.25). This enables us to construct approximate confidence intervals/bounds for and test hypotheses about the difference in means $\mu_Y - \mu_X$ for populations with finite variances. When both of the sample sizes, m and n, are sufficiently large (at least 25), this standard normal approximation is quite adequate. However, for moderate sample sizes (between 5 and 25) the sampling distribution of Z^* (9.25) can be better approximated by the t-distribution with 'degrees of freedom'.

$$df = \frac{\left(\frac{S_X^2}{m} + \frac{S_Y^2}{n}\right)^2}{\frac{\left(\frac{S_X^2}{m}\right)^2}{m-1} + \frac{\left(\frac{S_Y^2}{n}\right)^2}{n-1}}. \tag{9.31}$$

From previous discussion in this text, we know that the degrees of freedom for a t-distribution must be a positive integer. However, the 'degrees of freedom' df defined by (9.31) will, in general, not be an integer. Several different approaches have been proposed to deal with this situation, including: (i) interpolation using the two t-distributions with degrees of freedom corresponding to the positive integers immediately above and below df, respectively, (ii) use of a computer algorithm to numerically approximate the required probabilities associated with the non-integer degrees of freedom df, and (iii) a conservative approach based on the single t-distribution with integer degrees of freedom equal to $[[df]]$, where $[[u]]$ is the greatest positive integer less than or equal to u. (Thus, for example, if we obtain the value $df = 7.45$ from Eq. (9.31), then we would use a t-distribution with $[[7.45]] = 7$ degrees of freedom in this conservative approach.) In this text we take the conservative approach (iii) based on use of the t-distribution with degrees of freedom equal to $[[df]]$ to improve the approximate inference procedures for the difference in means $\mu_Y - \mu_X$ based on Z^* (9.25).

For the approximate 100*CL*% confidence interval for $\mu_Y - \mu_X$ this simply involves replacing the standard normal percentile $z_{(1\text{-CL})/2}$ in expression (9.26) by the corresponding percentile $t_{[[df]],(1\text{-CL})/2}$ for the t-distribution with $[[df]]$ degrees of freedom. Similarly, for the 100*CL*% confidence bounds for $\mu_Y - \mu_X$, $z_{1\text{-CL}}$ is replaced by $t_{[[df]],1\text{-CL}}$.

To use this approach to improve the approximation for any of the P-values in (9.28) - (9.30) when the sample sizes are only moderately large (between 5 and 25), we simply evaluate the relevant probabilities using the t-distribution with $[[df]]$ degrees of freedom, rather than the standard normal distribution. Thus, using this t-distribution approach, when the observed value of Z^* (9.27) is z^*, the approximate P-value for testing H_0: $[\mu_Y = \mu_X]$ against the alternative H_A: $[\mu_Y > \mu_X]$, for example, is $P(T \geq z^*)$, where T has a t-distribution with $[[df]]$ degrees of freedom.

Example 9.9. Can Goggled Green Turtle Hatchlings Find Their Way To the Sea? Almost immediately upon hatching in their beach nests, green turtle (*Chelomia mydas*) young begin moving toward the sea. One of the mechanisms suggested for this instinctive ability of green turtle hatchlings to find the sea is that they react positively to light sources (called positive phototropotaxis), since for many natural nesting beaches the open seaward horizon is much brighter than the darker landward vegetation. Mrosovsky and Shettleworth (1974) studied some of the details surrounding the mechanism of this reaction to light by hatchling green turtles. In particular, they were interested in whether the direction of the source of visual inputs (thus affecting which parts of the retina are stimulated) had an effect on the orientation and sea-finding ability of such hatchlings. In one of their experiments, Mrosovsky and Shettleworth selected 36 green turtle hatchlings on a beach in Bigisanti, Surinam and randomly assigned them to two groups. The 18 turtles in one of the groups (the control group) were fitted with goggles that covered the nasal fields of both eyes. The light input for this control group of turtles was thus restricted, but the restriction was symmetric with respect to the retinas of a turtle's two eyes. The 18 turtles in the second group were fitted with what are known as "harlequin goggles". They covered the nasal field for one eye and the temporal field for the other eye, corresponding to asymmetric restrictions on the light input to the two retinas for a turtle in this group. After fitting with the appropriate goggles, each of the 36 turtles was placed (one at a time), facing away from the sea in the center of an arena of 46 meters radius above the high tide zone of the beach. The arena sloped slightly upward in the direction of the sea, which was itself not directly visible at turtle eye level from within the arena. The measurement recorded for each turtle was the number of times it circled (i. e., crossed its own path) in the first 2 min after it had begun to crawl. The number of these 'circles' for each of the 36 turtles in the study is given in Table 9.6.

Clearly, these data are not from normal populations. In fact, the measurements are counts and, therefore, not even continuous random variables. However, the two-sample central limit theorem that led to the sampling distribution of Z^* (9.25) being approximately standard normal requires only that the two population variances are finite. This is clearly the case for the circle measurement data in Table 9.6. Thus the approximate inference procedures based on the *t*-distribution with degrees of freedom $[[df]]$, with *df* given by (9.31), can be applied to analyze the circling data.

We can use the **R** function *apply*( ) with the argument $MARGIN = 2$ (telling R to apply the function by column instead of by row; see *?apply* for more detail) on the data.frame *goggled_green_turtles* to calculate the results of

Table 9.6 Number of circles in the two-minute crawl period

Nasal field goggles (*X*)	Harlequin goggles (*Y*)
0	0
0	0
0	1
0	2
0	2
0	2
0	2
1	3
1	3
1	3
2	4
2	5
2	6
2	8
2	10
4	11
6	12
10	14

Source: Mrosovsky and Shettleworth (1974)

various functions applied by column. Using the *mean*( ) and *sd*( ) functions and the sample labels (X and Y) designated in Table 9.6, we see that the sample averages and standard deviations for the circling data are $\bar{x} = 1.83$, $\bar{y} = 4.89$, $s_X = 2.60$, and $s_Y = 4.31$

```
> apply(goggled_green_turtles, 2, mean)
nasal_field   harlequin
   1.833333    4.888889

> apply(goggled_green_turtles, 2, sd)
nasal_field   harlequin
   2.595245    4.309891
```

Evaluating expression (9.31), we see that.

$$df = \frac{\left(\frac{(2.60)^2}{18} + \frac{(4.31)^2}{18}\right)^2}{\frac{\left(\frac{(2.60)^2}{18}\right)^2}{18-1} + \frac{\left(\frac{(4.31)^2}{18}\right)^2}{18-1}} = \frac{1.981}{.0083 + .0626} = 27.94.$$

Taking the conservative approach to this approximation, we see that inferences about the difference in the means, $\mu_Y - \mu_X$, for these turtle circlings data can be based on the t-distribution with degrees of freedom [[df]] = {largest positive integer less than or equal to 27.94} = 27.

Thus, we estimate the difference, $\mu_Y - \mu_X$, between the mean number of circles for turtles with harlequin goggles and the mean number for turtles with symmetric nasal goggles to be $\bar{y} - \bar{x} = 4.89 - 1.83 = 3.06$ circles. An approximate 95% confidence interval for $\mu_Y - \mu_X$ is obtained from (9.26) with $z_{.025}$ replaced by $t_{27,.025}$, yielding

$$(\bar{y} - \bar{x}) \pm t_{27,.025}\sqrt{\frac{s_Y^2}{n} + \frac{s_X^2}{m}} = 3.06 \pm 2.052\sqrt{\frac{(2.60)^2}{18} + \frac{(4.31)^2}{18}}$$
$$= 3.06 \pm 2.435 = (0.625, 5.495) \text{ circles.}$$

Thus we are approximately 95% confident that the average number of circles for turtles with harlequin goggles will be somewhere between .625 and 5.495 circles higher than the average number of circles for turtles with symmetric nasal goggles. This provides a clear indication that the asymmetric distortion of the light source by the harlequin goggles has a greater negative effect on a green turtle hatchling's natural sea-finding instinct than does the symmetric blocking of the light by the nasal goggles. We note that since the value 0 is not contained in our approximate 95% confidence interval for $\mu_Y - \mu_X$, we would also reject H_0: [$\mu_Y = \mu_X$] in favor of H_A: [$\mu_Y \neq \mu_X$] at approximate significance level $\alpha = 1 - .95 = .05$. (You are asked in Exercise 9.B.18 to use the t-distribution with $df = 27$ degrees of freedom to find the approximate P-value for a test of H_0: [$\mu_Y = \mu_X$] versus H_A: [$\mu_Y \neq \mu_X$] for the green turtle hatchling data in Table 9.6.)

Section 9.3 Practice Exercises

9.3.1. *House Sizes as Related to Lot Sizes in North Carolina.* The dataset *house_lot_sizes* contains information about house and lot sizes for a random sample of 100 properties in Wake County, North Carolina, as collected by Woodard and Leone (2008). Consider two subsets of this dataset corresponding to the smallest 25 lot sizes (in acreage) and the largest 25 lot sizes (in acreage), respectively. Viewing these two subsets as representative samples of "*small*" and "*large*" lot sizes in Wake County, North Carolina, conduct the following analyses under the assumption that the populations of house sizes (in square feet) in Wake County, North Carolina for these two categories are normally distributed with mean house sizes μ_{small} and μ_{large} and variances σ^2_{small} and σ^2_{large}, respectively.

(a) Estimate the difference in mean house sizes, $\mu_{\text{large}} - \mu_{small}$.

(b) Find a confidence interval for $\mu_{\text{large}} - \mu_{small}$. Choose your own reasonable confidence level.

(c) Find the *P*-value for a test of $\text{H}_0 : \mu_{\text{large}} = \mu_{small}$ against the one-sided alternative H_A: $\mu_{\text{large}} > \mu_{small}$. What is your decision at significance level .02?

(d) Compare your findings in parts (a)–(c) with those obtained in Exercise 9.2.17 without the normality assumption.

9.3.2. *Diamonds—Does Color Matter?* Carry out statistical analyses similar to those prescribed in Exercise 9.2.19 but now under the more stringent assumption that the populations for the two color categories are normally distributed with mean values (in Singapore dollars) $\mu_{D,E}$ and $\mu_{F\ or\ lower}$ and variances $\sigma^2_{D,E}$ and $\sigma^2_{F\ or\ lower}$, respectively. Compare your results with those obtained in Exercise 9.2.19.

9.3.3. *Movie Ratings and Running Times.* The *Movie and Video Guide* is a ratings and information guide to movies prepared annually by Leonard Maltin. Moore (2006) selected a random sample of 100 movies from the 1996 edition of the *Guide*. He compiled the dataset *movie_facts* containing relevant information about the selected movies. Two of the pieces of information provided are the rating that Maltin gave each of the movies on a rising (worst to best) scale of 1, 1.5, 2, 2.5, 3, 3.5, 4 and the running time of the movie in minutes. Divide the 100 movies in the sample into two subsets, corresponding to movies with running times less than 90 minutes and those with running times of at least 90 minutes. Complete the following statistical analyses under the assumption that the populations of ratings for shorter running time (< 90 minutes) movies and longer running time (≥ 90 minutes) movies are normally distributed with means $\mu_{<90}$ and $\mu_{\geq 90}$ and variances $\sigma^2_{<90}$ and $\sigma^2_{\geq 90}$, respectively.

(a) Estimate the difference in mean ratings $\mu_{\geq 90} - \mu_{<90}$.

(b) Find an upper confidence bound for $\mu_{\geq 90} - \mu_{<90}$. Choose your own reasonable confidence level.

(c) Find the *P*-value for a test of H_0: $\mu_{\geq 90} = \mu_{<90}$. against the one-sided alternative H_A: $\mu_{\geq 90} < \mu_{<90}$. What is your decision at significance level .0250?

9.3.4. *Are Some Dung Piles Better Than Others? Onthophagus lecontei* is an American dung beetle that feeds on the dung of a number of different animal species. But would it, in fact, be more beneficial if they concentrated more heavily on wild rabbit (*Silvilagus cunicularius*) dung? Arellano et al. (2015) studied the effect of a variety of dung sources on the number, mass, and volume of the brood masses for dung beetles. The authors randomly paired adult beetles in the laboratory for breeding purposes and assigned them to horse, goat, or wild rabbit dung for brood development. They were interested in a number of factors, including which type of dung leads to longer (supposedly more competitive) offspring. The 44 offspring reared in wild rabbit dung had a mean length of $\bar{x}_{WR} = 5.45cm$ with standard deviation $s_{WR} = 0.05cm$, while the 23 offspring reared in horse dung had a mean length of $\bar{x}_H = 4.82cm$ with standard deviation $s_H = 0.16cm$. Complete the following statistical analyses under the assumption that the populations of lengths for adult dung beetles reared in wild rabbit dung and those reared in horse dung are normally distributed with means μ_{WR} and μ_H and variances σ^2_{WR} and σ^2_H, respectively.

(a) Estimate the difference in mean lengths $\mu_{WR} - \mu_H$.

(b) Find a lower confidence bound for $\mu_{WR} - \mu_H$. Choose your own reasonable confidence level.

(c) Find the *P*-value for a test of H_0: $\mu_{WR} = \mu_H$ against the one-sided alternative H_A: $\mu_{WR} > \mu_H$. What is your decision at significance level .010?

9.3.5. *Movie Cast Sizes and Ratings.* The *Movie and Video Guide* is a ratings and information guide to movies prepared annually by Leonard Maltin. Moore (2006) selected a random sample of 100 movies from the 1996 edition of the *Guide.* He compiled the dataset *movie_facts* containing relevant information about the selected movies. Two of the pieces of information provided are the rating that Maltin gave each of the movies on a rising (worst to best) scale of 1, 1.5, 2, 2.5, 3, 3.5, 4 and the listed number of cast members in each movie. Divide the 100 movies in the sample into two subsets, corresponding to movies with 6 or fewer listed cast members ($m = 53$ movies) and those with more than 6 listed cast members ($n = 47$ movies). Complete the following statistical analyses under the assumption that the populations of ratings for movies with smaller casts (6 or fewer listed cast members) and movies with larger casts (more than 6 listed cast members) are normally distributed with means $\mu_{\leq 6}$ and $\mu_{> 6}$ and variances $\sigma^2_{\leq 6}$ and $\sigma^2_{> 6}$, respectively.

(a) Estimate the difference in mean ratings $\mu_{>6} - \mu_{\leq 6}$.

(b) Find a lower confidence bound for $\mu_{>6} - \mu_{\leq 6}$. Choose your own reasonable confidence level.

(c) Find the *P*-value for a test of H_0: $\mu_{>6} = \mu_{\leq 6}$. against the two-sided alternative H_A: $\mu_{>6} \neq \mu_{\leq 6}$. What is your decision at significance level .015?

9.3.6. *National League Salaries Compared to American League Salaries.* Assume that the 2014 salaries of American League baseball players and the 2014 salaries of National League baseball players are each normally distributed with means μ_A and μ_N and variances σ_A^2 and σ_N^2, respectively. In Tables 1.17 and 1.23 we listed the 2014 baseball salaries for members of the New York Yankees (American League) and Cincinnati Reds (National League) baseball teams, respectively. Viewing these salaries as random samples from the larger populations of 2014 salaries for all baseball players in the American and National Leagues, respectively, conduct the following statistical analyses.

(a) Estimate the difference in average salaries $\mu_A - \mu_N$.

(b) Find a lower confidence bound for $\mu_A - \mu_N$. Choose your own reasonable confidence level.

(c) Find the *P*-value for a test of H_0: $\mu_A = \mu_N$ against the one-sided alternative H_A: $\mu_A > \mu_N$. What is your decision at significance level .010?

(d) Discuss any concerns you might have about these statistical "conclusions".

9.3.7. *Clearing Ultrasound Probes of Bacterial Infections.* One of the major sources for spreading nosocomial (hospital-acquired) infections from patient to patient is through the use of ultrasound probes at tertiary care facilities and it is essential that hospitals use effective ultrasound probe cleaning procedures. Ali et al. (2015) presented data for comparing different cleaning techniques, including the use of sterilized paper towels versus treatment with a 0.9% saline solution. The Colony Forming Unit (CFU) of bacterial counts using a standard agar plate were obtained from culture swabs for 50 probes conducted at the Radiology Department of the Aga Khan University Hospital in Karachi, Pakistan. Twenty-five of these probes were then wiped with sterilized paper towels and twenty-five of them were treated with a 0.9% saline solution. The CFU bacterial counts were then obtained again for each of the 50 probes after treatment. The before and after treatment CFU counts for the 50 probes are given in Table 9.7.

Table 9.7 Number of colony forming units (CFU) of bacterial counts for ultrasound probes before and after treatment with sterilized paper towels or 0.9% saline solution

Probe number	Number of colony forming units (CFU)	
	Before paper towel wipe	**After paper towel wipe**
1	350	136
2	142	62
3	190	106
4	300	190
5	409	211
6	390	192
7	159	61
8	198	101
9	302	192
10	296	136
11	322	166
12	172	72
13	104	78
14	151	91
15	133	71
16	202	131
17	102	89
18	109	79
19	167	99
20	79	59
21	107	78
22	89	55
23	202	121
24	197	101
25	106	79
	Before saline solution	**After saline solution**
26	292	51
27	302	42
28	261	49
29	302	97
30	192	39
31	201	32

(continued)

Table 9.7 (continued)

Probe number	**Number of colony forming units (CFU)**	
32	192	62
33	289	67
34	290	81
35	233	89
36	209	41
37	289	53
38	301	89
39	189	39
40	161	39
41	231	61
42	142	29
43	190	58
44	203	81
45	297	52
46	219	51
47	161	21
48	232	41
49	171	36
50	193	71

Source: Ali et al. (2015)

Complete the following statistical analyses under the assumption that the populations of numerical reduction in CFU counts from wiping with sterile paper towels and numerical reduction in CFU counts from being treated with a 0.9% saline solution are normally distributed with means μ_{SPT} and μ_{Saline} and variances σ^2_{SPT} and σ^2_{Saline}, respectively.

(a) Estimate the difference in mean numerical reductions $\mu_{SPT} - \mu_{Saline}$.

(b) Find a confidence interval for $\mu_{SPT} - \mu_{Saline}$. Choose your own reasonable confidence level.

(c) Find the *P*-value for a test of H_0: $\mu_{SPT} = \mu_{Saline}$ against the one-sided alternative $H_A : \mu_{SPT} > \mu_{Saline}$. What is your decision at significance level .040?

9.3.8. *Clearing Ultrasound Probes of Bacterial Infections Part Two.* Consider once again the data on clearing bacteria from ultrasound probes presented in Exercise 9.3.7. Conduct the exact same statistical analyses prescribed in that exercise but using the *percentage reduction* in CFU units rather than the numerical reduction in CFU units resulting from the two methods of treatment. Discuss how these two approaches to analyzing the data in Table 9.7 differ.

9.3.9. *Are All Euros Minted Equal?* In Exercises 7.B.4 and 7.B.6 you were asked to statistically compare the weights of Euro coins contained in two separate Packages 1 and 8, each containing 250 brand new coins. Now we have the machinery to make a formal simultaneous comparison of those two collections of Euros. Assume that the weights of Euros from the mint(s) that produced Packages 1 and 8 are normally distributed with means μ_1 and μ_8 and variances σ_1^2 and σ_8^2, respectively. Viewing the 250 Euro coins in each of these two packages as simple random samples from the mint(s) that produced them, carry out the following statistical analyses.

(a) Estimate the difference in mean weights $\mu_1 - \mu_8$.

(b) Find a confidence interval for $\mu_1 - \mu_8$. Choose your own reasonable confidence level.

(c) Find the P-value for a test of H_0: $\mu_1 = \mu_8$ against the two-sided alternative H_A: $\mu_1 \neq \mu_8$. What is your decision at significance level .027?

9.3.10. *Angioplasty Balloons—To Coat or Not To Coat—That Is the Question.* Atherosclerotic disease is a disease in which plaque builds up inside one's arteries, which can, among other problems, compromise blood flow to the legs and feet. Percutaneous transluminal coronary angioplasty (PTCA) is a minimally invasive procedure that uses a tiny balloon inserted into a blood vessel to open blocked coronary arteries and allows blood to more freely circulate to the legs and feet. Many times, however, the blood vessels once again narrow within a year after PTCA, requiring that additional measures, such as surgically inserting stints, to reduce this narrowing effect. Rosenfield et al. (2015) reported on the results of a study to investigate whether coating the angioplasty balloon during PTCA with the antineoplastic compound *paclitaxel* could extend the benefits from the procedure. The authors enrolled 476 patients in the study, of which 160 received standard PTCA and 316 received PTCA with a balloon coated with *paclitaxel*. All 476 patients were followed for a year post-PTCA. Among other diagnostics, the Walking Impairment Questionnaire (higher scores demonstrate greater mobility) was administered to the patients one year after the procedure. As part of this questionnaire, each patient achieved a walking distance score. The average walking distance score for the 160 patients who received the standard PTCA was $\bar{x}_{standard} = 22.2$ with standard deviation $s_{standard} = 35.4$. The corresponding average walking distance score for the 316 patients receiving PTCA coated with *paclitaxel* was $\bar{x}_{coated} = 31.5$ with standard deviation $s_{coated} = 35.4$. Assume that the one-year post-procedure walking distance scores for patients receiving standard PTCA and for patients receiving PTCA coated with *paclitaxel* are normally distributed with means $\mu_{standard}$ and μ_{coated} and variances $\sigma^2_{standard}$ and σ^2_{coated}, respectively.

(a) Estimate the difference in mean walking distance scores $\mu_{coated} - \mu_{standard}$.

(b) Find a lower confidence bound for $\mu_{coated} - \mu_{standard}$. Choose your own reasonable confidence level.

(c) Find the *P*-value for a test of $H_0 : \mu_{coated} = \mu_{standard}$ against the one-sided alternative H_A: $\mu_{coated} > \mu_{standard}$. What is your decision at significance level .056?

9.3.11. *Unintended Effects of Pesticides on Earthworms*. Insect growth regulators (IGRs) are advanced insecticides designed to mitigate the negative effects of harmful insects by preventing them from reaching maturity. They are labeled 'reduced risk' by the Environmental Protection Agency, meaning that they specifically target harmful insects while having minimal effects on beneficial insects. As the concentration of IGRs accumulates in a soil environment, however, they could lead to unintended negative consequences for other non-insect soil organisms, such as earthworms, that play important roles in the enrichment and improvement of soil for plants and other animals, including humans. In a designed laboratory study using artificially prepared soil, Badawy et al. (2013) investigated the effects that three different IGRs, namely, *buprofezin*, *lufenuron*, and *triflumuron*, had on growth changes and biochemical activity of the earthworm *Aporrectodea caliginosa*, commonly found in Egypt. Groups of four earthworms each were exposed to various dose levels of the three IGRs and a separate set of four earthworms served as a control in the same soil environment without any IGR infusion. One of the biochemical attributes studied was Acetylcholinesterase (AChE) activity, which plays an important role in neuromuscular junctions and brain synapses. AChE activity measurements were taken on the four control earthworms after 4 weeks in the pure artificial soil, as well as on the four earthworms that were exposed for 4 weeks to the same artificial soil infused with a concentration of 5 $mg\ a.i/kg$ soil of the IGR *lufenuron*. The average AChE activity for the four control earthworms at the end of the 4 weeks was $\bar{x}_{control} = 39.83\ \Delta OD_{412} \cdot \text{min}^{-1} \cdot mg\ protein^{-1}$ with standard deviation $s_{control} = 0.481\ \Delta OD_{412} \cdot \text{min}^{-1} \cdot mg\ protein^{-1}$. The corresponding average AChE activity for the four earthworms in the soil containing *lufenuron* was $\bar{x}_{lefenuron} = 3.70\ \Delta OD_{412} \cdot \text{min}^{-1} \cdot mg\ protein^{-1}$ with standard deviation $s_{lefenuron} = 0.741\ \Delta OD_{412} \cdot \text{min}^{-1} \cdot mg\ protein^{-1}$. Assume that the four week AChE activity measurements for the control earthworms and for the earthworms exposed to *lufenuron* are normally distributed with means $\mu_{control}$ and $\mu_{lufenuron}$ and variances $\sigma^2_{control}$ and $\sigma^2_{lufenuron}$, respectively.

(a) Estimate the difference in mean AChE activity $\mu_{control} - \mu_{lufenuron}$.

(b) Find a lower confidence bound for $\mu_{control} - \mu_{lufenuron}$. Choose your own reasonable confidence level.

(c) Find the P-value for a test of H_0: $\mu_{control} = \mu_{lufenuron}$ against the one-sided alternative H_A: $\mu_{control} > \mu_{lufenuron}$. What is your decision at significance level .001?

9.3.12. *Bird Diversity During the Wet and Dry Seasons in Tugu Wetland, Ghana.* Nsor and Obodai (2014) conducted a study to assess the effect of environmental factors on the diversity of bird populations in the wetlands areas of the Northern Region of Ghana. They observed birds from 7-11 am GMT once a week for a period of two years at each of six wetlands areas (Kukobila, Wuntori, Tugu, Adayili, Nabogo, and Bunglung) and the number of different bird species observed were compiled monthly during that period of time for each of these six areas. One aspect of their study was to compare the monthly numbers of bird species observed in each wetlands area during the 7 months of the dry season (November through May) and the 5 months of the wet season (June through October) over the two-year period. For one particular portion of the Tugu wetland (classified as a closed shallow marsh), they found that the average of the 14 monthly numbers of bird species observed during the dry season (over the two-year period) was $\bar{x}_{dry} = 7.4$ species with standard deviation $s_{dry} = 2.1$ species. For the same region the average of the 10 monthly numbers of bird species observed during the wet season (over the two-year period) was $\bar{x}_{wet} = 9.6$ species with standard deviation $s_{wet} = 3.6$ species. Assume that the monthly numbers of species observed in this portion of the Tugu wetland during dry and wet seasons are normally distributed with means μ_{dry} and μ_{wet} and variances σ^2_{dry} and σ^2_{wet}, respectively.

(a) Estimate the difference in mean number of bird species $\mu_{wet} - \mu_{dry}$.

(b) Find a lower confidence bound for $\mu_{wet} - \mu_{dry}$. Choose your own reasonable confidence level.

(c) Find the P-value for a test of H_0: $\mu_{wet} = \mu_{dry}$ against the one-sided alternative H_A: $\mu_{wet} > \mu_{dry}$. What is your decision at significance level .015?

9.3.13. *Bird Diversity During the Wet and Dry Seasons in Bunglung Wetland, Ghana.* In the same study discussed in Exercise 9.3.12, Nsor and Obodai (2014) also collected data from the Bunglung Wetland in Ghana. For one particular portion of the Bunglung wetland (classified as an artificial wetland), they found that the average of the 14 monthly numbers of bird species observed during the dry season (over the two-year period) was $\bar{x}_{dry} = 6.2$ species with standard deviation $s_{dry} = 2.2$ species. For the same region the average of the 10 monthly numbers of bird species observed during the wet season (over the two-year period) was $\bar{x}_{wet} = 4.8$ species with standard deviation $s_{wet} = 1.6$ species. Assume that the monthly numbers of species observed in this portion of the Bunglung wetland during dry and wet seasons are normally distributed with means μ_{dry} and μ_{wet} and variances σ^2_{dry} and σ^2_{wet}, respectively.

(a) Estimate the difference in mean number of bird species $\mu_{wet} - \mu_{dry}$.

(b) Find an upper confidence bound for $\mu_{wet} - \mu_{dry}$. Choose your own reasonable confidence level.

(c) Find the P-value for a test of H_0: $\mu_{wet} = \mu_{dry}$ against the one-sided alternative H_A: $\mu_{wet} < \mu_{dry}$. What is your decision at significance level .032?

(d) Compare and contrast these results with those you obtained in Exercise 9.3.12.

9.3.14. *Bird Diversity During the Wet Season in the Adayili and Nabogo Wetlands, Ghana.* Nsor and Obodai (2014) conducted a study to assess the effect of environmental factors on the diversity of bird populations in the wetlands areas of the Northern Region of Ghana. They observed birds from

7-11 am GMT once a week for a period of two years at each of six wetlands areas (Kukobila, Wuntori, Tugu, Adayili, Nabogo, and Bunglung) and the number of different bird species observed were compiled monthly during that period of time for each of these six areas. One aspect of their study was to compare the monthly numbers of bird species observed in different wetlands during the 5 months of the wet season (June through October) over the two-year period. For comparable portions of the Adayili and Nabogo wetlands (both classified as riparian wetlands), they found that the averages of the 10 monthly numbers of bird species observed during the rainy season (over the two-year period) for the Adayili and Nabogo wetlands were $\bar{x}_{Adayili} = 5.4$ species and $\bar{x}_{Nabogo} = 7.2$ species, respectively, with standard deviations $s_{Adayili} = 1.4$ species and $s_{Nabogo} = 2.6$ species, respectively. Assume that the monthly numbers of species observed in these portions of the Adayili and Nabogo wetlands during the wet season are normally distributed with means $\mu_{Adayili}$ and μ_{Nabogo} and variances $\sigma^2_{Adayili}$ and σ^2_{Nabogo}, respectively.

(a) Estimate the difference in mean number of bird species $\mu_{Adayili} - \mu_{Nabogo}$.

(b) Find a confidence interval for $\mu_{Adayili} - \mu_{Nabogo}$. Choose your own reasonable confidence level.

(c) Find the *P*-value for a test of $H_0 : \mu_{Adayili} = \mu_{Nabogo}$ against the two-sided alternative H_A: $\mu_{Adayili} \neq \mu_{Nabogo}$. What is your decision at significance level .021?

9.3.15. *How Well Does Your Beer Hold Its Foam?* Carry out statistical analyses similar to those prescribed in Exercise 9.2.21 but now under the more stringent assumption that the populations for the two beer production lots are normally distributed with mean percentage head collapse four minutes after pouring μ_1 and μ_2 and variances σ_1^2 and σ_2^2, respectively. Compare your results with those obtained in Exercise 9.2.21.

9.3.16. *You Can Lead a Slug to Food But Will It Eat?* Consider the woodland and waste site slug study discussed in Example 9.3. Assume that the Acceptability Indices (AI) to the plant *Allium Ursinum* (wild garlic) for *Arion Subfuscus* from woodland and waste sites are normally distributed with mean AI values $\mu_{woodland}$ and μ_{waste} and variances $\sigma^2_{woodland}$ and σ^2_{waste}, respectively.

(a) Estimate the difference in mean AI values $\mu_{woodland} - \mu_{waste}$.
(b) Find a confidence interval for $\mu_{woodland} - \mu_{waste}$. Choose your own reasonable confidence level.
(c) Find the *P*-value for a test of H$_0$: $\mu_{woodland} = \mu_{waste}$ against the two-sided alternative H$_A$: $\mu_{woodland} \neq \mu_{waste}$. What is your decision at significance level .060?

9.3.17. *Meniscal Repair—Vertical FasT-Fix Sutures or Horizontal FasT-Fix Sutures?* Surgery is most often the only option when faced with a torn medial meniscus—but what is the best surgical method for the repair? Kocabey et al. (2006) studied the performance characteristics of three different meniscal repair techniques, namely, the Vertical FasT-Fix Meniscal Repair Suture System (VFasT-Fix), the Horizontal FasT-Fix Meniscal Suture System (HFasT-Fix), and the RapidLoc Device (RLD) approach. Human cadaveric knees were used in the study, with six randomly assigned to each of the three surgery techniques. Each repaired meniscus was loaded into a servohydraulic device and cycled between 5 and 50 Newtons (*N*) at 1 *Hz* for 500 cycles. After cycle testing, the meniscus was subjected to tension loading at a slow rate of 5 *mm*/*min* (similar to the type of stresses that a meniscus might have to deal with during early rehabilitation) until failure of the repair occurred. One aspect of their study was to compare the load to failure test results for the VFasT-Fix and HFasT-Fix repair techniques. The authors found that the average load to failure for the six menisci repaired using the VFasT-Fix approach and the six repaired using the HFasT-Fix approach were $\bar{x}_{VFasT-Fix}$

$= 125.3N$ and $\bar{x}_{HFasT-Fix} = 89.7N$, respectively, with standard deviations $s_{VFasT-Fix} = 39N$ and $s_{HFasT-Fix} = 14$ N, respectively. Assume that the load to failure values for VFasT-Fix and HFastT-Fix meniscus repairs are normally distributed with means $\mu_{VFasT-Fix}$ and $\mu_{HFasT-Fix}$ and variances $\sigma^2_{VFasT-Fix}$ and $\sigma^2_{HFasT-Fix}$, respectively.

(a) Estimate the difference in mean load to failure values, $\mu_{VFasT-Fix} - \mu_{HFasT-Fix}$.

(b) Find a lower confidence bound for $\mu_{VFasT-Fix} - \mu_{HFasT-Fix}$. Choose your own reasonable confidence level.

(c) Find the *P*-value for a test of H_0: $\mu_{VFasT-Fix} = \mu_{HFasT-Fix}$ against the - one-sided alternative H_A: $\mu_{VFasT-Fix} > \mu_{HFasT-Fix}$. What is your decision at significance level .042?

9.3.18. *Meniscal Repair—Vertical FasT-Fix Sutures or RapidLoc Device?* Consider the study of meniscal repair techniques by Kocabey et al. (2006) discussed in Exercise 9.3.17. The authors were also interested in comparing the stiffness of the menisci during cycle loading for the VFasT-Fix and RapidLoc Device repair techniques. They found that the average stiffness measurement for the six menisci repaired using the VFasT-Fix approach and the six repaired using the RapidLoc approach were $\bar{x}_{VFasT-Fix} = 14.4N/mm$ and $\bar{x}_{RapidLoc} = 9.7N/mm$, respectively, with standard deviations $s_{VFasT-Fix} = 2.1\ N/mm$ and $s_{RapidLoc} = 0.44N/mm$, respectively. Assume that the average stiffness measurements for VFasT-Fix and RapidLoc meniscus repairs are normally distributed with means $\mu_{VFasT-Fix}$ and $\mu_{RapidLoc}$ and variances $\sigma^2_{VFasT-Fix}$ and $\sigma^2_{RapidLoc}$, respectively.

(a) Estimate the difference in mean stiffness measurements, $\mu_{VFasT-Fix} - \mu_{RapidLoc}$.

(b) Find a confidence interval for $\mu_{VFasT-Fix} - \mu_{RapidLoc}$. Choose your own reasonable confidence level.

(c) Find the *P*-value for a test of H_0: $\mu_{VFasT-Fix} = \mu_{RapidLoc}$ against the two-sided alternative H_A: $\mu_{VFasT-Fix} \neq \mu_{RapidLoc}$. What is your decision at significance level .057?

9.3.19. *Does Believing Help Make It So?* Does having a "pro attitude" (an attitude towards a favorable outcome, a concept considered by Thalbourne 2004) help individuals in a paranormal task? Storm and Thalbourne (2005) studied this question using a paranormal symbol-guessing experiment involving Zener symbols (star, waves, square, circle, and cross). A total of 131 participants volunteered for this experiment and each of them was classified as either a "believer" or "skeptic", depending on whether they scored at least 17 or less than 17, respectively, on the Australian Sheep-Goat Scale (ASGS) construct proposed by Thalbourne (1995). Eighty-seven of the participants were classified as "believers", while the remaining 44 were labeled as "skeptics". On each of 50 trials, participants were required to guess a computer's pre-selected Zener symbol by clicking the button under the symbol he/she thought would be the computer's pre-selected symbol for that trial. The number of correct guesses by each of the 131 participants is given in the dataset *believers_skeptics*, along with the information about whether a given participant was a "believer" or a "skeptic".

(a) How many of the symbols would you expect a participant to predict correctly if she/he were simply guessing on each trial?
(b) What are the averages and standard deviations for the number of correct predictions for the 87 "believers"? for the 44 "skeptics"?
(c) What are the null and alternative hypotheses of interest in this study?
(d) Find the *P*-value for an appropriate test of these hypotheses. What is your conclusion about "believers" versus "skeptics" at significance level .01?

9.4 Inference for the Difference in Means for Two Normal Populations with Equal Variances–Procedures Based on the Two Sample Averages and a Pooled Sample Standard Deviation

When conditions are such that the two continuous measurements X and Y are known to have underlying normal distributions with equal, but unknown, variances, we can utilize this information to construct procedures that are more effective in making inferences about the difference in the population means $\Delta = \mu_Y - \mu_X$. (Note that Δ also corresponds to the difference in population medians, since the mean and median coincide for normal distributions.)

Let $\bar{X}$ and $\bar{Y}$ be the sample averages and S_X and S_Y the sample standard deviations for independent random samples $X_1, \ldots, X_m$ and $Y_1, \ldots, Y_n$, from $N(\mu_X, \sigma)$ and $N(\mu_Y, \sigma)$ populations, respectively, where the population means μ_X, μ_Y, and the <u>common</u> population standard deviation σ are all unknown. The natural estimator of the difference $\mu_Y - \mu_X$ is the analogous observed difference in the sample averages $\bar{Y} - \bar{X}$. It follows from properties of expectation for a difference between two variables that the mean for $\bar{Y} - \bar{X}$ is

$$\mu_{\bar{Y}-\bar{X}} = \mu_{\bar{Y}} - \mu_{\bar{X}} = \mu_Y - \mu_X,$$

since the sample averages are unbiased estimators of the corresponding population parameters. Moreover, since the sample averages $\bar{X}$ and $\bar{Y}$ are based on independent random samples from the X and Y populations, the variance for $\bar{Y} - \bar{X}$ corresponds to the sum of the variances for the separate sample averages; that is,

$$\sigma^2_{\bar{Y}-\bar{X}} = \sigma^2_{\bar{Y}} + \sigma^2_{\bar{X}} = \frac{\sigma^2_Y}{n} + \frac{\sigma^2_X}{m} = \sigma^2\left(\frac{1}{n} + \frac{1}{m}\right)$$

and the corresponding standard deviation for $\bar{Y} - \bar{X}$ is

$$\sigma_{\bar{Y}-\bar{X}} = \sigma\sqrt{\frac{1}{m} + \frac{1}{n}}.$$

Since the underlying populations are normal, it follows that the standardized variable

$$Z^* = \frac{(\bar{Y} - \bar{X}) - (\mu_Y - \mu_X)}{\sigma\sqrt{\frac{1}{m} + \frac{1}{n}}} \tag{9.32}$$

has a standard $N(0, 1)$ distribution. However, since the common population standard deviation σ is unknown, we must estimate it in order to be able to make statistical inferences about the difference in means $\mu_Y - \mu_X$. Since both the X and Y samples carry information about their common population variance, it is natural to use both sample variances, S_X^2 and S_Y^2, to estimate σ^2 (and, hence, σ). The optimal way to combine these separate sample variances is to weight them by their respective sample sizes, thereby allowing the larger sample to have greater emphasis in this process. The resulting estimator for the common variance, called the *pooled estimator of* σ^2, is given by.

$$S_p^2 = \frac{(m-1)S_X^2 + (n-1)S_Y^2}{m+n-2}. \tag{9.33}$$

The corresponding pooled estimator of the standard deviation σ is then

$$S_p = \sqrt{\frac{(m-1)S_X^2 + (n-1)S_Y^2}{m+n-2}}. \tag{9.34}$$

Using this pooled estimator of σ in the standardized expression (9.32) yields the variable

$$T = \frac{(\bar{Y} - \bar{X}) - (\mu_Y - \mu_X)}{S_p\sqrt{\frac{1}{m} + \frac{1}{n}}}. \tag{9.35}$$

The probability distribution for T (9.35) is a t-distribution with $m + n - 2$ degrees of freedom, corresponding to the sum of the degrees of freedom (m - 1) and (n-1) associated with the X and Y sample variances, respectively.

This fact provides the basis for interval estimation of the difference in means $\mu_Y - \mu_X$. Such interval estimates for $\mu_Y - \mu_X$ are centered at the point estimator $\bar{Y} - \bar{X}$. The length of the interval is determined by $S_p\sqrt{\frac{1}{m}+\frac{1}{n}}$, which is the standard error of $\bar{Y} - \bar{X}$, and by the confidence level we want. The desired confidence level is attained through use of the appropriate percentiles of the t-distribution with $m + n - 2$ degrees of freedom. Proceeding as in other normality settings by taking the point estimator and adding and subtracting the appropriate margin of error, we obtain a confidence interval for the difference in means $\mu_Y - \mu_X$ based on the pooled t-statistic when the two populations have a common, but unknown, variance σ^2.

Point and Interval Estimation of the Difference in Population Means, $\Delta = \mu_Y - \mu_X$, for Two Normal Distributions with Common Variance The point estimator for the difference in means $\mu_Y - \mu_X$ is $\bar{Y} - \bar{X}$. The associated exact $100CL\%$ confidence interval for $\mu_Y - \mu_X$ is provided by the interval

$$\left(\bar{Y} - \bar{X} - t_{m+n-2,(1-CL)/2}\, S_p\sqrt{\frac{1}{m}+\frac{1}{n}},\ \bar{Y} - \bar{X} + t_{m+n-2,(1-CL)/2}\, S_p\sqrt{\frac{1}{m}+\frac{1}{n}}\right), \tag{9.36}$$

where $t_{m+n-2,(1-CL)/2}$ is the upper $((1 - CL)/2)$th percentile for the t-distribution with $m + n - 2$ degrees of freedom. The corresponding $100CL\%$ lower and upper confidence bounds for $\mu_Y - \mu_X$ are given by

$$\bar{Y} - \bar{X} - t_{m+n-2,1-CL}\, S_p\sqrt{\frac{1}{m}+\frac{1}{n}} \quad \text{and} \quad \bar{Y} - \bar{X} + t_{m+n-2,1-CL}\, S_p\sqrt{\frac{1}{m}+\frac{1}{n}},$$

respectively.

Example 9.10. Will Active-Exercise of a Newborn Infant Lead to an Earlier Onset of Walking Alone? If a newborn infant is held under her arms with her bare feet touching a flat surface, she will automatically attempt to make ordinary walking movements with her feet. In addition, if the backs of her feet are placed against the edge of a flat surface, she will make placing movements similar to those of a kitten. These automatic walking and placing reflexes are present in all normal infants and generally disappear by the time an infant reaches the age of eight weeks--but, while present, are they of any use for the development of the infant? How important is it during these first eight weeks for these two reflexes to be encouraged and stimulated? Could such encouragement and stimulation have any effect on the time at which an infant begins walking alone? Zelazo et al. (1972) conducted a study of newborn infants specifically designed to address this question. As part of their study, they enlisted twelve 1-week old male infants from middle-class and upper-middle-class families in the Boston area. For the study, six of these infants were randomly assigned to a control group and six were assigned to an active-exercise group. Infants in the active-exercise group received stimulation of both the walking and placing reflexes each day (four 3-minute sessions daily) from the beginning of the second week though the eighth week of age. Infants in the control group received no special training during this period of time. Mothers of these 12 infants then reported the ages at which their infants first walked alone. These data (in months) are presented in Table 9.8.

Here the X and Y sample sizes are $m = 6$ and $n = 6$, respectively, and the corresponding sample averages and sample variances are $\bar{x} = 11.708$,

Table 9.8 **Age (in months) at which infants first walked alone**

No-exercise group (X)	Active-exercise group (Y)
11.50	9.00
12.00	9.50
9.00	9.75
11.50	10.00
13.25	13.00
13.00	9.50

Source: Zelazo et al. (1972)

$\bar{y} = 10.125, s_x^2 = 2.310$, and $s_y^2 = 2.094$. There is nothing in these data to suggest non-normality for the populations of first walking times for either the no-exercise group or the active-exercise group. In addition, based on these 12 observations, it seems reasonable to assume equal variances for the two populations. Therefore, we will construct our confidence interval for $\mu_Y - \mu_X$ based on the pooled estimator of the common standard deviation σ given by S_p (9.34). For the walking age data of Table 9.7, we have.

$$s_p = \sqrt{\frac{(6-1)(2.310) + (6-1)(2.094)}{6+6-2}} = 1.484.$$

With pooled degrees of freedom $m + n - 2 = 6 + 6 - 2 = 10$, we find that $t_{10,.02} = 2.359$. Hence, we estimate the difference in mean first-walking times, $\mu_Y - \mu_X$, to be $\bar{y} - \bar{x} = 10.125 - 11.708 = -1.583$ months and the 96% pooled-t confidence interval for $\mu_Y - \mu_X$ is given by (9.36) to be

$$\bar{y} - \bar{x} \pm t_{10,.02} s_p \sqrt{\frac{1}{6} + \frac{1}{6}} = -1.583 \pm 2.359(1.484)(0.577)$$
$$= -1.583 \pm 2.020 = (-3.603, 0.437) \text{ months.}$$

Thus we are 96% confident that the mean first-walking time for newborn infants provided the active-exercise stimulation is between 3.603 months earlier (faster) and 0.437 months later (slower) for the active-exercise population than for the no-exercise population.

These results can also be obtained from a number of different software packages. For example, using the data.frame *infant_walking* based on the data from Table 9.8, the **R** function *t.test*( ) provides the following output, which includes estimates of μ_X and μ_Y and the 96% confidence interval (since we've specified the *conf.level* argument to be 0.96) for $\mu_X - \mu_Y$. (Be careful to note that the confidence interval reported by **R** is $\mu_X - \mu_Y$ not $\mu_Y - \mu_X$!)

```
> t.test(x = infant_walking$no_exercise, y = infant_walking$exercise,
conf.level = 0.96)

        Welch Two Sample t-test

data:  infant_walking$no_exercise and infant_walking$exercise
t = 1.8481, df = 9.9759, p-value = 0.09442
alternative hypothesis: true difference in means is not equal to 0
96 percent confidence interval:
 -0.4387392  3.6054059
sample estimates:
mean of x mean of y
 11.70833  10.12500
```

To conduct hypothesis tests about the difference in means μ_Y - μ_X for normal populations with common variance, we once again make use of the variable T (9.35).

Hypothesis Tests about the Difference in Population Means, $\Delta = \mu_Y - \mu_X$, for Two Normal Populations with Common Variance To test the null hypothesis H_0: $[\mu_X = \mu_Y]$ with two-sample data from normal populations $N(\mu_X, \sigma)$ and $N(\mu_Y, \sigma)$ with common variance, compute the statistic T (9.35) under the null hypothesis condition that $\mu_X = \mu_Y$, namely,

$$T = \frac{\bar{Y} - \bar{X}}{S_p\sqrt{\frac{1}{m} + \frac{1}{n}}} \tag{9.37}$$

and let t_{obs} be the attained value of T. Then, the exact P-value for normal populations with common variance for a test of H_0: $[\mu_Y - \mu_X = 0]$ against the alternatives H_A are:

H_A	P-value	
$\mu_Y > \mu_X$	$= P(T \geq t_{obs})$	(9.38)
$\mu_Y < \mu_X$	$= P(T \leq t_{obs})$	(9.39)
$\mu_Y \neq \mu_X$	$= 2P(T \geq t_{\text{obs}})$, if $t_{\text{obs}} \geq 0$	
	$= 2P(T \leq t_{\text{obs}})$, if $t_{\text{obs}} < 0$,	(9.40)

where $T \sim t(m + n\text{-}2)$, the t-distribution with $m + n - 2$ degrees of freedom.

Example 9.11. Active-Exercise and Onset of Walking Alone. Zelazo et al. (1972) were interested in assessing whether their active-exercise stimulation for newborn infants during the period 2-8 weeks of age would shorten the average time to onset of walking alone. Thus, using the notation of Example 9.10, we are interested in testing H_0: $[\mu_Y = \mu_X]$ versus the one-sided alternative H_A: $[\mu_Y < \mu_X]$. From Example 9.10 we know that the difference in sample averages is $\bar{y} - \bar{x} = -1.583$ months and the sample pooled standard deviation is $s_p = 1.484$. Computing the pooled *t*-statistic *T* (9.35), we see that

$$t_{obs} = \frac{-1.583}{1.484\sqrt{\frac{1}{6}+\frac{1}{6}}} = -1.848.$$

Hence, from (9.39) the *P*-value for our test of H_0: $[\mu_Y = \mu_X]$ against H_A: $[\mu_Y < \mu_X]$ is $P(T \leq -1.848) = P(t(10) \leq -1.848) = .0472$. With this *P*-value we would reject H_0: $[\mu_Y = \mu_X]$ in favor of H_A: $[\mu_Y < \mu_X]$ at significance level $\alpha = .05$, but not at significance level .025. Thus the data provide a moderate strength of support for the Zelazo-Zelazo-Kolb conjecture that their active-exercise stimulation for newborn infants during the period 2-8 weeks of age does shorten the average time to onset of walking alone.

As with the confidence interval for μ_Y - μ_X, the **R** function *t.test*( ) can be used to obtain the following output that includes the observed value of *T* and the associated one-sided *P*-value (since we specify *alternative* = *"greater"*) for this first-walking time hypothesis test. (As with the confidence interval, be careful to note that the value of *t* reported by **R** differs in sign from that in (9.35).)

```
> t.test(infant_walking$no_exercise, infant_walking$exercise,
alternative = "greater")

        Welch Two Sample t-test

data:  infant_walking$no_exercise and infant_walking$exercise
t = 1.8481, df = 9.9759, p-value = 0.04721
alternative hypothesis: true difference in means is greater than 0
95 percent confidence interval:
 0.03011819        Inf
sample estimates:
mean of x mean of y
 11.70833  10.12500
```

Large Sample Sizes m and n. When the sample sizes m and n are large, the t $(m + n\text{-}2)$ density curve can be well-approximated by the $N(0, 1)$ density curve. This fact can be used to provide approximations (for large m and n) for the confidence interval/bounds and hypothesis tests for $\Delta = \mu_Y - \mu_X$ described by Eqs. (9.36) and (9.38) - (9.40), respectively, by simply replacing the $t(m + n\text{-}2)$ percentiles and probabilities wherever they occur by the corresponding $N(0,1)$ percentiles and probabilities. However, remember that there are alternative procedures (see Sect. 3) available for such settings where both sample sizes are large that do not rely on the rather stringent condition (which underlies the procedures in this section) that the X and Y population variances are equal. The large sample procedures in Sect. 3 are the preferred methods of statistical inference when both sample sizes are large and we are interested in the difference of the two population means, $\Delta = \mu_Y - \mu_X$.

Section 9.4 Practice Exercises

9.4.1. *Are Men Hotter than Women?* It is a widely held belief that the normal body temperature for humans is 98.6 °*F*. Mackowiak et al. (1992) provide a critical evaluation of this statement through the collection of data from 148 individuals aged 18 through 40 years. The dataset *body_temperature_and_-heart_rate* contains body temperature and heart rate values (artificially generated by Shoemaker 1996, to closely recreate the original data considered by Mackowiak et al.) for 65 male and 65 female subjects. Conduct the following analyses under the assumption that the populations of human body temperatures for females and males are normally distributed with means μ_F and μ_M, respectively, and common variance σ^2.

(a) Estimate the difference in mean body temperatures, $\mu_M - \mu_F$.

(b) Find a confidence interval for $\mu_M - \mu_F$. Choose your own reasonable confidence level.

(c) Find the P-value for a test of H_0: $\mu_M = \mu_F$ against the one-sided alternative H_A: $\mu_F < \mu_M$. What is your decision at significance level .045?

9.4.2. *House Sizes as Related to Lot Sizes in North Carolina.* The dataset *house_lot_sizes* contains information about house and lot sizes for a random sample of 100 properties in Wake County, North Carolina, as collected by Woodard and Leone (2008). Consider two subsets of this dataset corresponding to the smallest 25 lot sizes (in acreage) and the largest 25 lot sizes (in acreage), respectively. Viewing these two subsets as representative samples of "*small*" and "*large*" lot sizes in Wake County, North Carolina, conduct the following analyses under the assumption that the populations of house sizes (in square feet) in Wake County, North Carolina for these two categories are normally distributed with mean house sizes μ_{small} and μ_{large}, respectively, and common variance σ^2.

(a) Estimate the difference in mean house sizes, $\mu_{\text{large}} - \mu_{small}$.

(b) Find a confidence interval for $\mu_{\text{large}} - \mu_{small}$. Choose your own reasonable confidence level.

(c) Find the P-value for a test of H_0: $\mu_{\text{large}} = \mu_{small}$ against the one-sided alternative H_A: $\mu_{\text{large}} > \mu_{small}$. What is your decision at significance level .02?

(d) Compare your results in parts (a)-(c) with those obtained in Exercise 9.3.1 without the common variance assumption.

9.4.3. *Diamonds—Does Color Matter?* Carry out statistical analyses similar to those prescribed in Exercises 9.2.19 and 9.3.2, but now under the assumption that the populations for the two color categories are normally distributed with mean values (in Singapore dollars) $\mu_{\text{D,E}}$ and $\mu_{\text{F} \textit{ or lower}}$ and common variance σ^2. Compare your results with those obtained in Exercises 9.2.19 and 9.3.2.

9.4.4. *Woodpecker Pads.* Woodpeckers play a major role as ecological engineers in forests as a consequence of their excavation of cavities that in turn create habitat for secondary cavity users. Tarbill et al. (2015) discuss the particular importance of this activity in recently burned forests as part of their study (over the two-year period 2009-2010) of the nesting characteristics of woodpecker species in a recently burned section of the Sierra Nevada Mountains in California. They examined the nest cavities of $m = 39$ black-backed woodpeckers and $n = 80$ hairy woodpeckers. The observed mean and standard deviation of the cavity heights for the 39 black-backed woodpeckers were $\bar{x}_{bb} = 4.72$ meters and $s_{bb} = 3.01$ meters, respectively. The corresponding data for the hairy woodpeckers were $\bar{x}_h = 7.23$ meters and $s_h = 4.10$ meters, respectively. Assuming that the populations of cavity heights for these two woodpecker species are normally distributed with means μ_{bb} and μ_h, respectively, and common variance σ^2, conduct the following statistical analyses.

(a) Find a confidence interval for $\mu_h - \mu_{bb}$. Choose your own reasonable confidence level.

(b) Find the *P*-value for a test of H_0: $\mu_h = \mu_{bb}$ against the one-sided alternative H_A: $\mu_h > \mu_{bb}$. What is your decision at significance level .037?

(c) Do you think your conclusions would be different if you dropped the common variance assumption and used the confidence interval and test procedures discussed in Sect. 9.3? Try it out!

9.4.5. *Movie Cast Sizes and Ratings*. Carry out statistical analyses similar to those prescribed in Exercises 9.3.5, but now under the assumption that the populations of ratings for movies with smaller casts (6 or fewer listed cast members) and movies with larger casts (more than 6 listed cast members) are normally distributed with means $\mu_{\le 6}$ and $\mu_{>6}$ and common variance σ^2. Compare your results with those obtained in Exercises 9.3.5.

9.4.6. *Angioplasty Balloons—To Coat or Not To Coat—That Is the Question.* Atherosclerotic disease is a disease in which plaque builds up inside one's arteries, which can, among other problems, compromise blood flow to the legs and feet. Percutaneous transluminal coronary angioplasty (PTCA) is a minimally invasive procedure that uses a tiny balloon inserted into a blood vessel to open blocked coronary arteries and allows blood to more freely circulate to the legs and feet. Many times, however, the blood vessels once again narrow within a year after PTCA, requiring that additional measures, such as surgically inserting stints, to reduce this narrowing effect. Rosenfield et al. (2015) reported on the results of a study to investigate whether coating the angioplasty balloon during PTCA with the antineoplastic compound *paclitaxel* could extend the benefits from the procedure. The authors enrolled 476 patients in the study, of which 160 received standard PTCA and 316 received PTCA with a balloon coated with *paclitaxel*. All 476 patients were followed for a year post-PTCA. Among other diagnostics, the Walking Impairment Questionnaire (higher scores demonstrate greater mobility/improvement) was administered to the patients one year after the procedure. As part of this questionnaire, each patient achieved a walking speed score. The average walking speed score for the 160 patients who received the standard PTCA was $\bar{x}_{standard} = 17.7$ with a standard deviation $s_{standard} = 31.1$. The corresponding average walking speed score for the 316 patients receiving PTCA coated with *paclitaxel* was $\bar{x}_{coated} = 21.2$ with a standard deviation $s_{coated} = 29.0$. Assume that the one-year post-procedure walking speed scores for patients receiving standard PTCA and for patients receiving PTCA coated with *paclitaxel* are normally distributed with means $\mu_{standard}$ and μ_{coated}, respectively, and common variance σ^2.

(a) Estimate the difference in mean walking speed scores $\mu_{coated} - \mu_{standard}$.

(b) Find a lower confidence bound for $\mu_{coated} - \mu_{standard}$. Choose your own reasonable confidence level.

(c) Find the P-value for a test of H_0: $\mu_{coated} = \mu_{standard}$ against the one-sided alternative H_A: $\mu_{coated} > \mu_{standard}$. What is your decision at significance level .067?

(d) In view of the observed data, does the assumption of common variance σ^2 seem reasonable? Justify your answer.

9.4.7. *Bird Diversity During the Wet Season in Kukobila and Wuntori Wetlands, Ghana.* Nsor and Obodai (2014) conducted a study to assess the effect of environmental factors on the diversity of bird populations in the wetlands areas of the Northern Region of Ghana. They observed birds from 7-11 am GMT once a week for a period of two years at each of six wetlands areas (Kukobila, Wuntori, Tugu, Adayili, Nabogo, and Bunglung) and the number of different bird species observed were compiled monthly during that period of time for each of these six areas. One comparison of interest in their study was the monthly number of bird species observed in different wetlands during the 5 months of the wet season (June through October) over the two-year period. For comparable portions of the Kukobila (open deep marsh) and Wuntori (closed shallow marsh) wetlands, they found that the averages of the 10 monthly numbers of bird species observed during the wet season (over the two-year period) for the Kukobila and Wuntori wetlands were $\bar{x}_{Kukobila} = 6.0$ species and $\bar{x}_{Wuntori} = 3.8$ species, respectively, with standard deviations $s_{Kukobila} = 1.1$ species and $s_{Wuntori} = 0.6$ species, respectively. Assume that the monthly numbers of species observed in these portions of the Kukobila and Wuntori wetlands during the wet season are normally distributed with means $\mu_{Kukobila}$ and $\mu_{Wuntori}$ and common variance σ^2.

(a) Estimate the difference in mean number of bird species $\mu_{Kukobila} - \mu_{Wuntori}$.

(b) Find a lower confidence bound for $\mu_{Kukobila} - \mu_{Wuntori}$. Choose your own reasonable confidence level.

(c) Find the P-value for a test of H_0: $\mu_{Kukobila} = \mu_{Wuntori}$ against the one-sided alternative H_A: $\mu_{Kukobila} > \mu_{Wuntori}$. What is your decision at significance level .045?

9.4.8. *Removing Spots and Stains From Works of Art on Paper.* When a work of art on paper becomes stained or spotted, great care must be taken in attempting to remove the blemish in order not to have a negative effect on the appearance of the work. Eirk (1972) studied a total of 31 possible methods for spot or stain removal, along with six separate controls involving treatment with distilled water only. She used rag paper obtained from two old ledgers, one with paper that was relatively white without disfiguring effects, while the sheets in the second ledger were degraded by foxing and mildew. One feature used for comparison of the treatments and controls was a brightness measure of the treated paper followed by a one-hour wash in running de-ionized water. Powdered sodium formaldehyde sulfoxylate (SFS) applied to damp sheets and washed after twenty minutes and 1:2 aqueous 5% hypochlorite/ 5% sodium metabisulfite (HSM) were two of the treatments studied by the author. Using the relatively white ledger paper without disfiguring effects, the average brightness score for ten replicates treated with SFS was $\bar{x}_{SFS} = 50.2$, with standard deviation $s_{SFS} = 0.82$, while the average brightness score for ten replicates with HSM was $\bar{x}_{HSM} = 55.4$, with standard deviation $s_{HSM} = 1.71$. Assume that the brightness scores for the SFS and HSM treatments are normally distributed with means μ_{SFS} and μ_{HSM}, respectively, and common variance σ^2.

(a) Estimate the difference in mean brightness scores $\mu_{SFS} - \mu_{HSM}$.

(b) Find an upper confidence bound for $\mu_{SFS} - \mu_{HSM}$. Choose your own reasonable confidence level.

(c) Find the P-value for a test of H_0: $\mu_{SFS} = \mu_{HSM}$ against the one-sided alternative H_A: $\mu_{SFS} < \mu_{HSM}$. What is your decision at significance level .015?

(d) Do you feel comfortable with the assumption of common variance for the SFS and HSM treatments? Why or why not? What alternative could you pursue if you are not comfortable with the assumption?

9.4.9. *Is Shakespeare Always Shakespeare?* Are great English authors consistent in the pattern of noun repetition they use in their works? In particular, does Shakespeare use a similar noun repetition pattern in his comedies as he does in his tragedies? Bennett (1957) thoroughly investigated this question for the Shakespearian comedy *As You Like It* and the Shakespearian tragedy *Julius Caesar*. He calculated the frequency of occurrence of all of the nouns in each of these plays and his tally is given in the dataset *Shakespeare_noun_use*.

(a) Assuming that *As You Like It* and *Julius Caesar* are reasonably representative of all the comedies (14) and tragedies (12), respectively, written by Shakespeare, find the *P*-value for an appropriate test of the hypothesis that Shakespeare uses the same pattern of noun repetition for his comedies as he does for his tragedies.

(b) Do you think it is reasonable to assume that *As You Like It* and *Julius Caesar* are representative of the 14 comedies and 12 tragedies? Why or why not? Do you want to add to the dataset by following Bennett's lead and count the noun repetitions in Merchant of Venice (comedy) and King Lear (tragedy)? Let us know what you find out......

9.4.10. *Goggled Green Turtles.* Consider the goggled green turtle hatchlings study discussed in Example 9.9. Assuming equal variances for the numbers of circles in the two–minute crawl period for hatchlings fitted with "nasal field" goggles and those fitted with "harlequin" goggles, find the *P*-value for a test of the hypotheses of interest. What is your decision at significance level .039? Do you believe that the assumption of equal variances is reasonable? Justify your answer.

9.5 Discussion of the Methods of Inference for the Difference Between the Centers of Two Populations with Independent Samples.

In Sects. 1, 2, 3, and 4 we discussed procedures for making inferences about the difference in the centers of two populations based on independent random samples from each. These procedures use different measures for the center of a population. The approximate procedures of Sect. 1 related to comparing appropriate proportions for the two populations. In Sect. 2 the emphasis is on the difference in medians for continuous populations. Such a difference is appropriate for all continuous populations, but particularly so for populations which are not necessarily symmetric. For such asymmetric populations, the median is often a better measure of center than is the mean, which can be heavily influenced by either the asymmetry or outliers in the population. The approximate procedures of Sect. 3 are quite broad in their applicability. They are appropriate for any populations where we have reasonably-sized samples from both populations and our primary interest is in the difference in the population means. The setting for Sect. 4 is the most restrictive of the four considered in this chapter, as it depends not only on underlying normality for the two populations but also on the condition that they have a common variance. An additional caveat for the procedures in Sects. 3 and 4 is that they might not be the best approach to consider for settings where the underlying populations are sharply skewed or heavy-tailed, since for such settings the sample means and sample standard deviations are not very reliable measures of center and variability for the two populations.

The important thing to remember is that when the assumptions underlying a particular set of procedures are reasonable for the two populations of interest, then those procedures will generally provide better inferences (i. e., shorter confidence intervals, more differentiating test procedures, etc.) for the difference in centers for the populations. However, as always, applying procedures to populations that do not satisfy the necessary assumptions for those procedures can often produce misleading inferential conclusions. Remember that diagnostic tools such as histograms, box plots, and normal probability plots can provide useful information about the shape and nature of the underlying populations.

Chapter 9 Comprehensive Exercises

9.A. Conceptual

9.A.1. Let $X_1, \ldots, X_m$ and $Y_1, \ldots, Y_n$ denote independent random samples from two distinct (X and Y) continuous populations. Let W and U (9.20) be the rank sum statistic and counting statistic, respectively, discussed in Sect. 2. Show that $W = U + \frac{n(n+1)}{2}$ when there are no tied values between the X's and/or Y's.

9.A.2. Let $X_1, \ldots, X_m$ and $Y_1, \ldots, Y_n$ be independent random samples from two distinct (X and Y) continuous populations. Let $S_1, \ldots, S_m$ and $R_1, \ldots, R_n$ denote the joint ranks of $X_1, \ldots, X_m$ and $Y_1, \ldots, Y_n$, respectively, among the combined sample of $N = (m + n)$ X and Y observations. Even though each individual S_i and R_j rank is random, explain why the total sum of the ranks, $\sum_{i=1}^{m} S_i + \sum_{j=1}^{n} R_j$, is not random. Show, either in general or for $m = 5$ and $n = 6$, that this total sum of ranks is always equal to the constant $N(N + 1)/2$.

9.A.3. The rank sum statistic W discussed in Sect. 2 uses only the sum of the combined samples ranks $R_1, \ldots, R_n$ of the Y observations. Wouldn't the statistic $V = \sum_{j=1}^{n} R_j - \sum_{i=1}^{m} S_i$, where $S_1, \ldots, S_m$ are the combined samples ranks of $X_1, \ldots, X_m$, respectively, be a more informative statistic to use in testing H_0: $[\eta_Y = \eta_X]$? Explain why this is not the case.

9.A.4. Consider the 100*CL*% confidence interval for the difference in population means, $\mu_Y - \mu_X$, as given in (9.26).

(a) For fixed sample sizes m and n and a given set of data, how does the length of this confidence interval vary as a function of the confidence level *CL*?

(b) For a fixed set of data and confidence level *CL*, how does the length of this confidence interval vary as a function of the two sample sizes, m and n?

9.A.5. Consider the 100*CL*% confidence interval for the difference in population means, $\mu_Y - \mu_X$, as given in (9.26) and the corresponding 100*CL*% upper confidence bound for $\mu_Y - \mu_X$. For fixed sample sizes m and n and a given set of data, compare the upper endpoint of the 100*CL*% confidence interval with the 100*CL*% upper confidence bound.

9.A.6. Notice that both of the 100*CL*% confidence intervals for the difference in population means, $\mu_Y - \mu_X$, given in (9.26) and (9.36) are centered at the point estimator for $\mu_Y - \mu_X$, namely, $\bar{Y} - \bar{X}$. Explain why this is not necessarily the case for the point estimator $\tilde{D}$ (9.14) for the difference in population medians $\eta_Y - \eta_X$ and the 100*CL*% confidence interval for $\eta_Y - \eta_X$ given in (9.18). What **can** be said about $\tilde{D}$ relative to the confidence interval in (9.18)?

9.B. Data Analysis/Computational

9.B.1. *Binge Drinking and Athletics.* Find an approximate 95% lower confidence bound for $p_Y - p_X$ for the athlete/non-athlete binge drinking data in Example 9.1.

9.B.2. *Driving Under the Influence and Athletics.* In Example 9.1, we used sample data from Leichliter et al. (1998) to compare the percentages of intercollegiate athletes and non-athletes who were involved in binge drinking

Table 9.9 Numbers of students reporting that they have driven under the influence in the last year prior to completing the core and alcohol survey

	Involvement in intercollegiate sports	
	Participants	Non-participants
Number completing survey	8749	42,734
Number who have driven under the influence during the prior year	3348	12,991

Source: Leichliter et al. (1998)

in the 2 weeks prior to completing a Core and Alcohol Survey. Those authors also reported sample data on whether the respondents had driven under the influence during the past year. These results are presented in Table 9.9 for participants and non-participants in intercollegiate sports.

Let p_{par} and p_{nonpar} denote the percentages of all participants and non-participants, respectively, in intercollegiate athletics who have driven under the influence during the prior year.

(a) Find a lower confidence bound for $p_{par} - p_{nonpar}$. Choose your own reasonable confidence level.

(b) Find the approximate P-value for an appropriate test of the conjecture that participants in intercollegiate athletics are more likely to have driven while under the influence in the prior year than non-participants. What is your decision at significance level .06?

(c) How do you think these results would compare with today's campuses?

9.B.3. *Gender and Musical Instrument Choice.* Consider the instrument opinion data in Table 9.2. Find a confidence interval for $p_Y - p_X$, where $p_X =$ [proportion of 9-11 year old boys who believe that girls should not play the trumpet] and p_Y = [proportion of 9-11 year old girls who believe that girls should not play the trumpet]. Choose your own reasonable confidence level.

9.B.4. *Gender and Musical Instrument Choice.* Consider the instrument opinion data in Table 9.2. Find the approximate *P*-value for a test of the hypothesis that both girls and boys agree that boys should not play the flute against the general alternative that they disagree on that issue.

9.B.5. *Insect Infection by Parasites.* Infection of an insect by a parasite can either lead directly to a lethal disease in the insect itself or it can be transmitted further to a vertebrate host by the insect. (Malaria is an example of the latter case, since it is a disease that is transmitted to man through various mosquito species carrying the infecting *Plasmodia* parasite.) An important part of such a host-parasite relationship is the defense system of the host. In the case of insects, the presence of the enzyme *phenoloxidase* has been suggested as a possible deterrent to infection by parasites. This enzyme produces quinones that react with proteins to produce a black pigment melanin, which is then deposited on a parasite by the insect as part of its defense against it. However, large amounts of active *phenoloxidase* can also kill insects. Hence, for generation of an adequate, but not lethal, supply of quinones to respond effectively to a parasite, close control of the *phenoloxidase* activity by the insect is essential. With this question in mind, Pye (1974) studied the activation of *prophenoloxidase* in the plasma of immune *Galleria mellonella* larvae in response to exposure to a variety of microbial products. For each product, ten *Galleria mellonella* larvae were involved, with five of them serving as controls (no immunization) and the other five being first immunized through injection of 1.0-μg doses of *Shigella flexneri lipopolysaccharide B* (Difco). Both the control and immune larvae were then exposed to the microbial product. Using a quick freezing method with acetone-dry ice, the level of *prophenoloxidase* activation was then obtained for all ten larvae by using a Guilford recording

Table 9.10 ***Prophenoloxidase*** **activation (units per .20** ***ml*** **plasma) for control larvae and immune larvae, both exposed to a 1** ***mg/ml*** **water mixture of the microbial product zymosan**

Control larvae	Immune larvae
79	381
64	361
82	425
13	353
174	339

Source: Pye (1974)

spectrophotometer. The data in Table 9.10 represent the results obtained for five control larvae and five immune larvae exposed to a 1 mg/ml water mixture of the microbial product Zymosan (a yeast polysaccharide). The measurements are in units of *prophenoloxidase* activity per .20 ml plasma of the larvae.

(a) Estimate the difference in the median *prophenoloxidase* activation levels for the control and immune larvae populations after exposure to the stipulated dose of Zymosan.

(b) Estimate the probability that a randomly selected control larva will exhibit a smaller *prophenoloxidase* activation level than a randomly selected immune larva after exposure of each to a 1 mg/ml water mixture of Zymosan.

(c) Find a confidence interval for the difference in the median *prophenoloxidase* activation levels for the control larvae population and the immune larvae population after exposure to the stipulated dose of Zymosan. Choose your own reasonable confidence level.

(d) Find the P-value for a test of the conjecture that immunization of the larvae leads to an increase in *prophenoloxidase* activation resulting from exposure to a 1 mg/ml water mixture of Zymosan. What is your decision at significance level .05?

9.B.6. *Anencephalus and Magnesium in Tap Water.* Anencephalus is a fatal, congenital birth anomaly where a child is born without an effectively functioning brain. Links between the occurrence of this disease and a number of environmental factors were investigated in Elwood (1977) and later by Archer (1979). One of the factors that Archer considered to be a possible influence on the anencephalus rate for a region was the magnesium content of its water. He obtained anencephalus rates (deaths from anencephalus / 1000 total births) for 36 cities in Canada for the period (1950-1969), as well as the average magnesium content of their water (parts per million) during that period of time. These two quantities are presented for these Canadian cities in Table 9.11.

For this exercise, we divide the Canadian cities into those considered to have unusually high magnesium tap water levels (≥ 7.6 ppm) and those with low magnesium levels (< 7.6 ppm) and search for potential differences in rates of death from anencephalus. (We consider an alternative approach to analyzing these same data without grouping by high or low magnesium level in Chap. 11.)

(a) Provide a list of the anencephalus death rates for the two samples created by this high/low magnesium criterion. What are the two sample sizes?

(b) Find a confidence interval for the differences in mean anencephalus death rate for areas with high magnesium tap water levels and those with low magnesium levels. Choose your own reasonable confidence level.

(c) Find the *P*-value for a test of the conjecture that cities with high magnesium tap water levels will have greater anencephalus death rates than those with low manesium levels. What is your decision at significance level .025?

Table 9.11 Rate of death from anencephalus per 1000 total births and magnesium in tap water (*ppm*) for thirty-six cities in Canada for the period (1950–1969)

Rate of death from anencephalus	Magnesium in tap water
1.47	23.9
0.97	8.6
0.77	15.1
0.80	0.5
0.93	10.4
1.39	26.0
1.23	0.6
1.41	8.3
1.32	7.5
1.03	7.9
1.10	26.3
1.51	7.6
1.17	7.5
1.08	8.6
0.74	8.3
1.17	8.4
1.55	2.2
1.60	3.1
1.79	1.7
0.61	27.5
1.24	8.6
2.12	0.7
1.56	0.6
0.90	7.5
1.37	7.7
1.54	6.7
0.77	11.6
1.46	3.0
1.69	2.8
1.28	5.0
1.45	8.7
2.04	0.7
0.79	0.2
1.44	7.5
1.11	7.8
1.28	6.5

Source: Archer (1979)

Table 9.12 Numbers of intercollegiate athletic team members and leaders reporting involvement in binge drinking in the 2 weeks prior to completing the core and alcohol survey

	Involvement in intercollegiate sports	
	Team member	**Leader**
Number completing survey	6651	2098
Number engaging in binge drinking in the previous two weeks	3618	1217

Source: Leichliter et al. (1998)

9.B.7. *Binge Drinking Athletes—Leaders or Not?* In Example 9.1, we used sample data from Leichliter et al. (1998) to compare the percentages of intercollegiate athletes and non-athletes who were involved in binge drinking in the 2 weeks prior to completing a Core and Alcohol Survey. Those authors also differentiated between whether an athlete was simply a member of the team or was considered a leader on the team. The binge drinking data for these two subgroups are presented in Table 9.12. Let p_{member} and p_{leader} denote the percentages of all participants in intercollegiate athletics who have engaged in binge drinking in the previous 2 weeks and who are team members only or leaders of teams, respectively.

(a) Find a confidence interval for $p_{leader} - p_{member}$. Choose your own reasonable confidence level.

(b) Find the approximate *P*-value for an appropriate test of the conjecture that leaders on intercollegiate athletic teams are more likely to have been involved in binge drinking in the previous 2 weeks than are athletes in lesser positions on their teams. What is your decision at significance level .045?

9.B.8. *Insect Infection by Parasites.* In his study of an insect's *prophenoloxidase* activation response to microbial products (see Exercise 9.B.5), Pye (1974) also

Table 9.13 ***Prophenoloxidase*** **activation (units per .20** ***ml*** **plasma) for control larvae and immune larvae, both exposed to .10** ***ml*** **aliquots of** ***Pseudomonas aeruginosa***

Control larvae	Immune larvae
33	225
90	139
32	287
23	217
130	211

Source: Pye (1974)

considered the microbial product *Pseudomonas aeruginosa*. The *prophenoloxidase* activation values (units per .20 *ml* plasma) for five control larvae and five immunized larvae after exposure to .10 *ml* aliquots of *Pseudomonas aeruginosa* are given in Table 9.13.

(a) Estimate the difference in median *prophenoloxidase* activation for the control and immune larvae populations after exposure to the stipulated dose of *Pseudomonas aeruginosa*.

(b) Estimate the probability that a randomly selected control larva will exhibit a smaller *prophenoloxidase* activation level than a randomly selected immune larva after exposure of each to .10 *ml* aliquots of *Pseudomonas aeruginosa*.

(c) Find a confidence interval for the difference in median *prophenoloxidase* activation for the control larvae population and the immune larvae population after exposure to the stipulated dose of *Pseudomonas aeruginosa*. Choose your own reasonable confidence level.

(d) Find the P-value for a test of the conjecture that immunization of the larvae leads to an increase in *prophenoloxidase* activation resulting from exposure to .10 *ml* aliquots of *Pseudomonas aeruginosa*. What is your decision at significance level .05?

9.B.9. *Anencephalus and Geomagnetic Flux.* In his study of factors affecting anencephalus death rates (see Exercise 9.B.6), Archer (1979) also considered the possible linkage between these rates and the horizontal geomagnetic flux of a region. The horizontal geomagnetic flux of a region has a strong influence on where incoming charged cosmic particles strike the earth's atmosphere, with higher flux regions diverting the particles to those with low flux. Since ionizing radiation is a known mutagen and carcinogen, it is possible that some of the geographical differences in congenital anomalies, such as anencephalus, could be accounted for by the differing intensities of cosmic radiation for the geographical regions. Dividing the 36 cities into those with high ($\geq .0162$) and low ($< .0162$) horizontal geomagnetic flux, respectively, the corresponding anencephalus death rates are given in Table 9.14.

Table 9.14 Rate of death from anencephalus per 1000 total births for thirty-six cities in Canada, divided into groups with high ($\geq .0162$) and low ($< .0162$) horizontal geomagnetic flux values, for the period (1950–1969)

Cities with high flux values	Cities with low flux values
1.47	0.77
0.97	0.93
0.80	1.32
1.39	1.51
1.23	1.17
1.41	1.55
1.03	1.60
1.10	1.79
1.08	0.61
0.74	0.90
1.17	1.37
1.24	0.77
2.12	1.46
1.56	1.69
1.54	1.28
1.45	2.04
0.79	1.44
1.11	1.28

Source: Archer (1979)

(a) Estimate the difference in mean death rates from anencephalus for cities with high horizontal geomagnetic flux values and those with low flux values.

(b) Find a lower confidence bound for the difference in mean anencephalus death rate for areas with high horizontal geomagnetic flux values and those with low flux values. Choose your own reasonable confidence level.

(c) Find the P-value for a test of the conjecture that cities with high horizontal geomagnetic flux values will have greater anencephalus death rates than those with low flux levels. What is your decision at significance level .030?

9.B.10. *Driving Under the Influence and Gender.* In Exercise 9.B.2, we used sample data from Leichliter et al. (1998) to compare the percentages of intercollegiate athletes and non-athletes who had driven while under the influence during the prior year. Those authors also reported the gender of the respondents. These results are presented in Table 9.15 for participants in intercollegiate sports.

Let $p_{femalepar}$ and $p_{malepar}$ denote the percentages of all female and male participants in intercollegiate athletics, respectively, who have driven under the influence during the prior year.

Table 9.15 Numbers of male and female intercollegiate athletes reporting that they have driven under the influence in the last year prior to completing the core and alcohol survey

	Male athletes	**Female athletes**
Number completing survey	4860	3.892
Number who have driven under the influence during the prior year	2181	1165

Source: Leichliter et al. (1998)

(a) Find a confidence interval for $p_{femalepar} - p_{malepar}$. Choose your own reasonable confidence level.

(b) Find the approximate P-value for an appropriate test of the conjecture that female participants in intercollegiate athletics are less likely to have driven while under the influence in the prior year than are male participants in intercollegiate athletics. What is your decision at significance level .06?

9.B.11. *If You Have Seen One Slug, Have You Seen Them All?* In Examples 9.3, 9.4, 9.5, and 9.6 we discussed statistical analyses of the data collected by Whelan (1982) on how the woodland site and waste site slugs responded to the toxic plant *Allium Ursinum*, commonly found in woodland but not waste sites, as the test gel. It would, of course, also be of interest to see how the two types of slugs responded to a toxic plant that was commonly found in waste, but not woodland, sites. In Table 9.16 we present precisely that data for the toxic waste site plant *Rumex obtusifolius*.

Conduct the same statistical analyses as in Examples 9.3, 9.4, 9.5, and 9.6 for the data on the toxic waste site plant *Rumex obtusifolius* in Table 9.16. Discuss

Table 9.16 Acceptability indices (AI) for *Arion Subfuscus* from woodland and waste sites with the toxic waste site plant *Rumex obtusifolius* as test gel

Woodland site slugs (X)	Waste site slugs (Y)
0.18	0.15
0.03	0.12
0.26	0.09
0.17	0.23
0.22	0.12
0.17	0.00
0.06	0.20
0.47	0.13
0.22	0.10
0.18	0.07

Source: Whelan (1982)

the similarities and differences between your findings and those obtained in Examples 9.3, 9.4, 9.5, and 9.6 for the toxic woodland plant *Allium Ursinum*.

9.B.12. *Hospital Admissions—Substance Abuse and/or Mental Illness.* In a study of hospital admissions and related costs, Salit et al. (1998) collected hospital discharge and admissions records from New York City public and private hospitals for the 2 years 1992 and 1993. Among other things, they found that 44,959 of the 244,345 public hospital admissions during that period were for substance abuse and/or mental illness. For private hospitals, 37,982 out of 139,641 admissions were for substance abuse and/or mental illness.

(a) Viewing these data from New York City as reasonably representative of data from all public and private hospitals, estimate the difference in the percentages of admissions due to substance abuse and/or mental illness for private and public hospitals.

(b) Find an approximate 95% confidence interval for the difference in the percentages of admissions due to substance abuse and/or mental illness for private and public hospitals.

9.B.13. *Baseball and Beer!* Baseball is the American pastime, but what goes with watching a baseball game? The well-known song says peanuts and crackerjack, but how about some beer to wash those snacks down? Wolfe et al. (1998) conducted a study to see just how much beer and baseball have become synonymous. Male spectators of drinking age were sampled over a three-game period—on a Friday night, a Saturday afternoon, and a Monday night—during the 1993 season at two major league ballparks. Wolfe et al. found that 65 out of 166 sampled spectators in the age group 20-35 had consumed alcohol immediately prior to entering the ballpark. For the age group 36-50, they found that 44 of the 145 sampled individuals had consumed alcohol immediately prior to entering the ballpark. Find the approximate P-value for a test of the conjecture that fans in the age group 20-35 are more likely to consume alcohol prior to going to a major league ball game than are fans in the age group 36-50.

9.B.14. *Baseball and Beer and Age.* In their study of beer and baseball (see Exercise 9.B.13), Wolfe et al. (1998) also found that 28 out of 212 sampled spectators in the age group 20-35 were legally intoxicated at the end of the fifth inning of the baseball game. The analogous sampling for the age group 51-65 yielded 4 out of 16 sampled spectators who were legally intoxicated at that stage of the ball game. Find an approximate 90% confidence interval for the difference in percentages of baseball fans in the age groups 20-35 and 51-65 who will be legally intoxicated at the end of the fifth inning of a baseball game.

9.B.15. *Did All Americans Have the Same Access to a Home Computer?* Internet usage is the norm for Americans today, but were there differences between groups within America in the 1990's as far as Internet access was concerned? Hoffman and Novak (1998) considered data provided by Nielsen Media Research from the Spring 1997 CommerceNet/Nielsen Internet Demographic Study (IDS), conducted from December 1996 through 1997. Among other things, the study found that 2173 of 4906 white respondents owned a home computer, while the corresponding figures for African Americans were 143 home computer owners out of 493 respondents. Find the approximate *P*-value for a test of the conjecture that the percentage of home computer owners was greater for white Americans than for African Americans in the 1990's.

9.B.16. *Buying a Personal Computer.* Hoffman and Novak (1998) considered data provided by Nielsen Media Research from the Spring 1997 CommerceNet/Nielsen Internet Demographic Study (IDS), conducted from December 1996 through 1997. One part of the data collected involved the number of respondents who plan to buy a personal computer in the next 6 months. Those figures for white Americans and African Americans were

819 out of 4906 and 134 out of 493, respectively. Find an approximate 97.5% confidence interval for the difference in percentages of white Americans and African Americans who plan to buy a personal computer in the 6 months following completion of the survey data collection in 1997. Comment on this finding in conjunction with the result of Exercise 9.B.15.

9.B.17. *Removing Spots and Stains From Works of Art on Paper.* Consider the study by Eirk (1972) in which she compared various approaches to removing stains or spots from works of art on paper, as previously discussed in Exercise 9.4.8. A second feature used for comparison of these treatments was the bursting strength in pounds per square inch of the dried paper following treatment. Again using the relatively white ledger paper without disfiguring effects, the observed average bursting strength for ten replicates of the powdered sodium formaldehyde sulfoxylate (SFS) treatment was $\bar{x}_{SFS} = 36.4$ pounds per square inch, with standard deviation $s_{SFS} = 5.17$ pounds per square inch, while the average bursting strength for ten replicates of the 1:2 aqueous 5% hypochlorite/5% sodium metabisulfite (HSM) treatment was $\bar{x}_{HSM} = 18.5$ pounds per square inch, with standard deviation $s_{HSM} = 1.63$ pounds per square inch. Assume that bursting strengths for the SFS and HSM treatments are normally distributed with means μ_{SFS} and μ_{HSM}, respectively, and common variance σ^2.

(a) Estimate the difference in mean bursting strengths $\mu_{SFS} - \mu_{HSM}$.

(b) Find a lower confidence bound for $\mu_{SFS} - \mu_{HSM}$. Choose your own reasonable confidence level.

(c) Find the P-value for a test of H_0: $\mu_{SFS} = \mu_{HSM}$ against the one-sided alternative H_A: $\mu_{SFS} > \mu_{HSM}$. What is your decision at significance level .001?

(d) Do you feel comfortable with the assumption of common variance for the SFS and HSM bursting strengths? Why or why not? What alternative could you pursue if you are not comfortable with the assumption?

9.B.18. *Will My Hair EVER Grow Again?* One of the major concerns for men as they age is whether they will lose some or all of their hair. While it is well known that much of male baldness can be blamed on genetic inheritance from mom (guys, look at the men on your mother's side for clues), hair restoration after initial loss of hair has become an important cosmetic industry for men. Kamimura et al. (2000) studied the effect that topical application of procyanidin B-2 (PB-2) isolated from apple juice might have on new hair growth. For 6 months they treated one group of 19 balding men twice a day with 1.8 *ml* of agent containing 1% PB-2, corresponding to 30 *mg* of PB-2 daily. A second group of 10 balding men served as a control group. They were treated in exactly the same way, except that the agent contained no PB-2. No other hair care products except shampoos and rinses were permitted during the study. Before and after the six-month period, hairs at a predetermined site were clipped from each participating subject and the diameters of the collected hairs were measured. The change in total hairs per .25 cm^2 and the change in terminal hairs (defined as >60 μm in diameter) for each of the participants was recorded and is presented in Table 9.17.

(a) Estimate the difference in the medians for total hair growth in the control and PB-2 treated populations.

(b) Estimate the probability that a randomly selected control individual will exhibit a smaller amount of total hair growth than a randomly selected individual treated with PB-2.

(c) Find a confidence interval for the difference in the medians in total hair growth for the control and PB-2 treated populations. Choose your own reasonable confidence level.

(d) Find the *P*-value for a test of the conjecture that treatment with PB-2 improves the amount of total hair growth for balding individuals. What is your decision at significance level .025?

Table 9.17 Total and terminal hair growth in each subject

Total hair change (hairs/.25cm^2)	Total terminal hair change (hairs/.25 cm^2)
Placebo controls	
0.3	0.94
1.4	1.27
3.0	−2.58
3.7	3.56
−1.5	−0.25
−2.0	−2.58
0	3.73
4.8	−2.62
2.4	−7.55
−11.3	−2.09
Treated with PB-2	
3.5	3.54
5.0	−2.74
7.3	5.87
18.3	3.12
14.5	1.53
6.7	−3.12
9.0	−0.17
−0.7	4.85
7.8	2.81
- 4.0	3.11
6.0	6.69
4.5	0.22
8.0	0.80
11.4	3.95
1.0	2.38
7.3	1.20
8.5	3.03
−0.7	0.71
13.5	0.00

Source: Kamimura et al. (2000)

9.B.19. *Will My Hair EVER Grow Again?* Consider the hair growth study by Kamimura et al. (2000) discussed in Exercise 9.B.18. Answer parts (a) through (d) of that Exercise again for total terminal hair growth.

9.B.20. *Will My Hair EVER Grow Again?* Consider the hair growth study by Kamimura et al. (2000) discussed in Exercise 9.B.18. Assume that total terminal hair growth for the Control and PB-2 treated populations are normally distributed with means $\mu_{Control}$ and μ_{PB-2} and variances $\sigma^2_{Control}$ and σ^2_{PB-2}, respectively.

(a) Estimate the difference in mean hair growth $\mu_{PB-2} - \mu_{Control}$.

(b) Find an upper confidence bound for $\mu_{Control} - \mu_{PB-2}$. Choose your own reasonable confidence level.

(c) Find the *P*-value for a test of $H_0 : \mu_{Control} = \mu_{PB-2}$ against the one-sided alternative $H_A : \mu_{PB-2} > \mu_{Control}$. What is your decision at significance level .010?

9.B.21. *Will My Hair EVER Grow Again?* Consider the hair growth study by Kamimura et al. (2000) discussed in Exercise 9.B.18. Assume that total hair growth for the Control and PB-2 treated populations are normally distributed with means $\mu_{Control}$ and μ_{PB-2} and variances $\sigma^2_{Control}$ and σ^2_{PB-2}, respectively.

(a) Estimate the difference in mean total hair growth $\mu_{PB-2} - \mu_{Control}$.

(b) Find an upper confidence bound for $\mu_{Control} - \mu_{PB-2}$. Choose your own reasonable confidence level.

(c) Find the *P*-value for a test of $H_0 : \mu_{Control} = \mu_{PB-2}$ against the one-sided alternative $H_A : \mu_{PB-2} > \mu_{Control}$. What is your decision at significance level .010?

9.C. Activities

9.C.1. *Are Female College Students More Liberal With Regard to Social Issues Than Male College Students?* Design an experiment (including the appropriate data to collect and how to collect it) that will enable you to statistically address this question. Collect the relevant data for samples of 10 men and

10 women, conduct an appropriate set of statistical analyses, and write a two-page report describing your experiment and statistical conclusions.

9.C.2. *Do College Science/Math Majors Spend Less Time Exercising Per Week than College Non-Science/Non-Math Majors?* Design an experiment (including the appropriate data to collect and how to collect it) that will enable you to statistically address this question. Collect the relevant data for samples of 10 college science/math majors and 10 college non-science/non-math majors, conduct an appropriate set of statistical analyses, and write a two-page report describing your experiment and statistical conclusions.

9.C.3. *Just For You!* Find a journal article in a field of your interest that presents the results of a study that involved independent samples from two distinct populations. Prepare a short (2-3 pages) summary report of the statistical findings in the article and attach a copy of the original paper with your summary.

9.C.4. *M&M Colors—Peanuts Versus Plain.* Mars, Inc. makes both M&M's Plain and M&M's Peanut candies. They claim that their production processes provide for the same percentage red pieces for both the plain and the peanut candies. Design an experiment (including the appropriate data to collect and how to collect it) that will enable you to statistically address this claim. Collect adequate relevant data, conduct an appropriate set of statistical analyses, and write a three-page report describing your experiment and statistical conclusions. (You can eat the M&M's upon completion of your report!)

9.C.5. *Lasting Power—Pennies or Nickels?* Is there a difference between the length of time that U. S. pennies and nickels stay in common circulation? Design an experiment (including the appropriate data to collect and how to collect it) that will enable you to statistically address this question. Collect adequate relevant data, conduct an appropriate set of statistical analyses, and write a three-page report describing your experiment and statistical conclusions.

9.C.6. *Do Female College Students Study More Than Male College Students?* Design an experiment (including the appropriate data to collect and how to collect it) that will enable you to statistically address this question. Collect the relevant data for samples of 10 male college students and 10 female college students, conduct an appropriate set of statistical analyses, and write a two-page report describing your experiment and statistical conclusions.

9.C.7. *Do Male College Students Get Better Grades Than Female College Students?* Design an experiment (including the appropriate data to collect and how to collect it) that will enable you to statistically address this question. Collect the relevant data for samples of 10 male college students and 10 female college students, conduct an appropriate set of statistical analyses, and write a two-page report describing your experiment and statistical conclusions.

9.C.8. *Does Smoking Participation Decrease with College Advancement?* Design an experiment (including the appropriate data to collect and how to collect it) that will enable you to statistically address this question. Collect the relevant data for samples of 10 underclassmen (freshmen or sophomores) and 10 upperclassmen (juniors or seniors), conduct an appropriate set of statistical analyses, and write a two-page report describing your experiment and statistical conclusions.

9.C.9. *Who Has More Friends on Facebook—Men or Women?* Design an experiment (including the appropriate data to collect and how to collect it) that will enable you to statistically address this question. Collect the relevant data for samples of 10 men and 10 women, conduct an appropriate set of statistical analyses, and write a two-page report describing your experiment and statistical conclusions.

9.C.10. *Do College Students Sleep Later on Weekends Than Their Parents?* Design an experiment (including the appropriate data to collect and how to collect it) that will enable you to statistically address this question. Collect the relevant data for samples of 20 college students and 20 parents (from different families), conduct an appropriate set of statistical analyses, and write a two-page report describing your experiment and statistical conclusions.

9.D. Internet Archives

9.D.1. *Surveys.* Identify three organizations that routinely collect survey data on current topics and locate the Internet sites where they periodically present the results of their surveys. Select one such survey of interest to you that involves comparison of percentages for at least two groups and prepare a brief report on its findings.

9.D.2. *Federal Government.* Identify three government agencies that routinely gather national data and locate the Internet sites where they periodically present the updates to their data collections. Select one specific data collection that is of interest to you and prepare a brief report using the data to compare two groups.

9.D.3. *Professional Societies.* Identify three professional societies that routinely gather information relevant to their membership and locate the Internet sites where they report their findings. Select one specific data collection that is of interest to you and prepare a brief report using the data to compare two groups.

9.D.4. *Nonprofit Organizations.* Identify three nonprofit organizations that routinely gather information relevant to their cause and locate the Internet sites where they report their findings. Select one specific data collection that is of interest to you and prepare a brief report using the data to compare two groups.

9.D.5. *Academic Organizations.* Identify three academic entities that routinely gather information relevant to their ongoing research projects and locate the Internet sites where they report their findings. Select one specific data collection that is of interest to you and prepare a brief report using the data to compare two groups.

9.D.6. *Medical Research.* Use the Internet to locate a paper published in a medical field within the past 2 years that presents a study involving data collection and comparison of two groups. If the data are not actually available in the published article, contact the authors to see if they will allow you to access the data. If you are successful, use the data to verify the statistical summary in the published article.

9.D.7. *Climate Change Research.* Use the Internet to locate a paper published within the past 2 years on a topic related to climate change that presents a study involving data collection and comparison of two groups. If the data are not actually available in the published article, contact the authors to see if they will allow you to access the data. If you are successful, use the data to verify the statistical summary in the published article.

9.D.8. *Social Science Research.* Use the Internet to locate a paper published in a social science field within the past 2 years that presents a study involving data collection and comparison of two groups. If the data are not actually available in the published article, contact the authors to see if they will allow you to access the data. If you are successful, use the data to verify the statistical summary in the published article.

9.D.9. *Humanities Research.* Use the Internet to locate a paper published in a humanities field within the past 2 years that presents a study involving data collection and comparison of two groups. If the data are not actually available in the published article, contact the authors to see if they will allow you to access the data. If you are successful, use the data to verify the statistical summary in the published article.

9.D.10. *STEM Research*. Use the Internet to locate a paper published in a STEM field (science, technology, engineering, or mathematics) within the past 2 years that presents a study involving data collection and comparison of two groups. If the data are not actually available in the published article, contact the authors to see if they will allow you to access the data. If you are successful, use the data to verify the statistical summary in the published article.

10 Statistical Inference for Two-Way Tables of Count Data

In Sect. 9.1 we discussed procedures designed for making statistical inference about the difference in the probabilities of a common event A for two populations. Those procedures are based on independent random samples of Bernoulli variables (i. e., either the event A occurs or it does not) from each of the two populations. One way to represent the observed outcomes of such Bernoulli random samples is in the following 2×2 table:

	Population 1	**Population 2**
Event A occurs	O_1	O_2
Event A does not occur	$m - O_1$	$n - O_2$

where m and n are the numbers of Bernoulli observations collected from Populations 1 and 2, respectively, and O_1 and O_2 are the numbers of these Bernoulli variables for which the event A occurred. For this setting and the statistical procedures of Sect. 9.1, we note that the numbers of observations,

© Springer International Publishing AG 2017

D.A. Wolfe, G. Schneider, *Intuitive Introductory Statistics*, Springer Texts in Statistics,
DOI 10.1007/978-3-319-56072-4_10

m and n, from Populations 1 and 2, respectively, are fixed in advance of the sampling, while O_1 and O_2 represent the random outcomes of this process.

This 2×2 table approach to representing the outcomes of two independent Bernoulli random samples can be generalized to deal with other settings where we have count data that can be categorized by two different criteria. Suppose that each measurement being collected can belong to one and only one of I categories, $C_1, \ldots, C_I$, for one attribute and to one and only one of J categories, $D_1, \ldots, D_J$, of a second defining attribute. Such data can be represented in the form of counts in an $I \times J$ two-way table (often referred to as a *contingency table*). In the setting above, for example, we have two population categories, $C_1 = \{\text{Population 1}\}$ and $C_2 = \{\text{Population 2}\}$, and two event categories, $D_1 = \{\text{event } A \text{ occurs}\}$ and $D_2 = \{\text{event } A \text{ does not occur}\}$. Thus $I = J = 2$ and we can represent the sample count data as noted in a 2×2 table.

In this chapter we discuss a number of test procedures designed to test a variety of appropriate hypotheses about general $I \times J$ two-way tables of count data. In Sect. 1, we consider an approximate procedure for testing equality of population proportions for an arbitrary number of populations, I, and an arbitrary number of categories, J. This procedure will be a direct extension of the test procedure discussed in Sect. 9.1 (for the special case of $I = 2$ populations and $J = 2$ categories) to the more general setting where either there are more than 2 populations or there are more than 2 categories or both. In Sect. 2 we present an approximate procedure for testing whether there is any association between two categorical attributes that can be used to classify observations; that is, do the two attributes occur independently in the population of interest? In Sect. 3 we discuss an exact procedure for testing equality of two population proportions. Although the procedure of Sect. 3 is more computationally intensive than the approximate hypothesis test for the same problem considered in Sect. 9.1, the P-values associated with the test in Sect. 3 are exact, rather than approximate. Finally, in Sect. 4, we present an approximate goodness-of-fit procedure for testing hypotheses about the probabilities associated with each of $I > 2$ possible categories for a multinomial count variable. This is a direct extension of the test procedure described in Sect. 6.3 for a single binomial variable, corresponding to $I = 2$ categories.

General Setting and Notation Consider a measurement that can belong to one and only one of I categories, $C_1, \ldots, C_I$, for one attribute and to one and only one of J categories, $D_1, \ldots, D_J$, of a second defining attribute. Collecting N such independent measurements, we record the numbers of these measurements that fall in the IJ different combinations of the two attribute-categories. Let O_{ij} denote the observed (random) number of these sample observations that belong to category C_i for the first attribute and category D_j for the second attribute, for $i = 1, \ldots, I$ and $j = 1, \ldots, J$. These counts can be represented in the form of the $I \times J$ configuration in Table 10.1.

Thus, the entry in the ith row (category C_i) and jth column (category D_j) of Table 10.1 is O_{ij} = [number of observations possessing both the attribute C_i and the attribute D_j], for $i = 1, \ldots, I$ and $j = 1, \ldots, J$. For this general setting, the O_{ij} entries represent the random outcomes for our sampling process and they will provide the basic data to test appropriate hypotheses about the probabilities of the various combinations of C and D attribute categories.

Table 10.1 Tabular representation of N sample observations categorized by two attributes

		Category for attribute D					
		D_1	D_2	$\cdots$	D_{J-1}	D_J	Total
Category for attribute C	C_1	O_{11}	O_{12}	$\cdots$	$O_{1,J-1}$	O_{1J}	$N_{1.}$
	C_2	O_{21}	O_{22}	$\cdots$	$O_{2,J-1}$	O_{2J}	$N_{2.}$
	$\vdots$	$\vdots$	$\vdots$		$\vdots$	$\vdots$	$\vdots$
	C_{I-1}	$O_{I-1,1}$	$O_{I-1,2}$	$\cdots$	$O_{I-1,J-1}$	$O_{I-1,J}$	$N_{(I-1).}$
	C_I	O_{I1}	O_{I2}	$\cdots$	$O_{I,J-1}$	O_{IJ}	$N_{I.}$
	Total	$N_{.1}$	$N_{.2}$	$\cdots$	$N_{.(J-1)}$	$N_{.J}$	N

The total $N_{i.}$ then represents the number of sample observations that belong to category C_i of the first attribute, for $i = 1, \ldots, I$. Similarly, the total $N_{.j}$ represents the number of sample observations that belong to category D_j of the second attribute, for $j = 1, \ldots, J$. How these data are used will depend on both the hypotheses of interest and the nature of the data collection (i. e., sampling) process.

10.1 General Test for Differences in Population Proportions

Consider a categorical variable for which the observed outcome belongs to one and only one of I possible categories $C_1, \ldots, C_I$. As a direct extension of the two population-two category setting of Sect. 9.1, we collect simple random samples independently from each of J separate populations and sort the observed outcomes into the I categories. Letting p_{ij} denote the probability that a random outcome from population j will belong to category C_i, for $i = 1, \ldots, I$ and $j = 1, \ldots, J$, the probability distributions for the J populations across the I categories can be represented as in Table 10.2. Since the I categories $C_1, \ldots, C_I$ completely partition each of the populations, the probability entries in Table 10.2 must satisfy the constraints that $p_{1j} + p_{2j} + \ldots + p_{Ij} = 1$ for every population $j = 1, \ldots, J$.

Table 10.2 Population probabilities for *J* populations partitioned by *I* common categories

		Population				
		1	2	…	J-1	J
Partitioning category	C_1	p_{11}	p_{12}	…	$p_{1,J-1}$	p_{1J}
	C_2	p_{21}	p_{22}	…	$p_{2,J-1}$	p_{2J}
	⋮	⋮	⋮	⋮	⋮	⋮
	C_{I-1}	$p_{I-1,1}$	$p_{I-1,2}$	…	$p_{I-1,J-1}$	$p_{I-1,J}$
	C_I	p_{I1}	p_{I2}	…	$p_{I,J-1}$	p_{IJ}

Table 10.3 Independent random samples from J populations partitioned by I common categories

		Population				
		1	2	...	J-1	J
Partitioning category	C_1	O_{11}	O_{12}	...	$O_{1,J\text{-}1}$	O_{1J}
	C_2	O_{21}	O_{22}	...	$O_{2,J\text{-}1}$	O_{2J}
	⋮	⋮	⋮	⋮	⋮	⋮
	$C_{I\text{-}1}$	$O_{I\text{-}1,1}$	$O_{I\text{-}1,2}$	...	$O_{I\text{-}1,J\text{-}1}$	$O_{I\text{-}1,J}$
	C_I	O_{I1}	O_{I2}	...	$O_{I,J\text{-}1}$	O_{IJ}
Sample size		n_1	n_2	...	$n_{J\text{-}1}$	n_J

A representation similar to Table 10.2 can be used to portray the sample data collected independently from each of the J populations and sorted into the I categories. Suppose we collect n_j sample observations from population j, for $j = 1, \ldots, J$. If we let O_{ij} denote the observed number of the sample observations from population j that belong to category i, then Table 10.3 provides such a representation of these J samples of categorical data.

One of the natural questions of interest in this setting is whether or not there are differences among the J populations with regard to their probabilities for the I categories. One way to address this question is test the null hypothesis

$$H_0 : \left[p_{i1} = p_{i2} = \ldots = p_{iJ}, \quad \text{for } i = 1 \ldots I\right] \tag{10.1}$$

against the general alternative that there are, in fact, differences of some kind in the category probabilities across the J populations. Notationally, this alternative corresponds to

$$H_A\text{: } \left[(p_{1s}, \ldots, p_{Is}) \neq (p_{1t}, \ldots, p_{It}), \quad \text{for at least one pair of populations } s, t \in \{1, \ldots, J\}\right]. \tag{10.2}$$

To test H_0 (10.1) against the general alternative H_A (10.2), we use the observed O_{ij} sample counts for the I categories from each of the populations, as given in Table 10.3. If the null hypothesis is true, the J populations do not differ in their probabilities for any of these I categories. Under H_0 (10.1), we have $p_{i1} = p_{i2} = \ldots = p_{iJ} =$ some common value, say p_i, for each $i = 1, \ldots, I$. Thus, when the null hypothesis is true, a natural estimator for the common probability p_i for the ith category C_i is the overall observed proportion in that category, pooled across all J samples; that is, if H_0 is true, we would estimate the common p_i by

$$\widehat{p_i} = \frac{O_{i1} + O_{i2} + \ldots + O_{iJ}}{n_1 + n_2 + \ldots + n_J}, \quad \text{for } i = 1, \ldots, I, \tag{10.3}$$

or, using the row totals notation of Table 10.1,

$$\widehat{p_i} = \frac{N_{i.}}{N}, \quad \text{for } i = 1, \ldots, I. \tag{10.4}$$

Now, when H_0 (10.1) is true, we would expect that each of the J populations would have something close to the pooled percentage, $\widehat{p_i}$ (10.4), of its sample observations falling in category C_i, for $i = 1, \ldots, I$. Thus, the number of sample observations from the jth population that we would expect to observe in category C_i when the null hypothesis, H_0 (10.1), is true is given by

$$E_{ij} = n_j \widehat{p_i} = \frac{n_j N_{i.}}{N}, i = 1, \ldots, I \text{ and } j = 1, \ldots, J, \tag{10.5}$$

or, using both the row and column totals notation of Table 10.1 (where, here, the column totals simply correspond to the fixed sample sizes from each of the J populations),

$$E_{ij} = \left[\text{expected number of sample observations from population } j \text{ in category } C_i \text{ when } H_0 \text{ is true}\right] = \frac{N_{i.} N_{.j}}{N}, i = 1, \ldots, I \text{ and } j = 1, \ldots, J. \tag{10.6}$$

Now, we have two quantities for each of the IJ category-population combinations in Table 10.3, namely, the observed counts, O_{ij}, and the expected counts, E_{ij} (10.6). If the null hypothesis H_0 is true, we would not expect these two counts to differ much across the category-population combinations. A natural statistic that is used to assess the magnitudes of these differences between the observed counts and the expected counts is the chi-square statistic

$$Q_1 = \sum_{all} \sum_{categories} \frac{(observed - \text{expected})^2}{\text{expected}} = \sum_{i=1}^{I} \sum_{j=1}^{J} \frac{(O_{ij} - E_{ij})^2}{E_{ij}}. \quad (10.7)$$

(Note that Q_1 (10.7) is similar to the Pearson goodness of fit statistic G (5.8) discussed in Sect. 5.4 to illustrate the simulation of sampling distributions.)

Evidence against the null hypothesis H_0 (10.1) in favor of the general alternative H_A (10.2) is clearly provided by large values of the chi-square statistic Q_1 (10.7). Moreover, when the null hypothesis is true, the sampling distribution for Q_1 can be well-approximated by the chi-square distribution with degrees of freedom = (I-1)(J-1), provided the sample sizes $n_1, \ldots, n_J$ are sufficiently large so that **each** of the IJ null expected counts E_{ij} (10.7) is at least 5.

To test the null hypothesis H_0 (10.1) versus the general alternative H_A (10.2) that the J populations do not totally agree on their probabilities for the I categories $C_1, \ldots, C_I$, let $q_{1(\text{obs})}$ be the observed value of the chi-square statistic Q_1 (10.7). Then, the approximate P-value for the test of H_0 versus H_A is

$$P\text{–}value \approx P\left(\chi^2_{(I-1)(J-1)} \geq q_{1(\text{obs})}\right), \quad (10.8)$$

where $\chi^2_{(I-1)(J-1)}$ has a chi-square distribution with (I-1)(J-1) degrees of freedom.

Example 10.1. Where Will a Yellow-Crowned Night Heron Build Her Nest? Yellow-crowned night herons nest in colonies that are generally circular in nature. Location of the individual nests within such a structure is an important feature of a heron colony. Are there criteria that determine where an individual female heron will build her nest? One possibility suggested by Bagley (1985) was that younger, less experienced hens and hens nearing the end of their reproductive years would be forced to build their nests near the edges of the colony circle where their nests (and eggs) would be less protected from predators than would those in the center of the colony. Such structuring of the nesting pattern would seem to be appropriate for optimal production of young herons during a breeding season, since the youngest and oldest female herons are also most likely to be the hens which lay fewer fertile eggs and, therefore, raise fewer young in the first place.

To study this question, Bagley (1985) observed a colony with forty-seven yellow-crowned night heron nests. Twenty-two of these nests were considered to be on the edges of the colony, while the other twenty-five were designated as interior nests. The numbers of hatched eggs for the 47 nests are presented in Table 10.4.

Here we have $J = 2$ populations corresponding to interior and edge nests, respectively, and $I = 3$ categories associated with the numbers of hatched eggs in the various nests. For $i = 3$, 4, and ≥ 5, let

Table 10.4 Numbers of nests with three, four, or five or more hatched eggs for twenty-five interior and twenty-two edge nests in a yellow-crowned night heron colony

	Number of hatched eggs			**Sample size**
	3	**4**	**5 or more**	
Interior nests	1	15	9	25
Edge nests	8	10	4	22

Source: Bagley (1985)

$$p_{i,\text{edge}} = \left[\text{proportion of all nests on the edges of yellow–crowned night heron colonies with } i \text{ hatched eggs}\right].$$

and

$$p_{i,\text{interior}} = \left[\text{proportion of all nests in the interior of yellow–crowned night heron colonies with } i \text{ hatched eggs}\right].$$

Then the null hypothesis H_0 (10.1) of interest here is given by

$$H_0 : \left[p_{i,\text{edge}} = p_{i,\text{interior}}, \text{ for } i = 3, 4, \text{ and } \geq 5\right]$$

and the alternative H_A (10.2) corresponds to

$$H_A : \left[\left(p_{3,\text{edge}}, p_{4,\text{edge}}, p_{\geq 5,\text{edge}}\right) \neq \left(p_{3,\text{interior}}, p_{4,\text{interior}}, p_{\geq 5,\text{interior}}\right)\right].$$

From (10.5) we see that the numbers of interior nests in our sample that we would expect to have 3, 4 or $\geq$5 hatched eggs, respectively, when H_0 (10.1) is true are:

$$E_{3,\text{interior}} = \frac{9(25)}{47} = 4.787, E_{4,\text{interior}} = \frac{25(25)}{47} = 13.298, \text{and } E_{\geq 5,\text{interior}} = \frac{13(25)}{47} = 6.915.$$

Similarly, the numbers of edge nests in our sample that we would expect to have 3, 4, or $\geq$5 hatched eggs, respectively, when H_0 (10.1) is true are:

$$E_{3,\text{edge}} = \frac{9(22)}{47} = 4.213, E_{4,\text{edge}} = \frac{25(22)}{47} = 11.702, \text{and } E_{\geq 5,\text{edge}} = \frac{13(22)}{47} = 6.085.$$

Combining these null expected nest numbers with the observed frequencies noted in Table 10.4, the value of the chi-square test statistic Q_1 (10.7) for the yellow-crowned night heron data becomes

$$\begin{aligned}
Q_1 &= \sum_{i=1}^{3}\sum_{j=1}^{2}\frac{(O_{ij}-E_{ij})^2}{E_{ij}} = \left\{\frac{(1-4.787)^2}{4.787}+\frac{(15-13.298)^2}{13.298}+\frac{(9-6.915)^2}{6.915}\right. \\
&\quad \left. +\frac{(8-4.213)^2}{4.213}+\frac{(10-11.702)^2}{11.702}+\frac{(4-6.085)^2}{6.085}\right\} \\
&= 2.996+0.218+0.629+3.404+0.248+0.714 \\
&= 8.209.
\end{aligned}$$

With $J = 2$ and $I = 3$, the null distribution of the test statistic Q_1 is approximately chi-square with (3–1)(2–1) = 2 degrees of freedom. Hence, from Eq. (10.8) and the **R** function *pchisq*( ), the approximate *P*-value for these yellow-crowned night heron hatchlings data is $P\text{–}value \approx P(\chi_2^2 \geq 8.209) = .016$.

```
> pchisq(8.209, df = 2, lower.tail = FALSE)
[1] 0.01649827
```

Thus there is some evidence from the sample data that the numbers of hatchlings do, indeed, differ from edge to interior night heron nests. This conclusion lends numerical support to Bagley's postulation that younger, less experienced hens and hens nearing the end of their reproductive years are forced to build their nests near the edges of a night heron colony circle.

Section 10.1 Practice Exercises

10.1.1. *College Athletes and Alcohol—Male/Female.* College athletes are exposed to a variety of personal demands and social influences that differ from those of their non-athlete colleagues. These circumstances can often result in the consumption of more alcohol by collegiate athletes than by non-athletes. Williams (2012) conducted a study designed to address this issue among college freshmen. He collected data from 263 freshmen college athletes at two NCAA Division I universities through the use of a modified version of the College Athlete Alcohol Survey. Using the North American

Table 10.5 College freshmen male and female athlete drinking habits

	Abstainer	Moderate	Heavy	Sample size
Male athletes	51	48	68	167
Female athletes	12	40	44	96

Source: Williams (2012)

Intercollegiate Athletic Association's guidelines for safe drinking, participants were assigned to one of three categories:

Abstainer (drinks no alcohol)

Moderate Drinker (male who consumes alcohol, but drinks fewer than 14 drinks per week and fewer than 4 drinks per occasion, or a female who consumes alcohol, but drinks fewer than 7 drinks per week and fewer than 3 drinks per occasion)

Heavy Drinker (male who consumes alcohol and drinks at least 14 drinks per week or 4 or more drinks per occasion, or a female who consumes alcohol and drinks at least 7 drinks per week or 3 or more drinks per occasion)

The breakdown of the 263 freshmen college athletes surveyed by Williams is provided in Table 10.5.

(a) Assuming these 263 athletes are representative of all NCAA Division I university athletes, state the null hypothesis of interest here.

(b) Construct the table of expected counts if the null hypothesis is true.

(c) Find the approximate P-value for an appropriate test of the null hypothesis. What is your conclusion at significance level .02?

(d) Comment on the implication of the analysis.

10.1.2. *College Athletes and Alcohol—Team/Individual Sport.* Consider the study of alcohol and freshmen athletes by Williams (2012) discussed in Exercise 10.1.1. He also classified each of the athletes by whether they participated in an individual or a team sport. The breakdown of the 263 athletes by individual versus team sport is provided in Table 10.6.

Table 10.6 College freshmen individual and team athlete drinking habits

	Abstainer	Moderate	Heavy	Sample size
Individual athletes	20	33	28	81
Team athletes	43	55	84	182

Source: Williams (2012)

(a) Assuming these 263 athletes are representative of all NCAA Division I university athletes, state the null hypothesis of interest here.
(b) Construct the table of expected counts if the null hypothesis is true.
(c) Find the approximate *P*-value for an appropriate test of the null hypothesis. What is your conclusion at significance level .02?
(d) Comment on the implication of the analysis.

10.1.3. *Where Are Students Carrying Weapons?* Meilman et al. (1998) reported on the regional distribution of students who carry weapons on college campuses, using data obtained from the nationwide Core Alcohol and Drug Survey of 28,253 students at 61 institutions during the 1994/95 academic years. Of those respondents, 9886 men and 14,659 women answered the following question: "During the last 30 days, how often have you carried a weapon (gun, knife, etc.), not including situations in which carrying the weapon occurred as "part of your job" or for "hunting purposes"? The responses are presented in Table 10.7. We are interested in assessing whether there are regional differences with respect to college students carrying weapons.

(a) Assuming these college students are representative of all US colleges, state the null hypothesis of interest here.
(b) Construct the table of expected counts if the null hypothesis is true.
(c) Find the approximate *P*-value for an appropriate test of the null hypothesis. What is your conclusion at significance level .045?
(d) Comment on the implication of the analysis.

Table 10.7 Number of US college students at institutions in different regions of the country who reported carrying a weapon at least once in the previous 30 days

	Carried weapon	Did not carry weapon	Total
Region			
Northeast	262	4335	4597
North central	559	7547	8106
South	451	4197	4648
West	446	6748	7194

Source: Meilman et al. (1998)

Table 10.8 Number of US college students at institutions of different sizes who reported carrying a weapon at least once in the previous 30 days

	Carried weapon	Did not carry weapon	Total
Campus size			
< 2500	454	7242	7696
2500–4999	409	5268	5377
5000–9999	236	2507	2743
10,000–19,999	330	3954	4284
$> 19{,}999$	293	4152	4445

Source: Meilman et al. (1998)

10.1.4. *Does Institution Size Matter where Students with Weapons Are Concerned?* Consider the study by Meilman et al. (1998) discussed in Exercise 10.1.3. They also looked at weapon carrying as a function of sizes of the college campuses. The breakdown of the survey responses by this criterion is given in Table 10.8.

(a) Assuming these college students are representative of all US colleges of similar sizes, state the null hypothesis of interest here.

(b) Construct the table of expected counts if the null hypothesis is true.

(c) Find the approximate P-value for an appropriate test of the null hypothesis. What is your conclusion at significance level .086?

(d) Comment on the implication of the analysis.

Table 10.9 Numbers of union and non-union employees who were completely satisfied with their employer's health insurance benefits

	Completely satisfied	Not completely satisfied
Union employees	181	212
Non-union employees	905	1681

Source: Gallup, Inc. (2015b)

10.1.5. *Does Belonging to a Union Lead to Better Health Insurance Benefits?* Gallup, Inc. conducted a survey of 2979 employed adults, aged 18 and older, about various aspects of their workplace environments. One of the questions asked whether a participant was completely satisfied with "the health insurance benefits your employer offers". The responses by the 393 employees who belonged to a union and 2586 employees who did not belong to a union are presented in Table 10.9.

(a) State the null hypothesis of interest here.

(b) Construct the table of expected counts if the null hypothesis is true.

(c) Find the approximate *P*-value for an appropriate test of the null hypothesis. What is your conclusion at significance level .033?

10.1.6. *Does Belonging to a Union Lead to Better Recognition at Work for Your Work Accomplishments?* Gallup, Inc. conducted a survey of 2979 employed adults, aged 18 and older, about various aspects of their workplace environments. One of the questions asked whether a participant was completely satisfied with "the recognition you receive at work for your work accomplishments". The responses by the 393 employees who belonged to a union and 2586 employees who did not belong to a union are presented in Table 10.10.

(a) State the null hypothesis of interest here.

(b) Construct the table of expected counts if the null hypothesis is true.

(c) Find the approximate *P*-value for an appropriate test of the null hypothesis. What is your conclusion at significance level .048?

Table 10.10 Numbers of union and non-union employees who were completely satisfied with the recognition they received at work for their work accomplishments

	Completely satisfied	Not completely satisfied
Union employees	138	255
Non-union employees	1293	1293

Source: Gallup, Inc. (2015b)

10.2 Test for Association (Independence) between Two Categorical Attributes

For the test procedure discussed in the previous section the count data in the $I \times J$ table corresponds to sample observations from each of J categorical populations sorted by I different categories $C_1, \ldots, C_I$. For that setting the numbers of observations from each of the populations are fixed and the random component of the $I \times J$ table of data is associated with how the observed sample items are distributed across the I categories $C_1, \ldots, C_I$. Thus, in Table 10.3, the column totals are fixed (corresponding the various sample sizes) and the row totals (representing the combined-samples counts in each of the categories $C_1, \ldots, C_I$) are random.

Another setting where the relevant sample data can be represented by an $I \times J$ table occurs when we are interested in assessing whether there is any statistical relationship between two different categorical variables for members of a single underlying population; that is, do the various combinations of the two categorical attributes vary independently across the population or is there some statistical pattern (i. e., dependence) between them? Thus, for example, is there a statistical relationship between self-esteem and problem drinking on campus? Between ethnicity and degree of cigarette smoking? Are there gender differences in what are viewed as the most important attributes for dates? For spouses? Is there a relationship between mental health and substance abuse disorders among homeless Americans?

Are there differences in how husbands and wives feel that their jobs are treated in a marriage?

All of these examples have a common data theme. Each subject under discussion can be classified into one and only one of at least two possible categorical levels for each of two different attributes. Thus each subject falls into one and only one simultaneous category for the two attributes. If we collect a random sample of N subjects from the population of interest and place each of them in the single two-attribute category that is appropriate, the resulting counts for the IJ two-attribute categories once again correspond to the general data representation in Table 10.1. However, unlike the setting for the previous section, here only the total number of subjects being distributed across the IJ two-attribute categories is fixed, since both the row and column totals are now random quantities that depend on the sample outcomes.

For this two-way categorical classification of sample data from a single population, one of the relevant questions to ask is whether or not classifications according to the two attributes are independent; that is, does knowledge of which category a subject belongs to for one of the attributes affect the probabilities associated with belonging to the various categories of the second attribute? To model this question, we add row and column totals to construct a probability table similar to that in Table 10.2, but now for the case where a single sample of data is categorized by two attributes. The resulting two-way classification probability table with row and column sums is given in Table 10.11.

Thus, for $i = 1, \ldots, I$ and $j = 1, \ldots, J$, the probability p_{ij} in Table 10.11 represents the probability that a randomly selected member of the population of interest belongs to both category C_i of attribute one and to category D_j of the second attribute; that is, $p_{ij} = P(C_i \text{ and } D_j) = P(C_i \cap D_j)$. The row total $p_{i.} = p_{i1} + p_{i2} + \ldots + p_{iJ} = P(C_i \text{ and } D_1) + P(C_i \text{ and } D_2) + \ldots + P(C_i \text{ and } D_J) = P(C_i)$ then corresponds to the probability that a randomly selected member of the population of interest belongs to category C_i of attribute one

Table 10.11 Population probabilities for categorization by two attributes

		Category for attribute D					
		D_1	D_2	...	D_{J-1}	D_J	$p_{i.} = P(C_i)$
Category for attribute C	C_1	p_{11}	p_{12}	...	$p_{1,J-1}$	p_{1J}	$p_{1.}$
	C_2	p_{21}	p_{22}	...	$p_{2,J-1}$	p_{2J}	$p_{2.}$
	$\vdots$	$\vdots$	$\vdots$		$\vdots$	$\vdots$	$\vdots$
	C_{I-1}	$p_{I-1,1}$	$p_{I-1,2}$	...	$p_{I-1,J-1}$	$p_{I-1,J}$	$p_{(I-1).}$
	C_I	p_{I1}	p_{I2}	...	$p_{I,J-1}$	p_{IJ}	$p_{I.}$
	$p_{.j} = P(D_j)$	$p_{.1}$	$p_{.2}$	...	$p_{.(J-1)}$	$p_{.J}$	1

without regard to which category of attribute two pertains. Similarly, the column total $p_{.j} = p_{1j} + p_{2j} + \ldots + p_{Ij} = P(C_1 \text{ and } D_j) + P(C_2 \text{ and } D_j) + \ldots + P(C_I \text{ and } D_j) = P(D_j)$ corresponds to the probability that a randomly selected member of the population of interest belongs to category D_j of attribute two without regard to which category of attribute one pertains.

With these representations for the probabilities of the individual C_i and D_j categories and their combinations $C_i \cap D_j$, we can use our knowledge about independence of two events (as discussed in Chap. 4) to write the hypothesis of independence between the classifications of a subject by the two attributes as

$$H_0 : \left[P(C_i \text{ and } D_j) = P(C_i) \times P(D_j), \text{for } i = 1, \ldots, I \text{ and } j = 1, \ldots, J\right], \quad (10.9)$$

or, equivalently,

$$H_0 : \left[p_{ij} = p_{i.} \times p_{.j} \text{ for } i = 1, \ldots, I \text{ and } j = 1, \ldots, J\right]. \quad (10.10)$$

The general dependence alternative to H_0 is naturally two-sided in nature, corresponding to

$$H_A : \left[P(C_i \text{ and } D_j) \neq P(C_i) \times P(D_j), \text{for at least one } (i, j) \text{ pair}\right], \quad (10.11)$$

or, equivalently, to

$$H_A: \left[p_{ij} \neq p_{i.} \times p_{.j} \text{ for at least one } (i, j) \text{ pair}\right]. \tag{10.12}$$

To test the null hypothesis of independence between classifications by the two attributes, H_0 (10.10), versus the general alternative of attribute dependence, H_A (10.12), we once again make use of the observed O_{ij} counts in the various two-way classification cells, as represented in Table 10.1. If the null hypothesis is true, we can first estimate $p_{i.}$ and $p_{.j}$ and then use the fact that $p_{ij} = p_{i.} \times p_{.j}$ to estimate p_{ij}. For each $i = 1, \ldots, I$ and $j = 1, \ldots, J$, the natural estimators for $p_{i.}$ and $p_{.j}$ are the proportions of observed outcomes in the ith row and jth column, respectively; that is,

$$\hat{p}_{i.} = \frac{N_{i.}}{N} \quad \text{for } i = 1, \ldots, I \tag{10.13}$$

and

$$\hat{p}_{.j} = \frac{N_{.j}}{N} \quad \text{for } j = 1, \ldots, J. \tag{10.14}$$

Hence, under the null H_0 hypothesis that $p_{ij} = p_{i.} \times p_{.j}$, the natural estimator for p_{ij} is then

$$\hat{p}_{ij} = \hat{p}_{i.}\hat{p}_{.j} = \frac{N_{i.}N_{.j}}{N^2}, \quad \text{for } i = 1, \ldots, I \text{ and } j = 1, \ldots, J. \tag{10.15}$$

Now, when H_0 (10.10) is true, we would expect to have somewhere close to the estimated percentage $\hat{p}_{ij} = \frac{N_{i.}N_{.j}}{N^2}$ of the N total sample observations falling in the crossed category $C_i \times D_j$, for $i = 1, \ldots, I$ and $j = 1, \ldots, J$. Thus, the number of the N sample observations that we would expect to observe in the crossed category $C_i \times D_j$, when the null hypothesis, H_0 (10.10), is true, is given by

$$E_{ij} = \left[\text{expected number of the } N \text{ sample observations in the crossed category } C_i \times D_j \text{ when } H_0(10.10) \text{ is true}\right]$$
$$= N\hat{p}_{i.}\hat{p}_{.j} = N\frac{N_{i.}N_{.j}}{N^2} = \frac{N_{i.}N_{.j}}{N}, \quad \text{for } i = 1, \ldots, I \text{ and } j = 1, \ldots, J. \tag{10.16}$$

As in Sect. 1, we now have two quantities for each of the IJ crossed categories $C_i \times D_j$ in Table 10.1, namely, the observed sample counts, O_{ij}, and the expected counts E_{ij} (10.16). If the null hypothesis H_0 is true, we would not expect these counts to differ by much across the IJ crossed categories $C_i \times D_j$.

We note that the expected counts in (10.16) for this independence setting are obtained from the $I \times J$ table in exactly the same way (but for different reasons) as the expected counts given in (10.5) for the J population problem. Thus, the general form of the chi-square statistic Q_1 given in eq. (10.7) can once again be used to assess the magnitudes of the attained differences between the observed and expected counts.

As in Sect. 1, evidence against the null hypothesis H_0 (10.10) in favor of the general alternative H_A (10.12) is provided by large values of the chi-square statistic Q_1 (10.7). Once again, when the null hypothesis H_0 (10.10) is true, the sampling distribution for Q_1 can be well-approximated by the chi-square distribution with degrees of freedom = $(I\text{-}1)(J\text{-}1)$, provided the number of sample observations, N, is sufficiently large so that each of the IJ expected counts E_{ij} (10.16) is at least 5.

To test the null hypothesis H_0 (10.10) of independence between the two attributes versus the general alternative H_A (10.12) that the two attributes are related in some fashion (i. e., they are dependent), let $q_{1(\text{obs})}$ be the observed value of the chi-square statistic Q_1 (10.7) with the expected counts given by (10.16). Then the approximate P-value for the test of H_0 (10.10) versus H_A (10.12) is

$$P\text{-}value \approx P\left(\chi^2_{(I-1)(J-1)} \geq q_{1(\text{obs})}\right), \tag{10.17}$$

where $\chi^2_{(I-1)(J-1)}$ has a chi-square distribution with (I-1)(J-1) degrees of freedom.

Example 10.2. Is Gender an Important Factor in Role Portrayals in Popular Magazines? When you look at an ad in a magazine, what do you notice about the people in the ad? Are they male or female? What ethnicity are the people? What roles are they portraying in the ad? Vigorito and Curry (1998) considered questions like these as part of their study into links between gender/ethnic identities and role portrayals in popular magazines. In this example we concentrate on the relationship between gender and role portrayal, leaving a similar discussion of the relationship between ethnicity and role portrayal for the exercises. Vigorito and Curry gathered data on magazine content from a cross section of 83 popular magazines during the summer of 1992. They coded 7935 individuals in ad illustrations in these magazines with regard to race, sex, and the roles they portrayed in the ads. Two hundred fifty of these individuals were infants for which the sex could not be determined from the ad illustration. The data in Table 10.12 represent the

Table 10.12 Gender-identified breakdown of role portrayals in magazine ad illustrations

	Gender		
Role portrayal	**Female**	**Male**	**Total**
Sport/fitness	114	166	280
Model/consumer	1386	526	1912
Occupational	588	1419	2007
Parent	222	142	364
Spouse/partner	319	287	606
Outdoor recreation	250	411	661
Other	880	975	1855
Total	3759	3926	7685

Source: Vigorito and Curry (1998)

breakdown of the remaining 7685 individuals for which gender could be identified into the following seven role categories: sport/fitness, model/consumer, occupational, parent, spouse/partner, outdoor recreation, and other.

Among the questions of interest to Vigorito and Curry was whether there is any relationship between gender and the roles that individuals play in magazine ad illustrations. Table 10.12 contains the observed frequencies in the fourteen gender-identified role portrayal combinations. In order to address the relationship between gender and role portrayal in magazine ads we must next use Expression (10.16) to construct the expected counts in each of these categories when the null hypothesis H_0 (10.10) of independence between gender and role portrayal in magazine ads is true. We illustrate the necessary calculations for the fourteen combinations and then summarize the null expected counts in Table 10.13.

From Expression (10.16) and the observed counts in Table 10.12, we find:

$$E_{\text{sport/fitness, female}} = \frac{(280)(3759)}{7685} = 136.96,$$
$$E_{\text{sport/fitness, male}} = \frac{(280)(3926)}{7685} = 143.04,$$

Table 10.13 Expected counts for gender-identified/role portrayal data in table 10.12 when the null hypothesis of independence is true

	Gender		
Role portrayal	**Female**	**Male**	**Total**
Sport/fitness	136.96	143.04	280
Model/consumer	935.23	976.77	1912
Occupational	981.69	1025.31	2007
Parent	178.05	185.95	364
Spouse/partner	296.42	309.58	606
Outdoor recreation	323.32	337.68	661
Other	907.34	947.66	1855
Total	3759	3926	7685

$$E_{\text{model/consumer, female}} = \frac{(1912)(3759)}{7685} = 935.23,$$

$$E_{\text{model/consumer, male}} = \frac{(1912)(3926)}{7685} = 976.77,$$

$$E_{\text{occupational, female}} = \frac{(2007)(3759)}{7685} = 981.69,$$

$$E_{\text{occupational, male}} = \frac{(2007)(3926)}{7685} = 1025.31,$$

$$E_{\text{parent, female}} = \frac{(364)(3759)}{7685} = 178.05,$$

$$E_{\text{parent, male}} = \frac{(364)(3926)}{7685} = 185.95,$$

$$E_{\text{spouse/partner, female}} = \frac{(606)(3759)}{7685} = 296.42,$$

$$E_{\text{spouse/partner, male}} = \frac{(606)(3926)}{7685} = 309.58,$$

$$E_{\text{outdoor recreation, female}} = \frac{(661)(3759)}{7685} = 323.32,$$

$$E_{\text{outdoor recreation, male}} = \frac{(661)(3926)}{7685} = 337.68,$$

$$E_{\text{other, female}} = \frac{(1855)(3759)}{7685} = 907.34,$$

$$E_{\text{other, male}} = \frac{(1855)(3926)}{7685} = 947.66.$$

Combining these expected counts under H_0 in Table 10.13 with the observed frequencies noted in Table 10.12 for the fourteen gender/role portrayal combinations, the value of the chi-square test statistic Q_1 (10.7) for these data becomes

$$
\begin{aligned}
Q_1 &= \sum_{i=1}^{7}\sum_{j=1}^{2}\frac{\left(O_{ij}-E_{ij}\right)^2}{E_{ij}} \\
&= \left\{\frac{(114-136.96)^2}{136.96}+\frac{(166-143.04)^2}{143.04}+\frac{(1386-935.23)^2}{935.23}\right. \\
&\quad +\frac{(526-976.77)^2}{976.77}+\frac{(588-981.69)^2}{981.69}+\frac{(1419-1025.31)^2}{1025.31} \\
&\quad +\frac{(222-178.05)^2}{178.05}+\frac{(142-185.95)^2}{185.95}+\frac{(319-296.42)^2}{296.42} \\
&\quad +\frac{(287-309.58)^2}{309.58}+\frac{(250-323.32)^2}{323.32}+\frac{(411-337.68)^2}{337.68} \\
&\quad \left.+\frac{(880-907.34)^2}{907.34}+\frac{(975-947.66)^2}{947.66}\right\} \\
&= \{3.849+3.685+217.266+208.026+157.883+151.166 \\
&\quad +10.849+10.388+1.720+1.647+16.627+15.920 \\
&\quad +0.824+0.789\} \\
&= 800.639.
\end{aligned}
$$

With $I = 7$ and $J = 2$, the null distribution of the test statistic Q_1 is approximately chi-square with (7–1)(2–1) = 6 degrees of freedom. Hence, from Eq. (10.17) and the **R** function *pchisq*( ), the approximate *P*-value for these gender-identified role portrayal magazine ad data is $P\text{-}value \approx P\left(\chi_6^2 \geq 800.639\right) \approx 0$.

```
> pchisq(800.639, df = 6, lower.tail = FALSE)
[1] 1.120478e-169
```

Thus there is very strong evidence in the Vigorito-Curry sample data that gender and magazine ad role portrayal are not independent attributes. Closer examination of the data shows clearly that the major contributors to the large observed value for the statistic Q_1 are the gender combinations with the model/consumer and occupational roles, with less sizable, but still important, contributions from the parent and outdoor recreation roles. Men are portrayed in occupational or outdoor recreation roles in magazine ads significantly more often than women, while the opposite is true for the model/consumer and parent roles.

We note that this chi-square analysis of the gender-identified role portrayal data can also be accomplished through use of the **R** function *chisq.test*( ) and the data.frame *gender_roles*.

```
> chisq.test(gender_roles)

        Pearson's Chi-squared test

data:  gender_roles
X-squared = 800.65, df = 6, p-value < 2.2e-16
```

Section 10.2 Practice Exercises

10.2.1. *Taking Care of Sick Parents.* Princeton Survey Research Associates of Princeton, New Jersey (1998) conducted an extensive series of surveys designed to assess American values about taking care of each other. One of the questions asked of the respondents in one of these surveys was:

Who should be responsible for taking care of parents if they become sick or disabled?

The respondents were also asked to self-classify themselves as Conservative, Moderate, or Liberal. The breakdown of the 1095 survey respondents with respect to both their political ideology and their answer to the stated question is:

	Numbers Who Gave This Response		
Question Response	**Conservatives**	**Moderates**	**Liberals**
People should feel entirely responsible	346	272	146
People should expect help from the government	339	409	221
It depends	60	61	28
Don't know/refused to answer	8	15	0

Is there a relationship between political ideology and viewpoint towards taking care of parents if they become sick or disabled?

(a) State the null hypothesis of interest here.

(b) Construct the table of expected counts if the null hypothesis is true.

(c) Find the approximate P-value for an appropriate test of the null hypothesis. What is your conclusion at significance level .01?

(d) Comment on the implication of the analysis.

10.2.2. *Is To Know Congress Really To Love Them?* Based on a survey by Gallup, Inc. (2015c), Americans in general do not have a very favorable view of the U. S. Congress, with 49% rating the way Congress is doing its job as poor or bad and only 15% rating it as excellent or good. Gallup chose to follow up and investigate whether these negative opinions might not be based at least partially on poorly informed public assumptions and impressions about Congress. They conducted a second poll in which they asked 1017 adult Americans, aged 18 and older, a set of five questions to ascertain their level of knowledge about Congress and its operations in addition to asking their opinion on how Congress is doing its job. The results from this second Gallup poll are presented in Table 10.14.

(a) State the null hypothesis of interest here.

(b) Construct the table of expected counts if the null hypothesis is true.

Table 10.14 Knowledge about Congress and opinion as to how well it is doing its job

	Number of five political questions answered correctly			
Rating of congress	**0**	**1**	**2–3**	**4–5**
Excellent/good	67	53	20	12
Fair	94	91	125	47
Poor/bad	73	131	190	114

Source: Gallup, Inc. (2015c)

(c) Find the approximate *P*-value for an appropriate test of the null hypothesis. What is your conclusion at significance level .059?
(d) Comment on the implication of the analysis.

10.2.3. *Birthright Citizenship*. Immigration was an important topic for debate during the 2016 presidential campaign. In particular, birthright citizenship for U. S.-born children of parents who are not legal residents was a major issue of discussion. The Pew Research Center (2015c) reported on the results of a poll that addressed a number of immigration issues, including birthright citizenship. One question they asked of respondents was: "Do you favor changing the Constitution to bar citizenship for U.S.-born children of parents who are not legal residents?" The responses from 1441 survey participants who identified themselves politically as Republican, Democrat, or Independent are given in Table 10.15.

(a) State the null hypothesis of interest here.
(b) Construct the table of expected counts if the null hypothesis is true.
(c) Find the approximate *P*-value for an appropriate test of the null hypothesis. What is your conclusion at significance level .088?

10.2.4. *Iran Nuclear Agreement*. The Joint Comprehensive Plan of Action (better known as the Iran Nuclear Agreement) is an international agreement on the nuclear program of Iran signed in Vienna, Austria on July 14, 2015 between Iran, the P5 + 1 (the five permanent members of the United Nations

Table 10.15 Should the Constitution be changed to bar birthright citizenship?

	Change constitution	Do not change constitution
Political identification		
Republican	223	198
Democrat	105	351
Independent	209	355

Source: Pew Research Center (2015c)

Table 10.16 U. S. public knowledge of and support for the Iran nuclear agreement

	Approve	Disapprove	Don't know
How much they had heard about the agreement			
A lot	158	191	19
A little	112	148	68
Nothing at all	40	111	157

Source: Pew Research Center (2015a)

Security Council—China, France, Russia, United Kingdom, United States—plus Germany), and the European Union. The Pew Research Center conducted a public opinion poll September 3–7, 2015 to assess the U. S. public's awareness and support for the agreement. The results of their poll are given in Table 10.16. The question of interest here is whether knowledge about the Iran Nuclear Agreement influenced the public's opinion about the pact.

(a) State the null hypothesis of interest here.

(b) Construct the table of expected counts if the null hypothesis is true.

(c) Find the approximate *P*-value for an appropriate test of the null hypothesis. What is your conclusion at significance level .022?

10.2.5. *Is College Worth It?* Gallup, Inc. and Purdue University conducted an extensive survey (from December 16, 2004 to June 29, 2015) of recent college graduates with a bachelor's degree or higher. The graduates were asked questions about student debt, employment, and opinions about the value of and satisfaction with their college educations, among other things. The results of this study were disseminated in the Gallup-Purdue Index Report (2015)—you might find the report interesting. One of the questions asked of the graduates was the amount of student loan debt they had incurred as undergraduates. The reported percentages of white, African-American, and

Table 10.17 Percentages of groups of alumni who graduated between 2006 and 2015 with various levels of student loan debt incurred as undergraduates

	White	African-American	Hispanic
Amount of student debt			
None	39	20	28
$1 – $25,000	27	30	38
$25,001 – $50,000	21	31	19
> $50,000	13	19	15

Source: Gallup-Purdue Index Report (2015)

Hispanic graduates with various levels of undergraduate student loan debt are presented in Table 10.17.

(a) What is the null hypothesis of interest here? Formally state it.

(b) For sake of this exercise, assume that these percentages are the result of survey responses from 5000 white graduates, 2000 African American graduates, and 2000 Hispanic graduates. Under this assumption, convert the percentages in Table 10.17 to the corresponding table with observed counts in each of the 12 debt level/ethnicity cells.

(c) Construct the associated table of expected counts for these cells if the null hypothesis is true.

(d) Find the approximate P-value for an appropriate test of the null hypothesis. What is your conclusion at significance level .050?

10.2.6. *Who Were the Nazis?* During World War II, the NSDAP (Nationalsozialistische Deutsche Arbeiter Partei), better known as the Nazi Party, was led by Adolph Hitler, but membership came from all walks of life and backgrounds. Jarausch and Arminger (1989) studied a number of factors that differentiated between those German teachers (members of the National Socialist Teachers League, NSLB) who were members of the NSDAP and those who chose not to join. One of the factors of interest to Jarausch and Arminger was the religion of the teacher. They examined the historical

Table 10.18 Religious preference and choice of membership in the Nazi Party for German teachers belonging to the National Socialist Teachers League

	Joined the Nazi party	Did not join the Nazi party	Total
Religion			
Protestant	1970	4273	6243
Catholic	787	3327	4114
None	327	2882	3209

Source: Jarausch and Arminger (1989)

records of 13,566 German teachers who belonged to the NSLB and recorded their religious preference (Protestant, Catholic, or none) and whether they had joined the Nazi Party. Their findings are reported in Table 10.18.

(a) State the null hypothesis that corresponds to the same proportion of Protestant teachers, Catholic teachers, and teachers indicating no religion choosing to join the Nazi Party during World War II.

(b) Construct the table of expected outcomes if this null hypothesis is true.

(c) Find the approximate *P*-value for an appropriate test of the null hypothesis . What is your conclusion at significance level .064?

10.3 Exact Procedure for Testing Equality of Two Population Proportions

In Sect. 9.1 we discussed an approximate hypothesis test for the equality of two population proportions. For that procedure based on the statistic V^*(9.8), the approximate *P*-value is obtained by using the fact that the sampling distribution for V^* can be well approximated by the standard normal distribution when both sample sizes m and n are large. In this section we consider a more computationally intensive test procedure for the same problem, but one that provides exact, rather than approximate, *P*-values. It will be the preferred procedure for testing equality of two population proportions when one or both of the sample sizes is not large enough for the standard normal distribution to provide a good approximation to the sampling distribution of V^*.

Consider independent random samples of sizes m and n Bernoulli variables (either an event A occurs or it does not) from two populations. Using the notation of Sect. 1, this setting corresponds to $J = 2$ populations and $I = 2$ categories and we can represent the observed outcomes of such Bernoulli random samples in the following 2×2 table, where $O_{11} + O_{21} = m$ and $O_{12} + O_{22} = n$.

Letting p_1 and p_2 denote the probability that the event A occurs for populations 1 and 2, respectively, we are interested here in testing the null hypothesis H_0: $[p_1 = p_2]$ versus either a one-sided or two-sided alternative. The test procedure, known as *Fisher's exact test*, is based on the total number of times the event A occurs, namely, $N_A = O_{11} + O_{12}$, and the observed proportions, $\frac{O_{11}}{N_A}$ and $\frac{O_{12}}{N_A}$, of these A outcomes that came from each of the two populations. (We note that an equivalent alternative test could be based on the total number of times that the event A *does not occur* and the corresponding proportions for each of the two populations.)

Let $N = m + n$ be the total number of sample observations from the two populations. Then the observed data situation presented in Table 10.19 corresponds to N_A of the N observations resulting in the occurrence of the event A and $N - N_A$ of them resulting in the non-occurrence of the event A. *Conditional* on the observed column and row totals (m, n, N_A, and $N - N_A$)

Table 10.19 Observed numbers of occcurrences of the event A for independent Bernoulli samples from two populations

	Population		
	1	**2**	**Total**
A occurs	O_{11}	O_{12}	N_A
A does not occur	O_{21}	O_{22}	$N - N_A$
Sample size	m	n	N

in Table 10.19, the sampling distribution of the statistic O_{11} = [number of occurrences of the event A from Population 1] when the null hypothesis H_0: $[p_1 = p_2]$ is true is given by the following expression:

$$P(O_{11} = x|N_A, N - N_A, m, n) = \frac{N_A!(N - N_A)!m!n!}{N!x!(m - x)!(N_A - x)!(n - N_A + x)!}, \quad (10.18)$$

for integer x values between max(0, N_A - n) and min(m, N_A).

Evidence against the null hypothesis H_0: $[p_1 = p_2]$ in favor of the general two-sided alternative H_A: $[p_1 \neq p_2]$ is clearly provided by either unusually large or unusually small observed values of O_{11}. For the one-sided alternative H_A: $[p_2 > p_1]$ only unusually small values of O_{11} are suggestive that the null hypothesis is not true, while for the one-sided alternative H_A: $[p_2 < p_1]$ rejection of H_0 is supported by unusually large values of O_{11}.

To test H_0: $[p_1 = p_2]$, let $o_{11(obs)}$denote the observed number of occurrences of the event A among the m sample items from Population 1. The exact P-value for a test of H_0: $[p_1 = p_2]$ against the alternative:

$$H_A : [p_2 > p_1] \text{ is } \quad P\text{–}value = P(O_{11} \leq o_{11(\text{obs})}|N_A, N - N_A, m, n) \quad (10.19)$$

$$H_A : [p_2 > p_1] \text{ is } \quad P\text{–}value = P(O_{11} \geq o_{11(\text{obs})}|N_A, N - N_A, m, n) \quad (10.20)$$

$$H_A : [p_2 \neq p_1] \text{ is } \quad P\text{–}value = 2 \times p^*, \quad (10.21)$$

where

$$p^* = \min\{P(O_{11} \leq o_{11(obs)}|N_A, N - N_A, m, n), P(O_{11} \geq o_{11(obs)}|N_A, N - N_A, m, n)\}$$

The exact P-values in (10.19)–(10.21) can be obtained using the **R** function *phyper*( ) (since this distribution is known as the *hypergeometric* distribution). They are also provided among the output of the **R** function*fisher.test*( ), which you can use (along with actual data) to conduct the hypothesis test.

Example 10.3. Does Political Affiliation Affect Our Views on Health Care Spending? Do all people have the right to receive the health care they need, regardless of how much it costs or are there limits on what our society can spend, even on health care? This is precisely one of the questions raised in a major national survey conducted by the Princeton Survey Research Associates for Americans Discuss Social Security and published in the report "Generation to Generation: American Values about Taking Care of Each Other" (1998). Among the groups compared on this question were the Republican and Democratic political parties. Each person in the survey sample was asked to select which of the following two statements came closer to the way he/she felt:

All people have the right to receive the health care they need, regardless of how much it costs.
There are limits on what our society can spend, even on health care.

Seven hundred six self-identified Democrats and 538 self-identified Republicans responded to this question and the summary of their responses is given in Table 10.20.

Letting p_{Dem} and p_{Rep} represent the proportion of all Democrats and Republicans, respectively, which are in favor of unlimited healthcare, we are interested in using the survey data in Table 10.20 to test the hull hypothesis H_0: $[p_{\mathrm{Dem}} = p_{\mathrm{Rep}}]$. The alternative of interest here is the one-sided alternative H_A: $[p_{\mathrm{Rep}} < p_{\mathrm{Dem}}]$, corresponding to the fact that Republican views generally tend to be a bit more fiscally conservative than those of Democrats. From Table 10.20 we see that the observed value of $O_{11} = O_{\text{Dem, Unlimited HC}}$ is

Table 10.20 Sample responses on limited versus unlimited health care by political party affiliation

	Democrats	Republicans
Unlimited healthcare	396	214
Limited healthcare	310	324
Sample size	706	538

Source: Princeton Survey Research Associates of Princeton, New Jersey (1998)

$o_{11(obs)} = o_{Dem\ ,\ Unlimited\ HC(obs)} = 396$. From (10.20), it follows that the exact *P*-value for testing H_0: $[p_{\text{Dem}} = p_{\text{Rep}}]$ versus the one-sided alternative H_A: $[p_{\text{Rep}} < p_{\text{Dem}}]$ is given by

$$P\text{–}value = P(O_{11} \geq 396 | N_{\text{UnlimHC}} = 610, N_{\text{LimHC}} = 634, m = 706, n = 538).$$

Using the **R** function *phyper*( ) we see that this *P*-value for these data is nearly 0, indicating that there is clearly a significant difference in how Republicans and Democrats view the issue of limited versus unlimited healthcare, no matter what the costs.

```
> phyper(396, m = 706, n = 538, k = 610, lower.tail = FALSE)
[1] 3.855707e-09
```

Note that we could have also used the **R** function *fisher.test*( ) along with the *health_care_by_affiliation* data.frame to conduct this hypothesis test.

Section 10.3 Practice Exercises

10.3.1. *Is Purple a Female Color?* Wypijewski (1997) reported on the results of a comprehensive scientific poll of American tastes in art, as commissioned by Vitaly Komar and Alexander Melamid in conjunction with the Nation Institute, a nonprofit of *The Nation* magazine. Random samples of 475 males and 526 females were asked to name their favorite color. Thirty-seven females and six males in the samples named purple as their favorite color. We are interested in exploring any gender differences among individuals who view purple as their favorite color.

(a) State the null hypothesis of interest here.

(b) Construct the appropriate table of counts for this setting.

(c) Find the exact *P*-value for an appropriate test of the null hypothesis. What is your conclusion at significance level .015?

(d) Compare your analysis with that in Exercise 9.1.11 using the approximate procedure based on a large sample approximation.

10.3.2. *Young of Year Gizzard Shad.* Consider the length of YOY gizzard shad data displayed in Table 1.38 in Exercise 1.B.13. Pool the ten observations from Sites 1 and 2 to constitute a single random sample of 20 observations from site "C" and do the same for the ten observations from Sites 3 and 4 to constitute a single random sample of 20 observations from Site "D". Let p_C and p_D denote the proportion of all the YOY gizzard shad from Sites "C" and "D", respectively, that were at least 30 *mm* in length at the time that Johnson (1984) conducted his study.

(a) State the null hypothesis that corresponds to Sites "C" and "D" being the same with respect to percentage of its YOY gizzard shad that were at least 30 *mm* in length at the time of the study.

(b) Construct the appropriate table of counts for this setting.

(c) Find the exact P-value for an appropriate test of the null hypothesis. What is your conclusion at significance level .037?

10.3.3. *Buyers Beware the Color!* The presence or absence of color in marketing communications (e.g., advertising, packaging) can affect the way a consumer digests information about a product. Lee et al. (2014) conducted a number of experiments to better understand this marketing feature. In one of their experiments, participants were asked to consider camping at a remote site where they could receive only a single radio station. The campsite manager offered two radios for rent: Option A was a simple analog radio, that appeared to be lighter and smaller, for $10 a day; Option B was a fancier digital radio, that appeared to be larger and heavier, but it had a more attractive display design along with high-precision tuner buttons and the ability to preset the radio to automatically receive numerous stations, for $18 a day. The participants were told that the two radios had the same sound quality. One group of 47 participants were presented black and white pictures of both Option A and Option B radios and asked to choose which one they wished to rent for the camping trip. A second group of 47 participants were presented color pictures of both Option A and Option B radios and asked to choose which one they wished to rent for the camping trip. Given the constraints of the camping trip, we would expect that participants would choose the more practical and cheaper Option A radio. The question of interest is whether the addition of color to the pictures of the radios led to unexpected choices by the participants. Of the 47 participants presented black and white pictures of the Option A and Option B radios, 35 chose the more practical Option A. Of the 47 participants presented color pictures of the Option A and Option B radios, only 23 chose the more practical Option A.

(a) State the null hypothesis of interest here.

(b) Construct the appropriate table of counts for this experiment.

(c) Find the exact P-value for an appropriate test of the null hypothesis. What is your conclusion at significance level .036?

10.3.4. *Army Rangers.* The Pentagon describes the Army Ranger School as "the Army's premier combat leadership course, teaching Ranger students how to overcome fatigue, hunger, and stress to lead soldiers during small unit combat operations". Students of the school are forced to train with minimal food and little sleep as they operate in difficult terrain, including woods, mountains, and swamplands. For the first time in history, two women successfully completed this rigorous and exhausting program in August 2015. Nineteen women and 381 men began the Army Ranger program in April 2015 and two women and 94 men successfully completed it and graduated in August 2015. What do these data say about the current success rates for male and female participants in the Army Ranger School?

(a) State the null hypothesis of interest here.

(b) Construct the appropriate table of counts for this experiment.

(c) Find the exact P-value for an appropriate test of the null hypothesis. What is your conclusion at significance level .080?

Table 10.21 Number of US college students at public and private institutions who reported carrying a weapon at least once in the previous 30 days

	Carried weapon	Did not carry weapon	Total
Type of institution			
Public	1205	12,801	14,006
Private	516	10,023	10,539

Source: Meilman et al. (1998)

10.3.5. *College Students Carrying Weapons—Private Versus Public Institutions.* Consider the study by Meilman et al. (1998) discussed in Exercise 10.1.3. They also compared weapon carrying at private and public institutions. The breakdown of the survey responses by this criterion is given in Table 10.21.

(a) Assuming these college students are representative of all public and private US colleges, state the null hypothesis of interest here.

(b) Find the exact *P*-value for an appropriate test of the null hypothesis. What is your conclusion at significance level .027?

(c) Comment on the implication of the analysis.

10.4 Goodness-of-fit Test for Probabilities in a Multinomial Distribution with I > 2 Categories

In Sect. 6.3 we discussed a procedure designed to use Bernoulli (binomial) sample data to test hypotheses about the value of $p_A = P(A) =$ [probability of the event A] for a population. Such a population is, of course, simply a categorical population with only $I = 2$ categories. In this section we discuss a more general procedure that will allow us to test hypotheses for categorical populations with an arbitrary number, $I \geq 2$, of categories. The test procedure utilizes multinomial categorical data collected from the population and the Pearson goodness of fit statistic previously encountered in Sect. 5.4 during our discussion of ways to use computer simulation to approximate the sampling distribution of a statistic.

Consider the setting where each measurement in a population can belong to one and only one of $I > 2$ categories, $C_1, \ldots, C_I$. Collecting N independent measurements from this population, we record the numbers of these sample observations, say $O_1, \ldots, O_I$, that fall in the I population categories $C_1, \ldots, C_I$, respectively. These observed category frequencies can be used to test the plausibility of a prescribed set of category proportions for the population.

Let $p_i = P(C_i)$ denote the proportion of the measurements in the population that belong to category C_i, for $i = 1, \ldots, I$. Let $p_{1,0}, \ldots, p_{I,0}$ be known numbers between 0 and 1, inclusive, such that $\sum_{i=1}^{I} p_{i,0} = 1$. Thus the I numbers $p_{1,0}, \ldots, p_{I,0}$ represent one possible way in which the population might be distributed across the categories $C_1, \ldots, C_I$. A natural question for this setting is whether or not the observed sample data agree with the division of the population associated with the set of categorical probabilities $p_{1,0}, \ldots, p_{I,0}$. Thus we are interested in testing.

$$H_0 : \left[p_i = p_{i,0}, \text{for } i = 1, \ldots, I\right] \tag{10.22}$$

against the general alternative

$$H_A : \left[p_i \neq p_{i,0}, \text{for at least one } i = 1, \ldots, I\right], \tag{10.23}$$

corresponding to at least two categories for which the population proportions do not agree with those prescribed by $p_{1,0}, \ldots, p_{I,0}$. This is generally known as the *goodness of fit testing problem*, since a test of the null hypothesis H_0 (10.22) can be thought of as a test designed to assess the "goodness of the fit" of the hypothesized categorical proportions $p_{1,0}, \ldots, p_{I,0}$ to the true breakdown of the population across the I categories.

In Example 5.10 we discussed a situation involving the distribution of colors claimed by Mars, Inc. for their M&M's Plain candy. There we had $I = 6$ categories, corresponding to the colors brown (C_1), yellow (C_2), red (C_3), orange (C_4), green (C_5), and blue (C_6), and the production proportions for these colors claimed by Mars, Inc. are $p_{\text{brown}} = .3$, $p_{\text{yellow}} = p_{\text{red}} = .2$ and $p_{\text{orange}} = p_{\text{green}} = p_{\text{blue}} = .1$. The relevant null hypothesis H_0 (10.22) corresponding to the company's claim for the M&M Plain color distribution is then given by

$$H_0 : \left[p_{\text{brown}} = .3, p_{\text{yellow}} = p_{\text{red}} = .2 \text{ and } p_{\text{orange}} = p_{\text{green}} = p_{\text{blue}} = .1\right].$$

In Sect. 5.4 we informally described a way to use the observed colors of randomly sampled M&M's Plain candy and the Pearson goodness of fit statistic G (5.8) to assess this color distribution claim by Mars, Inc. In what follows we formalize that discussion to deal with an arbitrary categorical population with any number of categories and hypothesized population proportions H_0 (10.22).

When the null hypothesis is true, we would expect that the proportion of sample observations falling in each of the categories would be relatively close to the corresponding population proportion prescribed in Expression (10.22). Thus, the number of sample observations that we would expect to observe in category C_i when the null hypothesis, H_0 (10.22), is true would be

$$E_i = Np_{i,0}, \quad \text{for } i = 1, \ldots, I. \tag{10.24}$$

Comparing these expected counts with the observed counts $O_1, \ldots, O_I$ through the Pearson goodness of fit statistic G (5.8), the test statistic for this setting is given by

$$Q_2 = \sum_{i=1}^{I} \frac{(O_i - E_i)^2}{E_i} = \sum_{i=1}^{I} \frac{\left(O_i - Np_{i,0}\right)^2}{Np_{i,0}}. \tag{10.25}$$

Evidence against the null hypothesis H_0 (10.22) in favor of the general alternative H_A (10.23) is provided by large values of the test statistic Q_2 (10.25). Once again, we can use an appropriate chi-square distribution to approximate the sampling distribution of Q_2. For this setting the appropriate chi-square degrees of freedom is $I - 1$ and the approximation is best if the sample size, N, is sufficiently large so that each of the expected counts E_i (10.24) is at least 5.

To test the null hypothesis that the population category proportions are as specified in H_0 (10.22) versus the general alternative H_A (10.23), let $q_{2(obs)}$ be the observed value of the chi-square statistic Q_2 (10.25). The approximate P-value for the test of H_0 versus H_A is then

$$P\text{–}value \approx P\left(\chi^2_{I-1} \geq q_{2(obs)}\right), \tag{10.26}$$

where χ^2_{I-1} has a chi-square distribution with $I - 1$ degrees of freedom.

Example 10.4. Are the Color Proportions Claimed by Mars, Inc. for M&M Plain Candy Correct? In Example 5.10 we discussed a situation involving the distribution of colors claimed by Mars, Inc. for their M&M's Plain candy. There we had $I = 6$ categories, corresponding to the colors brown (C_1), yellow (C_2), red (C_3), orange (C_4), green (C_5), and blue (C_6), and the production proportions of these colors claimed by Mars, Inc. are $p_{\text{brown}} = .3$, $p_{\text{yellow}} = p_{\text{red}} = .2$ and $p_{\text{orange}} = p_{\text{green}} = p_{\text{blue}} = .1$. The null hypothesis H_0 (10.22) corresponding to the company's claim for the M&M Plain color distribution is then given by.

$$H_0 : \left[p_{\text{brown}} = .3, p_{\text{yellow}} = p_{\text{red}} = .2 \text{ and } p_{\text{orange}} = p_{\text{green}} = p_{\text{blue}} = .1\right], \tag{10.27}$$

Table 10.22 Observed color counts for a bag of 800 Plain M&M candies

	Brown	Yellow	Red	Orange	Green	Blue
Observed counts	251	168	145	71	91	74

and we are interested in using the observed color counts from a sample of M&M Plain candy to test this null against the general alternative H_A (10.23) that at least one of the color production proportions for M&M Plain candy is different from those stipulated in (10.27).

Suppose we observe the color counts for a bag containing $N = 800$ Plain M&M's to be as specified in Table 10.22. From (10.24) we compute the expected color counts under the null hypothesis stipulation in (10.27) to be:

$$E_{\text{brown}} = 800(.3) = 240, \quad E_{\text{yellow}} = E_{\text{red}} = 800(.2) = 160,$$
$$E_{\text{orange}} = E_{\text{green}} = E_{\text{blue}} = 800(.1) = 80.$$

Combining these expected color counts, as stipulated by the goodness of fit null hypothesis H_0 (10.27), with the observed sample color counts given in Table 10.22, the value of the chi-square goodness of fit statistic Q_2 (10.25) for these M&M data is given by

$$\begin{aligned} Q_2 &= \sum_{i=1}^{6} \frac{(O_i - E_i)^2}{E_i} \\ &= \frac{(251-240)^2}{240} + \frac{(168-160)^2}{160} + \frac{(145-160)^2}{160} \\ &\quad + \frac{(71-80)^2}{80} + \frac{(91-80)^2}{80} + \frac{(74-80)^2}{80} \\ &= 0.504 + 0.4 + 1.406 + 1.013 + 1.513 + 0.45 = 5.286. \end{aligned}$$

Since there are $I = 6$ color categories, the null distribution of the goodness of fit test statistic Q_2 is approximately chi-square with $I - 1 = 5$ degrees of freedom.

Hence from Eq. (10.26) and the **R** function *pchisq*( ), the approximate *P*-value for these M&M Plain color data is $P\text{–}value \approx P(\chi_5^2 \geq 5.286) = .382$.

```
> pchisq(5.286, df = 5, lower.tail = FALSE)
[1] 0.3819824
```

Thus there is virtually no evidence in the observed color counts for our sample of 800 M&M Plain candies to warrant questioning the color proportions claimed by Mars, Inc.

Section 10.4 Practice Exercises

10.4.1. *Healthy Heart? Does the Answer Depend on Your Birth Month?* Cardiovascular disease, in particular acute myocardial infarction (AMI), is one of the major causes of death in developed countries. But is your susceptibility to this disease a function of which month you were born in? Stoupel et al. (2011) conducted a study of 22,047 patients diagnosed with AMI and admitted in the cardiology departments of the tertiary hospital at the Lithuanian University of Medical Sciences in Kaunas, Lithuania between the years 1990–2010. The month of birth for each of these patients is recorded in Table 10.23.

(a) It is natural *a priori* to assume that the month of birth should have no relationship to the incidence of AMI. State the null hypothesis that corresponds to this assumption.

(b) Construct the table of expected counts for each month if the null hypothesis in (a) is true.

(c) Assuming the data in Table 10.22 are representative of all patients with AMI, find the approximate *P*-value for an appropriate test of the null hypothesis in (a). What is your conclusion at significance level .045?

(d) What do you think could be the factor(s) that led to the result in part (c)? Do you think the results would be the same if we analyzed women and men separately? If you are curious, read more about it in Stoupel et al. (2011).

Table 10.23 Birth month for AMI patients, Kaunas, Lithuania

Month	Number of AMI patients born in month
January	2498
February	1934
March	2014
April	1893
May	1999
June	1783
July	1821
August	1694
September	1703
October	1601
November	1488
December	1619
Total	22,047

Source: Stoupel et al. (2011)

10.4.2. *Fair Dice.* Let X be the sum of the outcomes for the roll of two ordinary six-sided dice.

(a) What are the possible values for X?

(b) If the two dice are fair (i.e., the probability is $1/6$ for each of the numbers $1, 2, \ldots, 6$), what is the probability distribution for X?

(c) Our friend brings a new pair of six-sided dice for use in a board game and we decide to check if the dice are fair. What follows are the observed sums for 72 rolls of the two dice:

$8, 6, 8, 3, 5, 7, 8, 6, 5, 6, 4, 7, 9, 9, 6, 7, 6, 5, 4, 6, 3, 10, 8, 9, 9, 4, 9, 6, 7$
$7, 7, 4, 6, 3, 7, 7, 11, 10, 7, 8, 7, 9, 7, 10, 5, 6, 3, 10, 4, 6, 6, 8, 6, 7, 3, 4, 8,$
$7, 7, 7, 9, 6, 5, 4, 10, 6, 10, 6, 10, 9, 9, 3.$

(a) What is the null hypothesis that corresponds to the two dice being fair?

(b) Construct the table of expected outcomes if the two dice are fair.

(c) Construct the corresponding table of observed outcomes for the 72 dice rolls.

(d) Find the approximate *P*-value for an appropriate test of the null hypothesis that the two dice are fair. What is your conclusion at significance level .025?

10.4.3. *Random Number Generation.* When prescribing sampling schemes for data collection from a population, it is common to use random number generators to select which items of the population to include in the sample. Many algorithms are available in the literature and online for generation of the necessary random numbers. In such algorithms, each item in the population must have an equal probability of being included in the sample; that is, if the population size is N, then the random number generator must have probability $1/N$ of including each of the members of the population in the sample. The following set of 200 numbers were obtained from the set of integers {0, 1, 2, ..., 97, 98, 99} using a purported random number generator.

98	69	54	86	66	11	76	62	41	19	58	32
24	24	30	49	75	25	20	45	13	66	50	93
4	61	39	55	77	61	29	90	84	88	77	16
11	72	68	72	62	0	4	38	77	78	59	52
6	20	2	75	57	79	34	21	95	66	58	39
85	18	32	45	21	89	87	9	59	82	93	67
75	49	43	92	96	47	77	65	19	22	67	48
15	43	68	95	66	5	58	49	84	59	85	98
66	81	32	69	26	64	66	79	28	90	39	52
73	95	1	9	59	62	37	65	77	13	16	27
5	14	45	97	2	22	84	46	57	53	81	67
23	60	67	6	39	90	6	45	7	98	78	52
38	33	70	79	64	65	50	67	66	7	6	71
99	77	63	63	18	24	52	68	86	37	43	99
24	56	83	3	7	42	95	15	50	84	2	24
8	54	92	86	44	91	24	21	57	44	46	92
21	88	0	93	85	83	53	88				

(a) What is the null hypothesis that corresponds to these numbers being generated by an acceptable random number algorithm?
(b) Using each integer as its own category, how many of the generated numbers should we expect for each integer if the algorithm is a true random number generator?
(c) Construct the corresponding table of observed integer outcomes for the 200 generated random numbers.
(d) Find the approximate *P*-value for an appropriate test of the null hypothesis that the algorithm is a true random number generator. What is your conclusion at significance level .046?

10.4.4. *Random Number Generation—Approach Two.* Consider the integers generated by the purported random number generator given in Exercise 10.4.3.

(a) Using the interval categories [0,9), [10, 19), [20, 29), [30, 39), [40, 49), [50, 59), [60, 69), [70, 79), [80, 89), [90, 99), how many of the generated numbers should we expect for each of these ten categories if the algorithm is a true random number generator?
(b) Construct the corresponding table of observed category outcomes for the 200 generated random numbers.
(c) Find the approximate *P*-value for an appropriate test of the null hypothesis that the algorithm is a true random number generator. What is your conclusion at significance level .057?
(d) Compare this result with the result in Exercise 10.4.3.

10.4.5. *Random Number Generation—Approach Three.* Consider the integers generated by the purported random number generator given in Exercise 10.4.3.

(a) Using only the two categories of even or odd integer, how many of the generated numbers should we expect in the two categories if the algorithm is a true random number generator?

(b) Construct the corresponding table of observed category outcomes for the 200 generated random numbers.

(c) Find the approximate P-value for an appropriate test of the null hypothesis that the algorithm is a true random number generator. What is your conclusion at significance level .033?

(d) Compare this result with the results in Exercises 10.4.3 and 10.4.4.

10.4.6. *Random Number Generation—Approach Four—Does It Matter?* Consider the integers generated by the purported random number generator given in Exercise 10.4.3.

(a) Using only the two categories of a prime number or not a prime number, how many of the generated numbers should we expect in each of the categories if the algorithm is a true random number generator?

(b) Construct the corresponding table of observed category outcomes for the 200 generated random numbers.

(c) Find the approximate P-value for an appropriate test of the null hypothesis that the algorithm is a true random number generator. What is your conclusion at significance level .015?

(d) Compare this result with the results in Exercises 10.4.3, 10.4.4, and 10.4.5. Does the method of categorization matter?

Chapter 10 Comprehensive Exercises

10.A. Conceptual

10.A.1. Below are two questions of potential research interest. For each of these questions, describe the appropriate data to collect in order to address the question and state which procedure discussed in this chapter would provide the proper statistical analysis of these data.

Question 1: Does participation in intercollegiate athletics effect the length of time to graduation or even the graduation rate itself for a college student?

Question 2: Are the percentages of individuals between the ages of 16 and 25 who do not smoke at all, smoke occasionally (less than a pack a week), smoke regularly, but not heavily (between one and three packs a week), or are heavy smokers (more than three packs a week) the same for males and females?

10.A.2. *Degrees of Freedom.* There are k categories for the goodness of fit test procedure discussed in Sect. 4. However, the degrees of freedom for the chi-square approximation associated with the test statistic Q_2 (10.25) is only k-1, despite the fact that there are k terms (one for each category) in the sum for Q_2. Discuss an intuitive reason why the degrees of freedom for this chi-square approximation should not be k.

10.A.3. Below are two questions of potential research interest. For each of these questions, describe the appropriate data to collect in order to address the question and state which procedure discussed in this chapter would provide the proper statistical analysis of these data.

- Question 1: Is there any relationship between an individual's religious preference and her tolerance of the religious preferences of others?
- Question 2: Are there differences in drinking habits between college students who belong to sororities, fraternities or neither?

10.A.4. The test procedures discussed in Sects. 1 and 3 are both designed for settings where fixed numbers of sample observations are collected from at least two different populations. The question of interest for both procedures is whether certain categorical proportions are the same for the populations. Thus, under certain conditions they are competing test procedures.

(a) Specify the setting(s) where both procedures are applicable. What are the advantages and disadvantages of each procedure for such setting(s)?

(b) Which of the two procedures is more broadly applicable and why?

10.A.5. Both the test for differences in category proportions for two or more populations discussed in Sect. 1 and the test for association (independence) between two categorical attributes discussed in Sect. 2 use the same chi-square statistic Q_1 (10.7). Moreover, the two test procedures use the data in the observed $I \times J$ tables of counts in exactly the same way to compute the expected counts (Eqs. (10.6) and (10.16), respectively) when the appropriate null hypotheses are true. However, the ways in which the observed tables of counts are obtained is quite different for the two procedures. Discuss such data collection differences for the two settings, particularly with respect to the sampling methods for the two procedures and the interpretations of the column and row totals for the observed tables of counts.

10.A.6. *Degrees of Freedom.* Consider the test procedure discussed in Sect. 1 that is designed to test for differences in population proportions. If there are J populations and I categories, then we have IJ category-population cross entries in the observed data count Table 10.2. However, the degrees of freedom associated with the chi-square approximation for the associated test statistic Q_1 (10.7) is only (I-1)(J-1), despite the fact that there are IJ terms (one for each category-population combination) in the sum for Q_1. Discuss an intuitive reason why the degrees of freedom for this chi-square approximation should not be IJ.

10.A.7. *Lotteries.* In a lottery each number between 0 and 9 is designed to have the same chance of being drawn.

(a) In 6000 draws for the lottery, how many times should we expect the number 6 to appear? How many times should we expect the number 9 to appear?

(b) Suppose we make two-digit numbers from each consecutive pair of numbers drawn. For 6000 draws from the lottery, how many times should we expect the number 45 to appear? How many times should we expect that the two-digit number will be at least as large as 60?

(c) Suppose we make three-digit numbers from each consecutive triple of numbers drawn. For 6000 draws from the lottery, how many times should we expect the number 369 to appear? How many of the three-digit numbers should we expect to be less than 250?

10.B. Data Analysis/Computational

10.B.1. The following is a 5×4 table of observed counts collected from an experiment involving two categorical attributes, A and B:

		Category for attribute A			
		1	2	3	4
Category for attribute B	1	13	9	15	9
	2	11	5	8	10
	3	6	4	2	16
	4	8	9	10	5
	5	3	9	7	11

Find the following expected cross-category counts if there is no association between the two attributes (that is, they are independent).

(a) [expected count in category 1 for both attribute A and B]

(b) [expected count in category 3 for attribute A and category 2 for attribute B]

(c) [expected count in category 4 for attribute A]

(d) [expected count in category 3 for attribute B]

10.B.2. Consider the 5×4 table of observed counts in Exercise 10.B.1.

(a) Construct the corresponding table of expected cross-category counts if there is no association between the two attributes (that is, they are independent).

(b) Find the approximate P-value for an appropriate test of the hypothesis that there is no association between Attributes A and B (that is, they are independent).

10.B.3. *Lotteries.* In a fair lottery it is supposed to be equally likely to draw each integer number between 0 and 9, inclusive. Suppose we draw 500 such numbers using the lottery specified method and observe the following counts for the ten possible outcomes:

Number:	0	1	2	3	4	5	6	7	8	9
Observed Count:	60	42	35	88	50	32	70	55	40	28

(a) If the lottery is fair, what should the expected counts be for the ten possible outcomes?

(b) State the hypothesis that corresponds to the lottery being fair. Be explicit about all terms and numerical values.

(c) Using the observed counts given above, find an approximate P-value for an appropriate test for the fairness of this lottery system.

10.B.4. *Alcohol Consumption and Severity of Assault Injuries.* Shepherd et al. (1988) were interested in the possibility of a link between the amount of alcohol consumed by a victim of an assault and the severity of the injuries suffered as a result of the assault. For this purpose, they classified the severity of injuries from an assault into five categories:

- I = one hematoma or one laceration
- II = multiple hematomas or lacerations

Table 10.24 Numbers of treated patients with various combinations of severity of injury and level of alcohol consumption

	Severity of injury				
Level of alcohol consumption	**I**	**II**	**III**	**IV**	**V**
None	54	54	14	20	5
Light	49	96	9	28	17
Heavy	23	63	6	23	9

Source: Shepherd et al. (1988)

- III = one fracture
- IV = one fracture and hematomas and/or lacerations
- V = more than one fracture.

A victim's alcoholic consumption was categorized as: none, light (1–10 units), or heavy (> 10 units), where a unit of alcohol corresponded to either 1/2 pint of beer or lager, one measure of spirits, or one glass of wine. Following interviews and examinations of 470 consecutive victims of assault who came to an inner city accident and emergency service in 1986, they obtained the cross-categorized counts given in Table 10.24.

(a) State the null hypothesis of interest here.

(b) Construct the table of expected counts if the null hypothesis in (a) is true.

(c) Find the approximate *P*-value for an appropriate test of the null hypothesis in (a). What is your conclusion at significance level .05?

10.B.5. *Smoking and Hearing Loss.* Smoking has been linked to the occurrence of a number of serious diseases, including lung cancer, emphysema, and various forms of heart disease. However, cigarette smoking may also lead to increased deterioration of other health functions, especially those that tend to worsen with age anyway. In particular, clinical studies have suggested that cigarette smoking may be associated with accelerated hearing loss as a person ages. Cruickshanks et al. (1998) conducted an extensive population-based

Table 10.25 Prevalence of hearing loss for smoking/non smoking groups of subjects ages 48–59 in the Beaver Dam, Wisconsin study

Smoking behavior	Number of subjects	
	With hearing loss	Without hearing loss
Nonsmokers	86	448
Ex-smokers	101	344
Current smokers	66	189

Source: Cruickshanks et al. (1998)

study related to this issue. During the years 1993–1995, they gathered relevant information from questionnaires and examinations on over 3500 residents of the city/township of Beaver Dam, Wisconsin. For purposes of their study, Cruickshanks et al. defined a hearing loss to be a pure-tone average (*PTA*) of thresholds at 500, 1000, 2000, and 4000 *Hz* greater than 25-*bB* hearing level (*dB HL*) in the worse ear. In this exercise we concentrate on the relevant data for comparison of 534 nonsmokers (smoked fewer than 100 cigarettes in their lifetime), 445 ex-smokers, and 255 current smokers (at the time of the study) in the age group 48–59. The smoking related breakdown of the subjects in this age group who were diagnosed as having hearing losses is presented in Table 10.25.

(a) State the null hypothesis of interest here.

(b) Construct the table of expected counts if the null hypothesis in (a) is true.

(c) Find the approximate *P*-value for an appropriate test of the null hypothesis in (a) What is your conclusion at significance level .01?

10.B.6. ***Intensity of Smoking and Hearing Loss.*** In Exercise 10.B.5 we discussed a portion of the study by Cruickshanks et al. (1998) dealing with the overall effect of smoking on acceleration of hearing loss as one ages. They also collected data on the impact of the number of cigarettes and duration of smoking time on hearing. They defined the *total pack-years smoked* for a subject

Table 10.26 Prevalence of hearing loss by total pack-years smoked for subjects ages 48–59 in the Beaver Dam, Wisconsin study

Total pack-years	Number of subjects	
	With hearing loss	Without hearing loss
< 10	33	151
10 – 24	47	132
25 – 39	39	132
≥ 40	45	113

Source: Cruickshanks et al. (1998)

to be the number of cigarettes smoked per day divided by 20 cigarettes per pack, then multiplied by the number of years that the subject had smoked. In Table 10.26 are recorded the total pack-years smoked for each current or ex-smoker in the study, along with whether or not that subject was diagnosed as having a hearing loss.

(a) State the null hypothesis of interest here.

(b) Construct the table of expected counts if the null hypothesis in (a) is true.

(c) Find the approximate *P*-value for an appropriate test of the null hypothesis in (a). What is your conclusion at significance level .01?

10.B.7. *Smoking and Hearing Loss.* In Exercise 10.B.5 we discussed a portion of the study by Cruickshanks et al. (1998) dealing with the overall effect of smoking on acceleration of hearing loss as one ages. There we discussed the association between smoking behavior and hearing loss for subjects in the age group 48–59. Similar data for the age group 70–79 are given in Table 10.27.

Find the approximate *P*-value for a test of the hypothesis that smoking behavior and hearing loss are independent for this age grouping. Compare and contrast this result with your conclusion in Exercise 10.B.5.

10.B.8. *Intensity of Smoking and Hearing Loss.* In Exercise 10.B.6 we discussed a portion of the study by Cruickshanks et al. (1998) involving the overall

Table 10.27 Prevalence of hearing loss for smoking/non smoking groups for subjects ages 70–79 in the Beaver Dam, Wisconsin study

Smoking behavior	Number of subjects	
	With hearing loss	Without hearing loss
Nonsmokers	258	171
Ex-smokers	267	105
Current smokers	47	19

Source: Cruickshanks et al. (1998)

Table 10.28 Prevalence of hearing loss by total pack-years smoked for subjects ages 70–79 in the Beaver Dam, Wisconsin study

Total pack-years	Number of subjects	
	With hearing loss	Without hearing loss
< 10	77	30
10–24	66	33
25–39	57	21
≥ 40	106	37

Source: Cruickshanks et al. (1998)

relationship between smoking and hearing loss as one ages. There we discussed the association between total pack-years smoking and hearing loss for smokers in the age group 48–59. Similar data for the age group 70–79 are given in Table 10.28.

Find the approximate P-value for a test of the hypothesis that total pack-years smoking and hearing loss are independent for subjects in the age group 70–79. Compare and contrast this result with your conclusion in Exercise 10.B.6.

10.B.9. *M & M Colors.* Mars, Inc. claims that the color mix for M&M's Peanut candy is 20% brown, 20% yellow, 20% red, 10% orange, 10% green, and 20% blue. Suppose we observe the color counts for a bag containing $N = 750$ Peanut M&M's to be as specified in Table 10.29.

Table 10.29 Observed color counts for a bag of 750 Peanut M&M candies

	Brown	Yellow	Red	Orange	Green	Blue
Observed counts	168	145	190	83	62	102

(a) If the company's claim is correct, what would be the expected observed counts for the six colors?

(b) State the hypothesis that corresponds to the company's claim. Be explicit about all terms and numerical values.

(c) Using the observed counts given above, find an approximate P-value for an appropriate test of the company's claim. What is your conclusion for significance level .10?

10.B.10. *Flicker Squawks and Keos.* The common flicker, *colapres auratus*, has a diverse vocal repertoire. Flicker nestlings, however, produce only two distinct calls, squawk and keo. The keo is a common vocalization used by adult birds, both male and female, as well as by older nestlings to attract the parent(s) to the nest cavity to feed them (or to express agitation when they do not receive sufficient food from a parent!). In a number of bird species it is known that the vocalizations of the young change with time until a final innate template for the vocalization is achieved. Rosen (1979) conducted a study with flicker nestlings to see if the duration of their keo vocalizations changed as they matured. She observed the keo vocalizations for a group of flicker nestlings on four different days, corresponding to the nestlings being 17, 21, 22, and 24 days old. Using a Kay Electronic Company 6061B Sonograph, she made sonograms (visual representations -- plotting frequency of the sound on the ordinate versus time on the abscissa) of the keo vocalizations. The length (in *mm*) of the strip it creates on the sonogram represents the duration of an individual keo vocalization. The durations for the 71 keo vocalizations recorded by Rosen over the four days of observation are given in Table 10.30.

Table 10.30 Observed counts for lengths (in *mm*) of keo durations for flicker nestlings of various ages

Nestling age	Numbers of nestlings with keo durations of lengths (*mm*)			
	24–26	27–29	30–32	33–36
17 days	5	1	0	7
21 days	5	8	1	1
22 days	1	0	8	11
24 days	1	4	7	11

Source: Rosen (1979)

(a) State the null hypothesis of interest here.

(b) Construct the table of expected counts if the null hypothesis in (a) is true.

(c) Find the approximate *P*-value for an appropriate test of the null hypothesis in (a). What is your conclusion at significance level .03?

10.B.11. *Saving or Not?* Developing a systematic approach to saving is important in order to provide sufficient funds to support one's retirement years. Not surprisingly, people with higher income households tend to be better at this than those in lower income households simply because of the economics of the matter. What other factors might affect whether a person has taken this step toward systematically putting money away for her retirement? Princeton Survey Research Associates (1998) addressed a number of possible important factors, including gender, age, education, and race/ethnicity, in a nationwide survey conducted for the organization Americans Discuss Social Security. The results of their survey that relate to the age of the respondent are given in Table 10.31.

Find the approximate *P*-value for a test of the hypothesis that age and saving approach are independent attributes.

10.B.12. *Freshman Party Schools.* The public often perceives certain institutions of higher education as 'party schools'. However, are all such

Table 10.31 Survey frequencies of saving approaches among various age groups of non-retired adults

Age group	Number of respondents using saving approach		
	Systematic savers	Casual savers	Non-savers
18–34	49	136	118
35–49	90	151	80
≥ 50	80	131	46

Source: Princeton Survey Research Associates (1998)

Table 10.32 Numbers of college freshmen indicating that they partied at least 6 hours per week, categorized by type of four-year institution

Type of institution	Number partying at least 6 h per week
Public	11,304 out of 36,942 respondents
Nonsectarian private	12,088 out of 42,865 respondents
Protestant private	8104 out of 33,767 respondents
Catholic private	5966 out of 17,243 respondents

Source: Sax (1997)

institutions created equal with regard to their students' participation in such activities? Sax (1997) addressed a number of these and related issues in a comprehensive survey among college freshman in 1995. Respondents were asked whether or not they partied at least 6 h per week. The numbers of respondents from universities categorized as either public, nonsectarian private, Protestant private, or Catholic private who answered yes to this question are provided in Table 10.32. Were there differences among the types of four-year institutions with respect to partying by their freshmen in 1995?

(a) Formally state the null hypothesis of interest here. Be sure to clearly identify all relevant parameters.

(b) Construct the table of expected counts if the null hypothesis in (a) is true.

(c) Find the approximate P-value for an appropriate test of the null hypothesis in (a). What is your conclusion at significance level .075?

Table 10.33 Union soldiers rank and wartime mortality in the American Civil War

	Cause of wartime mortality		
	Illness	**Injury**	**Other**
Rank			
Private	335	75	10
Higher rank	15	11	1

Source: Lee (1999)

10.B.13. *Danger for Union Soldiers Based on Rank During the American Civil War*. The American Civil War was a deadly conflict for both Confederate and Union soldiers, but was it more deadly for some Union Army soldiers than for others? Lee (1999) investigated the pattern and causes of fatalities among Union soldiers by rank and placement in the battlefields. Based on a sample of 4295 recruits who enlisted in 45 companies organized in Ohio for whom information on both rank and duty was available, Lee compiled the wartime mortality data presented in Table 10.33 categorized by the soldiers' ranks.

(a) Formally state the null hypothesis of interest here.

(b) Construct the table of expected counts if the null hypothesis in (a) is true.

(c) Find the approximate P-value for an appropriate test of the null hypothesis in (a). What is your conclusion at significance level .061?

10.B.14. *Danger for Union Soldiers Based on Battlefield Placement During the American Civil War*. The American Civil War was a deadly conflict for both Confederate and Union soldiers, but was it more deadly for some Union Army soldiers than for others? Lee (1999) investigated the pattern and causes of fatalities among Union soldiers by rank and placement in the battlefields. Based on a sample of 4295 recruits who enlisted in 45 companies organized in Ohio for whom information on both rank and duty was available, Lee compiled the wartime mortality data presented in Table 10.34 categorized by the soldiers' battlefield placements.

Table 10.34 Union soldiers battlefield placement and wartime mortality in the American Civil War

	Cause of wartime mortality		
	Illness	**Injury**	**Other**
Battlefield placement			
Infantry	298	67	9
Non-infantry	52	19	2

Source: Lee (1999)

(a) Formally state the null hypothesis of interest here.

(b) Construct the table of expected counts if the null hypothesis in (a) is true.

(c) Find the approximate P-value for an appropriate test of the null hypothesis in (a). What is your conclusion at significance level .022?

10.B.15. *Danger for Union Soldiers Based on Combination of Rank and Battlefield Placement During the American Civil War*. The American Civil War was a deadly conflict for both Confederate and Union soldiers, but was it more deadly for some Union Army soldiers than for others? Lee (1999) investigated the pattern and causes of fatalities among Union soldiers by rank and placement in the battlefields. Based on a sample of 4295 recruits who enlisted in 45 companies organized in Ohio for whom information on both rank and duty was available, Lee compiled the wartime mortality data presented in Table 10.35 categorized by the combination of soldiers' ranks and battlefield placements.

(a) Formally state the null hypothesis of interest here.

(b) Construct the table of expected counts if the null hypothesis in (a) is true.

(c) Find the approximate P-value for an appropriate test of the null hypothesis in (a). What is your conclusion at significance level .061?

(d) Compare your findings with those in Exercises 10.B.13 and 10.B.14.

Table 10.35 Union soldiers rank and battlefield placement and wartime mortality in the American Civil War

	Cause of wartime mortality		
	Illness	**Injury**	**Other**
Rank/Battlefield Placement			
Private/infantry	285	60	8
Private/non-infantry	50	15	**2**
Higher rank/infantry	13	7	1
Higher rank/non-infantry	2	4	0

Source: Lee (1999)

Table 10.36 Participation in religious services and view of impact of science on society

	Impact of science on society		
	Mostly positive	**Mostly negative**	**No opinion**
Attend religious services			
Weekly or more	619	78	77
Monthly/yearly	570	33	67
Seldom/never	463	21	48

Source: Pew Research Center (2009)

10.B.16. *Science and Religion.* Does the degree of a person's religious participation affect their outlook on the impact of science on society? The Pew Research Center (2009) asked 1976 adults, 18 years of age or older, about their participation in religious services and whether they saw science as having a mostly positive or mostly negative effect on society. The results of the survey are given in Table 10.36.

(a) Formally state the null hypothesis of interest here.

(b) Construct the table of expected counts if the null hypothesis in (a) is true.

(c) Find the approximate *P*-value for an appropriate test of the null hypothesis in (a). What is your conclusion at significance level .025?

10.C. Activities

10.C.1. *Hair Color and Educational Level.* Discuss how to design an experiment to ascertain if there is any relationship between (true!) hair color and the highest educational degree a person attains. Specify both the numbers and types of categories for each of these attributes.

(a) State the hypothesis of interest here.

(b) Collect the necessary data from 20 students, 20 non-faculty employees and 20 faculty members at your university and test the hypothesis in (a).

10.C.2. *Hair Color and Educational Level.* In Question 10.C.1 you were asked to collect data from 20 students, 20 non-faculty employees and 20 faculty members at your university to assess whether there is any relationship between true hair color and a person's highest educational degree. Couldn't you have addressed the same issue more easily by simply sampling 60 students (or 60 staff employees or 60 faculty members)? Why or why not?

10.C.3. *Who Studies More?* Do freshmen, sophomore, junior, and senior college students spend about the same amount of time studying (on average) per week?

(a) Discuss how to design an experiment to address this question.

(b) State the hypothesis of interest here.

(c) Conduct the experiment and use your data to test the hypothesis in (b).

10.C.4. *Bridge.* In the card game of bridge, an ordinary deck of 52 cards is dealt to four players, each receiving a 13-card hand. One of the important features of the game of bridge is the number of honor cards (ace, king, queen, jack, and ten) that an individual holds in her hand.

(a) If a bridge hand is dealt at random to a player, what is the probability that the hand will not contain any honor cards? one honor card? two honor cards? more than two honor cards?

(b) Using your answer to (a), state an appropriate null hypothesis regarding honor cards that corresponds to dealing a 13-card hand at random from an ordinary deck of 52 cards.

(c) Using an ordinary (shuffled) deck of 52 cards, deal a hand of 13 cards and record whether the hand contains zero, one, two, or more than two honor cards.

(d) Repeat the experiment in (c) 160 times, with reshuffling between the dealing of each hand, and use the obtained counts to test the fairness of your deals.

10.C.5. *Bridge Again.* In part (d) of Exercise 10.C.4 you were asked to deal 160 *separate* 13-card hands from an ordinary deck of 52 cards, with reshuffling between the dealing of each hand. It would have been much easier to simply deal four complete 13-card hands each time you shuffled the deck. You would then have had to reshuffle the deck only 40 times, rather than the 160 required in part (d) of Exercise 10.C.4.

(a) Discuss why this proposed short-cut method is not equivalent to the more lengthy approach of part (d) of Exercise 10.C.4.

(b) Repeat part (d) of Exercise 10.C.4, but this time deal all four 13-card hands each time you shuffle the deck and repeat the complete process only 40 times. Record the number of hands that contain zero, one, two, or more than two honor cards.

(c) Compare the table of counts you obtained in part (b) with the table of counts you found in part (d) of Exercise 10.C.4. Are there substantial differences in the two tables? Discuss your finding.

10.C.6. *Religious Preference and Political Affiliation.* Is there a relationship between religious preference and political affiliation?

(a) Using five religious categories and three political preferences, discuss how to design an experiment to address this question.
(b) State the hypothesis of interest here.
(c) Conduct the experiment and use your data to test the hypothesis in (b).

10.C.7. *Fair Die?* Roll a six-sided die 120 times and record the frequencies with which each of the numbers $1, 2, 3, 4, 5$, and 6 occur. Using your data, find the *P*-value for a test of the hypothesis that the die is fair (i.e., that the outcomes are equally likely).

10.C.8. *Does Your Life Expectancy Depend on Your Month of Birth?* Doblhammer and Vaupel (2001) studied the relationship between month of birth and adult life expectancy. They concluded that people born in the Northern Hemisphere in autumn (October–December) live longer than those born in spring (April–June), but that the opposite is true in the Southern Hemisphere. Suppose you were asked to provide the statistical support for these conclusions.
(a) State the null and alternative hypotheses of interest here.
(b) What data would you need to test the null hypothesis?
(c) How would you design an experiment to collect the necessary data?
(d) Choose one of the states in the United States to serve as the data source and use public birth and death records to collect a small sample of the needed data.
(e) Using your sample data, obtain the approximate *P*-value for an appropriate test of the null hypothesis.

10.D. Internet Archives

10.D.1. *Original Skittles Flavors.* In Exercise 10.B.9 we discussed the proportions of the various colors of Peanut M & M's manufactured by

Mars, Inc. Skittles is another popular candy brand produced and marketed by the Wrigley Company, a division of Mars, Inc. Search the Internet to discover what flavors (colors) make up the original Skittles and in what proportions the Wrigley Company claims they are produced? Buy ten individual packages of the original Skittles and count the numbers of pieces of each color in these ten packages combined. Using these counts, test the hypothesis that the flavor proportions claimed by the Wrigley Company are correct—then you can enjoy your Skittles!

10.D.2. *Tropical Skittles Flavors.* Repeat Activity 10.D.1 for the variety of Tropical Skittles.

10.D.3. *Starburst Flavors.* Starburst is another candy brand produced and marketed by the Wrigley Company, a division of Mars, Inc. Search the Internet to discover what flavors make up the original Starburst and in what proportion the Wrigley Company claims they are produced? Buy ten individual packages of the original Starburst and count the numbers of pieces of each color in these ten packages combined. Using these counts, test the hypothesis that the flavor proportions claimed by the Wrigley Company are correct—then you can enjoy your Starburst!

10.D.4. *Freshman Party Schools—Updated.* In Exercise 10.B.12 you were asked to assess if there were any differences in 1995 among the types of four-year institutions with respect to partying by their freshmen. Search the Internet to find a published article that addresses a similar question for a more recent year. Discuss the findings from this update and compare it to the results for 1995.

10.D.5. Search the Internet to find a journal article that reports on a research study in which the data collected were used to test for differences in population proportions, as discussed in Sect. 1. Prepare a brief summary of the study and the associated statistical analyses carried out by the authors.

10.D.6. Search the Internet to find a journal article that reports on a research study in which the data collected were used to test for association (independence) between two categorical attributes, as discussed in Sect. 2. Prepare a brief summary of the study and the associated statistical analyses carried out by the authors.

10.D.7. Search the Internet to find a journal article that reports on a research study in which the data collected were used to test for goodness-of-fit for probabilities in a multinomial distribution with $I > 2$ categories, as described in Sect. 4. Prepare a brief summary of the study and the associated statistical analyses carried out by the authors.

10.D.8. Search the Internet to find a journal article that reports on a research study in which the data collected were used to test for differences in two population proportions, as described in Sect. 3. Prepare a brief summary of the study and the associated statistical analyses carried out by the authors. If they used the approximate test procedure discussed in Sect. 1, repeat their analyses using the exact test procedure presented in Sect. 3. Compare the results of the exact and approximate tests.

10.D.9. Gallup, Inc., is an American research-based global performance-management consulting company that "provides data-driven news based on U. S. and world polls, daily tracking and pubic opinion research". Their website www.gallup.com contains information about current and past public opinions on education, politics, the economy, and wellbeing, as well as other topics. Go to their website and find a report on a topic of interest to you that involves categorical data as discussed in this chapter. Prepare a short summary of Gallup's sampling methods, data collection, and statistical analyses as described in the report.

10.D.10. The Pew Research Center is a nonpartisan American organization that uses "public opinion polling, demographic research, content analysis, and other empirical social science research" to inform the public about "the issues, attitudes and trends shaping America and the world". Their interests include U. S. politics and policy; Internet, science and technology; religion and public life; and social and demographic trends. The website www.pewresearch.org contains past and current information they have gathered about these topics, as well as others. Go to their website and find a report on a topic of interest to you that involves categorical data as discussed in this chapter. Prepare a short summary of the Pew Research Center's sampling methods, data collection, and statistical analyses as described in the report.

11 Statistical Inference for Bivariate Populations

Many questions of interest (both in research and in applications) involve the relationship between two simultaneously collected variables (bivariate observations). For example, is there a relationship between the size of alumni donations to the general fund of a university and the performance of its basketball and football teams? How does the amount of annual rainfall affect the wheat yield in the United States? Does the amount of fracking wastewater injected into deep wells have an effect on the number and severity of earthquakes in the region? Is there any relationship between CO_2 production and sea levels? How does a prescribed diet-medication regimen affect blood pressure levels in subjects with severe high blood pressures? Is there any relationship between pine needle length and diameter of a pine tree? Does smoking or excessive drinking have an impact on mortality? Problems such as these are addressed statistically through the use of correlation or regression analyses.

© Springer International Publishing AG 2017
D.A. Wolfe, G. Schneider, *Intuitive Introductory Statistics*, Springer Texts in Statistics,
DOI 10.1007/978-3-319-56072-4_11

In Chap. 2 we focused on graphical and numerical methods to describe and display the relationship between two numerical or categorical variables. In this chapter we expand our understanding of such relationships by discussing statistical inference procedures for bivariate populations.

General Setting and Notation Let $(X_1, Y_1), \ldots, (X_n, Y_n)$ denote the items of a random sample from a bivariate population. The most appropriate statistical procedures for assessing the relationship between the X and Y variables depends on what information is available or can reasonably be assumed about the form of the underlying bivariate population.

Section 1 discusses correlation procedures for bivariate normal populations, while Sect. 2 is devoted to competitor rank-based correlation procedures that do not require bivariate normality. Section 3 introduces the least squares methodology for fitting straight lines to bivariate data. Section 4 presents linear regression inference methods based on least squares for bivariate normal populations, and Sect. 5 provides linear regression inference procedures that do not require bivariate normality.

11.1 Correlation Procedures for Bivariate Normal Populations

In this section we assume that the underlying population for our data $(X_1, Y_1), \ldots, (X_n, Y_n)$ follows what is known as a bivariate normal distribution. In particular, this assumption implies (among other things) that each of the individual variables X and Y follow univariate normal distributions as well. Moreover, for this bivariate normal distribution, the two random variables X and Y are independent (no relationship) if and only if they are uncorrelated. Thus to assess the strength of a relationship between two bivariate normal variables X and Y we can concentrate directly on the correlation between them.

In Sect. 2.2 we introduced the Pearson correlation coefficient r (2.1) as a measure of the association between two continuous random variables X and Y. In particular, this statistic can be used to assess the potential correlation between pairs of variables that follow a bivariate normal distribution. Applying our formula for r (2.1) to the data in our random sample $(X_1, Y_1), \ldots, (X_n, Y_n)$ leads to the Pearson sample correlation coefficient

$$R = \frac{\sum_{i=1}^{n}(X_i - \bar{X})(Y_i - \bar{Y})}{\sqrt{\sum_{i=1}^{n}(X_i - \bar{X})^2 \sum_{j=1}^{n}(Y_j - \bar{Y})^2}} = \frac{n\sum_{i=1}^{n} X_i Y_i - \left(\sum_{j=1}^{n} X_j\right)\left(\sum_{k=1}^{n} Y_k\right)}{\sqrt{n\sum_{i=1}^{n} X_i^2 - \left(\sum_{j=1}^{n} X_j\right)^2}\sqrt{n\sum_{i=1}^{n} Y_i^2 - \left(\sum_{j=1}^{n} Y_j\right)^2}}. \tag{11.1}$$

Recalling our discussion about the Pearson correlation coefficient from Chap. 2, we are reminded that $-1 \leq R \leq 1$ and large positive (negative) values of R near 1 (near -1) correspond to positive (negative) association between the X and Y variables. Values of R near zero are indicative of little or no association between the variables, which, under the bivariate normality assumption, corresponds to independence between X and Y. Other useful properties of R are discussed in Sect. 2.2.1 and Exercise 2.A.4.

To use the sample data $(X_1, Y_1), \ldots, (X_n, Y_n)$ to test hypotheses about the relationship between the variables X and Y we first compute the test statistic T given by

$$T = \frac{R\sqrt{n-2}}{\sqrt{1-R^2}}. \tag{11.2}$$

The sampling distribution for T when X and Y are independent is a t-distribution with $n - 2$ degrees of freedom. Moreover, we note that large (small) values of T correspond to large (small) values of R and values of T near zero correspond to values of R near zero. Thus P-values for hypothesis tests about the degree of association (correlation) between X and Y can be obtained from this sampling distribution for T when X and Y are independent.

Hypothesis Tests for the Independence of Bivariate Normal Variables To test the null hypothesis H_0: (X and Y are independent) with data $(X_1, Y_1), \ldots, (X_n, Y_n)$ from a bivariate normal population, compute the statistic T (11.2) and let t^* be the attained value of T. Then the P-values for a test of H_0 against the alternatives H_A are:

H_A	P-value	
X and Y positively correlated	$= P(T \geq t^*)$	(11.3)
X and Y negatively correlated	$= P(T \leq t^*)$	(11.4)
X and Y are not independent	$= 2P(T \geq \lvert t^* \rvert)$,	(11.5)

where $T \sim t(n\text{-}2)$, the t-distribution with $n - 2$ degrees of freedom.

Example 11.1. Swinging for Power—All or Nothing? Home runs can be a major contributor to success for major league baseball teams, but they often come at the expense of a substantial number of whiffs (strikeouts) as well. Table 11.1 contains the 2016 home run and strikeout statistics for a subset of major league players

We are interested in using these data to test the conjecture that swinging for power can lead to strikeouts, corresponding to the statistical statement that the number of home runs and the number of strikeouts are positively correlated. Letting X denote the number of home runs and Y the number of strikeouts, we summarize the 15 data pairs in Table 11.1 as follows:

$$\sum_{i=1}^{15} X_i = 496 \quad \sum_{i=1}^{15} X_i^2 = 16,848 \quad \sum_{i=1}^{15} Y_i = 1787 \quad \sum_{i=1}^{15} Y_i^2 = 232,543$$

$$\sum_{i=1}^{15} X_i Y_i = 60,007.$$

Table 11.1 2016 home run and strikeout statistics for a subset of major league players

Player	Number of home runs	Number of strikeouts
Mookie Betts	31	80
Robinson Cano	39	100
Yoenis Cespedes	31	108
Chris Bryant	39	154
Andrew McCutchen	24	143
Miguel Cabrera	38	116
Albert Pujols	31	75
Todd Frazier	40	163
Ryan Braun	30	98
Charlie Blackmon	29	102
Josh Donaldson	37	119
Chris Davis	38	219
Justin Turner	27	107
David Ortiz	38	86
Bryce Harper	24	117

Source: rotowire.com (2016)

Using the computational form of the expression for R (11.1), we see that

$$R = \frac{15(60,007) - (496)(1787)}{\sqrt{15(16,848) - (496)^2}\sqrt{15(232,543) - (1787)^2}} = \frac{13,753}{\sqrt{6704}\sqrt{294,776}}$$
$$= 0.3094.$$

Thus the value of T (11.2) becomes

$$T = \frac{.3094\sqrt{15-2}}{\sqrt{1-(.3094)^2}} = \frac{1.1156}{.9509} = 1.1732.$$

It follows from (11.3) that the P-value for testing the conjecture that the number of home runs and the number of strikeouts are positively correlated is given by $P(T \geq 1.1732)$, where T has a t-distribution with $n - 2 = 13$ degrees of freedom. We can find this P-value to be roughly 0.13 using the **R** function *pt*( ) as follows.

```
> pt(q = 1.1732, df = 13, lower.tail = FALSE)
[1] 0.1308819
```

Hence, there is not sufficient evidence to conclude that major league home runs and strikeouts are positively correlated.

It is important to emphasize two features of the test procedures based on the Pearson correlation coefficient R. First, they require that the underlying bivariate data follow a normal distribution. Properties of the test procedures in (11.3), (11.4), and (11.5) can be seriously affected if the joint distribution for the bivariate data differs significantly from the bivariate normal distribution. In Sect. 2 we consider rank-based correlation procedures that are appropriate even when the underlying distribution is not bivariate normal. Second, it is important to note that the test procedures in (11.3), (11.4), and (11.5) are specifically designed to detect only linear relationships between the two variables X and Y. They can be quite ineffective at detecting non-linear relationships between two variables. For the bivariate normal distribution, zero correlation between the two variables X and Y is equivalent to X and Y being independent. This is not the case for other bivariate distributions, as you are asked to consider in Exercises 11.A.5 and 11.A.6.

Section 11.1 Practice Exercises

11.1.1. Consider the following set of $n = 20$ (x, y) bivariate observations.

$(11, 78), (2, 88), (-2, 100), (-11, 83), (-5, 100), (2, 90), (-6, 87), (22, 82), (21, 92), (8, 90), (25, 85), (9, 93), (7, 92), (8, 96), (18, 100), (-14, 96), (-21, 86), (-26, 89), (-7, 93), (5, 80).$

Compute the value of the Pearson sample correlation coefficient R (11.1). What does this say about the relationship between the x and y observations?

11.1.2. Consider the following set of $n = 12$ (x, y) bivariate observations.

$$(1.3, 8), (1.8, 6.9), (0.9, 8.1), (1.6, 7), (2.6, 6.3), (1.5, 6.5),$$
$$(2.1, 6.4), (3, 5.8), (0.8, 8.3), (2.4, 8.3), (2.5, 6.6), (2.6, 6.6).$$

Compute the value of the Pearson sample correlation coefficient R (11.1). What does this say about the relationship between the x and y observations?

11.1.3. Consider the following set of $n = 8$ (x, y) bivariate observations.

$$(1, 1), (2, 4), (3, 9), (4, 16), (5, 25), (6, 36), (7, 49), (8, 64).$$

Compute the value of the Pearson sample correlation coefficient R (11.1). What does this say about the relationship between the x and y observations?

11.1.4. *Does Concurrent Temperature Affect Roost Time Departures for Snow Geese?* Wildlife science often nvolves trying to understand how environmental conditions affect wildlife habits. Freund et al. (2010) report data on such a study to assess how a variety of environmental conditions influence the time that lesser snow geese leave their overnight roost sites to fly to their feeding areas. In particular, they were interested in the effect that the air temperature might have on the time that the geese leave their roosts. Table 11.2 contains the times of roost departure in minutes before (−) or after (+) sunrise and the concurrent air temperature in degrees Centigrade for 36 different observation days in the 1987–1988 winter season.

(a) Assuming bivariate normality for the data, use the procedure based on the Pearson sample correlation coefficient to find the P-value for a test of the conjecture that departure time and concurrent temperature are not independent variables.

(b) Are there any concerns about the bivariate normality assumption in part (a)? Explain.

11.1.5. *Birth and Death Rates for the United States.* Birth and death rates play major roles over time (in addition to emigration and immigration rates) in the

Table 11.2 Snow geese roost departure times in minutes before (–) or after (+) sunset and concurrent air temperatures in degrees centigrade (°C)

Date	Departure time	Concurrent temperature (°C)
11/10/87	11	11
11/13/87	2	11
11/14/87	−2	11
11/15/87	−11	20
11/17/87	−5	8
11/18/87	2	12
11/21/87	−6	6
11/22/87	22	18
11/23/87	22	19
11/25/87	21	21
11/30/87	8	10
12/05/87	25	18
12/14/87	9	20
12/18/87	7	14
12/24/87	8	19
12/26/87	18	13
12/27/87	−14	3
12/28/87	−21	4
12/30/87	−26	3
12/31/87	−7	15
01/02/88	−15	15
01/03/88	−6	6
01/05/88	−14	2
01/07/88	−8	2
01/08/88	−19	0
01/10/88	−23	−4
01/11/88	−11	−2
01/12/88	5	5
01/14/88	−23	5
01/15/88	−7	8
01/16/88	9	15
01/20/88	−27	5
01/21/88	−24	−1
01/22/88	−29	−2
01/23/88	−19	3
01/24/88	−9	6

Source: Freund et al. (2010)

population stability of cities, states, and countries. One would expect (assuming emigration and immigration rates are not major contributors) that the population stability of such entities would be dependent on the relationship between their birth and death rates. Table 11.3 contains the birth and death rates per 1000 population for each of the 50 states in the United States.

(a) Assuming bivariate normality for the data, use the procedure based on the Pearson sample correlation coefficient to find the P-value for a test of the hypothesis that statewide birth rate and death rate are independent variables. Comment on the result.

(b) The overall birth rate for the entire United States in 2006 was 14.3 per 1000 population. Find the five states with the highest birth rates and, separately, the five states with the lowest birth rates. Comment on your findings.

(c) The overall death rate for the entire United States in 2006 was 8.1 per 1000 population. Find the six states with the highest death rates and, separately, the five states with the lowest death rates. Comment on your findings.

11.1.6. *How Effective is State Spending on Secondary Education?* Spending for secondary education is always a matter of concern for state legislatures across the United States. Merline (1991) used data from the Department of Education, National Center for Education Statistics in assessing the relationship between the amount of money spent on secondary education and various performance criteria for high-school seniors. Table 11.4 contains the spending ($) per high-school senior and the percentage of those seniors that graduated for each of the 50 states in the 1987–1988 school year.

(a) Assuming bivariate normality for the data, use the procedure based on the Pearson sample correlation coefficient to find the P-value for a test of the conjecture that spending level per pupil and high-school graduation rate are positively related. Comment on the result.

Table 11.3 Birth and death rates per 1000 population in 2006 for the 50 states in the United States

State	Birth rate	Death rate
Alabama	13.8	10.2
Alaska	16.2	5.0
Arizona	16.5	7.5
Arkansas	14.6	9.9
California	15.6	6.5
Colorado	14.9	6.2
Connecticut	12.0	8.4
Delaware	14.1	8.4
Florida	13.1	9.4
Georgia	15.9	7.3
Hawaii	14.9	7.4
Idaho	16.5	7.2
Illinois	14.2	8.0
Indiana	14.1	8.8
Iowa	13.7	9.2
Kansas	14.9	8.9
Kentucky	13.8	9.5
Louisiana	14.9	9.4
Maine	10.8	9.3
Maryland	13.8	7.8
Massachusetts	12.0	8.3
Michigan	12.6	8.5
Minnesota	14.3	7.2
Mississippi	15.9	9.9
Missouri	13.9	9.4
Montana	13.2	8.9
Nebraska	15.2	8.4
Nevada	16.1	7.6
New Hampshire	11.0	7.7
New Jersey	13.3	8.1
New Mexico	15.4	7.9
New York	12.9	7.7
North Carolina	14.4	8.4
North Dakota	13.5	9.2
Ohio	13.1	9.3
Oklahoma	15.1	9.9
Oregon	13.2	8.5

(continued)

Table 11.3 (continued)

State	Birth rate	Death rate
Pennsylvania	12.0	10.1
Rhode Island	11.7	9.1
South Carolina	14.3	9.0
South Dakota	15.1	9.0
Tennessee	13.9	9.4
Texas	17.1	6.7
Utah	20.7	5.3
Vermont	10.5	8.1
Virginia	14.1	7.6
Washington	13.6	7.2
West Virginia	11.6	11.4
Wisconsin	13.0	8.3
Wyoming	15.0	8.4

Source: United States National Center for Health Statistics, National Vital Statistics Reports (NVSR) (2009)

(b) Find the nine states with graduation rates greater than 80% and the nine states with the lowest graduation rates. Comment on factors other than spending level per pupil that might contribute to these clusters.

11.1.7. *County Birth and Death Rates*. In Exercise 11.1.5 we considered the possible relationship between statewide birth and death rates across the United States. It is also of interest to see if this relationship occurs across counties within a given state. Table 11.5 provides the 2004 birth and death rates per 1000 population for the 114 counties in the State of Missouri.

Assuming bivariate normality for the data, use the procedure based on the Pearson sample correlation coefficient to find the *P*-value for a test of the hypothesis that Missouri county birth rate and death rate are independent variables.

Table 11.4 Spending per high-school senior and the percentage of those seniors who graduated during the 1987–1988 school year

State	$ per pupil	% Graduated
Alaska	7971	65.5
New York	7151	62.3
New Jersey	6564	77.4
Connecticut	6230	84.9
Massachusetts	5471	74.4
Rhode Island	5329	69.8
Vermont	5207	78.7
Maryland	5201	74.1
Wyoming	5051	88.3
Delaware	5017	71.7
Pennsylvania	4989	78.4
Oregon	4789	73.0
Wisconsin	4747	84.9
Michigan	4692	73.6
Colorado	4462	74.7
New Hampshire	4457	74.1
Minnesota	4386	90.9
Illinois	4369	75.6
Maine	4246	74.4
Montana	4246	87.3
Washington	4164	77.1
Virginia	4149	71.6
Iowa	4124	85.8
Florida	4092	58.0
Kansas	4076	80.2
Ohio	3998	79.6
Nebraska	3943	85.4
Hawaii	3919	69.1
West Virginia	3858	77.3
California	3840	65.9
Indiana	3794	76.3
Missouri	3786	74.0
Arizona	3744	61.1
New Mexico	3691	71.9
Nevada	3623	75.8
Texas	3608	65.3
North Dakota	3519	88.3

(continued)

Table 11.4 (continued)

State	$ per pupil	% Graduated
Georgia	3434	61.0
South Carolina	3408	64.6
North Carolina	3368	66.7
South Dakota	3249	79.6
Louisiana	3138	61.4
Oklahoma	3093	71.7
Tennessee	3068	69.3
Kentucky	3011	69.0
Arkansas	2989	77.2
Alabama	2718	74.9
Idaho	2667	75.4
Mississippi	2548	66.9
Utah	2454	79.4

Source: Merline (1991)

11.2 Rank-Based Correlation Procedures

As noted in the previous section, the correlation procedures based on the Pearson correlation coefficient R can be ineffective at detecting relationships between the variables (X, Y) when they do not have a bivariate normal distribution. Fortunately, however, procedures based on the rankings of X_i among $X_1, \ldots, X_n$, for $i = 1, \ldots, n$ and the separate rankings of Y_j among $(Y_1, \ldots, Y_n)$ can be used to develop rank-based correlation procedures that are effective at detecting associations between the variables X and Y even when they do not follow a bivariate normal distribution.

In Sect. 2.2.2 we introduced the Spearman rank correlation coefficient r_S (2.3) as a measure of the association between two random variables X and Y based on these rankings. Applying this formula for r_S (2.3) to the data in our random sample $(X_1, Y_1), \ldots, (X_n, Y_n)$ leads to the Spearman sample correlation coefficient

Table 11.5 Birth and death rates per 1000 population in 2004 for the 114 counties in the state of Missouri

County	Birth rate	Death rate	County	Birth rate	Death rate
1	10.9	8.5	58	11.5	13.2
2	10.5	10.6	59	12.0	13.1
3	9.3	13.6	60	16.4	9.2
4	14.6	10.8	61	13.1	13.1
5	14.8	10.6	62	13.2	12.2
6	12.1	11.1	63	11.7	9.5
7	13.1	12.4	64	13.3	12.4
8	9.3	13.7	65	12.1	11.8
9	11.4	12.9	66	13.9	10.7
10	13.8	5.8	67	15.9	14.3
11	13.6	10.9	68	14.9	10.2
12	14.0	12.3	69	10.5	9.7
13	12.1	10.8	70	13.9	11.5
14	12.6	9.0	71	11.9	12.3
15	10.2	10.9	72	13.5	12.2
16	12.9	9.7	73	13.4	9.2
17	10.2	13.0	74	11.5	8.9
18	14.7	12.2	75	10.9	14.0
19	13.5	7.6	76	13.6	8.0
20	12.4	12.9	77	11.0	14.4
21	11.6	14.1	78	15.5	11.2
22	14.8	7.3	79	14.4	9.3
23	12.0	12.9	80	14.8	10.2
24	14.6	7.0	81	12.5	9.8
25	12.0	11.7	82	12.7	11.5
26	14.1	8.1	83	13.3	6.0
27	10.9	10.4	84	13.9	10.0
28	13.2	10.4	85	14.6	5.9
29	12.1	17.5	86	12.6	14.0
30	14.4	11.0	87	12.2	9.1
31	14.2	11.4	88	14.0	9.8
32	8.8	9.6	89	11.1	11.5
33	11.8	11.6	90	9.3	12.3
34	11.2	11.3	91	14.0	12.5
35	14.2	13.7	92	14.1	5.8
36	13.4	8.7	93	9.9	14.9
37	10.4	13.3	94	9.8	7.8

(continued)

Table 11.5 (continued)

County	Birth rate	Death rate	County	Birth rate	Death rate
38	13.3	15.4	95	12.5	11.0
39	13.2	9.3	96	12.1	9.3
40	13.7	12.1	97	12.9	14.2
41	12.8	13.9	98	11.4	13.0
42	12.7	13.0	99	13.8	10.2
43	8.0	15.4	100	14.1	9.0
44	11.5	11.7	101	13.1	11.4
45	11.4	11.3	102	14.0	13.9
46	13.9	12.4	103	13.1	12.2
47	13.8	15.4	104	9.2	10.1
48	15.6	9.0	105	16.2	13.6
49	15.6	10.0	106	13.5	10.4
50	13.1	7.3	107	12.2	12.7
51	14.4	7.0	108	14.1	11.4
52	11.4	10.2	109	13.9	9.4
53	14.4	10.2	110	12.4	10.8
54	12.5	10.3	111	10.3	16.1
55	13.0	10.8	112	14.8	7.9
56	10.7	11.2	113	13.8	12.5
57	15.1	8.0	114	14.7	11.1

Source: United States Census Bureau (2010)

$$R_S = \frac{\sum_{i=1}^{n}\left(R_i - \bar{R}\right)\left(S_i - \bar{S}\right)}{\sqrt{\sum_{i=1}^{n}\left(R_i - \bar{R}\right)^2 \sum_{j=1}^{n}\left(S_j - \bar{S}\right)^2}} = \frac{12\sum_{i=1}^{n}\left\{\left(R_i - \frac{n+1}{2}\right)\left(S_i - \frac{n+1}{2}\right)\right\}}{n(n^2-1)}, \qquad (11.6)$$

where $\bar{R} = \bar{S} = \dfrac{n+1}{2}$ and $\sum_{i=1}^{n}\left(R_i - \bar{R}\right)^2 = \sum_{j=1}^{n}\left(S_j - \bar{S}\right)^2 = \dfrac{n(n^2-1)}{12}$, as you are asked to show in Exercise 11.A.1. (In case of ties among the X's or separately ties among the Y's, assign the average of the involved ranks to each of the tied observations.) Note that R_S can also be obtained by using the computationally simpler expression (see Exercise 11.A.3)

$$R_S = 1 - \frac{6\sum_{i=1}^{n} D_i^2}{n(n^2-1)}, \tag{11.7}$$

where $D_i = S_i - R_i$, for $i = 1, \ldots, n$.

As with the Pearson correlation coefficient, $-1 \leq R_S \leq 1$, and large positive (negative) values of R_S near 1 (near -1) correspond to positive (negative) association between the X and Y variables. Values of R_S near zero are indicative of little or no association between the variables, corresponding to independence between X and Y. The sampling distribution of R_S when X and Y are independent can be used to obtain P-values for hypothesis tests about the degree of association (correlation) between X and Y.

Rank-Based Hypothesis Tests for the Independence of Bivariate Variables To test the null hypothesis H_0: (X and Y are independent) with data (X_1, Y_1) , ... , (X_n, Y_n) from a bivariate population, compute the Spearman sample correlation coefficient R_S (11.6) and let r_S^* be the attained value of R_S. Then the P-values for a test of H_0 against the alternatives H_A are:

H_A	P-value	
X and Y are positively correlated	$= P_0(R_S \geq r_S^*)$	(11.8)
X and Y are negatively correlated	$= P_0(R_S \leq r_S^*)$	(11.9)
X and Y are not independent	$= 2P_0(R_S \geq \lvert r_S^* \rvert)$,	(11.10)

where $P_0(R_S \geq r_S^*)$, $P_0(R_S \leq r_S^*)$, and $P_0(R_S \geq \lvert r_S^* \rvert)$ are obtained from the sampling distribution of R_S when H_0 is true.

Example 11.2. Does Your Baby Lamb Look Like Ewe? One of the focuses of the Research Farm at Ataturk University, Erzurum, Turkey, is on increasing meat quality and production in sheep. As part of this research, the sheep are

Table 11.6 Sheep weight (*kg*) from the research farm at Ataturk University

Mother's mating weight (X)	Lamb weight at seven months(Y)
50.5	25.0
44.3	26.5
47.7	23.5
44.8	27.2
51.9	26.6
56.8	31.0
58.4	34.5
51.5	27.5
50.0	22.7
54.9	27.1
52.3	27.9
58.5	29.0
52.6	30.3
55.8	28.4
50.9	23.7
48.0	21.6
55.2	35.5
52.1	26.3
53.3	31.4
50.2	34.4

Source: Özturk et al. (2005)

sampled periodically in order to monitor their biological growth. In particular, the ewe's weight at time of mating and her lamb's offspring 7 months after birth are routinely recorded. Table 11.6 (and the **R** dataset *sheep_weight*) contains the mother's mating weight (in *kg*) and her lamb offspring's weight (also in *kg*) at age 7 months for a subset of twenty ewe-lamb pairs from the Research Farm.

We are interested in testing whether there is a positive correlation between a ewe's mating weight (X) and the weight of her lamb offspring (Y) at seven months. The separate rankings of the X's and Y's (from least to greatest, of course) are as follows:

i	R_i	S_i	$D_i = S_i - R_i$
1	7	5	−2
2	1	7	6
3	3	3	0
4	2	10	8
5	10	8	−2
6	18	16	−2
7	19	19	0
8	9	11	2
9	5	2	−3
10	15	9	−6
11	12	12	0
12	20	14	−6
13	13	15	2
14	17	13	−4
15	8	4	−4
16	4	1	−3
17	16	20	4
18	11	6	−5
19	14	17	3
20	6	18	12

Thus, we have

$$\begin{aligned}\sum_{i=1}^{20} D_i^2 &= (-2)^2 + 6^2 + 0^2 + 8^2 + (-2)^2 + (-2)^2 + 0^2 + 2^2 + (-3)^2 + (-6)^2 \\ &\quad + 0^2 + (-6)^2 + 2^2 + (-4)^2 + (-4)^2 + (-3)^2 + 4^2 + (-5)^2 + 3^2 + (12)^2 \\ &= 436.\end{aligned}$$

It follows from (11.7) that

$$r_S^* = 1 - \frac{6(436)}{20\left\{(20)^2 - 1\right\}} = 1 - .3278 = .6722.$$

We can verify this result (reported as *rho*) using the **R** function *cor.test*( ) as follows.

```
> cor.test(x = sheep_weight$mother_weight,
           y = sheep_weight$lamb_weight,
           method = "spearman",
           alternative = "greater")

        Spearman's rank correlation rho

data:  sheep_weight$mother_weight and sheep_weight$lamb_weight
S = 436, p-value = 0.0007789
alternative hypothesis: true rho is greater than 0
sample estimates:
      rho
0.6721805
```

From the output of this command, we also see that the P-value for testing H_0: (X and Y are independent) against the one-sided alternative H_A: (X and Y are positively correlated) is given by $P_0(R_S \geq .6722) = 0.0008$, providing strong (not surprisingly) evidence that the mother's weight at mating is positively correlated with the offspring lamb's weight at seven months.

Section 11.2 Practice Exercises

For the exercises in this section, you can use the **R** function *cor.test*( ) with *method* = *"spearman"* to avoid performing the necessary calculations by hand.

11.2.1. Consider the following set of $n = 20$ (x, y) bivariate observations.

$(11, 78)$, $(2, 88)$, $(-2, 100)$, $(-11, 83)$, $(-5, 100)$, $(2, 90)$, $(-6, 87)$, $(22, 82)$, $(21, 92)$, $(8, 90)$, $(25, 85)$, $(9, 93)$, $(7, 92)$, $(8, 96)$, $(18, 100)$, $(-14, 96)$, $(-21, 86)$, $(-26, 89)$, $(-7, 93)$, $(5, 80)$.

Compute the value of the Spearman sample correlation coefficient R_S (11.6). What does this say about the relationship between the x and y observations?

11.2.2. Consider the following set of $n = 12$ (x, y) bivariate observations.

$(1.3, 8)$, $(1.8, 6.9)$, $(0.9, 8.1)$, $(1.6, 7)$, $(2.6, 6.3)$, $(1.5, 6.5)$, $(2.1, 6.4)$, $(3, 5.8)$, $(0.8, 8.3)$, $(2.4, 8.3)$, $(2.5, 6.6)$, $(2.6, 6.6)$.

Compute the value of the Spearman sample correlation coefficient R_S (11.6). What does this say about the relationship between the x and y observations?

11.2.3. Consider the following set of $n = 8$ (x, y) bivariate observations.

$$(1, 1), (2, 4), (3, 9), (4, 16), (5, 25), (6, 36), (7, 49), (8, 64).$$

Compute the value of the Spearman sample correlation coefficient R_S (11.6). What does this say about the relationship between the x and y observations? Contrast your answer with the result obtained in Exercise 11.1.3.

11.2.4. *Does Concurrent Temperature Affect Roost Time Departures for Snow Geese?* Consider the snow geese roost departure time data in Table 11.2, as discussed in Exercise 11.1.4. Using the procedure based on the Spearman sample correlation coefficient, find the P-value for a test of the hypothesis that departure time and concurrent temperature are independent variables. Compare your findings with those obtained in Exercise 11.1.4.

11.2.5. *Birth and Death Rates for the United States.* Consider the state birth and death rate data in Table 11.3, as discussed in Exercise 11.1.5. Using the procedure based on the Spearman sample correlation coefficient, find the P-value for a test of the hypothesis that birth rate and death rate are independent variables. Compare your findings with those obtained in Exercise 11.1.5.

11.2.6. *Psychological Relationships for Dizygous Twins.* Clark et al. (1961) investigated a variety of characteristics for dizygous (i.e., non-identical) twins. The data in Table 11.7 gives the test sores (totals of a number of different psychological tests) for 13 dizygous male twins.

Using the procedure based on the Spearman sample correlation coefficient, find the P-value for a test of the conjecture that the psychological test scores for dizygous twins are positively related.

11.2.7. *Does Voter Turnout in Presidential Elections Affect the Winning Percentage?* Voter turnout in presidential elections has varied over the years, depending at least partly on the intensity of the election campaigns. Generally, closer voter preferences between the Democratic and Republican

Table 11.7 Psychological test scores for dizygous male twins

Pair i	Twin X_i	Twin Y_i
1	277	256
2	169	118
3	157	137
4	139	144
5	108	146
6	213	221
7	232	184
8	229	188
9	114	97
10	232	231
11	161	114
12	149	187
13	128	230

Source: Clark et al. (1961)

candidates is associated with larger turnouts on Election Day, but is this also associated with smaller popular vote percentage by the winning candidate? Table 11.8 contains the total popular vote and percentage for the winning candidate for each of the elections since 1940.

(a) Using the procedure based on the Spearman sample correlation coefficient, find the P-value for a test of the conjecture that the total popular vote and winning percentage in presidential elections are negatively related.

(b) Is there something unusual about the 1968 and 1992 elections? Explain your answer.

11.2.8. *What's In That Hot Dog?* Researchers at *Consumer Reports* analyzed the caloric and sodium content of poultry based hot dogs. Table 11.9 contains their reported results on caloric content and milligrams (mg) of sodium for a test sample of seventeen poultry based hot dogs.

Table 11.8 Popular vote (in thousands) and winning percentage for presidential elections, 1940–2012

Year	Total popular vote (in thousands)	Winning percentage (%)
1940	49,900	54.7
1944	47,977	53.7
1948	48,834	49.6
1952	61,552	54.9
1956	62,027	57.4
1960	68,836	49.7
1964	70,098	61.1
1968	73,027	43.4
1972	77,625	60.2
1976	81,603	50.0
1980	86,497	50.5
1984	92,653	58.8
1988	91,595	53.4
1992	104,427	43.0
1996	96,278	49.2
2000	105,405	47.9
2004	122,295	50.7
2008	131,314	52.9
2012	129,085	51.1

Source: ropercenter.cornell.edu (2016)

Using the procedure based on the Spearman sample correlation coefficient, find the P-value for a test of the conjecture that there is a positive relationship between calories and sodium content in poultry based hot dogs.

11.3 Fitting a Least Squares Line to Bivariate Data

In Fig. 2.3 we display a scatterplot of height (Hgt97) and diameter (Diam97) from the dataset *pines_1997* for the pine trees at the Kenyon Center for Environmental Study (KCES) in 1997. From this scatterplot, it appears that tree height and tree diameter for the pines at KCES in 1997 are linearly related.

Table 11.9 Calories and sodium content (*mg*) for tested poultry hot dogs

Calories	Sodium content (*mg*)
129	430
132	375
102	396
106	383
94	387
102	542
87	359
99	357
107	528
113	513
135	426
142	513
86	358
143	581
152	588
146	522
144	545

Source: *Consumer Reports* (1986)

In such situations, it is often of interest to use a reasonable criterion to select a straight line that "best" represents this relationship.

The equation for a straight line passing through the point (x, y) is given by $y = a + bx$, where b is the slope of the line and a is the intercept of the line at $x = 0$. Our goal, then, in selecting a straight line that "best" represents this linear relationship between the x and y variables is equivalent to using our sample (x, y) data pairs to arrive at our "best" estimates $\hat{a}$ and $\hat{b}$of the slope b and intercept a, respectively. There are many criteria that can be used to obtain such a line to represent or model the linear association between x and y, but the most commonly used statistical criterion for fitting a straight line to bivariate data is known as the *least squares principle*. Simply stated, the least squares principle leads to the line that minimizes the sum of all vertical

Fig. 11.1 Least squares principle: minimizing the vertical squared distances of observed bivariate data from a line

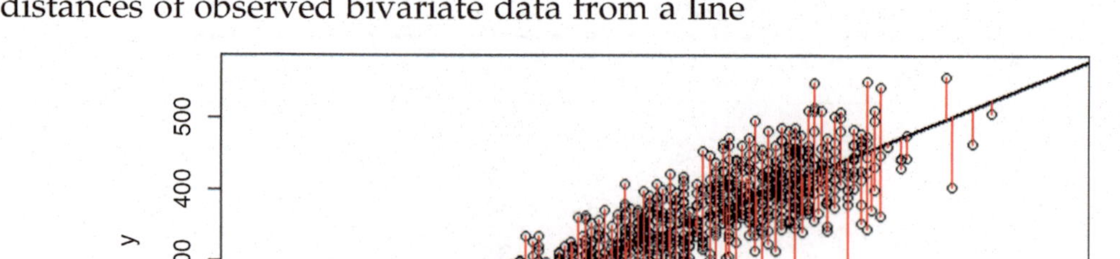

squared distances of the data pairs (x_i, y_i), $i = 1, \ldots, n$, from the line. This approach to obtaining the least squares line is illustrated in Fig. 11.1.

Fortunately, the slope and intercept estimates associated with the least squares line $y = \hat{a} + \hat{b}x$ fitted to the data pairs (x_i, y_i), $i = 1, \ldots, n$, can be easily calculated from the sample data. Expressions for calculating these estimates are provided in Eqs. (11.11) and (11.12).

Fitting the Least Squares Line to Bivariate Data Let $(x_1, y_1), \ldots, (x_n, y_n)$ be n bivariate observations. The least squares fitted line for these data is $y = \hat{a} + \hat{b}x$, where

$$\hat{b} = \frac{\sum_{i=1}^{n} (x_i - \bar{x})(y_i - \bar{y})}{\sum_{j=1}^{n} (x_j - \bar{x})^2} = \frac{n\sum_{i=1}^{n} x_i y_i - \left(\sum_{j=1}^{n} x_j\right)\left(\sum_{k=1}^{n} y_k\right)}{n\sum_{i=1}^{n} x_i^2 - \left(\sum_{j=1}^{n} x_j\right)^2}. \quad (11.11)$$

and

$$\hat{a} = \bar{y} - \hat{b}\bar{x} = \frac{\sum_{i=1}^{n} y_i - \hat{b} \sum_{j=1}^{n} x_j}{n}, \tag{11.12}$$

with $\bar{x} = \sum_{i=1}^{n} x_i/n$ and $\bar{y} = \sum_{i=1}^{n} y_i/n$.

Example 11.3. Least Squares Fitted Line Consider the following set of $n = 33$ bivariate (x, y) observations.

(3, 5), (7, 11), (11, 21), (15, 16), (18, 16), (27, 28), (29, 27), (30, 25), (30, 35), (31, 30), (31, 40), (32, 32), (33, 34), (33, 32), (34, 34), (36, 37), (36, 38), (36, 34), (37, 36), (39, 37), (39, 36), (39, 45), (40, 39), (41, 41), (42, 40), (42, 44), (43, 37), (38, 38), (44, 44), (45, 46), (46, 46), (47, 49), (50, 51)

Then we have

$$\sum_{i=1}^{33} x_i = 1104, \quad \sum_{i=1}^{33} y_i = 1124, \quad \sum_{i=1}^{33} x_i^2 = 41{,}086, \quad \sum_{i=1}^{33} x_i y_i = 41{,}355,$$

and it follows that

$$\hat{b} = \frac{(33)(41{,}355) - (1104)(1124)}{(33)(41{,}086) - (1104)^2} = 0.9036$$

and

$$\hat{a} = \frac{1124 - (0.903643)(1104)}{33} = 3.8296.$$

Thus the least squares line fit to these 33 observations is given by $y = 3.8296 + 0.9036\ x$. These data and the least squares fitted line obtained using the **R** function *lm*( ) are graphically depicted in Fig. 11.2.

Fig. 11.2 **Plot of $n = 33$ observations from Example 11.3 and fitted least squares line**

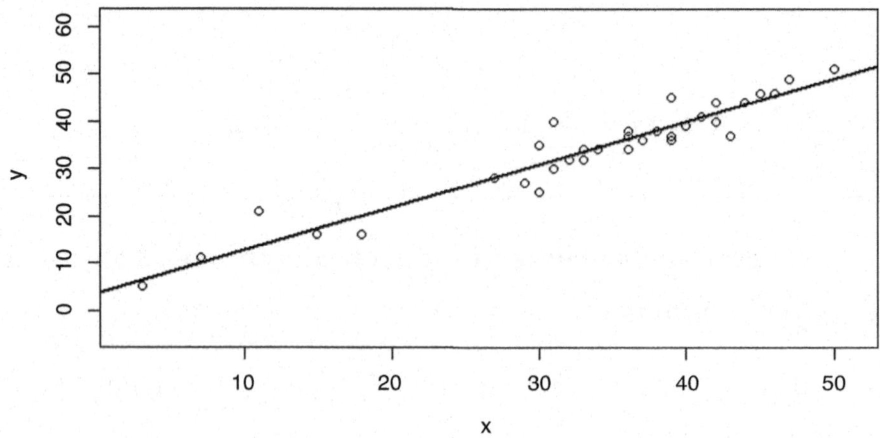

Section 11.3 Practice Exercises

11.3.1. Consider the following set of $n = 20$ (x, y) bivariate observations.

(11, 78), (2, 88), (−2, 100), (−11, 83), (−5, 100), (2, 90), (−6, 87), (22, 82), (21, 92), (8, 90), (25, 85), (9, 93), (7, 92), (8, 96), (18, 100), (−14, 96), (−21, 86), (−26, 89), (−7, 93), (5, 80).

(a) Plot the 20 bivariate observations.

(b) Does it look like a straight line might be a good way to represent the relationship between the x and y values? Why?

(c) Find the least squares fitted line for the 20 observations and plot it with the data.

(d) Comment on how well (or not) the least squares fitted line represents the data.

(e) Use the fitted least squares line to predict the y value for $x = 12$.

(f) Comment on the advisability of using the fitted least squares line to predict the y value for $x = 300$.

11.3.2. Consider the following set of $n = 12$ (x, y) bivariate observations.

(1.3, 8), (1.8, 6.9), (0.9, 8.1), (1.6, 7), (2.6, 6.3), (1.5, 6.5),
(2.1, 6.4), (3, 5.8), (0.8, 8.3), (2.4, 8.3), (2.5, 6.6), (2.6, 6.6).

(a) Plot the 12 bivariate observations.

(b) Does it look like a straight line might be a good way to represent the relationship between the x and y values? Why?

(c) Find the least squares fitted line for the 12 observations and plot it with the data.

(d) Comment on how well (or not) the fitted line represents the data.

(e) Use the fitted least squares line to predict the y value for $x = 1.7$.

(f) Comment on the advisability of using the fitted least squares line to predict the y value for $x = -16$.

11.3.3. Consider the following set of n = 14 (x, y) bivariate observations.

(17.3, 71.7), (19.3, 48.3), (19.5, 88.3), (19.7, 75), (22.9, 91.7), (23.1, 100), (26.4, 73.3), (26.8, 65), (27.6, 75), (28.1, 88.3), (28.2, 68.3), (28.7, 96.7), (29, 76.7), (29.6, 78.3).

(a) Plot the 14 bivariate observations.

(b) Does it look like a straight line might be a good way to represent the relationship between the x and y values? Why?

(c) Find the least squares fitted line for the 14 observations and plot it with the data.

(d) Comment on how well (or not) the fitted line represents the data.

(e) Use the fitted least squares line to predict the y value for $x = 27$.

(f) Comment on the advisability of using the fitted least squares line to predict the y value for $x = 70$ or $x = 9$.

11.3.4. Consider the following set of $n = 8$ (x, y) bivariate observations.

$(1, 1), (2, 4), (3, 9), (4, 16), (5, 25), (6, 36), (7, 49), (8, 64)$.

(a) Find the least squares fitted line for the 8 observations.

(b) Plot the 12 bivariate observations and the fitted least squares line.

(c) Comment on how well (or not) the fitted regression line represents the data.

(d) Do you think there is a better function (than a straight line) to represent the relationship between the x and y values? Explain.

11.3.5. Let $\hat{y} = \widehat{\alpha} + \widehat{\beta}x$ be the least squares fitted line for the n bivariate observations $(x_1, y_1), \ldots, (x_n, y_n)$.

(a) Is the least squares fitted line changed if we add the constant d to each of the y values while leaving the x values unchanged? How?

(b) Is the least squares fitted line changed if we multiply each of the y values by the constant d, while leaving the x values unchanged? How?

(c) What do your answers to parts (a) and (b) imply about the relationship between a fitted least squares line and the data $(x_1, y_1), \ldots, (x_n, y_n)$?

11.3.6. Let $\hat{y} = \hat{a} + \hat{b}x$ be the least squares fitted line for the n bivariate observations $(x_1, y_1), \ldots, (x_n, y_n)$.

(a) Is the least squares fitted line changed if we add the constant d to each of the x values while leaving the y values unchanged? How?

(b) Is the least squares fitted line changed if we multiply each of the x values by the constant d, while leaving the y values unchanged? How?

(c) What do your answers to parts (a) and (b) imply about the relationship between a fitted least squares line and the data $(x_1, y_1), \ldots, (x_n, y_n)$?

11.3.7. Let $\hat{y} = \hat{a} + \hat{b}x$ be the least squares fitted line for the n bivariate observations $(x_1, y_1), \ldots, (x_n, y_n)$.

(a) Is the least squares fitted line changed if we add the constant d to each of the y values and the constant k to each of the x values? How? What happens if $d = k$?

(b) Is the least squares fitted line changed if we multiply each of the y values by the constant d and multiply each of the x values by the constant k? How? What happens if $d = k$?

(c) What do your answers to parts (a) and (b) imply about the relationship between a fitted least squares line and the data $(x_1, y_1), \ldots, (x_n, y_n)$?

11.4 Linear Regression Inference for Normal Populations

Many statistical applications involve stochastic relationships between a dependent (response) variable and one or more independent (predictor) variables. Such statistical procedures are commonly referred to as *regression analyses*. These regression models can vary from the simplest linear relationship between the dependent variable and a single independent variable to very complex nonlinear relationships involving numerous predictor variables. In this section we present procedures for evaluating a possible linear relationship between a dependent random variable Y and a single predictor variable x when the underlying probability distribution is normal.

The linear model for this setting can be represented as follows:

$$Y = \alpha + \beta x + \varepsilon, \tag{11.13}$$

where Y is the dependent random variable, x is the known independent (predictor) variable, α and β are unknown parameters, and ε is assumed to have a normal distribution with mean 0 and unknown variance σ^2. Another way to state this model is that, for fixed x, the random variable Y has mean $E[Y \mid x] = \mu_{Y|x} = \alpha + \beta x$; that is, the expected value of Y is linearly related to the value of the predictor variable x. The line $\alpha + \beta x$ is called the *regression line* and it represents the assumed relationship between the dependent variable Y and the predictor variable x. In this section we concentrate on procedures for testing the null hypothesis H_0: $\beta = 0$, corresponding to no influence from the predictor variable x on the dependent variable Y, against appropriate ($\beta >$, $<$, or $\neq 0$) alternatives.

At each of n fixed values, $x_1, \ldots, x_n$, of the independent (predictor) variable x, we observe the value of the response variable Y. Thus, we collect a set of n paired observations $(x_1, y_1), \ldots, (x_n, y_n)$, where y_i is the observed value of the response variable Y_i when $x = x_i$. We then fit a least squares line to these paired observations by estimating the slope β and intercept α using the expressions in (11.11) and (11.12) respectively, to obtain the fitted least squares line $\widehat{\mu}_{Y|x} = \widehat{\alpha} + \widehat{\beta}x$.

To assess whether the independent (predictor) variable x provides any useful information about the dependent variable Y, we test the null hypothesis H_0: $\beta = 0$, corresponding to no statistically significant impact on Y by x against appropriate non-zero β alternatives. To simplify construction of the appropriate test statistic, we need some additional notation. Let

$$S_{xx} = \sum_{i=1}^{n} (x_i - \bar{x})^2, \quad S_{yy} = \sum_{j=1}^{n} (y_j - \bar{y})^2 \quad S_{xy} = \sum_{k=1}^{n} (x_k - \bar{x})(y_k - \bar{y}), \quad (11.14)$$

where $\bar{x} = \sum_{i=1}^{n} x_i/n$ and $\bar{y} = \sum_{j=1}^{n} y_j/n$.

Hypothesis Tests for a Linear Regression Effect of the Independent (Predictor) Variable To test the null hypothesis H_0: $\beta = 0$, corresponding to no linear relationship between the independent variable x and the dependent variable Y, against appropriate ($\beta >, <$, or $\neq 0$) alternatives, construct the test statistic

$$T = \frac{\sqrt{S_{xx}}\ \widehat{\beta}}{\sqrt{\dfrac{S_{yy} - \widehat{\beta}\, S_{xy}}{n-2}}}, \tag{11.15}$$

and let t_{obs} be the observed value of T. Then the P-values for a test of H_0 against the alternatives H_A are:

H_A	P-value	
$\beta > 0$	$= P(T \geq t_{obs})$	(11.16)
$\beta < 0$	$= P(T \leq t_{obs})$	(11.17)
$\beta \neq 0$	$= 2P(T \geq \|t_{obs}\|)$,	(11.18)

where $T \sim t(n\text{-}2)$, the t-distribution with $n - 2$ degrees of freedom. Note that the alternative $\beta > 0$ corresponds to a positive linear relationship between the independent and dependent variables x and Y so that larger Y values are associated with larger x values. Under the alternative $\beta < 0$ there is a negative linear relationship for which larger Y values are associated with smaller x values. The two-sided alternative $\beta \neq 0$ represents a general, non-directional, relationship between the predictor and dependent variables.

Confidence intervals can also be constructed for the slope parameter β (see Exercise 11.A.11). In addition, the fitted least squares regression line $\widehat{\mu}_{Y|x} = \widehat{\alpha} + \widehat{\beta}x$ can be used with care to predict Y outcomes for other potential values of the independent variable x within the range of the x-values used to obtain the least squares fit.

As previously noted in our discussion for the Pearson correlation coefficient in Sect. 1, always keep in mind that a failure to reject H_0: $\beta = 0$ is not equivalent to saying that there is no relationship between the independent (predictor) variable x and the dependent variable Y. It is simply indicative that there is no significant *linear* relationship between them, leaving open the possibility of a non-linear relationship.

The least squares approach presented in this section can also be used to test for a linear relationship with fixed slope $\beta_0 \neq 0$; that is, testing H_0: $\beta = \beta_0$ against alternatives $\beta >, <,$ or $\neq \beta_0$. See Exercise 11.A.10 for details.

Example 11.4. We can use the **R** dataset *pines_1997* to investigate whether there is a linear relationship between tree height (Hgt97) and tree diameter (Diam97) for the pine trees at the Kenyon Center for Environmental Study in 1997. We begin by using the **R** function *lm*( ) (short for "linear model") to fit the model in (11.13) and store it as the local variable *height_by_diameter_model*.

```
> height_by_diameter_model <- lm(Hgt97 ~ Diam97, data = pines_1997)
```

We can inspect the variable to see the estimated model of tree height regressed on tree diameter.

```
> height_by_diameter_model

Call:
lm(formula = Hgt97 ~ Diam97, data = pines_1997)

Coefficients:
(Intercept)       Diam97
      126.8         37.2
```

From this output, we see that $\widehat{\beta} = 37.2$, but is this enough to conclude that there is some linear relationship between height and diameter? To answer this, we use the **R** function *summary*( ) on our model, which will give us the information we need (and much more information that we don't currently need but can be useful in other settings!).

```
> summary(height_by_diameter_model)

Call:
lm(formula = Hgt97 ~ Diam97, data = pines_1997)

Residuals:
     Min       1Q   Median       3Q      Max
-276.623  -23.594    0.259   24.724  125.573

Coefficients:
            Estimate Std. Error t value Pr(>|t|)
(Intercept) 126.8025     4.4891   28.25   <2e-16 ***
Diam97       37.2031     0.7199   51.68   <2e-16 ***
---
Signif. codes:  0 '***' 0.001 '**' 0.01 '*' 0.05 '.' 0.1 ' ' 1

Residual standard error: 39.85 on 857 degrees of freedom
  (141 observations deleted due to missingness)
Multiple R-squared:  0.7571,	Adjusted R-squared:  0.7568
F-statistic:  2671 on 1 and 857 DF,  p-value: < 2.2e-16
```

From the table under the header "Coefficients", we see in the column titled "t value" that, at 51.68, the value of T is quite large. In the column after that, we are given the p-value defined in (11.18), which is practically zero. We can verify that 51.68 is a very extreme value for the $t(n\text{-}2)$ distribution with the **R** function *pt*( ).

```
> pt(51.68, nrow(pines_1997) - 2, lower.tail = FALSE)
[1] 1.089775e-284
```

Section 11.4 Practice Exercises

11.4.1. *Tree Height and Needle Length* Consider the 1997 data for the pine trees at the Kenyon Center for Environmental Study (KCS) contained in the dataset *pines_1997*. Find the *P*-value for a hypothesis test of the conjecture that the needle length (Needles97) of a pine tree is linearly related with a positive slope to the diameter (Diam97) of the tree.

11.4.2. *Does Size Really Matter?* In the February 18, 2000 edition of Singapore's *Business Times*, an advertisement (discussed in Chu, 2001) listed data, including weight in carats and value in Singapore dollars, for 308 round diamond stones. These data are provided in the dataset *diamonds_carats_color_cost*. Find the *P*-value for a hypothesis test of the conjecture that the value (cost) of a round diamond stone is linearly related (with positive slope) to the weight of the stone. Does size really matter?

11.4.3. *Does Your Baby Lamb Look Like Ewe?* Consider the mother and lamb weight data presented in Table 11.6 and discussed in Example 11.2. Find the *P*-value for a hypothesis test of the conjecture that the offspring lamb's weight at seven months is linearly related (with positive slope) to the parent ewe's weight at time of mating.

11.4.4. *Space Shuttle Challenger* After many previous successful launches, disaster struck on January 28, 1986. Only 73 seconds into its flight, the Space Shuttle Challenger exploded and all seven crewmembers died. In subsequent post launch analyses, it became clear that the cause of the disaster was the failure of an O-ring (used to help seal the joints of different components of the solid rocket booster) on the right solid rocket booster. Lighthall (1991) addressed some of the issues surrounding this O-ring failure in the Challenger, including the temperature at the time of the rocket launch. Table 11.10 contains the temperatures (°F) and depths of erosion (in *mils*, one *mil* equals .001 inch) for the O-rings in 22 previous successful space shuttle launches.

(a) Use the procedure based on the Pearson sample correlation coefficient to find the *P*-value for a test of the conjecture that O-Ring temperature at time of launch and depth of O-Ring erosion are negatively correlated.

(b) Obtain the fitted least squares line for these data, treating temperature as the independent variable and depth of erosion as the dependent variable.

(c) Find the *P*-value for a hypothesis test of the conjecture that the depth of O-ring erosion is linearly related (with negative slope) to the temperature of the O-ring at launch.

Table 11.10 Temperature (°F) at time of launch and depth of erosion (*mils*) for O-rings from 22 successful space shuttle launches

Temperature of O-ring	Depth of erosion
66	0
70	53
69	0
68	0
67	0
72	0
73	0
70	0
57	40
63	0
70	28
78	0
67	0
53	48
67	0
75	0
70	0
81	0
76	0
79	0
75	0
76	0

Source: Lighthall (1991)

(d) At the time of the fateful Challenger launch on January 28, 1986, the temperature of the O-rings at the launch site was an unusually low 29 °F. Using your fitted least squares line from part (b), what would you have predicted the O-ring erosion to be for that launch?

(e) Discuss the appropriateness of your statistical analyses in parts (a)-(d) and the decision to go ahead and proceed with the launch in spite of the low temperature.

11.4.5. *Arts Participation Across the States—Does Reading Lead to Attendance?* The National Endowment for the Arts periodically collects survey information from residents of the United States about their participation in the arts. A portion of the statewide results from the 2015 survey is presented in the publication *Arts Profile #11 (August 2016)*. In particular, the publication includes the percentage of each state's residents who attended a live music, theater, or dance performance in 2015 and the percentage of each state's residents who read literature in 2015. These two percentages are given in Table 11.11 for 20 of the states.

(a) Obtain the fitted least squares line for these data, treating percentage reading literature in 2015 as the independent variable and percentage attending a live event in 2015 as the dependent variable.

(b) Find the P-value for a hypothesis test of the conjecture that the percentage attending a live event in 2015 is linearly related (with positive slope) to the percentage reading literature in 2015.

(c) Use the fitted least squares line from part (a) to predict the percentage attending a live event in 2015 for the state of Ohio, where 46.0% of the residents read literature in 2015. Compare your prediction with the true value (from the survey) of 32.7% Ohioans who attended a live event in 2015.

(d) Use the fitted least squares line from part (a) to predict the percentage attending a live event in 2015 for the state of West Virginia, where 34.1% of the residents read literature in 2015. Compare your prediction with the true value (from the survey) of 21.5% West Virginians who attended a live event in 2015.

11.5 Rank-Based Linear Regression Inference

When there is concern that the underlying probability distribution for the dependent variable Y is not necessarily normal, inferences associated with the

Table 11.11 Percentage of state residents who attended a live music, theater, or dance performance in 2015 and percentage of state residents who read literature in 2015

State	Percentage attending live event	Percentage reading literature
Alabama	16.2	34.8
Alaska	40.6	59.3
Arizona	27.9	46.8
Arkansas	20.5	39.3
California	32.2	38.9
Colorado	44.4	59.0
Connecticut	42.0	52.0
Florida	24.4	30.5
Georgia	20.8	36.8
Hawaii	29.2	54.7
Illinois	34.7	47.6
Indiana	35.9	48.2
Kansas	37.1	49.4
Louisiana	25.3	36.6
Maine	29.8	45.8
Minnesota	40.5	49.3
Mississippi	17.8	21.7
Montana	40.8	57.8
Texas	27.2	37.5
Utah	51.0	57.0

Source: National Endowment for the Arts (2016)

least squares regression line presented in Sect. 4 can be unreliable. In such settings an approach that relies instead on the Kendall correlation coefficient (see Exercise 11.A.8) provides a useful alternative.

As in Sect. 4, we consider the linear model given in (11.13) but without the assumption of normality for the distribution of the dependent variable Y, and, as before, at each of n values of the independent variable x_i, $i = 1, \ldots, n$, we observe the value of the dependent variable Y_i, $i = 1, \ldots, n$. Here we also assume, without loss of generality, that the x's are labeled so that $x_1 \leq x_2 \leq \ldots \leq x_n$. (When you re-label the x values in increasing order, be sure to keep each Y value linked with its corresponding x value.). Compute the Kendall correlation coefficient K (see Exercise 11.A.8) for the (x_i, Y_i) pairs, namely,

$$K = \frac{\sum_{i=1}^{n-1} \sum_{j=i+1}^{n} c\left[(Y_j - Y_i)(x_j - x_i)\right]}{\left[\frac{n(n-1)}{2}\right]}, \tag{11.19}$$

where $c(t) = -1, 0, 1$ if $t <, =, > 0$. Thus, for each pair of subscripts (i, j), with $1 \le i < j \le n$, score 1 if $Y_j > Y_i$ and $x_j > x_i$, score -1 if $Y_j < Y_i$, and $x_j > x_i$, and score 0 if either $x_j = x_i$ or $Y_j = Y_i$.

Note that a pair of pairs (x_i, Y_i) and (x_j, Y_j) with $i < j$ receives a score of 1 if and only if $x_i < x_j$ and $Y_i < Y_j$. Such a pair of pairs is called a *concordant pair* since Y increases along with the increase in x. Moreover, a pair of pairs (x_i, Y_i) and (x_j, Y_j) with $i < j$ receives a score of -1 if and only if $x_i < x_j$ and $Y_i > Y_j$. Such a pair of pairs is called a *discordant pair* since Y decreases with the increase in x. The numerator of K (11.19) is then simply the number of concordant pairs minus the number of discordant pairs for all $\binom{n}{2} = \frac{n(n-1)}{2}$ pairings of the (x, y) pairs.

Hypothesis Tests for a Monotonic Regression Effect of the Independent (Predictor) Variable Without the Assumption of Normality To test the null hypothesis H_0: $\beta = 0$, corresponding to no linear relationship between the independent variable x and the dependent variable Y, against appropriate ($\beta >, <$, or $\neq 0$) alternatives, compute the test statistic

$$K = \frac{\sum_{i=1}^{n-1} \sum_{j=i+1}^{n} c\left[(Y_j - Y_i)(x_j - x_i)\right]}{\left[\frac{n(n-1)}{2}\right]}$$

and let k_{obs} be the observed value of K. Then the P-values for a test of H_0 against the alternatives H_A are:

H_A	P-value	
$\beta > 0$	$=P_0(K \geq k_{obs})$	(11.20)
$\beta < 0$	$=P_0(K \leq k_{obs})$	(11.21)
$\beta \neq 0$	$=2P_0(K \geq \lvert k_{obs} \rvert)$,	(11.22)

where $P_0(K \geq k_{obs})$, $P_0(K \leq k_{obs})$, and $2P_0(K \geq |k_{obs}|)$ are obtained from the sampling distribution of K when H_0 is true.

Note that the alternative $\beta > 0$ corresponds to a positive linear relationship between the independent and dependent variables x and Y so that larger Y values are associated with larger x values. Under the alternative $\beta < 0$ there is a negative linear relationship for which larger Y values are associated with smaller x values. The two-sided alternative $\beta \neq 0$ represents a general, non-directional, relationship between the predictor and dependent variables.

Example 11.5. How Hot Is It?—Ask the Crickets Crickets are cold-blooded animals, so the rates for their physiological processes are directly affected by temperature. This includes the rate at which they chirp. Pierce (1948) mechanically measured the frequency (in wing vibrations per second) of chirps made by a striped ground cricket at a variety of temperatures. His findings are displayed in Table 11.12.

For this experiment, temperature is the independent variable (x) and the corresponding number of chirps is the dependent (Y) variable and we will use the Kendall correlation coefficient K (11.19) to assess whether there is a positive linear relationship (slope $\beta > 0$) between these two variables. Ordering the pairs by temperature, we obtain:

Table 11.12 Cricket chirps per second and concurrent air temperatures in degrees Fahrenheit (°F)

Chirps per second	Concurrent temperature (°F)
20.0	88.6
16.0	71.6
19.8	93.3
18.4	84.3
17.1	80.6
15.5	75.2
14.7	69.7
17.1	82.0
15.4	69.4
16.2	83.3
15.0	79.6
17.2	82.6
16.0	80.6
17.0	83.5
14.4	76.3

Source: Pierce (1948)

Ordered temperatures	Chirps per second
69.4	15.4
69.7	14.7
71.6	16.0
75.2	15.5
76.3	14.4
79.6	15.0
80.6	16.0
80.6	17.1
82.0	17.1
82.6	17.2
83.3	16.2
83.5	17.0
84.3	18.4
88.6	20.0
93.3	19.8

The associated $c((Y_j - Y_i)(x_j - x_i))$ values for these $n = 15$ data pairs are as given in the following chart. You can generate these values for yourself using the **R** function *ConDis.matrix*( ) from the *asbio* package.

$c((Y_j - Y_i)(x_j - x_i))$ values for cricket chirps data

$j \backslash i$	1	2	3	4	5	6	7	8	9	10	11	12	13	14
2	−1													
3	1	1												
4	1	1	−1											
5	−1	−1	−1	−1										
6	−1	1	−1	−1	1									
7	1	1	0	1	1	1								
8	1	1	1	1	1	1	0							
9	1	1	1	1	1	1	1	0						
10	1	1	1	1	1	1	1	1	1					
11	1	1	1	1	1	1	1	−1	−1	−1				
12	1	1	1	1	1	1	1	−1	−1	−1	1			
13	1	1	1	1	1	1	1	1	1	1	1	1		
14	1	1	1	1	1	1	1	1	1	1	1	1	1	
15	1	1	1	1	1	1	1	1	1	1	1	1	1	−1

Summing the $c((Y_j - Y_i)(x_j - x_i))$ values in the chart, we obtain

$$\sum_{i=1}^{14} \sum_{j=i+1}^{15} c\left[\left(Y_j - Y_i\right)\left(x_j - x_i\right)\right] = 86 - 16 = 70,$$

so that

$$k_{obs} = \frac{70}{\frac{15(14)}{2}} = .6667.$$

Using the **R** function *cor.test*( ) with *method* = *'kendall'* and *alternative* = *'greater'* we see that the *P*-value for testing H_0: $\beta = 0$ against the alternative H_A: $\beta > 0$ is 0.0002537. Thus, there is very strong evidence that the rate of cricket chirps per minute has a positive ($\beta > 0$) linear relationship with concurrent temperature.

```
> cor.test(x = cricket_chirps$Chirps,
           y = cricket_chirps$Temperature,
           method = 'kendall',
           alternative = 'greater')

        Kendall's rank correlation tau

data:  cricket_chirps$Chirps and cricket_chirps$Temperature
z = 3.4768, p-value = 0.0002537
alternative hypothesis: true tau is greater than 0
sample estimates:
      tau
0.6763364

Warning message:
In cor.test.default(cricket_chirps$Chirps, cricket_chirps$Temperature,
:
  Cannot compute exact p-value with ties
```

Note that the tau statistic reported by *cor.test*( ) is slightly different than the k_{obs} that we calculated above. This has to do with a slight difference in the denominator that the **R** function is using as an adjustment for ties in the data. Although the details are beyond the scope of this book, the interested reader can learn more by searching the Internet for information about Kendall's Tau-a and Kendall's Tau-b.

We need to point out that while the conclusions based on the tests in (11.20), (11.21), and (11.22) based on K are stated in terms of the slope parameter for the linear model in (11.13), they are actually capable of detecting **any** monotonic (not necessarily linear) relationship between x and Y. This is not the case for the tests in (11.16), (11.17), and (11.18) associated with the least squares line and based on the assumption of normality.

Tests for Trend In the special case when the x-values are the time order (see, for example, Exercises 11.5.3 and 11.5.4), the procedures in (11.20), (11.21), and (11.22) can be viewed as tests against a time trend for the Y variable.

Section 11.5 Practice Exercises

11.5.1. *Does Lighting the Roost Give Snow Geese a Boost?* Wildlife science often involves trying to understand how environmental conditions affect wildlife habits. Freund et al. (2010) report data on such a study to assess how a variety of environmental conditions influence the time that lesser snow geese leave their overnight roost sites to fly to their feeding areas. In particular, they were interested in the effect that light intensity might have on the time that the geese leave their roosts. Table 11.13 contains the times of roost departure in minutes before (−) or after (+) sunrise and the concurrent light intensity for 36 different observation days in the 1987–1988 winter season.

Find the P-value for a test of the conjecture that there is a positive linear relationship between light intensity and roost departure time.

11.5.2. *Does "Strike 'Em Out" Equate with "Shut 'Em Down"?* A major part of many major league pitchers' arsenal is their ability to strike opponents out—but does that ability correlate with a low overall earned run average? Table 11.14 contains the nine-inning strikeout rates and the earned run averages for fifteen major leaguers who pitched at least 140 innings during the 2016 baseball season.

Use these data to test the conjecture that a major league pitcher's earned run average is negatively correlated with his nine-inning strikeout rate.

11.5.3. *Median Weekly Earnings for Men in Service Occupations.* The median weekly earnings in dollars for men in service occupations for the period of time from the first quarter of 2005 through the fourth quarter of 2015 are given in Table 11.15.

Find the P-value for testing the conjecture that there has been an increasing trend in median weekly salaries for men over the period of time from the first quarter of 2005 through the fourth quarter of 2015.

Table 11.13 Snow geese roost departure times in minutes before (–) or after (+) sunset and concurrent light intensity

Date	Departure time	Concurrent light intensity
11/10/87	11	12.6
11/13/87	2	10.8
11/14/87	–2	9.7
11/15/87	–11	12.2
11/17/87	–5	14.2
11/18/87	2	10.5
11/21/87	–6	12.5
11/22/87	22	12.9
11/23/87	22	12.3
11/25/87	21	9.4
11/30/87	8	11.7
12/05/87	25	11.8
12/14/87	9	11.1
12/18/87	7	8.3
12/24/87	8	12.0
12/26/87	18	11.3
12/27/87	–14	4.8
12/28/87	–21	6.9
12/30/87	–26	7.1
12/31/87	–7	8.1
01/02/88	–15	6.9
01/03/88	–6	7.6
01/05/88	–14	9.0
01/07/88	–8	7.1
01/08/88	–19	3.9
01/10/88	–23	8.1
01/11/88	–11	10.3
01/12/88	5	9.0
01/14/88	–23	5.1
01/15/88	–7	7.4
01/16/88	9	7.9
01/20/88	–27	3.8
01/21/88	–24	6.3
01/22/88	–29	6.3
01/23/88	–19	7.8
01/24/88	–9	9.5

Source: Freund et al. (2010)

Table 11.14 Nine-Inning strikeout rate and earned run average for 15 major leaguers who pitched at least 140 innings in the 2016 season

Player	Earned run average	Nine-inning strikeout rate
David Price	3.99	8.92
Max Scherzer	2.96	11.20
Justin Verlander	3.04	10.04
Madison Bumgarner	2.74	9.96
Jon Lester	2.44	8.75
Cole Hamels	3.32	8.97
Adam Wainwright	4.62	7.29
Bartolo Colon	3.43	6.01
Noah Syndergaard	2.60	10.68
Michael Pineda	4.82	10.60
Trevor Bauer	4.26	7.96
Chris Sale	3.34	9.25
Clayton Kershaw	1.69	10.39
Stephen Strasburg	3.60	11.15
Rick Porcello	3.15	7.63

Source: rotowire.com (2016)

11.5.4. *Median Weekly Earnings for Women in Service Occupations*. The median weekly earnings in dollars for women in service occupations for the period of time from the first quarter of 2005 through the fourth quarter of 2015 are given in Table 11.16.

Find the P-value for testing the conjecture that there has been an increasing trend in median weekly salaries for women over the period of time from the first quarter of 2005 through the fourth quarter of 2015. Compare your finding with that obtained in Exercise 11.5.3.

Chapter 11 Comprehensive Exercises

11.A. Conceptual

11.A.1. Let $R_1, \ldots, R_n$ be the ranks (from least to greatest) of the variables $X_1, \ldots, X_n$, respectively.

Table 11.15 Median weekly earnings (dollars) for men for each quarterly period between 2005 and 2015

Quarter	Median weekly salary	Quarter	Median weekly salary
Q1—2005	477	Q1—2011	565
Q2—2005	473	Q2—2011	544
Q3—2005	464	Q3—2011	528
Q4—2005	493	Q4—2011	578
Q1—2006	500	Q1—2012	563
Q2—2006	492	Q2—2012	529
Q3—2006	494	Q3—2012	530
Q4—2006	488	Q4—2012	550
Q1—2007	516	Q1—2013	576
Q2—2007	521	Q2—2013	534
Q3—2007	503	Q3—2013	562
Q4—2007	520	Q4—2013	546
Q1—2008	529	Q1—2014	581
Q2—2008	539	Q2—2014	580
Q3—2008	545	Q3—2014	585
Q4—2008	539	Q4—2014	588
Q1—2009	516	Q1—2015	575
Q2—2009	520	Q2—2015	587
Q3—2009	515	Q3—2015	571
Q4—2009	566	Q4—2015	607
Q1—2010	558		
Q2—2010	533		
Q3—2010	511		
Q4—2010	585		

Source: United States Department of Labor, Bureau of Labor Statistics (2016)

(a) Use algebra to show that $\bar{R} = \frac{1}{n}\sum_{i=1}^{n} R_i = \frac{n+1}{2}$.

(b) Use algebra to show that $\sum_{i=1}^{n} (R_i - \bar{R})^2 = \frac{n(n^2-1)}{2}$.

11.A.2. Use algebra to show that the two expressions for the slope estimator in (11.11) are equivalent. That is, show that

Table 11.16 Median weekly earnings (dollars) for women for each quarterly period between 2005 and 2015

Quarter	Median weekly salary	Quarter	Median weekly salary
Q1—2005	381	Q1—2011	431
Q2—2005	371	Q2—2011	439
Q3—2005	383	Q3—2011	427
Q4—2005	383	Q4—2011	440
Q1—2006	382	Q1—2012	450
Q2—2006	389	Q2—2012	435
Q3—2006	391	Q3—2012	440
Q4—2006	397	Q4—2012	420
Q1—2007	395	Q1—2013	447
Q2—2007	404	Q2—2013	461
Q3—2007	408	Q3—2013	447
Q4—2007	415	Q4—2013	452
Q1—2008	408	Q1—2014	459
Q2—2008	416	Q2—2014	452
Q3—2008	416	Q3—2014	467
Q4—2008	441	Q4—2014	470
Q1—2009	411	Q1—2015	461
Q2—2009	419	Q2—2015	457
Q3—2009	426	Q3—2015	465
Q4—2009	418	Q4—2015	471
Q1—2010	420		
Q2—2010	433		
Q3—2010	425		
Q4—2010	421		

Source: United States Department of Labor, Bureau of Labor Statistics (2016)

$$\frac{\sum_{i=1}^{n}(x_i-\bar{x})(y_i-\bar{y})}{\sum_{j=1}^{n}(x_j-\bar{x})^2}=\frac{n\sum_{i=1}^{n}x_iy_i-\left(\sum_{j=1}^{n}x_j\right)\left(\sum_{k=1}^{n}y_k\right)}{n\sum_{i=1}^{n}x_i^2-\left(\sum_{j=1}^{n}x_j\right)^2}.$$

11.A.3. Use algebra to show that the value of the Spearman sample correlation coefficient R_S (11.6) can be obtained by using the computationally simpler expression

$$R_S = 1 - \frac{6\sum_{i=1}^{n} D_i^2}{n(n^2 - 1)},$$

where $D_i = S_i - R_i$, for $i = 1, \ldots, n$.

11.A.4. Construct a set of bivariate observations $(X_1, Y_1), \ldots, (X_n, Y_n)$ for which the Pearson sample correlation coefficient R (11.1) is 0 (least indicative of a linear relationship between X and Y) but for which Y can be expressed as an explicit function of X (so their true relationship is perfect).

11.A.5. Construct a set of bivariate observations $(X_1, Y_1), \ldots, (X_n, Y_n)$ for which the Pearson correlation coefficient R (11.1) is 0 (least indicative of a linear relationship between X and Y) but for which the Spearman rank correlation coefficient R_S (11.6) and the Kendall correlation coefficient K (11.18) are both 1 (most indicative of a monotone positive relationship between X and Y).

11.A.6. Let (X, Y) be bivariate random variables. The population correlation between X and Y is defined by

$$\rho_{X,Y} = \frac{E[(X - \mu_X)(Y - \mu_Y)]}{\sigma_X \sigma_Y},$$

where $\mu_X = E(X)$, $\mu_Y = E(Y)$, $\sigma_X^2 = Var(X)$, and $\sigma_Y^2 = Var(Y)$.

(a) Show that $E[(X - \mu_X)(Y - \mu_Y)] = E[XY] - \mu_X\mu_Y$.

(b) Show that $\rho_{X,Y} = 0$ if X and Y are independent random variables.

(c) Is the converse to the statement in part (b) also true? That is, does $\rho_{X,Y} = 0$ also imply that X and Y are independent random variables?

Justify your answer.

[Hint: Consider the joint probability distribution given by

$$P((X,Y) = (-1,0)) = P((X,Y) = (0,0)) = P((X,Y) = (1,0)) = \frac{1}{3}.]$$

11.A.7. Let X be a positive random variable (i.e., $P(X > 0) = 1$). Define a second random variable $Y = X^2$. Clearly X and Y are dependent random variables, but they are not linearly related. As a result, the test procedures in (11.3), (11.4), and (11.5) based on the Pearson sample correlation coefficient R will not be particularly effective in detecting this dependence.

(a) Will the test procedures in (11.8), (11.9), and (11.10) based on the Spearman sample correlation coefficient R_S be capable of detecting this dependence? Justify your answer.

(b) Compare and contrast the linear regression and monotonic regression procedures discussed in Sects. 4 and 5, respectively, for this setting.

11.A.8. Let (X_1, Y_1), ... , (X_n, Y_n) be a random sample from a bivariate distribution. The *Kendall sample correlation coefficient* for these data is defined by

$$K = \frac{\sum_{i=1}^{n-1} \sum_{j=i+1}^{n} c[(Y_j - Y_i)(X_j - X_i)]}{\left[\frac{n(n-1)}{2}\right]}$$

where $c(t) = -1, 0, 1$ if $t <, =, > 0$. Thus, for each pair of subscripts (i, j), with $1 \le i < j \le n$, score 1 if $Y_j > Y_i$ and $X_j > X_i$, score -1 if $Y_j < Y_i$ and $X_j > X_i$, and score 0 if either $X_j = X_i$ or $Y_j = Y_i$.

(a) Construct a data set for which $K = -1$.

(b) Construct a data set for which $K = 1$.

(c) Construct a data set for which $K = 0$.

11.A.9. Let R_S be the Spearman sample correlation coefficient defined in (11.6).

(a) Construct a data set for which $R_S = -1$.
(b) Construct a data set for which $R_S = 1$.
(c) Construct a data set for which $R_S = 0$.

11.A.10. Suppose you were interested in testing for a linear or monotonic regression with fixed null hypothesis slope value $\beta_0 \neq 0$. Discuss how you might use the shifted sample data $Y_1 - \beta_0 x_1, \ldots, Y_n - \beta_0 x_n$ in conjunction with either the linear regression procedures based on T (11.14) or the monotonic regression procedures based on K (11.18) to test the null hypothesis $H_0 : \beta = \beta_0$ against appropriate alternatives $\beta > \beta_0$, $\beta < \beta_0$, or $\beta \neq \beta_0$.

11.A.11. *Confidence Interval for the Slope Parameter β for a Normal Population.* Let $\widehat{\beta}$ be the slope estimator given in (11.11) and let S_{xx}, S_{xy}, and S_{yy} be as defined in (11.14). When the underlying distribution is bivariate normal, a $100CL\%$ confidence interval for the slope parameter β is then given by

$$\widehat{\beta} \pm t_{n-2,\frac{(1-CL)}{2}} \frac{\sqrt{\dfrac{S_{yy} - \widehat{\beta} S_{xy}}{n-2}}}{\sqrt{S_{xx}}},$$

where $t_{n-2,\frac{(1-CL)}{2}}$ is the upper $\frac{(1-CL)}{2}$ percentile for the t-distribution with $n - 2$ degrees of freedom. The **R** functions *confint*( ) and *lm*( ) can be used together to obtain this confidence interval for β.

Use the **R** dataset *pines_1997* to obtain a 95% confidence interval for the slope parameter associated with a linear regression of the 1997 height (Hgt97) on the 1997 diameter (Diam97) for the pines at the Kenyon Center for Environmental Study (KCES).

11.A.12. *Confidence Interval for the Slope Parameter β for an Arbitrary Population.* Consider the linear model setting in (11.13) and let Y_i denote the value of the dependent variable Y at fixed value x_i of the independent variable x, for $i = 1, \ldots, n$. Suppose that all n x's are distinct. Compute the $N = \binom{n}{2} = n(n-1)/2$ individual sample slopes

$$S_{ij} = \frac{(Y_j - Y_i)}{(x_j - x_i)}, \quad 1 \leq i < j \leq n;$$

and let $S^{(1)} \leq \cdots \leq S^{(N)}$ be the ordered sample slope values. By properly specifying the *probs* argument, the **R** function *quantile*( ) can be used to obtain a 100CL% confidence interval for β based directly on these ordered sample slopes $S^{(1)} \leq \cdots \leq S^{(N)}$. For example, given a vector of data **x**, we can construct a 90% confidence interval using the following command.

```
> quantile(x, probs = c(0.05, 0.95))
```

Use the S_{ij} sample slope values for the cricket chirp rate data from Table 11.12 in Example 11.5 (without the (x, y) pair (80.6, 17.1) to avoid ties among the temperature values) to find a 95% confidence interval for the slope parameter associated with a linear regression of cricket chirp rate on concurrent temperature.

11.B. Data Analysis/Computational

11.B.1. *Do Strikeouts Affect Batting Averages?* In Example 11.1 we found that there was insufficient evidence to link the number of strikeouts with the number of home runs hit by major league ballplayers. A related question of interest is whether the number of strikeouts might be negatively related to the overall batting average for major leaguers. The data in Table 11.17 contains the number of strikeouts and final batting average from the 2016 season for the same 15 major league ballplayers considered in Example 11.1.

Table 11.17 2016 batting average and strikeout statistics for a subset of major league players

Player	Final batting average	Number of strikeouts
Mookie Betts	.318	80
Robinson Cano	.298	100
Yoenis Cespedes	.280	108
Chris Bryant	.292	154
Andrew McCutchen	.256	143
Miguel Cabrera	.316	116
Albert Pujols	.268	75
Todd Frazier	.225	163
Ryan Braun	.305	98
Charlie Blackmon	.324	102
Josh Donaldson	.284	119
Chris Davis	.221	219
Justin Turner	.275	107
David Ortiz	.315	86
Bryce Harper	.243	117

Source: rotowire.com (2016)

Use these data to test the conjecture that a major leaguer's batting average is negatively correlated with his number of strikeouts.

11.B.2. *Do Golf Handicaps "Drive" Stock Prices?* An investment compensation expert, Graef Crystal, undertook a study to investigate whether there is any link between the golf handicap for a company's CEO and the value of the company's publicly traded stock. He reported his findings in the May 31, 1998 issue of The New York Times under the heading "Investing It: Duffers Need Not Apply". Table 11.18 contains the golf handicaps (based on data obtained from the journal *Golf Digest*) and stock ratings (compiled by Crystal using data on the stock market performance of the companies) for 51 CEO's.

Find the P-value for a test of the conjecture that CEO handicap and Stock Rating are negatively correlated.

11.B.3. *Voter Turnout in Presidential Elections*. The population of the United States has steadily grown over the years since it became a nation, so we might

Table 11.18 Golf handicaps and stock ratings for major company CEOs

CEO handicap	Stock rating	CEO handicap	Stock rating
3.2	97	12.8	49
23.9	95	13.0	48
18.0	95	15.6	46
22.0	92	19.2	45
34.0	89	13.7	44
25.0	89	22.0	43
11.0	85	18.6	41
10.1	83	11.9	40
20.0	82	22.0	38
21.1	79	10.0	37
3.8	77	27.1	35
13.1	75	16.6	35
7.1	74	8.0	33
17.2	73	15.5	31
13.0	72	14.8	29
10.1	67	12.8	29
10.1	66	24.2	25
11.0	64	18.1	24
12.6	64	18.0	22
10.9	58	10.0	22
7.6	58	16.0	20
10.6	55	23.0	15
16.1	54	19.0	13
10.9	54	18.0	12
12.6	51	11.7	3
17.6	49		

Source: New York Times (1998)

expect that voter turnout in presidential elections would also have grown consistently over time from election to election. Table 11.19 contains the total popular vote (in thousands) for each of the elections from 1940 through 2012.

Find the P-value for an appropriate procedure to test if there is, indeed, an increasing trend in popular vote turnout for presidential elections over the period 1940–2012.

Table 11.19 Popular vote (in thousands) for presidential elections, 1940–2012

Year	Total popular vote (in thousands)
1940	49,900
1944	47,977
1948	48,834
1952	61,552
1956	62,027
1960	68,836
1964	70,098
1968	73,027
1972	77,625
1976	81,603
1980	86,497
1984	92,653
1988	91,595
1992	104,427
1996	96,278
2000	105,405
2004	122,295
2008	131,314
2012	129,085

Source: ropercenter.cornell.edu (2016)

11.B.4. *Do Caution Flags Really Slow Races?* The National Association for Stock Car Auto Racing (NASCAR) was founded in December 1947. It sponsors the Winston Cup, currently comprised of 36 races per year, with up to 43 cars competing in each race. One of the issues surrounding these races is the potential impact on fan enjoyment from slowing the race due to caution flags required when an incident (usually an accident) has occurred. The number of caution flags and winning time (in minutes) for 82 races over the period of time from 1975 through 2003 are given in Table 11.20. Each of these 82 races was held on a 2.5 mile track and the winner of each race completed the full 200 laps. (Thus, the winning times can be compared fairly.)

Table 11.20 Number of caution flags and winning times (in minutes) for 82 NASCAR races during 1975–2003

# of caution flags	Winning time	# of caution flags	Winning time
3	195.25	5	269.83
1	213.20	7	197.13
7	258.90	2	218.82
6	195.80	4	233.68
5	233.83	5	187.82
1	210.47	4	217.73
7	208.37	7	260.40
6	225.87	5	168.92
5	241.17	6	231.77
4	176.87	5	194.82
7	264.13	6	259.75
6	251.87	6	192.33
6	233.22	5	261.28
7	198.68	3	217.13
9	247.22	5	174.15
3	215.87	6	223.87
8	202.53	9	264.83
4	170.20	9	245.57
9	246.42	7	218.13
6	237.82	5	244.17
7	202.07	6	228.45
9	254.57	3	180.98
13	248.75	10	241.80
9	202.50	7	244.57
4	187.20	3	208.30
3	223.78	7	193.58
6	217.38	8	224.98
4	191.17	5	232.92
5	220.47	10	211.70
6	217.83	5	223.82
6	194.82	4	215.67
4	207.05	8	202.30
4	214.55	4	211.17
3	173.70	9	254.65
5	222.78	4	185.70
11	252.32	9	256.45
7	199.72	5	214.68

(continued)

Table 11.20 (continued)

# of caution flags	Winning time	# of caution flags	Winning time
7	229.60	3	185.43
7	223.23	6	222.90
9	209.83	5	209.17
5	222.40	8	234.92

Source: Winner (2006); Sporting News Books (2004)

(a) Obtain the fitted least squares line for these data, treating number of caution flags as the independent variable and winning time as the dependent variable.

(b) Find the P-value for a test of the null hypothesis that the winning time for a NASCAR race of 200 laps on a 2.5 mile track is linearly related to the number of caution flags. Which alternative hypothesis do you think is appropriate for this setting?

(c) Use the fitted least squares line from part (a) to predict the winning time for a race of this type with 6 caution flags. Compare this predicted value with the observed winning times for those races with 6 caution flags in our sample data.

(d) How would you feel about using the fitted least squares line in (a) to predict winning times for races with no caution flags? races with 15 caution flags?

11.B.5. *Come On—I Just Walked Him!* There is a general conception in baseball that walks seem to somehow come back to haunt a pitcher by scoring. But are walks really a major contributor to a pitcher's earned run average? Table 11.21 contains the nine-inning walk rates and the earned run averages for fifteen major leaguers who pitched at least 140 innings during the 2016 baseball season.

Use these data to test the conjecture that a major league pitcher's earned run average is positively correlated with his nine-inning walk rate.

Table 11.21 Nine-inning walk rate and earned run average for 15 major leaguers who pitched at least 140 innings in the 2016 season

Player	Earned run average	Nine-inning walk rate
David Price	3.99	1.96
Max Scherzer	2.96	2.21
Justin Verlander	3.04	2.25
Madison Bumgarner	2.74	2.14
Jon Lester	2.44	2.31
Cole Hamels	3.32	3.45
Adam Wainwright	4.62	2.67
Bartolo Colon	3.43	1.50
Noah Syndergaard	2.60	2.11
Michael Pineda	4.82	2.71
Trevor Bauer	4.26	3.32
Chris Sale	3.34	1.79
Clayton Kershaw	1.69	0.66
Stephen Strasburg	3.60	2.68
Rick Porcello	3.15	1.29

Source: rotowire.com (2016)

11.B.6. *Careful When You Chirp.* Consider the cricket chirp rate data from Table 11.12 in Example 11.5.

(a) What is the equation of the least squares line fit to these data (consider temperature as the independent variable x and cricket chirp rate as the dependent variable y)?

(b) Using this least squares fitted line, what would you estimate the chirp rate to be when the concurrent temperature is 70 ° F?

(c) Using this least squares fitted line, what would you estimate the chirp rate to be when the concurrent temperature is 100 ° F?

(d) Using this least squares fitted line, what would you estimate the chirp rate to be when the concurrent temperature is 32 ° F?

(e) Discuss your answers to parts (c) and (d) in the context of these data.

Table 11.22 Finish position and starting position for each of 42 drivers in a NASCAR race in 1987

Finish position	Start position	Finish position	Start position
1	1	2	4
3	11	4	7
5	13	6	6
7	3	8	5
9	31	10	22
11	19	12	15
13	21	14	8
15	37	16	35
17	24	18	14
19	36	20	42
21	27	22	18
23	28	24	40
25	38	26	33
27	2	28	39
29	29	30	10
31	12	32	17
33	23	34	16
35	20	36	34
37	9	38	30
39	26	40	25
41	32	42	41

Source: Winner (2006); Sporting News Books (2004)

11.B.7. *How Important Is a Good Starting Position?* The National Association for Stock Car Auto Racing (NASCAR) was founded in December 1947. It sponsors the Winston Cup, currently comprised of 36 races per year, with up to 43 cars competing in each race. Is it important to have a good starting position in these races? Table 11.22 contains the starting position and finish position for 42 drivers in a NASCAR race held in 1987.

(a) Using the procedure based on the Spearman sample correlation coefficient, find the *P*-value for a test of the hypothesis that starting position and finish position are independent variables against an appropriate alternative.

Table 11.23 Finish position and starting position for each of 42 drivers in a NASCAR race in 1997

Finish position	Start position	Finish position	Start position
1	4	2	6
3	20	4	22
5	9	6	11
7	14	8	34
9	17	10	15
11	10	12	5
13	40	14	7
15	32	16	29
17	38	18	37
19	27	20	23
21	1	22	25
23	24	24	13
25	26	26	16
27	33	28	36
29	31	30	12
31	8	32	28
33	35	34	18
35	21	36	19
37	39	38	3
39	42	40	2
41	41	42	30

Source: Winner (2006); Sporting News Books (2004)

(b) Repeat part (a) using the Kendall sample correlation coefficient. Are your findings similar?

11.B.8. *How Important Is a Good Starting Position?—Part II.* Table 11.23 contains the starting position and finish position for 42 drivers in a second NASCAR race held in 1997.

(a) Using the procedure based on the Spearman sample correlation coefficient, find the P-value for a test of the hypothesis that starting position and finish position are independent variables against an appropriate alternative.

(b) Repeat part (a) using the Kendall sample correlation coefficient. Are your findings similar?

(c) Compare your results for this race with those obtained for the 1987 race in Exercise 11.B.7.

11.B.9. *How Fast Is the Arctic Sea Ice Melting?* While the cause of climate change is bantered about in the popular media, the fact that it is occurring is not in question. Table 11.24 contains the extent of Arctic Sea Ice (in millions of square kilometers) in September for the years 1979 through 2012.

(a) Plot the Arctic Sea Ice data versus the years of measurement.

(b) Compute the Kendall sample correlation coefficient for the data in Table 11.24.

(c) Find the P-value for testing the conjecture that there is a decreasing trend in the extent of Arctic Sea Ice in September over the period of time from 1979 through 2012.

(d) Obtain the fitted least squares line for the data in Table 11.24. Plot this line on the plot of Arctic Sea Ice versus year of measurement. Does it look like a good fit?

(e) Use this fitted least squares line to find the P-value for a hypothesis test of the conjecture that there is a linear decline in September Arctic Sea Ice over the period of time from 1979 through 2012.

11.B.10. *Carbon Dioxide and Global Warming*. One aspect of climate change that has received a lot of attention from the scientific community is the effect of atmospheric CO_2 (carbon dioxide) concentration on global temperature. Table 11.25 contains the atmospheric CO_2 concentration in parts per million and the Global Land-Ocean Temperature Index from the Goddard Institute of Space Studies (GISTEMP) for the years 1979–2010. GISTEMP is reported in units of 1/100 of a degree Centigrade increase above the 1950–1980 mean and is known in the literature as the global surface temperature anomaly.

Table 11.24 **Extent of Arctic Sea ice (millions of square kilometers) in September for the years 1979–2012**

Year	Extent of arctic sea ice in september
1979	7.20
1980	7.85
1981	7.25
1982	7.45
1983	7.52
1984	7.17
1985	6.93
1986	7.54
1987	7.48
1988	7.49
1989	7.04
1990	6.24
1991	6.55
1992	7.55
1993	6.50
1994	7.18
1995	6.13
1996	7.88
1997	6.74
1998	6.56
1999	6.24
2000	6.32
2001	6.75
2002	5.96
2003	6.15
2004	6.05
2005	5.57
2006	5.92
2007	4.30
2008	4.68
2009	5.36
2010	4.90
2011	4.61
2012	3.61

Source: Witt (2013); Fetterer et al. (2016)

Table 11.25 Atmospheric CO_2 concentration (parts per million) and GISTEMP (1/100 °C), 1979–2010

Year	CO_2 concentration	GISTEMP
1979	336.67	8
1980	338.57	19
1981	339.92	26
1982	341.30	4
1983	342.71	25
1984	344.24	9
1985	345.81	4
1986	347.11	12
1987	348.72	27
1988	351.04	31
1989	352.68	19
1990	353.97	36
1991	355.37	35
1992	356.18	13
1993	356.69	13
1994	358.14	23
1995	360.02	37
1996	361.95	29
1997	363.18	39
1998	365.19	56
1999	367.86	31
2000	368.83	33
2001	370.43	47
2002	372.01	56
2003	374.45	55
2004	376.77	48
2005	378.30	63
2006	380.83	55
2007	382.56	58
2008	384.39	44
2009	386.34	57
2010	388.13	63

Source: Witt (2013); NASA Goddard Institute for Space Studies (GISS); Earth System Research Laboratory of the National Oceanic and Atmospheric Administration (NOAA)

(a) Compute both the Pearson correlation coefficient and the Spearman rank correlation coefficient between the atmospheric CO_2 concentration and GISTEMP.

(b) Using the Pearson correlation coefficient, find the P-value for a hypothesis test of the conjecture that atmospheric CO_2 concentration and GISTEMP are positively correlated. Do the same for a hypothesis test using the Spearman rank correlation coefficient.

(c) Obtain the fitted least squares line for the CO_2 concentration and GISTEMP data in Table 11.25.

(d) Using the Kendall correlation coefficient, find the P-value for a hypothesis test of the conjecture that atmospheric CO_2 concentration was increasing over the period of time from 1979 to 2010.

11.C. Activities

11.C.1. *Sodium and Calories in Canned Food.* Go to your favorite supermarket and randomly select ten different canned food items from the shelves. For each of these items, record the amount of grams of sodium per serving and total calories per serving. Using these data, find the P-value for an appropriate test of the conjecture that grams of sodium and calorie content are positively correlated for canned foods.

11.C.2. *Heart Rate and Blood Pressure.* Obtain heart rate (in beats per minute) and systolic blood pressure (in mm. Hg) values for eight women and eight men.

(a) Using all of the combined data for women and men, find the P-value for an appropriate test of the conjecture that heart rate and systolic blood pressure are positively correlated. Then carry out the same analysis separately for women and men. Discuss your findings.

(b) Find the least squares fitted lines for the combined data and then separately for women and men. Discuss your findings.

(c) Using the combined data, find the P-value for a test of the conjecture that there is a positive linear relationship between heart rate and systolic blood pressure.

11.C.3. *Coffee and Bedtime.* Survey at least 15 of your friends and/or classmates to obtain the following information from each of them: (i) average number of cups of coffee they drink in a 24 h day and (ii) their average bedtime, in minutes past ten p.m.

(a) Using these data, find the P-value for the conjecture that these two variables are positively correlated.

(b) Find the least squares fitted line for a linear regression of average bedtime on average daily cups of coffee. Find the P-value for an appropriate test of the significance of the linear regression.

11.D. Internet Archives

11.D.1. *Grip Strength and Fraility.* Chronological age is a natural marker of frailty. However, it is not a perfect marker, as there is wide variability in frailty between individuals of the same age. Numerous scientific studies have been conducted to investigate the possible connection between grip strength as a more reliable marker of frailty in older individuals. Use the Internet to find a scientific paper that addresses this association between grip strength and frailty. Summarize the findings discussed in the paper, particularly how the authors used correlation and regression to support their conclusions.

11.D.2. *Passing Yardage and College Football Victories.* Winning a college football game is dependent on a lot of performance variables. One of these variables is passing yardage. Use the Internet to find the following information for each of the Division I football games played on the most recent first Saturday in November:

(i) Passing Yardage for the Winning Team
(ii) Total Points Scored by the Winning Team
(iii) Total Points Scored by the Losing Team.

(a) Find the P-value for a test of the conjecture that total points scored by the winning team is positively correlated with the passing yardage for the winning team.
(b) Find the P-value for a test of the conjecture that total points scored by the losing team is negatively correlated with the passing yardage of the winning team.
(c) Find the least squares fitted line for the regression of total points scored by the winning team on the passing yardage for the winning team. Obtain the P-value for a test of the conjecture that there is a positive linear relationship between total points scored by the winning team and their passing yardage.

11.D.3. *Rushing Yardage and College Football Victories*. Use the Internet to find the rushing yardage for the winning team in each of the Division I football games played on the most recent first Saturday in November. Complete the following statistical analyses.

(a) Find the P-value for a test of the conjecture that total points scored by the winning team is positively correlated with the rushing yardage for the winning team.
(b) Find the P-value for a test of the conjecture that total points scored by the losing team is negatively correlated with the rushing yardage of the winning team.

(c) Find the least squares fitted line for the regression of total points scored by the winning team on the rushing yardage for the winning team. Obtain the P-value for a test of the conjecture that there is a positive linear relationship between total points scored by the winning team and their rushing yardage.
(d) Compare the results obtained in this exercise with those obtained in Exercise 11.D.2.

11.D.4. *How Important are Three-Point Shots in College Basketball?* One of the more recent changes to the rules of college basketball has been the addition of the three-point arc, beyond which a made field goal counts for three points rather than the standard two points. While this is clearly an exciting option for the fans attending games, how much effect has it actually had on winning or losing basketball games? Use the Internet to find the following information for each of the Division I basketball games played on the most recent third Saturday in January:

(i) Number of Made Three-Point Shots for the Winning Team
(ii) Total Points Scored by the Winning Team.

(a) Find the P-value for a test of the conjecture that total points scored by the winning team is positively correlated with the number of three-point shots they make.
(b) Find the least squares fitted line for the regression of total points scored by the winning team on the number of three-point shots they make. Obtain the P-value for a test of the conjecture that there is a positive linear relationship between total points scored by the winning team and the number of three-point shots they make.

11.D.5. *Trends in Never-Married Americans.* The share of never-married Americans has been on the rise for the past five decades and men are more likely than women to have never been married. Use the Internet to find one or more reports that provide data to support this statement. Using these data, find the P-value for a test of the conjecture that there is a positive trend in the percentage of never-married American women over the period of time from 1960 through 2012. Do the same for never-married American men.

11.D.6. *Explosion of Social Networking Sites.* As of October 2015, nearly two-thirds of American adults were using at least one social networking site. How fast has been the rise in this acceptance of social networking? Use the Internet to find one or more reports that provide data to address this question. Using these data, find the P-value for a test of the conjecture that the use of social networking sites has been increasing since it stood at 7% in 2005. Are there differences in the rate of this rise due to age, gender, education, and income? Discuss the relevant findings from your report(s).

11.D.7. *Shortage of Marriageable Men?* A Pew Research Center report found that over three-quarters of women surveyed cited that having a partner with a stable job was a very important attribute that they look for in someone to marry. But is that criterion becoming more difficult to satisfy? Use the Internet to find one or more reports that address this question over time for college-educated women who are 25–35 years old.

12 Statistical Inference for More Than Two Populations

Do generic versions of a drug do as well as the brand name version? Are there differences in blood pressure levels across ethnic groups? Which of three approaches to meniscal repair leads to the most effective recovery time? Are there differences between our major cities in the amount of time motorists spend in traffic congestion? How do competing car brands fare with regard to miles per gallon in city driving? Are there differences among types of higher education institutions with regard to the level of remaining student debt ten years after graduation?

Questions such as these require statistical inference procedures for analyzing independent random sample data from more than two populations. In Chap. 9 we discussed statistical inference procedures for analyzing independent random sample data from two populations. In this chapter we extend those results to accommodate $k > 2$ populations, commonly known as one-way Analysis of Variance (ANOVA) procedures.

© Springer International Publishing AG 2017

D.A. Wolfe, G. Schneider, *Intuitive Introductory Statistics*, Springer Texts in Statistics,
DOI 10.1007/978-3-319-56072-4_12

Section 1 presents one-way rank-based ANOVA methodology for analyzing independent random sample data from more than two populations that does not require normality of the populations. Section 2 details a one-way ANOVA procedure for more than two normal populations. In Section 3 we discuss a rank-based ANOVA procedure specifically designed to address settings where it is anticipated that the responses across the populations will be monotonically ordered.

General Setting and Notation Let $\{X_{11}, \ldots, X_{n_1 1}\}$, $\{X_{12}, \ldots, X_{n_2 2}\}, \ldots,$ $\{X_{1k}, \ldots, X_{n_k k}\}$ be k mutually independent random samples of sizes $n_1, \ldots,$ n_k from populations $1, 2, \ldots, k$, respectively. Let $N = \sum_{j=1}^{k} n_j$ be the total number of sample observations from the k populations. We assume that these populations are related to one another through the following one-factor ANOVA model:

$$X_{ij} = \tau_j + e_{ij}, \qquad i = 1, \ldots, n_j, j = 1, \ldots k,$$

where the N e's have the same probability distribution with median θ and τ_j is known as the " treatment j " effect, for $j = 1, \ldots, k$. Thus the only possible differences between the k populations are through their "treatment" effects $\tau_1, \ldots, \tau_k$. Throughout this chapter we will be interested in testing the null hypothesis

$$H_0 : [\tau_1 = \tau_2 = \cdots = \tau_k], \tag{12.1}$$

corresponding to no differences between the probability distributions for the k populations, against either general alternatives

$$H_1 : [\tau_1, \ldots, \tau_k \ \textit{not all equal}] \tag{12.2}$$

or ordered alternatives

$$H_2 : [\tau_1 \leq \tau_2 \leq \cdots \leq \tau_k, \; \textit{with at least one strict inequality}], \qquad (12.3)$$

corresponding to a monotonically non-decreasing "treatment" effect across the populations.

12.1 One-way Rank-Based General Alternatives ANOVA for More Than Two Populations

In Sect. 9.2 we introduced the idea of using joint rankings for independent random samples from two populations to make inferences about their medians without requiring the assumption of population normality. Here we take the same approach tor independent random samples from $k > 2$ populations.

First we combine all $N = \sum_{j=1}^{k} n_j$ observations from the k independent random samples and order (rank) them from least to greatest. In case of a group of 2 or more tied observations, use the average rank assigned to that group for each of the observations in the group. For $i = 1, \ldots, n_j$ and $j = 1, \ldots, k$, let R_{ij} denote the rank of X_{ij} in this joint ranking. For $j = 1, \ldots, k$, let R_j and $R_{.j}$ be the sum and average, respectively, of the joint ranks for the observations from the jth population; that is,

$$R_j = \sum_{i=1}^{n_j} R_{ij} \text{ and } R_{.j} = \frac{R_j}{n_j}, \quad \text{for} j = 1, \ldots, \text{k}. \qquad (12.4)$$

You are asked to show in Exercise 12.A.1 that the average of all of the ranks $1, \ldots, \text{N}$ assigned in this joint ranking is $R_{..} = \frac{1+2+\cdots+N}{N} = \frac{N+1}{2}$. If the null hypothesis $H_0 : [\tau_1 = \tau_2 = \cdots = \tau_k]$ is true, we would expect the k sample average ranks $R_{.1}, \ldots, R_{.k}$ to be similar and "close" to the combined average rank $R_{..} = \frac{N+1}{2}$. To assess possible deviations from H_0, we use the Kruskal-Wallis test statistic

$$Q = \frac{12}{N(N+1)} \sum_{j=1}^{k} n_j (R_{.j} - R_{..})^2 = \frac{12}{N(N+1)} \sum_{j=1}^{k} n_j \left(R_{.j} - \frac{N+1}{2} \right)^2. \quad (12.5)$$

We note that large values of Q provide evidence that the null hypothesis $H_0 : [\tau_1 = \tau_2 = \cdots = \tau_k]$ is not tenable. We can use the sampling distribution of Q when the null hypothesis is true to assess whether this evidence is sufficient to reject H_0 in favor of the general alternatives $H_1 : [\tau_1, \ldots, \tau_k \textit{ not all equal}]$ that the k population medians $\tau_1, \ldots, \tau_k$ are not the same.

One-way Rank-Based General Alternatives ANOVA for More Than Two Populations To test the null hypothesis $H_0 : [\tau_1 = \tau_2 = \cdots = \tau_k]$ against the general alternative $H_1 : [\tau_1, \ldots, \tau_k \textit{ not all equal}]$ using the k mutually independent random samples $\{X_{11}, \ldots, X_{n_1 1}\}, \{X_{12}, \ldots, X_{n_2 2}\}, \ldots, \{X_{1k}, \ldots, X_{n_k k}\}$ from populations 1, 2, ..., k, respectively, compute the Kruskal-Wallis statistic Q (12.5) and let q_{obs} be the attained value of Q. Then the P-value for a test of H_0 (12.1) against the general alternatives H_1 (12.2) is given by

$$P\text{–}value = P_0(Q \geq q_{obs}), \quad (12.6)$$

where $P_0(Q \geq q_{obs})$ is obtained from the sampling distribution of Q when H_0 is true.

Often it is easier to compute the Kruskal-Wallis statistic Q (12.5) using the equivalent representation (see Exercise 12.A.2)

$$Q = \left(\frac{12}{N(N+1)} \sum_{j=1}^{k} \frac{R_j^2}{n_j} \right) - 3(N+1). \tag{12.7}$$

Example 12.1. Meniscal Repair—Which Will It Be: FasT-Fix Sutures, Biodegradable Meniscus Arrows, or Vertical Mattress Sutures? Surgery is most often the only option when faced with a torn medial meniscus—but what is the best surgical method for the repair? Borden et al. (2003) studied the performance characteristics of three different meniscal repair techniques, namely, the FasT-Fix Meniscal Repair Suture System (FasT-Fix), the use of biodegradable Meniscus Arrows (MA), and the Vertical Mattress Sutures (VMS) approach. Human cadaveric knees were used in the study, with six randomly assigned to each of the three meniscus surgery techniques. Each repaired meniscus was loaded into a servohydraulic device and tension-loaded (similar to the type of stresses that a meniscus might have to deal with during early rehabilitation) until failure of the repair occurred. Table 12.1 contains the data for load at failure (Newtons (N)) for the eighteen meniscus surgeries in the study.

Table 12.1 Load (Newtons (N)) at failure for meniscal repairs

FasT-Fix	Meniscus arrows	Vertical mattress
88.0	44.9	97.3
119.8	46.1	106.4
65.8	59.3	118.2
82.9	35.5	99.7
149.9	50.7	106.5
117.1	56.8	84.2

Source: Borden et al. (2003)

Jointly ranking the eighteen failure times in Table 12.1 from least to greatest, we find the following ranks for the three repair mechanisms:

FasT-Fix	Meniscus arrows	Vertical mattress
10	2	11
17	3	13
7	6	16
8	1	12
18	4	14
15	5	9

.

Summing these ranks within repair mechanisms, we obtain:

$$R_1 = R_{FasT-Fix} = 10 + 17 + 7 + 8 + 18 + 15 = 75$$
$$R_2 = R_{Arrows} = 2 + 3 + 6 + 1 + 4 + 5 = 21$$
$$R_3 = R_{Mattress} = 11 + 13 + 16 + 12 + 14 + 9 = 75.$$

Combining these rank sums with the sample sizes $n_1 = n_2 = n_3 = 6$ and $N = 18$, we find from equation (12.7) that the observed value of the Kruskal-Wallis test statistic Q is

$$q_{obs} = \left(\frac{12}{18(18+1)}\left[\frac{(75)^2}{6} + \frac{(21)^2}{6} + \frac{(75)^2}{6}\right]\right) - 3(18+1) = 68.310 - 57$$
$$= 11.37.$$

Using the **R** function *pKW*( ) (from the *NSM3* package) along with the **R** dataset *meniscal_repairs_load_at_failure*, it follows that the associated *P*-value, $P_0(Q \geq 11.37)$, for a test of $H_0 : [\tau_1 = \tau_2 = \tau_3]$ versus the general alternative $H_1 : [\tau_1, \tau_2, \tau_3 \textit{ not all equal}]$ is approximately 0.0006.

```
> pKW(meniscal_repairs_load_at_failure)

Group sizes: 6 6 6
Kruskal-Wallis H Statistic: 11.3684
Monte Carlo (Using 10000 Iterations) upper-tail probability: 6e-04
```

Hence, there is strong evidence that the median failure time load is different for the three meniscal repair techniques FasT-Fix, Meniscus Arrows, and Vertical Mattress.

This example introduces two new concepts in **R** that we have not yet encountered. The first is that the dataset *meniscal_repairs_load_at_failure* is stored as a "list" rather than a "data.frame". The reason for this is that a data.frame is required to have an equal number of rows in every column. (In our case this means that each of the k samples would need to be the same size. Although this is the case in Example 12.1, it won't necessarily be that way throughout the rest of the chapter.)

The second new concept is that of a "Monte Carlo" sample. This is an approximation method that relies on taking a large sample from the null distribution of Q to estimate the P-value (rather than using the exact distribution to calculate it directly) because the calculation of the exact distribution becomes computationally infeasible when the sample sizes are moderately large.

Often in ANOVA settings when we reject the null hypothesis $H_0 : [\tau_1 = \tau_2 = \cdots = \tau_k]$, it is of interest to know which particular medians $\tau_1, \ldots, \tau_k$ are not equal. For instance, it would be of interest to know if there is sufficient statistical evidence from the data in Example 12.1 to conclude that the Meniscus Arrows technique is inferior to both the FasT-Fix and Vertical Mattress techniques with regard to median load to failure of the repaired menisci. Such inferential techniques are called *multiple comparison procedures*. For more information on the multiple comparison procedures associated with the Kruskal-Wallis test, see, for example, Chapter 6 of Hollander et al. (2014).

Section 12.1 Practice Exercises

12.1.1. *Arts Participation Across the States—Are There Regional Differences in Reading Literature?* The National Endowment for the Arts periodically collects survey information from residents about their participation in the arts.

Table 12.2 Percentages of state residents who read literature in 2015, stratified by regions of the United States

	State	Percentage Reading Literature (%)
Midwest	Illinois	47.6
	Kansas	49.4
	Minnesota	49.3
	Ohio	46.0
	South Dakota	52.1
Northeast	Delaware	42.6
	Maine	45.8
	Massachusetts	52.3
	New York	48.4
	Virginia	42.7
South	Alabama	34.8
	Florida	30.5
	Louisiana	36.6
	South Carolina	39.3
	Texas	37.5
West	Alaska	59.3
	California	38.9
	Idaho	56.4
	New Mexico	50.4
	Utah	57.0

Source: National Endowment for the Arts (2016)

A portion of the statewide results from the 2015 survey is presented in the publication *Arts Profile #11 (August 2016)*. In particular, the publication includes the estimated percentage of each state's residents who read literature in 2015. A subset of these statewide reading percentages is presented in Table 12.2, stratified by regions of the United States.

Find the *P*-value for a hypothesis test of the conjecture that the median percentage of state residents reading literature in 2015 differs across these four graphical regions of the United States.

12.1.2. *Are Nonprofit Hospitals Equally Charitable?* All nonprofit hospitals are dedicated to providing charity care (uncollectable debt) for those individuals who cannot afford it. The extent of this charity, however, can vary

Table 12.3 Charity care as a percentage of total expenses for a subset of twenty nonprofit hospitals in Colorado, Michigan, Ohio, and Virginia for the two-year period 2012–2013

Colorado	Michigan	Ohio	Virginia
6.00%	3.26%	1.41%	2.46%
1.90%	9.76%	2.86%	13.68%
0.02%	7.68%	8.48%	2.20%
8.03%	4.73%	10.77%	2.88%
3.29%	8.42%	2.19%	4.45%

Source: Watchdog.org (2016)

Table 12.4 Number of mountain bushtail possum observed as a function of the availability of hollow-bearing trees in various sites in the Central Highlands of Victoria, Southeastern Australia

Number of available hollow-bearing trees			
0–4	4–8	8–12	More than 12
0	0	1	4
0	2	3	5
1	2	4	5
1	4	7	6
2			11

Source: Lindenmayer et al. (2014)

considerably from state to state. Table 12.3 contains the uncollectable debt (as a percentage of expenses) for a subset of nonprofit hospitals in the states of Colorado, Michigan, Ohio, and Virginia over the two-year period 2012-2013.

Find the P-value for a test of the conjecture that the median charity care (as a percentage of total expenses) differs across these four states of Colorado, Michigan, Ohio, and Virginia.

12.1.3. *Want Possums?—Hollow Out a Tree*. One of the critical features for the presence of possums is the availability of large, old, hollow-bearing trees to provide appropriate habitat for the critters. Lindenmayer et al. (2014) reported on the results of a study conducted in the montane ash forests of the Central Highlands of Victoria, Southeastern Australia. Table 12.4 contains a subset of the data they collected.

Find the P-value for the Kruskal-Wallis test of the conjecture that there is a difference in the median number of possums for these different levels of available hollow-bearing trees.

12.2 One-way General Alternatives ANOVA for More Than Two Normal Populations

In this section we make the additional assumption that all k of the underlying populations are normally distributed. Thus, we assume that $\{X_{11},\ldots,X_{n_1 1}\}$, $\{X_{12},\ldots,X_{n_2 2}\},\ldots,$ $\{X_{1k},\ldots,X_{n_k k}\}$ are k mutually independent random samples of sizes $n_1, \ldots, n_k$ from normal populations with means $\tau_1,\ldots,\tau_k$, respectively, and common variance σ^2. We are still interested in testing the null hypothesis H_0 (12.1) against the general alternative H_1 (12.2), but here we will be able to design our procedure to take advantage of the assumed normality.

The k-sample test for equality of means under the normality assumption is based on a comparison of (i) the observed variability among the observations *within* the separate 'treatment' samples and (ii) the differences in variability *between* the treatment groups. (This is how the term analysis of variance (ANOVA) came to be used to describe procedures designed to test for differences in means from k populations.) To quantify this notion, we first need to define some sample measures of variability.

Let

$$\bar{X}_{.j} = \frac{1}{n_j}\sum_{i=1}^{n_j} X_{ij} \qquad (12.8)$$

be the sample mean for the observations from the jth population, for $j = 1, \ldots, k$, and let

$$\bar{X}_{..} = \frac{1}{N}\sum_{j=1}^{k}\sum_{i=1}^{n_j} X_{ij} = \frac{1}{N}\sum_{j=1}^{k} n_j \bar{X}_{.j} \tag{12.9}$$

be the *grand mean*, corresponding to the average of all $N = \sum_{j=1}^{k} n_j$ sample observations. The *total variation in the sample data* corresponds to the total sum of squares SST given by

$$SST = \sum_{j=1}^{k}\sum_{i=1}^{n_j} (X_{ij} - \bar{X}_{..})^2. \tag{12.10}$$

This total variation can be broken into two components, one component corresponding to the *between-treatments variation* SSB (the sum of squares attributable to any treatment differences) and a second component corresponding to the *within-treatments variation* SSE (the sum of squares due simply to randomness or sample error) as follows:

$$\text{SST} = \text{SSB} + \text{SSE}, \tag{12.11}$$

where

$$SSB = \sum_{j=1}^{k} n_j (\bar{X}_{.j} - \bar{X}_{..})^2 \tag{12.12}$$

and

$$SSE = \sum_{j=1}^{k}\sum_{i=1}^{n_j} (X_{ij} - \bar{X}_{.j})^2. \tag{12.13}$$

(You are asked to verify the relationship in (12.11) in Exercise 12.A.3.)

If the null hypothesis $H_0 : [\tau_1 = \tau_2 = \cdots = \tau_k]$ is true, each of the k samples comes from the same normal population with mean μ and variance σ^2. Under that condition, all k of the 'treatment' sample means, $\bar{X}_{.1}, \ldots, \bar{X}_{.k}$, should be similar and close to the grand mean, $\bar{X}_{..}$, leading to a small value for the between-treatments sum of squares SSB relative to the overall within-treatments error sum of squares SSE. Comparison of these two quantities will provide information about the validity of the null hypothesis H_0. This comparison is facilitated through the test statistic

$$F = \frac{SSB/(k-1)}{SSE/(N-k)}, \tag{12.14}$$

where SSB/(*k*-1) is commonly referred to as the mean square error due to 'treatment' and denoted by MSB, and SSE/(*N*-*k*) is commonly referred to as the mean square due to error (randomness) and denoted by MSE. Large values of *F*, corresponding to greater sample variability between treatments than within treatments, will be indicative of the alternative hypothesis $H_1 : [\tau_1, \ldots, \tau_k \textit{ not all equal}]$ and lead to rejection of the null hypothesis $H_0 : [\tau_1 = \tau_2 = \cdots = \tau_k]$.

One-way General Alternatives ANOVA for More Than Two Normal Populations To test the null hypothesis $H_0 : [\tau_1 = \tau_2 = \cdots = \tau_k]$ against the general alternative $H_1 : [\tau_1, \ldots, \tau_k \textit{ not all equal}]$ using the *k* mutually independent random samples $\{X_{11}, \ldots, X_{n_1 1}\}$, $\{X_{12}, \ldots, X_{n_2 2}\}, \ldots,$ $\{X_{1k}, \ldots, X_{n_k k}\}$ from normal populations 1, 2, …, *k* with common variance σ^2, compute the statistic *F* (12.14) and let f_{obs} be the attained value of *F*. Then the exact *P*-value for a test of H_0 (12.1) against the general alternatives H_1 (12.2) for normal populations with common variance σ^2 is given by

$$P\text{-}value = P_0(F \geq f_{obs}), \tag{12.15}$$

where *F* has an *f*-distribution with numerator degrees of freedom $k-1$ and denominator degrees of freedom $N-k$.

Often it is easier to calculate sample values for SST and SSB through the more computationally friendly expressions (which you are asked to verify in Exercises 12.A.4 and 12.A.5):

$$SST = \sum_{j=1}^{k} \sum_{i=1}^{n_j} X_{ij}^2 - \frac{\left(\sum_{j=1}^{k} \sum_{i=1}^{n_j} X_{ij} \right)^2}{N} \tag{12.16}$$

and

$$SSB = \left(\sum_{j=1}^{k} \frac{\left(\sum_{i=1}^{n_j} X_{ij} \right)^2}{n_j} \right) - \frac{\left(\sum_{j=1}^{k} \sum_{i=1}^{n_j} X_{ij} \right)^2}{N}. \tag{12.17}$$

The value of SSE can then be calculated from (12.11) to be SSE = SST – SSB.

Example 12.2. Where are For-Profit Hospitals? Hospitals can be categorized as either nonprofit, for-profit, or government (all levels). All three categories are represented in each of the fifty states—but are the relative proportions the same across the country? Table 12.5 contains data on the proportion of a state's hospitals that are for-profit for a subset of states from each of four regions of the country.

We are interested in using this subset of data to assess whether there are any overall differences in median proportions of for-profit hospitals across these four regions of the United States. Letting $\tau_{MW}, \tau_{NE}, \tau_S, \tau_W$ denote the median state percentage of for-profit hospitals in the Midwest, Northeast, South, and West, respectively, we are interested in testing the null hypothesis $H_0 : [\tau_{MW} = \tau_{NE} = \tau_S = \tau_W]$ against the general alternative $H_1 : [\tau_{MW}, \tau_{NE}, \tau_S, \tau_W \text{ not all equal}]$.

Here we have $k = 4$, $n_{MW} = n_{NE} = n_S = n_W = 5$, and $N = 20$. We use the more computationally friendly expressions in (12.16) and (12.17) to compute

Table 12.5 Proportion of for-profit hospitals for a subset of states in four regions of the country

	State	Proportion for-profit hospitals (%)
Midwest	Illinois	11
	Kansas	13
	Minnesota	1
	Ohio	13
	South Dakota	8
Northeast	Delaware	14
	Maine	3
	Maryland	4
	Massachusetts	14
	New York	0
South	Alabama	39
	Florida	50
	Louisiana	30
	South Carolina	39
	Texas	39
West	Alaska	9
	California	21
	Idaho	12
	New Mexico	39
	Utah	34

Source: American Hospital Association (2012)

SST, SSB, and SSE. First, we obtain the sum of all 20 observations and the sum of the squares of all 20 observations, as follows:

$$\sum_{j=1}^{4}\sum_{i=1}^{5}X_{ij} = [11 + 13 + 1 + 13 + 8 + 14 + 3 + 4 + 14 + 0 + 39 + 50 + 30$$
$$+39 + 39 + 9 + 21 + 12 + 39 + 34]$$
$$= 393$$

and

$$\sum_{j=1}^{4}\sum_{i=1}^{5}X_{ij}^2 = \left[(11)^2 + (13)^2 \ + \ (1)^2 + \cdots + (12)^2 + (39)^2 + (34)^2\right]$$
$$= 524 + 417 + 7963 + 3343 = 12,247.$$

Next, we need to calculate the sum of the observations separately within each of the four regions. Thus, we have:

$$\sum_{i=1}^{5} X_{iMW} = 11 + 13 + 1 + 13 + 8 = 46,$$

$$\sum_{i=1}^{5} X_{iNE} = 14 + 3 + 4 + 14 + 0 = 35,$$

$$\sum_{i=1}^{5} X_{iS} = 39 + 50 + 30 + 39 + 39 = 197,$$

and

$$\sum_{i=1}^{5} X_{iW} = 9 + 21 + 12 + 39 + 34 = 115.$$

From (12.16) we obtain

$$SST = 12,247 - \frac{(393)^2}{20} = 12,247 - 7722.45 = 4524.55$$

and from (12.17)

$$SSB = \left[\frac{(46)^2}{5} + \frac{(35)^2}{5} + \frac{(197)^2}{5} + \frac{(115)^2}{5}\right] - \frac{(393)^2}{20} = 11,075 - 7722.45$$
$$= 3352.55.$$

It follows from (12.11) that

$$\text{SSE} = \text{SST} - \text{SSB} = 4524.55 - 3352.55 = 1172.$$

The observed value of the test statistic $F(12.14)$ is then given by

$$f_{obs} = \frac{3352.55/(4-1)}{1172/(20-4)} = \frac{1117.52}{73.25} = 15.256.$$

The associated P-value for our test is then $P_0(F \geq 15.256)$, where F has an F-distribution with numerator degrees of freedom $k - 1 = 4 - 1 = 3$ and denominator degrees of freedom $N - k = 20 - 4 = 16$. Using the **R** function *pf*( ) we find that the P-value $= P_0(F \geq 15.256) = 0.00059$.

```
> pf(q = 15.256, df1 = 3, df2 = 16, lower.tail = FALSE)
[1] 5.936274e-05
```

Thus, there is very strong evidence in support of the alternative hypothesis $H_1 : [\tau_{MW}, \tau_{NE}, \tau_S, \tau_W \text{ not all equal}]$.

While it is important to understand the details involved in calculating f_{obs}, as the number of these groups and/or their sizes grow increasingly large these calculations can become quite tedious. Fortunately, the **R** function *oneway.test*( ) can be used to calculate f_{obs} and $P_0(F \geq f_{obs})$ for a particular dataset. We demonstrate its use (and verify the P-value above) with the **R** dataset *proportion_for_profit_hospitals*. Note that we also specify that the groups come from populations with equal variances using the *var.equal* argument. Although the *oneway.test*( ) function can handle situations when the variances are unequal, such a setting is beyond the scope of this text.

```
> oneway.test(formula = Proportion ~ Region,
              data = proportion_for_profit_hospitals,
              var.equal = TRUE)

        One-way analysis of means

data:  Proportion and Region
F = 15.256, num df = 3, denom df = 16, p-value = 5.936e-05
```

A discussion of multiple comparison procedures designed to assess which means $\tau_1, \ldots, \tau_k$ are not equal when the F-test leads to rejection of H_0 are discussed in Chapter 6 of Hollander et al. (2014).

Section 12.2 Practice Exercises

12.2.1. *Where Do Mayflies Call Home?* The absence of the common mayfly species *Stenacron interpunctatum* is sometimes used as an indicator of pollution conditions in a stream. One of the factors of interest in this regard is the age distribution of mayflies in different areas of a stream. In particular, do the mayflies migrate to different parts of the stream as they grow in size? Lamp (1976) studied the age distribution of *Stenacron interpunctatum* among four different habitats in Big Darby Creek, Ohio. One of the measurements he obtained was head width (in micrometer divisions, 1 division = .0345 *mm*). A subset of the data from his study is presented in Table 12.6.

Use the *F*-test to assess whether there are differences in median mayfly head widths among the four habitats in Big Darby Creek.

12.2.2. *Can Fish Be TOO Big?* To determine the number of game fish to stock in a given system and to set appropriate catch limits, it is important for fishery managers to be able to assess potential growth and survival of game fish in that system. Such growth and survival rates are closely related to the availability of appropriately sized prey. Young-of-year (YOY) gizzard shad (*Dorosoma cepedianum*) are the primary food source for game fish in many Ohio environments. However, because of their fast growth rate, YOY gizzard shad can quickly become too large for predators to swallow. Thus to be able to predict predator growth rates in such settings, it is useful to know both the density and the size structure of the resident YOY shad populations. With this in mind, Johnson (1984) sampled the YOY gizzard shad populations at four different sites in Kokosing Lake (Ohio) in summer 1984. The data in Table 12.7 are lengths (*mm*) for a subset of the YOY gizzard shad sampled by Johnson.

Use the *F*-test to assess whether there are any differences in median lengths for the YOY gizzard shad populations in these four Kokosing Lake sites.

Table 12.6 Head widths (in micrometer divisions, 1 division = .0345 *mm*) of *stenacron interpunctatum* nymphs in four habitats of Big Darby Creek, Ohio

Habitats						
A	B		C		D	
36	36	20	21	27	27	21
31	20	41	29	28	38	18
30	19	21	24	26	21	24
27	28	19	27	19	20	27
20	23	18	26	29	22	
33	28	46	18	26	30	
27	31	28	27	44	22	
18	25	22	27	23	18	
19	26	21	20	20	27	
28	29	30	46	24	30	
32	19	19	20	22	21	
22	20		38	31	24	
44	22		27	28	34	
37	24		28	20	28	
34	20		28	28	31	
37	19		18	28	22	
28	28		30	34	31	
44	24		19	25	31	
26	18		23	19	18	
73	19		49	30	22	
34	22		27	18	35	
21	19		32	36	22	
42	18		28		30	
54	27		27		23	
50	29		21		24	
25	25		34		23	
45	24		34		36	

Source: Lamp (1976)

12.2.3. *Meniscal Repair Techniques and Displacement*. In Example 12.1, we considered the differing effects of three meniscal repair options (FasT-Fix Sutures, Biodegradable Meniscus Arrows, and Vertical Mattress Sutures) on the load bearing ability of the repaired meniscus. Another issue of concern

Table 12.7 Length (in *mm*) of YOY gizzard shad from Kokosing Lake, Ohio

Sites			
I	II	III	IV
46	42	38	31
28	60	33	30
46	32	26	27
37	42	25	29
32	45	28	30
41	58	28	25
42	27	26	25
45	51	27	24
38	42	27	27
44	52	27	30

Source: Johnson (1984)

Table 12.8 Displacement (*mm*) at time of failure for meniscal repairs

FasT-Fix	Meniscus arrows	Vertical mattress
18.0	7.9	16.9
18.5	12.5	20.2
9.2	15.5	20.1
18.8	10.2	15.7
22.8	8.9	13.9
17.5	13.3	14.9

Source: Borden et al. (2003)

following such surgery is the degree of displacement of the repaired meniscus outside the joint. In the study by Borden et al. (2003) discussed in Example 12.1, they also measured the amount of displacement (*mm*) for the repaired meniscus under loadbearing. Table 12.8 contains the displacement amounts (*mm*) at time of failure for each of the eighteen surgically repaired cadaveric knees in the study.

Find the P-value for an F-test of the conjecture that the median displacement at time of failure is different for these three meniscal repair techniques.

12.3 One-way Rank-Based Ordered Alternatives ANOVA for More Than Two Populations

In many practical settings, the 'treatments" are such that the most appropriate alternative to the null (H_0) hypothesis of no "treatment" effects corresponds to increasing "treatment" effects for a natural labeling of the treatments. We emphasize that this natural ordering of the treatment labels must be anticipated *a priori* and cannot depend in any way on the observed sample data. Examples of such settings include "treatments" associated with severity of disease, drug dosage levels, intensity of a stimulus, and temperature. The proper alternative to H_0 for these situations is given by H_2 (12.3), corresponding to a monotonically non-decreasing "treatment" effect across the populations. Such alternatives represent a natural "one-sided" alternative to H_0. (Note that a monotonically non-increasing "treatment" effect alternative can be addressed as well by simply reversing the order of the treatment labels.)

Neither the Kruskal Wallis test procedure in (12.6) nor the F test in (12.15) utilizes this prior information regarding an anticipated ordered alternative, since values of both the Q (12.5) statistic and the F (12.14) statistic remain the same for any permutation of the treatment labels. Thus we must design a statistic that incorporates this anticipated *a priori* ordering into its evaluation of the sample data. A natural way to do this is to base our overall test statistic on individual pairwise two-sample statistics designed to detect one-sided alternatives, as discussed previously in Sect. 9.2.

For every pair of treatments u and v, with $1 \leq u < v \leq k$, define the statistic U_{uv} by

$$U_{uv} = \sum_{s=1}^{n_u} \sum_{t=1}^{n_v} \phi(X_{su}, X_{tv}), 1 \le u < v \le k, \tag{12.18}$$

where $\phi(a, b) = 1, 1/2, 0$ if $a <, =, > b$. Thus U_{uv} is the number of sample u before sample v precedences plus ½ of the tied observations across the two samples. It follows that U_{uv} will be large if the sample v observations tend to be larger than the sample u observations. Notice that U_{uv} is just the statistic U (9.20) that we used in Chap. 9 to test for equality of medians for two populations. Here, however, we have sample data from $k > 2$ populations and there are $\binom{n}{2} = \frac{n(n-1)}{2}$ pairwise U_{uv} statistics. Thus, we need to combine the pairwise sample information contained in the U_{uv} into a single test statistic. The natural way to accomplish this is to simply add them to obtain the Jonckheere-Terpstra statistic

$$J = \sum_{u=1}^{v-1} \sum_{v=2}^{k} U_{uv}. \tag{12.19}$$

Clearly large values of J will be associated with a generally increasing trend in sample values across the k treatments and provide evidence for rejecting the null hypothesis H_0 in favor of the ordered alternatives $H_2 : [\tau_1 \le \tau_2 \le \cdots \le \tau_k,\ \textit{with at least one strict inequality}]$.

One-Way Rank-Based Ordered Alternatives ANOVA for More Than Two Populations To test the null hypothesis $H_0 : [\tau_1 = \tau_2 = \cdots = \tau_k]$ against the ordered alternatives $H_2 : [\tau_1 \le \tau_2 \le \cdots \le \tau_k,\ \textit{with at least one strict inequality}]$ using the k mutually independent random samples $\{X_{11}, \ldots, X_{n_1 1}\}$, $\{X_{12}, \ldots, X_{n_2 2}\}$, …, $\{X_{1k}, \ldots, X_{n_k k}\}$ from populations 1, 2, …, k, compute the Jonckheere-Terpstra statistic J (12.19) and let j_{obs} be the attained value of J. Then the P-value for a test of H_0 (12.1) against the ordered alternatives H_2 (12.3) is given by

$$P\text{-}value = P_0(J \geq j_{obs}), \tag{12.20}$$

where $P_0(J \geq j_{obs})$ is obtained from the sampling distribution of J when H_0 is true.

Example 12.3. Fasting Metabolic Rate of White-Tailed Deer. Seasonal energy demands for deer and the nutritional quality of the range must be taken into account to prevent starvation of the animals during critical seasons. Silver et al. (1969) studied the fasting metabolic rate (FMR) of white-tailed deer. One of the questions of concern to the investigators was whether or not FMR is an increasing function of environmental temperature, for which they collected the data in Table 12.9.

For this setting, we are interested in testing the null hypothesis H_0 of no "treatment" effects against the ordered alternatives H_2 in (12.3) with $k = 4$, corresponding to

$$H_2 : [\tau_1 \leq \tau_2 \leq \tau_3 \leq \tau_4, \ \textit{with at least one strict inequality}].$$

Table 12.9 Fasting metabolic rate (FMR) for white-tailed deer (*kcal*/*kg*/day)

Two-month period			
January–February	**March–April**	**May–June**	**July–August**
36.0	39.9	44.6	53.8
33.6	29.1	54.4	53.9
26.9	43.4	48.2	62.5
35.8		55.7	46.6
30.1		50.0	
31.2			
35.3			

Source: Silver et al. (1969)

Here we have $n_1 = 7$, $n_2 = 3$, $n_3 = 5$, and $n_4 = 4$. To compute the test statistic J (12.19), we first must compute the $\binom{4}{2} = \frac{4(3)}{2} = 6$ U_{uv} (12.18) statistics, for $1 \leq u < v \leq 4$. Using the data in Table 12.9, we see that

$U_{12} = 7 + 1 + 7 = 15$	$U_{13} = 7 + 7 + 7 + 7 + 7 = 35$
$U_{14} = 7 + 7 + 7 + 7 = 28$	$U_{23} = 3 + 3 + 3 + 3 + 3 = 15$
$U_{24} = 3 + 3 + 3 + 3 = 12$	$U_{34} = 3 + 3 + 5 + 1 = 12$

From (12.19), it follows that $j_{obs} = 15 + 35 + 28 + 15 + 12 + 12 = 117$. Using the **R** function *pJCK( )* with the dataset *fmr_white_tailed_deer* we verify this and find that the P-value for our test of H_0 against the ordered alternatives $H_2 : [\tau_1 \leq \tau_2 \leq \tau_3 \leq \tau_4,\ \textit{with at least one strict inequality}]$ is $P_0 (J \geq j_{obs}) = P_0(J \geq 117) = 0.000023$. Thus, there is strong evidence that the FMR for white-tailed deer is, indeed, an increasing function of environmental temperature.

```
> pJCK(fmr_white_tailed_deer)
Group sizes: 7 3 5 4
Jonckheere-Terpstra J Statistic: 117
Exact upper-tail probability: 2.29768225124262e-05
```

Note that the **R** dataset *fmr_white_tailed_deer* is a list as discussed earlier, which allows for unequal sample sizes between groups. Here, the exact distribution of J is computed because the **R** function *pJCK( )* recognizes that the sample sizes are small enough to make such computations practical. If we had chosen a larger dataset (or explicitly specified "Monte Carlo" using the *method* argument), the Monte Carlo approximation would have been used, similar to Example 12.1.

For information on one-sided multiple comparison procedures associated with rejection of H_0 by the Jonckheere-Terpstra test, see, for example, Chapter 6 of Hollander et al. (2014).

Table 12.10 Value (in Singapore dollars) for round diamond stones of varying levels of purity and no larger than .75 carat in size

Color purity					
D	E	F	G	H	I
1302	1327	1427	1202	1126	1098
1641	1510	1468	1260	1222	1126
3490	1555	1551	1410	1316	1299
3921	1738	1593	1447	1420	1572
6372	3501	1956	2532	1485	2892
7368	4138	3480			
		3635			

Source: Chu (2001); Singapore Business Times (2000)

Section 12.3 Practice Exercises

12.3.1. *How Important is Color Purity in the Price of Diamonds?* In the February 18, 2000 edition of Singapore's *Business Times*, an advertisement (discussed in Chu, 2001) listed data, including purity of color and value in Singapore dollars, for 308 round diamond stones. A subset of these data is provided in Table 12.10 for stones no larger than .75 carat in size. The top color purity is coded as D, with decreasing level of purity down the alphabet from E through I (the most impure).

Find the *P*-value for a test of the conjecture that the value (in Singapore dollars) of diamond stones no larger than .75 carat is a non-decreasing function of the color purity of the diamond.

12.3.2. *Clearing Ultrasound Probes of Bacterial Infections.* One of the major sources for spreading nosocomial (hospital-acquired) infections from patient to patient is through the use of ultrasound probes at tertiary care facilities and it is essential that hospitals use effective ultrasound probe cleaning procedures. Ali et al. (2015) presented data for comparing three different probe cleaning procedures: (i) using sterilized paper towels; (ii) treatment with a 0.9% saline solution; and (iii) cleaning with a soap solution. The Colony Forming Unit (CFU) of bacterial counts using a standard agar plate were obtained from culture swabs for 75 probes conducted at the Radiology Department of the Aga Khan University Hospital in Karachi, Pakistan. Twenty-five of these probes were then wiped with sterilized paper towels, twenty-five of them were treated with a 0.9% saline solution, and the final twenty-five were cleaned with a soap solution. The CFU bacterial counts were then obtained again for each of the 75 probes after treatment. The before and after treatment CFU counts for the 75 probes are given in Table 12.11.

(a) Let $\tau_{paper\ towels}$, τ_{saline}, and τ_{soap} denote the median reduction in CFU bacterial counts from using paper towels, saline, and soap, respectively. Find the P-value for a test of the null hypothesis $H_0 : [\tau_{paper\ towels} = \tau_{saline} = \tau_{soap}]$ against the natural ordered alternatives $H_2 : [\tau_{paper\ towels} \leq \tau_{saline} \leq \tau_{soap}$ *with at least one strict inequality*].

(b) Using the Kruskal-Wallis procedure from Sect. 1, find the P-value for a test of the null hypothesis $H_0 : [\tau_{paper\ towels} = \tau_{saline} = \tau_{soap}]$ against the general alternatives $H_1 : [\tau_{paper\ towels}, \tau_{saline},$ *and* τ_{soap} *not all equal*]. Compare this P-value with your results from part (a).

12.3.3. *Effect of Heat Stress on Stocked Tiger Muskellunge.* Survival of stocked tiger muskellunge (*Esox masquinongy*) is sometimes unreliable in Ohio reservoirs. Previous research had shown that one of the factors affecting stocked muskellunge survival is stress-related mortality associated with the stocking process itself, including temperature increases during the process. Mather (1984) studied the glucose response of the fish to the stress of an increase in temperature. A sample of 24 tiger muskellunge were transferred from a 15 °C holding tank into a test tank (also held at 15 °C) and allowed 24 hours to recover. (This is the period of time that previous experimenters had found to be necessary for the fish's plasma glucose level to return to normal after a dipnet stressor.) A random sample of eight fish was then

Table 12.11 Number of colony forming units (CFU) of bacterial counts for ultrasound probes before and after treatment with sterilized paper towels, treatment with a 0.9% saline solution, or cleaning with a soap solution

Probe number	Number of colony forming units (CFU)	
	Before paper towel wipe	**After paper towel wipe**
1	350	136
2	142	62
3	190	106
4	300	190
5	409	211
6	390	192
7	159	61
8	198	101
9	302	192
10	296	136
11	322	166
12	172	72
13	104	78
14	151	91
15	133	71
16	202	131
17	102	89
18	109	79
19	167	99
20	79	59
21	107	78
22	89	55
23	202	121
24	197	101
25	106	79
	Before saline solution	**After saline solution**
26	292	51
27	302	42
28	261	49
29	302	97
30	192	39
31	201	32

(continued)

Table 12.11 (continued)

Probe number	**Number of colony forming units (CFU)**	
32	192	62
33	289	67
34	290	81
35	233	89
36	209	41
37	289	53
38	301	89
39	189	39
40	161	39
41	231	61
42	142	29
43	190	58
44	203	81
45	297	52
46	219	51
47	161	21
48	232	41
49	171	36
50	193	71
	Before cleaning with soap	**After cleaning with soap**
51	213	11
52	296	13
53	312	9
54	268	7
55	202	5
56	312	4
57	257	8
58	361	2
59	301	6
60	331	6
61	296	3
62	326	2
63	396	6
64	307	2
65	256	1
66	303	3

(continued)

Table 12.11 (continued)

Probe number	Number of colony forming units (CFU)	
67	309	2
68	268	8
69	292	7
70	302	2
71	368	6
72	317	1
73	314	1
74	316	2
75	309	5

Source: Ali et al. (2015)

Table 12.12 Plasma glucose for tiger muskellunge (*mg*%)

Hours after 12 °C temperature increase		
0 (baseline)	**1**	**4**
61.08	95.45	205.96
86.21	169.19	82.55
90.15	216.16	116.60
72.91	141.92	107.23
83.74	116.16	103.83
76.35	172.22	96.60
91.63	126.26	112.77
56.65	177.78	140.85

Source: Mather (1984)

removed from the tank, anesthetized, blood collected and plasma glucose determined. These data serve as a baseline or control sample. Next, the stressor (a 12 °C temperature increase) was applied to the test tank and blood samples were collected (in the way previously described) for random samples of eight additional fish at each of the time periods 1 and 4 hours after the temperature increase. These 24 plasma glucose measurements ($mg\%$) are given in Table 12.12.

It was expected that the temperature increase of 12°C would have an immediate effect of raising the $mg\%$ of plasma glucose for the fish and then it would gradually return to normal over a 24 hour period of time after the temperature increase. Find the P-value for a test of an appropriate hypothesis test of the conjecture that the 12°C temperature increase led to an increase in the plasma glucose for the fish within the first hour, but that this impact had begun to decline already by the time four hours had passed.

12.3.4. *Possums and Hollow-Bearing Trees Revisited.* Consider the hollow-bearing tree and number of bushtail possums data given in Table 12.4. Find the P-value for a test of the conjecture that the median number of bushtail possums living in a section of montane ash forest is a non-decreasing function of the number of available hollow-bearing trees. Compare your result with that obtained in Exercise 12.1.3.

Chapter 12 Comprehensive Exercises

12.A. Conceptual

12.A.1. Show that the average of the ranks 1, ..., N is $R_{..} = \frac{1+2+\cdots+N}{N} = \frac{N+1}{2}$.

12.A.2. Show that the Kruskal-Wallis statistic Q (12.5) can also be computed from the equivalent expression

$$Q = \left(\frac{12}{N(N+1)} \sum_{j=1}^{k} \frac{R_j^2}{n_j} \right) - 3(N+1).$$

12.A.3. Let SST, SSB, and SSE be as defined in (12.10), (12.12), and (12.13), respectively. Verify the fundamental identity in (12.11) relating these variance components.

12.A.4. Verify the computational formula for SST given in (12.16).

12.A.5. Verify the computational formula for SSB given in (12.17).

12.B. Data Analysis/Computational

12.B.1. *Meniscal Repair Techniques and Stiffness.* In Example 12.1, we considered the differing effects of three meniscal repair options (FasT-Fix Sutures, Biodegradable Meniscus Arrows, and Vertical Mattress Sutures) on the load bearing ability of the repaired meniscus. Another issue of concern following such surgery is the degree of stiffness in the repaired meniscus. In the study by Borden et al. (2003) discussed in Example 12.1, they also measured the degree of stiffness (Newtons(N)/mm) for the repaired meniscus under loadbearing. Table 12.13 contains the stiffness measurements (N/mm) at time of failure for each of the eighteen surgically repaired cadaveric knees in the study.

Find the P-value for an appropriate test of the conjecture that the median degree of stiffness at time of failure is different for these three meniscal repair techniques.

12.B.2. *Arts Participation Across the States—Are There Regional Differences in Performing or Creating Artworks?* The National Endowment for the Arts

Table 12.13 Stiffness (*N*/*mm*) at time of failure for meniscal repairs

FasT-Fix	Meniscus arrows	Vertical mattress
8.0	4.7	8.3
8.3	6.1	7.2
7.6	5.0	6.3
6.4	5.8	7.3
8.2	6.6	8.7
7.7	8.4	8.7

Source: Borden et al. (2003)

Table 12.14 Percentages of state residents who performed or created artworks in 2015, stratified by regions of the United States

	State	Percentage performing or creating artworks (%)
Midwest	Indiana	53.5
	Iowa	47.1
	Kentucky	35.0
	Nebraska	52.9
	Wisconsin	53.4
Northeast	Connecticut	56.7
	Pennsylvania	48.3
	Rhode Island	49.2
	Vermont	64.0
South	Arkansas	43.6
	Georgia	34.2
	Mississippi	38.5
	Tennessee	42.9
West	Arizona	41.0
	Hawaii	34.8
	Montana	59.9
	Nevada	47.0
	Washington	52.1

Source: National Endowment for the Arts (2016)

periodically collects survey information from residents about their participation in the arts. A portion of the statewide results from the 2015 survey is presented in the publication *Arts Profile #11 (August 2016)*. In particular, the publication includes the percentage of each state's residents who performed or created artworks in 2015. A subset of these statewide art creation/performance percentages is presented in Table 12.14, stratified by regions of the United States.

Find the *P*-value for a hypothesis test of the conjecture that the median percentage of state residents performing or creating artworks in 2015 differs across these four graphical regions of the United States.

Table 12.15 Calories for tested beef, meat, and poultry hot dogs

Beef	Meat	Poultry
186	173	129
181	191	132
176	182	102
149	190	106
184	172	94
190	147	102
158	146	87
139	139	99
175	175	107
148	136	113
152	179	135
111	153	142
141	107	86
153	195	143
190	135	152
157	140	146
131	138	144
149		
135		
132		

Source: Consumer Reports (1986)

12.B.3. *Caloric Content in Hot Dogs—Beef, Meat, and Poultry.* Researchers at *Consumer Reports* analyzed the caloric content of beef, meat, and poultry hot dogs. Table 12.15 contains their reported results for test samples of each type of hot dog.

Find the *P*-value for a hypothesis test of the conjecture that the median caloric content differs for the three types of hot dogs.

12.B.4. *How Tall Are You and How High Do You Sing?* Choral ensembles are composed of four basic singing groups: basses, altos, tenors, and sopranos. Are there general differences in heights for singers from these four groups? Chambers et al. (1983) consider the self-reported heights of singers in the

Table 12.16 Heights (inches) for members of the 1979 New York Choral Society

Sopranos		Altos		Tenors		Basses		
64	62	65	66	69	67	72	75	72
62	65	62	62	72	64	70	68	70
66	66	68	70	71		72	71	69
65	62	67	65	66		69	70	
60	65	67	64	76		73	74	
61	63	63	63	74		71	70	
65	65	67	65	71		72	75	
66	66	66	69	66		68	75	
65	65	63	61	68		68	69	
63	62	72	66	67		71	72	
67	65	62	65	70		66	71	
65	66	61	61	65		68	70	
62	65	66	63	72		71	71	
65	61	64	64	70		73	68	
68	65	60	67	68		73	70	
65	66	61	66	73		70	75	
63	65	66	68	66		68	72	
65	62	66		68		70	66	

Source: Chambers et al. (1983)

New York Choral Society in 1979. Table 12.16 contains the self-reported heights (in inches) for 130 members of the 1979 choral ensemble.

(a) Find the P-value for a test of the conjecture that the median heights for these four singing groups are not the same.

(b) Perhaps the result in part (a) is not surprising, since the soprano and alto groups are comprised of female singers, while the tenor and bass groups are comprised of male singers. Considering only the two male groups (tenors and basses), find the P-value for an appropriate procedure from Chap. 9 to test the hypothesis that median heights are different for tenor and bass singers. Do the same separate analysis for female soprano and alto singers.

(c) Discuss the implication of your results from parts (a) and (b).

Table 12.17 Cost (in millions of 2006 U. S. dollars) for a subset of nuclear accidents prior to 2010

Prior to 1990	1990–2000	2000–2010
267	62	30
78	254	3
32	384	41
2400	5	700
56		98
6700		

Source: Wikipedia (2016)

12.B.5. *Cost of Accidents at Nuclear Power Plants*. While nuclear reactors provide an alternative mechanism for supplying global power that has minimal direct effect on global warming and climate change, they do present other risks associated with radiation contamination from accidents. Table 12.17 contains the costs (in millions of 2006 U. S. dollars) for a subset of the nuclear accidents that occurred prior to 2010, broken down by the time periods: (i) Prior to 1990, (ii) 1990 – 2000, and (iii) 2000 – 2010.

(a) Find the *P*-value for an appropriate hypothesis test of the conjecture that the median cost of nuclear accidents (in constant 2006 U. S. dollars) differs for the three time periods.

(b) You might have noticed that there were two extremely expensive nuclear accidents prior to 1990. These outliers undoubtedly have a disproportionate effect on the analysis in part (a). Repeat part (a) without the two extreme values 2400 and 6700 for accidents prior to 1990. Compare your findings with the results obtained in part (a).

(c) Do a bit of exploring to learn about the specific nuclear accidents that led to the two extreme cleanup costs prior to 1990. Has there been another such extreme nuclear accident since 2010?

Table 12.18 Sodium content (mg) for tested beef, meat, and poultry hot dogs

Beef	Meat	Poultry
495	458	430
477	506	375
425	473	396
322	545	383
482	496	387
587	360	542
370	387	359
322	386	357
479	507	528
375	393	513
330	405	426
300	372	513
386	144	358
401	511	581
645	405	588
440	428	522
317	339	545
319		
298		
253		

Source: Consumer Reports (1986)

12.B.6. *Sodium Content in Hot Dogs—Beef, Meat, and Poultry.* Researchers at *Consumer Reports* analyzed the sodium content (mg) of beef, meat, and poultry hot dogs. Table 12.18 contains their reported results for test samples of each type of hot dog.

Find the P-value for a hypothesis test of the conjecture that the median sodium content differs for the three types of hot dogs. Compare your findings with those from Exercise 12.B.3 for the caloric content of hot dogs.

12.B.7. *Singer Heights Revisited.* In Exercise 12.B.4 you were asked to compare the median heights for the four singing groups: sopranos, altos, tenors, and basses. It might also be of interest to incorporate information about the pitch level for these four groups in our analysis. The lowest pitch is, of course, from the basses, followed in increasing pitch level by the tenors, altos, and, finally, the sopranos, with the highest pitch. Using the data from Table 12.16, find the P-value for an appropriate test of the conjecture that median heights in these four singing groups is an increasing function of the associated pitch levels. Compare your findings with those obtained in part (a) of Exercise 12.B.4.

12.B.8. *Bison, Burning, and Botany*. Plant species diversity is an important component of the health of native grassland prairie in the United States. Collins et al. (1998) conducted long-term field experiments at the Konza Prairie Long-Term Ecological Research site in northeastern Kansas to assess the effects of fire and bison grazing on plant species diversity. In one set of their experiments they recorded the total number of distinct species (grasses, forbs (non-grass flowering plants), and woody species) for twelve plots under each of four experimental conditions (total of 48 plots): control, burning only, bison grazing only, and the combination of burning and bison grazing. These species counts are given in Table 12.19.

(a) Find the P-value for an appropriate test of the null hypothesis that there is no difference in median species diversity under these four experimental conditions against the general alternative in (12.2).

(b) One possible conjecture might be that the species diversity would be least under the burning only setting, followed in increasing order by the control, burning and grazing, and, finally, bison grazing only. Find the P-value for an appropriate hypothesis test of this conjectured ordering. Compare your result with that found in part (a).

12.C. Activities

12.C.1. *Candle Color and Burning Stamina*. Purchase a package of ordinary birthday candles containing at least six red, six white, and six blue candles.

Table 12.19 Total number of species present in each of the Konza Prairie experimental plots under the four experimental conditions

Control	Burning only	Bison grazing only	Burning and bison grazing
34	31	38	41
46	30	55	53
51	36	59	59
55	48	76	69
34	35	55	36
40	37	63	54
46	40	65	66
55	49	81	76
34	30	63	54
39	32	69	60
49	33	71	71
62	35	88	81
36	32	48	49
41	33	51	52
54	43	55	64
63	47	90	89

Source: Collins et al. (1998)

Light each of the six red, six white, and six blue candles one at a time and record the length of time (in seconds) it takes for each of them to burn out. Find the P-value for an appropriate test of the null hypothesis that there are no differences in median burning time for the three colored candles against the general alternative that there is some difference in the medians.

12.C.2. *How Long Are the Commercials*? Time the length of at least four separate commercials for each of the following types of television shows:

(i) News Program
(ii) Drama Series
(iii) Comedy Series
(iv) College Football Game.

Find the P-value for an appropriate test of the null hypothesis that there are no differences in median commercial length across these four types of television shows against the general alternative that there is some difference in the medians.

12.C.3. *How Much Is It Going to Cost?* Pick a future, but arbitrary, Saturday night and find the price to stay for that Saturday night in a standard room with two double beds at a Hilton Garden Inn, Hyatt Place, Holiday Inn Express, and Courtyard by Marriott for each of the following cities:

Chicago Dallas Denver Pittsburgh St.Louis.

Find the P-value for an appropriate hypothesis test of the null hypothesis that there are no differences in median cost among these four hotel chains for a room with two double beds on a Saturday night in these five cities against the general alternative that there is some difference in the medians.

12.D. Internet Archives

12.D.1. *Only Child, Only Younger Siblings, Only Older Siblings, or Both Younger and Older Siblings—Does It Matter?* Use the Internet to locate a scientific paper that discusses differences in behavior or achievements between an only child, those with only younger brothers/sisters, those with only older brothers/sisters, and those with both older and younger brothers/sisters (i.e., middle children). Summarize the findings discussed in the paper, particularly how the authors used ANOVA to support their results.

12.D.2. *More About Nonprofit Hospitals.* In Exercise 12.1.2 we compared the charity care (uncollectable debt) provided for those individuals who cannot afford it by nonprofit hospitals in Colorado, Michigan, Ohio, and Virginia. Use the Internet to locate similar information for nonprofit hospitals in

California, Montana, Alabama, and Maine. Compare the information for these four states with the results in Exercise 12.1.2.

12.D.3. *Sense of Smell and Longevity*. Scientific researchers have found evidence to link one's sense of smell with longevity of life; that is, the better you smell (not how much you smell!), the more likely you are to live longer. Use the Internet to locate one or more published research papers that address this issue by comparing the longevity of individuals who have normal smelling abilities with (i) individuals who have moderate smelling loss and (ii) individuals who have more acute smelling loss. Discuss the findings in those publications. Which of the statistical procedures discussed in this chapter would be most appropriate for addressing the conjecture that loss of smelling has a negative impact on longevity of life?

Appendix A: Datasets Usage Throughout *IIS*

agricultural_chargeoff_rates_by_quarter: Example 2.6;

airline_arrivals: Example 2.7;

american_league_salary_2014: Example 1.6; Page 56, Chapter 1; Exercise 1.3.7; Exercise 1.3.19; Exercise 4.7.13; Exercise 4.B.5;

arion_subfuscus: Example 9.3;

average_HDL_levels: Example 7.1;

beer_head: Exercise 5.B.17; Exercise 9.2.21;

body_temperature_and_heart_rate: Exercise 5.B.13; Exercise 5.B.14; Exercise 7.6.4; Exercise 7.6.5; Exercise 7.B.7; Exercise 7.B.12; Exercise 9.2.18; Exercise 9.4.1;

chargeoff_rates: Example 2.5; Exercise 2.1.8;

college_rankings_2012: Exercise 2.B.2;

delinquency_rates: Exercise 2.B.1;

desimipramine: Example 8.2; Exercise 8.B.10;

diamonds_carats_color_cost: Exercise 7.B.1; Exercise 9.2.19;

emissions: Example 2.3; Exercise 2.1.7;

engineering_drawing_hours: Page 131, Chapter 1; Exercise 1.B.6;

female_amerindians: Example 7.6;

© Springer International Publishing AG 2017

D.A. Wolfe, G. Schneider, *Intuitive Introductory Statistics*, Springer Texts in Statistics,
DOI 10.1007/978-3-319-56072-4

fmr_white_tailed_deer: Example 12.3;

gender_roles: Example 10.2;

goggled_green_turtles: Example 9.9;

health_care_by_affiliation: Example 10.3;

homes_prices: Exercise 7.6.2;

house_lot_sizes: Exercise 7.B.8; Exercise 7.B.10; Exercise 9.2.17; Exercise 9.3.1; Exercise 9.4.2;

infant_walking: Example 9.10;

interstitial_lengths: Example 1.21;

kentucky_derby_2012: Exercise 2.1.9; Exercise 4.B.5;

meniscal_repairs_load_at_failure: Exercise 9.2.22; Example 12.1;

mother_smoking_age: Exercise 2.3.1;

mother_smoking_education: Example 2.9;

mother_smoking_education_1989_1993: Exercise 2.3.2;

mother_smoking_education_2010: Example 2.9;

motor_vehicle_death_rate_2012: Example 1.10; Example 1.17;

movie_facts: Exercise 5.B.12; Exercise 7.B.13; Exercise 7.B.14; Exercise 7.B.15; Exercise 9.2.20; Exercise 9.3..3; Exercise 9.3.5;

national_league_salary_2014: Example 1.20; Exercise 1.3.7; Exercise 1.3.19; Exercise 4.7.13;

nba_2015_2016: Exercise 2.B.5;

osu_math_salaries_2015: Example 3.1; Exercise 4.B.5;

pennies_age: First paragraph, Section 7.6; Example 7.8;

percentage_hatched_eggs: Example 8.1; Exercise 8.B.1; Exercise 8.B.9;

pew_science_survey_by_age_group: Exercise 2.3.5;

pew_science_survey_data_by_party: Exercise 2.3.4;

pines_1997: Exercise 1.B.16; Example 2.2; Exercise 2.1.5; Example 3.4; Exercise 3.1.10; Exercise 4.7.14; Exercise 4.B.1; Exercise 4.B.5; Exercise 6.C.9; Page 870, Chapter 11; Example 11.4; Exercise 11.A.11;

pmn_migration: Example 7.3;

population_estimates_2015: Exercise 2.B.3;

presidential_election_polls: Exercise 6.D.7;

proportion_for_profit_hospitals: Example 12.2;

q2_q4_agricultural_chargeoff_rates: Example 2.6;

reading_habits_2011: Example 3.5;

school_report_cards_2014: Example 3.2; Exercise 3.1.7;

sheep_weight: Example 11.2;

state_cdi: Exercise 2.3.3;

state_poverty_levels_2013: Exercise 1.1.11;

tiaa_cref: Exercise 2.1.1; Exercise 2.1.2; Exercise 4.B.5;

traffic_accidents: Example 1.2;

weekly_salaries: Example 2.1; Exercise 2.1.3; Exercise 2.1.4;

weight_of_Euros: Exercise 5.B.15; Exercise 7.B.4;

Appendix B: *R* Functions Usage Throughout *IIS*

aggregate( ): Example 3.5;

all( ): Example 4.12;

apply( ): Example 9.9;

as.numeric( ): Exercise 3.1.11;

barplot( ): Exercise 1.1.16; Exercise 1.1.27; Example 2.9; Example 3.5;

binconf( ): Page 652, Chapter 8; Example 8.3;

binom.test( ): Page 440, Chapter 6;

boxplot( ): Example 1.12;

chisq.test( ): Example 5.11;

choose( ): Example 5.8;

ConDis.matrix( ): Example 11.5;

confint( ): Exercise 11.A.11;

cor( ): Section 2.2.1; Example 2.8; Exercise 2.2.3;

cor.test( ): Example 11.2; Section 11.2, Practice Exercises; Example 11.5;

cumsum( ): Example 4.2;

dbinom( ): Example 4.10; Exercise 4.3.3; Exercise 4.3.9; Example 5.5; Example 5.8;

dgeom( ): Page 272, Chapter 4;

pJCK( ): Example 12.3;

© Springer International Publishing AG 2017

D.A. Wolfe, G. Schneider, *Intuitive Introductory Statistics*, Springer Texts in Statistics,
DOI 10.1007/978-3-319-56072-4

dmultinom( ): Example 5.6; Page 353, Chapter 5;

FindTriples( ): Exercise 1.B.6; Exercise 1.B.12;

fisher.test( ): Page 803, Chapter 10; Example 10.3;

hist( ): Pages 18, 21, Chapter 1; Example 4.2; Example 5.12; Example 7.3;

legend( ): Example 2.6;

lm( ): Example 11.3; Example 11.4; Exercise 11.A.11;

mean( ): Example 9.9;

median( ): Example 8.1;

outer( ): Example 1.21; Example 9.3;

plot( ): Example 2.1; Example 2.2; Example 2.3; Section 2.1.2; Example 2.6; Example 2.7; Example 2.9; Exercise 4.3.3;

pbinom( ): Exercise 4.3.9; Example 5.5; Figures 5.2, 5.3, and 5.4; Example 5.10; Exercise 5.A.1; Exercise 5.B.2; Exercise 5.B.3; Exercise 5.B.8; Exercise 5.B.9; Page 406, Chapter 5; Page 477, Chapter 6; Example 6.9; Page 497, Chapter 6;

pchisq( ): Page 386, Chapter 5; Example 10.1; Example 10.2; Example 10.4;

pf( ): Example 12.2;

phyper( ): Page 803, Chapter 10; Example 10.3;

pKW( ): Example 12.1;

dmultinom( ): Exercise 5.B.8; Exercise 5.B.9;

pnorm( ): Example 4.21; Page 299, Chapter 4; Example 4.22; Example 4.23; Example 4.25 (twice); Page 341, Chapter 5; Example 5.2; Example 5.3; Exercise 5.B.2; Exercise 5.B.3; Example 9.2;

ppoints( ): Page 307, Chapter 4;

psignrank( ): Page 556, Chapter 7; Page 557, Chapter 7; Exercise 7.2.3; Exercise 7.2.4; Example 8.1;

pt( ): Page 585, Chapter 7; Example 7.7; Example 11.1;

pwilcox( ): Example 5.7; Example 9.3; Page 693, Chapter 9; Example 9.4; Example 9.5; Page 697, Chapter 9;

Bibliography

AAA. (2016). *Foundation for traffic safety: Prevalence of self-reported aggressive driving behavior: United States, 2014*. Report Issued July 2016. www.aaafoundation.org

ABC News. (1994). *Telephone interview survey conducted by ABC News concerning the court system in the United States and the rights of crime victims.* February 2, 1994.

Abeles, H. F., & Porter, S. Y. (1978). The sex-stereotyping of musical instruments. *Journal of Research in Music Education, 26*, 65–75.

Al Jarad, N., Gellert, A. R., & Rudd, R. M. (1993). Bronchoalveolar lavage and ^{99m}*Tc-DTPA* clearance as prognostic factors in asbestos workers with and without asbestosis. *Respiratory Medicine, 87*, 365–374.

Ali, A., Rasheed, A., Siddiqui, A. A., Naseer, M., Wasim, S., & Akhtar, W. (2015). Non-parametric test for ordered medians: The Jonckheere Terpstra test. *International Journal of Statistics in Medical Research, 4*, 203–207.

American Association of University Professors. (2013). Here's the news: The annual report on the economic status of the profession, 2012–2013. *Academe, 99*(2), 4–19.

American Heritage College Dictionary. (1993). *Third edition*. Boston: Houghton Mifflin.

American Hospital Association. (2012). *AHA hospital statistics 2012 edition (2010 data)*. www.aha.org/. Accessed 2 Nov 2016.

American Orthopaedic Foot and Ankle Society. (1998). Telephone survey, as reported by Dr. Dean Edell, *Insight from America's Doctor at HealthCentral.com*. February 10, 1999.

American Orthopaedic Foot and Ankle Society. (1999). *Healthy trend continues: Low heels score high with working women. Autumn 1998 telephone survey*. Report issued on February 7, 1999.

Anderson, N. L. (1999). *Personal communication for report in Statistics 661*. Columbus: Ohio State University.

Annenberg Public Policy Center of the University of Pennsylvania. (2000). *Media in the home. Fifth annual survey of parents and children*. Philadelphia: Annenberg Public Policy Center of the University of Pennsylvania.

Archer, V. E. (1979). Anencephalus, drinking water, geomagnetism and cosmic radiation. *American Journal of Epidemiology, 109*, 88–97.

© Springer International Publishing AG 2017

D.A. Wolfe, G. Schneider, *Intuitive Introductory Statistics*, Springer Texts in Statistics,
DOI 10.1007/978-3-319-56072-4

Arellano, L., Castillo-Guevara, C., Huerta, C., Germán-García, A., & Lara, C. (2015). Effect of using different types of animal dung for feeding and nesting by the dung beetle *Onthophagus lecontei* (Coleoptera: Scarabaeinae). *Canadian Journal of Zoology, 93*, 337–343.

Ariel, B., Sutherland, A., Henstock, D., Young, J., Drover, P., Sykes, J., Megicks, S., & Henderson, R. (2016). Contagious accountability: A global multisite randomized controlled trial on the effect of police body-worn cameras on citizens' complaints against the police. *Criminal Justice and Behavior*. doi:10.1177/0093854816668218, published online September 22, 2016, 24 pages.

Atkinson, J., & Egeth, H. (1973). Right hemisphere superiority in visual orientation matching. *Canadian Journal of Psychology, 27*(2), 152–158.

Ault, R. G., Hudson, E. J., Linehan, D. J., & Woodward, J. D. (1967). A practical approach to the assessment of head retention of bottled beers. *Journal of the Institute of Brewing, 73*(6), 558–566.

Badawy, M. E. I., Kenawy, A., & El-Aswad, A. F. (2013). Toxicity assessment of Buprofezin, Lufenuron, and Triflumuron to the earthworm *Aporrectodea caliginosa*. *International Journal of Zoology*, Article ID 174523, 9 pages.

Bagley, F. (1985). *Personal communication for report in Statistics 661*. Columbus: Ohio State University.

Baxley, F., & Miller, M. (2006). Parental misperceptions about children and firearms. *Archives of Pediatric & Adolescent Medicine, 160*(5), 542–547.

Bennett, P. E. (1957). The statistical measurement of a stylistic trait in *Julius Caesar* and *As You Like It*. *Shakespeare Quarterly, 8*(1), 33–50.

Borden, P., Nyland, J., Caborn, D. N. M., & Pienkowski, D. (2003). Biomechanical comparison of the FasT-Fix meniscal repair suture system with vertical mattress sutures and meniscus arrows. *The American Journal of Sports Medicine, 31*(3), 374–378.

Brambilla, F., Cavagnini, F., Invitti, C., Poterzio, F., Lampertico, M., Sali, L., Maggioni, M., Candolfi, C., Panerai, A. E., & Müller, E. E. (1985). Neuroendocrine and psychopathological measures in *Aneroxia Nervosa*: Resemblances to primary affective disorders. *Psychiatry Research, 16*, 165–176.

Brondolo, E., Karlin, W., Alexander, K., Bobrow, A., & Schwartz, J. (1999). Workday communication and ambulatory blood pressure: Implications for the reactivity hypothesis. *Psychophysiology, 36*, 86–94.

Brust, J. (1984). *Personal communication for report in Statistics 661*. Columbus: Ohio State University.

Business Insider. (2012). *Jay Yarow, BUSINESS INSIDER, August 30, 2012. 51% of people think stormy weather affects 'cloud computing'*. http://www.businessinsider.com/. Accessed 21 June 2016.

Cable News Network (CNN). (2009). *Survey: Support for terror suspect torture differs among the faithful*. April 30, 2009. http://edition.cnn.com/2009/US/04/30/religion.torture

Cable News Network (CNN). (2015). *History in the making: 2 women will graduate from Army Ranger course*. www.cnn.com/2015/08/18/politics

Catalyst. (1998). *Two careers, one marriage: Making it work in the workplace*. Report issued January 20, 1998.

CBS News/New York Times Poll. (2014). September 10–14, 2014, as reported in PollingReport.com, 2015.

Centers for Disease Control and Prevention. (1995). United States National Center for Health Statistics, Health.

Centers for Disease Control and Prevention. (2003). *Behavioral risk factor surveillance system summary data quality report*. Atlanta: Centers for Disease Control and Prevention. www.cdc.gov

Centers for Disease Control and Prevention. (2016a, February). *WONDER online database*. http://wonder.cdc.gov/natality-current.html. Accessed 20 May 2016.

Centers for Disease Control and Prevention. (2016b, May). http://catalog.data.gov/dataset/u-s-chronic-disease-indicators-cdi-ff843. Accessed 21 May 2016.

Chambers, J. M., Cleveland, W. S., Kleiner, B., & Tukey, P. A. (1983). *Graphical methods for data analysis*. Belmont: Wadsworth.

Christenfeld, N., Phillips, D. P., & Glynn, L. M. (1999). What's in a name: Mortality and the power of symbols. *Journal of Psychosomatic Research, 47*(3), 241–254.

Chu, S. (2001). Pricing the C's of diamond stones. *Journal of Statistics Education, 9*(2), 12 pages online.

Clark, P. J., Vandenberg, S. G., & Proctor, C. H. (1961). On the relationship of scores on certain psychological tests with a number of anthropometric characters and birth order in twins. *Human Biology, 33*, 163–180.

Clawson, R. L., Hartman, G. W., & Fredrickson, L. H. (1979). Dump nesting in a Missouri wood duck population. *The Journal of Wildlife Management, 43*(2), 347–355.

Clay, S. W., & Conatser, R. R. (2003). Characteristics of physicians disciplined by the State Medical Board of Ohio. *Journal of the American Osteopathic Association, 103*(2), 81–88.

Cleveland.com. (2016). http://www.cleveland.com/datacentral/index.ssf.html. Accessed 30 July 2016.

College and University Professional Association for Human Resources. (2012). *2011–2012 National faculty salary survey executive summary*. Released March 2012. www.cupahr.org

Collins, S. L., Knapp, A. K., Briggs, J. M., Blair, J. M., & Steinauer, E. M. (1998). Modulation of diversity by grazing and mowing in native tallgrass prairie. *Science, 280*, 745–747.

Columbus Dispatch. (1998). Report on the results of the Buckeye State Poll conducted during the period February 17–26 for the Ohio State University College of Social and Behavioral Sciences. April 9, 1998 issue.

Consumer Reports (1986). June Issue, 366–367.

Cruickshanks, K. J., Klein, R., Klein, B. E. K., Wiley, T. L., Nondahl, D. M., & Tweed, T. S. (1998). Cigarette smoking and hearing loss: The epidemiology of hearing loss study. *Journal of the American Medical Association, 279*(21), 1715–1719.

Dean, A., & Voss, D. (1999). *Design and analysis of experiments*. New York: Springer.

Dearwater, S. R., Coben, J. H., Campbell, J. C., Nah, G., Glass, N., McLoughlin, E., & Bekemeier, B. (1998). Prevalence of intimate partner abuse in women treated at community hospital emergency departments. *Journal of the American Medical Association, 280*(5), 433–438.

DeGiorgio, C. M., Miller, P. R., Harper, R., Gornbein, J., Schrader, L., Soss, J., & Meymandi, S. (2014). Fish oil (n-3 fatty acids) in drug resistant epilepsy: A randomised placebo-controlled crossover study. *Journal of Neurology, Neurosurgery & Psychiatry, 86*, 65–70, published online on September 8, 2014, 6 pages.

Department of Biology, Kenyon College. (1998). Personal communication.

Depew, B. (1999). *Personal communication for report in Statistics 661*. Columbus: Ohio State University.

Devanand, D. P., Lee, S., Manly, J., Andrews, H., Schupf, N., Masurkar, A., Stern, Y., Mayeux, R., & Doty, R. L. (2015). Olfactory identification deficits and increased mortality in the community. *Annals of Neurology, 78*(3), 401–411.

Doblhammer, G., & Vaupel, J. W. (2001). Lifespan depends on the month of birth. *Proceedings of the National Academy of Sciences, 98*(5), 2934–2939.

Earth System Research Laboratory of the National Oceanic and Atmospheric Administration (NOAA). (2016). Available at. http://www.esrl.noaa.gov/gmd/ccgg/trends/

Eirk, K. G. (1972). An experimental evaluation of accepted methods for removing spots and stains from works of art on paper. *Bulletin of the American Group International Institute for Conservation of Historic and Artistic Works, 12*(2), 82–87.

Elwood, J. M. (1977). Anencephalus and drinking water composition. *American Journal of Epidemiology, 105*(5), 460–468.

Frantz, R. L., Ordy, J. M., & Udelf, M. S. (1962). Maturation of pattern vision in infants during the first six months. *Journal of Comparative and Physiological Psychology, 55*(6), 907–917.

Fetterer, F., Knowles, K., Meier, W., & Savoie, M. (2016, updated daily). *Sea Ice Index, Version 2.* [September 1979–2012] Boulder: National Snow and Ice Data Center. Available at http://nsidc.org/

Freund, R. J., Mohr, D., & Wilson, W. J. (2010). *Statistical methods* (3rd ed.). New York: Academic.

Gallup, Inc. (2012a). *Business Journal: Your employees don't "Get" your brand. J. H. Fleming, and D. Witters.* Report issued July 26, 2012. www.gallup.com

Gallup, Inc. (2012b). *Economy: U. S. workers least happy with their work stress and pay.* Report issued November 12, 2012. www.gallup.com

Gallup, Inc. (2014). *Well-being: Americans favor ban on smoking in public, but not total ban.* R. Riffkin; report issued July 30, 2014. www.gallup.com

Gallup, Inc. (2015a). *Final Gallup voter opinion polls and eventual results for Presidential Elections between 1936 and 2012.* www.gallup.com

Gallup, Inc. (2015b). *Workplace: Union members less content with safety, recognition at work.* A. Dugan; report issued September 11, 2015. www.gallup.com

Gallup, Inc. (2015c). *Politics: The more Americans know congress, the worse they rate it.* F. Newport, L. Saad, and M. Traugott; report issued September 21, 2015. www.gallup.com

Gallup, Inc. (2015d). *Politics: More republicans now prefer one-party government.* J. M. Jones; report issued October 12, 2015. www.gallup.com

Gallup, Inc. (2016, September). *Gallup Poll—Presidential and congressional approval ratings.* www.gallup.com. Accessed 4 September 2016.

Gallup-Purdue Index Report. (2015). *Great Jobs, great lives. The relationship between student debt, experiences and perceptions of college worth.* Report issued September 29, 2015. www.gallup.com

GfK Mediamark Research & Intelligence, LLC. (2010). *Top-line reports.* New York. www.gfkmri.com

GfK Mediamark Research & Intelligence, LLC. (2015a). *44% of US adults live in households with cell phones but no landlines.* Press release, April 2, 2015. New York. www.gfk.com/us

GfK Mediamark Research & Intelligence, LLC. (2015b). *Two-thirds of U. S. consumers believe companies need to be environmentally responsible.* Press release, April 21, 2015. New York. www.fgkmri.com

Groom, J.D. (1999). *Personal communication for report in Statistics 661.* Columbus: Ohio State University.

Hartlaub, B. A. (1997). *Personal communication. Student report.* Gambier: Kenyon College.

Hines, C. (1999). *Personal communication for report in Statistics 661.* Columbus: Ohio State University.

Ho, S. S., Radavelli-Bagatini, S., Dhaliwal, S. S., Hills, A. P., & Pal, S. (2012). Resistance, aerobic, and combination training on vascular function in overweight and obese adults. *The Journal of Clinical Hypertension, 14*(12), 848–854.

Hoffman, D. L., & Novak, T. P. (1998). Bridging the racial divide on the internet. *Science, 280*, 390–391.

Hollander, M., Wolfe, D. A., & Chicken, E. (2014). *Nonparametric Statistical Methods* (3rd ed.). New York: Wiley.

Honey, K. (2007). Microbicide trial screeches to a halt. *Journal of Clinical Investigation, 117*(5), 3 pages.

Hunter, J. T. (1997). *Personal communication for course project in experimental design course*. Gambier: Kenyon University.

Intellicast. (2016). http://www.intellicast.com/Local/History.aspx?month=9. Accessed 30 June 2016.

Jarausch, K. H., & Arminger, G. (1989). The German teaching profession and Nazi party membership: A demographic Logit model. *The Journal of Interdisciplinary History, 20*(2), 197–225.

Jobvite. (2014). *2014 Jobvite job seeker nation study: An authoritative survey of the social, mobile job seeker*. www.jobvite.com

Johnson, B. (1984). *Personal communication for report in Statistics 661*. Columbus: Ohio State University.

Kamimura, A., Takahashi, T., & Watanabe, Y. (2000). Investigation of topical application of procyanidin B-2 from apple to identify its potential use as a hair growing agent. *Phytomedicine, 7*(6), 529–536.

Kaptchuk, T. J., Stason, W. B., Davis, R. B., Legedza, A. R. T., Schnyer, R. N., Kerr, C. E., Stone, D. A., Nam, B. H., Kirsch, I., & Goldman, R. H. (2006). Sham device v inert pill: Randomized controlled trial of two placebo treatments. *British Medical Journal*. doi:10.1136/bmj.38726.603310.55, published 1 February 2006.

Kayle, K. A. (1984). *Personal communication for report in Statistics 661*. Columbus: Ohio State University.

Kentucky Derby. (2016). www.kentuckyderby.ag. Accessed 10 May 2016.

Kenyon Center for Environmental Study. (1997). Gambier, Ohio. Personal communication.

Kerr, H. (1983). *Personal communication for report in Statistics 661*. Columbus: Ohio State University.

Kocabey, Y., Chang, H. C., Brand, J. C., Jr., Nawab, A., Nyland, J., & Caborn, D. N. M. (2006). A biomechanical comparison of the FasT-Fix meniscal repair suture system and the RapidLoc device in cadaver meniscus. *Arthroscopy: The Journal of Arthroscopic and Related Surgery, 22*(4), 406–413.

Koga. (1999). *Personal communication for report in Statistics 661*. Columbus: Ohio State University.

Komar, V. & Melamid, A. (1997). *Painting by numbers: Komar and Melamid's scientific guide to art*. In J. Wypijewski (Ed.), New York: Farrar, Straus, & Giroux, Inc.

Kramer, A. D. I., Guillory, J. E., & Hancock, J. T. (2014). Experimental evidence of massive-scale emotional contagion through social networks. *Proceedings of the National Academy of Sciences of the United States of America, 111*(24), 8788–8790.

Lamp, W. O. (1976). *Statistical treatment of a study on the distribution of a stream insect by age*. Master's thesis, Ohio State University.

Lane, J. (1999). Morning coffee boosts blood pressure, stress hormones all day. Duke University Research News. *Science Daily* March 5.

Larson, D. W. (1999). *Personal communication*.

Larson, D. W., Matthes, U., Gerrath, J. A., Larson, N. W. K., Gerrath, J. M., Nekola, J. C., Walker, G. L., Porembski, S., & Charlton, A. (2000). Evidence for the widespread occurrence of ancient forests on cliffs. *Journal of Biogeography, 27*, 319–331.

Lee, C. (1999). Selective assignment of military positions in the Union Army: Implications for the impact of the Civil War. *Social Science History, 23*(1), 67–97.

Lee, H., Deng, X., Unnava, H. R., & Fujita, K. (2014). Monochrome forests and colorful trees: The effect of black-and-white versus color imagery on construal level. *Journal of Consumer Research, 41*(4), 1015–1032.

Lee, I.-M., & Paffenbarger, R. S., Jr. (1998). Life is sweet: Candy consumption and longevity. *British Medical Journal, 317*(7174), 1683–1684.

Leichliter, J. S., Meilman, P. W., Presley, C. A., & Cashin, J. R. (1998). Alcohol use and related consequences among students with varying levels of involvement in college athletics. *Journal of American College Health, 46*, 257–262.

Lighthall, F. F. (1991). Launching the space shuttle Challenger: Disciplinary deficiencies in the analysis of engineering data. *IEEE Transactions on Engineering Management, 38*(1), 63–74.

Lindenmayer, D. B., Barton, P. S., Lane, P. W., Westgate, M. J., McBurney, L., Blair, D., Gibbons, P., & Likens, G. E. (2014). An empirical assessment and comparison of species-based and habitat-based surrogates: A case study of forest vertebrates and large old trees. *PLOS ONE, 9*, e89807, February 24, 2014, 15 pages online.

Los Angeles Times. (2014). *Salvador Rodriguez, BUSINESS/Technology, March 4, 2014*. http://www.latimes.com/business/technology/. Accessed 21 June 2016.

Mackowiak, P. A., Wasserman, S. S., & Levine, M. M. (1992). A critical appraisal of 98.6°F, the upper limit of the normal body temperature, and other legacies of Carl Reinhold August Wunderlich. *Journal of the American Medical Association, 268*(12), 1578–1580.

Maltin, L. (1996). *Leonard Maltin's 1996 movie and video guide*. New York: Penguin Books.

March, G. L., John, T. M., McKeown, B. A., Sileo, L., & George, J. C. (1976). The effects of lead poisoning on various plasma constituents in the Canada goose. *Journal of Wildlife Diseases*, **12**, 14–19.

Mather, M. (1984). *Personal communication for report in Statistics 661*. Columbus: Ohio State University.

Matsunaga, E. (1962). The dimorphism in human normal cerumen. *Annals of Human Genetics, 25*, 273–286.

Meier, D. E., Emmons, C.-A., Wallenstein, S., Quill, T., Morrison, R. S., & Cassel, C. K. (1998). A national survey of physician-assisted suicide and euthanasia in the United States. *New England Journal of Medicine, 338*, 1193–1201.

Meilman, P. W., Leichliter, J. S., & Presley, C. A. (1998). Analysis of weapon carrying among college students, by region and institution type. *Journal of American College Health, 46*(6), 291–299.

Merline, J. W. (1991). Will more money improve education? *Journal of Consumer Research, 74*, 26–27.

Moore, T. L. (2006). Paradoxes in film ratings. *Journal of Statistics Education, 14*(1), 8 pages online.

Morrison, J., & Wickersham, P. (1998). Physicians disciplined by a state medical board. *Journal of the American Medical Association, 279*, 1889–1893.

Mrosovsky, N., & Shettleworth, S. J. (1974). Further studies of the sea-finding mechanism in green turtle hatchlings. *Behaviour, 51*, 195–208.

NASA Goddard Institute for Space Studies (GISS). (2016). Available at http://data.giss.nasa.gov/gistemp/

NASCAR Record & Fact Book. (2004). St. Louis: Sporting News Books.

National Basketball Association. (2016). http://stats.nba.com/league/team/

National Endowment for the Arts. (2016). *Arts data profile #11 (August 2016)— State-Level estimates of arts participation patterns (2012–2015)*. https://www.arts.gov. Accessed 18 Oct 2016

National Highway Traffic Safety Administration. (1998). *Survey on drivers' attitudes toward helping victims at traffic accidents*. www.nhtsa.gov.

National Highway Traffic Safety Administration. (2013). *Fatality analysis reporting system*. Released November 2013. www.nhtsa.gov

National Public Radio. (2010). *Health: The growing power of the sugar pill by Alix Spiegel*. Issued March 8, 2010.

National Public Radio. (2014). Scott Neuman, February 14, 2014. http://www.npr.org/sections/thetwo-way/2014/02/14/277058739/. Accessed 21 June 2016.

National Pubic Radio. (2015). *Sports and health in America. Prepared in conjunction with the Robert Wood Johnson Foundation and the Harvard T. H. Chan School of Public Health*. Released June 2015. www.npr.org

National Safety Council. (1996). *Accident Facts, 1996 Edition*. Itasca. www.nsc.org.

National Safety Council. (2014). *Injury Facts®, 2014 Edition*. Itasca. www.nsc.org (as reported by Insurance Information Institute (2015). www.iii.org)

National Science Foundation. (2014). *Unemployment among doctoral scientists and engineers remained below the national average in 2013*. Released September 2014. www.nsf.gov

National Women's Law Center. (2011). *State-by-State poverty data from the 2010 census*. Updated September 2011. www.nwlc.org

Navigation Technologies. (1999). What do you do when you're lost while driving? Telephone survey report released on January 8, 1999, as reported by M. Precker in *The Dallas Morning News*, January 28, 1999.

Neuman, M. D., & Werner, R. M. (2015). Marital status and postoperative functional recovery. *Journal of the American Medical Association Surgery, 151*, 194, Published online October 28, 2015, 3 pages.

New York City Department of Education. (2016). http://schools.nyc.gov/Accountability/tools/report/default.htm. Accessed 30 July 2016.

New York Times. (1989). *HEALTH; tests find 1 prisoner in 24 has AIDS virus*. September 21, 1989 issue.

New York Times. (1998). *Investing it; duffers need not apply*. May 31, Section 3, Page 1.

Nsor, C. A., & Obodai, E. A. (2014). Environmental determinants influencing seasonal variations of bird diversity and abundance in wetlands, Northern Region (Ghana). *International Journal of Zoology, 2014*, 1–10, Article ID 548401, 10 pages.

Ohio Department of Transportation. (1980). *Evaluation of the prepare to stop when flashing sign*. Report Issued February 1980. www.dot.state.oh.us

Okoro, C. A., Nelson, D. E., Mercy, J. A., Balluz, L. S., Crosby, A. E., & Mokdad, A. H. (2005). Prevalence of household firearms and firearm-storage practices in the 50 states and the District of Columbia: Findings from the Behavioral Risk Factor Surveillance System, 2002. *Pediatrics, 116*(3), e370–e376.

O'Neill, S. A., & Boulton, M. J. (1996). Boys' and girls' preferences for musical instruments: A function of gender? *Psychology of Music, 24*, 171–183.

Ostler, W. K., & Harper, K. T. (1978). Floral ecology in relation to plant species diversity in the Wasatch Mountains of Utah and Idaho. *Ecology, 59*(4), 848–861.

Özturk, Ö., Bilgin, Ö. C., & Wolfe, D. A. (2005). Estimation of population mean and variance in flock management: A ranked set sampling approach in a finite population setting. *Journal of Statistical Computation and Simulation, 75*, 905–919.

Perez, H. D., Horn, J. K., Ong, R., & Goldstein, I. M. (1983). Complement (C5)- derived chemotactic activity in serum from patients with pancreatitis. *The Journal of Laboratory and Clinical Medicine, 101*, 123–129.

Pérez-Stable, E. J., Herrera, B., Jacob, P., III, & Benowitz, N. L. (1998). Nicotine metabolism and intake in black and white smokers. *Journal of the American Medical Association, 280*, 152–156.

Petchesky, B. (2014). http://deadspin.com

Petrakis, N. L., Molohon, K. T., & Tepper, D. J. (1967). Cerumen in American Indians: Genetic implications of sticky and dry types. *Science, 158*, 1192–1193.

Pew Internet & American Life Project. (2016). *The rise of e-reading*. http://libraries.pewinternet.org/2012/04/04/. Accessed 24 Sept 2016.

Pew Research Center. (2009). *U.S. politics & policy: Public praises science; scientists fault public, media*. Report issued July 9, 2009. www.pewresearch.org

Pew Research Center. (2010). *Social and demographic trends: Women, men and the new economics of marriage, by Richard Fry*. Report issued January 19, 2010. www.pewresearch.org

Pew Research Center. (2012). *U.S. politics & policy: Partisan polarization surges in Bush, Obama years*. Report issued June 4, 2012. www.pewresearch.org

Pew Research Center. (2014). *2014 Global Attitudes Survey: Global opposition to U. S. surveillance and drones, but limited harm to America's image*. Report issued July 14, 2014. www.pewresearch.org

Pew Research Center. (2015a). *U. S. politics & policy: Support for Iran nuclear agreement falls*. Report issued September 8, 2015. www.pewresearch.org

Pew Research Center. (2015b). *Internet, science & tech: Teens, technology and romantic relationships*. Report issued October 1, 2015. www.pewresearch.org

Pew Research Center. (2015c). *U. S. politics & policy: On immigration policy, wider partisan divide over border fence than path to legal status*. Report issued October 8, 2015. www.pewresearch.org

Pew Research Center. (2016, May). *Internet & American life project*, science issues. http://www.pewinternet.com/datasets/2014-science-issues/. Accessed 21 May 2016.

Pierce, G. W. (1948). *The songs of insects*. Cambridge, MA: Harvard University Press.

Pierce, J. P., Choi, W. S., Gilpin, E. A., Farkas, A. J., & Berry, C. C. (1998). Tobacco industry promotion of cigarettes and adolescent smoking. *Journal of the American Medical Association, 279*(7), 511–515.

Pressman, R. M., Sugarman, D. B., Nemon, M. L., Desjarlais, J., Owens, J. A., & Schettini-Evans, A. (2015). Homework and family stress: With consideration of parents' self confidence, educational level, and cultural background. *The American Journal of Family Therapy, 43*(4), 297–313.

Princeton Survey Research Associates of Princeton, New Jersey (1998). *Generation to generation: American values about taking care of each other*. Report prepared for Americans Discuss Social Security; Released June 17, 1998. www.americansdiscuss.org

Public Religion Research Institute. (2013). http://publicreligion.org/research/2013/01/january-2013-tracking-poll-2/

Pye, A. E. (1974). Microbial activation of prophenoloxidase from immune insect larvae. *Nature, 251*, 610–613.

Rief, W., Nestoriuc, Y., Weiss, S., Welzel, E., Barsky, A. J., & Hofmann, S. G. (2009). Meta-analysis of the placebo response in antidepressant trials. *Journal of Affective Disorders, 118*, 1–8.

Ropercenter. (2016). www.ropercenter.cornell.edu. Accessed 17 Oct 2016.

Rosen, M. (1979). *Personal communication for report in Statistics 661*. Columbus: Ohio State University.

Rosenfield, K., Jaff, M. R., White, C. J., Rocha-Singh, K., Mena-Hurtado, C., Metzger, D. C., Brodmann, M., Pilger, E., Zeller, T., Krishnan, P., Gammon, R., Müller-Hülsbeck, S., Nehler, M. R., Benenati, J. F., & Scheinert, D. (2015). Trial of a paclitaxel-coated balloon for femoropopliteal artery disease. *New England Journal of Medicine, 373*(2), 145–153.

Rotowire. (2016). www.rotowire.com/baseball/player_stats.htm. Accessed 10 Oct 2016.

Salit, S. A., Kuhn, E. M., Hartz, A. J., Vu, J. M., & Mosso, A. L. (1998). Hospitalization costs associated with homelessness in New York City. *New England Journal of Medicine, 338*, 1734–1740.

Sax, L. J. (1997). Health trends among college freshmen. *Journal of American College Health, 45* (6), 252–264.

Schneider, G. (2014). **R** package *NSM3* for the third edition of *Nonparametric Statistical Methods*.

Schneider, L. S., Olin, J. T., & Pawluczyk, S. (1993). A double-blind crossover pilot study of *l*-Deprenyl (Selegiline) combined with Cholinesterase Inhibitor in Alzheimer's disease. *American Journal of Psychiatry, 150*, 321–323.

Schortman, E. M., & Urban, P.A. (1998). Personal communication.

Sciulli, P. W., & Carlisle, R. (1975). Analysis of the dentition from three Western Pennsylvania late woodland sites. I. Descriptive statistics, partition of variation and asymmetry. *Pennsylvania Archaeologist, 45*(4), 47–55.

Shepherd, J., Irish, M., Scully, C., & Leslie, I. (1988). Alcohol intoxication and severity of injury in victims of assault. *British Medical Journal (Clinical Research Education), 296*(6632), 1299.

Shkedy, Z., Aerts, M., & Callaert, H. (2006). The weight of Euro coins: Its distribution might not be as normal as you would expect. *Journal of Statistics Education, 14*(2), 15 pages online.

Shoemaker, A. L. (1996). What's normal?—Temperature, gender, and heart rate. *Journal of Statistics Education, 4*(2), 4 pages online.

Silberg, J., Pickles, A., Rutter, M., Hewitt, J., Simonoff, E., Maes, H., Carbonneau, R., Murrelle, L., Foley, D., & Eaves, L. (1999). The influence of genetic factors and life stress on depression among adolescent girls. *Archives of General Psychiatry, 56*(3), 225–232.

Silver, H., Colovos, N. F., Holter, J. B., & Hayes, H. H. (1969). Fasting metabolism of white-tailed deer. *Journal of Wildlife Management, 33*, 263–274.

Singapore Business Times. (2000). February 18, 2000 edition.

Smyth, J. M., Stone, A. A., Hurewitz, A., & Kaell, A. (1999). Effects of writing about stressful experiences on symptom reduction in patients with asthma or rheumatoid arthritis. *Journal of the American Medical Association, 281*(14), 1304–1309.

Society for Human Resource Management. (2014). *Report on employee job satisfaction and engagement: The road to recovery*. Issued May 2014. www.shrm.org

Storm, L., & Thalbourne, M. A. (2005). The effect of a change in pro attitude on paranormal performance: A pilot study using naïve and sophisticated skeptics. *Journal of Scientific Exploration, 19*(1), 11–29.

Stoupel, E., Tamoshiunas, A., Radishauskas, R., Bernotiene, G., Abramson, E., & Israelevich, P. (2011). Acute myocardial infarction (AMI) in context with the paradigm—Month of birth and longevity. *Health, 3*(12), 732–736.

Tarbill, G. L., Manley, P. N., & White, A. M. (2015). Drill, baby, drill: The influence of woodpeckers on post-fire vertebrate communities through cavity excavation. *Journal of Zoology, 296*, 95–103.

Thalbourne, M. A. (1995). Further studies of the measurement and correlates of belief in the paranormal. *Journal of the American Society for Psychical Research, 89*, 234–237.

Thalbourne, M. A. (2004). *The common thread between ESP and PK*. New York: The Parapsychology Foundation.

TIAA-CREF. (2016). www.tiaa-cref.org. Accessed 8 May 2016.

TIME Magazine. (1999). *Real knife, fake surgery, article by D. Thompson*. Issued February 22, 1999.

Turner, C. F., Ku, L., Rogers, S. M., Lindberg, L. D., Pleck, J. H., & Sonenstein, F. L. (1998). Adolescent sexual behavior, drug use, and violence: Increased reporting with computer survey technology. *Science, 280*(5365), 867–873.

United States Bureau of Transportation Statistics. (2016). *U. S. air carrier traffic statistics*. www.rita.dot.gov/bts/acts. Accessed 15 May 2016.

United States Census Bureau. (2010). *Library: Births, deaths, marriages, & divorces*. http://www.census.gov/library/publications/2010.html. Accessed 3 Oct 2016.

United States Census Bureau. (2012). *Arts, recreation, and travel: Statistical Abstract of the United States*. www.census.gov

United States Census Bureau. (2014a). *2012 and 2013 American Community Surveys*. Released September 2014. www.census.gov

United States Census Bureau. (2014b). *Population division, annual estimates of the resident population: April 1, 2010 to July 1, 2014*. Released December 2014. www.census.gov

United States Census Bureau. (2016). http://www.census.gov/popest/data/datasets.html. Accessed 30 May 2016.

United States Department of Education. (2016). *College scorecard data*. https://collegescorecard.ed.gov/data/. Accessed 30 May 2016.

United States Department of Energy. (2015). *Energy efficiency & renewable energy*. Released March 2015. www.fueleconomy.gov

United States Department of Labor. (2014). *Bureau of Labor Statistics. Injuries, illnesses, and fatalities*. Updated September 2014. www.bls.gov

United States Department of Labor. (2016). *Bureau of Labor Statistics. Labor force statistics (CPS)*. www.bls.gov/cps. Accessed 18 Apr 2016.

United States Department of Transportation. (2014). *Air travel consumer report*. Released February 2014. www.dot.gov

United States Department of Transportation. (2016). https://www.transportation.gov/airconsumer. Accessed 12 May 2016.

United States Environmental Protection Agency. (2016, May). *Greenhouse Gas Inventory Data Explorer*. Washington, DC: United States Environmental Protection Agency

United States Federal Reserve. (2016a). http://www.federalreserve.gov/releases/chargeoff/chgallnsa.htm. Accessed 8 May 2016.

United States Federal Reserve. (2016b). http://www.federalreserve.gov/releases/chargeoff/delallnsa.htm. Accessed 8 May 2016.

United States National Center for Health Statistics. (2009). *National Vital Statistics Reports (NVSR)*. http://www.cdc.gov/nchs/nvss.htm. Accessed 2 Oct 2016.

USA Today. (2013). *Chris Chase, USA TODAY Sports*. January 30, 2013. http://www.usatoday.com/story/gameon/2013/01/30. Accessed 21 June 2016.

Valkenburg, P. M., & Peter, J. (2007). Who visits online dating sites? Exploring some characteristics of online daters. *CyberPsychology & Behaviour, 10*(6), 849–852.

Vegetarian Times. (2008). *Vegetarianism in America*. www.vegetariantimes.com

Vigorito, A. J., & Curry, T. J. (1998). Marketing masculinity: Gender identity and popular magazines. *Sex Roles, 39*(1/2), 135–152.

VisitingAngels. (2013). *National survey reveals children choose mom over dad*. Released June 7, 2013. www.visitingangels.com

Wall Street Journal. (2012). *TECH: On Orbitz, Mac users steered to pricier hotels, by D. Mattioli.* http://www.wsj.com/articles/. Updated August 23, 2012.

Watchdog.org. (2016). http://watchdog.org/series/measuring-uncompensated-care. Accessed 2 Nov 2016.

Weinsberg, U., Adamic, L., & Develin, M. (2015). *The not-so-universal language of laughter.* Facebook Blog Post.

Whelan, R. J. (1982). An artificial medium for feeding choice experiments with slugs. *Journal of Applied Ecology, 19*(1), 89–94.

Wikipedia. (2016). www.en.wikipedia.org/. Accessed 2 Nov 2016.

Williams, R. D., Jr. (2012). Alcohol consumption and policy perception among college freshman athletes. *American Journal of Health Sciences, 3*(1), 17–22.

Wilson, K., Stoohs, R. A., Mulrooney, T. F., Johnson, L. J., Guilleminault, C., & Huang, Z. (1999). The snoring spectrum: Acoustic assessment of snoring sound intensity in 1,139 individuals undergoing polysomnography. *Chest, 115*(3), 762–770.

Winner, L. (2006). NASCAR Winston Cup race results for 1975–2003. *Journal of Statistics Education, 4*(3), 18 pages online.

Witt, G. (2013). Using data from climate science to teach introductory statistics. *Journal of Statistics Education, 21*(1), 24 pages online.

Wolfe, J., Martinez, R., & Scott, W. A. (1998). Baseball and beer: An analysis of alcohol consumption patterns among male spectators at major-league sporting events. *Annals of Emergency Medicine, 31*, 629–632.

Woodward, W. F. (1970). *A comparison of base running methods in baseball.* MSc thesis, Florida State University.

Woodard, R., & Leone, J. (2008). A random sample of Wake County, North Carolina residential real estate plots. *Journal of Statistics Education, 16*(3), 3 pages online.

Wypijewski, J. (1997). *Painting by numbers: Komar and Melamid's scientific guide to art.* Farrar, Straus, & Giroux, Inc.

Yoshizawa, K., Willett, W. C., Morris, S. J., Stampfer, M. J., Spiegelman, D., Rimm, E. R., & Giovannucci, E. (1998). Study of prediagnostic selenium level in toenails and the risk of advanced prostate cancer. *Journal of the National Cancer Institute, 90*(16), 1219–1224.

Yust, B. L. (1982). *Personal communication for report in Statistics 661.* Columbus: Ohio State University.

Zelazo, P. R., Zelazo, N. A., & Kolb, S. (1972). "Walking" in the newborn. *Science, 176*, 314–315.

Index

© Springer International Publishing AG 2017
D.A. Wolfe, G. Schneider, *Intuitive Introductory Statistics*, Springer Texts in Statistics,
DOI 10.1007/978-3-319-56072-4

Printed by Printforce, the Netherlands

NOTES

Notes